MASS SPECTROMETRY
of NUCLEOSIDES
and NUCLEIC ACIDS

MASS SPECTROMETRY
of NUCLEOSIDES
and NUCLEIC ACIDS

Edited by
Joseph H. Banoub
Patrick A. Limbach

CRC Press
Taylor & Francis Group
Boca Raton London New York

CRC Press is an imprint of the
Taylor & Francis Group, an **informa** business

CRC Press
Taylor & Francis Group
6000 Broken Sound Parkway NW, Suite 300
Boca Raton, FL 33487-2742

First issued in paperback 2019

© 2010 by Taylor & Francis Group, LLC
CRC Press is an imprint of Taylor & Francis Group, an Informa business

No claim to original U.S. Government works

ISBN-13: 978-1-4200-4402-7 (hbk)
ISBN-13: 978-0-367-38468-5 (pbk)

Library of Congress Cataloging-in-Publication Data

Mass spectrometry of nucleosides and nucleic acids / editors, Joseph H. Banoub, Patrick A. Limbach.
    p. cm.
  Includes bibliographical references and index.
  ISBN 978-1-4200-4402-7 (hardcover : alk. paper)
  1. Nucleic acids--Spectra. 2. Nucleosides--Spectra. 3. Mass spectrometry. I. Banoub, Joseph H. II. Limbach, Patrick A. III. Title.

QD433.M37 2010
572.80284--dc22

               2009040943

Visit the Taylor & Francis Web site at
http://www.taylorandfrancis.com

and the CRC Press Web site at
http://www.crcpress.com

Contents

Preface

Nucleic acids and their components play an important role in a variety of fundamental biological processes. The elucidation of the molecular structure of deoxyribonucleic acid (DNA) in 1953 by Watson and Crick led to the so-called *central dogma of molecular biology*. This dogma proclaims that DNA is transcribed to form messenger ribonucleic acid (mRNA) of a complementary sequence that directs its own replication and the sequence of nucleobases in the RNA is then translated to the corresponding sequence of amino acids to form a protein. In addition to mRNA, two other RNA species, transfer RNA (tRNA) and ribosomal RNA (rRNA), are required for this translation process.

The field of genomics has evolved over the last few years, culminating in the sequencing and characterization of the entire human genome. Furthermore, genetic theory contributed to the concept of coding by genes. A gene is defined, biochemically, as a segment of DNA (or, in a few cases, RNA) that encodes the information required to produce a functional biological product—a protein. Current challenges include the unraveling of the relationship between the genome and the proteome. The genome can be used to predict the total potential proteome, including all of the modifications that can be carried out after the initial translation of mRNA to protein by tRNAs and rRNAs.

Consequently, nucleic acid research has had a most profound impact on molecular biology. The study of noncovalent interactions between nucleic acids and peptides or proteins is a highly active research field. Such interactions are involved in many cellular processes, including replication, regulation of gene expression, and DNA packaging. Nearly all the functions of nucleic acids are accomplished by interacting with proteins. Recognition and selectivity are achieved through noncovalent contacts between virtually all biopolymers (proteins, DNA, RNA, polysaccharides, and membrane lipids) that are present in living organisms. The formation and dissociation of these weak interactions are crucial in a vast number of biochemical events. In addition, weak interactions between nucleases and DNA or RNA are central to the recombination, repair, and replication of these molecules in cells.

It is not possible to overestimate the role of metal ions in determining the three-dimensional architecture of nucleic acids, their helices, and higher-order molecular assemblies. Protein–protein, protein–nucleic acid, protein–ligand, and antibody–antigen interactions are other examples that involve noncovalent interactions, which are also of fundamental importance for the pharmaceutical industry in the evaluation of potential drug candidates.

The ability of DNA adducts to induce mutagenesis and carcinogenesis is dependent on their chemical structure, stability, and ability to be recognized by specialized DNA repair proteins. The local DNA sequence context plays a major role in mediating the rate of nucleobase adducts repair and their mispairing potency. Furthermore, the exact location of a DNA adduct within the gene determines whether the resulting mutation affects the structure and function of the gene product.

Among the biophysical techniques employed to study the large biomolecular assemblies involved in these tasks, mass spectrometry presents the greatest potential based on its intrinsic characteristics and accessible information. Therefore, mass spectrometry has had a major role in elucidating the regulatory components and mechanisms involved in the progression from gene to functional protein. Furthermore, analyses of nucleic acids is required to establish their size, purity, and sequence as a prerequisite to their use as molecular probes or therapeutics in biomedical science.

The development of soft-ionization techniques for mass spectrometry has, without question, transformed the field of nucleic acid chemistry. Significant progress in the area of accurate mass determination, sequencing, and the study of noncovalent interactions has been made possible by the use of ionization techniques such as matrix-assisted laser desorption ionization coupled to time-of-flight (MALDI-TOF) mass analyzers and electrospray ionization (ESI) coupled to conventional analyzers and to tandem mass spectrometric (MS/MS) instruments. Mass spectrometry is now one of the definitive methods for the characterization of isolated and synthetic DNA, RNA, and related products. In addition to accurately determining an intact mass, one can obtain information on the molecular structure and biopolymer sequence using a number of different mass spectrometry approaches. These approaches have provided valuable information on structurally specific bio-molecular DNA and RNA interactions. They can also enable the direct determination of stoichiometry for post-translational modifications of specific DNA–drug, DNA–protein, RNA–protein, and DNA–RNA noncovalent and covalent associations. Furthermore, these mass spectrometry approaches can be applied to DNA and RNA sequencing analyses, single-nucleotide polymorphisms, genotyping, mutation detection, genetic diagnosis, and probing of viral structures.

The editors acknowledge the special efforts of the contributors to this CRC/Taylor & Francis publication: Diethard K. Bohme (Canada), William Buchmann (France), Ed Dudley (UK), Daniele Fabris (USA), Valérie Gabelica (Belgium), Kristina Håkansson (USA), Filip Lemière (Belgium), Russell P. Newton (UK), Natalia Y. Tretyakova (USA), Antonio Triolo (Italy), Jean-Jacques Vasseur (France), and Yinsheng Wang (USA). Each contributor, an expert in their particular discipline, has shared their knowledge and understanding of the field of nucleic acid mass spectrometry.

Our aim is to expose you, the reader, to the latest developments in the field of nucleic acid mass spectrometry. The past 15 years have seen dramatic progress in this field. Thus, in addition to contributions that review and summarize these developments and advances, this book also contains chapters describing the next generation of mass spectrometry analyses of nucleic acids and their complexes that have been enabled by such past accomplishments.

This book should be equally accessible to the nucleic acids expert who is a mass spectrometry novice as well as to those with expertise in mass spectrometry but a minimal appreciation of nucleic acids. The exciting developments in mass spectrometry technology have fueled incredible advances in our understanding of nucleic acids and their complexes. We believe these contributions have captured these advances in outstanding fashion, serving to inspire new findings and developments of interest to all in this field.

Finally, no work of this scope is possible without the guidance and support of the publisher. The editors wish to especially acknowledge Ms. Jill J. Jurgensen for her expert help in preparing and producing this book.

Joseph H. Banoub (Canada)
Patrick A. Limbach (USA)

Editors

Joseph H. Banoub was born in 1947 and obtained his BS degree from the University of Alexandria (Egypt) in 1969. In 1977 he obtained his PhD degree in organic chemistry from the University de Montreal (Quebec) under the supervision of Stephen Hanessian. He spent three years as a research associate with the MDS Health Group at the National Research Council of Canada, Ottawa, synthesizing human neoglycoconjugate vaccines. In 1979 he moved to St. John's, Newfoundland, as a research scientist with the Federal Government of Canada, Department of Fisheries and Oceans. During this period he worked on the structural elucidation of various bacterial antigens and the formation of artificial vaccines for aquatic Gram-negative bacteria. He is currently principal research scientist and head of the special project program of the Department of Fisheries and Oceans and his research focus has evolved into the mass spectrometry and tandem mass spectrometry study of biomolecules (bacterial lipopolysaccharides, nucleic acids, and proteomics). He has been an adjunct professor of biochemistry at Memorial University of Newfoundland since 1984 and an adjunct professor of chemistry since 2006. His present research interests include the uses of tandem mass spectrometry for the structural elucidation of biologically active molecules.

Patrick A. Limbach was born in 1966 and obtained his BS degree from Centre College (Kentucky, USA) in 1988. In 1992, he obtained his PhD degree in analytical chemistry from The Ohio State University under the supervision of Alan G. Marshall. He spent two years as a postdoctoral researcher in the laboratory of James A. McCloskey at the University of Utah. In 1995, he accepted his first faculty position in the Department of Chemistry at Louisiana State University in Baton Rouge, Louisiana. In 2001 he moved to his current location in the Department of Chemistry at the University of Cincinnati. He is a professor of chemistry, serves as department head, and is an Ohio Eminent Scholar. His research interests include the development of new mass spectrometry methods for analyzing RNAs, the identification of unknown modified nucleosides, the characterization of RNA–protein complexes, and investigating the role of modified nucleosides in biological systems.

Contributors

Janna Anichina
York University
Toronto, ON, Canada

Joseph H. Banoub
Fisheries and Oceans Canada
Science Branch
Newfoundland, Canada

and

Department of Chemistry
Memorial University of Newfoundland
Newfoundland, Canada

Diethard K. Bohme
York University
Toronto, ON, Canada

William Buchmann
Analysis and Modeling Laboratory for
    Biology and Environment
University of Evry-Valley of the Essonne
Evry, France

C. Callie Coombs
Department of Chemistry
University of Cincinnati
Cincinnati, Ohio

Ed Dudley
Department of Biological Sciences
Swansea University
Singleton Park, Swansea,
    United Kingdom

Daniele Fabris
Department of Chemistry and
    Biochemistry
University of Maryland Baltimore
    County
Baltimore, Maryland

Valérie Gabelica
Mass Spectrometry Laboratory
University of Liège and FNRS
Liège, Belgium

Chunang Gu
Environmental Toxicology Graduate
    Program
University of California
Riverside, California

Nathan A. Hagan
Department of Chemistry and Biochemistry
University of Maryland Baltimore County
Baltimore, Maryland

Kristina Håkansson
Department of Chemistry
University of Michigan
Ann Arbor, Michigan

Haizheng Hong
Environmental Toxicology Graduate
    Program
University of California
Riverside, California

Mahmud Hossain
Department of Chemistry
University of Cincinnati
Cincinnati, Ohio

Farid Jahouh
Fisheries and Oceans Canada
Science Branch
Newfoundland, Canada

and

Department of Chemistry
Memorial University of Newfoundland
Newfoundland, Canada

Nicolas Joly
Department of Chemistry
Rue de l'Université
Béthune, France

Filip Lemière
Department of Chemistry
University of Antwerp
Antwerp, Belgium

Patrick A. Limbach
Department of Chemistry
University of Cincinnati
Cincinnati, Ohio

Patrick Martin
Department of Chemistry
Rue de l'Université
Béthune, France

Judith Miller-Banoub
Fisheries and Oceans Canada
Science Branch
Newfoundland, Canada

and

Department of Chemistry
Memorial University of Newfoundland
Newfoundland, Canada

Russell P. Newton
Biological Sciences, School of the
    Environment and Society
Swansea University
Singleton Park, Swansea, United Kingdom

Keiichiro Ohara
Max Mousseron Institute of Biomolecules
University of Montpellier
Montpellier, France

Michael Smietana
Max Mousseron Institute of
    Biomolecules
University of Montpellier
Montpellier, France

Natalia Y. Tretyakova
Department of Medicinal
    Chemistry
University of Minnesota
Minneapolis, Minnesota

Antonio Triolo
Menarini Ricerche Spa
Florence, Italy

Kevin B. Turner
Department of Chemistry and
    Biochemistry
University of Maryland Baltimore
    County
Baltimore, Maryland

Jean-Jacques Vasseur
Max Mousseron Institute of
    Biomolecules
University of Montpellier
Montpellier, France

Yinsheng Wang
Department of Chemistry
University of California
Riverside, California

Jiong Yang
Department of Chemistry
University of Michigan
Ann Arbor, Michigan

1 Overview of Recent Developments in the Mass Spectrometry of Nucleic Acid and Constituents

Joseph H. Banoub, Judith Miller-Banoub, Farid Jahouh,
Nicolas Joly, and Patrick Martin

CONTENTS

1.1   INTRODUCTION

Nucleic acids and their components play an important role in a variety of fundamental biological processes.[1] The elucidation of the molecular structure of DNA in 1953 by Watson and Crick led to the so-called central dogma of molecular biology.[2] This dogma proclaims that DNA is transcribed to form a messenger RNA (mRNA) of a complementary sequence that directs its own replication, and the sequence of nucleobases in the RNA is then translated to the corresponding sequence of amino acids to form a protein.[2] In addition to mRNA, two other RNA species, transfer RNA (tRNA) and ribosomal RNA (rRNA), are required for this translation process. Nucleic acid research has consequently had a most profound impact on molecular biology, initiating an avalanche of new disciplines. The field of genomics has evolved over the last few years, culminating in the sequencing and characterization of the entire human genome.[3] Current challenges include the unraveling of the relationship between the genome and the proteome. The genome can be used to predict the total potential proteome, including all of the modifications that can be carried out after the initial translation from mRNA to protein. Mass spectrometry (MS) has had a major role in elucidating the regulatory mechanism involved in the progression from gene to functional protein.[4]

Nucleotides represent the primary level for the storage and transmission of genetic information. They are not only crucial to the organization and execution of protein synthesis but also function as secondary messengers, metabolic regulators, and components of vitamins.[5] They also act as high-energy intermediates driving thermodynamically unfavorable enzyme-catalyzed reactions. Two classes of heterocyclic nitrogenous bases are found in nucleic acids; the two purine derivatives, adenine (A) and guanine (G), and the three pyrimidine derivatives, uracil (U), thymine (T), and cytosine (C). There have been many studies devoted to the separation, isolation, and purification of nucleic acids from biological samples as well as the separation and identification of constituents of hydrolyzed RNA and DNA at the nucleotide, nucleoside, and nucleobase levels.[6] Elucidation of the MS fragmentation patterns for simple nucleosides and nucleotides is a key step in developing an understanding of the MS fragmentation patterns of more complex oligonucleotides and nucleic acids. Seminal reviews by McCloskey and collaborators have established the fundamental importance of MS techniques and applications to nucleic acid research and other related biotechnological fields.[7–9]

The MS analysis of oligonucleotides is of primary importance since establishing the size, purity, and sequence of nucleic acids is a prerequisite to their use as molecular probes in biomedical science. Significant progress in the area of accurate mass determination, sequencing, and study of noncovalent interactions has been made possible by the use of MS ionization techniques such as matrix-assisted laser desorption/ionization coupled to time-of-flight (MALDI-TOF) and electrospray ionization (ESI) coupled to conventional analyzers and to tandem mass spectrometric (MS/MS) instruments.

Nucleic acids and their constituents, due to their polarity and thermal lability, have presented a considerable challenge for analysis by MS. Mixtures of a variety of nucleobases and nucleosides can be analyzed by combined gas chromatography-mass spectrometry (GC-MS), although it is usually necessary to convert them to more volatile derivatives. The GC-MS analysis of more complex nucleosides, with polar modifications in either the glycone (glycosyl portion) or the nucleobase (aglycone), had not been very successful.[10] However, with the development of newer ionization methods, it has become feasible to perform, without any derivatization, MS analysis on nucleosides, nucleotides, and other large related oligonucleotides (DNA and RNA).

Since the 1960s, it has been known that the mutagenic activity of many substances is linked to the degree with which they covalently bind to DNA. The analysis and measurement of these carcinogen–DNA adducts have been proposed as a means of evaluating the extent of human exposure to specific carcinogens and of monitoring the presence and effects of these specific carcinogens in the environment.[6,11a–e,12] Highly sensitive and structurally informative analytical techniques are thus needed for the general characterization of DNA modifications. MS has the potential to provide structural information and, as such, has played an important role in the structural elucidation of covalently modified nucleic acids.

Many ionization modes, such as electron ionization (EI) and chemical ionization (CI), have been reviewed and employed for the structural elucidation of nucleic acids, their constituents, and their covalently modified nucleic acids.[7–10,12] All of these techniques provide, when coupled with MS/MS, structural information for either single nucleobase-adducts or derivatized single nucleobase adducts. In general, these methods suffer from either a lack of sensitivity necessary for the analysis of trace levels of components derived from small amounts of DNA, or the need to use chemical derivatization techniques to increase volatility and decrease thermal lability. The new technologies developed in the last century, such as fast atom bombardment-mass spectrometry (FAB-MS), MALDI-TOF-MS, ESI-TQ-MS (triple quadrupole, TQ), quadrupole-ion-trap (QIT), and Fourier transform ion cyclotron resonance (FT-ICR), hybrid instruments such as the quadrupole orthogonal time-of-flight (Q-TOF), quadrupole-ion-trap-time-of-flight (QIT-TOF), and quadrupole-ion-trap-Fourier transform-ion cyclotron resonance-tandem mass spectrometry (QIT-FT-ICR-MS/MS) instruments have allowed the analysis of nonderivatized adducts of nucleosides and nucleotides. In addition, capillary electrophoresis (CE) coupled to MS/MS has been used for adduct detection.[6,11a–e,12]

Thus, the intense interest in the MS analysis of nucleic acids and their components has produced a wealth of published work. The major objectives of this chapter are to put into perspective the progress in current biotechnological research in this area, highlight the most popular ionization methods, and illustrate the diversity of strategies employed in the characterization and sequencing of DNA and RNA oligomers, nucleosides, nucleotides, and adducts of biomedical importance.

1.2 GAS-PHASE TECHNIQUES

The analysis of nucleic acids and derived products isolated from complex biological matrixes is synonymous with analysis at very low concentrations. Thus, tremendous efforts have been made to develop and implement MS methodologies for the study of the monomeric constituents of nucleic acids.[4,13–15a–c] Biemann and McColskey elucidated detailed fragmentation pathways for the major ions present in both electron ionization (EI) and chemical ionization (CI) mass spectra of simple purine and pyrimidine nucleosides.[16–18] Derivatization, which is of value in many cases, requires an additional sample manipulation step, and may produce unwanted and unexpected side products.[19–22]

1.2.1 ELECTRON IMPACT IONIZATION-MS FOR THE DETECTION OF DNA ADDUCTS

So far the biological significance of cytosine methylation has not been fully understood, but there is growing evidence suggesting a link between the development of human cancer and the perturbation of the methylation patterns due to the lack of DNA cytosine methyltransferase (Mtase) substrate specificity. A novel *in vitro* assay called mass tagging was developed, which permitted quantification of the DNA substrate preferences of cytosine Mtases.[23] This approach involved the labeling of target cytosine residues in the synthetic DNA duplexes with stable isotopes such as [15]N. Methylation was then measured by GC-EI-MS to determine the formation of 5-methylcytosine (5 mC) and DNA. Substrate selectivity was then determined through MS by the absence or presence of the label in 5 mC. This mass-tagging approach has proven to be a powerful tool for examining the substrate selectivity of cytosine DNA Mtases and could easily be adapted to the study of other kinds of selectivity.

1.2.2 ELECTRON-CAPTURE IONIZATION-MS FOR THE DETECTION OF DNA ADDUCTS

Electron capture (EC) is a sensitive ionization technique for MS analysis, providing selectivity toward electrophoric compounds. Advances in instrumentation have led to a more widespread application of this method in biomedical and pharmaceutical analysis.

Fedtke and Swenberg contributed a short review of MS techniques for DNA adducts, with emphasis on electron-capture negative ion chemical ionization-mass spectrometry (EC-MS) detection, especially with regard to their method for N^2,3-ethenoguanine.[24] They included in their work, a summary of earlier and more general reviews on the detection of DNA adducts as well as the normal constituents of DNA. Similarly, Giese et al. emphasized their own work in a short review on the same general subject, but with more attention to method development and instrumental aspects of EC-MS.[25] Detection of DNA adducts by means of different types of ionization techniques has also been reviewed.[26-29]

Electrophores are compounds that can be readily detected by the two closely related gas-phase detection techniques of EC-MS and electron-capture detection (ECD). These two analyses require that the electrophore be both thermally stable and volatile.[30,31] Usually, the compound is delivered to one of these detectors as a peak eluting from a GC column, although a direct insertion probe and a liquid chromatography (LC) belt interface can also be linked to the EC-MS instrument.[30,31]

In ECD, a gas-phase current of electrons is monitored, and the electrophore reduces the current by capturing one of the electrons. Similar EC takes place in the EC-MS, but under partial vacuum conditions, and the ion(s) formed from the analyte are measured instead. Once an electrophore captures an electron, it becomes an anion radical and it can either exist long enough to reach the detector or immediately fragment.

For standards, the two detectors are similarly sensitive, but the greater specificity of the EC-MS makes it considerably more powerful for biological samples. Aside from special analytes that are unusually volatile or nonvolatile, it is difficult to reach the low picogram level by GC-ECD when real samples requiring electrophore derivatization are tested.[32,33] Largely, this is because electrophoric contaminants are present or formed in the reagents and solvents during sample preparation.[34]

DNA adducts, due to their lack of volatility and thermal stability, are not good electrophores for detection by GC-EC-MS. Therefore, as part of sample preparation, one or more chemical reactions are necessary to make the DNA adducts available for detection by GC-EC-MS.

It must be kept in mind that the overall electrophore response in the GC-EC-MS is a combination of the recovery of the compound from the GC part into the EC ion source; the inherent EC properties of the compound; other events (e.g., wall effects) in the ion source; and the transmission efficiency (including the lifetime) of the electrophore anion to the detector.[35] A strong electrophore can be detected as a pure standard at the low attomole level on a modern instrument.[36]

Typically, the DNA adduct is converted into an electrophore, which is detected as either a parent anion radical (no dissociative EC) or as an exclusive anion fragment (dissociative EC), which is formed along with the corresponding neutral. Either way, the electrophore that is typically selected forms a single ion (aside from the associated isotope peak) maximizing the sensitivity. Ideally, the electrophore-labeled DNA adduct would give two or three anion fragments in similar amounts, and their ratio would be checked for each sample as an extra guard against interferences. This would be worth the slightly higher detection limit.

Although some standards of derivatized DNA adducts can be detected by GC-ECD (Table 1.1), DNA adducts that are derived from biological samples are usually only detectable by GC-EC-MS.

TABLE 1.1
DNA Adducts and Monomers Detected by Electrophore-Labeling GC-MS or GC-ECD (Detection is by GC-EC-MS Unless Indicated Otherwise)

Type of Adduct	Target on DNA	Sample Preparation Procedure	Yield (%)	Reported Detection Limit Standard	Reported Detection Limit DNA	Reference
4-Aminobiphenyl	C8-G (mostly)	DNA/NaOH/hexane/PFP	—	—	0.32 in 10^8 (100 g of DNA)	38,39
2-Aminofluorene	C8-G	NH_2NH_2/PFBzal	—	25 amol (S/N = 10)	—	40,41
Benzo[a]pyrene	N2-G (mostly)	DNA/HCl/heat/KO_2/PFBz/Si-SPE	22	14 amol (S/N = 20)	5 in 10^7 (100 g of DNA)	36,42
Cytosine	—	DNA/HCl/heat/CH_3I/C_6H_6	—	1 fmol	—	43
	—	PFB/Me/GC-ECD	35	7 fg	—	44–46
Etheno	$N^2,3$-G	DNA/HCl/IEC/C18-Si-SPE/PFBz/Si-SPE	—	190 amol (S/N = 10)	60 fmol/mol of G for 1 mg of DNA	24,47
Ethyl	O^2-T	PFBz/TLC/Si-FC	56	25 pg (S/N > 100)	—	48
Ethyl	O^4-T	PFBz/TLC	57	—	—	48,49
Ethyl	O^4-Tdn	PFBz/TLC/GC-ECD	55	0.44 fmol	—	30,50
Hydroxy	C8-dG	DNA/nucleases/TFA-acid/hydrazine/C18-SPE/acetyl/hydrolysis/ PFBz/Si-SPE	—	1.8 fmol	4 in 10^6 for 30 g of DNA	51,52
Hydroxy	C8-dG	Urine/C18(OH)-Si-SPE/acetyl/PFBz/HPLC/ EtOAc	—	1.8 fmol	1.8 pmol in 50 L of urine	51,52
Hydroxy	CH_3 on T	PFBz/TLC/Si-FC/PFBz/TLC	—	200 zmol (S/N = 3)	—	53
4'-Hydroxybutyl	O^6-G	DNA/HCl/Ab/PFBz/TMS	—	—	0.12–3 mol/mol of G	54
2'-Hydroxyethyl	N7-G	DNA/heat/HCl/HNO_2/PFBz/Si-SPE	9.7	1.3 amol (S/N = 10)	100 pg in 100 g of DNA	36,55,56

Source: Adapted from Kellersberger, K.A. et al., *J. Am. Soc. Mass Spectrom.* 16, 199, 2005. With permission.

However, it has been found that the measurement of 5-methylcytosine in DNA is usually performed by GC-ECD, but this is not an exception, since this naturally occurring, minor nucleobase is not actually a DNA adduct.[37] It is present in DNA in high concentrations (about 1 in every 100 cytosines in human DNA), relative to the dose of DNA adducts generally encountered in "natural" biological samples (1 adduct in $\geq 10^6$ normal nucleobases).

1.2.3 PHOTOIONIZATION DETECTION OF NUCLEIC ACID BASE CONSTITUENTS OF DNA FROM SPACE

Photoionization is the physical process in which an incident photon ejects one or more electrons (photoelectrons) from an atom, ion, or molecule. These ejected photoelectrons carry information about their preionized states. Photons with energies less than the electron binding energy may be absorbed or scattered, but will not photoionize the atom or ion.

The photoionization mode gives rise to a photoionization pattern, which has very specific characteristics with respect to spatial distribution, density, and relative yields of photolytic species. A particular photoionization mode is also very specific in terms of the ultimate chemical and structural effects induced to a given dielectric material. The concept of photoionization *mode* refers to very distinct interaction regimes of a high power laser pulse with a dielectric material; these are governed by a specific set of laws, and controlled by a specific set of parameters. For this reason, this concept is very well defined. It has been established that there are four fundamental photoionization modes based on four fundamental optical effects. These give rise to four very distinct interaction regimes: single-photon (SP) mode, filamentary (F) mode, optical breakdown (OB) mode, and below optical breakdown threshold (B/OB) mode.

The vacuum ultraviolet (VUV) photophysics and photochemistry of the pyrimidine and purine nucleic acid base constituents of DNA are of considerable interest in view of the possible delivery of these molecules from space to early Earth, and the role they could have played in the origin and development of life on our planet.[38]

It has been reported that nucleobases have been found in meteorites.[39–42] In an astrophysical context, the observation of the important nucleobases by radioastronomy requires initial laboratory studies on their gas-phase microwave spectra. These studies have indeed been carried out for the major tautomers of the three nucleic acid bases studied: adenine, thymine, and uracil.[43–45] Another possibility for astrophysical observation is the infrared spectra. This has also been measured in the laboratory in the gas phase and in low-temperature matrices, both phases being relevant to possible astrophysical measurements.[46–50] Gas-phase electronic spectra are also known, but the observed UV absorption or fluorescence bands would be difficult to measure and identify in astrophysical contexts.[51–55] All of these spectroscopic studies have enriched our knowledge of the structure of these nucleobases and their tautomeric variants. MS measurements of cometary grains were attempted on Halley's comet in 1986, and there have been some specific attempts to detect purine nucleobases in other space missions to comets.[56,57]

Apart from several photoelectron spectral studies, there have been very few studies of the photoionization phenomena of adenine, thymine, and uracil.[58–66] Dudek and a few others have carried out a small number of EI-MS studies in the 20–70 eV photon energy range on unmodified nucleobases; however, these studies showed that the adiabatic ionization energies measured were uncertain and no dissociative photoionization studies reported on these nucleobases.[17,67–75] Hans-Werner Jochims et al., as well as other researchers, have studied the photoionization MS of adenine, thymine, and uracil in the 6–22 eV photon energy range.[76–78] They have indicated that the energetic fragments resulting from dissociation of the nucleobases can cause subsequent damage in biological systems. DeVries et al. have studied the fragmentation of uracil and thymine induced by collision with slow multiply charged Xe (q^+, q = 5–25) ions.[79] Complete breakdown into atomic and diatomic fragments occurs via Coulomb dissociation under these conditions.

1.3 DESORPTION TECHNIQUES

Volatility restrictions for nucleic acid analysis have been circumvented by the use of desorption techniques that allow the production of molecular ions from thermally fragile biomolecules. The greatly increased mass range that has been made accessible by new ionization techniques has marked much of the progress in the field of MS. These key ionization techniques include field desorption (FD), Cf-plasma desorption (PD), desorption chemical ionization (DCI), secondary ion mass spectrometry (SIMS), MALDI, and FAB, with significant contributions being made by each of those methods.[80–91]

1.3.1 MALDI-TOF-MS TECHNIQUE

Concurrent with the initiation of the human genome-sequencing project, the research groups of Karas and Hillenkamp, and Tanaka and coworkers independently developed MALDI-TOF-MS.[95,96] Since its inception in 1987, it has been used for rapid DNA and RNA oligonucleotide sequencing and sizing studies, and has the potential to replace conventional gel electrophoresis.[92–94] For MALDI, the samples are usually prepared by cocrystallization of the analyte with a large molar excess of a photoabsorbing organic compound. Laser desorption and ionization of the embedded analyte provide a relatively gentle method of generating large intact molecular ions in the gas phase.[95–98] The ions produced in this single desorption event are then separated and detected effectively by simple TOF-MS. Therefore, MALDI-TOF-MS has the advantage of both speed and accuracy for the determination of the mass of large biomolecules such as proteins, DNA, and RNA oligomers.[98,99] Structural information deduced from ion fragmentation is also important, and two different types of fragmentation events have been observed in MALDI-TOF mass spectrometers. The first is the metastable decomposition occurring in the field-free region of the TOF flight tube, called post-source decay (PSD).[97,98] The second type is the in-source collision-induced dissociation (CID), and can be performed on MALDI-TOF-MS by employing higher potential differences in the source to induce energetic collisions. However, induction of fragmentation of ions prior to entering the flight tube decreases the potential for mixture analysis because individual components are dissociated prior to an ion selection stage.[100]

There has been a huge research effort directed to at the MS-based characterization of oligodeoxynucleotides and oligoribonucleotides, including the sequencing of DNA by direct MALDI-MS analysis, RNA sequencing, analysis of DNA tandem repeats, detection of single nucleotide polymorphisms (SNPs) using peptide nucleic acid (PNA) probes, genetic diagnosis, genotyping, mutation detection, and probing virus structures.[101–107]

Initially, the MALDI-MS method was predominantly applied in the analysis of proteins and peptides, with most studies using ion extraction, usually in the negative ion mode. To extend the MALDI-MS technique to the sequencing of DNA, it became clear that the choice of the matrix and the laser wavelength was crucial.[108,109] The difficulties encountered in the early application of MALDI-MS to nucleic acids have been reviewed in detail.[110] In 1995, MALDI instrumentation took a quantum leap when Vestal integrated the principle of delayed extraction into the MALDI process.[111] This step dramatically improved the resolution of the spectra, and the use of MALDI instruments began to move from the hands of a few specialists into the wider research community.

1.3.1.1 MALDI-TOF-MS Characterization of Nucleosides

The techniques used in the analysis of nucleosides include their isolation from urine by phenylboronate affinity gel chromatography, which takes advantage of the reaction of phenylboronate with the vicinal OH groups of the ribonucleosides.[112] Separation of the nucleosides is achieved equally well by reverse phase (RP)-HPLC or capillary electrophoresis in the micellar electrokinetic chromatography (MEKC) mode.[112–114] Identification of the modified nucleosides is now possible by MALDI-TOF-MS.

The MS analysis of urinary-modified nucleosides as tumor markers was performed by Liebich et al.[115] The analysis includes the isolation of the nucleosides from urine with phenylboronate gel and their separation and quantitation by high-performance liquid chromatography (HPLC) on C18 columns or by CE on uncoated columns applying a sodium dodecyl sulfate–borate–phosphate buffer. Identification of the nucleosides is performed with MALDI-TOF-MS including PSD spectra.

It has been shown that RP-HPLC and CE are equally suited for separation and quantitation of the nucleosides.[113,114] For the modified nucleosides, the reproducibilities are usually expressed by the coefficients of variation. For clinical studies on children with leukemia (as well as other malignancies) and on women with breast cancer, the nucleosides are measured by RP-HPLC. A typical example of urinary nucleosides separated by HPLC is shown in Figure 1.1.[115]

With the MALDI-TOF-MS and PSD spectra, the identities of normal and modified urinary nucleosides are proven. Examples are given in Figures 1.2 and 1.3 showing the first analysis of nucleosides from urine by MALDI-TOF-MS and PSD-MALDI. The PSD spectra of the nucleosides showed a characteristic cleavage between the nucleic base and the sugar moiety.[115] The resulting nucleic base peak is most informative for the identification of the nucleoside. Other fragments enhance the selectivity as they are characteristic for different nucleosides, except those with isomeric nucleic bases that produce similar PSD spectra.

1.3.1.2 MALDI-TOF-MS Characterization of Nucleotides

Methods involving either anion exchange or ion-pairing reagents are very suitable for separating polar nucleotides. However, the use of high salt concentrations and involatile ion-pairing reagents in these methods precludes the use of MS for detection. UV detection usually results in relatively poor sensitivity and selectivity. Both direct and indirect approaches for nucleotide analysis have been applied to circumvent the incompatibility with MS. Indirect methods consist of a separation of the nucleotides on an ion-exchange cartridge or using HPLC. The fractions containing the nucleotides are subsequently dephosphorylated using alkaline or acidic phosphatase. After dephosphorylation, the remaining nucleoside can be quantified using a radioimmunoassay (RIA), UV, or MS detection.[114–121] The main drawback of these indirect methods is the labor-intensive extra step in sample pretreatment, which is a possible source of error. Claire et al. have used eluents with low concentrations of tetrabutylammonium, the involatile ion-pairing reagent, in combination with phosphate on a microbore column.[122] Other direct methods have been developed using the volatile ion-pairing reagent dimethylhexylamine (DMHA).[123,124] Separation of nucleosides and nucleotides using a porous graphitic carbon column has also been described.[125] Weak anion-exchange (WAX) LC is another approach for the separation and direct determination of nucleotides with MS detection.[126] This method uses a pH gradient instead of a salt gradient to separate and elute the nucleotides from a WAX column. Veltkamp et al. successfully used this approach for the quantification of gemcitabine triphosphate.[127]

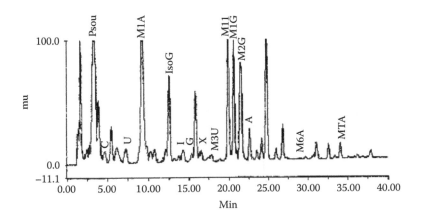

FIGURE 1.1 HPLC separation of urinary nucleosides. UV detection at 260 nm.

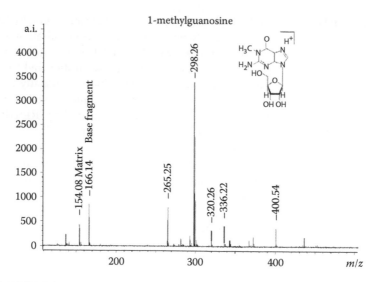

FIGURE 1.2   MALDI-TOF-MS spectrum of 1-methylguanosine.

1.3.1.3   Characterization of Oligonucleotides

1.3.1.3.1   Practical Aspects of Oligonucleotides Analysis

A systematic study of the ionization of the four standard nucleosides in a MALDI-TOF-MS has been carried out by Zhang and coworkers.[128] They confirmed, using several common matrixes, that ionization of DNA by the MALDI procedure was usually dominated by protonation and deprotonation of the bases and did not significantly involve the backbone. It was shown that the low ion signal, previously reported for poly-G DNA in MALDI-MS, was due to the difficult ionization of guanine. Protonation was dominated by a preprotonation step before laser ablation, and deprotonation was controlled by the thermal reaction.[128]

The analysis of oligonucleotides by MALDI-TOF-MS is usually performed under negative ionization conditions, following ammonium ion-exchange chromatography and with the addition of

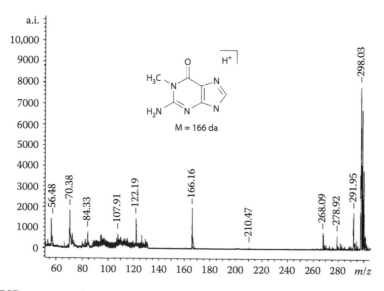

FIGURE 1.3   PSD spectrum of mass *m/z* 298 of 1-methylguanosine.

ammonium buffers to the MALDI matrix. The molecular ion region can be very complex due to the variable degree of ammoniation of the phosphate backbone. This causes a decrease in sensitivity and introduces ambiguity into the assignment of the relative molecular mass of the sample.

It was found that the use of an H+ ion-exchange resin *in situ* permitted the removal of alkali-metal ions from the oligonucleotide phosphate backbone, and this gave a gain in sensitivity of 1–2 orders of magnitude over previous MALDI-TOF methods.[129]

In other cases, when DNA was not significantly contaminated with alkali metals, NH_4^+ organic salts, as matrix additives, were used to replace alkali-metal ions in the polyphosphate backbone chain. Oligonucleotides can be converted into the ammonium form by the addition of either ammonium citrate or ammonium tartarate.[129] In-line microdialysis has also been used for rapid desalting prior to MALDI-MS.[130]

A novel approach for sample preparation was developed by Smirnov and collaborators.[131] This method was based on extracting the DNA out of the solution, onto a solid surface chip containing an attached binding site. Thus, desalting and concentration can be performed rapidly in a single step. The addition of the MALDI matrix material, after the removal of the supernatant solution, released the DNA from the surface, allowing the required cocrystallization on the support. The MS analysis was then performed directly from this support.[131] A similar method for the desalting of oligonucleotides (and peptides) involved the use of a 96-well solid-phase extraction plate packed with a reverse-phase sorbent that retained the biopolymer analytes, and any unretained inorganic ions can be washed away with deionized water.[132]

A study by Bentzley and coworkers revealed a variation in the robustness of oligonucleotide strands when subjected to repeated freeze–thaw cycles.[133] They showed, by using MALDI-MS to monitor purity, that oligonucleotide strands can suffer significant degradation when taken through repeated freeze–thaw cycles. The extent of degradation was dependent on base composition, solute concentration, strand length, and thawing conditions.

Acceptable mass resolution and sensitivity for DNA oligomers of more than 200 bases have been a challenge for MALDI-MS. Routine detection of these large DNA oligomers is complicated by both adduct formation and fragmentation, and therefore the development of new matrixes and MALDI experimental strategies has received much attention.

The matrix, 3-hydroxypicolinic acid (HPA), introduced in 1993, showed significant improvement, in terms of mass range, signal-to-noise ratio, and ability to analyze mixed-base oligomers up to 67 nucleotides, when compared directly with a set of 47 other compounds including most of the matrices previously used.[134] There is evidence that picolinic acid is an even better matrix than HPA, as it permitted the detection of oligonucleotides with up to 190 bases.[135]

Compounds that contain an intramolecular hydrogen bond such as 2-hydroxyacetophenones, 2-hydroxybenzophenones, salicylamide, and related species are another group with matrix potential.[136,137]

A unique matrix system, consisting mostly of 4-nitrophenol, was shown by Lin and collaborators to be very effective for the analysis of large DNA oligomers, especially when a cooled sample stage was used to prevent the sublimation of this matrix under vacuum.[138] UV laser desorption from this matrix allowed routine detection, at the picomole level, of DNA oligomers containing up to about 800 nucleotides. The effectiveness of this matrix was demonstrated when a double-stranded (ds) DNA oligomer, having more than 1000 base pairs (bp), was seen as a denatured single-stranded species with a molecular ion mass greater than 300 kDa.[138]

Lack of homogeneity in a sample makes it difficult to routinely optimize the laser fluence, and some spots can receive excess energy that results in a loss of sensitivity due to the formation of metastable ions.

It was found that addition of a sugar such as fructose or fucose to the matrix minimizes the effect of excess laser energy and allows detection, for example, of the 9 Da mass difference, which characterizes an A–T mutation.[139]

The use of a liquid matrix, such as glycerol, has been known to give better reproducibility, but since glycerol is nonabsorbing in the UV region, a second component must be included with it. The binary mix of 4-nitroaniline and glycerol is effective for both positive- and negative-ion formation of oligonucleotides and other biomolecules.[140] Hillenkamp and coworkers showed that IR-MALDI with a glycerol matrix gave excellent results for large dsDNA.[141–143] They observed, in the high-mass range, very little fragmentation with an IR laser compared with UV-MALDI, thus extending the mass limit for IR-MALDI to ca. 500 kDa with sensitivity in the subpicomole range and with good reproducibility.

An ingenious approach to desalting involved the use of 3,4-diaminobenezene as a matrix material dispersed in a sol–gel, the combined material acting as the sample support. Sodium ion adduction is effectively suppressed, allowing the detection of a 24-mer oligonucleotide at the 20 fmol level.[144] Similar results were reported for a sol–gel sample substrate doped with a crown ether.[145]

Another advance in matrix design was the use of a mixture of HPA and pyrazinecarboxylic acid (PCA). Deng and coworkers used this matrix with model synthetic oligonucleotides to obtain MALDI-TOF mass spectra with isotopic resolution for DNA segments up to 10.5 kDa in mass.[146] Equally impressive is the detection of separate ion peaks for both components of a mixture of two 23-mer DNA segments with a 7-Da difference. With respect to reproducibility, resolution, signal-to-noise ratio, and tolerance to metal salts in DNA analysis, this mixed HPA–PCA matrix is superior to HPA alone.[146]

1.3.1.3.2 Analysis of Synthetic Oligonucleotides

Ferris and collaborators have demonstrated that oligomerization of activated nucleic acids can be achieved by using the naturally occurring catalyst, montmorillonite.[147–151] Montmorillonite is a clay mineral, containing several chemical elements (Si, O, Na, K, Al, and Mg), and its molecular structure is a highly organized layer type. It has been used successfully for the structural characterization and catalytic oligomerization of a variety of activated nucleotides.[147–152] Analysis of the reaction products involved their extraction from the montmorillonite followed by chromatographic separation and characterization. The latter was usually achieved by gel electrophoresis or by the HPLC method using a UV detector.[148,150,151] This approach was not ideal, because neither method could "cleanly" separate oligonucleotides with similar molecular weights (MWs), and the UV detector had low sensitivity toward large oligomers formed in very small quantities or concentrations. Therefore, this approach did not provide definitive evidence for oligomers of more than ~12 nucleotides.

Zagorevskii et al. studied the development of the direct MALDI analysis of synthetic oligonucleotides formed by the starting reagents catalysis on minerals.[153] Adenosine 5′-phosphorimidazolide (ImpA; I) and uridine 5′-phosphorimidazolide (ImpU; II) were synthesized via a published procedure.[154] A simplified diagram of a typical chemical experiment and reaction components are shown in Scheme 1.1.[153] The MALDI-MS (– ion mode) of the desalted mixture of oligonucleotides from ImpA was recorded in both linear and reflectron modes. The reflectron mode MALDI-MS displayed several families of peaks separated by 329 Da due to a (pA) unit in which two series of ions dominated the mass spectrum; these series correspond to deprotonated linear, (pA)n, and cyclic, c-(pA) n, fragment ions, with a characteristic mass difference of 18 Da. The analysis of the samples R in the linear mode allowed for the detection of oligomers containing more than 30 nucleotides. It should be noted that high-molecular-weight oligonucleotides generally have a lower probability of detection by MALDI than their lower-molecular-weight analogs.[110]

1.3.1.3.3 Structural Studies of DNA and RNA

Compared with peptides, the overall sensitivity of DNA analysis by MALDI-TOF-MS is about 100 times lower due to the subsequent fragmentation of the ionized molecules, adduct formation, and low ionization efficiency.

An improvement in the detectability and stability of DNA analysis by MALDI-TOF-MS has been reported by Berlin and Gut.[155] They modified the oligonucleotides to neutralize all but one of the

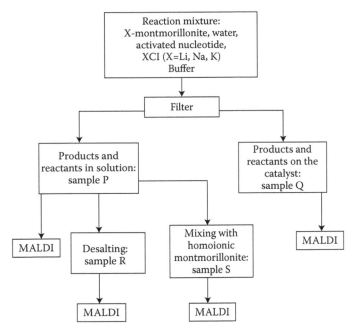

SCHEME 1.1 Diagram of a typical chemical experiment and reaction components.

backbone phosphates, thus creating a species with a single negative charge. Using the same nonprotonating matrix, the enhancement in sensitivity achieved by this relatively simple modification was comparable to that observed in the earlier work for DNA carrying a single positive charge.

DNA mapping studies have shown that careful matrix selection and technical advances such as time-lag focusing have provided excellent results for mixtures of different oligonucleotides.[156] For example, Figure 1.4 shows the quality of the MALDI-TOF spectra obtained for a mixture of oligonucleotides derived from the plasmid pBR322 (10, 11, 12, 13, 14, 15, 19, 20, and 50 nucleotides long), particularly when time-lag focusing was employed.

Since the early 1980s, solid-phase synthesis [phosphoramidite (PA) approach], using controlled-pore glass (CPG) as the solid support, was the method of choice for producing oligonucleotides.

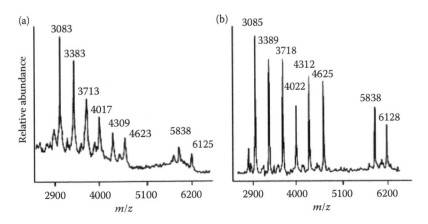

FIGURE 1.4 MALDI-TOF-MS spectra of a mixture of species (containing 10, 11, 12, 13, 14, 15, 19, and 20 nucleotides) derived from plasmid pBR322. Spectra were obtained (a) as negative ions without time-lag focusing and (b) as positive ions with time-lag focusing.

It was standard practice to have the growing oligonucleotide attached to the solid phase by a succinyl linker group.

Vasseur and collaborators tried to evaluate the efficiency of the solid-phase synthesis by MALDI-TOF-MS by employing a photolabile linker group that could be cleaved from the solid support by laser irradiation, thereby permitting simultaneous oligonucleotide cleavage and ionization.[157] This procedure, however, has proven less effective for the analysis of longer oligonucleotides.

MALDI analysis of succinyl-linked oligonucleotides generated fragment ions in both negative and positive modes, which allowed direct access to the nucleotide sequence and identification of the internucleosidic linkage.

The mass ladder generated in the MALDI-TOF-MS of 3′-CPG-supported cyanoethylphosphotriester oligonucleotide has been used to sequence the complete oligonucleotide (except the last two residues at the 5′ end) (Figure 1.5).[157] It has also been used to detect and identify modified residues and abnormal internucleosidic linkages.[158,159]

Hillenkamp and colleagues, using both static and delayed ion extraction, investigated IR-MALDI-MS with a CO_2 laser (10.6 μm).[141] Compared with an Er:YAG infrared laser ($\lambda = 2.94$ μm, $t = 90$ ns) and a frequency-tripled Nd:YAG UV laser ($\lambda = 355$ nm, $t = 15$ ns), fewer metastable ions were observed with the CO_2 laser, especially operating in the reflectron mode of the spectrometer. Thus, the mass spectra of large biomolecules (up to several hundred kilodaltons) could be better analyzed with a reflectron TOF-MS. This technique has been used in the analysis of gel-separated and electroblotted proteins, which were desorbed directly from a polyvinylidene fluoride (PVDF) membrane, and a dsDNA of 515 bp.[141]

Taranenko et al. demonstrated that UV-MALDI-TOF-MS could be used to determine the MW of polymerase chain reaction (PCR) products of intact 16S rRNA regions that were up to 1600 nucleotides in length and also profile their restriction digests.[160]

It has been established that the MALDI-TOF-MS of DNA multiplex assays using a photocleavable peptide-oligonucleotide conjugate was an attractive method, given that the unique mass of the peptide portion could serve as a mass marker label. The synthesis and characterization of three photocleavable peptide–DNA conjugates have been described by Olejnik and colleagues.[161] These conjugates were evaluated on the basis that the DNA part acts as the hybridization probe and the peptide part acts as a marker for the target DNA sequence. This approach has initiated further studies such as target-binding properties, cellular uptake, and exonuclease stability, including the properties of antisense or therapeutic oligonucleotides.

MALDI-MS can be used to monitor the efficiency of nucleotide coupling during the production-scale synthesis of oligonucleotides.[162,163] The failure species (N-1, N-2, N-3, etc.) were separated from the crude synthetic material by solid-phase extraction and were analyzed by MALDI-TOF-MS. The sequence information was obtained for a series of failure ions, each of which varied by the molecular mass of the specific nucleotide not coupled.[162]

1.3.1.3.4 DNA Sequencing

Pieles et al. first demonstrated DNA sequencing by controlled exonuclease digestion followed by mass spectral analysis of the resulting ladders.[164] Exonucleases, such as snake venom phosphodiesterase (PDase I) and bovine spleen phosphodiesterase (PDase II), sequentially remove mononucleotides from DNA oligodeoxynucleotides by hydrolyzing the phosphodiester bonds. PDase I attacks DNA at the 3′ end and releases 5′-mononucleotides, and PDase II attacks the oligodeoxynucleotide at the 5′ end and releases 3′-mononucleotides. Both of these PDases are nonprocessive enzymes that dissociate from the DNA molecule after the removal of each nucleotide, and bind DNA again to catalyze the release of the next monomer. Time-controlled exonuclease digestion of DNA thus generates a complete set of oligodeoxynucleotide fragments differing from each other by one mononucleotide, which is ideal for MS analysis. The introduction of delayed extraction in MALDI has made it possible to expand this approach to longer DNA fragments due to improved resolution and reduced fragmentation.

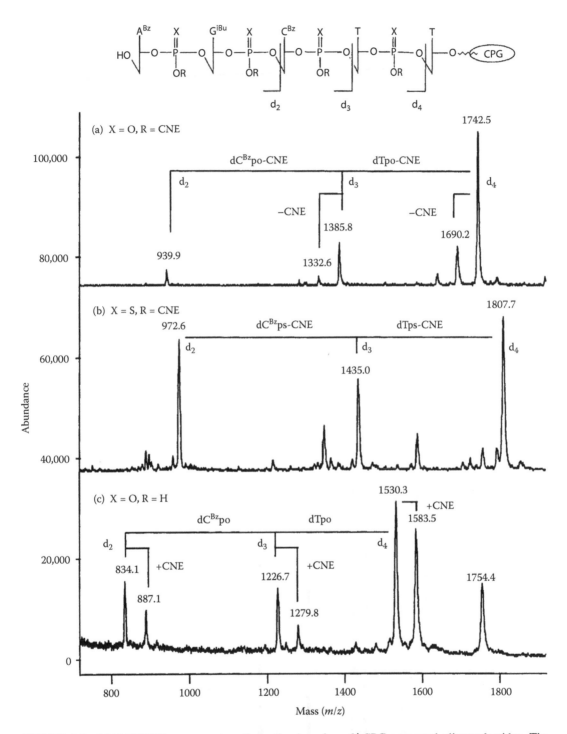

FIGURE 1.5   MALDI-TOF mass spectra of negative ions from 3′-CPG-supported oligonucleotides. The spectrum (a) of a cyanoethyl phosphotriester (poCNE) of 5′-AGCTT-3′ shows successive cleavage at the phosphate from the 3′ end with partial loss of CNE. (b) After the treatment of the 5-mer with 90 mM DBU solution in THF, the CNE group is mostly removed, but the same pattern of fragmentation is observed.

Oligodeoxynucleotides up to 50 bases in length were routinely sequenced using the information from separate digestions with 3′ to 5′ and 5′ to 3′ PDases and analysis by MALDI-TOF-MS.[164]

The relative intensities of ladder peaks in the mass spectra of a series of 5- and 7-mer oligonucleotides showed that the rate of PDase II digestion was influenced by the sequence of bases in the DNA strand.[165] Thus, it was shown that the 5′-terminal A or G nucleotides were cleaved 2–3 times faster than sequences terminating in C or T. This sensitivity of PDase II activity to base type extended to at least the third base in the sequence. Reaction rates could also be determined quantitatively from the time evolution of digestion as illustrated in Figure 1.6 for a typical 7-mer oligonucleotide. The initial MALDI spectrum shows the molecular ion before digestion, and the spectra taken at 3 and 6 min show the clear evolution of ladder peak intensities for each of the species generated by sequential loss of the 5′ residue.[165]

Wada found that the determination of the molecular mass of a PCR-amplified DNA by MALDI-TOF-MS was facilitated by treatment with restriction endonucleases.[166] Using this method, a 664-bp region from the *FAS* gene was analyzed and a two-nucleotide deletion in the *L1CAM* gene of a restriction fragment of 105 nucleotides was detected. Furthermore, the analysis of smaller fragments allowed separate detection of single-stranded oligonucleotides comprising individual digested fragments. This mixture analysis of restriction enzyme digests improved the resolution, sensitivity, and accuracy of MALDI-TOF-MS of DNA, facilitating its application in genetic diagnosis. For illustrative purposes, the calculated masses for a 664-bp segment of the *FAS* gene are shown in Figure 1.7. The molecular masses that were calculated for the sense and antisense strands (both with

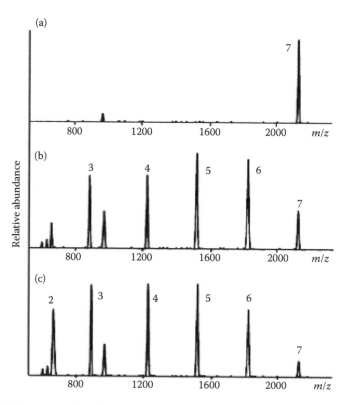

FIGURE 1.6 MALDI spectra of the 7-mer oligonucleotide 5′-AAAGTAA-3′ as a function of the time of digestion with PDase. (a) Obtained before digestion, showing only the molecular ion. (b and c) Obtained after 3 and 6 min digestion, respectively, clearly showing development of the sequence ladder by progressive removal of the 5′-nucleotide.

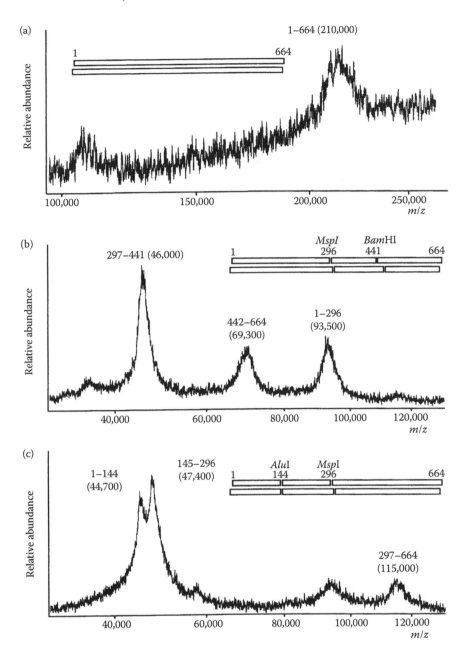

FIGURE 1.7 Positive-ion MALDI-TOF mass spectra of a 664 bp DNA amplified from the FAS gene. The ion m/z values are given with the sequence numbers of the digest fragment in parentheses. (a) The undigested PCR product. The doubly charged ions are observed at m/z 110,000. (b) Digest produced by restriction endonucleases *Msp*I and *Bam*HI. (c) Digest produced by *Alu*I and *Msp*I. The peak at m/z 93,000 is probably uncleaved DNA corresponding to sequence 1–296.

dA overhanging at the 3′ ends) were 205,966 and 204,780 Da, respectively. In the MALDI-TOF-MS, a broad signal corresponding to these DNA molecules was detected at m/z 210,000 with low intensity (Figure 1.7a). The PCR product was then digested with restriction endonucleases *Msp*I and *Bam*HI, cleaving at positions 296 and 441, respectively. The MALDI-TOF-MS of the digest revealed discrete molecular ion signals for three fragments, although the complementary strands were not resolved in each peak (Figure 1.7b). The signal for the fragment observed at m/z 46,000 was sharp

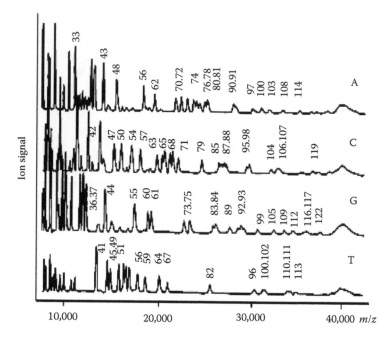

FIGURE 1.8 Negative-ion mass spectra of Sanger sequencing ladders produced by dideoxynucleotide chain termination of the PCR product of a double-strand 130 bp template (+ strand) with a 20 nt reverse primer.

compared with the larger one at *m/z* 93,500, and their peak widths at half-height were 2500 and 4000 units, respectively. The peak at *m/z* 93,000 was probably an intact DNA corresponding to sequences 1–296 (Figure 1.7c).[166]

Sequencing by the Sanger enzymatic method using gel electrophoresis for sequence detection is a relatively slow process, especially for genomic research. Early MS studies were restricted to short oligomers, but the sequencing of longer DNA fragments using MALDI-TOF DNA ladder detection has been demonstrated by Taranenko et al.[167] Using 20-mer primers, they successfully obtained DNA ladders from dsDNA templates of 130 bp (Figure 1.8). Both forward and reverse primers were

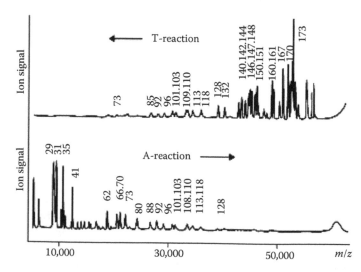

FIGURE 1.9 Negative-ion mass spectra of Sanger sequencing ladders produced by dideoxynucleotide chain termination of the PCR product of a double-stranded 200 bp template.

used, and the MALDI-TOF-MS of both modes could be combined to give a full 200 bp sequence (Figure 1.9). With improved resolution and sensitivity, mass spectrometric DNA sequencing has thus become a very valuable tool for both *de novo* sequencing and resequencing of known genes for mutation detection.

Nordhoff and collaborators reported the application of MALDI-TOF-MS to the rapid sequencing of DNA composed of 15–20 nucleotides using a protocol based on the Sanger enzymatic extension reactions with a ds template DNA.[168] All four sequencing reactions were performed simultaneously in one reaction vial. The sequencing products were separated and detected by MALDI-TOF-MS. The sequence was determined by comparing the measured molecular mass difference with the expected values. One reaction mixture typically includes 300 fmol of dsDNA template, 10 pmol of primer, and 200 pmol of each nucleotide monomer. It should be noted that neither the primer nor any of the nucleotide monomers were labeled. Solid-phase purification, concentration, and mass spectrometric sample preparation of the sequencing products were accomplished in a few minutes, and parallel processing of 96 samples was possible.[168] This work confirmed the increasing trend towards high-throughput MS analysis of DNA.

One mechanism of fragmentation that limits mass range and resolution in MALDI-MS analysis of DNA is the protonation of the nucleobase leading to base loss and subsequent backbone cleavage. Smith and colleagues established that the replacement of a 2′-hydrogen with a fluorine atom stabilized the *N*-glycosidic linkage, thus either reducing or blocking loss of the nucleobase.[169] The practical usefulness of this approach to improving MALDI-MS sensitivity was demonstrated by the successful incorporation of 2′-fluoronucleoside triphosphates into a growing chain by several DNA polymerases.[170] Furthermore, the presence of these modified oligonucleotides did not affect the incorporation of a dideoxy terminator step in a Sanger sequencing reaction. The UITma DNA polymerase gave the best results in the development of novel 2′-fluoronucleic acids suitable for MALDI analysis using the matrix 2,5-dihydroxybenzoic acid. Fragmentation of these modified nucleic acids during MALDI-TOF-MS was much less when compared with the analogous unmodified species.[170]

1.3.1.3.5 Enzymatic DNA Synthesis

It was shown that the in-house production of uniformly labeled dNTPs could be a complicated multi-step procedure that is both time consuming and material consuming with a relatively low yield.[171–175] The methodology requires that the DNA is extracted from the bacteria grown on labeled minimum medium; the DNA is then digested to deoxynucleoside monophosphates (dNMPs); kinases then convert the dNMPs to deoxynucleoside triphosphates (dNTPs); and this is followed by purification. This procedure is usually performed right before the reaction takes place and before any processing step is carried out. In addition, it is advisable to confirm that the reaction has ceased and that the complex has been cleaved prior to running the preparative gel for the separation of the cleavage products. There are several reasons for the careful monitoring of the progress of the reaction: First, the template–primer complex may be degraded by the time the synthesis is initiated; second, the unlabeled test confirms the specific activity of the enzyme preparation and determines the actual reaction time that is required; third, the labeled and unlabeled dNTP preparations may be of unequal purity, which affects the rate of the polymerase reaction in different ways; however, testing with unlabeled dNTPs confirms that the template–primer complex is available for the polymerase reaction.

DNA polymerases are also capable, by their "template-independent polymerase activities," of elongating the sequence that is being synthesized further than the template dictates.[176] Therefore, longer incubation can result in uncontrolled extensions of the DNA sequence.

Ambrus et al. carried out the monitoring of the progress of enzymatic ^{13}C/^{15}N-labeled DNA syntheses using MALDI-TOF-MS.[177] The analyses of three templates for a human telomeric 22-mer (a wild type) and a mutant human c-MYC promoter (18- and 22-mer) DNA and their reactions with the 3′ to 5′ exo$^-$ Klenow fragment of DNA polymerase demonstrated the usefulness of this protocol. Small amounts of samples (ca. 1–2 μL each) were taken from the reaction mixtures at different times and analyzed promptly by MALDI-TOF-MS. The progress of the reaction was monitored by

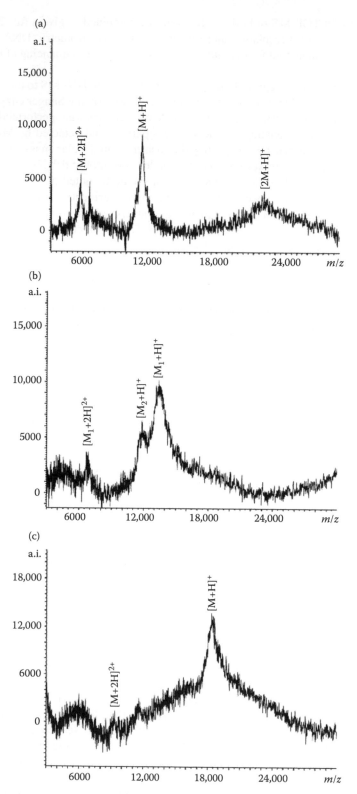

FIGURE 1.10 MALDI-TOF spectra of different stages in the enzymatic DNA synthesis, performed by a two-step on-plate successive washing for (a) the template–primer complex of the human telomeric DNA (MW(calc) D11,464 Da).

measuring the relative intensity ratios of ions corresponding to the desired products and the primer–template complexes. Insensitivity of the MALDI technique, due to a high salt content, was eliminated by the application of successive on-plate-desalting methods. Figure 1.10a shows the MALDI-TOF-MS of the template–primer complex used for the amplification of a human telomeric 22-mer oligonucleotide $(AG_3(T_2AG_3)_3)$. This oligonucleotide was dissolved in a solution (buffer) with high ionic strength, and the spectrum in Figure 1.10a was obtained after two on-plate washing steps. Although the S/N ratio was still low after the desalting steps, the original DNA complex can easily be identifiable. The spectrum in Figure 1.10a was also useful for the fast verification of the purity and possible degradation of the template–primer complex right before the labeled material is added. Figure 1.10b shows the MALDI-TOF-MS of a sample taken out of the reaction mixture for the human telomeric sequence after 30 min; in addition to the original peak at ca. 11.6 kDa, there is another peak at ca. 13.4 kDa, indicating that the initiation of the reaction was successful. Finally, Figure 1.10c shows the MALDI-TOF-MS that was obtained for a sample taken out after the fully completed reaction.

1.3.1.3.6 Study of Human Genetic Variations

SNP, the most common type of human genetic variation, is defined as a single-base change that occurs at a specific position in a genome, and different sequence alternatives (alleles) for these changes may exist in some individuals within a population. There are a number of known inherited diseases due to the SNP occurring at a particular locus and there are many complex disorders, such as cancers, that may result from the cumulative effect of several deleterious SNPs. The pharmaceutical industry, in their quest to design drugs customized for individual patient-specific responses, has an interest in the analysis of such variations.[178] By definition, an SNP is referred to as a mutation only when the less frequent allele has an abundance of 1% or greater. On average, one SNP is found in every 500–1000 bases in humans. Only a small portion lies within coding regions, and an even smaller percentage is responsible for amino acid changes in expressed proteins. Tost and Gut recently presented an excellent review on the major contributions in the field of genotyping SNP by MS.[179]

Fei and Smith described a method for typing SNPs by MALDI-TOF-MS where, instead of an unmodified dideoxynucleoside triphosphate (ddNTP), a mass-tagged ddNTP was used in a primer extension reaction.[180] The increased mass difference, due to the presence of the mass tag, greatly facilitated the accurate identification of the added nucleotide and was particularly useful for typing heterozygous samples. Further validation of the method was demonstrated in the analysis of five single-base mutations in exon IV of the human tyrosinase gene. Three plasmids and three human genomic DNA samples were amplified, and single nucleotide variations, within 182 bp PCR amplicons, were genotyped at five variable positions. The results were in 100% concordance with the conventional sequencing, and the genotypes were determined accurately at five sequence-tagged sites.[181]

Seichter and colleagues reported the genotyping of short tandem repeat (STR) polymorphisms.[182] Enhanced resolution MALDI-MS is required for the characterization of dinucleotide repeats, which is essential for animal genome studies. This mass resolution increase in MALDI-TOF-MS was achieved by modifying the RNA and DNA to discourage any intramolecular reactions that could result in fragmentation. RNA transcripts were synthesized enzymatically from PCR products containing a promoter sequence and did not require any special reagents or primer labels. The ssRNA transcripts produced were a prerequisite for high-resolution MS of nucleic acids. The MALDI-TOF-MS resolution obtained from RNA transcripts was found to be superior to that obtained for DNA molecules of a comparable size. It is interesting to note that although the resolution (full-width, half-maximum) of DNA dropped significantly with increasing fragment length (50-mer, 330; 75-mer, 115; and 100-mer, 40), it remained ~200 for all the RNA sizes observed.[182] These researchers have also characterized STRs in endonuclease-cleaved RNA transcripts by MALDI-TOF-MS.[183]

Wada and collaborators reported the measurement of polymorphic trinucleotide repeats in the androgen gene receptor.[184] Among the four different triplet sequences known to cause diseases (CAG, CCG, CTG, and GAA), it was established that the CAG repeats were responsible for fewer cases of pathological consequences than the other repeats. The molecular mass of amplified DNA

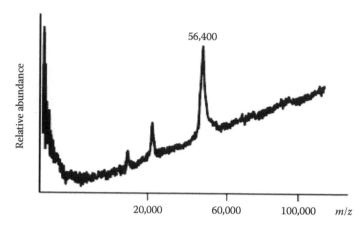

FIGURE 1.11 MALDI-TOF mass spectrum of an amplified DNA product from the androgen receptor gene containing 17 CAG repeats.

(200 bp) from the region containing the CAG repeat of the androgen receptor gene was measured by MALDI-TOF-MS. The single-charged molecular ion was detected using 0.1 pmol of DNA sample, and the number of repeats was determined from the molecular mass (Figure 1.11). It was established that the length of the CAG repeat sequence in the first exon of the androgen receptor gene was normally polymorphic, and the sequence ranged from 12 to 36 repeats.

Hung et al., using a MALDI-TOF-MS, developed a minisequencing method for the screening of the 1691G → A substitution in coagulation factor V.[80] This point mutation in the factor V molecule resulted in the amino acid substitution of an Arg[506] residue for a Gln residue. This mutation, also defined as factor Vleiden, is also responsible for the activation of protein C (APC) resistance, which is the most common genetic risk factor for familial thrombophilia. In this method, the authors amplified a fragment of genomic DNA containing the 1691st base, followed by minisequencing in the presence of dGTP and ddATP, ddCTP, and ddTTP. In this manner, the primer was extended by one base from one allele and two bases from the other allele. The extended products were analyzed using MALDI-TOF-MS. The base at position 1691 was identified based on the number of nucleotides added.

Wolfe et al. developed a novel genotyping method termed "incorporation and complete chemical cleavage" (ICCC).[185] This method consists of the incorporation, by PCR, of a chemically labile nucleotide (7-deaza-7-nitro-dATP, 7-deaza-7-nitro-dGTP, 5-hydroxy-dCTP, or 5-hydroxy-dUTP) followed by the specific oxidative chemical cleavage at every site containing the modified base in the resulting amplicon. The modified nucleotides have increased chemical reactivity and are capable of forming standard Watson–Crick base pairs with the identity of the cleaved fragments usually determining the genotype of the DNA. MALDI-TOF-MS analysis was used to determine the molecular mass of the cleaved fragments. Compared with other MS-based genotyping methods, the ICCC strategy provides greater flexibility as well as the unique opportunity to generate two useful sets of genotype-specific DNA fragments within a single assay, thus enabling the possibility of the internal confirmation of a genotype assignment.[185]

Clinical specimens often contain only a small fraction of mutated cells, and assays to detect mutations need to be highly sensitive, specific, and amenable to automated high-throughput methods, thereby permitting large-scale screening. Sun and coworkers described a screening method, the PNA-directed PCR primer extension MALDI-TOF (PPEM), which effectively addressed the requirements of assays for mutation screening.[186] DNA samples were first amplified using peptide PNA-directed PCR clamping reactions in which mutated DNA was preferentially enriched. The PCR-amplified DNA fragments were then sequenced through primer extension to generate diagnostic products. Finally, mutations were identified using MALDI-TOF-MS. This method was shown to detect as few as three copies of mutant alleles in the presence of a 10,000-fold excess of normal alleles in a robust and specific manner. Furthermore, it can be adapted for simultaneous detection of multiple mutations and is amenable to high-throughput automation.[186]

MALDI-TOF-MS has been successfully used to detect an SNP developed in the precore/basal core promoter region of the Hepatitis B virus and has also been used for high-throughput SNP analysis with fully integrated platforms.[187,188] MALDI-MS has also been used for the determination of allele frequencies in the postgenome sequencing era. Although there were different strategies devised for allele discrimination combined with MALDI, in practice only primer extension methods are nowadays routinely used. This combination enables the rapid, quantitative, and direct detection of several genetic markers simultaneously in a broad variety of biological samples.

In the field of molecular diagnostics, MALDI has been applied to the discovery of genetic markers that are associated with the expression of a disease susceptibility or drug response, as well as an alternative means for the diagnostic testing of a range of diseases for which the responsible mutations are already known. It is one of the first techniques where, based on SNPs, whole-genome scans were carried out. It is equally well suited for pathogen identification and the detection of emerging mutant strains, as well as for the characterization of the genetic identity and quantitative trait loci mapping in farm animals. MALDI can also be used as a detection platform for a range of novel applications that are more demanding than standard SNP genotyping such as mutation/polymorphism discovery, molecular haplotyping, analysis of DNA methylation, and expression profiling. Tost and Gut have reviewed the use of SNPs as genetic markers and the MALDI-MS detection related to clinical applications and molecular diagnostics.[188]

A novel approach was reported, using solid-phase capturable biotinylated dideoxynucleotides (biotin-ddNTPs) in a single-base extension for multiplex genotyping. In this method, oligonucleotide primers, with different MWs, and specific to polymorphic sites in the DNA template were extended with biotin-ddNTPs by DNA polymerase to generate 3′-biotinylated DNA products. Four biotin-ddNTPs with distinct MWs were selected to generate the extension products. These products were then complexed with streptavidin-coated solid-phase magnetic beads, and the unextended primers and other components in the reaction were washed away. The pure extension DNA products were subsequently released from the solid phase and analyzed by MALDI-TOF-MS. The mass of the extension products was determined using a stable oligonucleotide as a common internal mass standard. Since only the pure extension DNA products are introduced to the MS for analysis, the accuracy and scope of multiplexing in the SNP analysis were increased as the resulting mass spectrum was free of nonextended primer peaks and their associated dimers. The increase in mass difference provides improved resolution and accuracy in detecting the heterozygotes in the mass spectrum. Using this method, six nucleotide variations on synthetic DNA templates were detected. These variations mimicked the mutations in the *p53* gene and two disease-associated SNPs in the human hereditary hemochromatosis gene.[189,190]

1.3.1.3.7   Use of Ionic Liquid Matrices

Room-temperature ionic liquid matrices (ILMs) have recently been investigated for use in MALDI-MS and proven to be advantageous. Literature accounts of ILM performance with biological samples document increased sensitivity and ionization efficiency.

Jones et al. investigated the ILM-induced metastable decay of peptides and oligonucleotides and the stabilization of phospholipids in MALDI-FT-ICR-MS analyses.[191] Their research clearly demonstrated enhanced ionization efficiencies and improved spot-to-spot reproducibility with the use of ILMs.[192–196] It was shown that the use of ILMs increased the sensitivity of MALDI-TOF and -FT-ICR-MS for both pure compounds and complex biological samples. In addition, when ILMs were used with peptide and oligonucleotide analyses, there was increased fragmentation, which yielded complementary information.

Reports of oligonucleotide analysis, using MALDI-FT-ICR-MS, indicated that the use of the matrix 3-HPA was required to observe the parent ion with minimal fragmentation.[191] It was shown that MALDI-TOF-MS analysis of the oligonucleotide 5′-GGATTC-3′ required the negative ion

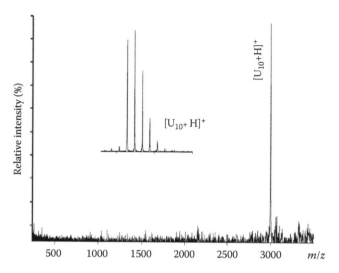

FIGURE 1.12 Analysis of standard 10-mer uridine (U_{10}) provided an experimental mass of 2998.328 Da (2998.294 Da calculated from sequence) and an accuracy of ~8 ppm. An initial amount of ~10 pmol was submitted to on-probe purification. CHCA was used as matrix in the positive ion mode.

mode to reveal abundant ion signals (Figure 1.12a). The behavior of oligonucleotides during MALDI-FT-ICR-MS with ILM was similar to that of peptides with respect to the metastable decay. The oligonucleotide used produced a significant amount of fragment ions using ILMs, with only a few fragments being the predictable and commonly observed metastable fragment ions. Furthermore, in the MALDI-MS, there were a number of peaks that could not be assigned, including some of the more abundant ions detected.[197] MALDI-FT-ICR-MS analysis using nBA-CHCA (n-butylamine part of 2-cyano-4-hydroxy-cinnamic acid) and DHB (2,5-dihydroxy benzoic acid) produces spectra very similar to the spectra obtained when using N-dimethylamine salt of CHCA (DEA-CHCA) as the matrix; however, the parent ion is present only in low abundance (*vide infra*). Moreover, it has been indicated that the substitution of a proton to a hydroxyl or ethyl group on the oligonucleotide can be exchanged with an alkali-metal cation (sodium, potassium, or cesium). This is only evident after the predicted fragments to be substituted are calculated with more than one alkali-metal cation.[197]

1.3.1.4 MALDI-TOF-MS Characterization of RNA–Protein Complexes

RNA–protein interactions are known to play important roles in various cellular processes. While several conserved amino acid sequence motifs for RNA binding have been identified, knowledge of the molecular mechanism of RNA–protein recognition is still limited. Thiede and Von Janta-Lipinski have used MALDI-TOF-MS (the − ion mode) to explore the noncovalent interactions between different peptides and RNAs.[198] They found that the best matrix for this work was 2,4,6-trihydroxyacetophenone, which showed approximately the same peak intensities for free peptides and RNA while allowing the detection of RNA–peptide complexes. The subject of noncovalent interactions between different peptides and RNAs is well detailed in Chapter 10 of this book.

1.3.1.5 Atmospheric Pressure MALDI-FT-ICR-MS Characterization
of RNA–Protein Complexes

Kellersberger et al. used atmospheric pressure (AP) MALDI-FT-ICR-MS to analyze both native RNA and chemically modified RNA.[199] This method was used to obtain the unambiguous characterization of RNA samples, which have been modified by solvent accessibility reagents used in the structural studies of RNA and protein–RNA complexes. The formation of cation adducts, was effectively reduced by extensive washing of the anionic analytes that were retained, by their strong interactions with the cationic layer of poly(diallyldimethylammonium chloride) (PADMAC), on the probe surface. This rapid desalting procedure resulted in the detection of DNA and RNA samples in high femtomole quantities distributed over a 4 × 4 mm sample well.

AP-MALDI-FT-ICR-MS was shown to provide high-resolution spectra for analytes as large as ~6.4 kDa with apparently very little of any metastable decomposition. Evidence for the low energy content typical of ions generated by AP-MALDI is the absence of any observable metastable decay for the selected precursor ions. This feature proved to be beneficial in the characterization of chemically modified RNA samples, which become particularly prone to base losses upon alkylation. The high resolution obtained by FT-ICR-MS enabled the application of a data-reduction algorithm capable of rejecting any signal devoid of plausible isotopic distribution, thus facilitating the analysis of complex analyte mixtures produced by the nuclease treatment of RNA substrates. Proper selection of nucleases and digestion conditions can ensure the production of hydrolytic fragments of manageable size, which could extend the range of applicability of this bottom-up strategy to the structural investigation of very large RNA and protein–RNA complexes.

1.3.1.6   MALDI-TOF-MS Characterization of DNA Adducts

Although most MS analyses of adducted bases or nucleosides have been carried out using ESI-MS/MS, several groups have applied MALDI-TOF-MS to the study of oxidative lesions and their stability, the role of repair mechanisms, and other biochemical features.[200]

Tretyakova and coworkers have shown that MALDI-TOF-MS analysis of exonuclease ladders can be successfully used to sequence DNA fragments containing a variety of carcinogen-induced nucleobase adducts.[201] This methodology, which is described in detail in Chapter 7 of this book, was used to determine the sequence of different synthetic 12–20-mer DNA strands containing both *O*-6-methylguanine and oxidative aromatic-guanine covalent adducts such as (**1**) and (**2**). These adducts were located without any prior information about the sequence or nature of the adducting agent. Sequence information was also obtained for DNA containing *O*-6-pyridyloxoguanine, even though this lesion normally blocks 3′-PDase.

D'Ham and coworkers synthesized oligonucleotides that contained modified pyrimidines, such as thymine glycol, 5,6-dihydrothymine, and 5-hydroxy-cytosine, in order to investigate the substrate specificity and excision mechanism of two *Escherichia coli* repair enzymes, endonuclease III and formamidopyrimidine DNA glycosylase (Fpg).[202] A GC-MS assay with HPLC prepurification was used to quantify the release of the modified base. The MALDI-TOF-MS analysis of the molecular masses of the repaired fragments from the synthetic oligonucleotides showed that the DNA backbone was cleaved by endonuclease III through a hydrolytic process with no β-elimination product being detected.

1.3.2   SURFACE-ENHANCED LASER DESORPTION/IONIZATION (SELDI)-TOF-MS

The need for methods to identify disease biomarkers is driven by the fact that the long-term survival rate for a patient is improved with the earliest possible detection of a diseased condition. Issaq et al. reviewed the technique of surface-enhanced laser desorption/ionization (SELDI) coupled to TOF-MS. It is a novel approach to biomarker discovery that combines two powerful techniques: chromatography and MS.[203a] SELDI-MS is similar to MALDI-MS in that biomacromolecular mass is determined. It is based on TOF-MS, where gaseous ions are created from the analytes in the solid state and a matrix is usually used as the energy receptacle to assist in the ionization process. However, by using surface-enhanced chips that are compatible with TOF-MS instrumentation, analytes can be captured on chemically modified chip surfaces and analyzed by MS directly from these affinity chips. This technique has been used for biomarker identification as well as the study of protein–protein and protein–DNA interaction. The versatility of SELDI-TOF-MS has permitted it to be used in projects ranging from the identification of potential diagnostic markers for prostate, bladder, breast, and ovarian cancers and Alzheimer's disease to the study of biomolecular interactions and the characterization of posttranslational modifications.[4,203]

It is known that ligand binding can be analyzed using a variety of common and time-consuming immunoassays. These methods may require secondary antibodies or gel electrophoresis, and the interpretation of results may be subjective.[204] However, SELDI-TOF-MS allows the immobilization of antigen, antibody, or oligo-DNA, on preactivated chips, with unambiguous identification of the binding antibodies, antigens, or DNA-binding proteins.

In an example of this technique, customized affinity DNA chip surfaces were designed to capture the sequence-specific protein–DNA interaction of LacI binding to the *lac* operon promoter sequence. By coupling this unique LacI DNA promoter sequence to a chip surface, the chip served to isolate LacI from the total protein lysate of *E. coli*. In addition, the chip surfaces utilized were compatible with the physiological buffers used to capture the LacI protein, and hence, binding conditions were optimized on the chips. Accordingly, the use of SELDI chip surfaces simplified molecular characterization of captured biomacromolecule analytes by MS.[205]

In addition, SELDI-TOF-MS has proven to be a powerful and versatile approach for the analysis of ligands. It eliminates the tracer-labeled secondary antibodies and allows the determination of the MWs of binding proteins and their ligands directly. This approach may also be considered for the detection of enzymes, receptors, or any other specific ligands.

The overall history and advances in SELDI-TOF-MS technology have also been reviewed by Merchant and Weinberger.[206] The fundamentals of SELDI-TOF analysis were presented and compared with other laser-based MS techniques. The application of SELDI-TOF-MS in functional genomics and biomarker discovery was discussed and exemplified by elucidating a biomarker candidate for prostatic carcinoma.[207]

SELDI-TOF-MS was performed on a Ciphergen Biosystems ProteinChip System, which was used in conjunction with DNA affinity capture (DACA), to study specific DNA–protein binding.[208] Using DNA molecules bound to a surface, the sequence-specific interactions were detected as a transcription factor mutation, which affected the binding profile for the TATA binding protein (TBP). A comparison between the methylated and unmethylated promoter-containing DNA fragments showed numerous binding profile differences over a mass range extending to >60 kDa.[207]

1.3.3 MS/MS Analysis of Oligonucleotides

Until recently, the ability to identify DNA- and RNA-based mass spectromertic sequencing was essentially limited to the use of ESI-MS/MS methods. This was mainly due to the fact that the majority of instruments using the MALDI ionization mode were limited to conventional MS instruments such as TOF-MS.

The development of new instruments with MALDI sources and true MS/MS capabilities, such as the Qq-TOF-MS and TOF-TOF-MS, has created the capacity to obtain high-quality tandem mass spectra of DNAs, RNAs, and other related nucleotides. In addition, the use of MALDI-FT-ICR-MS^n for high-resolution analyses of DNAs and RNAs has provided a powerful motivation for the development of strategies aimed at minimizing gas-phase decomposition and improving sensitivity and resolution. The utilization of selected matrix additives, refined trapping and axialization methods, and collisional cooling of trapped ions was initially proposed to increase the performance of instruments in which ions were generated within the FT-ICR-MS^n cell.[200,209–217]

1.3.3.1   AP-MALDI-FT-ICR-MS[n] of Normal and Chemically Modified RNA

Later, the introduction of external MALDI sources enabled the decoupling of the ion production from the analysis step and afforded a more efficient way to control the cell vacuum through multiple stages of differential pumping.[218–220] This feature has gained further significance since the demonstration that low-energy ions could be effectively produced by injecting inert gas into the external source during the desorption event, which was in agreement with the initial results that were obtained with a MALDI-TOF instrument.[221–225] AP-MALDI has been combined with FT-ICR-MS to obtain the unambiguous characterization of RNA samples modified by solvent accessibility reagents used in structural studies of RNA and protein–RNA complexes.

The formation of cation adducts, typical of MS analysis of nucleic acids, was effectively reduced by extensive washing of the anionic analytes that were retained on the probe surface. This rapid desalting procedure, using a cationic layer of PADMAC, permitted dthe etection of DNA and RNA samples in high femtomole quantities (Table 1.2).[226] AP-MALDI-FT-ICR-MS was shown to provide high-resolution spectra for analytes as large as ca. 6.4 kDa. Under optimal conditions, there were no significant metastable products observed in the AP-MALDI-FT-ICR-MS analysis of diverse sequences of DNA and RNA oligonucleotides, as shown for the uridine decamer (U10) (Figure 1.12).[226] The product ion scan of the 12-mer deoxyoligonucleotide (D12) obtained by sustained off-resonance irradiation collision-induced dissociation (SORI-CID) is shown in Figure 1.14 and Table 1.2. Consistent with the normally low energy content of ions generated by AP-MALDI, there was no metastable decomposition observed in the isolation spectrum, even after extended trapping (up to 5 s), and despite the intrinsic fragility of nucleic acid ions.[226] Subsequent collisional activation provided a characteristic series of product ions, which properly matched the analyte sequence (Figure 1.13).[226] In spite of the fact that singly charged ions of relatively large species tend to produce less extensive fragmentation than the corresponding multiply charged ions under normal CID conditions, the product ions obtained from collisional activation of D12 at m/z 3628.65 allowed nearly complete sequence coverage. The observation of the lack of any significant metastable decay of the precursor ions selected for tandem experiments offers further evidence of the low energy

---

TABLE 1.2
Summary of Standard DNA and RNA Samples Employed in the Study

Name	Sequence	Calc. Monoisotopic Mass (Da)
AA	AA[a]	564.16
TT	TT[a]	546.14
AG	AG[b]	612.14
G_4	GGG G[a]	1254.25
QZ11	ATC GAT C[a]	2079.39
U_{10}	$(U)_{10}$[b]	2998.29
T_{10}	$(T)_{10}$[a]	2978.50
D10	ATC GGC CAC A[a]	2995.55
D12	CAG TCA GCT CAG[a]	3628.65
U_{20}	$(U)_{20}$[b]	6018.96
T_{20}	$(T)_{20}$[a]	6058.54
QZ2	TAT TGC TTT AAA AAC TCA AAA[a]	6393.10
QZ1	TTT TGA GTT TTT AAA GCA ATA[a]	6446.13

Source: Adapted from Kellersberger, K.A. et al., *J. Am. Soc. Mass Spectrom.* 16, 199, 2005. With permission.

[a]   Oligodeoxyribonucleotides.

[b]   Oligoribonucleotides.

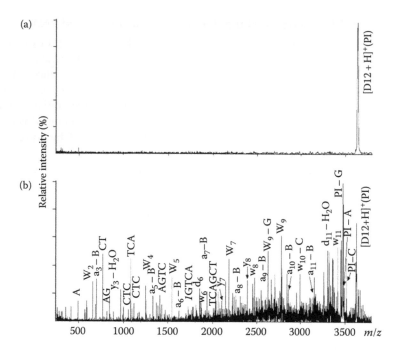

FIGURE 1.13 Isolation spectrum (a) and sustained off-resonance irradiation collision-induced dissociation (SORI-CID) product ion spectrum (b) obtained from singly charged D12 (Table 1.2). An initial amount of ~45 pmol was submitted to on-probe purification. CHCA was used as matrix in positive ion mode. No metastable decomposition could be observed after trapping the precursor ion for up to 5 s. P.I. indicates the precursor ion observed at 3628.65 *m/z*.

content typical of ions generated by AP-MALDI. This feature proves to be very beneficial in the characterization of chemically modified RNA samples, which become particularly prone to base losses upon alkylation. The high resolution offered by FT-ICR-MS enabled the application of a data-reduction algorithm capable of rejecting any signal devoid of plausible isotopic distribution, thus facilitating the analysis of complex analyte mixtures produced by nuclease treatment of RNA substrates.

Proper selection of nucleases and digestion conditions can ensure the production of hydrolytic fragments of manageable size, which could extend the range of applicability of this bottom-up strategy in the structural investigation of very large RNA and protein–RNA complexes.

1.3.3.2 RNA Fragmentation in MALDI-TOF-TOF-MS/MS Studied by H/D Exchange

Several H/DX-MS studies of DNA fragmentation have been conducted on nucleotide tetramers with one or two high proton affinity bases (adenine, guanine, and cytosine) in a low proton affinity base (thymine) environment.[226–230a] To obtain comparable data sets, DNA tetramers of similar compositions were analyzed. It has been shown that during the analysis of RNA tetramers (UCUU) containing a high number of proton affinity base(s), the obtained backbone fragmentation data were not very informative, because the mass of the H/D-exchanged cytosine equaled that of the H/D-exchanged uracil. MS/MS fragmentation data obtained from a (UCUU) tetramer were evaluated and this showed that it occurred by loss of only neutral or charged cytosine. It was also noted that uracil was rarely lost in the fragmentation process.

Anderson et al. performed hydrogen–deuterium exchange on a series of RNA and DNA tetranucleotides and studied their gas-phase chemistry and fragmentation patterns using a high-resolution MALDI-TOF-TOF-MS/MS instrument (the + ion mode). The authors were specifically interested in elucidating the remarkably different fragmentation behavior of RNA and DNA, specifically, the characteristic and abundant production of c and y ions from RNA versus a dominating generation of $(a - B)^-$ and w^- ions from DNA analytes.[203b] The analysis yielded important information on all significant backbone cleavages as well as on the nucleobase losses. It has been suggested that the common fragmentation mechanisms for RNA and DNA, as well as an important RNA-specific reaction, required the presence of the 2-hydroxyl group leading to c and y ions. A rigorous comparison of the fragmentation of H/D-exchanged oligonucleotides in the CID and PSD only modes was also performed. The data were qualitatively identical in these two modes, which strongly suggested that identical fragmentation mechanisms were in operation. The only difference was an increased signal from low-mass fragments in the CID mode. The previous studies using H/DX-MS for oligonucleotide fragmentation analysis showed that H/DX did not affect the characteristic fragmentation pattern of the oligonucleotide. Figure 1.14 shows the full MS/MS spectra of deuterated UGUU and UCUA. Fragmentation of the H/D-exchanged RNA tetramers yielded information on the mechanism of N-glycosidic bond cleavage. The peak at m/z 1063.22 (Figure 1.14a) corresponded to the loss of neutral guanine from the UGUU tetramer. The mass difference of 155 Da, as compared with the intact precursor, indicated that guanine was eliminated as a completely exchanged species. Therefore, it was concluded that this loss of the neutral nucleobase was observed consistently for all RNA tetramers, a result which is similar to that previously reported for DNA by Gross et al.[227] A crucial point to mention is that, upon the departure of a fully deuterated neutral nucleobase, the hydrogens in the remaining ion are rearranged in a way that inevitably transferred a nonmobile proton to a mobile site. Thus, as hydrogen became mobile, intramolecular H/DX in the gas phase was proven to be possible.[227]

1.3.4 SECONDARY ION MASS SPECTROMETRY

Patterning DNA onto surfaces has recently received considerable attention due to its applications in fundamental biology and biomedical research as genomic arrays, diagnostics, and biosensors.[228a–d] The diverse methods of preparing DNA microarrays have developed into a flexible tool that has a number of applications including gene mapping, drug discovery, biosensor development, and sequence analysis. The importance of standardizing the methodology for "reading" the array becomes apparent when the number of immobilized DNA probes in an array (which can reach tens of thousands), combined with the number of types of DNA, and the number of ways of fixing the probe material to the surface are considered.[228a–d]

Castner and his group made a preliminary assessment with TOF-SIMS as a method for the analysis of DNA arrays.[228e] The spectra of the arrays obtained by TOF-SIMS were complex; however, a detailed table of all positive and negative ions was built up for all the common DNA components. This information was then used to support the analysis of oligomer spectra through the application of principal component analysis.[228e]

Castner et al. have indicated that for enhancing the performance of DNA-modified surfaces for microarray and biosensor applications, it was crucial to accurately characterize the chemistry and structure of immobilized DNA molecules on micropatterned surfaces.[228e] Several methods have been developed for fabricating micron-scale DNA patterns, including contact and noncontact printing of presynthesized DNA onto substrates, and *in situ* synthesis of microarrays using electrochemistry[228c] and photolithography.[228f] Microprinting techniques are widely used for DNA microarray fabrication on commercial array slides containing hundreds to thousands of spotted features. The resulting immobilized DNA density and distribution within individual microarray spots have profound influences on the subsequent target-capture performance.[228g,229,230] It was shown that spot-to-spot variations in DNA surface density and distribution can lead to inconsistent target capture, inaccurate data quantification, and misleading results.[228g,229,230] Thus, accurate quantitative analysis of printed DNA microarray diagnostics is possible only if controlled and reliable spot uniformity, with respect to spot density, spot size and shape, and repeatability, is achieved.

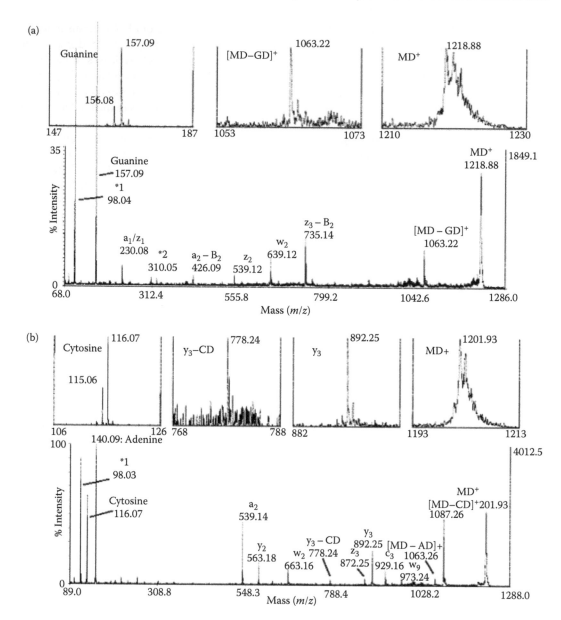

FIGURE 1.14 Tandem mass spectra of H/D-exchanged UGUU (a) and UCUA (b), with important regions enlarged above each spectrum. The precursor ions (UGUU: m/z 1218.88 and UCUA: m/z 1201.93) show the high degree of deuteration required to perform the H/DX MS/MS analysis.

Innovations in imaging x-ray photoelectron spectroscopy (XPS) and TOF-SIMS now permit more detailed studies of micropatterned surfaces.[228g,229,230]

1.3.5 LIQUID SECONDARY ION MASS SPECTROMETRY

Nucleoside phosphoramidites (PAs) are the most widely used building blocks in contemporary solid-phase synthesis of oligonucleotides. The accurate MW measurements of such molecules, which contain acid-labile moieties, may be easily determined by MS using a matrix system [triethanolamine (TEOA)–NaCl] on liquid secondary ion mass spectrometry (LSIMS) equipped with a double focusing.

FAB-MS and LSIMS are effective tools for analyzing PAs. However, it was found that it was not always possible to detect the molecular-related ions (MRIs) of the various PAs.[207,231] On the other hand, little attention has been paid to the study of suitable matrices for LSIMS or FAB-MS measurements of the PAs. A greatly enhanced reliability of PA-mass measurements would be obtained if the accurate MWs of PAs were to be measured. A suitable matrix system, such as TEOA [N(CH$_2$CH$_2$OH)$_3$, (TEOA)]–NaCl, for the detection of the MRIs of a variety of PAs has been reported by Fujitake et al.[232] Their method measures the accurate MWs of various PAs as adduct ions [M + Na]$^+$ with an average mass error of less than 0.4 ppm, thus providing the formulae of various PAs. For example, the LSIMS spectrum (low MS level) of an RNA derivative (Figure 1.15) using the TEOA–NaCl matrix characterized a structure where the fragment ions at m/z 839 and 651 correspond to the Na$^+$C-adducted ions generated by the removal of the pivaloyloxymethyl (POM) or dimethoxytrityl (DMT) groups from the RNA derivative, together with a base peak due to DMTC ion at m/z 303.

1.4   SPRAY IONIZATION TECHNIQUES

1.4.1   Electrospray Ionization

ESI is a well-established, robust technique when interfaced with liquid chromatography (LC-MS) or with similar techniques such as capillary electrophoresis (CE-MS).[233,234] ESI-MS allows rapid, accurate, and sensitive analysis of a wide range of analytes from low-mass polar compounds (<200 Da) to biopolymers (>200 kDa).[235,236] A useful feature of ESI is its ability to generate multiply charged quasimolecular ions in the gas phase without fragmentation. These ions can have a high mass, but the m/z values are sufficiently low to allow their analysis in quadrupole and QIT instruments; these instruments are usually limited to a mass range of less than 6000 Da for singly charged ions. It is an excellent means of characterization of DNA varying in size, from around 70- to 106-mer oligonucleotides.[236,237] ESI appears to be almost unlimited in the size of molecules that can be ionized, and recent reports describe macromolecular ions with masses varying from 4.0×10^5 Da to polymer ions ranging up to 5.0×10^6 Da and with nearly 5000 charges.[238] These megadalton

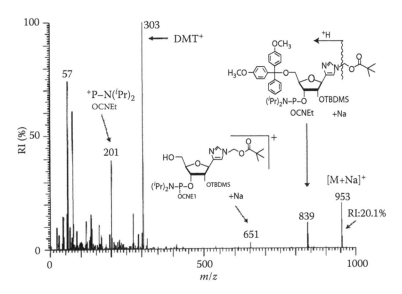

FIGURE 1.15   Positive ionization LSIMS spectrum of the RNA derivative having an M_r 930 Da using the TEOA–NaCl matrix.

macromolecules are now routinely analyzed in commercial mass spectrometers with a wide range of different analyzers that can be interfaced to an ESI source.

1.4.1.1 ESI-MS Characterization of Nucleosides

1.4.1.1.1 Analysis of Medicinal Nucleosides

Several methods have been developed to analyze nucleosides and nucleobases that are found in Chinese medicine. HPLC with a diode array detector (DAD) and ESI-MS/MS of adenine, guanosine, uridine, cytidine, and adenosine in the *Banlangen* injection (a Chinese patent drug) was reported by Fujitake et al.[232,239] In a separate experiment, the detection of amounts of uracil, adenine, and adenosine in Lingzhi (a Chinese patent drug) was effected by HPLC-DAD-MS.[240] In two species of Lingzhi, the dried sporophore of *Ganoderma lucidum* and *Ganoderma sinense*, the qualitative analysis of six nucleosides: adenosine, cytidine, guanosine, inosine, thymidine, uridine, and five nucleobases: adenine, guanine, hypoxanthine, thymine, and uracil, was determined.[241] Quantitative analyses showed that uridine was the most abundant nucleoside in the Lingzhi samples and nine target analytes were found to be different in the pileus and stipes of the fruiting bodies and among the different species of *Ganoderma* spp.

Capillary electrophoresis, coupled with mass spectrometry, has been used for the qualitative and quantitative analysis of 12 nucleosides and nucleobases in natural and cultured *Cordyceps* (traditional Chinese medicine).[240] A method combining HPLC with photodiode array detection and MS has been developed for the simultaneous separation, identification, and quantification of nucleosides in *Cordyceps sinensis* (Cs), a traditional Chinese medicine, and *Cordyceps mycelia* (Cm), a cultured *Cordyceps*.[242] LC-ESI-MS using the selective ion monitoring (SIM) mode allowed the quantification of nucleosides in Cs and Cm. The limits of detection (LOD) and limits of quantification (LOQ) for nucleosides were in the order of 0.1–0.6 and 0.5–2.0 µg mL^{-1}, respectively.

1.4.1.1.2 Analysis of Modified Nucleosides

McCloskey and collaborators have reported a detailed study of several species of *Methanococci*, a lineage of methanogenic marine *Euryarchaea* that grows over an unusual temperature range. They attempted to determine if the posttranscriptional modifications of the tRNAs were reflected in the phylogenetic relationships.[243] The tRNAs from *Methanococci vannielii*, *Methanococci maripaludis*, the thermophile *Methanococci thermolithotrophicus*, the hyperthermophiles *Methanococci jannaschii* and *Methanococci igneus* were digested and analyzed by LC-ESI-MS. McCloskey and collaborators revealed up to 24 modified nucleosides, including the complex tricyclic nucleoside "wyosine" that is characteristic of position 37 in tRNA^{Phe}, and two other members of this family with unknown structures.

The hypermodified nucleoside 5-methylaminomethyl-2-thiouridine (**3**), previously reported only in bacterial tRNA, was identified by LC-ESI-MS in four of the five organisms. The methylated nucleosides, 2′-O-methyladenosine (**4**), N^2,2′-O-dimethylguanosine (**5**), and N^2,N^2,2′-O-trimethylguanosine (**6**), were found only in hyperthermophile tRNA, consistent with their proposed roles in thermal stabilization of tRNA.[243] An important facet of these studies has been the greater accuracy in nucleoside assignments resulting from the combined use of HPLC relative retention times and ESI-MS.[244]

3   **4**   **5** $R_1 = Me_1 R_2 = H$
                                                                      6 $R_1 = R_2 = Me$

1.4.1.1.3   Analysis of Urinary Nucleosides

Chung and coworkers reported the analysis of nucleosides in urine by LC-ESI-MS using direct (i.e., no extra purification step) sample injection.[245] This methodology appears to have great potential for the detection of urinary nucleosides down to 0.2 nmol mL^{-1}.

Wang et al. developed a simple, rapid, and efficient CE-MS method to analyze urinary nucleosides.[246] This method depended on the high resolution obtained by CE and the MS potential for quantification, which provided a suitable method for the study of urinary nucleoside profiles.

1.4.2   ESI-MS CHARACTERIZATION OF NUCLEOTIDES

1.4.2.1   Analysis of Nucleotides in Champagne

In the food industry, mononucleotides, particularly 5′-nucleotides, such as guanosine-5′-monophosphate (5′-GMP) and inosine-5′-monophosphate (5′-IMP), have been used as flavoring ingredients and can be produced by enzymatic RNA degradation during yeast autolysis.[247,248] RNA represents more than 95% of the total nucleic acid content of yeast cells and is known to degrade more than DNA during autolysis.[249,250] One aspect of the traditional production of Champagne is long aging on lees, during which yeast autolysis occurs.[251] This results in intracellular yeast constituents, such as the degradation products of nucleic acids, being released into the champagne.[259] Formation of such nucleotides in Champagne will clearly affect its quality.[252–254] Aussenac and coworkers established a method for the isolation, separation, and identification of monophosphate nucleotides that they applied to the analysis of Champagne that had been aged on lees for 8 years.[255] For the first time, using ESI-MS analysis, 5′-IMP, 5′-AMP, 5′-CMP, 5′GMP, 5′UMP, and the 3′- and/or 2′-isomers were identified in Champagne.

1.4.2.2   Determination of Nucleotides in Biomimicking Prebiotic Synthesis

It has been found that N-(O,O-diisopropyl) phosphoryl amino acid (Dipp-aa) (Figure 1.16) can induce some interesting biomimicking reactions that could lead to the formation of polypeptides and oligonucleotides.[256–260]

Liu et al. carried out simultaneous measurement of trace monoadenosine (AMP) and diadenosine monophosphate (ApA) in biomimicking prebiotic synthesis using HPLC-UV-ESI-MS.[261] Using this method, the positional isomers 2′-, 3′-, and 5′-AMP, and 3′ to 5′ ApA were simultaneously analyzed by a gradient elution with 10 mM NH$_4$Ac–CH$_3$OH mobile phase. The specific ions, [AMP+H]$^+$ at *m/z* 348 and [ApA+H]$^+$ at *m/z* 597, were employed for the identification of the four nucleotides. The reliability, sensitivity, and accuracy of the method ensured that the trace target nucleotides could be easily observed and accurately determined, whereas these trace nucleotides could not be distinguished and determined merely by the ESI-MS technique. This method was found to be suitable for further investigation on the reaction between Dipp-aa and single adenosine, cytidine, guanosine, uridine, or the mixed nucleosides.

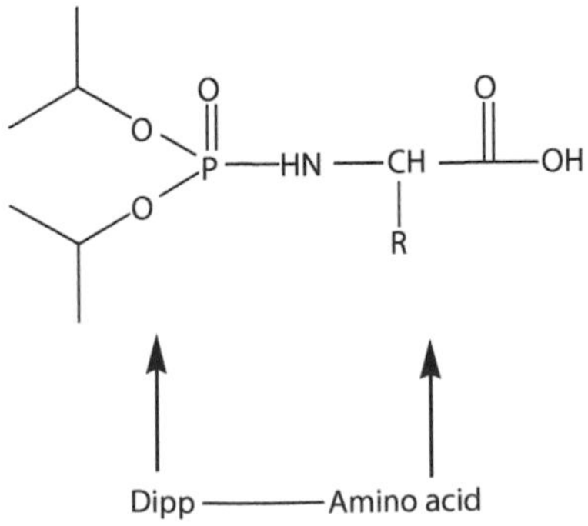

FIGURE 1.16   Diagram of Dipp-aa.

1.4.2.3   Determination of Nucleotides on Pressure-Assisted Capillary Electrophoresis

Another method for the determination of nucleotides is based on pressure-assisted capillary electrophoresis (PACE) coupled to ESI-MS. Cao and Moini, who had developed a PACE-MS method for the analysis of peptides, applied a similar PACE-MS method for the characterization of nucleotides using a noncharged polymer coated capillary to prevent sample adsorption.[262,263] This method enabled the simultaneous determination of nucleotides and CoA compounds. They indicated that the fluctuating current during electrophoresis resulted in time-shift migration of the analytes that clogged the capillary inlet.

Soga et al. proposed a new PACE-MS method for the analysis of nucleotides and CoAs using a normal fused-silica capillary where silanols were masked with phosphates to prevent the nucleotides from interacting with the capillary wall.[264] Nucleotides were driven toward the cathode by both Electroosmotic flow (EOF) and applied air pressure, followed by MS detection.[265] This approach is reproducible, quantitative, and robust, and it was readily applied to the simultaneous analysis of nucleotides from *E. coli* samples. Compared with other techniques, this method has several advantages: (1) various types of phosphorylated species such as nucleotides, nicotinamide–adenine dinucleotides, and CoA compounds are simultaneously analyzed; (2) sensitivity and selectivity were sufficient for quantification; and (3) good reproducibility, linearity, and robustness are obtained. The utility of this method was demonstrated by the analysis of other phosphorylated species in *E. coli* wild type, *pfkA* or *pfkB* knockout mutants, and the results provided valuable information about the enzyme activity.

1.4.2.4   HPLC-ESI-MS Quantification of Nucleotides

HPLC-ESI-MS methods that were able to quantify and/or separate nucleotides have been developed.[122,124,266–269] However, some of these methods used phosphate buffer; although known to improve chromatographic performance, it also results in ion suppression and, therefore, poor reproducibility.[122,124] When using phosphate buffer, automation is not possible because the salt precipitation in the ESI source necessitates very frequent cleanings. It is well known that cyclic nucleoside monophosphates also have ion-suppressive properties.[270] Surprisingly, in most published nucleotide LC-MS methods, the potential negative effects of ion suppression on sensitivity, accuracy, and reproducibility have not been addressed.[267]

Klawitter et al. studied the development and the validation of an assay for the quantification of 10 nucleotides AMP, ADP, ATP, CDP, CTP, FAD, GDP, GTP, UDP, and UTP.[271] They compared different extraction procedures and evaluated, using LC-ESI-MS analysis, the effect of ion suppression by nucleotides, the mobile phase, and the matrix. They also systematically assessed the influence of ischemia during the tissue collection process.[271] They showed that ATP and ADP concentrations and the resulting energy charge in tissues are very sensitive to hypoxia (Figure 1.17).

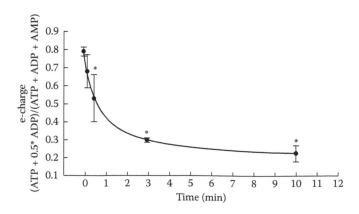

FIGURE 1.17   Time-dependent changes of cellular energy charge during collection of kidney tissue from rats.

Freeze clamping of the perfused organ is considered to be the "gold standard." Their results indicated that it was important to have the collection procedures appropriately validated and compared with the "gold standard" before meaningful conclusions could be drawn from nucleotide concentration data.

A major concern was the ion-suppressive properties of nucleotide compounds and the possible ion-suppressive properties of other compounds in the matrix of the PCA extracts. These were systematically investigated following the procedure described by Müller and coworkers, where $NADP^+$ (m/z 742) was monitored in the single ion mode, resulting in a constant signal.[272] When matrix extracts or other solutions were injected and analyzed, suppression of the $NADP^+$ signal and the appearance of "negative peaks" were noted (Figure 1.18).[271] The injection of 20 µL of PCA extract from 200 mg of rat kidney tissue resulted in an $NADP^+$ signal that decreased with the gradient and showed negative peaks at the positions where ion suppression occurred (injection peak and nucleotides). The decrease caused by the gradient was also observed when methanol or water was injected, and this was caused by the ion-pairing reagent in the mobile phase. Although ion suppression by the mobile phase reduced sensitivity, it was reproducible between runs and did not negatively affect the quantitation and reproducibility of the assay. The nucleotides themselves turned out to be the major source of ion suppression under the chosen conditions (Figure 1.18).

1.4.3 ESI-MS Characterization of DNA Adducts

1.4.3.1 Enediyne-DNA Adducts

The enediyne-DNA adducts were evaluated by ESI-MS. This study indicated that both the position and mode of attachment of a drug or ligand to a DNA sequence could be determined.[273] In addition to confirming the information previously obtained by gel electrophoresis, the analysis provided both the identification and stoichiometry of the adducts. They concluded that the analysis of the carefully selected oligonucleotides that were incubated with the novel enediyne ligand revealed several important clues about the interactions between ligand and DNA. Mass analysis confirmed a direct interaction that was not mediated by metals or other compounds. CID of adducts revealed that, in many cases, the ejection of the ligand and the cleavage of the DNA backbone occurred to a similar extent. This indicated that the DNA–ligand bond is of a comparable strength to the bonds within the oligonucleotide, thereby suggesting that the DNA–ligand adduct is formed by a covalent linkage.

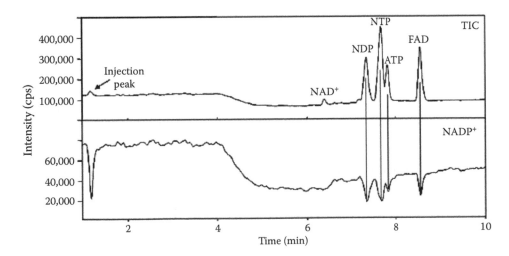

FIGURE 1.18 Representative ion-suppression experiment.

1.4.3.2 Etheno-DNA Adducts

The etheno derivatives, 1, N^2- and $N^{2,3}$-etheno-2′-deoxyguanosine (**7**, **8**), 1,N^6-etheno-2′-deoxy-adenosine (**9**), and 3,N^4-etheno-2′-deoxycytidine (**10**), are promutagenic lesions in DNA. These adducts are produced by either endogenous metabolic processes (e.g., lipid peroxidation) or metabolic activation of exogenous chemicals (e.g., vinyl chloride or urethane).

Doerge et al. developed a method for the quantification of trace (one adduct in 10^8 normal nucleotides from 100 μg of DNA) levels of etheno-DNA adducts using on-line sample preparation LC-ESI-MS.[274] They used automated solid-phase extraction and stable labeled internal standards for the determination of (**9**), which was contained in crude DNA hydrolysates from untreated rodent and human tissues. The method was also applied in the analysis of liver DNA from untreated and urethane-treated B6C3F1 mice, untreated rat liver, human placenta, and several commercial DNA preparations.[274]

| 7 | 8 | 9 | 10 |

1.4.3.3 Base-Alkylated Adducts

Melphalan (**11**) is a bifunctional alkylating agent that covalently binds with intracellular nucleophilic sites. Esmans and coworkers developed a methodology, using ESI-MS, to detect and identify DNA adducts, including the alkylation sites, within a nucleotide.[275,276] This subject is dealt with in more detail in Chapter 6 of this book.

11

1.4.3.4 LC-MS Analysis of Oxidized Damaged Bases on Nucleotides

It has been proposed that reactive oxygen species, produced by inflammatory cells, contribute to inflammation-mediated carcinogenesis, in both animals and humans, by inducing oxidative DNA damage.[277–280] 8-Hydroxydeoxyguanosine (8-OHdG), a form of guanine oxidized at the C-8 position, is considered to be fairly stable and the most abundant oxidative lesion among many oxidized nucleosides.[281] The development of HPLC coupled with ultraviolet/electrochemical detection (HPLC–UV–ECD) for the sensitive and precise analysis of 8-OHdG has furthered our understanding of its biological function.[282] 8-OHdG can induce GC:TA transversion, which is frequently found in the tumor genes of a variety of cancers.[283,284]

Under inflammatory conditions, both 8-nitroguanine (NO_2Gua) and 8-hydroxydeoxyguanosine (8-OHdG) are found in tissues. Measurement of both these types of damaged bases in nucleotides may be able to provide evidence of the possible correlation between inflammation and carcinogenesis. For the establishment of an *in vivo* model, Yuji Ishii's group developed a sensitive and precise *in vitro* method for the determination of NO_2Gua, which uses LC-MS and 6-methoxy-2-naphthyl glyoxal (MTNG) derivatization.[285] The procedure for DNA digestion in their method is identical to that widely used for 8-OHdG measurement, enabling them to detect the two damaged bases in the same DNA sample. In order to validate their method, NO_2Gua levels were measured in a DNA sample using LC-MS. The buffer solutions, at the high concentration used in the DNA digestion process, cause ion suppression in MS analyses, and can damage the MS detector; therefore, isolation of the NO_2Gua from the buffer by its retention on a column is a required step. The derivatization of NO_2Gua by MTNG was able to improve its polarity (Figure 1.19a). ESI-MS (the − ion mode) was set up with SIM at m/z 391 and 394 for NO_2Gua-MTNG and $[^{13}C, ^{15}N2]$-NO_2Gua-MTNG as surrogate standard, respectively (Figure 1.19). The average recoveries from DNA samples spiked with 25, 50, and 250 nM NO_2Gua were 99.4%, 99.8%, and 99.1% with correction, using the added surrogate standard, respectively. To ascertain the applicability of this method to DNA samples harboring the two damaged bases, NO_2Gua and 8-OHdG levels were measured in calf thymus DNA treated with ONOO⁻. As a result, both NO_2Gua and 8-OHdG levels were clearly increased with ONOO⁻ dose dependency, the amount of NO_2Gua at the high dose ONOO⁻ being almost the same as those of 8-OHdG. LC-MS was able to determine NO_2Gua in a small amount of DNA sample, and is therefore expected to be a very powerful tool for the evaluation of DNA damage induced by reactive nitrogen species.

1.4.4 ESI-MS GENOTYPING OF SNPS

A large number of methods have been devised for determining the allelic state of SNPs and they have been comprehensively reviewed.[286–288] The exact determination of allele frequency in polyploidy organisms, and eventually in populations, is essential in association studies. Thus, there is growing interest in methodologies that are suitable for the quantitative determination of allele abundances.[289–291] Enzymatic amplification of DNA sequences by PCR is usually the first step in the genotyping of SNPs.[292] MALDI-MS and ESI-MS have become valuable methods that are suitable for allele determination after amplification by PCR.[293–298] While MALDI-MS is predominantly utilized for the high-throughput analysis of short products of primer extension minisequencing reactions, ESI-MS is applicable to the direct mass analysis of single- and double-stranded PCR amplicons up to 500 bp or more.[294,299,300]

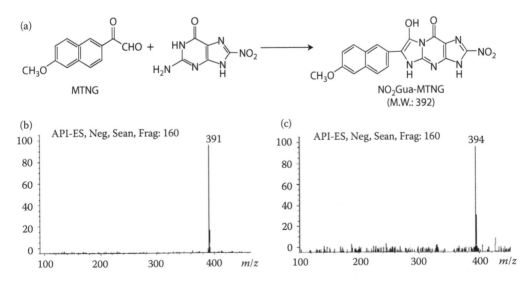

FIGURE 1.19 (a) The reaction of NO_2Gua with the MTNG. MS spectra of NO_2Gua-MTNG (b) and $[^{13}C, ^{15}N2]$-NO_2Gua-MTNG (c).

Oberacher et al. optimized the suppression of adducts in PCR products for semiquantitative SNP genotyping by LC-MS.[301a] It was shown that partially denaturing during HPLC separation was the method of choice for the physical mutation scanning approach. This method used an ion-pair reversed-phase HPLC with electrospray ionization quadrupole ion trap mass spectrometry (ICEMS). This ICEMS method obviates the need for the selection of appropriate temperatures for resolving the heteroduplices. It also allows the discrimination, due to distinct mass differences between nucleobases, of the different alleles even when they coelute.[301b] The optimization experiments have shown that the suppression of cation adduction of nucleic acids in ESI-MS was strongly supported by the chromatographic separation at temperatures above 60°C, the eluent having concentrations of 20–30 mmol L^{-1} of butyldimethylammonium bicarbonate, and the addition of 10 mmol L^{-1} EDTA to the sample. This cation suppression was essential for the correct genotyping of both diploid and polyploid genomes by ICEMS. The strategy applied here for the semiquantitative determination of allele frequencies was based on the chromatographic purification, under denaturing conditions, of PCR fragments containing the polymorphic sites, followed by subsequent MS detection of the single-stranded species that are characteristic of the different alleles. Since the allelic ratio in the genome is conserved during PCR, LC, and MS, the relative signal intensities in the deconvoluted mass spectra were utilized to deduce the relative allelic ratios.[301a,301b]

PCR products of SNP 44–174, containing a T > C polymorphism, were amplified and analyzed by ICEMS for the determination of allelic ratios of SNPs in the tetraploid potato genome. The series of multiply charged ions were deconvoluted to yield the molecular masses of the single strands that allowed the identification and the relative quantitation of the different alleles (Figure 1.20 and Table 1.3).[301a] It was shown that differentiation between homo- and heterozygous samples, using an

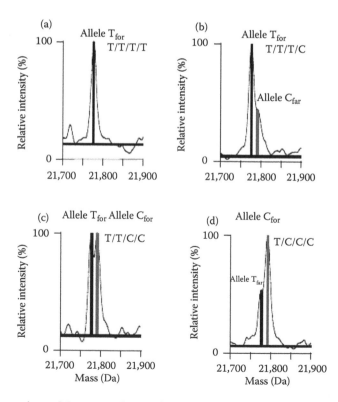

FIGURE 1.20 Comparison of the genotyping results obtained for four different allelic combinations of the potato SNP 44–174 by measuring the relative peak intensities of the forward strands of the two possible alleles in the deconvoluted mass spectra.

TABLE 1.3
Semiquantitative Genotyping of Four Different Allelic Combinations of the Potato SNP 44–174 by Measuring the Relative Peak Intensities of the Forward and Reverse Single Strands of the Two Possible Alleles T and C in the Deconvoluted Mass Spectrum

Genotype	T/T/T/T		T/T/T/C		T/T/C/C		C/C/C/T	
Allele	T	C	T	C	T	C	T	C
Theoretical frequency (%)	100	0	75	25	50	50	25	75
Average measured allele frequency (%)	100	0	72	28	52	48	34	66
Standard deviation (%)	0.0	4.7	3.0	1.6				
Number of measurements	3	3	10	3				
Absolute deviation from expected	—	±3	±2	±9				

Source: Adapted from Oberacher, H. et al., *J. Am. Soc. Mass Spectrom.* 15, 1897, 2004. With permission.

ion trap MS, could be effected by measuring the molecular masses of the single strands; these differed by 16 Da, and corresponded to the two alleles.[301a,301b]

Extracting the signal intensities from the deconvoluted mass spectra of PCR amplicons facilitates the rapid, semiquantitative determination of relative allele ratios by ICEMS with less than 10% error. Nevertheless, the use of high-resolution MS could further improve quantitative accuracy. In contrast to most other established methods for allele quantitation, ICEMS can analyze PCR amplicons in less than 15 min and without any other postPCR sample treatment.[301a,302b] Moreover, there is potential for a further increase in sample throughput with available rapid column regeneration and equilibration protocols that are compatible with the monolithic separation media.[302] A major advantage of MS investigation of SNP is the ability to simultaneously investigate individual both strands of the PCR product, enhancing both confidence in the identification and quantitative accuracies.

The advantages of ion-pair reversed-phase HPLC combined with ESI-Q-TOF-MS for the semiquantitative genotyping of SNP has been evaluated.[303] To do this they used, as a reference sample, mitochondrial DNA mixtures that have different levels of heteroplasmy at nucleotide position 16,519. The observed assay performance regarding accuracy, reproducibility, and sensitivity suggests that the described technique represents one of the most powerful assays available today for the determination of allelic contents.

1.4.5 ESI-MS CHARACTERIZATION OF OLIGONUCLEOTIDES

1.4.5.1 Practical Aspects of Oligonucleotides Analysis

The factors affecting the ESI-MS analysis of oligonucleotides that are crucial to understanding their analytical requirements have been discussed previously by Crain.[11a–e,15] Native and modified synthetic oligonucleotides have been traditionally purified by reverse-phase HPLC using volatile ion-pairing mobile phases. Gilar et al. demonstrated the purification of 10–90 nmol of oligonucleotides in a single injection on a 4.6 × 75 mm HPLC column packed with porous 2.5 μm C18 sorbent.[304,305] This group has also reported the analysis of desalted oligodeoxynucleotides using carefully optimized LC-ESI-TOF-MS, which allows a throughput of nearly 1000 samples in 24 h.[306,307] The average mass accuracy was 80 ppm for oligonucleotides up to 110-mer length.

Stults and Masters reported a method for the rapid purification of synthetic oligodeoxynucleotides to remove sodium counterions prior to ESI-MS.[236] This method entailed purefying oligomers by gel electrophoresis, and precipitating them by ammonium acetate (where ammonium ions replaced the sodium ions), and then dissolving these oligomers in water. Negative-ion ESI-MS scans were presented for oligomers with up to 48 residues in which the most intense peaks correspond to the $[M-nH]^{n-}$ ions.

Cavanagh et al. described the use of a simple in-line gel cartridge for desalting of solutions prior to ESI-MS, which is valuable for the investigation of protein–DNA interactions.[308]

Muddiman and his group have carried out an evaluation of sample preparation methods.[309] They point out that prior to any MS analysis, PCR products require a cleanup protocol that is both rapid and amenable to automation with the potential to process hundreds of sample per day. They compare various combinations of Geneclean (a commercial product), ethanol, or 2-propanol precipitation and microdialysis, characterizing each method by examining the quality of the ESI-FT-ICR spectra of an 82 base-pair PCR product. Although ethanol precipitation followed by microdialysis is the best procedure, other less time-consuming alternatives can give results that are almost as good (Geneclean with ethanol precipitation or tandem ethanol precipitation).[309]

The accuracy of the determination of the mass of n-mer fragments was illustrated by the work of Deroussent et al.[310] An ESI-MS method for the analysis of desalted oligonucleotides used for pharmacological studies was evaluated for sensitivity and accuracy with two antisense ODN sequences. The transformed and raw negative-ion ESI-MS for a 25-mer phosphorothioate (Figure 1.21) showed the $[M+H]^+$ of the 24-mer sequence (M_r 7456.25 Da), indicating that thymidine phosphorothioate (M_r 320.2 Da) has been lost at the 3′ terminal end. Mass analysis of the 25-mer phosphorothioate (M_r 7776.58 ± 1.44 Da) was performed to within 0.001% accuracy

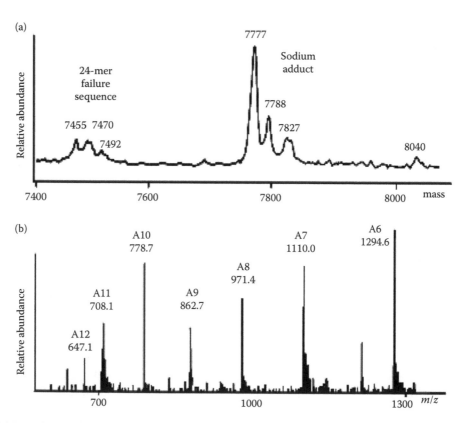

FIGURE 1.21   Transformed (a) and negative-ion spectra (b) of a 25-mer phosphorothioate oligonucleotide that targets the GAG region of the human immunodeficiency virus (HIV)-1 genome. Thr-transformed spectrum shows the intact oligonucleotide whose mass was determined as 7777.58 (1.44 Da and the 24-mer species (M_r 7456.25) derived from the loss of thymidine phosphorothioate from the 3′ end.

(standard error of 0.05 Da) for a sample concentration of 12 pmol μL^{-1}. In contrast, the ESI-MS of a 77-mer (M_r 24,039 ± 1.21 Da) showed the [M+Na]$^+$ adduct as the predominant peak, indicating the retention of only a single sodium ion.

Another method for the suppression of signals from alkali-adducted ions involved the addition of millimolar concentrations of a series of organic bases with solution values that ranged from 11.5 to 5.5 pKb.[311] Stronger bases, such as triethylamine and piperidine, were most effective in reducing the signals from bound sodium, but also decreased the total ion counts from the oligonucleotide. However, imidazole (pH ≈ 8.0) provided a modest suppression of sodium/potassium adduct ions with a fourfold improvement in sensitivity. Coaddition of imidazole and triethylamine or piperidine produced high ion abundance and good suppression of cation-adducted species, especially for samples of phosphodiester or phosphorothioate oligomers which were not desalted via preliminary precipitation or by HPLC. The addition of high concentrations of imidazole generated a bimodal distribution of charge states, reflecting the different gas-phase conformations for single-stranded oligomers.[311] The introduction of 0.1 M imidazole in acetonitrile after the HPLC column is also reported to afford increased sensitivity for the ESI-MS analysis of 20–50-mer oligonucleotides.[312]

A mixed-mode stationary phase has been developed by Van Breemen et al., particularly for LC-ESI-MS of oligonucleotides.[313] The potential affinity ligand, 3-(thymid-1-yl)propanoic acid, was immobilized on an aminopropyl reverse-phase HPLC column to give a system with a combination of reverse-phase, ion exchange, and affinity properties. Optimum column performance was achieved using gradients that utilized the potential affinity properties.

1.4.5.2   ESI-MS Analysis of Synthetic Oligonucleotides

A versatile synthetic route for the synthesis of purine and pyrimidine nucleosides modified as the 2′-O-{2-[(N,N-dimethylamino)oxy]ethyl} derivatives was developed by Prakash et al.[314,315] The synthetic oligonucleotides with this 2′-O modification were characterized by HPLC, capillary gel electrophoresis, and ESI-MS. They found that this modification enhanced the binding affinity of the oligonucleotides for the complementary RNA but not for DNA, and produced excellent nuclease stability.

1.4.5.3   ESI-MS Oligonucleotides Sequencing

Janning and coworkers describe an approach for the enzymatic digestion of DNA, yielding oligonucleotides ranging from dinucleoside monophosphate to octanucleoside heptaphosphates.[316] Calf thymus DNA was digested by means of the benzon nuclease, an unspecific nuclease, and alkaline phosphatase to remove the terminal phosphate. The mixture of oligonucleotides was separated using capillary-zone electrophoresis (CZE) with a buffer system with a strong electro-osmotic flow. The oligomers were separated into groups with nucleotides of the same chain length; typically 10 dimers (mass 516–596 Da), 20 trimers (mass 805–925 Da), 35 tetramers (mass 1094–1254 Da), and 56 pentamers (mass 1383–1583 Da). The CZE-ESI-MS were recorded in the negative-ion mode using a multisector-field MS.

The combination of exonuclease digestion and MS is another excellent sequencing method for oligonucleotides of greater than 25-mer. However, during an exonuclease digestion, the rapid buildup of nucleotide concentration produces strong signals of nucleotide cluster ions in ESI-MS. Wu and Aboleneen reported a method where alkaline phosphatase was added with snake venom phosphodiesterase to the oligonucleotide solution to convert the interfering nucleotides into noninterfering nucleosides, a procedure that eliminated the effect of the cluster ions.[317] With this approach, the signal/noise ratio of the reconstructed molecular weight spectrum for relatively large oligonucleotides was greatly improved, and only a single digestion was needed for sequencing.

1.4.5.4   Combined ESI-MS Analyses of DNA Oligonucleotides and DNA Adducts

There have been a number of MS and non-MS detection techniques used in the analysis of DNA adducts.[318–327] It has been established that the MS-based detection techniques provide excellent and reliable structural information on the DNA adducts compared with the majority of published DNA-adduct studies. Yong-Lai Feng et al. developed the technique of pressure-assisted electrokinetic injection (PAEKI) for measurements of DNA oligonucleotides and their adducts using CE-ESI-MS.[328] The identification and measurement of negatively charged DNA oligonucleotides and their benzo[a]pyrene-7,8,9,10-tetrahydro-7,8-dihydrodiol-9,10-epoxide (BDE) adducts were carried out by using a CE-MS system employing an on-line enrichment technique, the constant PAEKI.

1.4.6   ESI-MS CHARACTERIZATION OF PCR PRODUCTS

Oberacher et al. developed an application for the analysis of genomic DNA by PCR-LC-ESI-MS.[329] PCR is a cyclic reaction where a temperature-resistant DNA polymerase is used to specifically amplify a certain DNA segment framed by a pair of oligonucleotide primers with binding sites on opposite strands.[330] PCR solutions contain core components that are indispensable for DNA replication (polymerase, primers, dNTPs, and buffer) as well as a number of different additives, such as bovine serum albumin (BSA), PCR enhancers, nonionic detergents, and/or small concentrations of solvents like glycerol, dimethyl sulfoxide (DMSO), or formamide. These auxiliary substances are added to either relieve amplification inhibition or to facilitate the efficient amplification of long or problematic templates like the GC-rich sequences. It has been shown that *Bacillus subtilis* and *Bacillus atrophaeus* are two closely related species with few distinguishing phenotypic characteristics. Also, *B. subtilis* can be divided into two subgroups, W23 (type strain, W23) and 168 (type strain, 168). Comparison of sequence variability in the intergenic spacer region (ISR) between the 16S and 23S rRNA genes has been a useful tool for differentiating between the bacterial species. This ISR region has been shown to vary in size and sequence even within closely related taxonomic groups.[329,330]

Johnson and coworkers reported the analysis by ESI-MS of the PCR products amplified from the ISR-PCR using a primer pair corresponding to conserved sequences of the 16S and 23 rRNA genes, including the 5'-terminal end of the 23S rRNA and a conserved portion of the ISR.[331] It was found that a 119 or 120 base-pair PCR product was produced for *B. atrophaeus* strains; however, the W23 and 168 subgroups of *B. subtilis* each produced products of 114 base pairs. The differentiation of *B. subtilis* and *B. atrophaeus* and the genetic similarity of *B. subtilis* subgroups W23 and 168 were confirmed by ESI-MS analysis. Accurate determination of the molecular weight of PCR products from the 16S-23S rRNA ISR using ESI-MS has shown potential as a general technique for characterizing closely related bacterial species.

1.4.7   ESI-MS CHARACTERIZATION OF COVALENT OR NONCOVALENT DNA
AND RNA COMPLEXES

ESI-MS has proved to be a useful tool for studying biomolecular structures and noncovalent interactions involving proteins with metals, ligands, peptides, oligonucleotides, and other proteins.[332] Since the first reports in 1991 by Ganem, Li, and Henion, the use of conventional ESI-MS for studying noncovalent complexes in solution has been successfully demonstrated and reviewed.[333,334] Analyses of noncovalent complexes of DNA and RNA by MS have been covered in a review by Hofstadler and Griffey.[335] This technique has been used to determine the stoichiometry and dissociation constants for a variety of biological noncovalent complexes by either titration or competition experiments. The advantages of ESI-MS over other techniques for noncovalent binding studies include high sensitivity, analysis speed, stoichiometric information capabilities, and the ability to identify unknown species.

1.4.7.1   ESI-MS Characterization of Covalent DNA Complexes

DNA methylation is a postreplicative process that is regulated by the DNA Mtases. This biologically important class of enzymes catalyzes the transfer of the activated methyl group from the cofactor

S-adenosyl-L-methionine to adenine N-6, cytosine N-4, or cytosine C-5 within their DNA recognition sequences. A particular DNA sequence may exist as its fully methylated, unmethylated, or transient hemimethylated form. DNA methylation can be regarded as an increase in the information content of the DNA serving a wide variety of biological functions. Mtases flip their target adenines or cytosines out of the DNA, double helix into a pocket within the cofactor-binding domains, where catalysis takes place. Biochemical evidence for a base-flipping mechanism of the N^6-adenine DNA Mtases M.*Eco*RI and M.*Taq*I was obtained using duplex oligodeoxyribonucleotides containing the fluorescent base analogue, 2-aminopurine, at the target positions.

Krauss and coworkers investigated the site of the covalent coupling of the N^6-adenine DNA Mtases M.*Taq*I and M.*Cvi*BIII with duplex DNA.[336] This was achieved by synthesizing a duplex DNA that had the adenine moiety in the recognition sequence 5′-TCGA-3′ replaced by 5-iodoracil, affording species (**12**), which was then cross-linked to the enzyme using a high-yield photochemical reaction. Analyses by anion-exchange chromatography revealed that almost all the duplex oligodeoxyribonucleotide (ODN) had reacted, forming the nucleopeptide (**13**). The total mixture from the photo-cross-linking reaction from M.*Cvi*BIII was treated with chymotrypsin, and the resulting nucleopeptide was purified by anion-exchange chromatography and analyzed by ESI in the negative ion mode using a double-focusing sector field instrument.

12   **13**

The first 13 residues of the peptide sequence were found to be DFIVGNPPXVVRP, where the residue at position 9 (X) was not recognized as one of the 20 natural amino acids. This peptide sequence corresponds to the amino acid residues 114–126 of M.*Cvi*BIII, which contains a tyrosine residue at position 122, strongly suggesting that this is the site of DNA coupling with the intact Mtase. The C-terminal part of the nucleopeptide and its overall structure were verified by ESI-MS. The DNA strands of the covalently linked nucleopeptide underwent dissociation in the electrospray interface, and the modified strand was detected as five to ninefold negatively charged ions. The modified strand's deconvoluted MS showed that the observed mass of the nucleopeptide was 5900.5 Da. This observed value was in close agreement with the mass of 5900.6 Da, which was calculated for the nucleopeptide containing the amino acid residues 114–129 of M.*Cvi*BIII and which has a chymotrypsin cleavage site at its C-terminal end.[336]

Wong and Reich described a similar, highly sensitive strategy that combined laser-induced photo-cross-linking and HPLC-ESI-MS for identification of the amino acid residues involved in Mtase-DNA recognition duplexes to the enzyme M.*Eco*RI.[337] The covalent linking of hemimethylated DNA duplexes to the enzyme M.*Eco*RI was achieved by photochemical coupling between a 5-iodouracil, incorporated in the recognition site, and the tyrosine 204 residue in the enzyme. Negative ionization ESI-MS analysis was utilized to obtain the accurate masses of these oligonucleotides. As shown in Table 1.4, the mass data provide unequivocal confirmation of the structure of the component species and the coupled complex.

TABLE 1.4
ESI-MS Data for Complex of M.*Eco*RI with a DNA Strand

Species	Calcd mass/Da	Exptl mass/Da
DNA (dGGCG GAATTCGCGG)[a]	4345.1	4344.7 ± 0.43
DNA-I (dGGCG GAAITCGCGG)[b]	4456.1	4455.3 ± 0.08
M.*Eco*RI (325 amino acids)	37,913.4	37,916.9 ± 30.52
DNA-M.*Eco*RI complex	42,241.5	42,242.9 ± 21.28

Source: Adapted from Wong, D.L. and Reich, N.O., *Biochemistry* 39, 15410, 2000.
          With permission.
[a] The M.*Eco*RI recognition site in the mer-14 oligonucleotide is underlined.
[b] The recognition site is modified by substitution of iodouracil (**I**) for thymine (**T**).

1.4.7.2   ESI-MS Characterization of Noncovalent DNA Complexes

DNA is capable of binding to many different types of molecules, and MS has become a crucial technique for the study of these relatively weak complexes. Thus, ESI-MS is a useful tool for the characterization of specific drug binding to DNA and can be used in many studies of protein–DNA interaction.

1.4.7.2.1   ESI-MS Characterization of Complexes of Double-Stranded Oligonucleotides and Drug Molecules

Gabelica et al. investigated, by using ESI-MS, the complex formation between double-stranded (ds) oligonucleotides and various antitumor drugs.[338] The ds oligonucleotides used were formed by a nonself-complementary oligonucleotide d(GGGGATATGGGG) and the complementary strand d(CCCCATATCCCC). Two types of antitumor drugs, intercalators (ethidium bromide, amsacrine, and ascididemin), and minor-groove binders (Hoechst 33258, netropsin, distamycin A, berenil, and DAPI) were used. This study assessed whether the relative intensities in the MS reflected the actual abundances of the species in the solution phase. This is dealt with in more detail in Chapter 8 of this book.

An elegant study of noncovalent binding of the antitumor drugs nogalamycin (**14**) and daunomycin (**15**) to duplex DNA has been reported by Sheil and co-workers.[339] The conditions for the drug–duplex complex formation were carefully optimized for the self-complementary oligonucleotides [5′-d(GGCTAGCC)-3′ or 5′-d(CGGCGCCG)-3′] and these formed the corresponding duplex DNA, which were bound to three molecules of the drug. There was also some evidence for dsDNA with four molecules of the drug bound to it. The assignments of the MS of the complexes of (**14**) with both 10-mer oligonucleotides are given in Table 1.5. For the larger 12-mer species, 5′-(TGAGCTAGCTCA)$_2$-3′, binding of up to four nogalamycin and six daunomycin molecules was observed. These data were consistent with the neighbor exclusion principle, whereby intercalation occurred between every other base pair to the extent where up to four bound drugs would be expected for the 8-mer and up to six for the 12-mer.[339] This group also reported, in another positive-ion ESI-MS study of a 16-mer dsDNA, no evidence of any of the ssDNA ions that had complicated many of the early ESI-MS studies of dsDNA and DNA–drug complexes.[340] They prevented the formation of ssDNA by carefully choosing the DNA sequences and by using a relatively high salt concentration (0.1 M ammonium acetate, pH 8.5) as well as a low desolvation temperature (40°C).[340]

In a different study, the ESI-MS of complexes of 16-mer dsDNA with cisplatin, daunomycin, and distamycin was obtained under conditions where the amounts of ssDNA produced were negligible.

TABLE 1.5
Assignments of Major Ions in the Optimized ESI Spectra of Complexes Formed by Mixing Nogalamycin (N) with Duplex DNA 5′-d(GGCTAGCC) (d1) or 5′-d(CGGCGCCG)-3′ (d2) in a Drug/DNA Ratio of 5:1 (Multiple Binding by Nogalamycin is Evident)

Fragment	Expl. Mass (d1)/Da	Exptl. Mass (d2)/Da
N	786.5	786.2
[M–5H]$^{5-}$	481.4	481.3
[M–4H]$^{4-}$	601.9	601.7
[M–3H]$^{3-}$	802.6	802.8
[M–2H]$^{2-}$		1204.8
[2M–5H]$^{5-}$		963.0
[2M+N–5H]$^{5-}$		1120.9
[2M+N–4H]$^{4-}$		1401.2
[M+N–3H]$^{3-}$/[2M+N–6H]$^{6-}$	1065.1	1065.4
[2M+2N–5H]$^{5-}$	1278.4	1278.1
[M+N–2H]$^{2-}$/[2M+2N–4H]$^{4-}$	1597.8	1598.3
[2M+3N–6H]$^{6-}$	1196.4	
[2M+3N–SH]$^{5-}$	1436.1^a	1436.4
[2M+3N–4H]$^{4-}$	1795.4	1795.3
[M+2N–3H]$^{3-}$/[2M+4N–6H]$^{6-}$	1327.9	1327.9
[2M+4N–5H]$^{5-}$	1593.7	1592.9
[M+2N–2H]$^{2-}$/[2M+4N–4H]$^{4-}$	1992.2	
[2M+5N–5H]$^{5-}$	1751.5	

Source: Adapted from Kapur, A. et al., *Rapid Commun. Mass Spectrom.* 13, 2489, 1999. With permission.
^{a} Most intense ion attributable to the complex.

In the gas phase, the dsDNA complexes with daunomycin and distamycin appeared to have less strand separation than the uncomplexed dsDNA.

1.4.7.3 ESI-MS Characterization of Noncovalent Complexes of Peptides and Double-Stranded DNA

The GCN4 peptides are a group of proteins that bind to dsDNA through a sequence-specific interaction and are important in the regulation of gene transcription in yeast. In addition to a basic DNA-binding domain, these proteins also contain a leucine zipper dimerization domain.[341,342] The protein dimers specifically bind to dsDNA containing the binding element 5′-ATGA(C/G)TCAT-3′ to form a tetramolecular noncovalent complex.[343] This noncovalent complex has been detected in the gas phase by ESI-MS. It is important to note that no ions were detected for the peptide dimer or for the DNA–peptide monomer complex, indicating that the tetrameric complex is an especially stable unit under the detection parameters. The observation of these important complexes by MS was extended to several types of dsDNA with up to 30 bp. The ability to observe these large complexes by MS introduces an analytical technique that can probe specific protein/DNA interactions and that can be applied to the study of other biological problems, including the study of the transcription processes.

1.4.7.4 ESI-MS Characterization of Complexes of Actinomycin D and Oligonucleotides

Zhou et al. studied the interactions between actinomycin D (ActD) and ssDNA 5′-CGTAACCA ACTGCAACGT-3′ and a duplex stranded DNA (dsDNA) with the same sequence using microchip-based, nongel sieving, electrophoresis and ESI-MS.[344] The ssDNA was designed according to the conserved regions of open reading frame 1b (replicase 1B) following the Tor 2 SARS genome sequence of 15,611–15,593. The binding constants and stoichiometries were obtained using Scatchard analysis. To validate the results provided by microchip electrophoresis, the stoichiometries of ActD and oligonucleotides were compared with ESI-MS data. The results from both microchip electrophoresis and ESI-MS showed that ActD bound much more tightly to the ssDNA-related severe acute respiratory syndrome associated coronavirus (SARSCoV) than its own complementary dsDNA.[344]

1.4.7.5 ESI-MS Characterization of Complexes of Ribonuclease A and Cytidylic Acid Ligands

Zhang and coworkers described a fully automated, chip-based nanoESI-MS system for the investigation of noncovalent interactions between ribonuclease A (RNase A, M_r 13,682) and cytidylic acid ligands (2′-CMP, M_r 323.2 Da; CTP, M_r 483.1 Da).[345] Titration and competitive binding approaches were both performed prior to the automated nanoESI-MS analysis with a Q-TOF-MS hybrid instrument. The measured Kd values for the complexes RNase A-2′-CMP and RNase A-CTP were found to be in agreement with the available published values obtained by standard spectroscopic techniques.

Several ESI sources in different commercial MS instruments were evaluated for use in the study of noncovalent interactions between ribonuclease A and cytidylic acid by Haskins et al.[346] It was observed that while the spectra from these various sources were very different, the structural information relating to the complex was, in all cases, identical. These studies also showed that the noncovalently bonded complexes between RNase A and various cytidylic acid substrates were most stable at lower charged states.

1.4.7.6 ESI-MS Characterization of Complexes of Platinum(II) and Nucleobases

ESI-MS study of platinum(II) complexes with nucleobases and DMSO has been investigated by Franska.[347] They showed that in ESI-MS (low cone voltage) adenine, guanine, and cytosine tend to form [M+PtCl+2DMSO]+ cationic complex ions.

At higher cone voltages, decomposition involving the loss of DMSO, HCl, and the sugar moiety (for nucleosides), was observed. Franska presumed that the nucleobase was situated in the *trans*-position to the chlorine and that platinum was attached to N-7 atom of guanine, N-3 atom of cytosine, and N-1 atom of adenine. The Na+ or K+ attachment was observed in ESI-MS as [M−H+PtCl+2DMSO+Na]+ and [M−H+PtCl+2DMSO+K]+ ions, which were characterized in low abundances. Uracil or thymine (or their nucleosides) form neutral complexes containing a deprotonated nucleobase (or nucleosides). The relatively higher acidity of uracil and thymine compared with that of the other nucleosides may explain the existence of complexes containing deprotonated molecules. It is difficult to propose structures for these complexes with the results obtained. The interactions of ruthenium(II)–DMSO complexes, shown to have anticancer properties, with nucleobases, nucleosides, and nucleotides have also been successfully studied using ESI-MS.[348,349]

1.4.7.7 ESI-MS Characterization of DNA Triplexes

Mariappan et al. performed the analysis of GAA/TTC DNA triplexes using NMR and ESI-MS.[350] They reported NMR and MS analysis for the pH and ligand-induced destabilization of triplexes using short intramolecular and intermolecular parallel GAA-TTC triplexes model systems. These assays could be applicable as general methods for characterizing structure, dynamics, and stability of DNA and DNA–ligand complexes. The authors have also investigated berenil, netropsin, distamycin, and acridine orange as model ligands. Berenil, distamycin, and netropsin are minor-groove-binding drugs, whereas acridine orange is an intercalator.

1.4.7.8 ESI-MS Characterization of RNA Noncovalent Interactions

RNA serves fundamental structural and catalytic roles in many cellular functions.[351,352] The product of gene transcription is mRNA, and its translation on ribosomes yields functional proteins. There are other functions for RNA, including the components (and perhaps the catalysis) of the RNA-splicing apparatus, the ribosome (including its peptidyl transferase activity), the signal recognition particle, RNA editing and other posttranscriptional RNA modifications, X-chromosome inactivation, telomere maintenance, and both natural and engineered regulation of gene expression through sense–antisense interactions.[353,354] RNA participates in noncovalent associations with metals, proteins, and DNA as well as other molecules of RNA. RNA is also the genetic material of many viruses and retroviruses such as HIV and HCV.[355,356]

Hoyne et al. highlighted the increasingly significant role of ESI-MS in the study of the noncovalent interactions involved in RNA–RNA complexes and also described the application of µESI-MS for the study of base-pairing interactions between separate RNA oligonucleotides.[357] A set of complementary RNA oligonucleotides (Table 1.6) was synthesized, and the RNA samples were desalted by sequential

TABLE 1.6
Comparison of Calculated and Measured M_r Values for Selected RNA Oligonucleotides and their Complexes

	RNA Type	Sequence and Structure	Calcd M_r/Da	Exptl. M_r/Da
Single strands	RNA1	*CAUCUCAGUCGCACUUA GCUCAGUCAUAGA	9494	9477[a]
				9592[b]
	RNA2	*GUGCGACUGAGAUG	4526	4510[a]
	RNA3	*UCUAUGACUGAGCU	4408	4406[c]
RNA complexes	RNA1 + RNA2	GUAGAGUCAGCGUG* *CAUCUCAGUCGCACUUA GCUCAGUCAUAGA	14,020	13,987[a] 14,102[d]
	RNA1 + RNA3	UCGAGUCAGUAUCU* *CAUCUCAGUCGCACUUA GCUCAGUCAUAGA	13,902	13,883[a] 13,996[b]
	RNA1 + RNA2 + RNA3	GUAGAGUCAGCGUG* UCGAGUCAGUAUCU* *CAUCUCAGUCGCACUUA GCUCAGUCAUAGA	18,428	18,395[a] 18,507[d]

Source: Adapted from Hoyne, P.R. et al., *Rapid Commun. Mass Spectrom.* 15, 1539, 2001. With permission.

Note: (*) denotes the 5′ terminus of the oligonucleotide.

[a] The mass discrepancy is due to replacement of a hydroxyl group by a hydrogen in one or both strands.

[b] The mass discrepancy is due to the presence of one tert-butyldimethylsilyl group in the RNA1 strand.

[c] Within 0.01% mass accuracy for the instrument.

[d] The mass discrepancy is due to a combination of the above modifications in the individual strands.

steps of gel filtration and cation-exchange chromatography prior to analysis by μESI-MS. The spectra of these desalted oligonucleotides (RNA1, RNA2, and RNA3) are shown in Figure 1.22.

Transformation of the multiply charged spectrum of RNA1 showed a prominent species with a mass of 9477 Da, which agreed with the calculated mass of the oligonucleotide with one OH → H modification (Table 1.6). A second peak with a mass of 9592 Da indicated a minor contamination

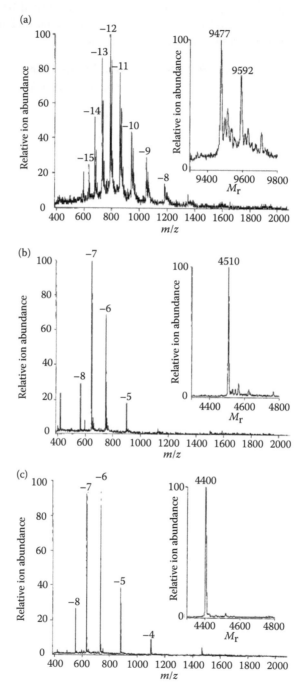

FIGURE 1.22   Negative-ion μESI-MS of three RNA oligonucleotides over the *m/z* range 440–2000. Samples were infused in water after desalting off-line suing a cation-exchange microcartridge. The inserts show transformed spectra and experimental M_r values: (a) RNA 1, (b) RNA 2, and (c) RNA 3.

due to the presence of a single remaining tert-butyldimethylsilyl protecting group on one of the 29 possible 2′-hydroxyl positions in the RNA1 oligonucleotide. The measured M_r for the oligonucleotide RNA2 again suggests an OH → H modification, whereas for oligonucleotide RNA3 there is agreement between the observed and calculated M_r values. Appropriate mixtures of these complementary RNA oligonucleotides produced the expected two- and three-component double-helical complexes as confirmed by the measured M_r values (Table 1.6).[357]

A unique member of the RNA-binding protein family is the translational regulator, Bacteriophage T4 regA protein. It consists of 122 amino acid residues and regulates the expression of as many as 30 early T4 genes, including the translation of its own mRNA. This is drastically different from the other translational repressors, which generally recognize only one or a few RNA sequences. Liu et al. used ESI-FT-ICR-MS to study the interactions of bacteriophage T4 regA protein with RNAs of various size and sequence. They were able to observe, by using very gentle interface conditions, the formation of 1:1 regA/RNA complexes for four different RNAs.[358] Filter binding experiments determined that each of the RNAs studied contained the sequence for the recognition element for regA. The high-resolution ESI-FT-ICR-MS of regA itself gave a charge state distribution that indicated an M_r value 14,618.4 Da, which is very close to the calculated value of 14,618.7 Da. When regA was mixed with a selected RNA at a molar ratio of 1:4 and analyzed by ESI-FT-ICR-MS, the 1:1 regA/RNA complexes (at higher m/z) and excess unbound RNA (at lower m/z) were detected (Figure 1.23). Numerous additional species, corresponding to other complexes and degradation products, were observed in the RNA profile. Interestingly, the corresponding regA complexes, as identified by accurate mass measurements, were also observed in many of the regA/RNA profiles. This correlation between complexed and free species allowed insight into the relative binding of different species with the protein. In some cases, the loss of a single nucleotide resulted in a dramatic change in binding affinity. The data obtained for the CID off-resonance irradiation directly revealed the information regarding the binding domain of regA in the regA/RNA complex.[358]

1.4.7.9   ESI-MS Characterization of RNA Duplexes

LC-ESI-MS with a hexafluoroisopropanol/triethylamine ion-pairing buffer and a methanol gradient was used to study a short interfering RNA (RNAi) duplex isolated from ocular metabolites.[359] Using ESI, the duplex was preserved in the gas phase for analysis by a triple quadrupole MS. The RNAi pathway utilizes short pieces (19–21 nucleotides) of dsRNA or small interfering RNA (siRNA) to interact with the RNA-induced silencing duplex, SIRNA-027. With this methodology, metabolites from rabbit ocular vitreous humor and retina/choroid tissue were identified and a pattern of siRNA

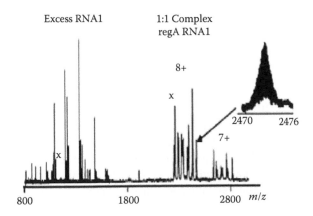

FIGURE 1.23 ESI mass spectrum of a mixture of RegA and oligonucleotide RNA1. Two sets of peaks at higher m/z correspond to a 1:1 complex with either 8+ or 7+ charge. Uncomplexed excess RNA1 species appear at lower m/z. (inset) Resolution of isotopic peaks. The peaks marked x indicate a particular RNA1 and its complex. Several such pairs can be identified.

degradation was established. Results showed that the duplex was metabolized predominantly from the end of the siRNA duplex with the weakest (calculated) binding energy, indicating that the ability of the siRNA to split into single strands is a factor in its degradation. It was rationalized that the nuclease preference to attack at one particular site or side of the duplex could be due to the nature of the single strand in the duplex. A duplex without the abasic and phosphorothioate groups showed a similar degradation profile, but without as clear a preference for one side of the duplex, indicating that chemical modifications might also play a role in directing the degradation.

1.4.7.10   Interaction of Cisplatin with DNA

The antineoplastic activity of cisplatin [*cis*-diaminedichloroplatinum(II)] is thought to derive from its tendency to form DNA adducts, especially through coordination with N7 of guanine. Various adducts are formed upon interaction of platinum complexes with nucleotides, but the extent of the contribution of individual adducts to antitumor activity is unknown. Adducts that were formed in the reaction of cisplatin with various nucleotides have been separated by CZE and identified by ESI-MS.[360] In the investigation of the cisplatin–DNA interaction mechanism, the outer-sphere association of several 20-mer oligonucleotides with $[Pt(NH_3)_4]^{2+}$ and $[Pt(py)_4]^{2+}$ has been measured by ESI-MS, and the affinity constants have been determined for some of these complexes. The results suggested that electrostatic effects and hydrogen-bonding potential were important factors in the outer-sphere complexation.[361]

1.4.8   Charge Detection ESI-MS

Fuerstenau and Benner investigated the potential of a charge detection (CD) ESI-TOF-MS instrument capable of determining the mass of multiply charged electrospray ions generated from samples of macromolecules in the megadalton size range.[237] The instrument utilized a sensitive amplifier that detected the charge on a single ion as it passed through a tube detector. A velocity measurement of an ion with known electrostatic energy provided the value of *m/z* for that ion. Thus, simultaneous determination of *z* and *m/z* for an ion permitted a mass assignment to be made in each case. ESI of DNA and polymer molecules with masses greater than 1×10^6 Da and charge numbers in excess of $425e^-$ were readily detected by MS. The on-axis single-ion detection configuration provided a duty cycle of nearly 100% and extended the practical application of ESI-MS to the analysis of very large molecules.

Schultz and co-workers used this CD-ESI-MS instrument for the rapid detection and analyses of PCR products (1525, 1982, and 2677 bp) with a minimum of postPCR sample cleanup.[362] A longer double-stranded, linear DNA sample has also been analyzed and both positive-ion mass spectra and ion charge-state distributions were obtained.

1.4.9   Atmospheric Pressure Photoionization MS

Atmospheric pressure photoionization (APPI) is based on the photoionization of a selected species with an ionization energy (IE) lower than the photon energy (*hv*). Usually it is a krypton discharge lamp that produces mostly 10 eV photons that is used. There has been a growing interest in the analysis of biomolecules by this technique, which permits the use of almost all kinds of solvents, even the most nonpolar ones, and is insensitive to the presence of salts.[363] Nucleic bases, ribonucleosides, and ribonucleotides have been investigated using APPI-MS.[364] In-source fragmentations have been studied, initially under thermospray conditions with the photoionization lamp switched off. In this mode, the fragmentations are minor and the compounds do not suffer from thermal degradation.

The fragmentation patterns of these biomolecules have been further studied both directly and with dopant-assisted photoionization using three different dopant molecules (toluene, anisole, and acetone). The relative ionization energies and proton affinities of the sample and dopant molecules play key roles in the relative abundance of the different precursor ions. The variety of precursor ions (radical molecular ions, or molecules protonated at various sites) that are generated under APPI conditions results in the versatility of this method. Some of the fragmentation patterns appear to be dependent on the nature of the dopant employed.

1.5　TANDEM MASS SPECTROMETRY

In addition to single-stage MS, tandem mass spectrometry (MS/MS) has become a particularly important analytical application in the structural characterization of nucleobases, 2′-deoxynucleosides, and 2′-deoxynucleotides.[365] The first example of an MS/MS of an oligonucleotide tetramer was reported by Sindona's group in 1983, and they also suggested the "zwitterionic structure" of gaseous oligonucleotides.[366] Subsequently, it was found that DNA and RNA oligonucleotide sequences were easier to generate using the MS/MS because the oligonucleotide ions were selectively fragmented in the MS collision cell.[366,367] These fragmentations are primarily induced by collision with neutral gas molecules. There have been a multitude of published reports describing the use of MS/MS for complete DNA sequencing.[368–370] These reported that MS/MS methods used several different kinetic MS/MS instruments, including metastable kinetic ion decomposition (MIKE), both low- and high-energy CID analyses of the deprotonated DNA anions, sector instruments, and lately, modern FT-ICR-MS^n.[304,371–377]

The availability of instruments (such as MALDI-TOF-TOF, MALDI-FT-ICR-, MALDI-Q-TOF, MALDI-QIT-TOF, and QIT-FT-ICR-MS^n) that provide the advantage of both high-sensitivity and comprehensive fragmentations, in either high- or low-energy CID-MS/MS, has resulted in a plethora of new technological developments in high-resolution hybrid MS/MS.[89,358,365–367,376]

1.5.1　FAB-MS/MS

1.5.1.1　FAB-MS/MS Characterization of Nucleosides

Sindona and coworkers investigated the synthesis of the isoxazolidinyl nucleosides by the 1,3-dipolar cycloaddition of nitrones to vinylnucleosides.[378] They obtained 1-(isoxazolidin-5-yl)thymine (AdT) (16) and the uracil (AdU) (17), which are analogues of 3′-deoxythymidine (ddT) (18) and 2′,3′-dideoxyuridine (ddU) (19). AdT and AdU are representatives of a new class of antiviral agents, which were characterized by FAB-MS/MS using a V6-ZAB-2F multisector instrument. The MS/MS analysis of (16) and (17) were similar to those of the dideoxyribose nucleosides, displaying fragment ions corresponding to the [BH$_2$]$^+$ and [M−BH]$^+$ species. They also show a [B+27]$^+$ species that appears to be formed from a retrocycloaddition process, which was obviously less evident in the dideoxyribose nucleosides (Figure 1.24). This behavior appeared to be consistent with protonation of the analytes at the pyrimidine rings. Model isoxazolidines, in which the nucleobase was replaced by a phenyl or a naphthyl moiety, displayed the expected behavior of species with a localized charge on the N−O moiety of the isoxazolidine ring.[378]

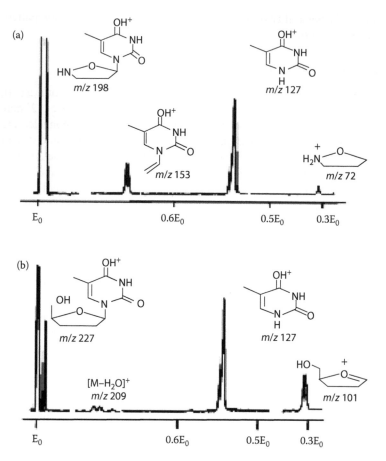

FIGURE 1.24 FAB tandem mass spectra (MIKE) of (a) the [M+H]⁺ species from AdT and (b) the [M+H]⁺ species from ddT. Glycerol was used as the matrix, and spectra were recorded in the absence of collision gas in the second field-free region of a B-E sector instrument.

1.5.1.2 FAB-MS/MS Characterization of Nucleotides

Newton and coworkers investigated the presence of cyclic nucleotides in the culture medium of *Corynebacterium murisepticum* by using FAB-MS/MS.[379] Earlier reports had suggested that the culture media contained cytidine-3′,5′-cyclophosphate (**20**) and inosine-3′,5′-cyclophosphate (**21**). These nucleotides were found within the bacterial cells as well as in the growth medium. The fragmentation of the [M+H]⁺ ion obtained by FAB-MS/MS for each of this series of cyclonucleosides was similar and is properly explained in Chapter 4 of this book.[379]

20   **21**

1.5.2   ESI-MS/MS

1.5.2.1   MS/MS Characterization of Nucleosides

1.5.2.1.1   MS/MS Characterization of Synthetic Nucleosides

ESI-MS/MS is an excellent technique for the separation and quantitation of various nucleosides; for example, Ribavirin (**22**), a synthetic nucleoside with a broad spectrum of antiviral activity. Ribavirin, in combination with pegylated interferons α-2b and α-2a, has been successfully used in the treatment of viral hepatitis C.[380] Shou and collaborators developed an LC-ESI-MS/MS method for the separation of ribavirin in human plasma and serum using MRM of the MS/MS transition [M+H]⁺ for the simultaneous determinations of total ribavirin (ribavirin, ribavirin monophosphate, ribavirin diphosphate, and ribavirin triphosphate) in monkey red blood cells.[381] The MS/MS is selected to monitor m/z 245 → 113 and 250 → 113, for ribavirin and [¹³C]ribavirin, respectively, using positive ESI.

22

1.5.2.1.2   MS/MS Characterization of Carbocyclic Nucleosides

Ziagen (abacavir) (**23**) is a carbocyclic nucleoside analogue possessing *in vitro* activity and *in vivo* efficacy against HIV.[382–384] Fung and coworkers reported an ion-pairing MS/MS method using the positive ion mode ESI to simultaneously quantify ziagen and three metabolites, carbovir monophosphate (**24**) and the corresponding di- and triphosphates.[385] *N,N*-Dimethylhexylamine (DMHA) was used as the ion-pairing agent, allowing excellent separation of the four compounds by reverse-phase HPLC. This methodology has also been applied successfully in the analysis of these compounds in human liver cells treated with ziagen.[385]

23   **24**

1.5.2.1.3   MS/MS Characterization of Halogenonucleosides

Nucleoside analogues, 2′-deoxy-2′-fluorouridine (2′-FU; **25**) and 2′-deoxy-2′-fluorocytidine (2′-FC; **26**), have been used in the development of therapeutic aptamers. It is important to understand the biological fate of these nucleoside analogues in order to evaluate and increase their efficacy as aptamers *in vivo* by reducing their degradation half-life by endogenous nucleases. Richardson et al. developed an ESI-MS/MS method to detect and quantify the nucleosides 2′-FU and 2′-FC in enzymatically hydrolyzed DNA, and using this methodology, they have shown that both nucleosides are incorporated into the DNA of rats following a chronic treatment program.[386]

An HPLC-ESI-MS/MS-based method was developed for the measurement of both chlorinated DNA lesions including 5-chloro-2′-deoxycytidine (5-CldCyd), 8-chloro-2′-deoxyguanosine (8-CldGuo), and 8-chloro-2′-deoxyadenosine (8-CldAdo), and chlorinated RNA nucleosides including 5-chlorocytidine (5-ClCyd), 8- chloroguanosine (8-ClGuo), and 8-chloroadenosine (8-ClAdo). Such a method has been used to measure the level of chlorinated DNA and RNA lesions in cells incubated with HOCl.[387] The assay has been applied for monitoring the extent of the different chlorinated nucleosides in the DNA of freshly isolated human white blood cells (Table 1.7).

1.5.2.1.4   MS/MS Characterization of Brain Basal Nucleosides

An LC-ESI-MS/MS method, using the positive ion mode, has been reported for the determination of brain basal nucleosides (inosine, guanosine, and adenosine) in microdialysates from the striatum and cortex of freely moving rats.[388] A microdialysis probe was surgically implanted into either the striatum or the cortex of individual rats, Ringer's was then solution was used as the perfusion medium, at a flow rate of either 0.3 or 0.5 μL min⁻¹, and direct analysis was performed off-line by LC-MS/MS experiments using MRM. The MRM transitions for the analytes were as follows: inosine (m/z 269 → 137), guanosine (m/z 284 → 152), and adenosine (m/z 268 → 136). The detection limits for inosine, guanosine, and adenosine were, respectively, 80, 80, and 40 pg.

1.5.2.1.5   MS/MS Characterization of Other Antiretroviral Nucleoside

Examples of antiretroviral nucleosides that have been approved for use in HIV therapy are zalcitabine (ddC), lamivudine (3TC), didanosine (ddI), stavudine, carbovir (CBV), zidovudine (AZT), and tenofovir (PMPA) and its administrated form [tenofovir diisoproxyl fumarate (TDF)]. The ESI-MS analysis of these nucleosides revealed intense protonated (PMPA, ddC, ddI, 3TC, CBV, and

---

TABLE 1.7
Mass Spectrometric and HPLC Features of the Different Chlorinated Ribo and 2-Deoxyribonucleosides Detected by HPLC-MS/MS under HPLC Conditions A or B Used for Optimal Separation of 2′-Deoxyribonucleosides and Ribonucleosides, respectively

Product	Molecular Weight	Retention Time (min)	Main Transition	Collision Energy (eV)	Limit of Quantification (fmol)
5-CldCyd	261	14.5 (A)	262 → 146	13	5
8-CldGno	301	19.1 (A)	302 → 186	16	25
8-CldAdo	285	22.5 (A)	286 → 170	19	2
5-ClCyd	277	11.7 (B)	278 → 146	15	5
8-ClGno	317	17.8 (B)	318 → 186	21	15
8-ClAdo	301	22.0 (B)	302 → 170	31	2

Source: Adapted from Badouard, K. et al., *J. Chromatogr. B* 827, 26, 2005. With permission.

TABLE 1.8
Selected Ion Transitions (*m/z* Values) and Optimized ESI-MS/MS Parameters for HPLC-ESI-MS/MS Analysis of NRTIs, in Multiple Reactions Monitoring (MRM) Mode

	Precursor Ion [M+H]+ (*m/z*)	Product Ion (*m/z*)	ESI Capillary Voltage (kV)	Cone Voltage (V)	Collision Energy (eV)
[M+H]+					
PMPA	288.3	176.0	+3.4	−50	−25
ddC	212.0	112.1	+2.0	−31	−15
3TC	230.2	112.1	+3.4	−40	−12
ddA	236	135.9	+3.4	−31	−15
CEV	248.2	152.1	+3.4	−31	−14
TDF	520.2	176.0	+3.4	−40	−50
[M−H]−					
ddI	235.1	135.1	−3.4	+50	+24
AZT	265.9	223.0	−3.4	+31	+11

Source: Adapted from Bezy, V. et al., *J. Chromatogr. B* 821, 132, 2005. With permission.

TDF) or deprotonated (ddI, ddA, and AZT) molecule signals. PMPA is negatively charged in the aqueous HPLC mobile phase, but gives positively charged ions during the ESI process (Table 1.8). The ESI parameters (capillary and cone voltages) were optimized for each compound to obtain the highest signal intensity. The MS was then switched to the product ion scan mode, and the precursor ions were fragmented by CID with argon. Collision energies were optimized to obtain the best intensity for the most abundant product ions.[389]

There is an ongoing search for novel nucleosides that are effective against the AIDS virus.[43] The systematic study of the structure–reactivity relationships of AZT analogues has shown a pronounced enhancement of antiHIV activity when either the 5-position of the pyrimidine ring is substituted by halogens or the O-4′ atom has been replaced by either a methylene group or a sulfur atom.[44,45]

ESI-MS has been used for the structural characterization of AZT (**27**) and a novel series of 3′-azido-2′,3′,4′-trideoxy-5-halogeno-4′-thio-β-D-uridine nucleosides and their respective α-anomers [**28–33**].[390] Low-energy CID-MS/MS analysis of the protonated molecules [M+H]+ confirmed the predicted fragmentation route for AZT and the series of uridines and also provided characteristic fingerprint patterns, permitting anomer differentiation. The mass spectra of the anomeric pair **28** and **31**, obtained from the protonated molecule [M+H]+ at *m/z* 348, illustrated clear differentiation between the anomers.

27

28 X = Br
29 X = Cl
30 X = F

31 X = Br
32 X = Cl
33 X = F

1.5.2.1.6   MS/MS Characterization of Antiretroviral Nucleoside Possessing Conformational Rigidity

The low-energy collision MS/MS analysis of the various protonated molecules [M+H]⁺ and cluster ions such as [M+Na]⁺, [2M+H]⁺, and [2M+Na]⁺ isolated from the aromatic analogs of the antiviral 2′,3′-dideoxy-2′,3′-didehydronucleosides has been reported.[390b] These analyzed 2′,3′-dideoxy-2′,3′-didehydronucleosides contained two chiral centers (C-1′ and C-3′), which possessed the α/β and D/L centers in a furanose sugar. The analytes were two isolated diastereoisomeric mixtures of 1-(3-benzoyl-oxymethyl-1,3-dihydrobenzo[c]furan-1-yl)thymine: **34** [(1′R,3′S) and (1′S,3′S)] and **35** [(1′R,3′R) and (1′S,3′R)], and the four optically pure stereoisomers of the analogous uracil (1′S,3′R) **36**, (1′R,3′S) **37**, (1′S,3′S) **38**, and (1′R,3′R) **39**.[390b] The MS/MS analysis of the precursor ion [M+H]⁺ at *m/z* 365 isolated from nucleosides **34** and **35** showed, as was expected, that there were no significant differences between product ion intensities within each enantiomeric pair. However, the scans for [2M+Na]⁺, [2M+H]⁺, and [M+Na]⁺ ions at *m/z* 751, 729, and 387, obtained from the four pure enantiomers **36–39**, indicated that there were discriminating differences in fingerprint ion intensities in each enantiomeric pair.[390b]

34   35   36

37   38   39

1.5.2.1.7   MS/MS Characterization of Ribonucleoside Analogues

5-Azacytidine (5AC) is a hypomethylating agent that has anticancer properties and is used to treat various malignancies. HPLC-ESI-MS/MS has been used to quantify it in plasma, even though it is unstable and is rapidly hydrolyzed to several by-products, including 5-azacytosine and 5-azaura-cil.[391] Reverse-phase LC-MS/MS has been used to achieve a sensitive and specific assay method allowing pharmacokinetic and pharmacodynamic studies on 5AC in combination therapy trials. The LC-MS/MS method developed was applied to monitor plasma samples from patients who had received 5AC in combination with phenylbutyrate.[392]

1.5.2.1.8   MS/MS Characterization of Pyrimidine Nucleoside Antiviral Agents

ESI-MS/MS has played a significant role in the characterization and analysis of pyrimidine nucleo-side, an antiviral agent that also has antitumor properties.[393–399]

Biemann and McCloskey were the first to report the mass spectra of compounds containing the pyrimidine ring.[16] The importance of nucleic acid constituents as therapeutic drugs and the signifi-cance of their role in biotechnology and human genome studies means that nucleic acid bases con-stitute an important target in studies of the dissociation of heterocyclic compounds. The ionic dissociation reactions of heterocyclic compounds have been regarded as some of the most complex processes in MS. In ESI-MS, the collision process is an effective means of introducing sufficient internal energy into either the protonated or deprotonated molecular ions and to promote extensive fragmentation. CID is a technique that is valuable for this type of analysis and has been employed in the structural characterization of many pyrimidine nucleosides and pyrimidine bases.[15a,18,67,400–409] The low-energy (5 eV) CID-MS of these four antiviral agents are given in Figure 1.25.[15c]

1.5.2.1.9   Quantitative Analysis of Therapeutic Oligonucleotides and their Metabolism by ESI-MS/MS

ESI-MS/MS has been extensively used to study therapeutic oligonucleotides (OGNs), leading to the discovery of novel agents designed to specifically inhibit target gene expression in animal and human disease models.[410] The full ESI (−mode) scan of the OGN showed that the major ions were a series of multiply deprotonated ions $(M-nH)^{n-}$ forming a charge state distribution, each of which represents a multiply charged ion of the intact molecule that had a specific number of protons removed from the sugar–phosphate backbone. The signals of the multiply charged series can be readily deconvoluted to yield the molecular mass. The pH of the mobile phase, additives in the mobile phase, and the MS tuning parameters have a strong influence on the charge distribution and

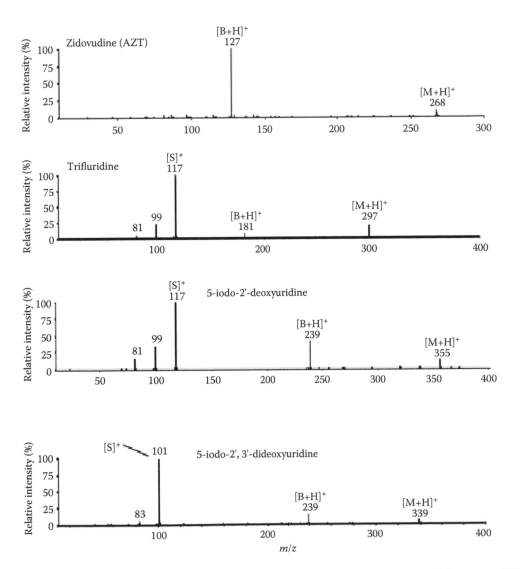

FIGURE 1.25   CID product ion spectra of $[M+H]^+$ ions of pyrimidine antiviral at low collision energy (CE) of 5 eV.

their detectability in the QIT and the triple quadrupole MS instruments.[311,312,410–417] It was found that the protons can be substituted with a variable number of other cations in the sugar–phosphate backbone, resulting in decreased sensitivity and highly complex spectra. The degree of cation adduction increased with the length of the OGNs and was more extensive with phosphorothioates than with phosphodiesters.[311] A high-quality mass spectrum is a fundamental requirement for positive metabolite identification. Therefore, effective removal of cations and careful control of instrument parameters are required for OGN metabolite studies using an LC–ESI-MS or MS/MS.

1.5.2.2 MS/MS Characterization of Ribonucleosides

ESI-MS has long provided an important means for the characterization of covalent modification of nucleosides. HPLC-ESI-MS has been used to examine the fragmentation of the ESI-produced [M+H]+ and [M–H]− ions of 2′- and 3′-*O*-methylribonucleosides, in order to establish a sensitive method for the differentiation of the isomeric ribonucleosides.[406]

Negative-ion ESI-MS/MS has allowed the differentiation of 2′-*O*-methylribonucleosides from their 3′-*O*-methylated counterparts. Several characteristics were common in the fragmentation chemistry of ribose-methylated nucleosides. The product ion spectra of the [M–H]− ions of the corresponding purine nucleosides showed that the most abundant ions were the deprotonated ions of the nucleobases (Figure 1.26). The product ions of *m/z* 190 and *m/z* 206 for 2′-*O*-methyladenosine and 2′-*O*-methylguanosine respectively, allowed the differentiation of 2′-*O*-methyladenosine and 2′-*O*-methylguanosine from their corresponding 3′-*O*-methylated analogs. These two ions are formed from the familiar cross-ring cleavage of the ribose moiety (loss of a $C_3H_6O_3$ fragment).

1.5.2.3 MS/MS Characterization of Geometric Deoxynucleoside Adducts

MS/MS studies of carcinogen–DNA adducts, derived from *in vitro* reactions, were carried out to determine their fragmentation pathways. These pathways were then used to help differentiate various adduct structures that had been isolated from *in vivo* sources.[418–421] Differentiating the

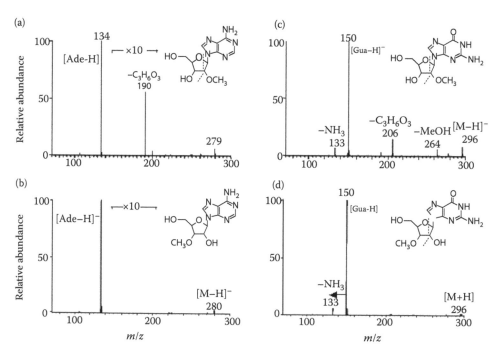

FIGURE 1.26 Product-ion spectra of the [M–H]− ions of 2′-*O*-methyladenosine (a), 3′-*O*-methyladenosine (b), 2′-*O*-methylguanosine (c), and 3′-*O*-methylguanosine (d).

structures of geometric isomers obtained from the reaction of a metabolized carcinogen with DNA has proven to be more challenging. However, studies of thymidine glycol, PAH-modified nucleosides, and PAH tetraols suggest that diastereomers may be differentiated based on the *cis/trans* arrangement of hydroxyl groups bound to a PAH.[378,422,423]

The product ion studies of diastereomeric benzo[ghi]fluoranthene-2-deoxynucleoside adducts by ESI and QIT-MS have been conducted by Li et al.[424] They examined the fragmentation characteristics of the protonated molecule ions derived from a total of 10 different B[ghi]F adducts (Scheme 1.2). These adducts include four diastereomers of B[ghi]F-5a-*N*2-dG and B[ghi]F-5a-*N*6-dA and two diastereomers of B[ghi]F-5a-*N*4-dC. The reaction of each diol-epoxide produces two adducts that differ in structure based on the *cis/trans* arrangement of the hydroxyl group and the nucleic acid bound to the carbons of B[ghi]F.

1.5.2.4 MS/MS Characterization of Nucleotides

Ge et al. reported the detection and quantification of *trans*-Zeatin riboside-5′-monophosphate in coconut water (*Cocos nucifera* L.) by CZE-ESI-MS/MS after SPE and on-line sample stacking.[425]

The MS/MS with multiple MRM detection was carried out to obtain sufficient selectivity and sensitivity for the cytokinin nucleotides. Example of representative fragmentation patterns of protonated DZMP under low-energy conditions is illustrated in Figure 1.27 as a selected example among the six cytokinin nucleotide standards.

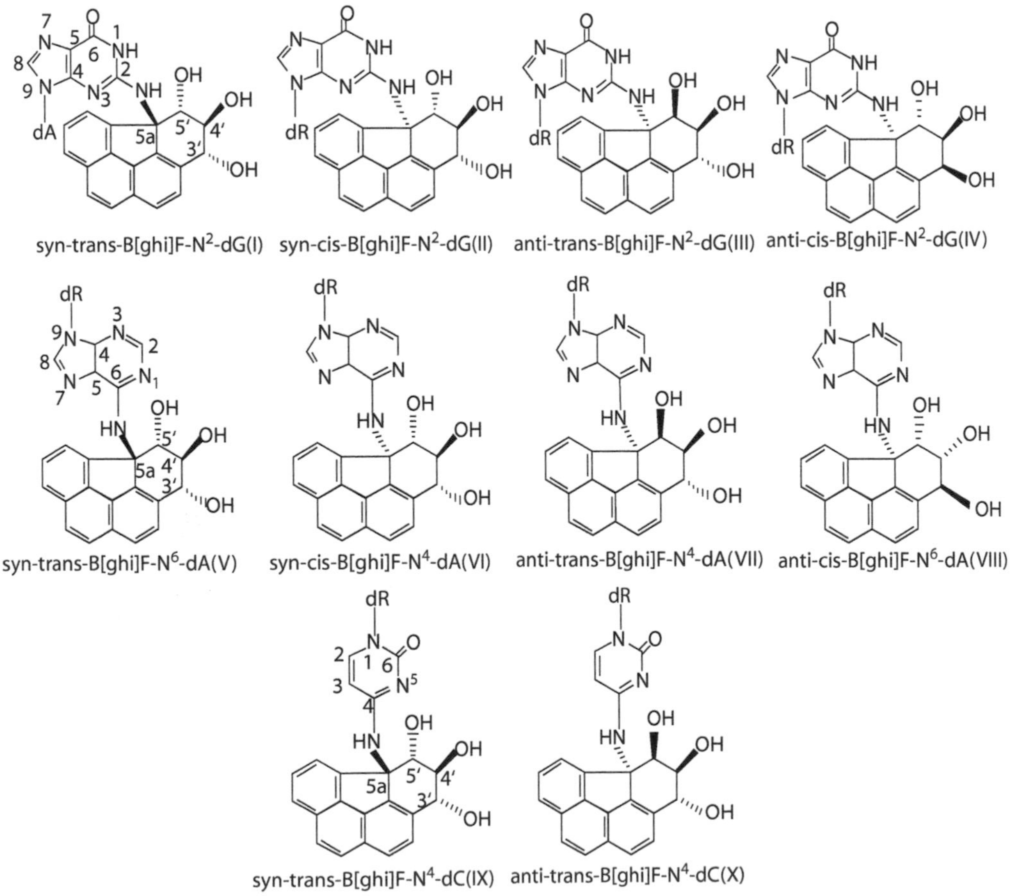

syn-trans-B[ghi]F-N²-dG(I)   syn-cis-B[ghi]F-N²-dG(II)   anti-trans-B[ghi]F-N²-dG(III)   anti-cis-B[ghi]F-N²-dG(IV)

syn-trans-B[ghi]F-N⁶-dA(V)   syn-cis-B[ghi]F-N⁴-dA(VI)   anti-trans-B[ghi]F-N⁴-dA(VII)   anti-cis-B[ghi]F-N⁶-dA(VIII)

syn-trans-B[ghi]F-N⁴-dC(IX)   anti-trans-B[ghi]F-N⁴-dC(X)

SCHEME 1.2 Different B[ghi]F adducts.

1.5.2.5 CID-MS/MS Sequencing of Small Oligonucleotides

Elucidation of the sequence of oligonucleotides by MS/MS of multiply charged DNA ions has been studied intensively. McLuckey and collaborators reported a neat study of the fragmentation of multiply charged oligonucleotides by ESI in a QIT-MS^n instrument, leading them to propose a nomenclature for different oligonucleotide fragment types, analogous to what had been developed for peptides.[375,376,426] The MS/MS of triply charged anions, (such as [d(A_4)]$^{3-}$), performed under gentle CID conditions with low-amplitude supplementary rf signals afforded product ions that resulted from the loss of adenine. They also established in the MS/MS of [d(A_4)]$^{3-}$, the concept of the formation of the [a_4–B_4(A)]$^{2-}$ ion and its complement ion w_1^- (using the McLuckey notation).[376] Under moderate CID conditions consecutive fragmentations occurred by further decompositions involving charge-separation reactions. The same authors also studied the ESI-QIT-MS/MS analysis of the [**M–7H**]$^{7-}$ precursor ion (indicated by an asterisk) of the 5'-d(TGCATCGT)-3' oligomer (Figure 1.28).[375] The authors observed the loss of an adenine leading to the formation of the [**M–7H–B(A)**]$^{6-}$ ion that dissociated to produce the following complementary ions: {(a_4–B_4(A))$^{2-}$ and w_4^{4-}} and {(a_4–B_4(A))$^{3-}$ and w_4^{3-}}. Spectral interpretation was greatly facilitated by the fact that complementary ions, arising from decompositions involving charge-separation reactions, were usually observed in either MS/MS or MS^n.[356,407,427]

ESI-MS/MS sequencing using a quadrupole–hexapole–quadrupole instrument for a series of synthetic, self-complementary, isomeric DNA hexamers, namely, d(CAGCTG), d(CGATCG), and d(CGTACG) (M_r 1791) and d(CATATG), d(TGATCA), and d(TGTACA) (M_r 1790), has been conducted.[428] Low-energy CID-MS/MS analysis of the multicharged oligonucleotide anions [M–3H]$^{3-}$ and [M–4H]$^{4-}$ provided distinct fingerprint patterns, which permitted the discrimination of individual isomeric DNA hexamers.[428]

Baker and coworkers verified the sequence of synthetic methylphosphonate oligodeoxynucleotides using ESI-MS/MS via flow injection.[429] The product ion scans from multiply protonated (4+ and 5+) were studied, the product ions were identified and the base sequence of the intact methylphosphonate oligodeoxynucleotide molecule was determined. Oligomers containing as many as 18 bases have been successfully characterized by this method.

Tabet and colleagues reported the study of low-energy collision MS/MS of multiply deprotonated products of single-strand oligonucleotides prepared in a nanospray external source coupled to a QIT-MS^n instrument.[430] The negative charge could be located, for a selected anion, by studying the fragmentation pattern. Multideprotonated 5'-dephosphorylated nucleotides undergo isomerization into ion–dipole complexes prior to dissociation, leading to competitive [BH] and [B]$^-$ losses. In contrast, 5'-phosphorylated nucleotides yielded competitive HPO_3 and nucleobase

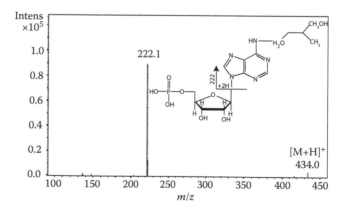

FIGURE 1.27 Positive product spectrum of DZMP with the protonated molecule ([M+H]$^+$) as a precursor ion.

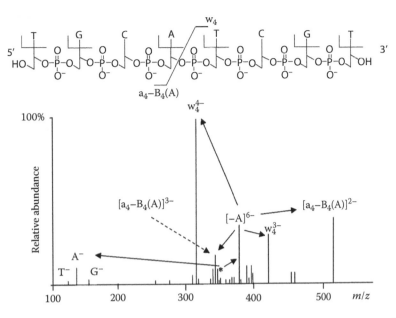

FIGURE 1.28 ESI-QIT-MS/MS analysis of 5′-d(TGCATCGT)-3′. The precursor ion $[M-7H^+]^{7-}$ is indicated by an asterisk. The solid arrows indicate pairs of complementary ions derived from the precursor ion and two different dissociations of the ion derived by initial loss of adenine $[M-7H^+-A^-]^{6-}$. (Adapted from McLuckey, S.A., Van Berkel, G.J., and Glish, G.L., *J. Am. Soc. Mass Spectrom.* 3, 60, 1992. With permission. Copyright 1993, American Chemical Society.)

eliminations. The proposed mechanisms were discussed in terms of the charge state of the precursor and the thermochemical gas-phase properties. This study established that the potential energy was minimal for distant charged phosphorylated groups. These findings suggested that a specific conformation is normally favored in order to minimize Coulombic repulsion during the desolvation process.[430]

1.5.2.5.1 Determination of Nearest Neighbors

The determination of nearest neighbors was a widely used method in early studies of DNA sequence and biosynthesis and for measurement of average dinucleotide frequencies.[431] This technique was based on polymerase-mediated incorporation of specific 5′-^{32}P-labeled nucleotides into the growing polynucleotide chain complementary to the DNA of interest, in four parallel experiments. Nuclease digestion to 3′-mononucleotides directed the label into the adjacent residue, which was then identified using 2D-TLC.

The method of Rozenski and McCloskey was based on the analysis of fragment ions of the nucleic acid formed in the ionization region by ESI-MS nozzle-skimmer fragmentation.[432] The fragmentation of the oligonucleotide included the formation of cyclic phosphate dimers ($N^1pN^2 > p$). The subsequent MS/MS analysis created product ions ($N^1 > p$ or $N^2 > p$) whose structures indicated whether the nearest neighbors of any selected nucleotide were modified. This method was applicable to RNA and DNA and directly to components of oligonucleotide mixtures through the use of LC-ESI-MS/MS.[432] In direct combination with an HPLC separation, components of complex mixtures could be successfully analyzed.[433]

1.5.2.5.2 Oligonucleotide Sequencing Programs

It is well known that the information content of product ion spectra depends critically on the type of MS/MS instrument and on the experimental conditions used. Furthermore, the complexity of the spectrum increases dramatically with the size of the fragmented oligonucleotide, which may result in considerable difficulties in spectrum interpretation.[434,435] Little et al. outlined a sequencing strategy for the manual interpretation of MS/MS spectra, but the deduction of sequence information was time-consuming and highly technical and could only be performed in laboratories where the researchers had extensive experience in MS/MS.[436] Hence, automated interpretation of product ion spectra is a prerequisite for the applicability of MS/MS in routine nucleic acids sequence analysis.

McCloskey and coworkers first broached the automation of the analysis of oligonucleotide mass spectra.[437] This group developed a computer algorithm, based on the MS/MS fragmentation scheme described by McLuckey et al., enabling the sequencing of oligonucleotides using the spectra from an ESI-CID triple quadrupole MS/MS.[376,377,427] The method was intended for the rapid sequencing of a 15-mer or less DNA or RNA oligonucleotide with a completely unknown structure. Identification of the sequence-relevant ions, produced from extensive fragmentation in the quadrupole collision cell, was based primarily on the recognition of 3'- and 5'-terminal residues as initial steps in mass ladder propagation. This method used the alignment of overlapping nucleotide chains, constructed independently from each terminus, and experimentally measured molecular mass for the rejection of incorrect sequence candidates. These algorithms for sequence derivation were embodied in a computer program that requires <2s for execution.[437] The efficacy of the procedure was demonstrated for sequence location of simple modifications in the base and sugar. The potential for direct sequencing of components of mixtures was shown using an unresolved fraction of unknown oligonucleotides from ribosomal RNA.[437]

Huber and coworkers developed a program for the computer-aided interpretation of product ion spectra obtained for CID of multiply charged oligodeoxynucleotide ions generated by ESI.[434,438] This algorithm, for the comparative sequencing (COMPASS) of oligonucleotides, was shown to be suitable for the sequence verification of nucleic acids ranging in length up to 80 nucleotides. The COMPASS program is based on matching the mass spectrum generated by CID-MS/MS to m/z values produced from a systematically varied reference sequence employing established fragmentation pathways. The potential and limits of the COMPASS program regarding the lengths, the charge state of the oligodeoxynucleotide, the selected collision energy, and the analyzed amount of sample using a QIT-MS^n instrument have been evaluated.[367] The COMPASS algorithm has proven to be reproducible and is applied for the genotyping of the polymorphic, Y-chromosomal locus M9 contained in a 62-bp PCR product.[302]

Another program worth mentioning is the mongo oligo mass calculater. The mongo oligo program was originally written by Limbach, Pomerantz, and Rozenski, and allowed the mass calculation of oligonucleotides, CID fragments, and endo- and exonuclease digestion.[439] It also contains extra output options and utilities for handling the input sequence and user residue assignment.

1.5.2.6 MS/MS Characterization of DNA Complexes

1.5.2.6.1   Leukotriene A4–Nucleoside Complexes

Leukotriene A4 (LTA4), a chemically reactive conjugated triene epoxide, is formed by the reaction of arachidonic acid with molecular oxygen catalyzed by 5-lipoxygenase. LTA4 is an intermediate in the formation of the biologically active eicosanoids leukotriene B4 and leukotriene C4. ESI-MS and MS/MS have been applied in the study of the covalent binding of LTA4 with uridine, cytidine, adenosine, and guanosine.[440] Competition reactions with nucleoside mixtures established that guanosine was most reactive toward LTA, yielding five major and at least six minor covalent adduct species. Data from CID-MS/MS of the molecular anion [M–H]⁻ at m/z 600.3 suggested that guanosine attacked LTA4 at either C-12 or C-6, with the opening of the epoxide at C-5 to yield a series of adducts. Observation of the ready covalent attachment of guanine to LTA4 raises the possibility that this intermediate of leukotriene biosynthesis formed on or near the cellular nuclear envelope may react with nucleosides and nucleotides present in RNA or DNA.[440]

1.5.2.6.2   MS/MS Characterization Metal–Oligonucleotide Complexes

Metal–oligonucleotide complexes investigated by positive mode nanoelectrospray MS/MS have revealed fundamental aspects of their gas-phase behavior.[441] The addition of transition metal ions, such as iron(II), iron(III), and zinc(II), leads to very stable metal–oligonucleotide complexes showing heavily altered fragmentation patterns, in contrast to the uncomplexed oligonucleotides. The site of metal ion complexation was located by CID experiments. It was found that all three metal ions investigated coordinate predominantly to the central phosphate groups of the oligonucleotides. Furthermore, it is demonstrated that the fragmentation of such complexes strongly depends on the metal ion complexed as well as on the sequence of the nucleobases in the oligonucleotide.

1.5.2.6.3   Duplex and Triplex DNA Binding

The mode by which ligand binds to duplex and triplex DNA was assessed by the group of Gabelica using a combination of ESI-MS/MS and molecular modeling.[442] This approach was used to analyze seven benzopyridoindole and benzopyridoquinoxaline drugs, bound to different duplex DNA and triple helical DNA. The ligands were ranked according to the collision energy necessary to dissociate 50% (CE50) of the complex with the duplex or the triplex in MS/MS. To determine the probable ligand-binding site and binding mode, molecular modeling was used to calculate relative ligand-binding energies in different binding sites and binding modes. The selectivity of formation and CID-MS/MS of the complexes with the duplex and triplex DNA are explained in Chapter 8 of this book.

1.5.2.7   MS/MS Characterization of DNA Adducts

ESI-MS/MS has proven to be a promising technique for the detection, identification, and quantitation of covalently attached DNA adducts and their derived products. Furthermore, MS/MS analyses readily provide structural characterizations as well as sequence information, permitting the rapid and accurate analysis of moderately sized oligonucleotides. This work has been previously reviewed by Esmans et al., Crain, and Apruzzese and Vouros.[10–12]

1.5.2.7.1   Analysis of Covalently Modified Nucleosides and Bases

Radiation or general oxidative damage to DNA is a complex problem, and the development of high-quality assays for damaged nucleosides or bases is of great interest. An LC-ESI-MS/MS method (an MRM mode) was developed for the assessment of six oxidized 2′-deoxyribonucleosides and two modified purine bases found within both isolated and cellular DNA, which are radiation-induced degradation products of DNA.[405] Stable isotopically labeled internal standards were prepared and used for isotope dilution MS measurements. This methodology was applied to the simultaneous assessment of 8-oxo-7,8-dihydro-2′-deoxyguanosine (**40**), 8-oxo-7,8-dihydro-2′-deoxyadenosine (**41**), 5-hydroxy-2′-deoxyuridine (**42**), 5,6-dihydroxy-5,6-dihydrothymidine (**43**), 5-formyl-2′-deoxyuridine (**44**), and 5-(hydroxymethyl)-2′-deoxyuridine (**45**). This sensitive and specific assay was also applied for the quantification of 2,6-diamino-4-hydroxy-5-formamidopyrimidine (**46**) and 4,6-diamino-5-formamidopyrimidine (**47**).[405]

Another highly sensitive quantitative HPLC-ESI-MS/MS method was developed by Brink et al. for the simultaneous determination of O-6-methyl-2-deoxyguanosine, 8-oxo-7,8-dihydro-2-deoxy-guanosine, and 1,N6-etheno-2-deoxyadenosine in DNA using on-line sample preparation.[443] The method developed is useful in the study of the biological significance of exogenous DNA adducts as an increment to background DNA damage and the role of modulating factors such as DNA repair.

1.5.2.7.2 Deoxyguanosine Adducts

LC-MS/MS has been used in the quantification of O-6-methyl and O-6-ethyl deoxyguanosine adducts as reported by Churchwell et al.[444] The carcinogenicity of many alkylating agents is derived from their ability to form persistent DNA adducts that induce mutations. Their work describes and validates the methodology, based on LC-ESI-MS/MS, for the separate or concurrent quantification by isotope dilution of O-6-methyl-2'-deoxyguanosine (O-6-methyl-dG) and O-6-ethyl-2'-deoxy-guanosine (O-6-Et-dG) DNA adducts. These methods are useful for the evaluation of DNA adducts, in relation to cellular processes that modify carcinogenic and toxicological responses in both experimental animals and humans.

1.5.2.7.3 Estrogen-DNA Adducts

The Esmans group reported the identification of estrogen-DNA adducts in both breast tumor and healthy breast tissue using nanoLC coupled to nano-ESI-MS/MS.[445] Estrogen-DNA adducts of 4-hydroxy-equilenin were detected by CID-MS/MS in patients who were using hormone replacement therapy (HRT), such as Premarin, but there was no significant difference detected between the two groups.[446]

1.5.2.7.4 Analysis of Covalently Modified Oligonucleotides

Marzilli et al. reported a novel method to map guanine bases in short oligonucleotides.[447] In this method, target oligonucleotide strands were subjected to guanine-specific methylation, and then the modified strand was analyzed by ESI-QIT-MS^n. The CID product ion scan of the monomethylated oligonucleotide strand indicated rapid depurination. Further collision (MS^3) of the apurinic oligo-nucleotide induced preferential cleavage of the backbone at the site of depurination. The mass of the resulting complementary product ions identifies the position of each guanine base in the sequence. This methodology has been demonstrated for oligonucleotide sequences up to 10 bases in length, confirming the utility of this selective fragmentation by QIT-MS^n for sequencing oligonucleotides.[447]

ESI-MS/MS has been used to determine the position of p-benzoquinone adducted to a 7-mer oligonucleotide.[448] The CID spectra of ions with different charge states led to ion profiles allowing full sequencing.

The mutagenicity of the tobacco carcinogen, benzo[a]pyrene, is believed to result from the reaction of a metabolite, benzo[a]pyrene diol epoxide with DNA. Early studies of the promutagenic lesions established that the guanine adduct (+)-$trans,anti$-7R,8S,9S-trihydroxy-10S-(N^2-2'-deoxy-guanosinyl)-7,8,9,10-tetrahydrobenzo[a]pyrene (**1**) was implicated. This work depended on the excision of the damaged nucleoside by repair enzymes, a process that is not always effective and does not easily allow the chemical structure and stereochemistry of the lesion to be determined.

Tretyakova and her group developed an elegant strategy for the direct quantitative analysis of (**1**) and its stereoisomers, originating from specific guanine nucleobases within DNA sequences derived from the p53 tumor suppressor gene and the K-ras proto-oncogene.[449,450] Guanosine, substituted with a stable isotope (^{15}N or ^{13}C), was incorporated into specific positions in a set of synthetic DNA strands corresponding to key sequences from these two genes (Table 1.9). Adduct formation following exposure to BDE was determined by HPLC-ESI-MS/MS by direct comparison of the labeled and unlabeled ions. The reaction occurred preferentially at the guanosine in methylated CG dinucle-otides with Adduct (**1**) being the main product. Small amounts of adduction were also observed from three other stereoisomers of BDE.

TABLE 1.9
Synthetic Oligonucleotides, Incorporating Isotopically Labeled Guanosine and Methylated Cytosine, Used to Locate and Quantify the Adduction of Benzo[a]pyrene Diol Epoxide to DNA by HPLC/ESI-MS/MS

Sequence	Calcd Mol. Mass	Exptl. Mol. Mass
CC^{Me}C[^{15}N3-G]GCACC^{Me}CG^{Me}CGTC^{Me}CG^{Me}CG	5784.7	5785.3
CC^{Me}CGGCACC^{Me}C [^{15}N3-G]^{Me}CGTC^{Me}CG^{Me}CG	5784.7	5785.2
CC^{Me}CGGCACC^{Me}CG^{Me}C[^{15}N3-G] TC^{Me}CG^{Me}CG	5784.7	5785.3
CC^{Me}CGGCACC^{Me}CG^{Me}CGTC^{Me}C[^{15}N3-G]^{Me}CG	5784.7	5785.2
^{Me}CG^{Me}CGGA^{Me}CG^{Me}CGGGTGC^{Me}CGGG	5981.9	5982.9
ATGGG^{Me}C[^{15}N3-G]GCATGAAC^{Me}CGGAGGCCCA	7773.1	7774.2
ATGGG^{Me}CGGCAT[^{15}N3-G]AAC^{Me}CGGAGGCCCA	7773.1	7774.4
ATGGG^{Me}CGGCATGAAC^{Me}C [^{15}N3-G]GAGGCCCA	7773.1	7774.1
ATGGG^{Me}CGGCATGAAC^{Me}CG[^{15}N3-G]AGGCCCA	7773.1	7774.3
ATGGG^{Me}CGGCATGAAC^{Me}CGGA[^{15}N3-G]GCCCA	7773.1	7774.1
TGGGCCTC^{Me}CGGTTCATGC^{Me}CGCCCA	7614.0	7612.0
GCTTT[^{15}N3-G]AGGTG^{Me}CGTGTTTGTG	6547.3	6546.8
GCTTGA[^{15}N3-G]GTG^{Me}CGTGTTTGTG	6547.3	5646.8
GCTTTGAG[^{15}N3-G]TG^{Me}CGTGTTTGTG	6547.3	6546.8
GCTTTGAGGTG^{Me}C [^{15}N3, ^{13}C-G]TGTTTGTG	6548.3	6548.5
GCTTTGAGGTG^{Me}C [^{15}N5-G]TGTTTGTG	6549.3	6549.4
GCTTTGAGGTG^{Me}CGT[^{15}N3-G]TTTGTG	6547.3	6546.7
CACAAACA^{Me}CGCACCTCAAAGC	6336.2	6336.4
ATGGGC [^{15}N3,^{15}C-G]GCATGAAC	4646.1	4645.4
ATGGG^{Me}C [^{15}N3, ^{13}C-G] GCATGAAC	4660.1	4659.5
GTTCATGCCGCCCAT	4504.0	4503.1
GTTCATGC^{Me}CGCCCAT	4518.0	4517.4
CATGAACC [^{15}N3-G]GAGGCCCATC	5785.8	5786.8
CATGAAC^{Me}C[^{15}N3-G]GAGGCCCATC	5799.8	5800.1
GATGGGCCTCCGGTTCATG	5835.8	5836.3
GATGGGCCTC^{Me}CGGTTCATG	5849.8	5850.3
GCTTTGAGGTGC[^{15}N3,^{13}C-G]TGTTTGTG	6534.3	6533.0
CACAAACACGCACCTCAAAGC	6322.2	6321.6

Source: Adapted from Matter, R. et al., *Chem. Res. Toxicol.* 17, 731, 2004. With permission.

The carcinogenic effect of sunlight is due in part to the formation of dimeric pyrimidine photo-products induced by short wavelength UV. The Cadet group obtained data for a group of such photodimerization products of the bases, nucleosides, dinucleoside monophosphates, and dinucleotides using ESI-MS/MS in the MRM mode.[451] The fragmentation patterns are sensitive to the mechanism of collisional activation (TQ compared to QIT), especially for diastereoisomeric cyclobutane dimers. Similar modified trinucleotides have been identified by LC-ESI-MS/MS in the nuclease P1 digests of DNA irradiated at 254 nm.[452] Tandem cross-linking of thymine and purine bases can be studied in synthetic DNA by incorporation of 5-(phenylthiomethyl)-2′-deoxyuridine in either a 3′ or 5′ position vicinal to adenosine or guanosine residues. UV irradiation induces the cross-link, and a sensitive assay was developed for the four possible linked dinucleoside monophosphates by HPLC-ESI-MS/MS.[453] These four lesions were found in both single-strand oligonucleotides and isolated DNA subjected to γ irradiation. The tandem G,T lesions were produced more efficiently than the corresponding A,T cross-linked species.

1.5.3   ESI-FT-ICR-MS[n] MAPPING NONCOVALENT LIGAND BINDING TO *STEMLOOP* DOMAINS OF THE HIV-1 PACKAGING SIGNAL

During the replication of human immunodeficiency virus type 1 (HIV-1), specific interactions between the packaging signal (ψ) of viral RNA and the nucleocapsid (NC) domain of the Gag polyprotein mediate the vital functions of genome recognition, dimerization, and packaging.[454–459] The relatively low mutation frequency of this ~120 nt stretch of genomic RNA makes the packaging signal a very promising target for the development of new therapeutic strategies to contend with the rapid emergence of drug-resistant strains.[450–453,460–467] For this reason, steps have been made toward the identification of possible inhibitors, which may bind specific ψ-RNA structures and disrupt their normal functions.[468–470]

Fabris et al. presented the stoichiometry and binding affinity of NC for hairpin SL2, SL3, and SL4 of HIV-1 ψ-RNA that were determined directly by ESI-FT-ICR-MS[n].[471] In 2006, Fabris et al. studied the binding modes and structural determinants of the noncovalent complexes formed by aminoglycoside antibiotics with conserved domains of the HIV-1 packaging signal (ψ-RNA) by ESI-FTICR-MS[n].[472]

Using the same approach, the aminoglycosidic antibiotics were found capable of forming stable noncovalent complexes with all three RNA constructs in the absence of NC, but their effects on the corresponding protein–RNA assemblies were shown to be remarkably different.[472–477] These studies are described in detail in Chapter 9 of this book.

1.6   NOVEL MASS ANALYZERS

The development of novel mass analyzers has been especially important for complex biomolecule analyses. The performance characteristics of FT-ICR-MS, TQ-MS, and QIT-MS have been thoroughly reviewed by Yates, and McLuckey and Wells.[100,367]

FT-ICR-MS[n] is emerging as the preferred platform for the investigation of the sequences of modified oligonucleotides based on the small amount of sample required and the unambiguous diagnostic fingerprints that are generated. Detailed studies by McLafferty and colleagues have provided a thorough understanding of the fragmentation of the multiply charged oligonucleotides using FT-ICR-MS.[181,435,454,455,478] They used nozzle-skimmer (NS) dissociation and CID-MS/MS and MS[n] experiments to investigate the dissociation mechanisms of multiply charged oligosaccharides up to 108 nucleotides in length. To correctly sequence oligonucleotides that may contain various chemical modifications, an understanding of how these modifications might influence the fragmentation pattern is essential. Sannes-Lowery and Hofstadler explored how glycone and backbone modifications affected fragmentation patterns observed in modified oligonucleotides that were fragmented by infrared multiple photon dissociation in the external reservoir of an ESI-FT-ICR-MS[n].[456] They observed that the chemical modifications influence which fragment types dominate (i.e., $a_n - B$ versus C_n), and the ease with which the oligonucleotides are fragmented and sequenced.

Substantive studies of oligonucleotide sequencing by CID-MS/MS with triple quadrupole or quadrupole–hexapole–quadrupole have also been performed.[435,456,478] Precursor ions with higher charges gave many low-mass, nonspecific fragment ions. In contrast, doubly and singly charged ions required higher collision energies owing to their greater stability, leading to an overall decrease in sensitivity. It has been generally accepted that the extent of fragmentation of the multiply charged precursor ion is dependent on the amount of energy applied to it. This was predictably greater in triple quadrupole instruments compared with to ion-trapping analyzers. It should also be recognized that the QIT and FT-ICR analyzers have the capability for multiple stages of CID and mass analysis, which is very useful for the understanding of the genesis of the product ions.[11a–e]

Makarov developed the Orbitrap mass analyzer, which operates by radially trapping ions about a central spindle electrode.[457] An outer barrel-like electrode is coaxial with the inner spindle-like electrode, and *m/z* values are measured from the frequency of harmonic ion oscillations of the orbitally trapped ions along the axis of the electric field. This axial frequency is independent of the energy and spatial spread of the ions. The ion frequencies are measured nondestructively by the acquisition of the time-domain image current transients, followed by Fourier transformation to obtain the mass spectra.

In the ESI-LTQ-Orbitrap hybrid MS, the ions are transferred from the ESI source through three stages of differential pumping using RF guide quadrupoles. Features of the LTQ-Orbitrap hybrid MS include high mass resolution (up to 150,000), large space charge capacity, high mass accuracy (2–5 ppm), an m/z range of at least 6000, and dynamic range greater than 10.[457]

1.6.1 CHARACTERIZATION OF DNA NONCOVALENT COMPLEXES USING LTQ-ORBITRAP HYBRID MASS SPECTROMETER

LC-MS^n analysis has been used to study the reaction of deoxynucleosides with the active metabolites of two heterocyclic aromatic amines (NHOH-PhIP and NHOH-IQ).[458] Sequential MS^3 experiments of the formed adducts were carried out with an ion trap MS. The accurate mass measurements acquired with an LTQ-Orbitrap mass analyzer supported the determination of the ion structures. Particular attention was given to the diagnostic ions formed by the linking of the heterocyclic aromatic amine (HAA) and the deoxynucleoside. A total of five adduct structures have been characterized, in which two are new compounds: dG-N7-IQ and dA-N^6-IQ. The C-8 and N-2 atoms of dG were found to be the most reactive, resulting in the formation of two different adducts with IQ and one adduct with PhIP. An unusual nondepurinating dG-N7-IQ adduct has also been characterized, and a mechanism was proposed for its formation on the basis of the reactivity of arylamines.

ESI-MS (the − ion mode) was used to study the interaction between DNA duplexes and peptides. It was shown that peptides, containing two adjacent basic residues, formed noncovalent complexes with both single-strand and duplex DNA. The CID analysis of the complexes formed showed unexpected fragmentation pathways. In the gas phase, the binary complexes dissociated without disruption of the noncovalent interaction, by fragmentation of either the covalent bonds of the peptide backbone and/or along the backbone of the ssDNA. This demonstrated the strong binding between the peptide and the DNA strand.

Sequential MS^n performed with an LTQ-Orbitrap mass analyzer was used to investigate the nature of the interaction of duplex DNA and peptide complexes. The CID spectra indicated a major fragmentation that resulted from the disruption of the noncovalent interactions and the formation of binary complexes and single-strand ions. A preferential formation of complexes with thymidine containing single strands was observed. An alternative pathway was also detected, in which complexes were dissociated along the covalent bond of the peptide and/or DNA according to the basicity. The experimental data obtained suggested the presence of strong salt bridge interactions between the DNA and the peptides containing basic residues.

1.6.2 ION MOBILITY SPECTROMETRY

The complexity of biological samples has encouraged the development of analytical techniques that achieved high resolving power and high accuracy measurement. Ion mobility spectrometry (IMS) has long been explored for fast separation and detection of gas-phase ions based on their differential travel time through a low homogeneous electric field. The size-to-charge (s/z) ratio separation mechanism of IMS is different from the m/z ratio measurement mechanism employed by mass spectrometry; thus the combination of these two techniques allows two-dimensional separation.

Various types of IMS-MS are available today and their advantages for application in a wide range of analytes have been described.[459,479–481] Separation of isomers, isobars, and conformers; reduction of chemical noise; and measurement of ion size become possible when IMS cells are added to MS. In addition, structurally similar ions and ions of the same charge state can be separated into families of ions that appear along a unique mass-mobility correlation line.

There are four methods of ion mobility separation currently used in mass spectrometry. These are drift-time ion mobility spectrometry (DTIMS), aspiration ion mobility spectrometry (AIMS), differential-mobility spectrometry (DMS) [also called field-asymmetric waveform ion mobility spectrometry (FAIMS)], and traveling-wave ion mobility spectrometry (TWIMS). DTIMS provides the highest IMS resolving power and is the only IMS method that can directly measure collision cross-sections. AIMS is a low-resolution mobility separation method that can monitor ions in a continuous manner. DMS and FAIMS offer continuous-ion monitoring capability and orthogonal ion mobility separation in which high-separation selectivity can be achieved. TWIMS is a method of IMS that has low resolving power but good sensitivity and is well intergrated into a commercial mass spectrometer.

1.6.2.1   Matrix-Assisted Laser-Desorption Ionization Followed by Ion Mobility Separation and Time-of-Flight Mass Analysis (MALDI-IM-TOF-MS)

MALDI-IM-TOF-MS has been used to characterize native and chemically modified DNA oligonucleotides up to eight bases in length.[482] Mobility resolution between 20 and 30 can be used to separate oligonucleotides of different lengths, but not to differentiate between isomers or even different compositions of the same length.[482] MALDI-IM-TOF-MS does, however, have additional utility in the analysis of mixtures of DNA oligonucleotides and peptides, because these classes of molecules can be distinguished on the basis of differences in their mobility. Oligonucleotide sequencing is also possible by MALDI-IM-TOF-MS.[482] Ion signals corresponding to nucleobase losses (w-type and y-type fragments) were identified by the use of differences in ion mobility. MALDI-IM-TOF-MS was also used to resolve DNA–platinum adducts from the corresponding unmodified oligonucleotides.[482] Finally, it is predicted that high-performance FT-ICR-MS, when combined with IMS, will enable superior resolving power for the analysis of complex biological samples.

1.7   SUMMARY

Nucleic acids and their components play an important role in a variety of fundamental biological processes. Their study has been at the heart of molecular biology and has initiated a wide range of new disciplines over the last two decades. Among these new disciplines has been mass spectrometry techniques for the production of quasi-molecular ions of thermally fragile biomolecules. These have been of such groundbreaking significance that they have led to the awarding of Nobel Prizes in Chemistry (2002) to John Fenn for the development of ESI and to Koichi Tanaka for MALDI.[483,484] These novel soft ionization methods are ideally suited to the accurate determination of the molecular weights of the nucleic acid biopolymers and their constituents. They have also provided valuable information on structurally specific biomolecular DNA and RNA interactions. These were essential in establishing the direct determinations of stoichiometry for posttranslational modifications for specific DNA–drug, DNA–protein, RNA–protein, and DNA–RNA noncovalent and covalent associations, including DNA and RNA sequencing analysis of SNPs, genotyping, mutation detection, genetic diagnosis, and probing of viral structures. In the past decade, dramatic progress in the field of MS has established MS as a vital bioanalytical tool in many fields such as functional genomics, proteomics, early drug discovery, and chemical diagnostics. These activities are currently fueled by resources being devoted to drug development and other biologically related activities. Progress in these applications is accelerated by exciting developments that has led to improved sensitivity, mass accuracy, mass-to-charge, and speed available in new mass spectrometer systems.[368]

REFERENCES

1. Dulbecco, R., Ginsberg, H.S., in *Microbiology, 3rd edn*, B.D. Davis, R. Dulbecco, H.N. Eisen, and H.S. Ginsberg, eds, Harper and Row, Philadelphia, 1980.
2. Watson, J.D. and Crick, F.H.C., Molecular structure of nucleic acids: A structure for deoxyribose nucleic acid, *Nature* 171, 737, 1953.

3. Jasny, B.R. and Roberts, L., Building on the DNA revolution, *Science* 300, 277, 2003.
4. Forde, C.E. and McCutchen-Maloney, S.L., Characterization of transcription factors by mass spectrometry and the role of SELDI-MS, *Mass Spectrom. Rev.* 21, 419, 2002.
5. Saenger, N., *Principles of Nucleic Acid Structure*, Springer, New York, 1984.
6. Kaneko, T., Katoh, K., Fujimoto, M., Kumagai, M., Tamaoka, J., and Katayama-Fujimura, Y.J., Determination of the nucleotide composition of a deoxyribonucleic acid by high-performance liquid chromatography of its enzymatic hydrolysate: A review, *Method. Microbiol.* 4, 229, 1986.
7. McCloskey, J.A., Techniques for the structure elucidation of complex nucleosides by mass spectrometry, in *Proc. 4th Int. Roundtable on Nucleosides, Nucleotides and their Biological Application*, F.C. Alderweineldt, and E.L. Esmans, eds, University of Antwerp, Antwerp, pp. 47–67, 1982.
8. McCloskey, J.A., Mass spectrometry of nucleic acid constituents and related compounds, in *Mass Spectrometry in the Health and Life Sciences*, A.L. Burlingame, N. Castagnoli, eds., Elsevier, Amsterdam, pp. 521–546, 1985.
9. McCloskey, J.A., Experimental approaches to the characterization of nucleic acid constituents by mass spectrometry, in *Mass Spectrometry in Biomedical Research*, S.J. Gaskell, ed., Wiley, New York, pp. 75–95, 1986.
10. Esmans, E.L., Broes, D., Hoes, I., Lemiere, F., and Vanhoutte, K., Liquid chromatography–mass spectrometry in nucleoside, nucleotide and modified nucleotide characterization, *J. Chromatogr. A* 794, 109, 1998.
11. (a) Crain, P.F., in *Electrospray Ionization Mass Spectrometry. Fundamentals, Instrumentation and Applications*, R.B. Cole, ed., Wiley, New York, 1997, p. 421. (b) Liguori, A., Napoli, A., Siciliano, C., and Sindona, G., Deacylation of 2-*N*-isobutyryl- and 2-*N*-isobutyryl-6-*O*-methyl-2′-deoxyguanosine in the condensed and gas phase. A kinetic investigation, *J. Chem. Soc., Perkin Trans. 2*, 1833, 1994. (c) Izatt, R.M., Christen, J.J., and Rytting, H., Sites and thermodynamic quantities associated with proton and metal ion interaction with ribonucleic acid, deoxyribonucleic acid, and their constituent bases, nucleosides, and nucleotides, *Chem. Rev.* 71, 439, 1971. (d) Petit, V.W., Zirah, S., Rebuffat, S., and Tabet, J.-C., Collision induced dissociation-based characterization of nucleotide peptides: Fragmentation patterns of microcin C7-C51, an antimicrobial peptide produced by *Escherichia coli*, *J. Am. Soc. Mass Spectrom.* 19, 1187, 2008. (e) Angelico, R., Ceglie, A., Cuomo, F., Cardellicchio, C., Mascolo, G., and Colafemmina, G., Catanionic systems from conversion of nucleotides into nucleo-lipids, *Langmuir* 24, 2348, 2008.
12. Apruzzese, W.A. and Vouros, P., Analysis of DNA adducts by capillary methods coupled to mass spectrometry: A perspective, *J. Chromatogr. A* 794, 97, 1998.
13. McCloskey, J.A., *Methods in Enzymology*, Academic Press, New York, Chapter 4, 1990.
14. McCloskey, J.A. and Crain P.F., Progress in mass spectrometry of nucleic acid constituents: Analysis of xenobiotic modifications and measurements at high mass, *Int. J. Mass Spectrom. Ion Process.* 118–119, 593, 1992.
15. (a) Crain, P.F., Mass spectrometric techniques in nucleic acid research, *Mass Spectrom. Rev.* 9, 505, 1990. (b) Crain, P.F. and McCloskey, J.A., Applications of mass spectrometry to the characterization of oligonucleotides and nucleic acids, *Curr. Opin. Biotechnol.* 9, 25, 1998. (c) Kamel A.M. and Munson B., Collisionally-induced dissociation of substituted pyrimidine antiviral agents: Mechanisms of ion formation using gas phase hydrogen/deuterium exchange and electrospray ionization tandem mass spectrometry, *J. Am. Soc. Mass Spectrom.* 18, 1477, 2007.
16. Bieman, K. and McCloskey, J.A., Application of mass spectrometry to structure problems.1 VI. nucleosides 2, *J. Am. Chem. Soc.* 84, 2005, 1962.
17. McCloskey, J.A., in *Basic Principles in Nucleic Acid Chemistry*, P.O.P. Ts'O, ed., Academic Press, New York, p. 209, 1974.
18. Wilson, M.S. and McCloskey, J.A., Chemical ionization mass spectrometry of nucleosides. Mechanisms of ion formation and estimations of proton affinity, *J. Am. Chem. Soc.* 97, 3436, 1975.
19. Schram, K.H. and McCloskey, J.A., in *GLC and HPLC Determination of Therapeutic Agents, Part 3*, K. Tsuji, ed., Marcel Dekker, New York, p. 1149, 1979.
20. Von Minden, D.L., Stillwell, R.N., Koenig, W.A., Lyman, K.J., and McCloskey, J.A., Mass spectrometry of 7-methylpurine nucleosides: Studies of a characteristic oxygen incorporation reaction that occurs during trimethylsilylation, *Anal. Biochem.* 50, 110, 1972.
21. Panzica, R.P., Townsend, L.B., von Minden, D.L., Wilson, M.S., and McCloskey, J.A., Formation of 3,5,5-trimethyl derivatives of dihydrouracil nucleosides by reaction with methylsulfinyl carbanion and methyl iodide, *Biochim. Biophys. Acta* 331, 147, 1973.
22. Schulten, H.R., Biochemical, medical, and environmental applications of field-ionization and field-desorption mass spectrometry, *Int. J. Mass Spectrom. Ion Phys.* 32, 97, 1979.

23. Rusmintratip, V., Riggs, A.D., and Sowers, L.C., Examination of the DNA substrate selectivity of DNA cytosine methyltransferases using mass tagging, *Nucleic Acids Res.* 28, 3594, 2000.

24. Fedtke, N. and Swenberg, J.A., Quantitative analysis of DNA adducts: The potential for mass spectral analysis, in *Monitoring Human Exposures to Carcinogens: Analytical Epidemiological and Ethical Considerations*, Skipper, P.L., Koshier, F., Groopman, J.D., eds, CRC Press, Boca Raton, pp. 171–188, 1991.

25. Giese, R.W., Saha, M., Abdel-Baky, S., and Allam, K., Measuring DNA adducts by gas chromatography-electron capture-mass spectrometry: Trace organic analysis, Method. *Enzymol.* 271, 504, 1996.

26. Norwood, C.B. and Vouros, P., DNA modifications: investigations by mass spectrometry, in *Mass Spectrometry: Clinical and Biomedical Applications,* Vol. 2, D.M. Desiderio, ed., Plenum Press, New York, pp. 89–133, 1994.

27. Talaska, G. and Roh, J.H., 32P-Postlabelling and mass spectrometric methods for analysis of bulky, polyaromatic carcinogen-DNA adducts in humans, *J. Chromatogr.* 580, 293, 1992.

28. Chiarelli, M.P. and Lay, J.O. Jr., Mass spectrometry for the analysis of carcinogen-DNA adducts, *Mass Spectrom. Rev.* 11, 447, 1992.

29. Farmer, P.B. and Sweetman, G.M.A., Mass spectrometric detection of carcinogen adducts, *J. Mass Spectrom.* 30, 1369, 1995.

30. Trainor, T.M., Giese, R.W., and Vouros, P., Mass spectrometry of electrophore-labeled nucleosides: Pentafluorobenzyl and cinnamoyl derivatives, *J. Chromatogr.* 452, 369, 1988.

31. Annan, R.S., Kresbach, G.M., Giese, R.W., and Vouros, P., Trace detection of modified dna bases via moving-belt liquid chromatography—mass spectrometry using electrophoric derivatization and negative chemical, *J. Chromatogr.* 465, 285, 1989.

32. Eklund, G., Josefsson, B., and Roos, C., Determination of volatile halogenated hydrocarbons in tap water, seawater and industrial effluents by glass capillary gas chromatography and electron capture detection, *J. High Res. Chromatogr. Chromatogr. Commun.* 7, 34, 1978.

33. Petersen, B.A., Hanson, R.N., Giese, R.W., and Karger, B.L., Picogram analysis of free triiodothyronine and free thyroxine hormones in serum by equilibrium dialysis and electron capture gas chromatography, *J. Chromatogr.* 126, 503, 1976.

34. Rogers, E.J. and Giese, R.W., Sample preparation for gas chromatography with electron capture detection: Determination of total and free thyroxin in serum, *Anal. Biochem.* 164, 439, 1987.

35. Culbertson, J.A., Sears, L.J., Knighton, W.B., and Grimsrud, E.P., Origin of adduct ions in the electron-capture mass spectrum of tetracyanoethylene, *Org. Mass Spectrom.* 27, 277, 1992.

36. Abdel-Baky, S. and Giese, R.W., Gas chromatography/electron capture negative-ion mass spectrometry at the zeptomole level, *Anal. Chem.* 63, 2986, 1991.

37. Fisher, D.H. and Giese, R.W., Determination of 5-methylcytosine in DNA by gas chromatography-electron-capture detection, *J. Chromatogr.* 452, 51, 1988.

38. *The Molecular Origins of Life*, A. Brack, ed., Cambridge University Press, Cambridge, UK, 1998.

39. Stocks, P.G. and Schwartz, A.W., Uracil in carbonaceous meteorites, *Nature* 282, 709, 1979.

40. Stocks, P.G. and Schwartz, A.W., Nitrogen-heterocyclic compounds in meteorites: Significance and mechanisms of formation, *Geochim. Cosmochim. Acta* 45, 563, 1981.

41. Hua, L.L., Kobayashi, K., Ochiai, E.I., Gerke, C.W., Gerhardt, K.O., and Ponnamperuma, C., Identification and quantification of nucleic acid bases in carbonaceous chondrites, *OLEB* 16, 226, 1986.

42. Shimoyama, A., Hagishita, S., and Harada, K., Search for nucleic acid bases in carbonaceous chondrites from Antarctica, *Geochem. J.* 24, 343, 1990.

43. Brown, R.D., Godfrey, P.D., McNaughton, D., and Pierlot, A.P., A study of the major gas-phase tautomer of adenine by microwave spectroscopy, *Chem. Phys. Lett.* 156, 61, 1989.

44. Brown, R.D., Godfrey, P.D., McNaughton, D., and Pierlot, A.P., Microwave spectrum of the major gas-phase tautomer of thymine, *J. Chem. Soc. Chem. Commun.* 37, 1989.

45. Brown, R.D., Godfrey, P.D., McNaughton, D., and Pierlot, A.P., Microwave spectrum of uracil, *J. Am. Chem. Soc.* 110, 2329, 1988.

46. Colarusso, P., Zhang, K.Q., Guo, B., and Bernath, P.F., The infrared spectra of uracil, thymine, and adenine in the gas phase, *Chem. Phys. Lett.* 269, 39, 1997.

47. Graindourze, M., Smets, J., Zeegers-Huyskens, T., and Maes, G., Fourier transform-infrared spectroscopic study of uracil derivatives and their hydrogen bonded complexes with proton donors: Part I. Monomer infrared absorptions of uracil and some methylated uracils in argon matrices, *J. Mol. Struct.* 222, 345, 1990.

48. Yu Ivanov, A., Plokhotnichenko, A.M., Radchenko, E.D., Sheina, G.G., and Blagoi, Y.P., FTIR spectroscopy of uracil derivatives isolated in Kr, Ar and Ne matrices: matrix effect and Fermi resonance, *J. Mol. Struct.* 372, 91, 1995.

49. Nowak, M.J., IR matrix isolation studies of nucleic acid constituents: the spectrum of monomeric thymine, *J. Mol. Struct.* 193, 35, 1989.

50. Nowak, M.J., Lapinski, L., Kwiatkowski, J., and Leszcynski, J., Molecular Structure and Infrared Spectra of Adenine. Experimental Matrix Isolation and Density Functional Theory Study of Adenine [15]N Isotopomers, *J. Phys. Chem.* 100, 3527, 1996.

51. Clark, L.B., Peschel, G.G., and Tinoco, I. Jr., Vapor Spectra and Heats of Vaporization of Some Purine and Pyrimidine Bases, *J. Phys. Chem.* 69, 3615, 1965.

52. Lührs, D.C., Viallon, J., and Fischer, I., Excited state spectroscopy and dynamics of isolated adenine and 9-methyladenine, *Phys. Chem. Chem. Phys.* 3, 1827, 2001.

53. Plützer, C., Nir, E., de Vries, M.S., and Kleinermanns, K., IR–UV double-resonance spectroscopy of the nucleobase adenine, *Phys. Chem. Chem. Phys.* 3, 5466, 2001.

54. Fujii, M., Tamura, T., Mikami, N., and Ito, M., Electronic spectra of uracil in a supersonic jet, *Chem. Phys. Lett.* 126, 583, 1986.

55. Kim, N.J., Jeong, G., Kim, Y.S., Sung, J., and Kim, S.K., Resonant two-photon ionization and laser induced fluorescence spectroscopy of jet-cooled adenine, *J. Chem. Phys.* 113, 10052, 2000.

56. Kissel, J. and Krueger, F.R., The organic component in dust from comet Halley as measured by the PUMA mass spectrometer on board Vega, *Nature* 326, 755, 1987.

57. Varmuza, K., Werther, W., Krueger, F.R., Kissel, J., and Schmid, E.R., Organic substances in cometary grains: comparison of secondary ion mass spectral data and californium-252 plasma desorption data from reference compounds, *Int. J. Mass Spectrom.* 189, 79, 1999.

58. Hush, N.S. and Cheung, A.S., Ionization potentials and donor properties of nucleic acid bases and related compounds, *Chem. Phys. Lett.* 34, 11, 1975.

59. Urano, S., Yang, X., and LeBreton, P.R., UV photoelectron and quantum mechanical characterization of DNA and RNA bases: Valence electronic structures of adenine, 1,9-dimethyl-guanine, 1-methylcytosine, thymine and uracil, *J. Mol. Struct.* 214, 315, 1989.

60. Lauer, G., Schäfer, W., and Schweig, A., Functional subunits in the nucleic acid bases uracil and thymine, *Tetrahedron Lett.* 45, 3939, 1975.

61. Yu, C., O_Donnell, T.J., and LeBreton, P.R., Ultraviolet photoelectron studies of volatile nucleoside models. Vertical ionization potential measurements of methylated uridine, thymidine, cytidine, and adenosine, *J. Phys. Chem.* 85, 3851, 1981.

62. Palmer, M.H., Simpson, I., and Platenkamp, R.J., The electronic structure of flavin derivatives: Part I. Ab initio calculations for 1H-alloxazine and 10H-isoalloxazine, their reduced derivatives and related compounds; assignments of photoelectron spectra, *J. Mol. Struct.* 66, 243, 1980.

63. Dougherty, D., Wittel, K., Meeks, J., and McGlynn, S.P., Photoelectron spectroscopy of carbonyls. Ureas, uracils, and thymine, *J. Am. Chem. Soc.* 98, 3817, 1976.

64. Kubota, M. and Kobayashi, T.J., Electronic structure of uracil and uridine derivatives studied by photoelectron spectroscopy, *Electron. Spectrosc. Rel. Phenom.* 82, 61, 1996.

65. Peng, S., Padva, A., and LeBreton, P.R., Ultraviolet photoelectron studies of biological purines: the valence electronic structure of adenine, *Proc. Natl. Acad. Sci. USA* 73, 2966, 1976.

66. Lin, J., Yu, C., Peng, S., Akiyama, I., Li, K., Kao Lee, L., and LeBreton, P.R., Ultraviolet photoelectron studies of the ground-state electronic structure and gas-phase tautomerism of purine and adenine, *J. Am. Chem. Soc.* 102, 4627, 1980.

67. Rice, J.M., Dudek, G.O., and Barber, M., Mass spectra of nucleic acid derivatives. pyrimidines, *J. Am. Chem. Soc.* 87, 4569, 1965.

68. Rice, J.M. and Dudek, G.O., Mass spectra of nucleic acid derivatives. II. Guanine, adenine, and related compounds, *J. Am. Chem. Soc.* 89, 2719, 1967.

69. Ulrich, J., Teoule, R., Massot, R., and Cornu, A., Etude de la fragmentation de derives de L'uracile et de la thymine par Spectrometrie de masse, *Org. Mass Spectrom.* 2, 1183, 1969.

70. Occolowitz, J.L., Carbon-14 as a label in mass spectrometry, *Chem. Commun.* 1226, 1968.

71. Barrio, M.G., Scopes, D.I.C., Holtwick, J.B., and Leonard, N.J., Syntheses of all singly labeled [15N] adenines: Mass spectral fragmentation of adenine, *Proc. Natl. Acad. Sci. USA* 78, 3986, 1981.

72. Sethi, S.K., Gupta, S.P., Jenkins, E.E., Whitehead, C.W., Townsend, L.B., and McCloskey, J.A., Mass spectrometry of nucleic acid constituents. Electron ionization spectra of selectively labeled adenines, *J. Am. Chem. Soc.* 104, 3349, 1982.

73. Brown, E.G. and Mangat, B.S., Structure of a pyrimidine amino acid from pea seedlings, *Biochim. Biophys. Acta* 177, 427, 1969.

74. Smith, K.C., Aplin, R.T., A mixed photoproduct of uracil and cysteine (5-S-cysteine-6-hydrouracil). A possible model for the in vivo cross-linking of deoxyribonucleic acid and protein by ultraviolet light, *Biochemistry* 5, 2125, 1966.

75. Hecht, S.M., Gupta, A.S., and Leonard, N.J., Position of uridine thiation: The identification of minor nucleosides from transfer RNA by mass spectrometry, *Biochim. Biophys. Acta* 182, 444, 1969.
76. Jochims, H.W., Schwell, M., Baumgartel, H., and Leach, S., Photoion mass spectrometry of adenine, thymine and uracil in the 6–22 eV photon energy range, *Chem. Phys.* 314, 263, 2005.
77. Gohlke, S. and Illenberger, E., Probing biomolecules: Gas phase experiments and biological relevance, *Europhys. News* 33, 207, 2002.
78. Abouaf, R., Pommier, J., and Dunet, H., Negative ions in thymine and 5-bromouracil produced by low energy electrons, *Int. J. Mass Spectrom.* 226, 397, 2003.
79. deVries, J., Hoekstra, R., Morgenstern, R., and Schlathölter, T., Charge Driven Fragmentation of Nucleobases, *Phys. Rev. Lett.* 91, 053401, 2003.
80. Hung, K., Sun, H., Ding, M., Kalafatis, P., Simioni, P., and Guo, B., A matrix-assisted laser desorption/ ionization time-of-flight based method for screening the 1691G -> A mutation in the factor V gene, *Blood Coagul. Fibrin.* 13, 117, 2002.
81. Schulten, H.R., Schiebel, H.M., Fresnius, Z., High resolution field desorption mass spectrometry, *Anal. Chem.* 280, 139, 1976.
82. Schulten, H.R. and Beckey, H.D., High resolution field desorption mass spectrometry-I: Nucleosides and nucleotides, *Org. Mass Spectrom.* 7, 861, 1973.
83. McNeal, C.J., Narang, S.A., Macfarlane, R.D., Hsiung, H.M., and Brousseau, R., Sequence determination of protected oligodeoxyribonucleotides containing phosphotriester linkages by californium-252 plasma desorption mass spectrometry, *Proc. Natl. Acad. Sci. USA* 77, 735, 1980.
84. Otake, N., Ogita, T., Miyazaki, Y., Yonehara, H., Macfarlane, R.D., and McNeal, C.J., [252]Cf plasma desorption mass spectrometry of Adenomycin (C19-97 substance), *J. Antibiot.* 34, 130, 1981.
85. McNeal, C.J., Ogilvie, K.K., Theriault, N.Y., and Nemer, M.J., A new method for the analysis of fully protected oligonucleotides by californium-252 plasma desorption mass spectrometry. 3. Positive ions, *J. Am. Chem. Soc.* 104, 981, 1982.
86. Cotter, R.J. and Fenselau, C., The effects of heating rate and sample size on the direct exposure/chemical ionization mass spectra of some biological conjugates, *Biomed. Mass Spectrom.* 6, 287, 1979.
87. Esmans, E.L., Freyne, E.J., Vanbroeckhoven, J.H., and Alderweireldt, F.C., Chemical ionization desorption mass spectrometry as an additional tool for the structure elucidation of nucleosides, *Biomed. Mass Spectrom.* 7, 377, 1980.
88. Hunt, D.F., Shabanowitz, J., Botz, F.K., and Brent, D.A., Chemical ionization mass spectrometry of salts and thermally labile organics with field desorption emitters as solids probes, *Anal. Chem.* 49, 1160, 1977.
89. Eicke, A., Sichtermann, W., and Benninghoven, A., Secondary ion mass spectrometry of nucleic acid components: Pyrimidines, purines, nucleosides and nucleotides, *Org. Mass Spectrom.* 15, 289, 1980.
90. Unger, S.E., Schoen, A.E., Cooks, R.G., Ashmorth, D.J., Gomes, J.D., and Chang, C.J., Identification of modified nucleosides by secondary-ion and laser-desorption mass spectrometry, *J. Org. Chem.* 46, 4765, 1981.
91. Clay, K.L., Wahlin, L., and Murphy, R.C., Interlaboratory reproducibility of relative abundances of ion currents in fast atom bombardment mass spectral data, *Biomed. Mass Spectrom.* 10, 489, 1983.
92. Marziali, A. and Akeson, M., New DNA sequencing methods, *Annu. Rev. Biomed. Eng.* 3, 195, 2001.
93. Murray, K.K., DNA sequencing by mass spectrometry, *J. Mass Spectrom.* 31, 1203, 1996.
94. Monforte, J.A. and Becker, C.H., High-throughput DNA analysis by time-of-flight mass spectrometry, *Nat. Med.* 3, 360, 1997.
95. Karas, M., Bachmann, D., Bahr, U., and Hillenkamp, F., Matrix-assisted ultraviolet laser desorption of non-volatile compounds, *Int. J. Mass Spectrom. Ion Processes* 78, 53, 1987.
96. Tanaka, K., Waki, H., Ido, Y., Akita, S., Yoshida, Y., and Yoshida, T., Protein and polymer analyses up to m/z 100.000 by laser ionization time-of-flight mass spectrometry, *Rapid Commun. Mass Spectrom.* 2, 151, 1988.
97. Spengler, B., Post-source decay analysis in matrix-assisted laser desorption/ionization mass spectrometry of biomolecules, *J. Mass Spectrom.* 32, 1019, 1997.
98. Nordhoff, E., Kirpekar, F. Hahner, St. Hillenkamp F. and Roepstorff, P. MALDI-MS as a new method for the analysis of nucleic acids (DNA and RNA) with molecular masses up to 150kDa, *Applications of Modern Mass Spectrometry in Plant Science Research,* (R.P. Newton and T.J. Walton, eds., pp. 86–101, *Clarendon Press,* London, 1996.
99. Spengler, B., Kirsch, D., Kaufmann, R., and Jaeger, E., Peptide sequencing by matrix-assisted laser-desorption mass spectrometry, *Rapid Commun. Mass Spectrom.* 6, 105, 1992.
100. Yates, J.R., Mass spectrometry and the age of the proteome, *J. Mass Spectrom.* 33, 1, 1998.

101. Koster, H., Tang, K., Fu, D.J., Braun, A., Van den Boom, D., Smith, C.L., Cotter, R.J., and Cantor, C.R., A strategy for rapid and efficient DNA sequencing by mass spectrometry, *Nat. Biotechnol.* 14, 1123, 1996.

102. Fu, D.J., Tang, K., Braun, A., Reuter, D., Darnhofer-Demar, B., Little, D.P., O'Donnell, M.J., Cantor, C. R., and Koster, H., Sequencing exons 5 to 8 of the p53 gene by MALDI-TOF mass spectrometry, *Natl. Biotechnol.* 16, 381, 1998.

103. Hahner, S., Ludemann, H.C., Kirpekar, F., Nordhoff, E., Roepstorff, P., Galla, H.J., and Hillenkamp, F., Matrix-assisted laser desorption/ionization mass spectrometry (MALDI) of endonuclease digests of RNA, *Nucleic Acids Res.* 25, 1957, 1997.

104. Faulstich, K., Worner, K., Brill, H., and Engels, W.J., A sequencing method for RNA oligonucleotides based on mass spectrometry, *Anal. Chem.* 69, 4349, 1997.

105. Ross, P.L. and Belgrader, P., Analysis of short tandem repeat polymorphisms in human DNA by matrix-assisted laser desorption/ionization mass spectrometry, *Anal. Chem.* 69, 3966, 1997.

106. Ross, P.L., Lee, K., and Belgrader, P., Discrimination of single-nucleotide polymorphisms in human DNA using peptide nucleic acid probes detected by MALDI-TOF mass spectrometry, *Anal. Chem.* 69, 4197, 1997.

107. Griffin, J.T., Tang, W., and Smith, L.M., Genetic analysis by peptide nucleic acid affinity MALDI-TOF mass spectrometry, *Nat. Biotechnol.* 15, 1368, 1997.

108. Nordhoff, E., Ingendoh, A., Cramer, R., Overberg, A., Stahl, B., Karas, M., Hillenkamp, F., and Crain, P.F., Matrix-assisted laser desorption/ionization mass spectrometry of nucleic acids with wavelengths in the ultraviolet and infrared, *Rapid Commun. Mass Spectrom.* 6, 771, 1992.

109. Nordhoff, E., Kipekar, F., Karas, M., Cramer, R., Hahner, S., Hillenkamp, F., Kristiansen, K., Roepstroff, P., and Lezius, A., Comparison of IR- and UV-matrix-assisted laser desorption/ionization mass spectrometry of oligodeoxynucleotides, *Nucleic Acids Res.* 22, 2460, 1994.

110. Nordhoff, E., Kirpekar, F., and Roepstorff, P., Mass spectrometry of nucleic acids, *Mass Spectrom. Rev.* 15, 67, 1996.

111. Vestal, M.L., Juhasz, P., and Martin, S.A., Delayed extraction matrix-assisted laser desorption time-of-flight mass spectrometry, *Rapid Commun. Mass Spectrom.* 9, 1044, 1995.

112. Gehrke, C.W., Kuo, K.C., Davis, G.E., Suits, R.D., Waalkes, T.P., and Borek, E., Quantitative high-performance liquid chromatography of nucleosides in biological materials, *J. Chromatogr.* 150, 455, 1978.

113. Liebich, H.M., Di Stefano, C., Wixforth, A., and Schmidt, H.R.J., Quantitation of urinary nucleosides by high-performance liquid chromatography, *Chromatogr. A* 763, 193, 1997.

114. Liebich, H.M., Xu, G., Lehmann, R., Haring, H.U., Lu, P., and Zhang, Y., Analysis of normal and modified nucleosides in urine by capillary electrophoresisa), *Chromatographia* 45, 396, 1997.

115. Liebich, H.M., Müller-Hagedorn, S., Klaus, F., Meziane, K., Kim, K.R., Frickenschmidt, A., and Kammerer, B.J., Chromatographic, capillary electrophoretic and matrix-assisted laser desorption ionization time-of-flight mass spectrometry analysis of urinary modified nucleosides as tumor markers, *Chromatogr. A* 1071, 271, 2005.

116. Slusher, J.T., Kuwahara, S.K., Hamzeh, F.M., Lewis, L.D., Kornhauser, D.M., and Lietman, P.S., Intracellular zidovudine (ZDV) and ZDV phosphates as measured by a validated combined high-pressure liquid chromatography-radioimmunoassay procedure, *Antimicrob. Agents Chemother.* 36, 2473, 1992.

117. Robbins, B.L., Waibel, B.H., and Fridland, A., Quantitation of intracellular zidovudine phosphates by use of combined cartridge-radioimmunoassay methodology, *Antimicrob. Agents Chemother.* 40, 2651, 1996.

118. Darque, A., Valette, G., Rousseau, F., Wang, L.H., Sommadossi, J.P., and Zhou, X.J., Quantitation of intracellular triphosphate of emtricitabine in peripheral blood mononuclear cells from human immunodeficiency virus-infected patients, *Antimicrob. Agents Chemother.* 43, 2245, 1999.

119. Solas, C., Li, Y.F., Xie, M.Y., Sommadossi, J.P., and Zhou, X.J., Intracellular nucleotides of (−)-2′,3′-deoxy-3′-thiacytidine in peripheral blood mononuclear cells of a patient infected with human immunodeficiency virus, *Antimicrob. Agents Chemother.* 42, 2989, 1998.

120. Moore, J.D., Valette, G., Darque, A., Zhou, X.J., and Sommadossi, J.P., Simultaneous quantitation of the 5′-triphosphate metabolites of zidovudine, lamivudine, and stavudine in peripheral mononuclear blood cells of HIV infected patients by high-performance liquid chromatography tandem mass spectrometry, *J. Am. Soc. Mass Spectrom.* 11, 1134, 2000.

121. King, T., Bushman, L., Anderson, P.L., Delahunty, T., Ray, M., and Fletcher, C.V.J., Quantitation of zidovudine triphosphate concentrations from human peripheral blood mononuclear cells by anion exchange solid phase extraction and liquid chromatography–tandem mass spectroscopy; an indirect quantitation methodology, *Chromatogr. B Analyt. Technol. Biomed. Life Sci.* 831, 248, 2006.

122. Claire, R.L. III., Positive ion electrospray ionization tandem mass spectrometry coupled to ion-pairing high-performance liquid chromatography with a phosphate buffer for the quantitative analysis of intracellular nucleotides, *Rapid Commun. Mass Spectrom.* 14, 1625, 2000.

123. Pruvost, A., Becher, F., Bardouille, P., Guerrero, C., Creminon, C., Delfraissy, J.F., Goujard, C., Grassi, J., and Benech, H., Direct determination of phosphorylated intracellular anabolites of stavudine (d4T) by liquid chromatography/tandem mass spectrometry, *Rapid Commun. Mass Spectrom.* 15, 1401, 2001.

124. Tuytten, R., Lemiere, F., Dongen, W.V., Esmans, E.L., and Slegers, H., Short capillary ion-pair high-performance liquid chromatography coupled to electrospray (tandem) mass spectrometry for the simultaneous analysis of nucleoside mono-, di- and triphosphates, *Rapid Commun. Mass Spectrom.* 16, 1205, 2002.

125. Xia, Y.Q., Jemal, M., Zheng, N., and Shen, X., Utility of porous graphitic carbon stationary phase in quantitative liquid chromatography/tandem mass spectrometry bioanalysis: Quantitation of diastereomers in plasma, *Rapid Commun. Mass Spectrom.* 20, 1831, 2006.

126. Shi, G., Wu, J.T., Li, Y., Geleziunas, R., Gallagher, K., Emm, T., Olah, T., and Unger, S., Novel direct detection method for quantitative determination of intracellular nucleoside triphosphates using weak anion exchange liquid chromatography/tandem mass spectrometry, *Rapid Commun. Mass Spectrom.* 16, 1092, 2002.

127. Veltkamp, S.A., Hillebrand, M.J., Rosing, H., Jansen, R.S., Wickremsinhe, E.R., Perkins, E.J., Schellens, J.H., and Beijnen, J.H., Quantitative analysis of gemcitabine triphosphate in human peripheral blood mononuclear cells using weak anion-exchange liquid chromatography coupled with tandem mass spectrometry, *J. Mass Spectrom.* 41, 1633, 2006.

128. Kong, Y., Zhu, Y., and Zhang, J.-Y., Ionization mechanism of oligonucleotides in matrix-assisted laser desorption/ionization time-of-flight mass spectrometry, *Rapid Commun. Mass Spectrom.* 15, 57, 2001.

129. Langley, G.J., Herniman, J.M., Davies, N.L., and Brown, T., Simplified sample preparation for the analysis of oligonucleotides by matrix-assisted laser desorption/ionisation time-of-flight mass spectrometry, *Rapid Commun. Mass Spectrom.* 13, 1717, 1999.

130. Liu, C., Wu, Q., Harms, A.C., and Smith, R.D., On-line microdialysis sample cleanup for electrospray ionization mass spectrometry of nucleic acid samples, *Anal. Chem.* 68, 3295, 1996.

131. Smirnov, I.P., Hall, L.R., Ross, P.L., and Haff, L.A., Application of DNA-binding polymers for preparation of DNA for analysis by matrix-assisted laser desorption/ionization mass spectrometry, *Rapid Commun. Mass Spectrom.* 15, 1427, 2001.

132. Gilar, M., Belenky, A., and Wang, B.H., High-throughput biopolymer desalting by solid-phase extraction prior to mass spectrometric analysis, *J. Chromatogr. A* 921, 3, 2001.

133. Davis, D.L., O'Brien, E.P., and Bentzley, C.M., Analysis of the degradation of oligonucleotide strands during the freezing/thawing processes using MALDI-MS, *Anal. Chem.* 72, 5092, 2000.

134. Wu, K.J., Steding, A., and Becker, C.H., Matrix-assisted laser desorption time-of-flight mass spectrometry of oligonucleotides using 3-hydroxypicolinic acid as an ultraviolet-sensitive matrix, *Rapid Commun. Mass Spectrom.* 7, 142, 1993.

135. Tang, K., Taranenko, N.I., Allman, S.L., Chen, C.H., Chang, L.Y., and Jacobson, K.B., Picolinic acid as a matrix for laser mass spectrometry of nucleic acids and proteins, *Rapid Commun. Mass Spectrom.* 8, 673, 1994.

136. Krause, J., Stoeckli, M., and Schlunegger, U.P., Studies on the selection of new matrices for ultraviolet matrix-assisted laser desorption/ionization time-of-flight mass spectrometry, *Rapid Commun. Mass Spectrom.* 10, 1927, 1996.

137. Zhu, Y.F., Chung, C.N., Taranenko, N.I., Allman, S.L., Martin, S.A., Haff, L., and Chen, C.L., The study of 2,3,4-Trihydroxyacetophenone and 2,4,6-trihydroxyacetophenone as matrices for DNA detection in matrix-assisted laser desorption/ionization time-of-flight mass spectrometry, *Rapid Commun. Mass Spectrom.* 10, 383, 1996.

138. Lin, H., Hunter, J.M., and Becker, C.H., Laser desorption of DNA oligomers larger than one kilobase from cooled 4-nitrophenol, *Rapid Commun. Mass Spectrom.* 13, 2335, 1999.

139. Shahgholi, M., Garcia, B.A., Chiu, N.H.L., Heaney, P.J., and Tang, K., Sugar additives for MALDI matrices improve signal allowing the smallest nucleotide change (A:T) in a DNA sequence to be resolved, *Nucleic Acids Res.* 29, e91, 2001.

140. Williams, T.L. and Fenselau, C., p-Nitroaniline/glycerol: A binary liquid matrix for MALDI analysis, *Eur. J. Mass Spectrom.* 4, 379, 1998.

141. Menzel, C., Berkenkamp, S., and Hillenkamp, F., Infrared matrix-assisted laser desorption/ionization mass spectrometry with a transversely excited atmospheric pressure carbon dioxide laser at 10.6 μm wavelength with static and delayed ion extraction, *Rapid Commun. Mass Spectrom.* 13, 26, 1999.

142. Kirpekar, F., Berkenkamp, S., and Hillenkamp, F., Detection of double-stranded DNA by IR- and UV-MALDI mass spectrometry, *Anal. Chem.* 71, 2334, 1999.

143. Berkenkamp, S., Menzel, C., Karas, M., and Hillenkamp, F., Performance of infrared matrix-assisted laser desorption/ionization mass spectrometry with lasers emitting in the 3 μm wavelength range, *Rapid Commun. Mass Spectrom.* 11, 1399, 1997.

144. Chen, W.Y. and Chen, Y.C., Reducing the alkali cation adductions of oligonucleotides using sol–gel-assisted laser desorption/ionization mass spectrometry, *Anal. Chem.* 75, 4223, 2003.

145. Weng, M.F. and Chen, Y.C., Using sol-gel/crown ether hybrid materials as desalting substrates for matrix-assisted laser desorption/ionization analysis of oligonucleotides, *Rapid Commun. Mass Spectrom.* 18, 1421, 2004.

146. Zhou, L.H., Deng, H.M., Deng, Q.Y., and Zhao, S.K., A mixed matrix of 3-hydroxypicolinic acid and pyrazinecarboxylic acid for matrix-assisted laser desorption/ionization time-of-flight mass spectrometry of oligodeoxynucleotides, *Rapid Commun. Mass Spectrom.* 18, 787, 2004.

147. Ferris, J.P., Mineral catalysis and prebiotic synthesis: Montmorillonite-catalyzed formation of RNA, *Elements* 1, 145, 2005.

148. Huang, W. and Ferris, J.P., Synthesis of 35-40 mers of RNA oligomers from unblocked monomers. A simple approach to the RNA world, *Chem. Commun.* 1458, 2003.

149. Wang, K.J. and Ferris, J.P., Catalysis and selectivity in prebiotic synthesis: Initiation of the formation of oligo(U)s on montmorillonite clay by adenosine-5′-methylphosphate, *Origins Life Evol. Biosphere* 35, 187, 2005.

150. Ferris, J.P., Montmorillonite catalysis of 30-50 mer oligonucleotides: Laboratory demonstration of potential steps in the origin of the RNA world, *Origins Life Evol. Biosphere* 32, 311, 2002.

151. Ferris, J.P., *Chem. Tracts Biochem. Mol. Biol.* 12, 419, 1999.

152. Rupert, J.P., Granquist, W.T., and Pannavaia, T.J., Catalytic properties of clay minerals. In: NEWMAN, A.C.D. (ed.), *Chemistry of Clays and Clay Minerals, Mineralogical Society Monograph*, 6, 275, 1987.

153. Zagorevskii, D.V., Aldersley, M.F., and Ferris., J.P. MALDI analysis of oligonucleotides directly from montmorillonite, *J. Am. Soc. Mass Spectrom.* 17, 1265, 2006.

154. Joyce, G.F., Inoue, T., and Orgel, L.E., Non-enzymatic template-directed synthesis on RNA random copolymers: Poly(C, U) templates, *J. Mol. Biol.* 176, 279, 1984.

155. Berlin, K. and Gut, I.G., Analysis of negatively 'charge tagged' DNA by matrix-assisted laser desorption/ionization time-of-flight mass spectrometry, *Rapid Commun. Mass Spectrom.* 13, 1739, 1999.

156. Lichtenwalter, K.G., Apffel, A., Bai, J., Chakel, J A., Dai, Y., Hahnenberger, K.M., Li, L., and Hancock, W.S., Approaches to functional genomics: Potential of matrix-assisted laser desorption ionization–time of flight mass spectrometry combined with separation methods for the analysis of DNA in biological samples, *J. Chromatogr., B* 745, 231, 2000.

157. Meyer, A., Spinelli, N., Imbach, J.-L., and Vasseur, J.-J., Analysis of solid-supported oligonucleotides by matrix-assisted laser desorption/ionization time-of-flight mass spectrometry, *Rapid Commun. Mass Spectrom.* 14, 234, 2000.

158. Meyer, A., Spinelli, N., Bres, J.-C., Dell'Aquila, C., Morvan, F., Lefebvre, I., Rayner, B., Imbach, J.-L., and Vasseur, J.-J., Direct MALDI-TOF MS analysis of oligonucleotides on solid support through a photolabile linker, *Nucleos. Nucleot. Nucleic Acids* 20, 963, 2001.

159. Guerlavais. T., Meyer, A., Depart, F., Imbach, J.-L., Morvan, F., and Vasseur, J.-J., Use of MALDI-TOF mass spectrometry to monitor solid-phase synthesis of oligonucleotides, *Anal. Bioanal. Chem.* 374, 57, 2002.

160. Taranenko, N.I., Hurt, R., Zhou, J.Z., Isola, N.R., Huang, H., Lee, S.H., and Chen, C.H.J., Laser desorption mass spectrometry for microbial DNA analysis, *Microbiol. Meth.* 48, 101, 2002.

161. Olejnik, J., Ludemann, H., Krzymanska-Olejnik, E., Berkenkamp, S., Hillenkamp, F., and Rothschild, K.J., Photocleavable peptide-DNA conjugates: Synthesis and applications to DNA analysis using MALDI-MS, *Nucleic Acids Res.* 27, 4626, 1999.

162. Alazard, D., Filipowsky, M., Raeside, J., Clarke, M., Majlessi, M., Russell, J., and Weisburg, W., Sequencing of production-scale synthetic oligonucleotides by enriching for coupling failures using matrix-assisted laser desorption/ionization time-of-flight mass spectrometry, *Anal. Biochem.* 301, 57, 2002.

163. Mouradian, S., Rank, D.R., and Smith, L.M., Analyzing sequencing reactions from bacteriophage M13 by matrix-assisted laser desorption/ionization mass spectrometry, *Rapid Commun. Mass Spectrom.* 10, 1475, 1996.

164. Pieles, U., Zurcher, W., Schar, M., and Moser, H.E., Matrix-assisted laser desorption ionization time-of-flight mass spectrometry: A powerful tool for the mass and sequence analysis of natural and modified oligonucleotides, *Nucleic Acids Res.* 21, 3191, 1993.

165. Bentzley, C.M., Johnston, M.V., and Larsen, B.S., Base specificity of oligonucleotide digestion by calf spleen phosphodiesterase with matrix-assisted laser desorption ionization analysis, *Anal. Biochem.* 258, 31, 1998.
166. Wada, Y., Separate analysis of complementary strands of restriction enzyme-digested DNA. An application of restriction fragment mass mapping by matrix-assisted laser desorption/ionization mass spectrometry, *J. Mass Spectrom.* 33, 187, 1998.
167. Taranenko, N.I., Allman, S.L., Golovlev, V.V., Taranenko, N.V., Isola, N.R., and Chen, C.H., Sequencing DNA using mass spectrometry for ladder detection, *Nucleic Acids Res.* 26, 2488, 1998.
168. Nordhoff, E., Leubbert, C., Thiele, G., Heiser, V., and Lehrach, H., Rapid determination of short DNA sequences by the use of MALDI-MS, *Nucleic Acids Res.* 28, e86, 2000.
169. Tang, W., Scalf, M., Smith, L.M., Controlling DNA fragmentation in MALDI-MS by chemical modification, *Anal. Chem.* 69, 302, 1997.
170. Ono, T., Scalf, M., and Smith, L.M., 2′-Fluoro modified nucleic acids: Polymerase-directed synthesis, properties and stability to analysis by matrix-assisted laser desorption/ionization mass spectrometry, *Nucleic Acids Res.* 25, 4581, 1997.
171. Smith, D.E., Su, J.Y., and Jucker, F.M., Efficient enzymatic synthesis of ^{13}C, ^{15}N-labeled DNA for NMR studies, *J. Biomol. NMR* 10, 245, 1997.
172. Phan, A.T., and Patel, D.J., A site-specific low-enrichment ^{15}N, ^{13}C isotope-labeling approach to unambiguous NMR spectral assignments in nucleic acids, *J. Am. Chem. Soc.* 124, 1160, 2002.
173. Kettani, A., Gorin, A., Majumdar, A., Hermann, T., Skripkin, E., Zhao, H., Jones, R., and Patel, D.J., A dimeric DNA interface stabilized by stacked A · (G · G · G · G) · A hexads and coordinated monovalent cations1, *J. Mol. Biol.* 297, 627, 2000.
174. Masse, J.E., Bortmann, P., Dieckmann, T., and Feigon, J., Simple, efficient protocol for enzymatic synthesis of uniformly ^{13}C/^{15}N-labeled DNA for heteronuclear NMR studies, *Nucleic Acids Res.* 26, 2618, 1998.
175. Zimmer, D.P. and Crothers, D.M., NMR of enzymatically synthesized uniformly ^{13}C^{15}N-labeled DNA oligonucleotides, *Proc. Natl. Acad. Sci. USA* 92, 3091, 1995.
176. Clark, J.M., Joyce, C.M., and Beardsley, G.P., Novel blunt-end addition reactions catalyzed by DNA polymerase I of *Escherichia coli*, *J. Mol. Biol.* 198, 123, 1987.
177. Ambrus, A., Chen D., Whatcott, C., Somogyi, A., and Yang, D., Matrix-assisted laser desorption/ionization time-of-flight mass spectrometry protocol for monitoring the progress of enzymatic ^{13}C/^{15}N-labeled DNA syntheses, *Anal. Biochem.* 342, 246, 2005.
178. Stoerker, J., Mayo, J.D., Tetzlaff, C.N., Sarracino, D.A., Schwope, I., and Richert, C., Rapid genotyping by MALDI-monitored nuclease selection from probe libraries, *Nat. Biotechnol.* 18, 1213, 2000.
179. Tost, J. and Gut, I.G., Genotyping single nucleotide polymorphisms by mass spectrometry, *Mass Spectrom. Rev.* 21, 389, 2002.
180. Fei, Z. and Smith, L.M., Analysis of single nucleotide polymorphisms by primer extension and matrix-assisted laser desorption/ionization time-of-flight mass spectrometry, *Rapid Commun. Mass Spectrom.* 14, 950, 2000.
181. Little, D.P. and McLafferty, F.W., Sequencing 50-mer DNAs using electrospray tandem mass spectrometry and complementary fragmentation methods, *J. Am. Chem. Soc.* 117, 6783, 1995.
182. Krebs, S., Seichter, D., and Forster, M., Genotyping of dinucleotide tandem repeats by MALDI mass spectrometry of ribozyme-cleaved RNA transcripts, *Nat. Biotechnol.* 19, 877, 2001.
183. Seichter, D., Krebs, S., and Forster, M., Rapid and accurate characterisation of short tandem repeats by MALDI-TOF analysis of endonuclease cleaved RNA transcripts, *Nucleic Acids Res.* 32, e16, 2004.
184. Wada, Y., Mitsumori, K., Terachi, T., and Ogawa, O., Measurement of polymorphic trinucleotide repeats in the androgen receptor gene by matrix-assisted laser desorption/ionization time-of-flight mass spectrometry, *J. Mass Spectrom.* 34, 885, 1999.
185. Wolfe, J.L., Kawate, T., Sarracino, D.A., Zillmann, M., Olson, J., Stanton, V.P., Jr., and Verdine, G.L., A genotyping strategy based on incorporation and cleavage of chemically modified nucleotides, *Proc. Natl. Acad. Sci. USA* 99, 11073, 2002.
186. Sun, X.Y., Hung, K., Wu, L., Sidransky, D., and Guo, B.B., Detection of tumor mutations in the presence of excess amounts of normal DNA, *Nat. Biotechnol.* 20, 186, 2002.
187. Lau, C.C., Yue, P.Y.K., Chui, S.H., Chui, A.K.K., Yam, W.C., and Wong, R.N.S., Detection of single nucleotide polymorphisms in hepatitis B virus precore/basal core promoter region by matrix-assisted laser desorption/ionization time-of-flight mass spectrometry, *Anal. Biochem.* 366, 93, 2007.
188. Sauer, S., Typing of single nucleotide polymorphisms by MALDI mass spectrometry: Principles and diagnostic applications, *Clin. Chim. Acta* 363, 95, 2006.

189. Kim, S., Ulz, M.E., Nguyen, T., Li, C.-M., Sato, T., Tycko, B., and Ju, J., Thirtyfold multiplex genotyping of the p53 gene using solid phase capturable dideoxynucleotides and mass spectrometry, *Genomics* 83, 924, 2004.

190. Tost, J. and Gut, I.G., Genotyping single nucleotide polymorphisms by MALDI mass spectrometry in clinical applications, *Clin. Biochem.* 38, 335, 2005.

191. Jones, J.J., Mariccor, S., Batoy, A.B., and Wilkins, C.L., Ionic liquid matrix-induced metastable decay of peptides and oligonucleotides and stabilization of phospholipids in MALDI FTMS analyses, *J. Am. Soc. Mass Spectrom.* 16, 2000, 2005.

192. Armstrong, D., Zhang, L.K., He, L., and Gross, M., Ionic liquids as matrixes for matrix-assisted laser desorption/ionization mass spectrometry, *Anal. Chem.* 73, 3679, 2001.

193. Carda-Broch, S., Berthod, A., and Armstrong, D.W., Ionic matrices for matrix-assisted laser desorption/ionization time-of-flight detection of DNA oligomers, *Rapid Commun. Mass Spectrom.* 17, 553, 2003.

194. Mank, M., Stahl, B., and Boehm, G., 2,5-Dihydroxybenzoic acid butylamine and other ionic liquid matrixes for enhanced MALDI-MS analysis of biomolecules, *Anal. Biochem.* 76, 2938, 2004.

195. Zabet-Moghaddam, M., Heinzle, E., and Tholey, A., Qualitative and quantitative analysis of low molecular weight compounds by ultraviolet matrix-assisted laser desorption/ionization mass spectrometry using ionic liquid matrices, *Rapid Commun. Mass Spectrom.* 18, 141, 2004.

196. Li, Y.L., Hsu, F., and Gross, M.L., Ionic-liquid matrices for improved analysis of phospholipids by MALDI-TOF mass spectrometry, *J. Am. Soc. Mass Spectrom.* 16, 679, 2005.

197. Banoub, J.H., Newton, R.P., Esmans, E., Ewing, D.F., and Mackenzie, G., Recent developments in mass spectrometry for the characterization of nucleosides, nucleotides, oligonucleotides, and nucleic acids, *Chem. Rev.* 105, 1869, 2005.

198. Thiede, B., Von Janta-Lipinski, M., Noncovalent RNA-peptide complexes detected by matrix-assisted laser desorption/ionization mass spectrometry, *Rapid Commun. Mass Spectrom.* 12, 1889, 1998.

199. Kellersberger, K.A., Yu, E.T., Merenbloom, S.I., and Fabris, D., Atmospheric pressure MALDI-FTMS of normal and chemically modified RNA, *J. Am. Soc. Mass Spectrom.* 16, 199, 2005.

200. Gasparutto, D., Saint-Pierre, C., Jaquinod, M., and Cadet, J., MALDI-TOF mass spectrometry as a powerful tool to study enzymatic processing of DNA lesions inserted into oligonucleotides, *Nucleos. Nucleot. Nucleic Acids* 22, 1583, 2003.

201. Tretyakova, N., Matter, B., Ogdie, A., Wishnok, J.S., and Tannenbaum, S.R., Locating nucleobase lesions within DNA sequences by MALDI-TOF mass spectral analysis of exonuclease ladders, *Chem. Res. Toxicol.* 14, 1058, 2001.

202. D'Ham, C., Romieu, A., Jaquinod, M., Gasparutto, D., and Cadet, J., Excision of 5,6-Dihydroxy-5,6-dihydrothymine, 5,6-dihydrothymine, and 5-hydroxycytosine from defined sequence oligonucleotides by *Escherichia coli* Endonuclease III and Fpg proteins: Kinetic and mechanistic aspects, *Biochemistry* 38, 3335, 1999.

203. (a) Issaq, H.J., Veenstra, T.D., Conrads, T.P., and Felschow, D., The SELDI-TOF MS approach to proteomics: protein profiling and biomarker identification, *Biochem. Biophys. Res. Commun.* 292, 587, 2002. (b) Andersen, T.E., Kirpekar, F., and Haselmann, K.F., RNA fragmentation in MALDI mass spectrometry studied by H/D-exchange: Mechanisms of general applicability to nucleic acids, *J. Am. Soc. Mass Spectrom.* 17, 1353, 2006.

204. Zhu, Y., Valdes, R. Jr., Simmons, C.Q., Linder, M.W., Pugia, M.J., and Jortani, S.A., Analysis of ligand binding by bioaffinity mass spectrometry, *Clin. Chim. Acta* 371, 71, 2006.

205. Forde, C.E., Gonzales, A.D., Smessaert, J.M., Murphy, G.A., Shields, S.J., Fitch, J.P., and McCutchen-Maloney, S.L., A rapid method to capture and screen for transcription factors by SELDI mass spectrometry, *Biochem. Biophys. Res. Commun.* 290, 1328, 2002.

206. Merchant, M. and Weinberger S.R., Recent advancements in surface-enhanced laser desorption/ionization-time of flight-mass spectrometry, *Electrophoresis* 21, 1164, 2000.

207. Kim, S., Edwards, J.R., Deng, L., Chung, W., and Ju, J., Solid phase capturable dideoxynucleotides for multiplex genotyping using mass spectrometry, *Nucleic Acids Res.* 30, e85, 2002.

208. Bane, T.K., LeBlanc, J.F., Lee, T.D., and Riggs, A.D., DNA affinity capture and protein profiling by SELDI-TOF mass spectrometry: Effect of DNA methylation, *Nucleic Acids Res.* 30, e69, 2002.

209. Castoro, J.A., Köster, C., and Wilkins, C.L., Matrix-assisted laser desorption/ionization of high-mass molecules by Fourier-transform mass spectrometry, *Rapid Commun. Mass Spectrom.* 6, 239, 1992.

210. Köster, C., Castoro, J.A., and Wilkins, C.L., High-resolution matrix-assisted laser desorption/ionization of biomolecules by Fourier transform mass spectrometry, *J. Am. Chem. Soc.* 114, 7572, 1992.

211. Rempel, D.L. and Gross, M.L., High pressure trapping in Fourier transform mass spectrometry: A radio-frequency-only-mode event, *J. Am. Soc. Mass Spectrom.* 3, 590, 1992.

212. Schweikhard, L., Guan, S., and Marshall, A.G., Quadrupolar excitation and collisional cooling for axialization and high pressure trapping of ions in Fourier transform ion cyclotron resonance mass spectrometry, *Int. J. Mass Spectrom. Ion Proc.* 120, 71, 1992.

213. Solouki, T. and Russell, D .H., Laser desorption studies of high mass biomolecules in Fourier-transform ion cyclotron resonance mass spectrometry, *Proc. Natl. Acad. Sci. USA* 89, 5701, 1992.

214. Guan, S., Whal, M.C., Wood, T.D., and Marshall, A.G., Enhanced mass resolving power, sensitivity, and selectivity in laser desorption Fourier transform ion cyclotron resonance mass spectrometry by ion axialization and cooling, *Anal. Chem.* 65, 1753, 1993.

215. Pasa-Tolic, L., Huang, S., Guan, S., Kim, H.S., and Marshall, A.G.J., Ultrahigh-resolution matrix-assisted laser desorption/ionization Fourier transform ion cyclotron resonance mass spectra of peptides, *J. Mass Spectrom.* 30, 825, 1995.

216. Pastor, S.J., Castoro, J.A., and Wilkins, C.L., High-mass analysis using quadrupolar excitation/ion cooling in a Fourier transform mass spectrometer, *Anal. Chem.* 67, 379, 1995.

217. Yao, J., Dey, M., Pastor, S.J., and Wilkins, C.L., Analysis of high-mass biomolecules using electrostatic fields and matrix-assisted laser desorption/ionization in a Fourier transform mass spectrometer, *Anal. Chem.* 67, 3638, 1995.

218. McIver, R.T.J., Li, Y., and Hunter, R.L., High-resolution laser desorption mass spectrometry of peptides and small proteins, *Proc. Natl. Acad. Sci. USA*. 91, 4801, 1994.

219. Wu, J., Fannin, S.T., Franklin, M.A., Molinski, T.F., and Lebrilla, C.B., Exact mass determination for elemental analysis of ions produced by matrix-assisted laser desorption, *Anal. Chem.* 67, 3788, 1995.

220. Heeren, R.M. and Boon, J.J., Rapid microscale analyses with an external ion source Fourier transform ion cyclotron resonance mass spectrometer, *Int. J. Mass Spectrom. Ion Proc.* 157/158, 391, 1996.

221. Baykut, G., Jertz, R., and Witt, M., Matrix-assisted laser desorption/ionization Fourier transform ion cyclotron resonance mass spectrometry with pulsed in-source collision gas and in-source ion accumulation, *Rapid Commun. Mass Spectrom.* 14, 1238, 2000.

222. O'Connor, P.B. and Costello, C.E., A high pressure matrix-assisted laser desorption/ionization Fourier transform mass spectrometry ion source for thermal stabilization of labile biomolecules, *Rapid Commun. Mass Spectrom.* 15, 1862, 2001.

223. O'Connor, P.B., Budnik, B.A., Ivleva, V.B., Kaur, P., Moyer, S.C., Pittman, J.L. and Costello, C.A., A high pressure matrix-assisted laser desorption ion source for Fourier transform mass spectrometry designed to accommodate large targets with diverse surfaces, *J. Am. Soc. Mass Spectrom.* 15, 128, 2004.

224. Krutchinsky, A.N., Loboda, A.V., Spicer, V.L., Dworschak, R., Ens, W., and Standing, K.G., Orthogonal injection of matrix-assisted laser desorption/ionization ions into a time-of-flight spectrometer through a collisional damping interface, *Rapid Commun. Mass Spectrom.* 12, 508, 1998.

225. Gauthier, J.W., Trautman, T.R., and Jacobson, D.B., Sustained off-resonance irradiation for collision-activated dissociation involving Fourier transform mass spectrometry. Collision-activated dissociation technique that emulates infrared multiphoton dissociation, *Anal. Chim. Acta* 246, 211, 1991.

226. Kellersberger, K.A., Yu, E.T., Merenbloom, S.I., and Fabris, D., Atmospheric pressure MALDI-FTMS of normal and chemically modified RNA, *J. Am. Soc. Mass Spectrom.* 16, 199, 2005.

227. Gross, J., Leisner, A., Hillenkamp, F., Hahner, S., Karas, M., Schafer, J., Lutzenkirchen, F., and Nordhoff, E., Investigations of the metastable decay of DNA under ultraviolet matrix-assisted laser desorption/ionization conditions with post-source-decay analysis and hydrogen/deuterium exchange, *J. Am. Soc. Mass Spectrom.* 9, 866, 1998.

228. (a) Bejjani, B.A. and Shaffer, L.G., Application of array-based comparative genomic hybridization to clinical diagnostics, *J. Mol. Diagn.* 8, 528, 2006. (b) Lamartine, J., The benefits of DNA microarrays in fundamental and applied bio-medicine, *Mater. Sci. Eng. C-Biomimetic Supramol. Syst.* 26, 354, 2006. (c) Egeland, R.D. and Southern, E.M., Electrochemically directed synthesis of oligonucleotides for DNA microarray fabrication, *Nucleic Acids Res.* 33, 125, 2005. (d) Barbulovic-Nad, I., Lucente, M., Sun, Y., Zhang, M.J., Wheeler, A.R., and Bussmann, M., Bio-microarray fabrication techniques-A review, *Crit. Rev. Biotechnol.* 26, 237, 2006. (e) Lee, C.-H., Harbers, G.M., Grainger, D.W., Gamble, L.J., and Castner, D.W., Fluorescence, XPS and ToF-SIMS surface chemical state image analysis of DNA microarrays, *J. Am. Chem. Soc.* 29(30), 9429, 2007. (f) Dufva, M., Fabrication of high quality microarrays, *Biomol. Eng.* 22, 173–184, 2005. (g) Pirrung, M.C., How to make a DNA chip, *Angew. Chem.-Int.* Ed. 41, 1276, 2002.

229. Gong, P., Harbers, G.M., and Grainger, D.W., Multi-technique comparison of immobilized and hybridized oligonucleotide surface density on commercial amine-reactive microarray slides, *Anal. Chem.* 78, 2342, 2006.

230. Peterson, A.W., Heaton, R.J., and Georgiadis, R.M., The effect of surface probe density on DNA hybridization, *Nucleic Acids Res.* 29, 5163, 2001.
231. Toren, P.C., Betsch, D.F., Weith, H.L., and Coull, J.M., Determination of impurities in nucleoside 3′-phosphoramidites by fast atom bombardment mass spectrometry, *Anal. Biochem.* 152, 291, 1986.
232. Fujitake, M., Harusawa, S., Araki, L., Yamaguchi, M., Lilley, D.M.J., Zhao Z.Y., and Kurihara, T., Accurate molecular weight measurements of nucleoside phosphoramidites: A suitable matrix of mass spectrometry, *Tetrahedron* 61, 4689, 2005.
233. Whitehouse, C.M., Dreyer, R.N., Yamashita, M., and Fenn, J.B., Electrospray interface for liquid chromatographs and mass spectrometers, *Anal. Chem.* 57, 675, 1985.
234. Von Brocke, A., Nicholson, G., and Bayer, E., Recent advances in capillary electrophoresis/electrospray-mass spectrometry, *Electrophoresis* 22, 1251, 2001.
235. Smith, R.D., Loo, J.A., Edmonds, G.G., Barinaga, C.J., and Udseth, H.R., New developments in biochemical mass spectrometry: Electrospray ionization, *Anal. Chem.* 62, 882, 1990.
236. Stults, J.T. and Marsters, J.C., Improved electrospray ionization of synthetic oligodeoxynucleotides, *Rapid Commun. Mass Spectrom.* 5, 359, 1991.
237. Fuerstenau, S.D. and Benner, W.H., Molecular weight determination of megadalton DNA electrospray ions using charge detection time-of-flight mass spectrometry, *Rapid Commun. Mass Spectrom.* 9, 1528, 1995.
238. Nohmi, T. and Fenn, J.B., Electrospray mass spectrometry of poly(ethylene glycols) with molecular weights up to five million, *J. Am. Chem. Soc.* 114, 3241, 1992.
239. Gao, J.L., Leung, K.S.Y., Wang, Y.T., Lai, C.M., Li, S.P., Hu, L.F., Lu, G.H., Jiang, Z.H., and Yu, Z.L., Qualitative and quantitative analyses of nucleosides and nucleobases in Ganoderma spp. by HPLC–DAD-MS, *J. Pharm. Biomed. Anal.* 44, 807, 2007.
240. Feng-Qing Yanga, Liya Geb, JeanWan Hong Yong, Swee Ngin Tanb, Shao-Ping Li, Determination of nucleosides and nucleobases in different species of *Cordyceps* by capillary electrophoresis–mass spectrometry, *J. Pharm. Biomed. Anal.* 50, 307, 2009.
241. Guo, F.Q., Li, A., Huang, L.F., Liang, Y.Z., and Chen, B.M., Identification and determination of nucleosides in Cordyceps sinensis and its substitutes by high performance liquid chromatography with mass spectrometric detection, *J. Pharm. Biomed. Anal.* 40, 623, 2006.
242. Fan, H., Li, S.P., Xiang, J.J., Lai, C.M., Yang, F.Q., Gao, J.L., Wang, Y.T., Qualitative and quantitative determination of nucleosides, bases and their analogues in natural and cultured *Cordyceps* by pressurized liquid extraction and high performance liquid chromatography-electrospray ionization tandem mass spectrometry (HPLC-ESI-MS/MS). *Anal. Chim. Acta* 567, 218, 2006.
243. McCloskey, J.A., Graham, D.E., Zhou, S.L., Crain, P.F., Ibba, M., Konisky, J., Soll, D., and Olsen, G.J., Post-transcriptional modification in archaeal tRNAs: Identities and phylogenetic relations of nucleotides from mesophilic and hyperthermophilic Methanococcales, *Nucleic Acids Res.* 29, 4699, 2001.
244. Pomerantz, S.C. and McCloskey, J.A., Analysis of RNA hydrolyzates by liquid chromatography-mass spectrometry, *Method. Enzymol.* 193, 796, 1990.
245. Lee, S.H., Jung, B.H., Kim, B.C., and Chung, B.C., A rapid and sensitive method for quantitation of nucleosides in human urine using liquid chromatography/mass spectrometry with direct urine injection, *Rapid Commun. Mass Spectrom.* 18, 973, 2004.
246. Wang, S., Zhao, X., Mao, Y., and Cheng, Y., Novel approach for developing urinary nucleosides profile by capillary electrophoresis–mass spectrometry, *J. Chromatogr. A* 1147, 254, 2007.
247. Benaiges, M.D., Lopez-Santin, J., and Sola, C., Production of 5′-ribonucleotides by enzymatic hydrolysis of RNA, *Enzyme Microb. Technol.* 12, 86, 1990.
248. Todd, B.E.N., in *Proc. 9th Australian Wine Industry Technical Conf.*, Adelaide, C.S. Stockley, A.N. Sas, R.S. Johnson, and T H. Lee, eds, Winetitles, Adelaide, p. 33, 1996.
249. Belem, M.A.F. and Lee, B.H., Production of RNA derivatives by Kluyveromyces fragilis grown on whey Producción de derivados de RNA mediante Kluyveromyces fragilis inoculada en suero de queso, *Food Sci. Technol. Int.* 3, 437, 1997.
250. Mounolou, J.C., *The Yeasts*, A.H. Rose and J.S. Harrison, eds, Vol. 2, Academic Press, London (The Physiology and Biochemistry of Yeasts), p. 309, 1971.
251. Trevelyan, W.E., Effect of procedures for the reduction of the nucleic acid content of SCP on the DNA content of Saccharomyces cerevisiae, *J. Sci. Food Agric.* 29, 903, 1978.
252. Feuillat, M. and Charpentier, C., Autolysis of yeasts in champagne, *Am. J. Enol. Viticult.* 33, 6, 1982.
253. Courtis, K., Todd, B., and Zhao, J., The potential role of nucleotides in wine flavour, *Aust. Grapegrower Winemaker* 409, 31, 1998.

254. Charpentier, C. and Feuillat, M., in *Wine Microbiology and Biotechnology*, G.H. Fleet (ed.), Harwood Academic Press, Lausanne, p. 225, 1993.
255. Aussenac, J., Chassagne, D., Claparols, C., Charpentier, M., Duteurtre, B., Feuillat, M., and Charpentier, C.J., Purification method for the isolation of monophosphate nucleotides from Champagne wine and their identification by mass spectrometry, *Chromatogr. A* 907, 155, 2001.
256. (a) Xue, C.B., Yin, Y.W., and Zhao, Y.F., Studies on phosphoserine and phosphothreonine derivatives: N-diisopropyloxyphosphoryl-serine and -threonine in alcoholic media, *Tetrahedron Lett.* 29, 1145, 1988. (b) Ji, G.J., Xue, C.B., Zeng, J.N., Li, L.P., Chai, W.G., and Zhao, Y.F., Synthesis of. N-(diisopropyloxyphosphoryl)amino acids and peptides, *Synthesis* 6, 444, 1988.
257. (a) Li, Y.M., Yin, Y.W., and Zhao, Y.F., Phosphoryl group participation leads to peptide formation from N-phosphorylamino acids, *Int. J. Pept. Protein Res.* 39, 375, 1992. (b) Li, C.X., Fu, H., Zhao, Y.F., and Cheng, C.M., Synthesis of nucleoside N-phosphoamino acids and peptide formation, *Origins Life Evol. Biosphere* 35, 11, 2005.
258. Gait, M.J., *Oligonucleotide Synthesis: A Practical Approach [M]*, IRL Press, Oxford, pp. 1–197, 1984.
259. Dugas, H. and Chem, B., *A Chemical Approach to Enzyme Action [M]*, Springer, New York, pp. 21–110, 1996.
260. Zhou, W.H., Ju, Y., and Zhao, Y.F., Simultaneous formation of peptides and nucleotides from n-phospho-threonine, *Origins Life Evol. Biosphere* 26, 547, 1996.
261. Liu, H., Zhao, C., Lu, J., Liu, M., Zhang, S., Jiang, Y., and Zhao, Y., Simultaneous measurement of trace monoadenosine and diadenosine monophosphate in biomimicking prebiotic synthesis using high-performance liquid chromatography with ultraviolet detection and electrospray ionization mass spectrometry characterization, *Anal. Chim.* Acta 566, 99, 2006.
262. Cao, P. and Moini, M., Pressure-assisted and pressure-programmed capillary electrophoresis/electrospray ionization time of flight—mass spectrometry for the analysis of peptide mixtures, *Electrophoresis* 19, 2200, 1998.
263. Soga, T., Ueno, Y., Naraoka, H., Matsuda, K., Tomita, M., and Nishioka, T., Pressure-assisted capillary electrophoresis electrospray ionization mass spectrometry for analysis of multivalent anions, *Anal. Chem.* 74, 6224, 2002.
264. Soga, T., Ishikawa, T., Igarashi, S., Sugawara, K., Kakazu, Y., and Tomita, M., Analysis of nucleotides by pressure-assisted capillary electrophoresis–mass spectrometry using silanol mask technique, *J. Chromatogr. A* 1159, 125, 2007.
265. Lukacs, K.D. and Jorgenson, J.W., Capillary zone electrophoresis: Effect of physical parameters on separation efficiency and quantitation, *J. High Res. Chromatogr.* 8, 407, 1985.
266. Von Ballmoos, C., Brunner, J., and Dimroth, P., The ion channel of F-ATP synthase is the target of toxic organotin compounds, *Proc. Natl. Acad. Sci. USA* 101, 11239, 2004.
267. Qian, T., Cai, Z. and Yang, M.S., Determination of adenosine nucleotides in cultured cells by ion-pairing liquid chromatography–electrospray ionization mass spectrometry, *Anal. Biochem.* 325, 77, 2004.
268. Nordstrom, A., Tarkowski, P., Tarkowska, D., Dolezal, K., Astot, C., Sandberg, G., and Moritz, T., Derivatization for LC-electrospray ionization-MS: A tool for improving reversed-phase separation and ESI responses of bases, ribosides, and intact nucleotides, *Anal. Chem.* 76, 2869, 2004.
269. Buchholz, A., Takors, R., and Wandrey, C., Quantification of intracellular metabolites in Escherichia coli K12 using liquid chromatographic-electrospray ionization tandem mass spectrometric techniques, *Anal. Biochem.* 295, 129, 2001.
270. Witters, E., Roef, L., Newton, R.P., Van Dongen, W., and Van Onkelen, H.A., Quantitation of cyclic nucleotides in biological samples by negative electrospray tandem mass spectrometry coupled to ion suppression liquid chromatography, *Rapid Commun. Mass Spectrom.* 10, 225, 1996.
271. Klawitter, J., Schmitz, V., Klawitter, J., Leibfritz, D., and Christians, U., Development and validation of an assay for the quantification of 11 nucleotides using LC/LC–electrospray ionization-MS, *Anal. Biochem.* 365, 230, 2007.
272. Müller, C., Schafer, P., Stortzel, M., Vogt, S., and Weinmann, W., Ion suppression effects in liquid chromatography–electrospray-ionisation transport-region collision induced dissociation mass spectrometry with different serum extraction methods for systematic toxicological analysis with mass spectra libraries, *J. Chromatogr. B* 773, 47, 2002.
273. Sherman, C.L., Pierce, S.E., Brodbelt, J.S., Tuesuwan, B., and Kerwin, S.M., Identification of the adduct between a 4-Aza-3-ene-1,6-diyne and DNA using electrospray ionization mass spectrometry, *J. Am. Soc. Mass Spectrom.* 17, 1342, 2006.
274. Doerge, D.R., Churchwell, M.I., Fang, J.L., and Beland, F.A., Quantification of etheno-DNA adducts using liquid chromatography, on-line sample processing, and electrospray tandem mass spectrometry, *Chem. Res. Toxicol.* 13, 1259, 2000.

275. Hoes, I., Lemiere, F., Van Dongen, W., Vanhoutte, K., Esmans, E.L., Van Bockstaele, D., Berneman, Z.N., Deforce, D., and Van den Eeckhout, E.G., Analysis of melphalan adducts of 2'-deoxynucleotides in calf thymus DNA hydrolysates by capillary high-performance liquid chromatography–electrospray tandem mass spectrometry, *J. Chromatogr. B* 736, 43, 1999.
276. Hoes, I., Van Dongen, W., Lemiere, F., Esmans, E.L., Van Bockstaele, D., and Berneman, Z.N., Comparison between capillary and nano liquid chromatography–electrospray mass spectrometry for the analysis of minor DNA–melphalan adducts, *J. Chromatogr. B* 748, 197, 2000.
277. Pinlaor, S., Ma, N., Hiraku, Y., Yongvanit, P., Semba, R., Oikawa, S., Murata, M., Sripa, B., Sithithawom, P., and Kawanishi, S., Repeated infection with Opisthorchis viverrini induces accumulation of 8-nitroguanine and 8-oxo-7,8-dihydro-2'-deoxyguanine in the bile duct of hamsters via inducible nitric oxide synthase, *Carcinogenesis* 25, 1535, 2004.
278. Ding, X., Hiraku, Y., Ma, N., Kato, T., Saito, K., Nagahama, M., Semba, R., Kuribayashi, K., and Kawanishi, S., Inducible nitric oxide synthase-dependent DNA damage in mouse model of inflammatory bowel disease, *Cancer Sci.* 96, 157, 2005.
279. Chaiyarit, P., Ma, N., Hiraku, Y., Pinlaor, S., Yongvanit, P., Jintakanon, D., Murata, M., Oikawa, S., and Kawanishi, S., Nitrative and oxidative DNA damage in oral lichen planus in relation to human oral carcinogenesis, *Cancer Sci.* 96, 553, 2005.
284. Bruner, S.D., Norman, D.P. and Verdine, G.L., Structural basis for recognition and repair of the endogenous mutagen 8-oxoguanine in DNA, *Nature* 403, 859, 2000.
285. Ishii, Y., Ogara, A., Okamura, T., Umemura, T., Nishikawa, A., Iwasaki, Y., Ito, R., Saito, K., Hirose, M. and Nakazawa, H., Development of quantitative analysis of 8-nitroguanine concomitant with 8-hydroxydeoxyguanosine formation by liquid chromatography with mass spectrometry and glyoxal derivatization, *J. Pharm. Biomed. Anal.* 43, 1737, 2007.
286. Wang, D.G., Fan, J.B., Siao, C.J., Berno, A., Young, P., Sapolsky, R., Ghandour, G. et al., Large-scale identification, mapping, and genotyping of single-nucleotide polymorphisms in the human genome, *Science* 280, 1077, 1998.
287. Gut, I.G., Automation in genotyping of single nucleotide polymorphisms, *Hum. Mutat.* 17, 475, 2001.
288. Kristensen, V.N., Kelefiotis, D., Kristensen, T. and Borresen-Dale, A.L., High-throughput methods for detection of genetic variation, *Biotechniques* 30, 318, 2001.
289. Rickert, A.M., Premstaller, A., Gebhardt, C. and Oefner, P.J., Genotyping of SNPs in a polyploid genome by pyrosequencing (TM), *Biotechniques* 32, 592, 2002.
290. Sham, P., Bader, J.S., Craig, I., O'Donovan, M. and Owen, M., DNA pooling: A tool for large-scale association studies, *Nature Rev. Gent.* 3, 862, 2002.
291. Null, A.P., Nepomuceno, A.I., and Muddiman, D.C., Implications of hydrophobicity and free energy of solvation for characterization of nucleic acids by electrospray ionization mass spectrometry, *Anal. Chem.* 75, 1331, 2003.
292. Reynolds, R., Sensabaugh, G. and Blake, E., Analysis of genetic-markers in forensic DNA samples using the polymerase chain-reaction, *Anal. Chem.* 63, 2, 1991.
293. Haff, L.A. and Smirnov, I.P., Single-nucleotide polymorphism identification assays using a thermostable DNA polymerase and delayed extraction MALDI-TOF mass spectrometry, *Genome Res.* 7, 378, 1997.
294. Li, J., Butler, J.M., Tan, Y., Lin, H., Royer, S., Ohler, L., Shaler, T.A., Hunter, J.M., Pollart, D.J., Monforte, J.A. and Becker, C.H., Single nucleotide polymorphism determination using primer extension and time-of-flight mass spectrometry, *Electrophoresis* 20, 1258, 1999.
295. Griffin, T.J., Hall, J.G., Prudent, J.R., and Smith, L.M., Direct genetic analysis by matrix-assisted laser desorption ionization mass spectrometry, *Proc. Natl. Acad. Sci. USA* 96, 6301, 1999.
296. Sauer, S., Lechner, D., Berlin, K., Lehrach, H., Escary, J.L., Fox, N., and Gut, I.G., A novel procedure for efficient genotyping of single nucleotide polymorphisms, *Nucleic Acids Res.* 28, 13, 2000.
297. Huber, C.G. and Oberacher, H., Analysis of nucleic acids by on-line liquid chromatography-mass spectrometry, *Mass Spectrom. Rev.* 20, 310, 2001.
298. Zhang, S., Van Pelt, C.K., Huang, X., and Schultz, G.A., Detection of single nucleotide polymorphisms using electrospray ionization mass spectrometry: Validation of a one-well assay and quantitative pooling studies, *J. Mass Spectrom.* 37, 1039, 2002.
299. Muddiman, D.C., Null, A.P., and Hannis, J.C., Precise mass measurement of a double-stranded 500 base-pair (309 kDa) polymerase chain reaction product by negative ion electrospray ionization Fourier transform ion cyclotron resonance mass spectrometry, *Rapid Commun. Mass Spectrom.* 13, 1201, 1999.
300. Oberacher, H., Oefner, P.J., Parson, W., and Huber, C.G., On-line liquid chromatography mass spectrometry: A useful tool for the detection of DNA sequence variation, *Angew. Chem. Int. Edit.* 40, 3828, 2001.

301. (a) Oberacher, H., Parson, W., Hölzl G., Oefner, P.J., and Huber, C.G., Optimized suppression of adducts in polymerase chain reaction products for semi-quantitative SNP genotyping by liquid chromatography-mass spectrometry, *J. Am. Soc. Mass Spectrom.* 15, 1897, 2004. (b) Oberacher, H., Huber, C.G. and Oefner, P.J., Mutation scanning by ion-pair reversed-phase high-performance liquid chromatography electrospray ionization mass spectrometry (ICEMS), *Hum. Mut.* 21, 86, 2003.

302. Berger, B., Hölzl, G., Oberacher, H., Niederstätter, H., Huber, C.G., and Parson, W., Single nucleotide polymorphism genotyping by on-line liquid chromatography-mass spectrometry in forensic science of the Y-chromosomal locus M9, *J. Chromatogr. B* 782, 89, 2002.

303. Niederstatter, H., Oberacher, H., and Parson, W., Highly efficient semi-quantitative genotyping of single nucleotide polymorphisms in mitochondrial DNA mixtures by liquid chromatography electrospray ionization time-of-flight mass spectrometry, *Int. Congr. Ser.* 1288, 10, 2006.

304. Phillips, D.R. and McCloskey, J.A., A comprehensive study of the low energy collision-induced dissociation of dinucleoside monophosphates, *Int. J. Mass Spectrom. Ion Process.* 128, 61, 1993.

305. Gilar, M., Analysis and purification of synthetic oligonucleotides by reversed-phase high-performance liquid chromatography with photodiode array and mass spectrometry detection, *Anal. Biochem.* 298, 196, 2001.

306. Fountain, K.J., Gilar, M., and Gebler, J.C., Analysis of native and chemically modified oligonucleotides by tandem ion-pair reversed-phase high-performance liquid chromatography/electrospray ionization mass spectrometry, *Rapid Commun. Mass Spectrom.* 17, 646, 2003.

307. Fountain, K.J., Gilar, M., and Gebler, J.C., Electrospray ionization mass spectrometric analysis of nucleic acids using high-throughput on-line desalting, *Rapid Commun. Mass Spectrom.* 18, 1295, 2004.

308. Cavanagh, J., Benson, L.M., Thompson, R., and Naylor, S., In-line desalting mass spectrometry for the study of noncovalent biological complexes, *Anal. Chem.* 75, 3281, 2003.

309. Null, A.P., George, L.T., and Muddiman, D.C., Evaluation of sample preparation techniques for mass measurements of PCR products using ESI-FT-ICR mass spectrometry, *J. Am. Soc. Mass Spectrom.* 13, 338, 2002.

310. Deroussent, A., LeCaer, J.P., Rossier, J., and Gouyette, A., Electrospray mass-spectrometry for the characterization of the purity of natural and modified oligonucleotides, *Rapid Commun. Mass Spectrom.* 9, 1, 1995.

311. Greig, M. and Griffey, R.H., Utility of organic bases for improved electrospray mass spectrometry of oligonucleotides, *Rapid Commun. Mass Spectrom.* 9, 97, 1995.

312. Deguchi, K., Ishikawa, M., Yokotura, T., Ogata, I., Ito, S., Mimura, T., and Ostrander, C., Enhanced mass detection of oligonucleotides using reverse-phase high-performance liquid chromatography/electrospray ionization ion-trap mass spectrometry, *Rapid Commun. Mass Spectrom.* 16, 2133, 2002.

313. Van Breemen, R.B., Tan, Y.C., Lai, J., Huang, C.R., and Zhao, X.M., Immobilized thymine chromatography mass spectrometry of oligonucleotides, *J. Chromatogr. A* 806, 67, 1998.

314. Prakash, T.P., Kawasaki, A.M., Fraser, A.S., Vasquez, G., and Manoharan, M.J., Synthesis of 2'-O-[2-[(N,N-dimethylamino)oxy]ethyl] modified nucleosides and oligonucleotides, *Org. Chem.* 37, 357, 2002.

315. Prakash, T.T., Kawasaki, A.M., Lesnik, E.A., Sioufi, N., and Manoharan, M., Synthesis of 2'-O-[2-[(N,N-dialkylamino)oxy]ethyl]-modified oligonucleotides: Hybridization affinity, resistance to nuclease, and protein binding characteristics, *Tetrahedron* 59, 7413, 2003.

316. Janning, P., Schrader, W., and Linscheid, M., A new mass-spectrometric approach to detect modifications in DNA, *Rapid Commun. Mass Spectrom.* 8, 1035, 1994.

317. Wu, H.Q. and Aboleneen, H., Improved oligonucleotide sequencing by alkaline phosphatase and exonuclease digestions with mass spectrometry, *Anal. Biochem.* 290, 347, 2001.

318. Worth, C.C.T., Schmitz, O.J., Lliem, H.C., and Wlebler, M., Synthesis of fluorescently labeled alkylated DNA adduct standards and separation by capillary electrophoresis, *Electrophoresis* 21, 2086, 2000.

319. Lewtas, J., Walsh, D., Williams, R., and Dobias, L., Air pollution exposure DNA adduct dosimetry in humans and rodents: Evidence for non-linearity at high doses, *Mutat. Res.* 378, 51, 1997.

320. Norwood, C.B., Jackim, E., and Cheer, S., DNA adduct research with capillary electrophoresis, *Anal. Biochem.* 213, 194, 1993.

321. Funk, M., Ponten, I., and Seidel, A., Jernstrom, B., Critical parameters for adduct formation of the carcinogen (+)-anti-benzo[a]pyrene-7,8-dihydrodiol 9,10-epoxide with oligonucleotides *Bioconjugate Chem.* 8, 310, 1997.

322. Mala, Z., Kleparnik, K., and Bocek, P., Highly alkaline electrolyte for single-stranded DNA separations by electrophoresis in bare silica capillaries, *J. Chromatogr. A* 853, 371, 1999.

323. Nguyen, A.L., Luong, J.H.T., and Masson, C., Determination of nucleotides in fish tissues using capillary electrophoresis, *Anal. Chem.* 62, 2490, 1990.

324. Sharma, M., Jain, R., Ionescu, E., and Slocum, H.K., Capillary electrophoretic separation and laser induced fluorescence detection of the major DNA-adducts of cisplatin and carboplatin, *Anal. Biochem.* 228, 307, 1995.

325. Duhachek, S.D., Kenseth, J.R., Casale, G.P., Small, G.J., Porter, M.D., and Jankowiak, R., Monoclonal antibody-gold biosensor chips for detection of depurinating carcinogen-DNA adducts by fluorescence line-narrowing spectroscopy, *Anal. Chem.* 72, 3709, 2000.

326. Wang, H., Lu, M., Weinfeld, M., and Le, X.C., Enhancement of immunocomplex detection and application to assays for DNA adduct of benzo[a]pyrene, *Anal. Chem.* 75, 247, 2003.

327. Roberts, K.P., Lin, C.H., Singhal, M., Casale, G.P., Small, G .J., and Jankowiak, R., On-line identification of depurinating DNA adducts in human urine by capillary electrophoresis—fluorescence line narrowing spectroscopy, *Electrophoresis* 21, 799, 2000.

328. Feng, Y.L., Lian, H., and Zhu, J., Application of pressure assisted electrokinetic injection technique in the measurements of DNA oligonucleotides and their adducts using capillary electrophoresis–mass spectrometry, *J. Chromatogr. A* 1148, 244, 2007.

329. Oberacher, H., Niederstätter, H., Casetta, B., and Parson, W., Some guidelines for the analysis of genomic DNA by PCR-LC-ESI-MS, *J. Am. Soc. Mass Spectrom.* 17, 124, 2006.

330. Erlich, H., *PCR Technology. Principles and Applications for DNA Amplification*, Stockton Press, New York, 1989.

331. Johnson, Y.A., Nagpal, M., Krahmer, M.T., Fox, K.F., and Fox, A., Precise molecular weight determination of PCR products of the rRNA intergenic spacer region using electrospray quadrupole mass spectrometry for differentiation of B-subtilis and B-atrophaeus, closely related species of bacilli, *J. Microbiol. Meth.* 40, 241, 2000.

332. Veenstra, T.D., Electrospray ionization mass spectrometry: A promising new technique in the study of protein/DNA noncovalent complexes, *Biochem. Biophys. Res. Commun.* 257, 1, 1999.

333. Ganem, B., Li, Y T., and Henion, J.D., Detection of noncovalent receptor-ligand complexes by mass spectrometry, *J. Am. Chem. Soc.* 113, 6294, 1991.

334. Ganem, B., Li, Y.T., and Henion, J.D., Observation of noncovalent enzyme-substrate and enzyme-product complexes by ion-spray mass spectrometry, *J. Am. Chem. Soc.* 113, 7818, 1991.

335. Hofstadler, S.A. and Griffey, R.H., Analysis of noncovalent complexes of DNA and RNA by mass spectrometry, *Chem. Rev.* 101, 377, 2001.

336. Holz, B., Dank, N., Eickhoff, J.E., Lipps, G., Krauss, G., and Weinhold, E., Identification of the binding site for the extrahelical target base in N-6-adenine DNA methyltransferases by photo-cross-linking with duplex oligodeoxyribonucleotides containing 5-iodouracil at the target position, *J. Biol. Chem.* 274, 15066, 1999.

337. Wong, D.L. and Reich, N.O., Identification of tyrosine 204 as the photo-cross-linking site in the DNA—EcoRI DNA methyltransferase complex by electrospray ionization mass spectrometry, *Biochemistry* 39, 15410, 2000.

338. Gabelica, V., De Pauw, E., and Rosu, F., Interaction between antitumor drugs and a double-stranded oligonucleotide studied by electrospray ionization mass spectrometry, *J. Mass Spectrom.* 34, 1328, 1999.

339. Kapur, A., Beck, J.L., and Sheil, M.M., Observation of daunomycin and nogalamycin complexes with duplex DNA using electrospray ionisation mass spectrometry, *Rapid Commun. Mass Spectrom.* 13, 2489, 1999.

340. Gupta, R., Kapur, A., Beck, J.L., and Sheil, M.M., Positive ion electrospray ionization mass spectrometry of double-stranded DNA/drug complexes, *Rapid Commun. Mass Spectrom.* 15, 2472, 2001.

341. Verheijen, J.C., van der Marel, G.A., van Boom, J.H., and Metzler-Nolte, N., Transition metal derivatives of peptide nucleic acid (PNA) oligomers-synthesis, characterization, and DNA finding, *Bioconjugate Chem.* 11, 741, 2000.

342. Veenstra, T.D., Benson, L.M., Craig, T.A., Tomlinson, A.J., Kumar, R., and Naylor, S., Metal mediated sterol receptor-DNA complex association and dissociation determined by electrospray ionization mass spectrometry, *Nat. Biotechnol.* 16, 262, 1998.

343. Deterding, L.J., Kast, J., Przybylski, M., and Tomer, K.B., Molecular characterization of a tetramolecular complex between dsDNA and a DNA-binding leucine zipper peptide dimer by mass spectrometry, *Bioconjugate Chem.* 11, 335, 2000.

344. Zhou, X., Shen, Z., Li, D., He, X., and Lin, B., Study of interactions between actinomycin D and oligonucleotides by microchip electrophoresis and ESI-MS, *Talanta* 72, 561, 2007.

345. Zhang, S., Van Pelt, C.K., and Wilson, D.B., Quantitative determination of noncovalent binding inter-actions using automated nanoelectrospray mass spectrometry, *Anal. Chem.* 75, 3010, 2003.

346. Haskins, N.J., Ashcroft, A.E., Phillips, A., and Harrison, M., The evaluation of several electrospray sys-tems and their use in noncovalent bonding studies, *Rapid Commun. Mass Spectrom.* 8, 120, 1994.

347. Franska, M., Electrospray ionization mass spectrometric study of platinum (II) complexes with nucleo-bases and dimethyl sulfoxide, *Int. J. Mass Spectrom.* 261, 86, 2007.

348. Beck, J.L., Humphries, A., Sheil, M.M., and Ralph, S.F., Electrospray ionisation mass spectrometry of ruthenium and palladium complexes with oligonucleotides, *Eur. J. Mass Spectrom.* 5, 489, 1999.

349. Davey, J.M., Moerman, K.L., Ralph, S.F., Kanitz, R., and Sheil, M.M., Comparison of the reactivity of cis-[RuCl$_2$(DMSO)$_4$] and *trans*-[RuCl$_2$(DMSO)$_4$] towards nucleosides, *Inorg. Chim. Acta* 281, 10, 1998.

350. Mariappan, S.V.S., Cheng, X., van Breemen, R.B., Silks, L.A., Gupta, G., Analysis of GAA/TTC DNA triplexes using nuclear magnetic resonance and electrospray ionization mass spectrometry, *Anal. Biochem.* 334, 216, 2004.

351. Hope, I.A. and Struhl, K., GCN4 protein, synthesize in vitro, binds HIS3 regulatory sequences: Implications for general control of amino acid biosynthetic genes in yeast, *Cell* 43, 177, 1985.

352. Hope, I.A. and Struhl, K., Functional dissection of a eukaryotic transcriptional activator protein, GCN4 of Yeast, *Cell* 46, 885, 1986.

353. Hope, I.A. and Struhl, K., GCN4, a eukaryotic transcriptional activator protein, binds as a dimmer to target DNA, *EMBO J.* 6, 2781, 1987.

354. Talanian, R.V., McKnight, C.J., and Kim, P.S., Sequence specific DNA-binding by a short peptide dimer, *Science* 249, 769, 1990.

355. Gallo, R.C., Sarin, P.S., Gelmann, E.P., Guroff, M.R., Richardson, E., Kalyanaraman, V.S., Mann, D. et al., Isolation of human T-cell leukemia-virus in acquired immune-deficiency syndrome (AIDS), *Science* 220, 865, 1983.

356. Morice, Y., Roulot, D., Grando, V., Stirneman, J., Gault, E., Jeantils, V., Bentata, M. et al., Phylogenetic analyses confirm the high prevalence of hepatitis C virus (HCV) type 4 in the Seine-Saint-Denis district (France) and indicate seven different HCV-4 subtypes linked to two different epidemiological patterns, *Gen. Virol.* 82, 1001, 2001.

357. Hoyne, P.R., Benson, L.M., Veenstra, T.D., Maher, L.J., and Naylor, S., RNA-RNA noncovalent interac-tions investigated by microspray ionization mass spectrometry, *Rapid Commun. Mass Spectrom.* 15, 1539, 2001.

358. Liu, C., Tolic, L.P., Hofstadler, S.A., Harms, A.C., Smith, R.D., Kang, C., and Sinha, N., Probing regA/RNA interactions using electrospray ionization Fourier transform ion cyclotron resonance-mass spectro-metry, *Anal. Biochem.* 262, 67, 1998.

359. Beverly, M., Hartsough, K., Machemer, L., Pavco, P., and Lockridge, J., Liquid chromatography electro-spray ionization mass spectrometry analysis of the ocular metabolites from a short interfering RNA duplex, *J. Chromatogr. B* 835, 62, 2006.

360. Warnke, U., Gysler, J., Hofte, B., Tjaden, U.R., van der Greer, J., Kloft, C., Schunack, W., and Jaehde, U., Separation and identification of platinum adducts with DNA nucleotides by capillary zone elec-trophoresis and capillary zone electrophoresis coupled to mass spectrometry, *Electrophoresis* 22, 97, 2001.

361. Carte, N., Legendre, F., Leize, E., Potier, N., Reeder, F., Chottard, J.C., and Dorsselaer, A.V., Determination by electrospray mass spectrometry of the outersphere association constants of DNA/platinum complexes using 20-mer oligonucleotides and ([Pt(NH$_3$)$_4$]$^{2+}$, 2Cl$^-$) or ([Pt(py)$_{(4)}$]$^{2+}$, 2Cl$^-$), *Anal. Biochem.* 284, 77, 2000.

362. Schultz, J.C., Hack, C.A., and Benner, W.H., Polymerase chain reaction products analyzed by charge detection mass spectrometry, *Rapid Commun. Mass Spectrom.* 13, 15, 1999.

363. Bos, S.J., van Leeuwen, S.M., and Karst, U., From fundamentals to applications: Recent developments in atmospheric pressure photoionization mass spectrometry, *Anal. Bioanal. Chem.* 384, 85, 2006.

364. Bagag, A., Giuliani, A., and Laprévote, O., Atmospheric pressure photoionization mass spectrometry of nucleic bases, ribonucleosides and ribonucleotides, *Int. J. Mass Spectrom.* 264, 1, 2007.

365. (a) McLafferty, F.W. (ed.), *Tandem Mass Spectrometry*, Wiley, New York, 1983. (b) Busch, K.L., Glish, G.L., and McLuckey, S.A., *Mass Spectrometry/Mass Spectrometry*, VCH Publishers, Germany, 1988.

366. Panico, M., Sindona, G., and Uccella, N., Bioorganic applications of mass spectrometry. Fast-atom-bombardment-induced zwitterionic oligonucleotide quasimolecular ions sequenced by MS/MS, *J. Am. Chem. Soc.* 105, 5607, 1983.

367. McLuckey, S.A. and Wells, M., Mass analysis at the advent of the 21st century, *Chem. Rev.* 101, 571, 2001.

368. Grosse-Herrenthey, A., Maier, T., Gessler, F., Schaumann, R., Böhnel, H., Kostrzewa, M., and Krüger M., Challenging the problem of clostridial identification with matrix-assisted laser desorption and ionization-time-of-flight mass spectrometry (MALDI-TOF MS), *Anaerobe* 14, 242, 2008.

369. Keto-Timonen, R., Nevas, M., and Korkeala, H., Efficient DNA fingerprinting of Clostridium botulinum types A, B, E, and F by amplified fragment length polymorphism analysis, *Appl Environ Microbiol* 71, 1148, 2005.

370. Vertes, A., Medical Applications of Mass Spectrometry, Chapter 8, in *Mass Spectrometry in Proteomics*, Elsevier, Amsterdam, pp. 173–194, 2008.

371. Cerny, R.L., Gross, M.L., and Grotjahn, L., Fast atom bombardment combined with tandem mass spectrometry for the study of dinucleotides, *Anal. Biochem.* 156, 424, 1986.

372. Liguori, A., Sindona, G., and Uccella, N., Sequence effect on the slow degradations of dinucleotides by fast atom bombardment tandem mass spectrometry, *Biomed. Environ. Mass Spectrom.* 16, 451, 1988.

373. Isern-Flecha, I., Jiang, X.Y., Cooks, R.G., Pfleiderer, W., Chae, W.G., and Chang, C., Characterization of an alkylated dinucleotide by desorption chemical ionization and tandem mass spectrometry, *Biomed. Environ. Mass Spectrom.*, 14, 17, 1987.

374. Neri, N., Sindona, G., and Uccella, N., Bio-organic applications of mass spectrometry 2. slow degradation of oligonucleotide diesterophosphate anions—the mass spectrometric method of fast-atom bombardment—MIKE CID, *Gazz Chim. Ital.* 113, 197, 1983.

375. McLuckey, S.A. and Habibi-Goudarzi, S., Decompositions of multiply charged oligonucleotide anions, *J. Am. Chem. Soc.* 115, 12085, 1993.

376. McLuckey, S.A., Van Berkel, G.J., and Glish, G.L., Tandem mass spectrometry of small, multiply charged oligonucleotides, *J. Am. Soc. Mass Spectrom.* 3, 60, 1992.

377. Medzihradszky, K.F., Campbell, J.M., Baldwin, M.A., Falick, A.M., Juhasz, P., Vestal, M.L., and Burlingame, A.L., The characteristics of peptide collision-induced dissociation using a high-performance MALDI-TOF/TOF tandem mass spectrometer, *Anal. Chem.* 71, 552, 2000.

378. Colacino, E., Giorgi, G., Liguori, A., Napoli, A., Romeo, R., Salvini, L., Siciliano, C., and Sindona, G., Structural characterization of isoxazolidinyl nucleosides by fast atom bombardment tandem mass spectrometry, *J. Mass Spectrom.* 36, 1220, 2001.

379. Newton, R.P., Kingston, E.E., and Overton, A., Mass spectrometric identification of cyclic nucleotides released by the bacterium *Corynebacterium murisepticum* into the culture medium, *Rapid Commun. Mass Spectrom.* 12, 729, 1998.

380. Walker, M.P., Appleby, T.C., Zhong, W., Lau, J.Y.N., and Hong, Z., Hepatitis C virus therapies: Current treatments, targets and future perspectives, *Antiviral Chem. Chemother.* 14, 1, 2003.

381. Shou, Z.W., Bu, H.Z., Addison, T., Jiang, X., and Weng, N.D., Development and validation of a liquid chromatography/tandem mass spectrometry (LC/MS/MS) method for the determination of ribavirin in human plasma and serum, *J. Pharmaceut. Biomed. Anal.* 29, 83, 2002.

382. Daluge, S.M., Good, S.S., Faletto, M.B., Miller, W.H., St. Clair, M.H., Boone, L.R., Tisdale, M. et al., 1592U89, A novel carbocyclic nucleoside analog with potent, selective anti-human immunodeficiency virus activity, *Antimicrob. Agents Chemother.* 41, 1082, 1997.

383. Faletto, M.B., Miller, W.H., Garvey, E.P., St. Clair, M.H., Daluge, S.M., and Good, S.S., Unique intracellular activation of the potent anti-human immunodeficiency virus agent 1592U89, *Antimicrob. Agents Chemother.* 41, 1099, 1997.

384. Staszewski, S., Katlama, C., Harrer, T., Massip, P., Yeni, P., Cutrell, A., Tortell, S.M. et al., A dose-ranging study to evaluate the safety and efficacy of abacavir alone or in combination with zidovudine and lamivudine in antiretroviral treatment naive subjects, *AIDS* 12, F197, 1998.

385. Fung, E.N., Cai, Z.W., Burnette, T.C., and Sinhababu, A.K., Simultaneous determination of Ziagen and its phosphorylated metabolites by ion-pairing high-performance liquid chromatography-tandem mass spectrometry, *J. Chromatogr. B* 754, 285, 2001.

386. Richardson, F.C., Zhang, C., Lehrman, S.R., Koc, H., Swenberg, J.A., Richardson, K.A., and Bendele, R.A., Quantification of 2′-fluoro-2′-deoxyuridine and 2′-fluoro-2′-deoxycytidine in DNA and RNA isolated from rats and woodchucks using LC/MS/MS, *Chem. Res. Toxicol.* 15, 922, 2002.

387. Badouard, K., Masudab, M., Nishino, H., Cadet, J., Favier, A., and Ravanat, J.L., Detection of chlorinated DNA and RNA nucleosides by HPLC coupled to tandem mass spectrometry as potential biomarkers of inflammation, *J. Chromatogr. B* 827, 26, 2005.

388. Zhu, Y.X., Wong, P.S.H., Zhou, Q., Sotoyama, H., and Kissinger, P.T., Identification and determination of nucleosides in rat brain microdialysates by liquid chromatography/electrospray tandem mass spectrometry, *J. Pharmaceut. Biomed. Anal.* 26, 967, 2001.

389. Bezy, V., Morin, P., Couerbe, P., Leleu, G., and Agrofoglio, L., Simultaneous analysis of several antiretro-viral nucleosides in rat-plasma by high-performance liquid chromatography with UV using acetic acid/hydroxylamine buffer Test of this new volatile medium-pH for HPLC–ESI-MS/MSJ. *Chromatogr. B* 821, 132, 2005.

390. (a) Banoub, J., Gentil, E., Taber, B., Fahmi, N.E., Ronco, G., Villa, P., and Mackenzie, G., Electrospray tandem mass spectrometry of 3′-azido-2′,3′-dideoxythymidine and of a novel series of 3′-azido-2′,3′,4′-trideoxy-4′-thio-5-halogenouridines and their respective alpha-anomers, *Spectroscopy* 12, 69, 1994. (b) Len, C., Mackenzie, G., Ewing, D.F. and Banoub, J.H., Electrospray tandem mass-spectrometric analysis of diastereo- and stereoisomeric pyrimidine nucleoside analogues based on the 1,3-dihydrobenzo[c]furan core, *Carbohydr. Res.* 338(22), 2311, 2003.

391. Zhao, M., Rudek, M.A., He, P., Hartke, C., Gore, S., Carducci, M.A., and Baker, S.D., Quantification of 5-azacytidine in plasma by electrospray tandem mass spectrometry coupled with high-performance liquid chromatography, *J Chromatogr. B Analyt. Technol. Biomed. Life Sci.* 813, 81, 2004.

392. Gore, S.D., Baylin, S.B., Carducci, M.A., Cameron, E.E., Gilbert, J., Miller, C.B., and Herman, J.G., Sequential DNA methyltransferase and histone deacetylase inhibition to re-express silenced genes: Pre-clinical and early clinical modeling, *Proc. Am. Assoc. Cancer Res.* 42, 681, 2001.

393. Kamel, A.M. and Munson, B., Collisionally-induced dissociation of substituted pyrimidine antiviral agents: Mechanisms of ion formation using gas phase hydrogen/deuterium exchange and electrospray ionization tandem mass spectrometry, *J. Am. Soc. Mass Spectrom.* 18, 1477, 2007.

394. Lefebvre, I., Pompon, A., Valette, G., Perigaud, C., Gosselin, G., and Imbach, J.L.L., On-line cleaning HPLC-UV-MS: A tool for analyzing new anti-HIV drugs in biological media, *Chromatogr. Gas Chromatogr.* 15, 868, 1997.

395. Salgado, A., Bidois, L., Naesens, L., De Clercq, E., and Balzarini, J., Phosphoramidate derivatives of d4T as inhibitors of HIV: The effect of amino acid variation, *Antivir. Res.* 35, 195, 1997.

396. Roberts, W.L., Buckley, T.J., Rainey, P.M., and Jatlow, P.I., Solid-phase extraction combined with radio-immunoassay for measurement of zalcitabine (2′,3′-dideoxycytidine) in plasma and serum, *Clin. Chem.* 40, 211, 1994.

397. Szinai, I., Veres, Z., Ganzler, K., Hegedus-Vajda, J., and De Clercq, E., Metabolism of anti-herpes agent 5-(2-chloroethyl)-2′-deoxyuridine in mice ad rats, *Eur. J. Drug Met. Pharmacok.* 16, 129, 1991.

398. Kreutzberger, A. and Sellheim, M., Antiviral Agents. 26. Synthesis of 4,6-disubstituted 2cyanaminopy-rimidine and mass spectrometric studies of their structure, *Chemiker. Zeitung.* 108, 253, 1984.

399. Kreutzberger, A. and Richter, B., Antiviral substances. 21. Perfluoroalkyl-2-(4-nitroanilino)pyrimidines, *J. Fluorine Chem.* 20, 227, 1982.

400. Wang, Y., Vivekananda, S., and Zhang, K., ESI-MS/MS for the differentiation of diastereomeric pyrimidine glycols in mononucleosides, *Anal. Chem.* 74, 4505, 2002.

401. Sakurai, T., Matsuo, T., Kusai, A., and Nojima, K., Collisionally activated decomposition spectra of normal nucleosides and nucleotides using a four-sector tandem mass spectrometer, *Rapid Commun. Mass Spectrom.* 3, 212, 1989.

402. Crow, F.W., Tomer, K.B., Gross, M.L., McCloskey, J.A., and Bergstrom, D.E., Fast atom bombardment combined with tandem mass spectrometry for the determination of nucleosides, *Anal. Biochem.* 139, 243, 1984.

403. Reddy, D. and Iden, C.R., Analysis of modified deoxynucleosides by electrospray ionization mass spectrometry, *Nucleos. Nucleot.* 12, 815, 1993.

404. Frelon, S., Douki, T., Ravanat, J.L., Pouget, J.P., Torabene, C., and Cadet, J., High-performance liquid chromatography-tandem mass spectrometry measurement of radiation-induced base damage to isolated and cellular DNA, *J. Chem. Res. Toxicol.* 13, 1002, 2000.

405. Hua, Y., Wainhaus, S.B., Yang, Y., Shen, L., Xiong, Y., Xu, X., Zhang, F., Bolton, J.L., and van Breemen, R.B., Comparison of negative and positive ion electrospray tandem mass spectrometry for the liquid chromatography tandem mass spectrometry analysis of oxidized deoxynucleosides, *J. Am. Soc. Mass Spectrom.* 12, 80, 2001.

406. Zhang, Q. and Wang, Y., Differentiation of 2′-O- and 3′-O-methylated ribonucleosides by tandem mass spectrometry, *J. Am. Soc. Mass Spectrom.* 17, 1096, 2006.

407. Ovcharenko, V., Szacon, E., Tkaczynski, T., Matosiuk, D., and Pihlajal, K., Electron impact mass spectra of substituted 1-aryl-2-arylsulphonylamino-1,4,5,6-tetrahydropyrimidines, *Rapid Commun. Mass Spectrom.* 11, 1407, 1997.

408. Nelson, C.C., and McCloskey, J.A., Collision induced dissociation of uracil and its derivatives, *J. Am. Soc. Mass Spectrom.* 5, 339, 1994.

409. Ramsey, R.S., Van Berkel, G.J., McLuckey, S.A., and Glish, G.L., Determination of pyrimidine cyclo-butane dimmers by electrospray ionization trap mass-spectrometry, *Biol. Mass Spectrom.* 21, 347, 1992.

410. Lin, Z.J., Li, W., and Dai, G., Application of LC–MS for quantitative analysis and metabolite identification of therapeutic oligonucleotides, *J. Pharmaceut. Biomed. Anal.* 44, 330, 2007.
411. Dai, G., Wei, X., Liu, Z., Liu, S., Marcucci, G., and Chan, K.K., Characterization and quantification of Bcl-2 antisense G3139 and metabolites in plasma and urine by ion-pair reversed phase HPLC coupled with electrospray ion-trap mass spectrometry, *J. Chromatogr. B* 825, 201, 2005.
412. Johnson, J.L., Guo, W., Zang, J., Khan, S., Bardin, S., Ahmad, A., Duggan, J.X., and Ahmad, I., Quantification of raf antisense oligonucleotide (rafAON) in biological matrices by LC-MS/MS to support pharmacokinetics of a liposome-entrapped rafAON formulation, *Biomed. Chromatogr.* 19, 272, 2005.
413. Muddiman, D.C., Cheng, X., Udseth, H.R., and Smith, R.D., Charge-state reduction with improved signal intensity of oligonucleotides in electrospray ionization mass spectrometry, *J. Am. Soc. Mass Spectrom.* 7, 697, 1996.
414. Huang, G. and Krugh, T.R., Large-scale purification of synthetic oligonucleotides and carcinogen-modified oligodeoxynucleotides on a reverse-phase polystyrene (PRP-1) column, *Anal. Biochem.* 190, 21, 1990.
415. Apffel, A., Chakel, J.A., Fischer, S., Lichtenwalter, K., and Hancock, W.S., Analysis of oligonucleotides by HPLC-electrospray ionization mass spectrometry, *Anal. Chem.* 69, 1320, 1997.
416. Coleman, R.S., Kesicki, E.A., Arthur, J.C., and Cotham, W.E., Analysis of post-synthetically modified oligodeoxynucleotides by electrospray ionization mass spectrometry. *Bioorg. Med. Chem. Lett.* 4, 1869, 1994.
417. Oberacher, H., Walcher, W., and Huber, C.G., Effect of instrument tuning on the detectability of biopolymers in electrospray ionization mass spectrometry, *J. Mass Spectrom.* 38, 108, 2003.
418. Embrechts, J., Lemiere, F., Van Dongen, W., and Esmans, E.L., Equilenin-2'-deoxynucleoside adducts: Analysis with nano-liquid chromatography coupled to nano-electrospray tandem mass spectrometry, *J. Mass Spectrom.* 36, 317, 2001.
419. Li, L., Chiarelli, M.P., Branco, P.S., Antunes, A.M., Marques, M.M., Goncalves, L.L., and Beland, F.A., Differentiation of isomeric C8-substituted alkylaniline adducts of guanine by electrospray ionization and tandem quadrupole ion trap mass spectrometry, *J. Am. Soc. Mass Spectrom.* 14, 1488, 2003.
420. Song, R., Zhang, W., Chen, H., Ma, H., Dong, Y., Sheng, G., Zhen, Z., and Fu, J., Site determination of phenyl glycidyl ether-DNA adducts using high-performance liquid chromatography with electrospray ionization tandem mass spectrometry, *Rapid Commun. Mass Spectrom.* 19, 1120, 2005.
421. Wang, Y., Taylor, J.S., and Gross, M.L., Fragmentation of photomodified oligodeoxynucleotides adducted with metal ions in an electrospray-ionization ion-trap mass spectrometer, *J. Am. Soc. Mass Spectrom.* 12, 1174, 2001.
422. Chiarelli, M.P., Chang, H.F., Olsen, K.W., Barbacci, D., Cho, B.P., and Huffer, D.M., Structural differentiation of diastereomeric benzo[ghi]fluoranthene adducts of deoxyadenosine by matrix-assisted laser desorption/ionization time-of-flight mass spectrometry and postsource decay, *Chem. Res. Toxicol.* 16, 1236, 2003.
423. Huffer, D.M., Chang, H.F., Cho, B.P., Zhang, L., and Chiarelli, M.P., Product ion studies of diastereomeric benzo[ghi]fluoranthene tetraols by matrix-assisted laser desorption/ionization time-of-flight mass spectrometry and post-source decay, *J. Am. Soc. Mass Spectrom.* 12, 376, 2001.
424. Li, L., Chang, H.F., Olsen, K.W., Cho, B.P., and Ciarelli, M.P., Product ion studies of diastereomeric benzo[ghi]fluoranthene-2'-deoxynucleoside adducts by electrospray ionization and quadrupole ion trap mass spectrometry, *Anal. Chim. Acta* 557, 191, 2006.
425. Ge, L., Yong, J.W.H.S., Tan, N., Yang, X.H., and Ong, E.S., Analysis of cytokinin nucleotides in coconut (*Cocos nucifera L.*) water using capillary zone electrophoresis-tandem mass spectrometry after solid-phase extraction, *J. Chromatogr. A* 1133, 322, 2006.
426. McLuckey, S.A., Vaidyanathan G., and Habibi-Goudarzi, S., Charged vs neutral nucleobase loss from multiply charged oligonucleotide anions, *J. Mass Spectrom.* 30, 1222, 1995.
427. Stephenson, J.L., Jr. and McLuckey, S.A., Simplification of product ion spectra derived from multiply charged parent ions via ion/ion chemistry, *Anal. Chem.* 70, 3533, 1998.
428. Gentil, E. and Banoub, J., Characterization and differentiation of isomeric self-complementary DNA oligomers by electrospray tandem mass spectrometry, *J. Mass Spectrom.* 31, 83, 1996.
429. Baker, T.R., Keough, T., Dobson, R.L.M., Riley, T.A., Hasselfield, J.A., and Hesselberth, P.E., Antisense DNA oligonucleotides. The use of ionspray tandem mass spectrometry for the sequence verification of methylphosphonate oligodeoxyribonucleotides, *Rapid Commun. Mass Spectrom.* 7, 190, 1993.
430. Favre, A., Gonnet, F., and Tabet, J.-C., Location of the negative charge(s) on the backbone of single-stranded deoxyribonucleic acid in the gas phase, *Eur. J. Mass Spectrom.* 6, 389, 2000.

431. Josse, J., Kaiser, A.D., and Kornberg, A., Enzymatic synthesis of deoxyribonucleic acids. VIII. Frequencies of nearest neighbor base sequences in deoxyribonucleic acids, *J. Biol. Chem.* 236, 864, 1961.

432. Rozenski, J. and McCloskey, J.A., Determination of nearest neighbors in nucleic acids by mass spectrometry, *Anal. Chem.* 71, 1454, 1999.

433. Rozenski, J. and McCloskey, J.A., Characterization of oligonucleotide sequence isomers in mixtures using HPLC/MS, *Nucleos. Nucleot. Nucleic Acids* 18, 1539, 1999.

434. Oberacher, H., Wellenzohn, B., and Huber, C.G., Comparative sequencing of nucleic acids by liquid chromatography-tandem mass spectrometry, *Anal. Chem.* 74, 211, 2002.

435. Premstaller, A. and Huber, C.G., Factors determining the performance of triple quadrupole, quadrupole ion trap and sector field mass spectrometer in electrospray ionization mass spectrometry. Suitability for de novo sequencing, *Rapid Commun. Mass Spectrom.* 15, 1053, 2001.

436. Little, D.P., Aaserud, D.J., Valaskovic, G.A., and McLafferty, F.W., Sequence information from 42-108-mer DNAs (complete for a 50-mer) by tandem mass spectrometry, *J. Am. Chem. Soc.* 118, 9352, 1996.

437. Ni, J.S., Pomerantz, S C., Rozenski, J., Zhang, Y., and McCloskey, J.A., Interpretation of oligonucleotide mass spectra for determination of sequence using electrospray ionization and tandem mass spectrometry, *Anal. Chem.* 68, 1989, 1996.

438. Oberacher, H., Parson, W., Oefner, P.J., Mayr, B.M., and Huber, C.G., Applicability of tandem mass spectrometry to the automated comparative sequencing of long-chain oligonucleotides, *J. Am. Soc. Mass Spectrom.* 15, 510, 2004.

439. Rozenski, J., http://library.med.utah.edu/masspec/mongohlp.htm

440. Reiber, D.C. and Murphy, R.C., Covalent binding of LTA(4) to nucleosides and nucleotides, *Arch. Biochem. Biophys.* 379, 119, 2000.

441. Monn, S.T.M. and Schürch, S., Investigation of metal-oligonucleotide complexes by nanoelectrospray tandem mass spectrometry in the positive mode, *J. Am. Soc. Mass Spectrom.* 16, 370, 2005.

442. Rosu, F., Nguyen, C.H., De Pauw, E., and Gabelica, V., Ligand binding mode to duplex and triplex DNA assessed by combining electrospray tandem mass spectrometry and molecular modeling, *J. Am. Soc. Mass Spectrom.* 18, 1052, 2007.

443. Brink, A., Lutz, U., Völkel, W., and Lutz, W.K., Simultaneous determination of O^6-methyl-2'-deoxyguanosine, 8-oxo-7,8-dihydro-2'-deoxyguanosine, and 1,N^6-etheno-2'-deoxyadenosine in DNA using on-line sample preparation by HPLC column switching coupled to ESI-MS/MS, *J. Chromatogr. B* 830, 255, 2006.

444. Churchwell, M.I., Beland, F.A., and Doerge, D.R., Quantification of O^6-methyl and O^6-ethyl deoxyguanosine adducts in C57BL/6N/Tk$^{+/-}$ mice using LC/MS/MS, *J. Chromatogr. B* 844, 60, 2006.

445. Haglund, J., Van Dongen, W., Lemiere, F., and Esmans, E.L., Analysis of DNA-phosphate adducts in vitro using miniaturized LC-ESI-MS/MS and column switching: Phosphotriesters and alkyl cobalamins, *J. Am. Soc. Mass Spectrom.* 15, 596, 2004.

446. Embrechts, J., Lemiere, F., Van Dongen, W., Esmans, E.L., Buytaert, P., Van Marck, E., Kockx, M., and Makar, A., Detection of estrogen DNA-adducts in human breast tumor tissue and healthy tissue by combined nano LC-nano ES tandem mass spectrometry, *J. Am. Soc. Mass Spectrom.* 14, 482, 2003.

447. Marzilli, L.A., Barry, J.P., Sells, T., Law, S.J., Vouros, P., and Harsch, A., Oligonucleotide sequencing using guanine-specific methylation and electrospray ionization ion trap mass spectrometry, *J. Mass Spectrom.* 34, 276, 1999.

448. Glover, R.P., Lamb, J.H., and Farmer, P.B., Tandem mass spectrometry studies of a carcinogen modified oligodeoxynucleotide, *Rapid Commun. Mass Spectrom.* 12, 368, 1998.

449. Tretyakova, N., Matter, R., Jones, R., and Shallop, A., Formation of benzo[a]pyrene diol epoxide-DNA adducts at specific guanines within K-ras and p53 gene sequences: Stable isotope-labeling mass spectrometry approach, *Biochemistry* 41, 9535, 2002.

450. Matter, R., Wang, G., Jones, R., and Tretyakova, N., Formation of diastereomeric benzo[alpha]pyrene diol epoxide-guanine adducts in p53 gene-derived DNA sequences, *Chem. Res. Toxicol.* 17, 731, 2004.

451. Douki, T., Court, M., and Cadet, J., Electrospray-mass spectrometry characterization and measurement of far-UV-induced thymine photoproducts, *J. Photochem. Photobiol. B Biol.* 54, 145, 2000.

452. Wang, Y.S., Taylor, J.S., and Gross, M.L., Nuclease P1 digestion combined with tandem mass spectrometry for the structure determination of DNA photoproducts, *Chem. Res. Toxicol.* 12, 1077, 1999.

453. Bellon, S., Ravanat, J.L., Gasparutto, D., and Cadet, J., Cross-linked thymine-purine base tandem lesions: Synthesis, characterization, and measurement in gamma-irradiated isolated DNA, *Chem. Res. Toxicol.* 15, 598, 2002.

454. Little, D.P., Speir, J.P., Senko, M.W., O'Connor, P.B., and McLafferty, F.W., Infrared multiphoton dissociation of large multiply charged ions for biomolecule sequencing, *Anal. Chem.* 66, 2809, 1994.

455. Little, D.P., Thannhauser, T.W., and McLafferty, F.W., Verification of 50-mer to 100-mer DNA and RNA sequences with high-resolution mass-spectrometry, *Proc. Natl. Acad. Sci. USA* 92, 2318, 1995.
456. Sannes-Lowery, K.A., and Hofstadler, S.A., Sequence confirmation of modified oligonucleotides using IRMPD in the external ion reservoir of an electrospray ionization Fourier transform ion cyclotron mass spectrometer, *J. Am. Soc. Mass Spectrom.* 14, 825, 2003.
457. (a) Hu, Q., Noll, R.J., Li, H., Makarov, A., Hardman, M., and Cooks, R.G., The orbitrap: A new mass spectrometer, *J. Mass Spectrom.* 40, 430, 2005 (b) Makarov, A., Denisov, E., Lange, O., and Horning, S., Dynamic range of mass accuracy in LTQ Orbitrap hybrid mass spectrometer, *J. Am. Soc. Mass. Spectrom.* 17, 977, 2006.
458. Jamin, E.L., Arquier, D., Canlet, C., Rathahao, E., Tulliez, J., and Debrauwer, L., New insights in deoxynucleoside the formation of adducts with the heterocyclic aromatic amines PhIP and IQ by means of ion trap MSn and accurate mass measurement of fragment ions, *J. Am. Soc. Mass Spectrom.* 18, 2107, 2007.
459. Furutani, H., Ugarov, M.V., Prather, K., and Schultz, J.A., US Patent 7170052, 2007.
460. Goh, W.C., Sodroski, J.G., Rosen, C.A., and Haseltine, W.A., Expression of the art gene protein of human T-lymphotropic virus type-III (HTLV-III/LAV) in bacteria, *J. Virol.* 63, 4085, 1987.
461. Linial, M.L. and Miller, A.D., Retroviral RNA packaging—sequence requirements and implications, *Curr. Top. Microbiol. Immunol.* 157, 25, 1990.
462. Darlix, J L., Gabus, C., Nugeyre, M.T., Clavel, F., and Barre-Sinussi, F., Cis elements and Trans-acting factors involved in the RNA dimerization ot the human-immunodeficiency-virus HIV-1, *J. Mol. Biol.* 216, 689, 1990.
463. Dickson, C., Eisenman, R., Fan, H., Hunter, E., and Reich, N., Protein biosynthesis and assembly. Weiss, R., ed., In *RNA Tumor Viruses*, pp. 513-648. Cold Spring Harbor Laboratory Press: Plainview, NY, 1985.
464. Darlix, J.L., Lapadat-Tapolsky, M., de Roquigny, H., and Roques, B.P., First glimpses at structure-function relationships of the nucleocapsid protein of retroviruses, *J. Mol. Biol.* 254, 523, 1995.
465. Coffin, J.M., Hughes, S.H., and Varmus, H. *Retroviruses*, Cold Spring Harbor Laboratory Press, Plainview, NY, 1997.
466. Johnson, V.A., Brun-Vezinet, F., Clotet, B., Conway, B., Kuritzkes, D.R., Pillay, D., Shapiro, J.M., Telenti, A., and Richman, D.D., Update of the drug resistance mutations in HIV-1: 2005, *Top. HIV Med.* 13, 51, 2005.
467. Yusa, K., Kavlick, M.F., Kosalaraksa, P., and Mitsuya, H., HIV-1 acquires resistance to two classes of antiviral drugs through homologous recombination, *Antivirus Res.* 36, 179, 1997.
468. McPike, M.P., Sullivan, J.M., Goodisman, J., and Dabrowiak, J.C., Footprinting, circular dichroism and UV melting studies on neomycin B binding to the packaging region of human immunodeficiency virus type-1 RNA, *Nucleic Acids Res.* 30, 2825, 2002.
469. Ennifar, E., Paillart, J.C., Marquet, R., Ehresmann, B., Ehresmann, C., Dumas, P., and Walter, P., HIV-1 RNA dimerization initiation site is structurally similar to the ribosomal A site and binds aminoglycoside antibiotics *J. Biol. Chem.* 278, 2723, 2003.
470. McPike, M.P., Goodisman, J., and Dabrowiak, J.C., Footprinting and circular dichroism studies on paromomycin binding to the packaging region of human immunodeficiency virus type-1, *Bioorg. Med. Chem.* 10, 3663, 2002.
471. Hagan, N., and Fabris, D., Direct mass spectrometric determination of the stoichiometry and binding affinity of the complexes between nucleocapsid protein and RNA stem-loop hairpins of the HIV-1 psi-recognition element, *Biochemistry* 42, 10736, 2003.
472. Turner, K.B., Hagan, N.A., and Fabris, D., Inhibitory effects of archetypical nucleic acid ligands on the interactions of HIV-1 nucleocapsid protein with elements of Psi-RNA, *Nucleic Acids Res.* 34, 1305, 2006.
473. Cerny, R.L., Tomer, K.B., Gross, M.L., and Grotjahn, L., Fast-atom bombardment combined with tandem-mass spectrometry for determining structures of small oligonucleotides, *Anal. Biochem.* 165, 175, 1987.
464. Kellersberger, K.A., Yu, E., Kruppa, G.H., Young, M.M., and Fabris, D., Top-down characterization of nucleic acids modified by structural probes using high-resolution tandem mass spectrometry and automated data interpretation, *Anal. Chem.* 76, 2438, 2004.
475. Kirpekar, F. and Krogh, T.N., RNA fragmentation studied in a matrix-assisted laser desorption/ionisation tandem quadrupole/orthogonal time-of-flight mass spectrometer, *Rapid Commun. Mass Spectrom.* 15, 8, 2001.
476. Tromp, J.M., and Schurch, S., Gas-phase dissociation of oligoribonucleotides and their analogs studied by electrospray ionization tandem mass spectrometry, *J. Am. Soc. Mass Spectrom.* 16, 1262, 2005.

477. Turner, K.B., Hagan, N.A., Kohlway, A.S., and Fabris D., Mapping noncovalent ligand binding to stem-loop domains of the HIV-1 packaging signal by tandem mass spectrometry, *J. Am. Soc. Mass Spectrom.* 17, 1401, 2006.
478. Little, D.P., Chorush, R.A., Speir, J.P., Senko, M.W., Kelleher, N L., and McLafferty, F.W., Rapid sequencing of oligonucleotides by high resolution mass spectrometry, *J. Am. Chem. Soc.* 116, 4893, 1994.
479. Woods, A.S., Ugarov, M., Egan, T., Koomen, J., Gillig, K.J., Fuhrer, K., Gonin, M., and Schultz, J.A., Lipid/peptide/nucleotide separation with MALDI-ion mobility-TOF MS, *Anal. Chem.* 76, 2187, 2004.
480. Kanu, A.B., Dwivedi, P., Tam, M., Matz, L., and Hill Jr. H.H., Ion mobility-mass spectrometry, *J. Mass Spectrom.* 43, 1, 2008.
481. Verbeck, G.F., Ruotolo, B.T., Gillig, K.J., and Russell, D.H., Resolution equations for high-field ion mobility, *J. Am. Soc. Mass Spectrom.* 15, 1320, 2004.
482. Koomen, J.M., Ruotolo, B.T., Gillig, K.J., McLean, J.A., Russell, D.H., Kang, M., Dunbar, K.R., Fuhrer, K., Gonin, M., and Schultz, J.A., Oligonucleotide analysis with MALDI-ion-mobility-TOF-MS, *Anal Bioanal Chem.* 373, 612, 2002.
483. Fenn, J.B., Mann, M., Meng, C.K., Wong, S.F., and Whitehouse, C.M., Electrospray ionization for mass spectrometry of large biomolecules, *Science* 246, 64, 1989.
484. Tanaka, K., The origin of macromolecule ionization by laser irradiation, *Angew. Chem., Int. Edit.* 42, 3860, 2003.

2 Mass Spectrometric Analysis of Deoxyinosines

Antonio Triolo

CONTENTS

2.1 INTRODUCTION

Deoxyinosines are a class of biologically important deoxynucleosides that have hypoxanthine as the nucleobase. They have been investigated as markers of cancer and other metabolic disorders. Some examples of these deoxynucleotides and their significance as markers are 2′-deoxyinosine (**1**, Chart 2.1), an essential intermediate in the catabolism of adenine deoxynucleotides, used as a marker of altered purine metabolism; 5′-deoxyinosine (**2**, Chart 2.1), which has been detected in the urine of leukemia patients; and 2′-deoxyinosine, its derivatives, and other modified purine and pyrimidine bases found in DNA hydrolysates that have been exposed to genotoxic carcinogens. These base changes in the composition of DNA can lead to mutations in critical genes, such as those involved in the regulation of cellular growth. The study of these modified bases, at the nucleoside level, is essential to understand the development of various proliferative diseases at the molecular level. There are deoxyinosines, and related structures (such as 3′-deoxyinosine (**3**, Chart 2.1), an antileishmaniasis agent, or didanosine (**4**, Chart 2.1), an antiretroviral drug used against AIDS, that are currently being used in the therapy of viral infections, microbial infections, and proliferative disorders and still other deoxyinosine-related compounds that are under investigation for their possible therapeutic value.

The study of nucleobases, nucleosides, nucleotides, and nucleic acids is an area in biology where the application of mass spectrometry (MS) is most relevant. MS has been mostly applied to the study of the "natural" constituents of nucleic acids, and only in minor part to the study of other purine and pyrimidine derivatives. This work summarizes some of the relevant aspects of MS studies performed on deoxyinosine and its analogues.

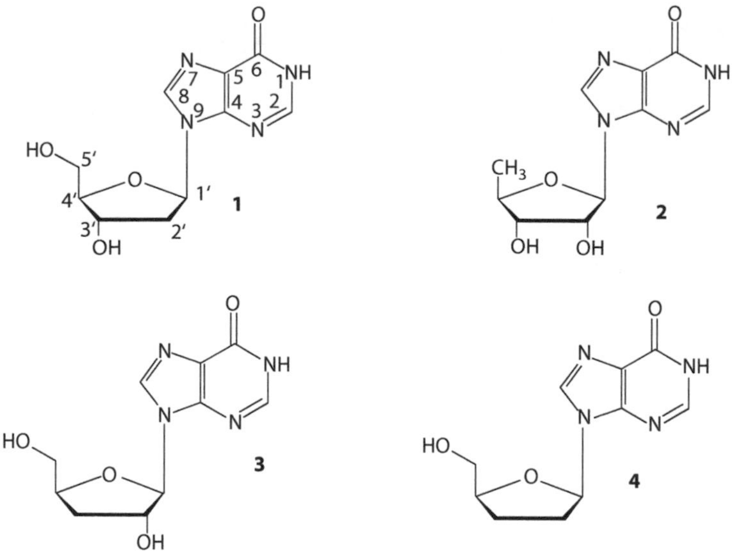

CHART 2.1   Structural formulae of compounds 1 to 4 (details in the text).

2.2   MASS SPECTRA OF 2′-DEOXYINOSINE

In 1986, Dizdaroglu[1,2] reported the electron ionization (EI) mass spectrum of the trimethylsilyl derivative of 2′-deoxyinosine in a gas chromatography (GC)-MS study aimed at identifying the radiation-
generated free hydroxy radical damage to DNA. The presence of 2′-deoxyinosine-$(Me_3Si)_3$ was
identified in the chromatogram and was ascribed to the deamination of 2′-deoxyadenosine under the
hydrolysis conditions used. The analogue 8-hydroxy-2′-deoxyinosine-$(Me_3Si)_4$, which was probably
formed by deamination of 8-hydroxy-2′-deoxyadenosine, a known product in γ-irradiated DNA,
was also identified. At this time, liquid-phase ionization techniques had not been developed and
EI-GC-MS of the trimethylsilyl derivative of the nucleoside was the primary method of analysis.[3,4]
This methodology was limited by the risk of introducing artifacts during the time-consuming, labor-
intensive sample preparation and derivatization procedures and by the excessive fragmentation of
the molecules; also, not every modified nucleoside, even after derivatization, had sufficient volatility
and stability to be analyzed by GC.

The pioneering work by Esmans et al. described the first direct liquid introduction (DLI) interfaced
liquid chromatography (LC)-MS spectrum of 2′-deoxyinosine, showing a strong protonated molecular
ion observed at m/z 253.[5] The analysis of labile compounds including bases, nucleosides, and nucleotides later became routine, with improvements in instrumentation sensitivity and the development of
atmospheric pressure ionization techniques, which allowed easy interfacing with LC.

Interest in 2′-deoxyinosine, a product of the catabolism of adenine deoxynucleotide, was driven by
its importance in the diagnosis of inherited purine metabolism disorders. Variations of the level
2′-deoxyinosine in biological fluids indicate errors in metabolism that can result in serious disorders.
The need for a sensitive, specific, and high-throughput determination of the levels of 2′-deoxyinosine
in clinical samples has led to the development of different methods based on high-performance liquid
chromatography (HPLC)-MS.[6] Frycák et al. has presented a complete description of the MS and
product ion MS/MS spectra of 2-deoxyinosine and of other purines and pyrimidines that are potential
markers of metabolic disorders.[7] With the aim of finding a high-throughput method for screening
metabolic errors in urine, the authors compared the fragmentation patterns with a series of nucleosides and bases (including 2′-deoxyinosine) by electrospray ionization (ESI) and atmospheric pressure chemical ionization (APCI), under positive and negative ionization modes. The authors found
that, under ESI conditions, there was clustering between the analytes and the sodium and potassium

ions that were present in the samples. They found that the APCI method provided a higher response for the analyte because there was less aggregation and it was less sensitive to the presence of trace metal ions. Therefore, APCI was chosen as the most suitable ionization method for the direct, high-throughput analysis of urine. MS/MS detection was used to ensure the selectivity and specificity for the analytes. Under either ESI or APCI conditions, in the positive ion mode, the protonated quasimolecular ion species of 2′-deoxyinosine was observed at m/z 253, whereas the negative ion mass spectrum showed the deprotonated quasimolecular ion species at m/z 251 (Scheme 2.1).

The product ion spectrum of 2′-deoxyinosine showed the base, hypoxanthine, to be either protonated (m/z 137) or deprotonated (m/z 135), depending on the ion charge. In each case, this is the result of the transfer of one hydrogen atom from deoxyribose to the base with a corresponding loss to the sugar residue of 116 Da. In the negative product ion scan, there is a minor fragment observed at m/z 161, which results from a neutral loss of 90 Da ($C_3H_6O_3$) caused by the cleavage of the O4–C1 and C2–C3 bonds in the furanose ring. In the product ion scan described by Frycák, obtained with an ion-trap instrument, the m/z 161 ion has 6.4% relative abundance. Deoxyribose product ions were not observed under the conditions he described. The fragmentation pattern of 2′-deoxyinosine is depicted in Scheme 2.1; the italicized numbers refer to the negative ion spectrum.

In summary, with respect to MS/MS analysis, the structural information is essentially limited to the mass of the base. The mass of the sugar is determined by the difference between the precursor and product ions. While this is sufficient for confirming the identity of a known compound, for related, unknown compounds, it may be necessary to further fragment the product ions by performing MS^n experiments with an ion-trap instrument.

An essential step for locating modifications on the unknown hypoxanthine-related moiety is elucidating the fragmentation pathways of the unmodified one. The MS^3 and MS^4 spectra of the hypoxanthine fragment ion have not been reported for 2′-deoxyinosine. However, in a study that was aimed at the identification of modified urinary nucleosides that could be potential biomarkers, Kammerer et al. reported the fragmentation pathway for ribonucleoside inosine.[8] Irrespective of the origin of the hypoxanthine ion, it is logical to expect that the fragmentation pattern, under identical MS conditions, would be the same. The authors demonstrate this for the base itself and for the corresponding ribonucleoside. For this reason, the fragmentation pattern of protonated hypoxanthine, as described by Kammerer in the MS^2 scan of the base and in the MS^3 scan of inosine, can

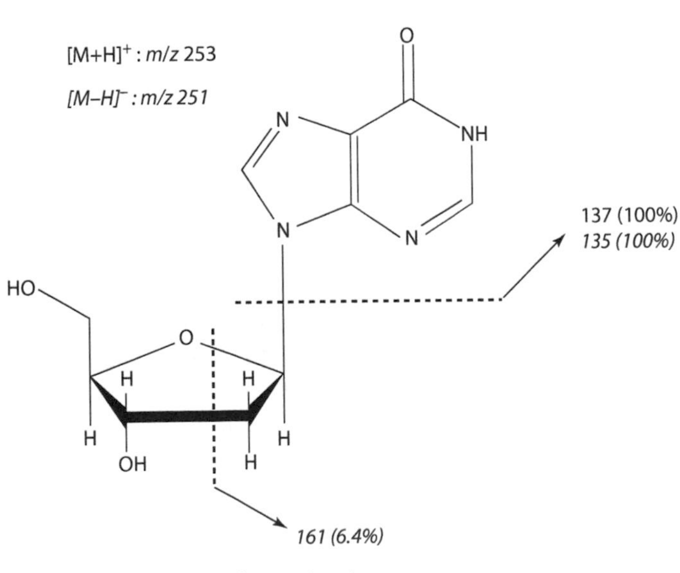

SCHEME 2.1 Fragmentation pattern of 2′-deoxyinosine.

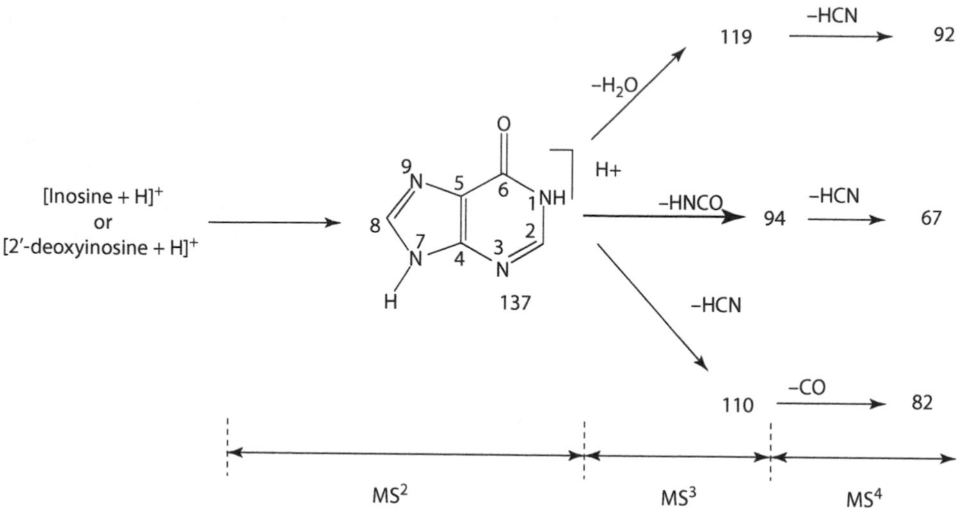

SCHEME 2.2   MS^n fragmentation pattern of the protonated hypoxanthine.

also be applied to 2'-deoxyinosine. This fragmentation is briefly described as follows: the protonated base, m/z 137, loses water (m/z 119), HCN (m/z 110), and HCNO (m/z 94) during the third stage of mass analysis or MS^3 spectrum; in the MS^4 spectra, m/z 119 loses HCN (m/z 92), m/z 110 loses CO (m/z 82), and m/z 94 loses HCN (m/z 67). The overall fragmentation pattern is depicted in Scheme 2.2.

In this work, the authors also describe the fragmentation of the base in the MS^3 scan of 1-methyl and 7-methylinosine. They show that, like hypoxanthine, the protonated 7-methylhypoxanthine (m/z 151) loses water (m/z 133), HCN (m/z 124), and HCNO (m/z 108), whereas 1-methylhypoxanthine loses water, CH_3CN (m/z 110), and CH_3NCO (m/z 94). It is apparent from the data that one can easily differentiate the two isomers, and that losses of HNCO and HCN in hypoxanthine involve N1–C6 and N1–C2 cleavages, respectively.

Ito et al. described a quantitative LC-MS/MS method, based on negative ESI and selected reaction monitoring (SRM), for the rapid screening of patients at risk of disorders of purine and pyrimidine metabolism.[9] The simultaneous analysis of 17 purines and pyrimidines, including 2'-deoxyinosine, from urine or urine-soaked filter paper strips has been accomplished with minimal sample preparation: urine was taken to pH = 7.5, diluted if necessary, centrifuged and injected into the LC-MS/MS; the paper strips were extracted with methanol, the methanol was evaporated, and the residue of dissolved acetonitrile centrifuged and injected into the LC-MS/MS. Stable isotope-labeled internal standards were used to improve the accuracy of the measurements; the internal standard for 2'-deoxyinosine was 8-^{13}C-adenine. A reverse-phase HPLC column with linear gradient elution (0.05 M ammonium formate and methanol) was used for the separation. Total analysis time was 15 min, which is about one-third of an equivalent HPLC method with UV detection. For the SRM analysis of 2'-deoxyinosine, the transition between the deprotonated adduct with formic acid (m/z 297) and the deprotonated analyte (m/z 251) was monitored with a detection limit of 0.5 μmol L^{-1}. Lu et al. developed an LC-MS/MS method for the simultaneous quantitation of nitrogen-containing metabolites of *Salmonella enterica* in which 2'-deoxyinosine was determined along with 90 other substances.[10] The task was challenging because of the presence of a large number of analytes, the majority of which were present only in trace amounts. These trace species had to be analyzed in the presence of other, potentially interfering compounds. The authors used a polar-embedded C18 column and a 55-min gradient elution (0.1% formic acid in water and 0.1% formic acid in methanol) in the positive mode ESI, and with SRM detection of the analytes.

In the case of the protonated molecule of 2′-deoxyinosine, the SRM transition monitored the decomposition of m/z 253 → m/z 137 the protonated base [BH_2]$^+$. The limit of detection for the 90 purified metabolites was approximately 5 ng mL^{-1} in solution.

However, only 36 of the metabolites extracted from either the growing or carbon-starved cultures could be reliably quantified. Unfortunately, 2′-deoxyinosine was not one of these 36 metabolites.

The study of DNA lesions following exposure to cytotoxic and mutagenic agents is another application that requires an accurate determination of the amount of 2′-deoxyinosine. Dong et al. investigated the effects of nitric oxide-mediated genotoxicity on plasmid pUC19 DNA.[11] The nitrosating derivatives of nitric oxide induce deamination of the DNA bases leading to the formation of hypoxanthine from adenine; xanthine from guanine; uracil from cytosine, as well as abasic sites and interstrand cross-links. The authors used ESI-MS with single-ion monitoring (SIM) to quantitate the deaminated bases after obtaining the corresponding nucleosides; detection limit was 100 fmol and sensitivity was six lesions per seven bases with 50 μg of DNA. In another study, they measured the levels of deaminated bases in DNA from human TK6 cells exposed to toxic levels of nitric oxide.[12]

2.3   MASS SPECTRUM OF 5′-DEOXYINOSINE

This nucleoside (**2**, Chart 2.1) was found in the urine of chronic myelogenes leukemia (CML) patients.[13,14] The nucleoside was isolated by HPLC and was identified by UV, NMR, and the GC-MS of the trimethylsilyl derivative. The assignment was confirmed by comparison with the chromatographic and spectroscopic properties of a standard. The EI-MS of 5′-deoxyinosine-(Me_3Si)$_3$ is depicted in Figure 2.1. It shows a very low-abundance molecular ion at m/z 468, and the ion from the loss of a methyl radical at m/z 453. The base peak, m/z 260, comes from cleavage of the glycosidic bond with transfer of a hydrogen on the base and charge retention on the sugar ion [S–H]$^+$. The differentiation by EI-MS of the trimethylsislyl (TMS) derivatives of 2′,3′- and 5′-deoxynucleotides has

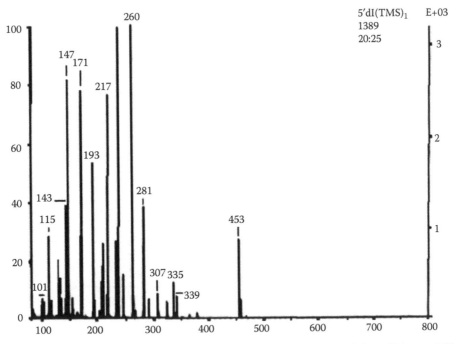

FIGURE 2.1   EI mass spectrum of 5′-deoxyinosine (ME$_3$Si)$_3$. (Reproduced from Schram, K.H., *Mass Spectrom. Rev.* 17, 131–251, 1998. Copyright 1999, John Wiley & Sons. With permission.)

been discussed by Reimer et al.[15] Briefly, the indicators of deoxygenation at the 5′ position are the ion [S–H]⁺ as the base peak, the absence of m/z 103 (ion $C_4H_{11}OSi^+$), and the occurrence of the ion [B + 128]⁺ at m/z 335. No other mass spectrum has been described for this compound.

According to Schram,[14] the origin of 5′-deoxyinosine may be ascribed to the liberation of 5′-deoxyadenosine from vitamin B12, and subsequent deamination by adenosine deaminase. Alternatively, 5′-deoxyribose may be cleaved from 5′-deoxyadenosine by the action of 5′-deoxy-5′-methylthioadenosine phosphorylase, and coupled as 1-phosphate to hypoxanthine by purine nucleoside phosphorylase.

2.4   MASS SPECTRA OF 2′,3′-DIDEOXYINOSINE (DIDANOSINE)

There are several deoxyinosine analogues that have been investigated as potential antiviral or antiprotozoarian agents, but there are no MS data available for them. An example of one of these analoques is 3′-deoxyinosine (**3**, Chart 2.1), which has been proposed as an antileishmanial agent.[16] However, there is a significant MS analysis of didanosine, with the structure 2′,3′-dideoxyinosine (**4**, Chart 2.1), an antiretroviral drug used in HIV therapy. Didanosine inhibits virus replication by blocking the viral inverse transcriptase, and is normally administered along with other reverse transcriptase inhibitors such as 3′-azidothymidine (zidovudine), 2′,3′-dideoxythymidine (stavudine), and 2′,3′-dideoxycytidine (zalcitabine). Didanosine is biotransformed into its active form, a triphosphorylated metabolite, once it is inside the cells. Mass spectrometry is an essential tool that can be used during all stages of drug development from the structural elucidation of metabolites and impurities to the quantitative determination of the drug or its metabolites in biological fluids for either pharmacokinetics or therapeutic drug monitoring. There have been many different MS studies performed on didanosine.[17,18] The ESI-MS of didanosine was described by Reddy and Iden in their investigation of a series of modified deoxynucleosides.[19] The positive ion MS afforded the protonated molecule at m/z 237, whereas the negative ion spectrum afforded the deprotonated molecule at m/z 235. Fragmentation of the molecular ions was achieved via in-source collision-induced dissociation by increasing the cone-skimmer voltage. This is a convenient way to obtain structural information when an MS/MS instrument is not available.[20] Although in-source fragmentation can generate the same fragment ions as MS/MS, it is limited in its application to biological samples because of chemical background interferences from the mobile-phase components and coeluting compounds.

The positive ion MS/MS spectrum of didanosine, obtained with a triple quadrupole instrument, is shown in Figure 2.2.

The two fragments observed arise from the cleavage of the C–N bond linking the base to 2′,3′-dideoxyribose. As with 2′-deoxyinosine, the base peak is the protonated hypoxanthine (m/z 137), whereas the low-abundance signal at m/z 101 is the oxonium ion of the sugar residue, as shown in

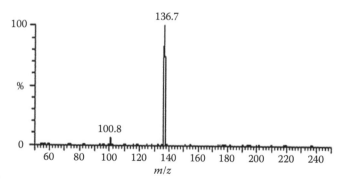

FIGURE 2.2   Positive ion MS/MS spectrum of the didanosine protonated quasimolecular ion, m/z 237. (Adapted from Estrela, R.C.E. et al., *J. Mass Spectrom.* 38, 378–385, 2003. With permission. Copyright 2003, Wiley–Blackwell.)

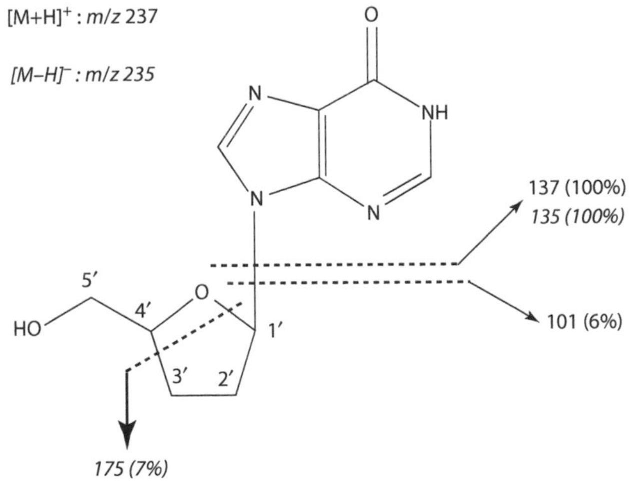

SCHEME 2.3 Fragmentation pattern of didanosine.

Scheme 2.3. Font et al. reported the negative ion MS/MS scan of didanosine, along with the spectra of other antiretroviral nucleosides, using different cone–skimmer voltages and collision energies.[21]

The fragmentation pathways of the endogenous purine nucleosides were found to be similar to that of the didanosine.

Figure 2.3 shows the negative MS/MS spectrum of didanosine deprotonated molecule obtained with a triple quadrupole instrument.

The main fragmentation process of the deprotonated quasimolecular ion involved the loss of the dideoxyribose (100 Da), yielding the deprotonated hypoxanthine at m/z 135. Another fragmentation path involves the loss of 60 Da, ascribed to the cleavage of the sugar ring at the C1–O4 and C3–C4 bonds, resulting in the loss of glycolic aldehyde. The main fragmentation processes of didanosine are summarized in Scheme 2.3; the italicized numbers indicate the masses observed in the negative ion MS/MS.

In the above spectrum, with respect to didanosine, an important observation is in the negative ion mode, where the fragmentation of the sugar ring involves cleavage of the C3–C4 bond, instead of the C2–C3 bond as is observed in 2′-deoxyinosine. Neutral losses of 60 and 90 Da, due to the fragmentation of the sugar ring in nucleosides, have been previously described, and a mechanism for the loss of 90 Da has been proposed.[7,22] The MS/MS (– ion mode) of deprotonated molecule of didanosine showed low-abundance ions at m/z 192, 148, and 108 (Figure 2.3). The product ion at m/z 192 was formed by the loss of HNCO from m/z 235, whereas m/z 148 and 108 arise from losses of HCN from m/z 175 and 135, respectively. It can be concluded that the sensitivity and specificity of the LC-MS/MS detection by SRM are particularly effective in meeting the analytical requirements associated with viral reverse transcriptase inhibitor therapy.

In fact, individual variations in efficacy and toxicity require that methods used to monitor antiretroviral drug levels in biological fluids have sensitivities in the low-ppb range, and HPLC-UV methods cannot easily achieve this. Moreover, the multidrug strategy in HIV infection requires a method capable of the simultaneous quantification of the range of drugs that are coadministered. This cannot be achieved using standard immunoassays, which are sufficiently sensitive, but specific for one drug only; however, it can be achieved using LC-MS/MS. There have been many published reports detailing LC-MS/MS methods for the quantitation of didanosine, alone or simultaneously with other antiretroviral drugs.

Volosov et al. developed a simple method for the determination of 15 anti-HIV drugs in humans, starting with only 80 μL of plasma.[23] After the addition of cimetidine as the internal standard,

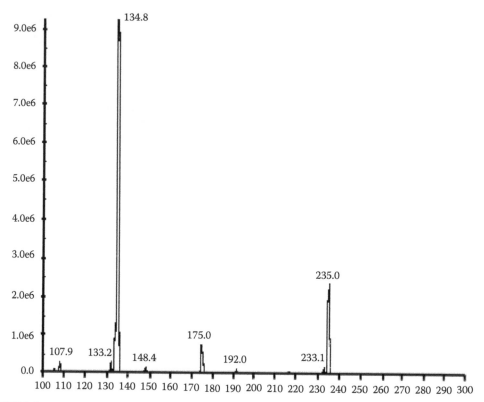

FIGURE 2.3 Negative ion MS/MS spectrum of the didanosine deprotonated quasimolecular ion, m/z 235. (Adapted from Compain, S. et al., *J. Mass Spectrom.* 40, 9–18, 2005. With permission. Copyright 2005, Wiley–Blackwell.)

plasma proteins were precipitated with acetonitrile and injected into an LC-MS/MS system. Didanosine was analyzed by SRM using positive ESI. Any combination of the administered drugs could be analyzed in 4.5 min.

Estrela et al. developed and validated an LC-MS/MS method that focused on didanosine to assess the bioequivalence of two different formulations.[24] Didanosine was determined in human serum by SRM with positive ESI using lamivudine as the internal standard.

Sample purification was performed by solid-phase extraction (SPE). The method was linear in the range 10–1500 ng mL^{-1}, with a detection limit of 10 ng mL^{-1}.

Clark et al. described another dedicated LC-MS/MS method for the determination of didanosine in maternal plasma, amniotic fluid, and fetal and placental tissues.[25] Analysis was performed using stavudine as the internal standard, in a 3-min run time and with detection limits of 1 ng mL^{-1}. An assay for the determination of didanosine and stavudine in several different matrixes such as plasma, peripheral blood mononuclear cells, alveolar cells, semen, and tonsils was described by Huang et al.[26] SPE was used to purify the samples prior to analysis by positive ESI with SRM. Depending on the matrix analyzed, isocratic or gradient HPLC elution was used to separate the analytes, and the limits of detection ranged from 0.5 to 10 ppb. Compain et al. described an LC-MS/MS assay for the quantitation of didanosine in the presence of other drugs (zidovudine, stavudine, zalcitabine, lamivudine, and abacavir) in plasma and peripheral blood mononuclear cells, using 2-chloroadenosine as the internal standard.[27] The aim of the work was to establish the highest sensitivity possible in order to quantitate intracellular drug levels, as well as any residual drug remaining in the plasma between two doses, a factor responsible for drug resistance. After purification by SPE, the drugs were analyzed by ESI-SRM using either positive or negative modes. These authors, contrary to most of the published results, reported a better response for didanosine in negative ESI, in which they obtained a quantitation limit of 0.5 ng mL^{-1} for plasma and 0.025 ng per pellet for intracellular samples. The method was fully validated for linearity, recovery, accuracy and precision, and stability in plasma and in extraction solvent. Using automated SPE, it was possible to analyze 72 samples for each analytical run. The authors also discuss the limits with respect to accuracy, sensitivity, and robustness of fast assays with simplified purification procedures and no or short chromatographic separations.

The development of nucleoside reverse transcriptase inhibitors as therapeutic agents requires the need to quantitate, accurately and with high sensitivity, the intracellular levels of the active triphosphate metabolite. It was found that the enzymatic dephosphorylation of the nucleoside was usually necessary to obtain sufficient chromatographic retention coupled to a high sensitivity of detection. Cahours et al. describes one such method for the measurement of didanosine 5′-triphosphate (ddATP) in CEM-T4 cells.[28] The ddATP was extracted from cells that were isolated by anion exchange SPE, dephosphorylated and analyzed by LC-ESI-MS/MS. Detection sensitivity was 0.02 ng mL^{-1} with this method. The method was also extended to the simultaneous detection of five other nucleoside reverse transcriptase inhibitors. The same group also published a method employing capillary electrophoresis (CE)-MS/MS. The detection limit of 2 µg L^{-1} for didanosine was obtained with this method.[29]

2.5 MASS SPECTROMETRY OF 2′-DEOXYINOSINE-RELATED NUCLEOSIDES FROM DNA ADDUCTS

The covalent binding of genotoxic carcinogens and their active metabolites to DNA is generally considered to be the starting point of a series of processes that can lead to cancer. Electrophilic groups within the structure of the genotoxic compounds react with nucleophilic sites on the DNA forming DNA adducts. These adducts may induce errors in the DNA sequence during replication or repair. Identification and quantitation of DNA adducts is essential in determining the genotoxic potential of a carcinogen and the extent of exposure to it. It is also important to determine what, if any, site specificity the carcinogen may possess. All of these imply a considerable analytical effort. DNA adducts are generally labile compounds, occurring *in vivo* with extremely low frequencies (1–10 adducts per 10^9 unmodified bases). Since the amount of DNA that might be available at the start of analysis could be as little as 50 µg, care must be taken in sample preparation and treatment, and very sensitive techniques are required. The techniques of ^{32}P-postlabeling, immunoassays, and fluorescence spectroscopy are some currently used procedures. A comprehensive summary of problems and techniques relating to DNA adduct analysis, with special emphasis on LC-ESI-MS, has been recently published by Singh and Farmer.[30] Even before the advent of the state-of-the-art atmospheric pressure ionization techniques [such as ESI, APCI, and atmospheric pressure photoionization (APPI)], mass spectrometry with use of traditional techniques (EI, CI), electron-capture chemical ionization (EC-CI) and soft ionization techniques (FAB, field desorption, or thermospray), were employed for the structural elucidation of DNA adducts.[31] It was with the introduction of techniques such as ESI that the analysis of underivatized labile DNA adducts became reliable.[32] Banoub et al. have reviewed selected examples of DNA adduct analyses.[33]

It is important to point out that the elucidation of the structure of DNA adducts is as such, a complex problem that requires MS to be integrated with other analytical techniques such as NMR, chemical derivatization, and even the synthesis of candidate structures. However, for the specific and sensitive quantitation of DNA adducts, LC-MS/MS with SRM has emerged as a valid alternative to more traditional techniques such as ^{32}P-postlabeling.[30]

The MS analyses of DNA adducts are commonly performed, after enzymatic hydrolysis of the DNA, on the corresponding modified deoxynucleosides. DNA adducts giving 2′-deoxyinosine-related nucleosides have been reported in a few cases, compared with modified deoxynucleosides derived from DNA bases. The deoxyinosine derivative being isolated and identified after enzymatic hydrolysis of DNA treated with nitrous acid (1 M, pH 4.2, 25°C for 24 h) was *N*2-(2-deoxyinosyl) deoxyguanosine (**5**, Chart 2.2).[34] Identification was performed by UV at pH 2.5, 7, and 13, proton

NMR, reduction with sodium dithionite, and MS of the TMS and *N,O*-permethylated derivatives. The main fragmentation pathway involved two consecutive losses of the sugars with hydrogen transfer to the cross-linked base. Exact mass measurements were performed to confirm the found structure through its molecular formula. The proposed mechanism for the formation of this cross-linked nucleoside occurred through the diazotization of a guanine residue, followed by attack at this position by the exocyclic amino group of a neighboring guanine on the opposite strand. Fifteen years later, Kirchner et al. gave further evidence of the covalent nature of this dimeric adduct. They also reported the ESI-MS of the underivatized substance, showing the in-source adducts of the molecule with proton (*m/z* 517), sodium (*m/z* 540), and potassium (*m/z* 556).[35]

Solomon and Segal alkylated calf thymus DNA, 2′-deoxyadenosine, 2′-deoxycytidine, 2′-deoxyguanosine, 2′-deoyinosine, and thymidine with acrylonitrile (pH 7, 37°C, 10–40 days).[36] From the original DNA, they isolated the 1-cyanoethyl adducts with guanine and thymine, and the 1-carboxyethyl adducts of adenine and cytosine. These adducts were formed by the hydrolysis of the

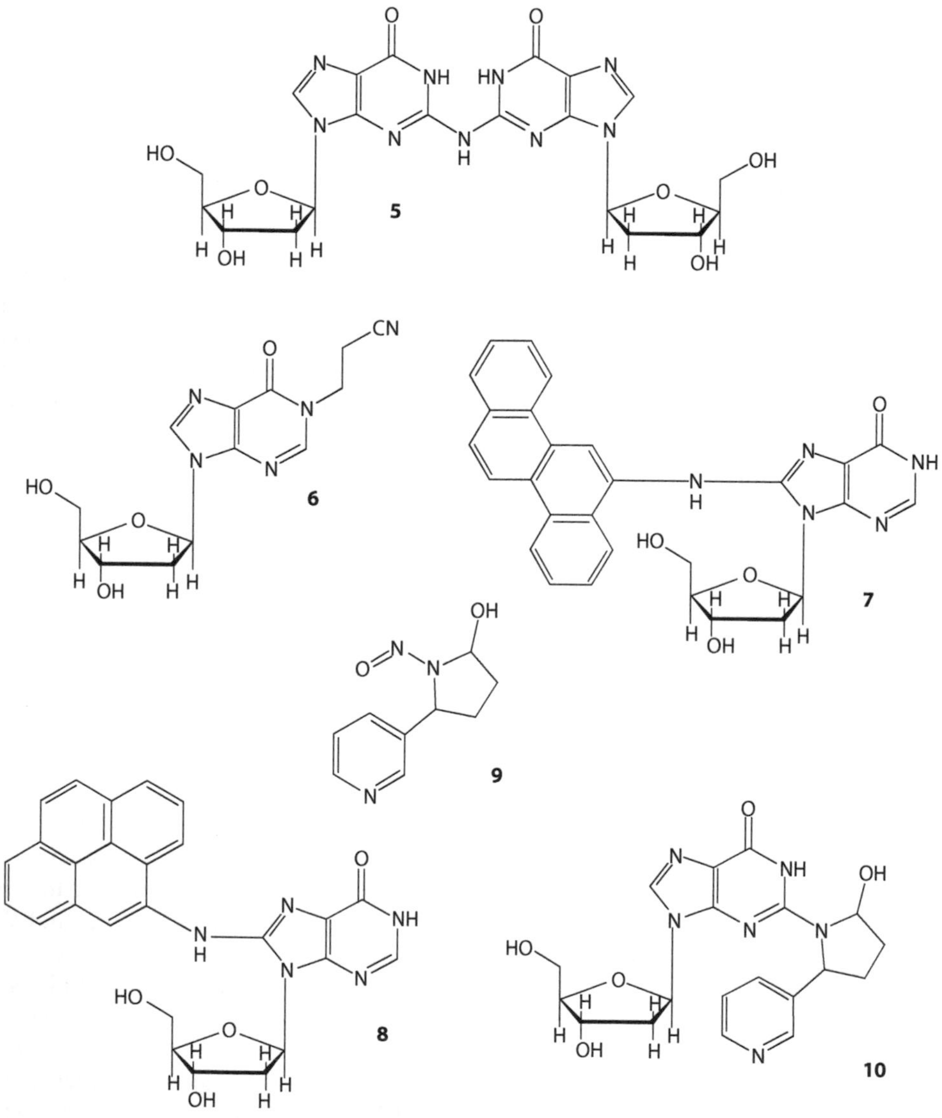

CHART 2.2 Structural formulae of compounds 5 to 10 (details in the text).

primitive cyanoethyl adduct, catalyzed intramolecularly by the adjacent exocyclic nitrogen, which forms a six-membered cyclic intermediate with the carbon of the CN group. There was no evidence that any 2′-deoxyinosine derivatives were isolated from the treated DNA; however, when acrylo- nitrile was reacted with the deoxynucleoside, a stable 1-cyanoethyl adduct (**6**, Chart 2.2) was formed. This was evidenced by the desorption CI-MS, showing a low-abundance protonated molecular ion at m/z 306, and fragment ions at m/z 190 (protonated base), 137 (m/z 190 minus acrilonitrile), and 117 (sugar oxonium ion).

Delclos et al. studied the reaction of DNA with *N*-hydroxy-6-aminochrysene, a metabolic activation product of 6-aminochrysene.[37] They were able to isolate and identify, by proton NMR and MS, *N*-(deoxyinosine-8-yl)-6-aminochrysene (**7**, Chart 2.2), 5-(deoxyguanosine-*N*2-yl)-6- aminochrysene, and *N*-(deoxyguanosine-8-yl)-6-aminochrysene from the hydolysate. The deoxy- inosinyl derivative was considered to be a product of the spontaneous oxidation of the corresponding deoxyadenosine adducts, which occurred during sample preparation. In the DNA isolated from rat hepatocytes and treated with radiolabeled 6-amino- and 6-nitrochrysene, the authors found that the main adducts were with 8-deoxyinosyl and 8-deoxyguanosyl.

Chae et al. investigated the reaction between DNA and 4-nitropyrene, another polycyclic aro- matic compound.[38] The reduction of the nitro group that was catalyzed by xanthine oxidase and analyzed by HPLC with either UV or radioactive detection gave three peaks putatively corre- sponding to the DNA adducts. The first peak was composed of two poorly separated substances, which decomposed to give the second peak, assigned to 4-nitropyrene-9,10-dione. Due to their instability, attempts to isolate the two adducts for MS analysis were unsuccessful. The third peak was analyzed by positive and negative ESI-MS, and it was characterized as a 2′-deoxyinosine adduct (Mr 467 Da) (putative structure: **8**, Chart 2.2). The same HPLC peaks were found in DNA hydrolysates from both rat liver and mammary glands after treatment with 4-nitropyrene.

Upadhyaya et al. reported the identification of the adducts of 5′-hydroxy-*N*′-nitrosonornicotine (5′-hydroxyNNN: **9**, Chart 2.2) with 2′-deoxyguanosine and DNA.[39] *N*′-NNN is a component of tobacco that has a metabolite, 5′-hydroxyNNN, which can form an electrophilic species that can react with DNA.

A stable precursor, 5′-acetoxyNNN, of the active species was reacted with 2′-deoxyguanosine or DNA in the presence of esterase (pH 7, 37°C, 1 h). The authors, using a strategy involving mul- tiple approaches, identified two adducted isomeric species, 2[2-hydroxy-5-(3-pyridyl)pyrrolidin-1 -yl]deoxyinosine (**10**, Chart 2.2) and *N*2[5-(3-pyridyl)tetrahydrofuran-2-yl]deoxyguanosine (structure not shown).

They detected the expected species using LC-MS/MS with SRM and monitoring the transition m/z 415 → 299 for the loss of deoxyribose from the protonated quasimolecular ion, in low abun- dance from the complex reaction mixture.

To confirm the identified structures, they synthesized them and compared those HPLC retention times and the peaks detected by SRM with the experimental data. They also characterized the syn- thetic standards by UV, 1H-NMR, ESI MS/MS, and exact mass determination and determined the stereochemistry of the pyridyl-substituted carbons by making the appropriate chemical transforma- tions and comparing the HPLC retention times with those of synthetic standards with defined ste- reochemistry. According to the authors, the sensitivity and specificity of the LC-MS/MS were of critical importance for the detection and identification of the adducts.

REFERENCES

1. Dizdaroglu, M., Characterization of free radical-induced damage to DNA by the combined use of enzy- mic hydrolysis and gas chromatography-mass spectrometry, *J. Chromatogr.* 367, 357–366, 1986.
2. Dizdaroglu, M., Free-radical-induced formation of an 8,5′-cyclo-2′-deoxyguanosine, *Biochem. J.* 238, 247–254, 1986.
3. McCloskey, J.A. (ed.), *Methods in Enzymology: Constituents of Nucleic Acids: Overview and Strategy*, Academic Press, San Diego, pp. 771–781, 1990.

4. Dizdaroglu, M., Gas chromatography-mass spectrometry of free radical-induced products of pyrimidines and purines in DNA, in *Methods in Enzymology*, J.A. McCloskey (ed.), Academic Press, San Diego, pp. 842–857, 1990.
5. Esmans, E.L., Luyten, Y., and Alderweireldt, F.C., Liquid chromatography-mass spectrometry of nucleosides using a commercially available direct liquid introduction probe, *Biomed. Mass Spectrom.* 10, 347–351, 1983.
6. Chace, D.H., Mass spectrometry in the clinical laboratory, *Chem. Rev.* 101, 445–478, 2001.
7. Frycák, P., Huskova, R., Adam, T., and Lemr, K., Atmospheric pressure ionization mass spectrometry of purine and pyrimidine markers of inherited metabolic disorders, *J. Mass Spectrom.* 37, 1242–1248, 2002.
8. Kammerer, B., Frickenschmidt, A., Mueller, C.E., Laufer, S., Gleiter, C.H., and Liebich, H., Mass spectrometric identification of modified urinary nucleosides used as potential biomedical markers by LC-ITMS coupling, *Anal. Bioanal. Chem.* 382, 1017–1026, 2005.
9. Ito, T., Van Kuilenburg, A.B.P., Bootsma, A.H., Haasnoot, A.J., Van Cruchten, A., Wada, Y., and Van Gennip, A.H., Rapid screening of high-risk patients for disorders of purine and pyrimidine metabolism using HPLC-electrospray tandem mass spectrometry of liquid urine or urine-soaked filter paper strips, *Clin. Chem.* 46, 445–452, 2000.
10. Lu, W., Kimball, E., and Rabinowitz, J.D., A high-performance liquid chromatography-tandem mass spectrometry method for quantitation of nitrogen-containing intracellular metabolites, *J. Am. Soc. Mass Spectrom.* 17, 37–50, 2006.
11. Dong, M., Wang, C., Deen, W.M., and Dedon, P.C., Absence of 2′-deoxyoxanosine and presence of abasic sites in DNA exposed to nitric oxide at controlled physiological concentrations, *Chem. Res. Toxicol.* 16, 1044–1055, 2003.
12. Dong, M. and Dedon, P.C., Relatively small increases in the steady-state levels of nucleobase deamination products in DNA from human TK6 cells exposed to toxic levels of nitric oxide, *Chem. Res. Toxicol.* 19, 50–57, 2006.
13. Chheda, G.B., Patrzyc, H.B., Bhargava, A.K., Crain, P.F., Sethi, S.K., McCloskey, J.A., and Dutta, S.P., Isolation and characterization of a novel nucleoside from the urines of chronic myelogenous leukaemia patients, *Nucleosides Nucleotides* 6, 597–611, 1987.
14. Schram, K.H., Urinary nucleosides, *Mass Spectrom. Rev.* 17, 131–251, 1998.
15. Reimer, M.L.J., McClure, T.D., and Schram, K.H., Differentiation of isomeric 2′-, 3′- and 5′-deoxynucleotides by electron ionization and chemical ionization-liked scanning mass spectrometry, *Biomed. Environ. Mass Spectrom.* 18, 533–542, 1989.
16. Wataya, Y. and Hiraoka, O., 3′-Deoxyinosine as an anti-leishmanial agent: The metabolism and cytotoxic effects of 3′-deoxyinosine in *Leishmania tropica promastigotes*, *Biochem. Biophys. Res. Commun.* 123, 677–683, 1984.
17. Lee, M.S. (ed.), *LC/MS Applications in Drug Development*, Wiley, New York, 2002.
18. Rossi, D.T. and Sinz, M.W. (eds), Mass *Spectrometry in Drug Discovery*, Marcel Dekker, New York, 2002.
19. Reddy, D. and Iden, C.R., Analysis of modified deoxynucleosides by electrospray ionization mass spectrometry, *Nucleosides Nucleotides* 12, 815–826, 1993.
20. Bruins, A.P., ESI source design and dynamic range consideration, in *Electrospray Ionization Mass Spectrometry. Fundamentals, Instrumentation and Applications*, R.B. Cole (ed.), John Wiley and Sons, New York, pp. 107–136, 1997.
21. Font, E., Lasanta, S., Rosario, O., and Rodriguez, J.F., Analysis of antiretroviral nucleosides by electrospray ionization mass spectrometry and collision induced dissociation, *Nucleosides Nucleotides* 17, 845–853, 1998.
22. Hua, Y., Wainhaus, S.B., Yang, Y., Shen, L., Xiong, Y., Xu, X., Zhang, F., Bolton, J.L., and Van Breemen, R.B., Comparison of negative and positive ion electrospray tandem mass spectrometry for the liquid chromatography-electrospray tandem mass spectrometry analysis of oxidized deoxynucleotides, *J. Am. Soc. Mass Spectrom.* 12, 80–87, 2001.
23. Volosov, A., Alexander, C., Ting, L., and Soldin, S.J., Simple rapid method for quantification of antiretrovirals by liquid chromatography-tandem mass spectrometry, *Clin. Biochem.* 35, 99–103, 2002.
24. Estrela, R.C.E., Salvadori, M.C., Raices, R.S.L., and Suarez-Kurtz, G., Determination of didanosine in human serum by on-line solid-phase extraction coupled to high-performance liquid chromatography with electrospray ionization tandem mass spectrometric detection: Application to a bioequivalence study, *J. Mass Spectrom.* 38, 378–385, 2003.
25. Clark, T.N, White, C.A., and Bartlett, M.G., Determination of didanosine in maternal plasma, amniotic fluid, fetal and placental tissues by high-performance liquid chromatography-tandem mass spectrometry, *Biomed. Chromatogr.* 20, 605–611, 2006.

26. Huang, Y., Zurlinden, E., Lin, E., Li, X., Tokumoto, J., Golden, J., Murr, A., Engstrom, J., and Conte, J., Liquid chromatographic-tandem mass spectrometric assay for the simultaneous determination of didanosine and stavudine in human plasma, bronchoalveolar lavage fluid, alveolar cells, peripheral blood mononuclear cells, seminal plasma, cerebrospinal fluid and tonsil tissue, *J. Chromatogr. B* 799, 51–61, 2004.
27. Compain, S., Schlemmer, D., Levi, M., Pruvost, A., Goujard, C., Grassi, J., and Benech, H., Development and validation of a liquid chromatographic/tandem mass spectrometric assay for the quantitation of nucleoside HIV reverse transcriptase inhibitors in biological matrices, *J. Mass Spectrom.* 40, 9–18, 2005.
28. Cahours, X., Tran, T.T., Mesplet, N., Kieda, C., Morin, P., and Agrofoglio, L.A., Analysis of intracellular didanosine triphosphate at sub-ppb level using LC-MS/MS, *J. Pharm. Biomed. Anal.* 26, 819–827, 2001.
29. Cahours, X., Dessans, H., Morin, P., Dreux, M., and Agrofoglio, L., Determination at ppb level of an anti-human immunodeficiency virus nucleoside drug by capillary electrophoresis-electrospray ionization tandem mass spectrometry, *J. Chromatogr. A* 895, 101–109, 2000.
30. Singh, R. and Farmer, P.B., Liquid chromatography-electrospray ionization-mass spectrometry: The future of DNA adduct detection, *Carcinogenesis* 27, 178–196, 2006.
31. Esmans, E.L., Broes, D., Hoes, I., Lemiere, F., and Vanhoutte, K., Liquid chromatography-mass spectrometry in nucleoside, nucleotide and modified nucleotide characterization, *J. Chromatogr. A* 794, 109–127, 1998.
32. Crain, P.F., ESI MS of nucleic acids and their constituents, in *Electrospray Ionization Mass Spectrometry. Fundamentals, Instrumentation and Applications*, R.B. Cole (ed.), John Wiley and Sons, New York, pp. 421–458, 1997.
33. Banoub, J.H., Newton, R.P., Esmans, E., Ewing, D., and Mackenzie, G., Recent developments in mass spectrometry for the characterization of nucleosides, nucleotides, oligonucleotides, and nucleic acids, *Chem. Rev.* 105, 1869–1915, 2005.
34. Shapiro, R., Dubelman, S.F., Aaron, M., Crain, P.F., and McCloskey, J.A., Isolation and identification of cross-linked nucleosides from nitrous acid treated deoxyribonucleic acid, *J. Am. Chem. Soc.* 99, 302–303, 1977.
35. Kirchner, J.J., Sigurdsson, S.T., and Hopkins, P.B., Interstrand cross-linking of duplex DNA by nitrous acid: Covalent structure of the dG-to-dG cross-link at the sequence 5'-CG, *J. Am. Chem. Soc.* 114, 4021–4027, 1992.
36. Solomon, J.J. and Segal, A., Direct alkylation of calf thymus DNA by acrylonitrile. Isolation of cyanoethyl adducts of guanine and thymine and carboxyethyl adducts of adenine and cytosine, *Environm. Health Perspect.* 62, 227–230, 1985.
37. Delclos, K.B., Miller, D.W., Lay, J.O., Jr., Casciano, D.A., Walker, R.P., Fu, P.P., and Kadlubar, F.F., Identification of C8-modified deoxyinosine and N2- and C8-modified deoxyguanosine as major products of the *in vitro* reaction of *N*-hydroxy-6-aminochrysene with DNA and the formation of these adducts in isolated rat hepatocytes treated with 6-nitrochrysene and 6-aminochrysene, *Carcinogenesis* 8, 1703–1709, 1987.
38. Chae, Y.-H., Ji, B.-Y., Lin, J.-M., Fu, P.P., Cho, B.P., and El-Bayoumi, K., Nitroreduction of 4-nitropyrene is primarily responsible for DNA adduct formation in the mammary gland of female CD rats, *Chem. Res. Toxicol.* 12, 180–186, 1999.
39. Upadhyaya, P., McIntee, E.J., Villalta, P.W., and Hecht, S.S., Identification of adducts formed in the reaction of 5'-acetoxy-*N*'-nitrosonornicotine with deoxyguanosine and DNA, *Chem. Res. Toxicol.* 19, 426–435, 2006.

3 Tandem Mass Spectrometry of Nucleic Acids

Jiong Yang and Kristina Håkansson

CONTENTS

3.1   INTRODUCTION

Tandem mass spectrometry, also termed MS/MS, refers to any mass spectrometric method that involves at least two stages of mass analysis.[1] The most generally implemented scheme involves the following: (1) mass selection of a precursor (parent) ion in the first stage, (2) energy deposition into this selected precursor ion to induce fragmentation via cleavage of covalent chemical bonds, and (3) mass spectrometric detection of the resulting product (fragment) ions. This fundamental operating principle can be achieved in two conceptually different implementations, as illustrated in Figure 3.1: tandem in space MS/MS utilizing at least two consecutive mass analyzers or tandem in time MS/MS involving a series of mass analyses in a single mass analyzer.

Several techniques, including traditional enzymatic sequencing methods, are available for structural characterization of nucleic acids. For example, nuclear magnetic resonance spectroscopy

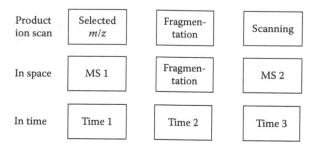

FIGURE 3.1 Comparison of a product ion scan by space-based and time-based tandem mass spectrometers. (Adapted from de Hoffmann, E. and Stroobant, V., *Mass Spectrometry Principles and Applications*, 2nd ed., Wiley, Chichester, 2002. With permission.)

is widely used for RNA structural analysis[2,3] but suffers from limited sensitivity. Another powerful approach is chemical labeling followed by fluorescence resonance energy transfer (FRET) measurements,[4] which can provide *single-molecule* sensitivity.[5] However, structural information from FRET is limited to distance constraints for the added fluorophores and does not provide a detailed picture. MS/MS is a well-established technique for characterizing oligonucleotides and PCR products,[6–12] including analysis of genetic markers, such as short tandem repeats and single nucleotide polymorphisms.[13] MS/MS fragmentation patterns provide oligonucleotide sequence information as well as information on the presence and location of chemical modifications, including modified nucleobases and alterations of sugar rings and the phosphate backbone. Chemically modified oligonucleotides often cannot be characterized by classical enzymatic techniques. MS/MS may also be used to probe gas-phase structures and folding of nucleic acids.[14–16]

A variety of MS/MS fragmentation strategies are available for nucleic acid structural characterization, with collision-activated dissociation (CAD)[17] being the most common. This chapter will provide a brief introduction to MS/MS instrumentation followed by a discussion of the pros and cons of the various fragmentation techniques. For further information, the reader is referred to excellent review articles by Wu and McLuckey,[12] Hofstadler et al.,[8] and Frahm and Muddiman.[18]

3.2 IONIZATION METHODS FOR MS/MS

The first step in any mass spectrometric analysis is the volatilization and ionization of the analyte of interest. For biomacromolecules such as nucleic acids, the two superior techniques for accomplishing this task are electrospray ionization (ESI)[19] and matrix-assisted laser desorption/ionization (MALDI).[20,21] The use of either technique is implicit in the following text.

3.2.1 ELECTROSPRAY IONIZATION

ESI is regarded as a very "soft" ionization method[19] and has been shown to successfully ionize oligonucleotides as large as 8000 base pairs, corresponding to a molecular mass of 5 MDa.[22] In ESI, the solution containing the analyte is pushed through a narrow inner diameter (~50 μm) capillary held at an electric potential of 3–6 kV with respect to the entrance of the mass spectrometer (located ~3–20 mm away), corresponding to an electric field of ~10^6 V/m. Analytes should preferably exist as ions in a solution, either in protonated/metal-adducted cationic forms, or as deprotonated anions. Depending on the polarity of the electric field, either cations or anions will accumulate at the tip of the capillary, resulting in increased Coulomb repulsion and electrostatic nebulization (electrospray) of the liquid. In order to assist nebulization, an organic solvent is usually added to the solution to decrease the surface tension compared to that of pure water. Furthermore, nebulizing gas (i.e., nitrogen) can also be added (the so-called ion spray or pneumatically assisted electrospray[23]). The resulting mist contains multiply charged droplets from which the solvent rapidly evaporates. Heating of the mass spectrometer inlet or addition of hot drying gas is often used to assist solvent evaporation, which results in an increase in droplet charge density, causing disruption of droplets into smaller offspring droplets due to Coulomb repulsion overcoming surface tension. According to the charge residue mechanism,[24] this process occurs in multiple steps until the analyte is free of solvent, thereby forming free gas-phase ions. In the ion evaporation mechanism,[25] the surface electric field at a certain droplet size is high enough to cause field desorption (i.e., ejection) of analyte ions from the droplets. ESI has a major advantage over MALDI in that it forms multiply charged ions from large molecules, thereby facilitating tandem mass spectrometric analysis through more efficient energy deposition (more highly charged ions are accelerated to higher kinetic energies at the same electric potential), and because precursor ions are less stable due to intramolecular Coulomb repulsion.

3.2.2 MATRIX-ASSISTED LASER DESORPTION/IONIZATION

MALDI is also a highly successful ionization method for mass spectrometric analysis of large molecules with advantages of its own, such as higher tolerance to salts and detergents in samples, and high sensitivity: the detection limit of MALDI coupled with time-of-flight (TOF) MS can be as low as 250 attomol.[26] The first MALDI mass spectra of small oligonucleotides were reported in 1990.[27,28] In MALDI, analyte molecules are mixed with an excess of the matrix, which usually consists of organic acids (e.g., 3,5-dimethoxy-4-hydroxycinnamic acid, α-cyano-4-hydroxycinnamic acid, or 2,5-dihydroxybenzoic acid) in an organic solvent (normally acetonitrile or ethanol) often containing some trifluoroacetic acid. This solution is spotted onto a MALDI plate and the solvent is allowed to evaporate, leaving analyte/matrix crystals. The MALDI spot is irradiated by an intense (1–10 μJ) pulse from either an ultraviolet (UV) or an infrared (IR) laser. In UV-MALDI, the matrix absorbs most of the laser energy, resulting in ejection of both matrix and analyte species into the gas phase. The precise ionization mechanism is not well understood, but is thought to involve gas-phase proton-transfer reactions.[29] In contrast to ESI, MALDI generally produces singly charged ions, but multiply charged ions ($[M + nH]^{n+}$) can also be observed for large species.

ESI and MALDI are complementary ionization techniques due to their very different mechanisms (volatilization/ionization from a liquid versus the solid state). For nucleic acids, the negative ion mode is typically used in ESI due to the high acidity of the sugar–phosphate backbone, which promotes deprotonation in solution. Both positive and negative ion modes are used in MALDI, in which ionization is believed to occur in the gas phase following laser desorption.

3.3 MS/MS INSTRUMENTATION

A variety of different instruments are commercially available for both tandem in space and tandem in time MS. A brief overview of the most popular variants is given in the following two sections.

3.3.1 TANDEM IN SPACE MASS SPECTROMETERS

As described above, tandem in space MS/MS requires at least two consecutive mass analyzers. These analyzers can be of the same type, or they can be of different types to form a so-called hybrid mass spectrometer. The classical tandem in space mass spectrometer is the triple quadrupole, which uses a first quadrupole (Q) as a mass filter to select the mass-to-charge (m/z) ratio of a precursor ion of interest, a second quadrupole as a collision cell to induce fragmentation of the chosen precursor ion and to confine the generated product ions, and a third quadrupole in a scanning fashion to detect the generated product ions. The triple quadrupole tandem mass spectrometer was first described by Yost and Enke.[30] A schematic diagram of their original instrument (with an electron ionization source) is shown in Figure 3.2. In many modern triple quadrupole instruments, the second quadrupole

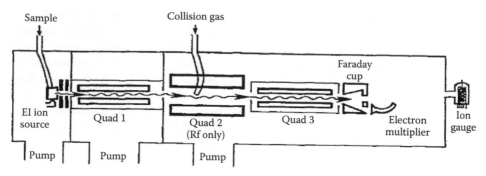

FIGURE 3.2 Schematic diagram of the original triple quadrupole (Q-Q-Q) tandem in space mass spectrometer developed by Yost and Enke.[30] (Reproduced from Yost, R.A. et al., *Int. J. Mass Spectrom. Ion Processes* 30, 127, 1979. With permission.)

(collision cell) has been replaced with a higher-order multipole, such as a hexapole or octopole, which provides improved confinement of product ions for better detection efficiency.[31]

Different kinds of mass analyzers have different performance characteristics. Advantages of the quadrupole include relatively high scan speed and large dynamic range, features highly suitable for on-line chromatographic coupling and quantitative experiments. However, the mass accuracy and resolution of quadrupole mass analyzers are moderate, and they also have limited m/z range (typically <4000 m/z). Thus, they are not suitable for coupling with MALDI, which produces mostly singly charged biomacromolecular ions of high m/z values.

The TOF mass analyzer[32,33] has the advantages of higher speed, mass accuracy, and resolution compared with the quadrupole. In addition, it has virtually unlimited m/z range. The latter feature makes it an ideal mass analyzer to interface with MALDI and the overall performance has rendered it a popular substitution for the third quadrupole in a triple quadrupole instrument to form a Q-Q-TOF configuration. Two commercial instruments with this general configuration are available: the Q-TOF™ from Waters Corporation and the QSTAR™ from Applied Biosystems. A schematic diagram of the original Q-TOF instrument is shown in Figure 3.3. Here, a hexapole is used as a collision cell and another hexapole with the main purpose of ion focusing precedes the first quadrupole. MS/MS in a Q-Q-TOF-type instrument is analogous to MS/MS in a Q-Q-Q-type instrument with the main difference being in how the product ions are detected.

Another commercially available hybrid tandem in space mass spectrometer is Bruker's Apex Ultra™, which uses a quadrupole for mass selection, a hexapole collision cell for fragmentation, and a Fourier transform ion cyclotron resonance (FT-ICR) mass analyzer[34,35] for product ion detection. Advantages of the FT-ICR mass analyzer include ultrahigh resolution and mass accuracy. FT-ICR mass analyzers can also be used for tandem in time MS/MS as described below.

For MS/MS of MALDI-generated ions, the tandem TOF or the TOF/TOF mass spectrometer is another choice. This type of instrument uses an electrostatic ion gate[36] to perform precursor ion selection in the first TOF analyzer. Because ions of different m/z ratios have different velocities when traveling through the TOF flight tube, they will reach the ion gate at different instants in time and ions can be m/z selected by opening this gate for a very brief period. The first TOF vacuum tube is followed by a collision cell for precursor ion fragmentation, and product ions are detected with the second TOF analyzer. Commercially available TOF/TOF instruments are currently manufactured by Applied Biosystems and by Bruker Daltonics.

MS/MS of MALDI-generated ions can also be achieved with a single TOF analyzer via the so-called postsource decay (PSD).[37] This implementation takes advantage of the metastable nature of MALDI-generated ions, that is, their probability of dissociating while traveling down the TOF tube due to elevated internal energies resulting from the MALDI process. Similar to MALDI TOF/TOF,

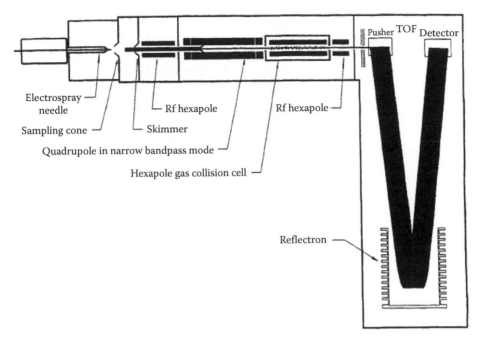

FIGURE 3.3 Schematic diagram of a hybrid quadrupole-time-of-flight (Q-h-TOF) tandem in space mass spectrometer. In this particular instrument, a hexapole (h) rather than a quadrupole serves as a collision cell. An additional hexapole located in front of the quadrupole serves as a focusing device. (Reproduced from Morris, H.R. et al., *Rapid Commun. Mass Spectrom.* 10, 889, 1996. With permission.)

an ion gate is used to perform precursor ion selection. However, a collision cell is redundant due to metastable decay. Product ions formed inside the TOF tube will have the same velocity as the corresponding precursor ion but different kinetic energies due to their lower mass. Thus, they can be separated with an electrostatic ion mirror located at the end of the flight tube. Such ion mirrors, or reflectrons, are frequently present in TOF mass analyzers due to their focusing ability, which improves resolution.[38] The advantage of PSD compared with TOF/TOF MALDI MS/MS is the less advanced instrumentation. However, PSD suffers from low fragmentation efficiency and difficulties in generating product ion spectra of uniform quality.

A final configuration for tandem in space MS/MS is the sector-type instrument, which uses various combinations of electromagnetic and electrostatic sectors to perform precursor ion selection (nonscanning operation) and to detect product ions (scanning operation). Similar to the TOF/TOF instrumentation, a collision cell is located between the two stages of mass analysis. Sector instruments provide high mass accuracy and resolution, but because of their bulky size, cost, and low scan speed, they are being replaced by other mass analyzers to a large extent.

3.3.2 TANDEM IN TIME MASS SPECTROMETERS

Tandem in time mass spectrometers are ion-trapping devices in which ions can be stored for extended periods of time. There are three types of ion-trap mass analyzers: the quadrupole ion trap, the FT-ICR mass analyzer, and the more recently developed orbitrap mass analyzer.[39] Only the first two former ones can currently be used for tandem in time MS/MS. For such experiments, ions are first injected into the device and allowed to assume stable ion trajectories (i.e., orbiting without hitting the walls of the trap). For MS/MS, an excitation waveform is first applied to a pair of electrodes "capping" the trap. This waveform selectively excites ions of all *m/z* ratios except the one of the ion

FIGURE 3.4 Geometries of linear (top) and hyperbolic (bottom) tandem in time quadrupole ion traps. (Reproduced from Hager, J.W., *Anal. Bioanal. Chem.* 378, 845, 2004. With permission.)

of interest. Consequently, all undesired ions assume larger trajectories and are either ejected from the trap or neutralized at the trap walls and, as a result, only the desired precursor ions remain. Secondly, selected precursor ions are fragmented through any of the fragmentation strategies described in Section 3.3. Thirdly, the ion trap is voltage scanned (for the quadrupole ion-trap mass analyzer), or subjected to a swift frequency sweep (an FT-ICR mass analyzer), to mass-selectively detect generated product ions. Thus, all events occur in the same space but at different time points, as described in Figure 3.1.

Quadrupole ion traps of two main designs are currently on the market: either a linear device with two endcaps such as the LTQ™ mass spectrometer from Thermo Fisher or a hyperbolically shaped design, such as the HCT™ from Bruker Daltonics. Both geometries are depicted in Figure 3.4. Similar to the quadrupole mass analyzer, quadrupole ion traps have moderate mass accuracy and resolution. Their main advantage compared with Q-Q-Q- or Q-Q-TOF-type instruments is the ability to perform MS^n analyses in which one specific first-generation product ion is further isolated and fragmented followed by detection of second-generation product ions and so on. Improved product ion detection performance can be accomplished by interfacing a quadrupole ion trap with a higher performing mass analyzer such as a TOF (e.g., Hitachi's NanoFrontier™ linear ion-trap-TOF liquid chromatograph/mass spectrometer), an FT-ICR (the LTQ-FT™ from Thermo Fisher), or an orbitrap (the LTQ-orbitrap™ from Thermo Fisher).

3.3.3 Nucleic Acid Product Ion Nomenclature

When subjecting an oligonucleotide ion to gas-phase fragmentation, eight types of fragments can be generated by cleaving the four possible sites of a phosphodiester bond. The most commonly used nomenclature for these product ions was introduced by McLuckey et al.[40] and is depicted in Scheme 3.1. a_n, b_n, c_n, and d_n are the symbols for fragments containing the 5′ end of the oligonucleotide and w_n, x_n, y_n, and z_n are the symbols for fragments containing the 3′ end. The subscript, n, denotes the number of nucleotide residues in that particular product ion.

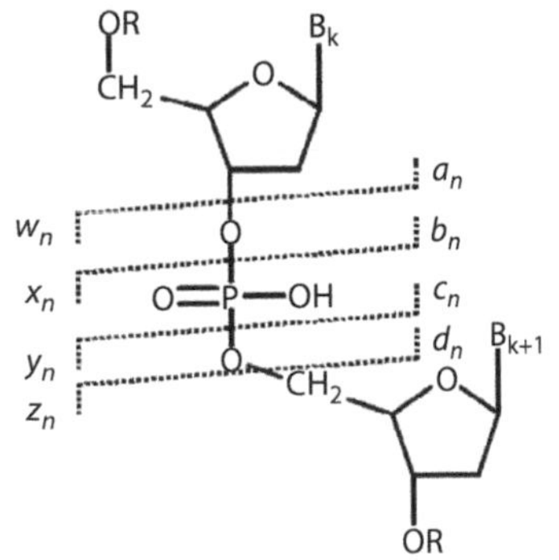

SCHEME 3.1 Nomenclature for oligonucleotide product ions. (Adapted from McLuckey, S.A., Van Berkel, G.J., and Glish, G.L., *J. Am. Soc. Mass Spectrom.* 3, 60, 1992.)

3.4 CAD OF NUCLEIC ACIDS

CAD is the most widely used tandem mass spectrometric fragmentation technique. There are two CAD regimes: high-energy collisions and low-energy collisions. The high-energy collision regime, which is achievable in TOF/TOF or sector-type instruments, is believed to involve electronic excitation.[41] In such instruments, precursor ions are accelerated through several kV, and helium is the most commonly used target gas. Low-energy collisions (accelerating voltage of <100 V), which mostly involve vibrational excitation of precursor ions,[42] are often divided into "beam-type CAD" and "ion-trap-type CAD" where the latter type involves lower collision energies. In both cases, heavier target gases such as argon and xenon can be used to transfer more energy into precursor ions.[43]

Beam-type CAD is achievable in, for example, Q-Q-Q- and Q-Q-TOF-type configurations. Here, the precursor ion kinetic energy is determined by the potential difference between the ion source and the collision cell. The fraction of this energy that can be imparted into precursor ions through inelastic collisions is determined by[17]

$$E = \frac{N}{m_p + N}E_k, \tag{3.1}$$

in which N is the atomic weight of the target gas, m_p is the molecular weight of precursor ion, and E_k is the precursor ion kinetic energy. In quadrupole ion-trap-type CAD, the achievable kinetic energy is limited due to the limited dimensions of the trap. Here, precursor ion kinetic energy is increased by applying a radiofrequency (rf) waveform in resonance with the axial frequency of selected precursor ions stored in the trap. The amplitude of this waveform has to be selected such that precursor ions are still confined within the trap. However, because the magnitude of this amplitude determines the so-called low mass cutoff, that is, the lowest m/z ratio that can undergo stable motion inside the ion trap, it will affect the m/z range of product ions that can be detected. This unavoidable phenomenon is often referred to as the "1/3 rule" because, practically, product ions with m/z ratios less than 1/3 of the precursor ion m/z cannot be detected in quadrupole ion-trap-type CAD MS/MS.

3.4.1 CAD OF DNA

McLuckey et al. were the first to apply CAD to characterize negatively charged oligodeoxynucleotide ions in a quadrupole ion-trap instrument.[40] Figure 3.5 shows a typical CAD spectrum of a DNA oligomer. In such experiments, neutral nucleobase (B) loss is a major fragmentation pathway, but sequence-specific backbone cleavage, mainly in terms of $(a-B)$ and w ions, is also observed.[40,44] Several groups have performed extensive research to explore oligodeoxynucleotide fragmentation pathways in CAD and several fragmentation mechanisms have been proposed. McLuckey and

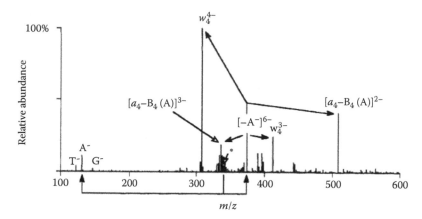

FIGURE 3.5 CAD MS/MS spectrum of the $[M-7H]^{7-}$ ion in a quadrupole ion-trap mass spectrometer, where $M = d(TGCATCGT)$. Abundant w- and $(a-B)$-type product ions are observed. The asterisk indicates the m/z location of the precursor ion. (Reproduced from McLuckey, S.A. and Habibi-Goudarzi, S., *J. Am. Chem. Soc.* 115, 12085, 1993. With permission.)

Habibi-Goudarzi proposed a two-step fragmentation scheme (nucleobase loss followed by backbone cleavage) via 1,2-elimination involving hydrogens from the sugar.[45] These authors suggested that a nucleobase can be lost either as a neutral or as an anion and found the following order of preference for loss of anionic base: $A^- > T^- > G^- > C^-$.[45] Similar conclusions were reached by Rodgers et al.[46] Barry et al. proposed a base-catalyzed two-step internal elimination scheme in which the 2′- proton of the sugar is attacked by a negatively charged 3′-phosphate oxygen and a base anion is lost.[47] Gross and Wan proposed that intramolecular proton transfer occurs from an adjoining 5′-phosphate to a nucleobase as shown in Scheme 3.2, thereby forming a zwitterionic intermediate that dissociates via loss of a neutral base followed by backbone cleavage.[48] The major proton source was verified by the CAD of DNA with a methylphosphonate backbone and hydrogen/deuterium exchange experiments.[48,49] This mechanism explains the reduced cleavage typically seen on the 3′ side of thymidine residues, which have low proton affinity.[50,51]

A major disadvantage of nucleic acid beam-type CAD is secondary fragmentation, such as water and additional nucleobase loss, which complicates spectral interpretation and reduces sensitivity. Enhanced sensitivity has been reported by Muddiman and Hannis through the incorporation of a 7-deaza purine analog, eliminating extensive depurination.[52] Additional and complementary information can be obtained through fragmentation of radical anions, as demonstrated by McLuckey et al. through ion–ion reactions[53] and by Hvelplund and coworkers through high-energy collisions with noble gas atoms.[54] Particularly, reduced nucleobase loss is seen.

3.4.2 CAD OF RNA

Oligoribonucleotides have been much less characterized with MS/MS than DNA oligonucleotides and their fragmentation pathways are not as well understood. Kirpekar and Krogh reported that low-energy CAD of oligoribonucleotide cations results primarily in abundant c-type ions and their complementary y-type ions as the major sequence ions.[55] Similar results were obtained by Schurch et al. for anionic RNA.[56] An example is shown in Figure 3.6. Thus, the CAD fragmentation patterns of DNA and RNA are drastically different in that a different backbone covalent bond is preferentially cleaved. This difference points to a role of the 2′ sugar position in the fragmentation mechanism. Experiments by Tromp and Schurch have shown that the nucleobases are unlikely to play a key role in RNA CAD fragmentation.[57] More recent work by Kirpekar and Krogh utilized hydrogen/deuterium exchange and high-energy CAD in a MALDI TOF/TOF mass spectrometer to further address the

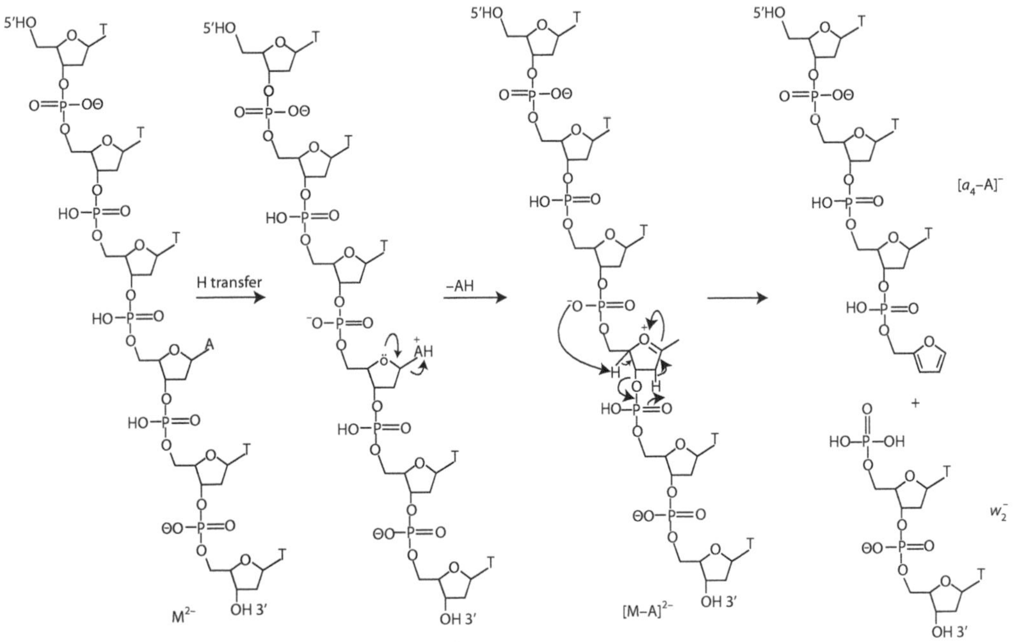

SCHEME 3.2 Mechanism for the formation of $(a_4{-}B)^-$ and the complementary w_2 ion from the doubly deprotonated hexadeoxynucleotide d(TTTATT). (Reproduced from Wan, K.X. and Gross, M.L., *J. Am. Soc. Mass Spectrom.* 12, 580, 2001. With permission.)

different fragmentation behavior of DNA and RNA.[58] These authors suggested that, in this energy regime, a common fragmentation mechanism exists for DNA and RNA, but that an RNA-specific reaction, requiring a 2'-hydroxyl group, also occurs that leads to generation of *c*- and *y*-type ions.

3.4.3 CAD OF CHEMICALLY MODIFIED OLIGONUCLEOTIDES

Chemically modified oligonucleotides play an important role in biomedical/pharmaceutical research, particularly for antisense applications. Many traditional drugs combat disease by targeting faulty proteins, but antisense oligonucleotides intervene at an earlier stage by preventing the production of these incorrect proteins. Antisense oligonucleotides are typically 13–25 nucleotides long and are designed to hybridize to messenger RNA by Watson–Crick base pairing.[59] Antisense compounds are widely used as therapeutics for malignant disease, and there is a rapid increase in the number of antisense molecules progressing past Phase I, II, and III clinical trials.[60,61] Many antisense compounds are modified oligonucleotides, such as analogs with unnatural nucleobases, modified sugars (particularly at the sugar 2' position), or altered phosphate backbones.[62] Traditional enzymatic sequencing techniques are not suitable for characterizing antisense compounds due to their altered chemical nature, and alternative analytical approaches are therefore desired.

MS/MS involving CAD of oligonucleotide anions has been shown to allow characterization of chemically modified oligomer DNA and RNA.[12,48,57,63] For example, Tromp and Schurch examined oligoribonucleotides with different 2' substituents, including 2'-methoxy-ribose and 2'-fluoro-ribose. These modifications resulted in the generation of all different types of backbone product ions in comparable abundance, contrary to unmodified RNA, which shows a strong preference for *c*/*y*-type ion formation.[57] However, because the major fragmentation channel in CAD of DNA involves neutral nucleobase loss,[48] this fragmentation technique may not be ideal for characterizing oligonucleotides with modified bases.

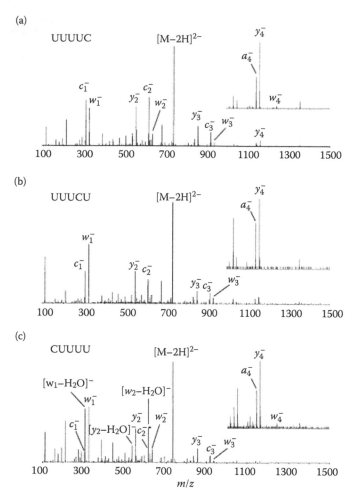

FIGURE 3.6 CAD MS/MS spectra of the RNA pentanucleotides UUUUC (a), UUUCU (b), and CUUUU (c) obtained by the dissociation of the $[M-2H]^{2-}$ precursor ions with a collision energy of 30 eV. The main sequence-defining fragment ions are the c, y, and w series. (Reproduced from Schurch, S., Bernal-Mendez, E., and Leumann, C.J., *J. Am. Soc. Mass Spectrom.* 13, 936, 2002. With permission.)

3.5 PHOTODISSOCIATION OF NUCLEIC ACIDS

3.5.1 INFRARED MULTIPHOTON DISSOCIATION

Infrared multiphoton dissociation (IRMPD)[64,65] is, like low-energy CAD, a vibrational excitation fragmentation method. IRMPD and CAD oligonucleotide fragmentation therefore likely proceed through a similar mechanism. The most common IRMPD implementation involves a continuous wave CO_2 laser at 10.6 μm. IRMPD is commonly used in FT-ICR mass analyzers because it eliminates the need to introduce a collision gas, thereby eliminating the pumping downtime required to restore ultrahigh vacuum conditions required for ultrahigh performance. IRMPD has also been implemented in quadrupole ion traps[66] where one major advantage is circumvention of the so-called 1/3 rule (see description in Section 3.3) because an rf waveform is not required. Instead, precursor ion internal energy is raised through multiple absorptions of IR photons. In general, IRMPD fragmentation patterns are similar to those from CAD. However, for nucleic acids, more efficient fragmentation can be seen due to the strong absorption at 10.6 μm (the typical wavelength used) by backbone phosphate groups.[67,68] Brodbelt and Keller have compared CAD and IRMPD of deprotonated and protonated oligonucleotides in a quadrupole ion trap (Figure 3.7) and found that IRMPD can minimize often uninformative [M–B] ions (in which "M" denotes the precursor ion and "B" denotes a nucleobase), which dominate in CAD spectra.[69] This phenomenon can be rationalized

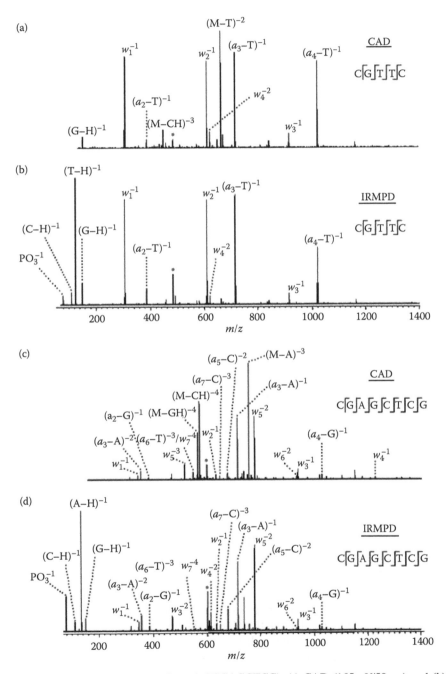

FIGURE 3.7   MS/MS data for d(CGTTC) and d(CGAGCTCG). (a) CAD (165 mV/50 ms) and (b) IRMPD (34 W/50 ms) of the 5-mer, −3 charge state. (c) CAD (215 mV/50 ms) and (d) IRMPD (27 W/50 ms) of the 8-mer, −4 charge state. Precursor ions are marked with asterisks. (Reproduced from Keller, K.M. and Brodbelt, J.S., *Anal. Biochem.* 326, 200, 2004. With permission.)

from the mechanism described in Scheme 3.2: the first step following activation is nucleobase loss and some further activation is needed to generate backbone cleavage. Further activation of first-generation product ions easily occurs in the IR laser beam (because product ions are generated "in beam") whereas minimum further activation occurs in CAD because m/z ratios (and therefore axial frequencies) of product ions are different from that of the precursor ions, and product ions are thereby unaffected by the applied rf waveform. Secondly, free phosphate and nucleobase ions are observed from IRMPD (an effect of the circumvention of the 1/3 rule), which can aid the identification of modified bases.[69]

McLafferty and coworkers have demonstrated IRMPD for the characterization of large biomolecules, including oligonucleotides with up to 108 residues.[64,70] Complete sequencing of a 50-mer was achieved.[70] Sannes-Lowery and Hofstadler implemented IRMPD in the external ion reservoir of an FT-ICR mass spectrometer to characterize modified oligonucleotides, resulting in increased sequence coverage and product ion yields compared with in-cell IRMPD because ions are exposed to a range of laser irradiation times, and metastable ions are stabilized by the high gas pressure in this region.[71] This approach was found to yield sequence-specific (a–B) and w-type ions that provide rapid and accurate sequencing of modified oligonucleotides containing phophorothioate backbones and 2′-methoxyethyl ribose.[71] However, as described above for CAD of DNA, reduced backbone cleavage on the 3′-side of thymidine residues was observed, thereby reducing sequence coverage.

3.5.2   UV PHOTODISSOCIATION

In addition to the relatively frequently utilized IR photons, a few examples involving significantly more energetic UV photons for gas-phase dissociation of nucleic acids have been reported. McLafferty and coworkers employed 193 nm photons from an argon-fluoride excimer laser to dissociate larger multiply charged biomolecules, including peptides, proteins, and DNA.[72] The main reaction observed in UV photodissociation (UVPD) of all-T 30-mer anionic DNA appeared to involve electron ejection, although low-abundance a- or z-type ions were also seen (these ions cannot be differentiated based on mass due to the symmetry of dT_{30}). In more recent work, Gabelica et al. have explored the gas-phase fragmentation behavior of single- and double-stranded anionic DNA at ~260 nm.[14] These authors found that electron photodetachment is the major reaction when guanine is present and that the photodetachment efficiency directly depends on the number of guanines. CAD of the resulting radical anions provide fragmentation patterns similar to electron detachment dissociation (EDD) (described below) and, similar to EDD, electron photodetachment dissociation (EPD) appears useful for probing noncovalent intra- and intermolecular interactions in nucleic acids. In the follow-up work, the sequence dependence of the photodetachment yield was determined for hexamer oligodeoxynucleotides and was found to follow the trend $dG_6 > dA_6 > dC_6 > dT_6$, which is inversely correlated with the base ionization potentials (G < A < C < T).[73]

3.6   GAS-PHASE ION–ELECTRON REACTIONS OF NUCLEIC ACIDS

3.6.1   ELECTRON-CAPTURE DISSOCIATION

Electron-capture dissociation (ECD) involves radical ion chemistry and has been shown to provide unique fragmentation patterns for a variety of biomolecules, including peptides and proteins,[74] peptide nucleic acids,[75] polymers,[76] lantibiotics,[77] and a siderophore.[78] ECD requires multiply charged cationic analytes, which are irradiated with low-energy (typically <0.5 eV) electrons to form a charge-reduced radical intermediate via electron capture. This radical intermediate undergoes facile dissociation, which in all cases investigated so far yields complementary structural information compared with MS/MS of even-electron ions (e.g., CAD or IRMPD). Scheme 3.3 shows the general ECD fragmentation pathway. ECD was originally implemented in an FT-ICR mass analyzer, which has the advantage of being able to simultaneously store free electrons and large biomolecular ions,

$$[M+nH]^{n+} + e^{-}_{(<1\,eV)} \rightarrow [M+nH]^{(n-1)+\bullet} \rightarrow \text{fragments}$$

SCHEME 3.3 ECD fragmentation route.

a criterion that facilitates the ECD reaction. However, more recent implementations in quadrupole ion traps have been demonstrated in which electron confinement is aided by a weak magnetic field (to partially overcome the low mass cutoff value, which is typically much higher than the electron mass).[79,80] ECD has also been reported in a digital ion-trap configuration.[81]

A particularly interesting feature of ECD is that it has been shown to yield extremely "soft" fragmentation for peptides and proteins in that backbone bonds can be cleaved without losing labile post-translational modifications, thereby allowing localization of modifications.[82,83] Also, backbone covalent bonds can be ruptured without breaking the noncovalent interactions of a protein's higher-order structure.[84] The latter feature has been exploited in the investigation of protein gas-phase folding and unfolding.[85,86] Hakansson et al. showed that oligodeoxynucleotides also display different fragmentation behavior in ECD compared with other MS/MS techniques.[87] However, in those experiments, the fragmentation was nucleobase specific and rather limited. More recent results with an improved electron injection system[88] and the possibility to mass-selectively accumulate precursor ions (improving sensitivity) demonstrate more extensive fragmentation, resulting in complete sequencing of small (5- to 7-mer) oligonucleotides,[89] as shown in Figure 3.8 for dA_6. The same group also found that the ECD fragmentation patterns of oligoribonucleotides, including A_6, C_6, and CGGGGC are nucleobase dependent, suggesting that backbone cleavage proceeds following electron capture at the nucleobases.[90] Overall, only limited backbone cleavage was observed in ECD of RNA. The main caveat for analyzing oligonucleotides by ECD is the requirement of positively charged precursor ions.

3.6.2 ELECTRON DETACHMENT DISSOCIATION

EDD, introduced by Zubarev et al. in 2001, involves higher-energy ($\geq$10 eV) electrons than ECD, and has been shown to provide unique radical-driven fragmentation for negatively charged peptide ions.[91–93] Electron irradiation results in electron detachment from precursor ions to form a charge-reduced radical intermediate that undergoes dissociation. It should be noted that this intermediate

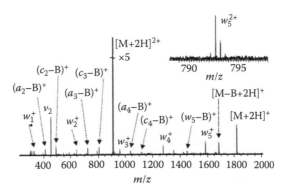

FIGURE 3.8 ECD MS/MS of dA_6 employing mass-selective external ion accumulation and an indirectly heated dispenser cathode as electron source. Rich fragmentation is observed, including a complete w/d ion series (only the w label is given), and several (a/z–B) and (c/x–B) ions (only a and c labels are given). The inset shows one out of three doubly charged product ions, possibly due to a zwitterionic precursor ion structure. (Reproduced from Schultz, K.N. and Hakansson, K., *Int. J. Mass Spectrom.* 234, 123, 2004. With permission.)

$$[M-nH]^{n-} + e^-_{fast(>10\ eV)} \rightarrow [M-nH]^{(n-)-\bullet} + 2e^- \rightarrow fragments$$

SCHEME 3.4   EDD fragmentation route.

is different from that in ECD, which contains an extra electron rather than a hole. Scheme 3.4 shows the EDD fragmentation pathway. Similar to ECD, labile modifications, including sulfate and phosphate groups, are retained in EDD,[91,93,94] allowing, for example, localization of post-translational modifications in acidic peptides.

EDD has so far been demonstrated in FT-ICR and ion-trap mass analyzers. For example, Yang et al. implemented EDD on a commercial FT-ICR mass spectrometer utilizing two different configurations: a heated filament electron source and an indirectly heated hollow dispenser cathode electron source.[16] Filament EDD was demonstrated for the first time although the dispenser cathode configuration provided higher EDD efficiency and richer fragmentation patterns for hexamer oligodeoxynucleotides. Indirectly heated dispenser cathodes provide higher flux electron beams with lower energy spread[95] and the hollow configuration allows simultaneous access to an IR laser for IRMPD.[96] Filament and dispenser cathode EDD spectra of the oligonucleotide dA_6 are shown in Figure 3.9. Similar to ECD of the same analyte (Figure 3.8), even-electron d/w ion series dominate the spectra but numerous a/z (both even-electron and radical species), $(a/z-B)$, c/x, $(c/x-B)$, and $(d/w-B)$ ions as well as minimal nucleobase losses are also observed. Comparison of EDD data with previous radical oligonucleotide anion dissociation experiments reveals similarities as well as differences. Electron transfer from CCl_3^+ to doubly deprotonated dA_3 followed by CAD of the resulting radical $[dA_3-2H]^{-\bullet}$ in an ion trap resulted mainly in d/w ions, as in EDD.[53] In addition, one a/z ion was observed although as an even-electron species rather than the radical a/z ion observed in EDD. Transfer of a hydrogen atom should be facilitated at the longer activation period during ion-trap CAD compared with EDD. In stark contrast to EDD, the ion-trap experiments resulted in sugar cross-ring cleavage. Similar cross-ring cleavage products were also observed in 100 keV collisions of $[dA_5-2H]^{2-}$ with He gas, which results in abundant charge-reduced radical species from electron loss.[54]

The application of EDD for RNA characterization has been shown by Yang and Håkansson.[90] EDD provided complete sequence coverage for the RNAs A_6, C_6, G_6, U_6, CGGGGC, and GCAUAC. The EDD fragmentation patterns were different from those observed with CAD and IRMPD in that the dominant product ions corresponded to d- and w-type ions rather than c- and y-type ions. The minimum differences between oligoribonucleotides suggest that EDD proceeds following direct electron detachment from the phosphate backbone.

Yang et al. also applied EDD to fragment a triply deprotonated hexamer DNA duplex.[16] Filament EDD resulted solely in charge reduction to form a radical species. The absence of any product ions is similar to ECD of large proteins[84] and some post-translationally modified peptides,[88,97] and can be explained through retention of intramolecular noncovalent interactions, preventing product ions from separating. These data constituted the first piece of evidence that EDD may allow characterization of nucleic acid gas-phase structure when combined with IRMPD or CAD. Of particular interest are the characterization of RNA structure and folding, and the characterization of nucleic acid-containing complexes. In the follow-up work, Mo and Håkansson applied a range of different MS/MS techniques, including EDD, IRMPD, activated ion (AI) EDD, and EDD/IRMPD MS^3, in an FT-ICR mass spectrometer to the characterization of three isomeric 15-mer oligodeoxynucleotides with different sequences and predicted solution-phase structures.[15] The goal of this work was to explore whether the structural differences between these oligodeoxynucleotides could be directly probed with MS/MS. It was found that all three 15-mers had higher-order structures in the gas phase, although preferred structures were only predicted for two of them in solution. Nevertheless, EDD, AI EDD, and EDD/IRMPD MS^3 experiments yielded different cleavage patterns with less backbone fragmentation for the more stable solution-phase structure as compared with the other two

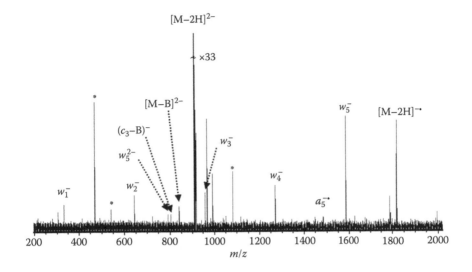

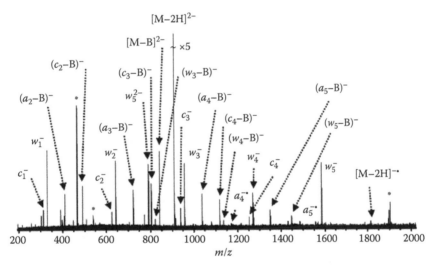

FIGURE 3.9 EDD MS/MS of dA_6 with a heated filament electron source (upper spectrum) and an indirectly heated hollow dispenser cathode electron source (lower spectrum). A complete d/w (only one label is given) ion series is observed, allowing full sequencing of the oligonucleotide. More extensive fragmentation is observed with the dispenser cathode configuration compared to the filament, resulting in increased fragmentation efficiency and higher sensitivity. Electronic noise spikes are labeled with asterisks. (Reproduced from Yang, J. et al., *Anal. Chem.* 77, 1876, 2005. With permission.)

15-mers. By contrast, no major differences were observed in IRMPD although the extent of backbone cleavage was higher with that technique for all three 15-mers. Thus, experiments utilizing the radical ion chemistry of EDD can provide complementary structural information compared with traditional slow heating methods, such as IRMPD, for structured nucleic acids. Backbone (including w, d, and $(a-B)$ ions and few a radical ions) and nucleobase cleavages (normalized to charge) observed following EDD/IRMPD MS^3 of the three investigated 15-mers are shown in Figure 3.10.

In order to apply EDD to the sequencing and structural characterization of unknown nucleic acids, rigorous workflows have to be developed. A sequencing strategy may include the following: (1) determination of the accurate molecular weight of the DNA/RNA to obtain an approximate

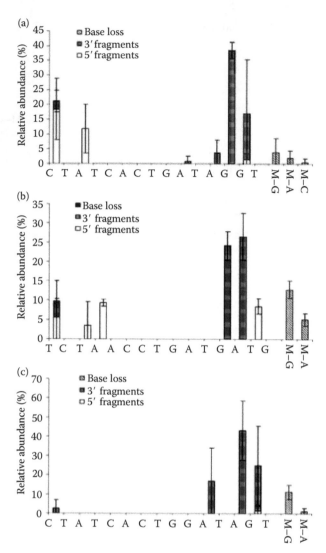

FIGURE 3.10 Backbone (including w, d, and (a–B) ions and few a radical ions) and nucleobase cleavages (normalized to charge) observed following EDD/IRMPD MS^3 of 15mer-1 (a), 15mer-2 (b), and 15mer-3 (c) where 15mer-1 and 3 have predicted hairpin structures with 15mer-3 being more stable. 15mer-2 does not have a predicted solution-phase structure. (Reproduced from Mo, J. and Hakansson, K., *Anal. Bioanal. Chem.* 386, 675, 2006. With permission.)

residue number; (2) applying EDD to fragment the DNA/RNA; (3) grouping of product ions into w and d ion series (other series, i.e., a, b, c, and x, y, z, have fixed mass differences with respect to the w and d series, e.g., 79.9663 for w and y ions); and (4) utilization of w and d ion series to determine the sequence, including possible modifications. For automation purposes, this procedure should be developed in collaboration with bioinformaticists.

The potential of EDD to probe nucleic acid gas-phase structures is valuable for elucidating RNA structure and folding in the absence of water without the need for large sample quantities. For example, metal ions such as Mg^{2+} are important for the structure and function of both DNA and RNA: Mg^{2+} stabilizes tRNA tertiary structure[98] and mediates interactions between DNA and drug molecules.[99] EDD may allow further insights into metal stoichiometry and specificity of metal-binding sites. Furthermore, for ligands, for example, drug molecules, the soft nature of EDD may allow direct inference of interaction sites with nucleic acids. The most popular technique for probing nucleic acid gas-phase structure (which may or may not be similar to the solution-phase structure) is gas-phase hydrogen/deuterium exchange.[100–106] This experimental strategy may also benefit from EDD in that hydrogen scrambling (which prevents detailed determination of deuteration sites) may be reduced compared with, for example, CAD and IRMPD. Low degrees of scrambling have been reported for ECD of peptides,[107,108] whereas CAD can cause complete deuterium randomization.[109–115] A remaining challenge with EDD is the low fragmentation efficiency compared with other MS/MS techniques, particularly for larger nucleic acids and for MS^n-type or AI EDD experiments involving a combination with IRMPD.[15] A partial explanation for the low efficiency of the latter experimental categories may be the currently employed experimental geometries, which do not provide maximum overlap between ions, photons, and electrons. Improved geometries with either an off-axis laser beam[88] or a pneumatic probe-guided laser beam[116] have been reported for AI ECD. Another option may be the utilization of quadrupolar axilization[117–121] of ions.

3.7 SUMMARY

MS/MS of ESI- or MALDI-generated ions can provide several levels of information in nucleic acid structural determination, including sequence information, identity and location of chemical modifications, and higher-order structure information. A number of different mass spectrometers are commercially available to perform MS/MS experiments, each with their own pros and cons that need to be carefully considered. For these instruments, several fragmentation methods employing different physical principles are available. Although low-energy CAD, involving vibrational excitation, is the most common manner of invoking gas-phase dissociation, alternative strategies are likely to provide complementary information. For example, IRMPD, which also is a vibrational excitation technique, can yield more efficient fragmentation and provide informative low-mass product ions that can, for example, identify modified nucleobases. Techniques involving electronic excitation and/or radical formation can yield higher sequence coverage because the fragmentation behavior is less affected by the identity of neighboring nucleobases. For example, RNA and DNA yield similar fragmentation patterns in EDD in contrast to the drastically different backbone cleavage patterns seen in CAD and IRMPD. Both EDD and EPD show promise for characterizing nthe ucleic acid higher-order structure.

ACKNOWLEDGMENTS

This work was supported by a NSF Career award (CHE-05–47699), a research award from the American Society for Mass Spectrometry (sponsored by Thermo Electron), a starter grant from the Petroleum Research Fund, and an Elisabeth Caroline Crosby Research Award from the University of Michigan.

REFERENCES

1. McLafferty, F.W., *Tandem Mass Spectrometry*, Wiley, New York, 1983.
2. Fürtig, B., et al., NMR spectroscopy of RNA, *ChemBioChem.* 4, 936, 2003.
3. Al-Hashimi, H.M., Dynamics-based amplification of RNA function and its characterization by using NMR spectroscopy, *ChemBioChem.* 6, 1506, 2005.
4. Selvin, P.R., Fluorescence resonance energy transfer, *Methods Enzymol.* 246, 6264, 1995.
5. Zhuang, X., Single-molecule RNA science, *Annu. Rev. Biophys. Biomol. Struct.* 34, 399, 2005.
6. Banoub, J.H., et al., Recent developments in mass spectrometry for the characterization of nucleosides, nucleotides, oligonucleotides, and nucleic acids, *Chem. Rev.* 105, 1869, 2005.
7. Benson, L.M., Null, A.P., and Muddiman, D.C., Advantages of *Thermococcus kodakaraenis* (KOD) DNA polymerase for PCR-mass spectrometry based analyses, *J. Am. Soc. Mass Spectrom.* 14, 601, 2003.
8. Hofstadler, S.A., Sannes-Lowery, K.A., and Hannis, J.C., Analysis of nucleic acids by FTICR MS, *Mass Spectrom. Rev.* 24, 265, 2005.

9. Muddiman, D.C. and Smith, R.D., Sequencing and characterization of larger oligonucleotides by electrospray ionization Fourier transform ion cyclotron resonance mass spectrometry, *Rev. Anal. Chem.* 17, 1, 1998.

10. Murray, K.K., DNA sequencing by mass spectrometry, *J. Mass Spectrom.* 31, 1203, 1996.

11. Ni, J., et al., Interpretation of oligonucleotide mass spectra for determination of sequence using electrospray ionization and tandem mass spectrometry, *Anal. Chem.* 68, 1989, 1996.

12. Wu, J. and McLuckey, S.A., Gas-phase fragmentation of oligonucleotide ions, *Int. J. Mass Spectrom.* 237, 197, 2004.

13. Null, A.P. and Muddiman, D.C., Perspectives on the use of electrospray ionization Fourier transform ion cyclotron resonance mass spectrometry for short tandem repeat genotyping in the post-genome era, *J. Mass Spectrom.* 36, 589, 2001.

14. Gabelica, V., et al., Electron photodetachment dissociation of DNA polyanions in a quadrupole ion trap mass spectrometer, *Anal. Chem.* 78, 6564, 2006.

15. Mo, J. and Hakansson, K., Characterization of nucleic acid higher order structure by high resolution tandem mass spectrometry, *Anal. Bioanal. Chem.* 386, 675, 2006.

16. Yang, J., et al., Characterization of oligodeoxynucleotides by electron detachment dissociation Fourier transform ion cyclotron resonance mass spectrometry, *Anal. Chem.* 77, 1876, 2005.

17. McLuckey, S.A., Principles of collisional activation in analytical mass spectrometry, *J. Am. Soc. Mass Spectrom.* 3, 599, 1992.

18. Frahm, J.L. and Muddiman, D.C., Nucleic acid analysis by Fourier transform ion cyclotron resonance mass spectrometry at the beginning of the twenty-first century, *Curr. Pharm. Design* 11, 2593, 2005.

19. Fenn, J.B., et al., Electrospray ionization for mass spectrometry of large biomolecules, *Science* 246, 64, 1989.

20. Tanaka, K., et al., Protein and polymer analyses up to m/z 100000 by laser ionization time-of-flight mass spectrometry, *Rapid Commun. Mass Spectrom.* 2, 151, 1988.

21. Karas, M. and Hillenkamp, F., Laser desorption ionization of proteins with molecular masses exceeding 10000 daltons, *Anal. Chem.* 60, 2299, 1988.

22. Schultz, J.C., Hack, A.C., and Benner, W.H., Mass determination of megadalton-DNA electrospray ions using charge detection mass spectrometry, *J. Am. Soc. Mass Spectrom.* 9, 305, 1998.

23. Bruins, A.P., Covey, T.R., and Henion, J.D., Ion spray interface for combined liquid chromatography/atmospheric pressure ionization mass spectrometry, *Anal. Chem.* 59, 2642, 1987.

24. Dole, M., et al., Molecular beams of macroions, *J. Chem. Phys.* 49, 2240, 1968.

25. Iribarne, J.V. and Thomson, B.A., On the evaporation of small ions from charged droplets, *J. Chem. Phys.* 64, 2287, 1976.

26. Zhang, Z.Y., et al., 3-Hydroxycoumarin as a new matrix for matrix-assisted laser desorption/ionization time-of-flight mass spectrometry of DNA, *J. Am. Soc. Mass Spectrom.* 17, 1665, 2006.

27. Börnsen, K.O., Schär, M., and Widmer, H.M., Matrix-assisted laser desorption and ionization mass spectrometry and its applications in chemistry, *Chimia* 44, 412, 1990.

28. Spengler, B., et al., Molecular weight determination of underivatized oligodeoxyribonucleotides by positive-ion matrix-assisted ultraviolet laser-desorption mass spectrometry, *Rapid Commun. Mass Spectrom.* 4, 99, 1990.

29. Karas, M. and Kruger, R., Ion formation in MALDI: The cluster ionization mechanism, *Chem. Rev.* 103, 427, 2003.

30. Yost, R.A. and Enke, C.G., Selected ion fragmentation with a tandem quadrupole mass spectrometer, *J. Am. Chem. Soc.* 100, 2274, 1978.

31. Tolmachev, A.V., Udseth, H.R., and Smith, R.D., Radial stratification of ions as a function of mass to charge ratio in collisional cooling radio frequency multipoles used as ion guides or ion traps, *Rapid Commun. Mass Spectrom.* 14, 1907, 2000.

32. Stephens, W.E., A pulse mass spectrometer with time dispersion, *Phys. Rev.* 69, 691, 1946.

33. Wiley, W.C. and McLaren, I.H., Time-of-flight mass spectrometry with improved resolution, *Rev. Sci. Instr.* 26, 1150, 1955.

34. Comisarow, M.B. and Marshall, A.G., Fourier transform ion cyclotron resonance spectroscopy, *Chem. Phys. Lett.* 25, 282, 1974.

35. Comisarow, M.B. and Marshall, A.G., Frequency-sweep Fourier transform Ion cyclotron resonance spectroscopy, *Chem. Phys. Lett.* 26, 489, 1974.

36. Piyadasa, C.K.G., et al., A high resolving power ion selector for post-source decay measurements in a reflecting time-of-flight mass spectrometer, *Rapid Commun. Mass Spectrom.* 12, 1655, 1998.

37. Spengler, B., Post-source decay analysis in matrix-assisted laser desorption/ionization mass spectrometry of biomolecules, *J. Mass Spectrom.* 32, 1019, 1997.
38. Mamyrin, B.A., et al., The mass-reflectron, a new nonmagnetic time-of-flight mass spectrometer with high resolution, *Sov. Phys. JETP* 37, 45, 1973.
39. Makarov, A., et al., Performance evaluation of a hybrid linear ion trap/orbitrap mass spectrometer, *Anal. Chem.* 78, 2113, 2006.
40. McLuckey, S.A., Van Berkel, G.J., and Glish, G.L., Tandem mass spectrometry of small, multiply-charged oligonucleotides, *J. Am. Soc. Mass Spectrom.* 3, 60, 1992.
41. Yamaoka, H., Dong, P., and Durup, J., Energetics of the collision-induced dissociations $C_2H_2^+ \rightarrow C_2H^+ + H$ and $C_2H_2^+ \rightarrow H^+ + C_2H$, *J. Chem. Phys.* 51, 3465, 1969.
42. Schwartz, R.N., Slawsky, Z.I., and Herzfeld, K.F., Calculation of vibrational relaxation times in gases, *J. Chem. Phys.* 20, 1591, 1952.
43. Vachet, R.W. and Glish, G.L., Effects of heavy gases on the tandem mass spectra of peptide ions in the quadrupole ion trap, *J. Am. Soc. Mass Spectrom.* 7, 1194, 1996.
44. Wang, Z., et al., Structure and fragmentation mechanisms of isomeric T-rich oligodeoxynucleotides: A comparison of four tandem mass spectrometric methods, *J. Am. Soc. Mass Spectrom.* 9, 683, 1998.
45. McLuckey, S.A. and Habibi-Goudarzi, S., Decompositions of multiply-charged oligonucleotide anions, *J. Am. Chem. Soc.* 115, 12085, 1993.
46. Rodgers, M.T., et al., Low-energy collision-induced dissociation of deprotonated dinucleotides: Determination of the energetically favored dissociation pathways and the relative acidities of the nucleic acid bases, *Int. J. Mass Spectrom. Ion Processes* 137, 121, 1994.
47. Barry, J.P., et al., Mass and sequence verification of modified oligonucleotides using electrospray tandem mass-spectrometry, *J. Mass. Spectrom.* 30, 993, 1995.
48. Wan, K.X. and Gross, M.L., Fragmentation mechanisms of oligodeoxynucleotides: Effects of replacing phosphates with methylphosphonates and thymines with other bases in T-rich sequences, *J. Am. Soc. Mass Spectrom.* 12, 580, 2001.
49. Wan, K.X., et al., Fragmentation mechanisms of oligodeoxynucleotides studied by H/D exchange and electrospray ionization tandem mass spectrometry, *J. Am. Soc. Mass Spectrom.* 12, 193, 2001.
50. Green-Church, K.B. and Limbach, P.A., Mononucleotide gas-phase proton affinities as determined by the kinetic method, *J. Am. Soc. Mass Spectrom.* 11, 24, 2000.
51. Podolyan, Y., Gorb, L., and Leszczynski, J., Protonation of nucleic acid bases. A comprehensive post-Hartree–Fock study of the energetics and proton affinities, *J. Phys. Chem. A* 104, 7346, 2000.
52. Hannis, J.C. and Muddiman, D.C., Tailoring the gas-phase dissociation and determining the relative energy of activation for dissociation of 7-deaza purine modified oligonucleotides containing a repeating motif, *Int. J. Mass Spectrom.* 219, 139, 2002.
53. McLuckey, S.A., Stephenson, J.L., and O'Hair, R.A.J., Decompositions of odd- and even-electron anions derived from deoxypolyadenylates, *J. Am. Soc. Mass Spectrom.* 8, 148, 1997.
54. Liu, B., et al., Electron loss and dissociation in high energy collisions between multiply charged oligonucleotide anions and noble gases, *Int. J. Mass Spectrom.* 230, 19, 2003.
55. Kirpekar, F. and Krogh, T.N., RNA fragmentation studied in a matrix-assisted laser desorption/ionization tandem quadrupole/orthogonal time-of-flight mass spectrometer, *Rapid Commun. Mass Spectrom.* 15, 8, 2001.
56. Schurch, S., Bernal-Mendez, E., and Leumann, C.J., Electrospray tandem mass spectrometry of mixed-sequence RNA/DNA oligonucleotides, *J. Am. Soc. Mass Spectrom.* 13, 936, 2002.
57. Tromp, J.M. and Schurch, S., Gas-phase dissociation of oligoribonucleotides and their analogs studied by electrospray ionization tandem mass spectrometry, *J. Am. Soc. Mass Spectrom.* 16, 1262, 2005.
58. Andersen, T.E., Kirpekar, F., and Haselmann, K.F., RNA fragmentation in MALDI mass spectrometry studied by H/D-exchange: Mechanisms of general applicability to nucleic acids, *J. Am. Soc. Mass Spectrom.* 17, 1353, 2006.
59. Dias, N. and Stein, C.A., Antisense oligonucleotides: Basic concepts and mechanisms, *Mol. Cancer Ther.* 1, 347, 2002.
60. Tamm, I., Antisense therapy in malignant diseases: Status quo and quo vadis? *Clin. Sci.* 110, 427, 2006.
61. Aboul-Fadl, T., Antisense oligonucleotides: The state of the art, *Curr. Med. Chem.* 12, 2193, 2006.
62. Cook, P.D., *Antisense Drug Technology: Principles, Strategies and Applications*, S.T. Crooke (ed.), Marcel Dekker, New York, 2001.
63. Monn, S.T.M. and Schurch, S., New aspects of the fragmentation mechanisms of unmodified and methylphosphonate-modified oligonucleotides, *J. Am. Soc. Mass Spectrom.* 18, 984, 2007.

64. Little, D.P., et al., Infrared multiphoton dissociation of large multiply-charged Ions for biomolecule sequencing, *Anal. Chem.* 66, 2809, 1994.

65. Woodin, R.L., Bomse, D.S., and Beauchamp, J.L., Multiphoton dissociation of molecules with low power continuous wave infrared laser radiation, *J. Am. Chem. Soc.* 100, 3248, 1978.

66. Stephenson, J.L., et al., Infrared multiple-photon dissociation in the quadrupole ion trap via a multipass optical arrangement, *J. Am. Soc. Mass Spectrom.* 5, 886, 1994.

67. Crowe, M.C. and Brodbelt, J.S., Differentiation of phosphorylated and unphosphorylated peptides by high-performance liquid chromatography-electrospray ionization-infrared multiphoton dissociation in a quadrupole ion trap, *Anal. Chem.* 77, 5726, 2005.

68. Flora, J.W. and Muddiman, D.C., Selective, sensitive, and rapid phosphopeptide identification in enzymatic digests using ESI-FTICR-MS with infrared multiphoton dissociation, *Anal. Chem.* 73, 3305, 2001.

69. Keller, K.M. and Brodbelt, J.S., Collisionally activated dissociation and infrared multiphoton dissociation of oligonucleotides in a quadrupole ion trap, *Anal. Biochem.* 326, 200, 2004.

70. Little, D.P., et al., Sequence information from 42-108-mer DNAs (complete for a 50 mer) by tandem mass spectrometry, *J. Am. Chem. Soc.* 118, 9352, 1996.

71. Sannes-Lowery, K.A. and Hofstadler, S.A., Sequence confirmation of modified oligonucleotides using IRMPD in the external reservoir of an electrospray ionization Fourier transform ion cyclotron mass spectrometer, *J. Am. Soc. Mass Spectrom.* 14, 825, 2003.

72. Guan, Z., et al., 193 nm photodissociation of larger multiply-charged biomolecules, *Int. J. Mass Spectrom. Ion Processes* 157/158, 357, 1996.

73. Gabelica, V., et al., Base-dependent electron photodetachment from negatively charged DNA strands upon 260-nm laser irradiation, *J. Am. Chem. Soc.* 129, 4706, 2007.

74. Zubarev, R.A., Kelleher, N.L., and McLafferty, F.W., Electron capture dissociation of multiply charged protein cations. A nonergodic process, *J. Am. Chem. Soc.* 120, 3265, 1998.

75. Olsen, J.V., et al., Comparison of electron capture dissociation and collisionally activated dissociation of polycations of peptide nucleic acids, *Rapid Commun. Mass Spectrom.* 15, 969, 2001.

76. Cerda, B.A., et al., Sequencing of specific copolymer oligomers by electron-capture-dissociation mass spectrometry, *J. Am. Chem. Soc.* 124, 9287, 2002.

77. Kleinnijenhuis, A.J., et al., Localization of intramolecular monosulfide bridges in lantibiotics determined with electron capture dissociation, *Anal. Chem.* 75, 3219, 2003.

78. Liu, H.C., et al., Collision activated dissociation, infrared multiphoton dissociation, and electron capture dissociation of the *Bacillus anthracis* siderophore petrobactin and its metal ion complexes, *J. Am. Soc. Mass Spectrom.* 18, 842, 2007.

79. Baba, T., et al., Electron capture dissociation in a radio frequency ion trap, *Anal. Chem.* 76, 4263, 2004.

80. Silivra, O.A., et al., Electron capture dissociation of polypeptides in a three-dimensional quadrupole ion trap: Implementation and first results, *J. Am. Soc. Mass Spectrom.* 16, 22, 2005.

81. Ding, L. and Brancia, F.L., Electron capture dissociation in a digital ion trap mass spectrometer, *Anal. Chem.* 78, 1995, 2006.

82. Cooper, H.J., Hakansson, K., and Marshall, A.G., The role of electron capture dissociation in biomolecular analysis, *Mass Spectrom. Rev.* 24, 201, 2005.

83. Zubarev, R.A., Reactions of polypeptide ions with electrons in the gas phase, *Mass Spectrom. Rev.* 22, 57, 2003.

84. Horn, D.M., Ge, Y., and McLafferty, F.W., Activated ion electron capture dissociation for mass spectral sequencing of larger (42 kDa) proteins, *Anal. Chem.* 72, 4778, 2000.

85. Horn, D.M., et al., Kinetic intermediates in the folding of gaseous protein ions characterized by electron capture dissociation mass spectrometry, *J. Am. Chem. Soc.* 123, 9792, 2001.

86. Breuker, K., et al., Detailed unfolding and folding of gaseous ubiquitin ions characterized by electron capture dissociation, *J. Am. Chem. Soc.* 124, 6407, 2002.

87. Hakansson, K., et al., Electron capture dissociation and infrared multiphoton dissociation of oligodeoxynucleotide dications, *J. Am. Soc. Mass Spectrom.* 14, 23, 2003.

88. Hakansson, K., et al., Combined electron capture and infrared multiphoton dissociation for multistage MS/MS in an FT-ICR mass spectrometer, *Anal. Chem.* 75, 3256, 2003.

89. Schultz, K.N. and Hakansson, K., Rapid electron capture dissociation of mass-selectively accumulated oligodeoxynucleotide dications, *Int. J. Mass Spectrom.* 234, 123, 2004.

90. Yang, J. and Hakansson, K., Fragmentation of oligoribonucleotides from gas-phase ion-electron reactions, *J. Am. Soc. Mass Spectrom.* 17, 1369, 2006.

91. Budnik, B.A., Haselmann, K.F., and Zubarev, R.A., Electron detachment dissociation of peptide di-anions: An electron–hole recombination phenomenon, *Chem. Phys. Lett.* 342, 299, 2001.

92. Haselmann, K.F., et al., Electronic excitation gives informative fragmentation of polypeptide cations and anions, *Eur. J. Mass Spectrom.* 8, 117, 2002.

93. Kjeldsen, F., et al., C(alpha)-C backbone fragmentation dominates in electron detachment dissociation of gas-phase polypeptide polyanions, *Chem. Eur. J.* 11, 1803, 2005.

94. Kweon, H.K. and Hakansson, K., Metal oxide-based enrichment combined with gas-phase ion–electron reactions for improved mass spectrometric characterization of protein phosphorylation, *J. Proteome Res.*, 7, 745, 2008.

95. Tsybin, Y.O., et al., Improved low-energy electron injection systems for high rate electron capture dissociation in Fourier transform ion cyclotron resonance mass spectrometry, *Rapid Commun. Mass Spectrom.* 15, 1849, 2001.

96. Tsybin, Y.O., et al., Combined infrared multiphoton dissociation and electron capture dissociation with a hollow electron beam in Fourier transform ion cyclotron resonance mass spectrometry, *Rapid Commun. Mass Spectrom.* 17, 1759, 2003.

97. Chalmers, M.J., et al., Protein kinase A phosphorylation characterized by tandem Fourier transform ion cyclotron resonance mass spectrometry, *Proteomics* 4, 970, 2004.

98. Stein, A. and Crothers, D.M., Conformational changes of transfer RNA. The role of magnesium(II), *Biochemistry* 15, 160, 1976.

99. Fan, J.Y., et al., Self-assembly of a quinobenzoxazine-Mg^{2+} complex on DNA: A new paradigm for the structure of a drug–DNA complex and implications for the structure of the quinolone bacterial gyrase–DNA complex, *J. Med. Chem.* 38, 408, 1995.

100. Chipuk, J.E. and Brodbelt, J.S., Gas-phase hydrogen/deuterium exchange of 5'- and 3'-mononucleotides in a quadrupole ion trap: Exploring the role of conformation and system energy, *J. Am. Soc. Mass Spectrom.* 18, 724, 2007.

101. Crestoni, M.E. and Fornarini, S., Gas-phase hydrogen/deuterium exchange of adenine nucleotides, *J. Mass Spectrom.* 38, 854, 2003.

102. Robinson, J.M., et al., Hydrogen-deuterium exchange of nucleotides in the gas phase, *Anal. Chem.* 70, 3566, 1998.

103. Griffey, R.H., et al., Gas-phase hydrogen-deuterium exchange in phosphorothioate d(GTCAG) and d(TCGAT), *Rapid Commun. Mass Spectrom.* 13, 113, 1999.

104. Hofstadler, S.A., Sannes-Lowery, K.A., and Griffey, R.H., Enhanced gas-phase hydrogen-deuterium exchange of oligonucleotide and protein ions stored in an external multipole ion reservoir, *J. Mass Spectrom.* 35, 62, 2000.

105. Gabelica, V., et al., Fast Gas-Phase Hydrogen/Deuterium Exchange Observed for a DNA G Quadruplex, *Rapid Commun. Mass Spectrom.* 19, 201, 2005.

106. Freitas, M.A. and Marshall, A.G., Gas phase RNA and DNA ions. 2. Conformational dependence on the gas-phase H/D exchange of nucleotide-5'-monophosphates, *J. Am. Soc. Mass Spectrom.* 12, 780, 2001.

107. Kweon, H.K. and Hakansson, K., Site-specific amide hydrogen exchange in melittin probed by electron capture dissociation Fourier transform ion cyclotron resonance mass spectrometry, *Analyst* 131, 275, 2006.

108. Rand, K.D., et al., Electron capture dissociation proceeds with a low degree of intramolecular migration of peptide amide hydrogens, *J. Am. Chem. Soc.* 130, 1341, 2008.

109. Johnson, R.S., Krylov, D., and Walsh, K.A., Proton mobility within electrosprayed peptide ions, *J. Mass Spectrom.* 30, 386, 1995.

110. McLafferty, F.W., et al., Gaseous conformational structures of cytochrome c, *J. Am. Chem. Soc.* 120, 4732, 1998.

111. Buijs, J., et al., Inter- and intra-molecular migration of peptide amide hydrogens during electrospray ionization, *J. Am. Soc. Mass Spectrom.* 12, 410, 2001.

112. Demmers, J.A.A., et al., Factors affecting gas-phase deuterium scrambling in peptide ions and their implications for protein structure determination, *J. Am. Chem. Soc.* 124, 11191, 2002.

113. Hagman, C., et al., Inter-molecular migration during collisional activation monitored by hydrogen/deuterium exchange FT-ICR tandem mass spectrometry, *J. Am. Soc. Mass Spectrom.* 15, 640, 2004.

114. Hoerner, J.K., et al., Is there hydrogen scrambling in the gas phase? Energetic and structural determinants of proton mobility within protein ions, *J. Am. Chem. Soc.* 126, 7709, 2004.

115. Jorgensen, T.J.D., et al., Intramolecular migration of amide hydrogens in protonated peptides upon collisional activation, *J. Am. Chem. Soc.* 127, 2785, 2005.

116. Mihalca, R., et al., Combined infrared multiphoton dissociation and electron capture dissociation using co-linear and overlapping beams in Fourier transform ion cyclotron resonance mass spectrometry, *Rapid Commun. Mass Spectrom.* 20, 1838, 2006.

117. Hendrickson, C.L., Drader, J.J., and Laude, D.A., Jr., Simplified application of quadrupolar excitation in Fourier transform ion cyclotron resonance mass spectrometry, *J. Am. Soc. Mass Spectrom.* 6, 448, 1995.
118. de Hoffmann, E. and Stroobant, V., *Mass Spectrometry Principles and Applications*, 2nd ed., Wiley, Chichester, 2002.
119. Yost, R.A., et al., High efficiency collision-induced dissociation in an RF-only quadrupole, *Int. J. Mass Spectrom. Ion Processes*, 30, 127, 1979.
120. Morris, H.R., et al., High sensitivity collisionally-activated decomposition tandem mass spectrometry on a novel quadrupole/orthogonal-acceleration time-of-flight mass spectrometer, *Rapid Commun. Mass Spectrom.* 10, 889, 1996.
121. Hager, J.W., Recent trends in mass spectrometer development, *Anal. Bioanal. Chem.* 378, 845, 2004.

4 Mass Spectrometric Application in the Study of Cyclic Nucleotides in Biochemical Signal Transduction

Russell P. Newton

CONTENTS

4.1 BACKGROUND

The discovery of a single molecule, adenosine 3′,5′-cyclic monophosphate (cyclic AMP, cAMP), by Nobel Prize winner Sutherland (1971) and his colleagues in 1958 constituted a seminal moment in biochemistry, heralding the birth of many of the current concepts of both biological signaling between cells and metabolic regulation.[1] The original second messenger concept was that hormones and neurotransmitters, the primary biochemical messengers, are recognized by, and bind to, specific cell membrane receptors.[2] This binding indirectly stimulates adenylyl cyclase that is responsible for the synthesis of the second(ary) messenger, cAMP, which is subsequently released into the cell (Figure 4.1). This concept is now extended to include two separate populations of such membrane receptors; R_s, which on binding of the agonist stimulates the cyclase, and R_i, which inhibits it. They each have a characteristic serpentine (or seven-pass) structure that crosses the membrane seven times.[3] G-proteins are sited between the receptors and adenylyl cyclase and are the means of transducing the signal from receptor to cyclase.[4] G-proteins are heterotrimeric GTP-binding proteins that undergo a sequential cycle of events: interaction with the stimulated receptor leads to the binding of GTP, followed by G-protein dissociation into α and βγ subunits; interaction of this released

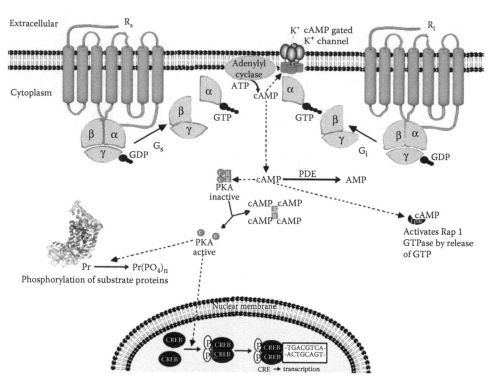

FIGURE 4.1   cAMP second messenger system.

the α subunit with the adenylyl cyclase, affecting the latter's activity, then GTP conversion to GDP, followed by G-protein reassociation into the trimer with consequent loss of interaction with the cyclase.

The nascent intracellular cAMP elicits physiological responses to the primary messenger by stimulating the phosphorylation of a wide variety of substrate proteins through the action of cAMP-dependent protein kinases (PKA). On activation, this PKA dissociates to release a regulatory dimer, which is composed of two subunits each binding two molecules of cAMP, and two free catalytic subunits. These catalytic subunits are active in this free form, but are inactive when associated with the regulatory subunits in the absence of cAMP.[5] The various substrate proteins change conformation on phosphorylation as a result of increased surface charge, with a consequent change in activity. cAMP action is switched off both through its hydrolysis by a number of intracellular cyclic nucleotide phosphodiesterases and by the reversal of the target protein phosphorylation by the action of phosphatases.[6]

In addition to extranuclear effects, migration of the active PKA catalytic units to the nucleus allows phosphorylation of a cAMP-response element-binding (CREB) protein; the nascent phosphorylated CREB dimerizes, and is then able to interact with DNA at a CRE, thereby effecting initiation of specific gene transcriptions.[7] Other direct actions of cAMP, not involving protein kinase, have also now been identified, including a cAMP-gated K+ channel in odorant cells and activation of Rap 1.[8,9] The latter is a GTPase functioning in cell proliferation, differentiation, and morphogenesis, which binds to Epac, an exchange protein directly activated by cAMP. This Epac–Rap 1 interaction facilitates the release of GDP from Rap 1 and its replacement with GTP.[9]

While cAMP appears omnipotent in its effects upon mammalian metabolic pathways, a second naturally occurring cyclic nucleotide, guanosine 3',5'-cyclic monophosphate (cyclic GMP, cGMP) has analogous but more restricted roles, functioning for example, in visual transduction in the eye,

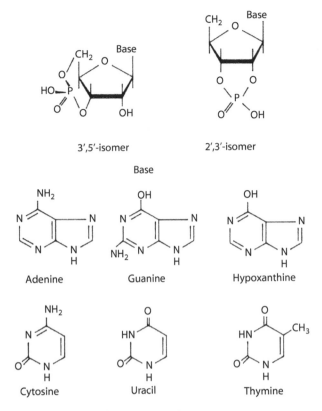

FIGURE 4.2 Structures of 3′,5′- and 2′,3′-cyclic nucleotides.

as a atrial natriuretic peptide mediator, and as a nitric oxide respondent. The cGMP metabolism and role are affected by the range of enzymes and proteins parallel to those of cAMP, that is, agonist receptors, guanylyl cyclases, cGMP phosphodiesterases, cGMP-dependent protein kinases, and cGMP-binding proteins.

Knowledge of these cyclic nucleotide second messenger systems has enabled their successful therapeutic manipulation by numerous pharmacological agents, acting at both enzymic and receptor targets, ranging from β-blockers to Viagra®, with over a thousand currently prescribed drugs eliciting their effects via cyclic nucleotide-mediated processes. Thus, in order to allow their more selective manipulation and thereby optimize benefit to risk more detailed understanding of these cyclic nucleotide systems is a continuing goal.

The characteristic structure of a cyclic nucleotide is a phosphodiester ring linked to a nucleoside; the 3′,5′-isomers are the active second messengers of interest and the 2′,3′-isomers are merely nucleic acid breakdown products of no significance in cyclic nucleotide signal transduction mechanisms (Figure 4.2). As described above, cAMP and cGMP have established functions as second messengers. The analogous biochemical functions of purine and pyrimidine nucleotides (e.g., in coenzyme function, in driving thermodynamically unfavorable reactions, and as components of DNA and RNA) suggest by extrapolation that pyrimidine cyclic nucleotides would have second messenger functions analogous to those of the purines, cAMP, and cGMP. This poses the questions as to whether cyclic nucleotides other than the latter two (Figure 4.2) are naturally occurring, and if so, whether they possess analogous regulatory functions and what their potentials are as further pharmacological targets. Furthermore, because of the emphasis on established and potential pharmacological applications, the focus of cyclic nucleotide research has been predominantly upon mammalian systems. Nevertheless, there is also huge potential in the manipulation of cyclic nucleotide systems in higher plants and microorganisms for agricultural and biotechnological applications, but as yet such nonmammalian systems are much less well elucidated than those of mammalian cells.

It can thus be seen that cyclic nucleotide biochemistry is one of the major areas of research that offers both short- and long-term commercial, medical, and humanitarian benefit. The contributions of mass spectrometry (MS) to these studies can be described as

a. Unequivocal identification of putative cyclic nucleotides in tissue extracts and as products of enzymic reactions.
b. Structural elucidation of naturally occurring cyclic nucleotide analogs.
c. Structural elucidation of synthetic cyclic nucleotide analogs produced for use in radioimmunoassay, cell permeation, or enzyme manipulation.
d. Quantitation of cyclic nucleotides in tissue extracts.
e. Quantitation of cyclic nucleotide-related enzyme activities, that is, nucleotidyl cyclases, cyclic nucleotide phosphodiesterases, and cyclic nucleotide-responsive protein kinases.
f. Identification and monitoring of proteins involved in cyclic nucleotide-regulated pathways.

4.2 INITIAL APPLICATIONS OF MS TO CYCLIC NUCLEOTIDE STUDIES

The elucidation of the structure of cAMP and the recognition of its second messenger role were achieved before MS was widely used in determining the molecular structure of biomolecules. While the first mass spectra of nucleosides and nucleotides were published in the late 1960s and early 1970s, the more "traditional and established" MS were relatively unhelpful in the study of cyclic nucleotides. The first detailed fragmentation study of cyclic nucleotides was carried out by McCloskey's group, using electron impact ionization of the trimethylsilyl (TMS) derivatives of cAMP.[10] We utilized this methodology to provide the first mass spectrum of putative cAMP in extracts from plants, a highly contentious issue at that time.[11–14] The spectrum obtained (Figure 4.3) contains the molecular ion of cAMP (TMS)$_3$ at m/z 545, together with the diagnostic ions at m/z 310 and 378, the latter being unique to the 3′,5′-cyclic nucleotide, thereby allowing its differentiation from the 2′,3′-isomer. However, acquisition of the TMS cyclic nucleotide molecular ions after derivatization of cyclic nucleotides in a complex biological tissue extract is difficult, and it was the advent of "soft" methods of ionization that facilitated the productive and regular use of MS analysis in studies of cyclic nucleotide biochemistry.

4.2.1 FAST-ATOM BOMBARDMENT MS OF CYCLIC NUCLEOTIDES

Fast-Atom Bombardment (FAB) ionization has proved to be an excellent means of generating mass spectra containing a prominent protonated molecule from cyclic nucleotides, without any need for derivatization.[15] FAB mass spectra, both from chemically synthesized standards and partially purified extracts containing putative cyclic nucleotides, had the predominant ions as the protonated cyclic nucleotide and its sodium and glycerol adducts, together with peaks derived from the glycerol matrix (Figure 4.4).[16,17] These peaks were, however, common to both 3′,5′- and 2′,3′-isomers, thus FAB-MS alone did not provide unambiguous identification of the cyclic nucleotide second messengers. Such unambiguous identification was obtained by collisionally induced dissociation (CID) of the protonated molecules, which yield mass-analyzed ion kinetic energy (MIKE) spectra that readily differentiate between the two isomers (Figure 4.4b and c). With both cAMP isomers, the major peak in the CID/MIKES spectrum of the protonated molecule (m/z 330) occurs at m/z 136, corresponding to the protonated base, $[BH_2]^+$, with the 3′,5′-cAMP spectrum containing diagnostic fragments at m/z 164 ($[BH_2+28]^+$) and 178 ($[BH_2+42]^+$) arising from the S_2 and S_1 cleavages (Figure 4.5a). Other significant fragments are seen at m/z 313 and 119, corresponding to the loss of NH_3 from the protonated molecule and the protonated base, respectively; the peak at m/z 238, corresponding to

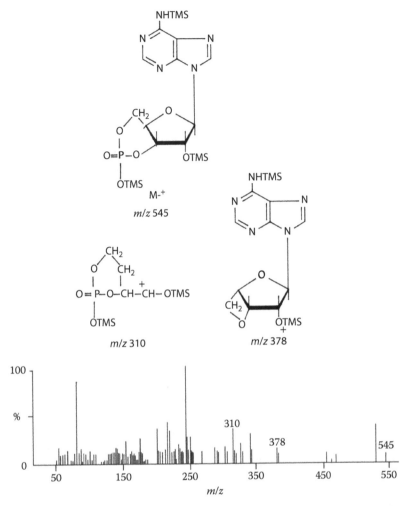

FIGURE 4.3 Electron impact mass spectrum of TMS derivative of 3′,5′-cAMP from plant tissue extract. (Modified from Newton, R.P. et al., *Phytochemistry* 19, 1909, 1980.)

[MH–92]+, provides an indication of the contribution of the glycerol matrix to the parent ion. The 2′,3′-cAMP does not yield the S$_1$ product of [BH$_2$+42]+ (*m/z* 178), as this cleavage is blocked by the substitution at the 2′-*O* position; hence the major peaks in the 2′,3′-cAMP CID/MIKE spectrum are at *m/z* 136 ([BH$_2$]+), at *m/z* 164 ([BH$_2$+28]+), and at *m/z* 202 ([BH$_2$+66]+), the latter fragment occurring as a result of three cleavages releasing the 5′-CH$_2$OH and the 2′,3′-cyclic phosphodiester moiety (Figure 4.5b).

The diagnostic fragmentation patterns illustrated for the two cAMP isomers above (Figure 4.5) have been shown to be replicated in the other cyclic nucleotides (Figure 4.2) with the CID/MIKE spectra from the protonated molecules of the 3′,5′-isomers containing [BH$_2$]+, [BH$_2$+28]+, and [BH$_2$+42]+ fragments while in the spectra from the 2′,3′-isomers [BH$_2$]+ and [BH$_2$+28]+ are present but [BH$_2$+42]+ is not.[16,17] This means of isomeric differentiation has been applied successfully to many partially purified extracts to solve controversial issues relating to cyclic nucleotides. There were debates over an extended period regarding whether cAMP and cGMP, as had already been proven to exist in animals, bacteria, and protozoa, occurred in higher plants. The occurrence in plants was questioned both on theoretical grounds, on the basis of plants having different control mechanisms, and on technical grounds, because of the lack of resolution that was inherent to the techniques

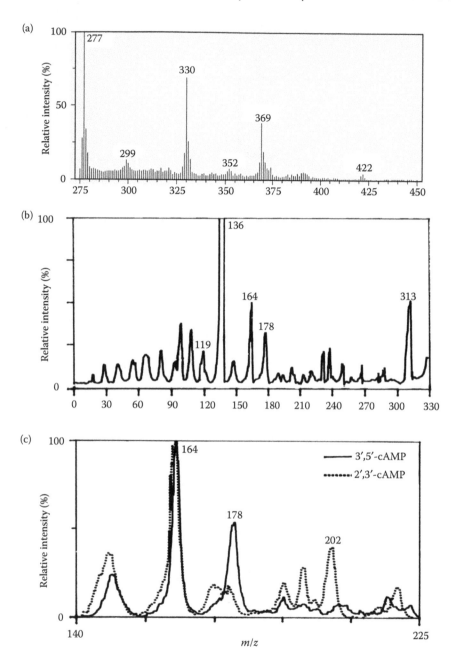

FIGURE 4.4 FAB mass spectra of cAMP. (a) FAB mass spectrum of 3′,5′-cAMP. (b) CID/MIKE spectrum of 3′,5′-cAMP. (c) Partial, diagnostic, CID/MIKE spectra of 3′,5′-cAMP and 2′,3′-cAMP. (Modified from Kingston, E.E., Beynon, J.H., and Newton, R.P., *Biomed. Mass Spectrom.* 11, 367, 1984; Kingston, E.E. et al., *Biomed. Mass Spectrom.* 12, 525, 1985.)

used in the investigation (e.g., in the interference of plant secondary metabolites in cyclic nucleotide immunoassays).[18] These arguments were unequivocally resolved by the application of FAB/MIKES analysis to the protonated molecules of putative 3′,5′-cAMP and 3′,5′-cGMP from partially purified plant extracts, and the demonstration of the relevant diagnostic $[BH_2]^+$ and $[BH_2+28]^+$ and $[BH_2+42]^+$ ions in both cases.[17,19,20] FAB/MIKES also provided the first unambiguous identification of the

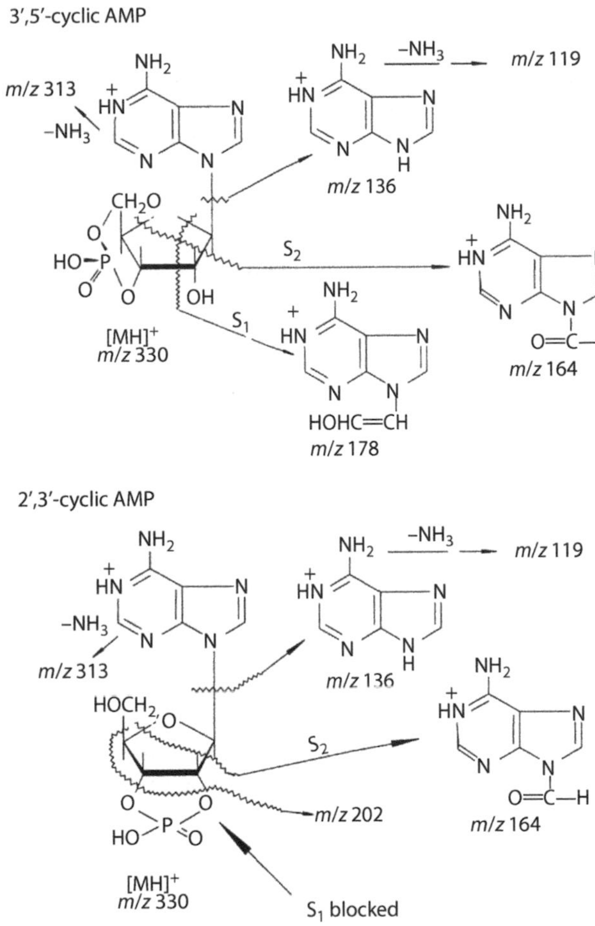

FIGURE 4.5 FAB/MIKES fragmentation profiles of [MH]$^+$ of 3′,5′-cAMP and 2′,3′-cAMP. (From Kingston, E.E., Beynon, J.H., and Newton, R.P., *Biomed. Mass Spectrom.* 11, 367, 1984; Kingston, E.E. et al., *Biomed. Mass Spectrom.* 12, 525, 1985.)

reaction product of putative adenylyl cyclase activity in plants, in *Pisum*, and cAMP formation by enzymic synthesis was similarly confirmed in *Medicago* cell cultures.[20–22] This successful strategy was then further employed in demonstrating the natural occurrence of 3′,5′-cyclic nucleotides other than cAMP and cGMP, showing the presence of cytidine 3′,5′-cyclic monophosphate (cyclic CMP) in several mammalian tissues, later followed by inosine 3′,5′-cyclic monophosphate (cyclic IMP), uridine 3′,5′-cyclic monophosphate (cyclic UMP), deoxythymidine 3′,5′-cyclic monophosphate (cyclic dTMP), and xanthosine 3′,5′-cyclic monophosphate (cyclic XMP).[23,24] This was then further extended to demonstrate the presence of the latter five cyclic nucleotides in plant tissues.[25] In the latter case, the fact that the spectra appeared to indicate differences in the relative distribution of these cyclic nucleotides in plants, with cyclic CMP and cyclic UMP elevated in the fast-growing meristems and cyclic IMP and cyclic dTMP elevated in the slower-growing nonmeristematic zones, was of particular interest, suggesting a possible role of at least some of the cyclic nucleotides in plant growth regulation.[25]

Application of the partial purification-FAB/MIKES strategy to the analysis of bacterial cyclic nucleotide profiles revealed the presence of 3′,5′-cAMP and 3′,5′-cGMP, which were anticipated to be present in cells of *Corynebacterium murisepticum*; 3′,5′-cyclic CMP, -IMP, and -UMP, cyclic nucleotides not previously reported in bacteria, and the cyclic deoxynucleotide, 2′-deoxy 3′,5′-cAMP, not previously even demonstrated to be naturally occurring.[26] Intriguingly, the cell culture fluid contained 3',5'-cAMP, 3',5'-cyclic CMP, 3',5'-IMP, and 2'-deoxy 3',5'-cAMP; as no other cyclic or noncyclic nucleotides were present in this culture medium, the implication was that these four cyclic nucleotides were being selectively released for a specific, but as yet undetermined, purpose. The established and hypothetical roles of cyclic nucleotides as metabolic regulators suggest that their regression from the bacterium has two possible purposes, either to rapidly switch off intracellular metabolic processes stimulated by the cyclic nucleotides, or to signal from one cell to another.[26]

4.3 STRUCTURAL ELUCIDATION OF NATURALLY OCCURRING CYCLIC NUCLEOTIDE ANALOGS

In addition to providing unambiguous identification of naturally occurring putative 3',5'-cyclic nucleotides, FAB-MS has made a major contribution to the structural elucidation of novel substituted cyclic nucleotides and cyclic nucleotide analogs in biological extracts. In many instances the MS analysis has been combined with simple chemical or enzymic modification, with the resultant combined MS data from the original compound and the derived products being sufficient to provide conclusive structural information without recourse to NMR or other analytical techniques. For example, the novel compound adenosine-2'-O-monophosphate-3',5'-cyclic pyrophosphate has been identified in extracts from the red seaweed *Porphyra* by this approach.[18] Although the analyte was found to contain an adenine to phosphate ratio of 1:3 by spectrophotometry, the mass spectrum contained an apparent protonated molecule at m/z 490, inconsistent with the previously assumed identity of adenosine triphosphate (M_r 507). The CID/MIKE spectrum obtained from [M+H]$^+$ contained peaks at m/z 136 and 164, corresponding to [BH$_2$]$^+$ and [BH$_2$+28]$^+$, respectively, but [BH$_2$+42]$^+$ was absent, indicating that the 2'-O position was blocked (Figure 4.6). Mild acid hydrolysis liberated a compound with [M+H]$^+$ at m/z 410, consistent with the loss of a phosphate, which was confirmed spectrophotometrically. The CID/MIKE spectrum now contained [BH$_2$+42]$^+$ at m/z 178, indicating that the phosphate moiety had been removed from the 2'-O position. A more extended acidic hydrolysis produced three further compounds, one with [M+H]$^+$ at m/z 428, indicating addition of water to the previous dephosphorylation product, and two with [M+H]$^+$ at m/z 348, indicating loss of a second phosphate. The CID/MIKE spectra of the latter two compounds were essentially identical to those of 3'-AMP and 5'-AMP, the data collectively reflecting successive hydrolyses of the parent 2'-phosphoadenosine-3',5'-cyclic pyrophosphate (Figure 4.6). By this strategy, 18 common nucleotides and two cyclic nucleotides plus 15 novel nucleotides were identified, comprising eight deoxynucleotides, two cyclic deoxycyclic nucleotides, 10 aminoacylnucleotides, and two nucleoside trisphosphates, together with N^6, N^6-dimethyladenosine 5'-monophosphate, the 2'-phosphoadenosine-3',5'-cyclic pyrophosphate described above, and 5'-phosphoadenosine-2',3'-cyclic pyrophosphate.[18] The presence of these compounds in *Porphyra*, which is considered to be one of the least differentiated eukaryotic plants, would seem to be indicative of a basal, as yet undetermined, role of evolutionary significance. Of relevance in a pharmacological context, one or more of these compounds may be responsible for the apparently potent effects that render the red seaweed extract a prized herbal medicine in some cultures.

An analogous strategy was used to resolve the initial debate surrounding the enzyme cytidylyl cyclase, reported to be a specific enzyme that functioned to catalyze the synthesis of cyclic CMP from CTP. Initial assays utilized methods previously applied to adenylyl and guanylyl cyclases, in which the conversion of radiolabeled triphosphate to a cyclic nucleotide is estimated after the separation of the cyclic nucleotide product from unused substrate and side products by the use of an ion-exchange resin. While initial reports demonstrated clear differences between the cytidylyl cyclases and the adenylyl and guanylyl cyclases, indicating that cyclic CMP was not being synthesized as a result of lack of specificity of one or both of the latter two enzymes, claims that this was a specific enzyme with the sole function of cyclic CMP synthesis were contested on the basis that after more extensive chromatography, the major radiolabeled product appeared to be composed of

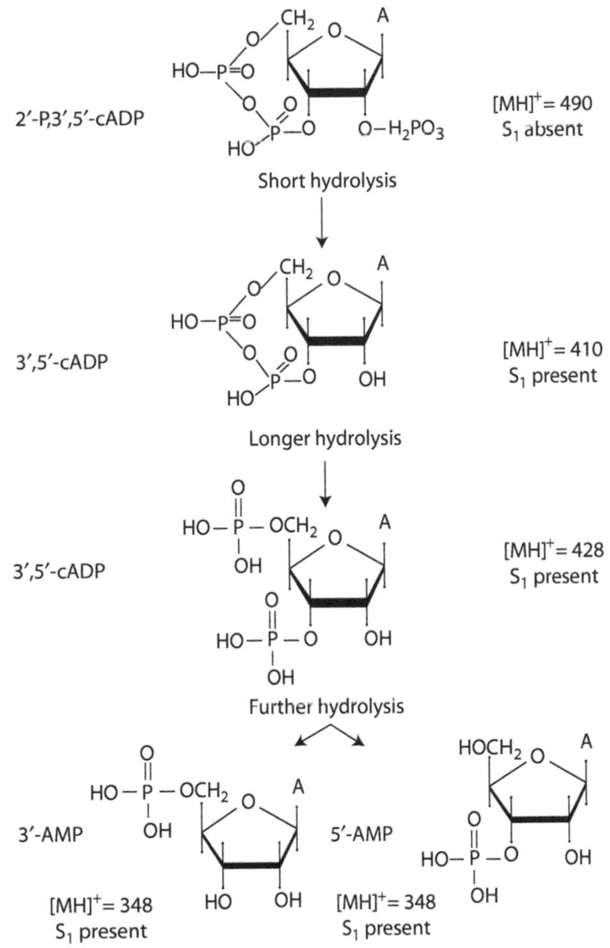

FIGURE 4.6 Identification of adenosine-2'-O-monophosphate-3',5'-cyclic pyrophosphate from the red seaweed *Porphyra*. (Modified from Newton, R.P., Kingston, E.E., and Overton, A., *Rapid Commun. Mass Spectrom.* 9, 305, 1995.)

two compounds, thus questioning the validity of the original identity of the enzyme product as cyclic CMP.[27] The skepticism over this identification was further supported by a report that the incubate of a putative cytidylyl cyclase matrix after the reaction contained more than one compound immunoreactive with anticyclic CMP serum. In order to determine conclusively whether cyclic CMP was produced by the putative cytidylyl cyclase activity, and to identify any side products formed from CTP during the course of the enzyme-catalyzed reaction, the incubation mixtures at the end of parallel reactions utilizing (a) unlabeled, (b) ^{14}C-single labeled, and (c) ^{32}P, ^{14}C-double labeled CTP as substrate were separately subjected to chromatographic analysis, spectrophotometry, and either FAB-MS with MIKE spectrometry of (a) or isotopic ratio determination for (b) and (c). Further data were acquired by subjecting the separated reaction products to selective hydrolysis, followed by re-examination by chromatography and then isotopic ratio determination or FAB/ MIKE spectrometry as appropriate.[28] Seven UV-absorbing and radioactively labeled products (where labeled substrates were employed) were observed, all with UV spectra indicative of cytidine-containing compounds. CTP, the original substrate plus its two products of hydrolysis, CDP and CMP, were identified on the basis of chromatographic comparison with standards, isotopic ratios, and the FAB/MIKE spectra obtained from their protonated molecules at *m/z* 484, 404, and 324, respectively (Table 4.1). The occurrence of these three compounds would have been predicted as a

TABLE 4.1

Identification of Components of Cytidylyl Cyclase Reaction

			Products of Hydrolysis by		
			(a) Nucleotidase	(b) Phosphodiesterase	(c) HCl
Compound	molP/mol	[MH]⁺	[MH]⁺	[MH]⁺	[MH]⁺
CTP	3	484	484, 404, 324	484	484, 404, 324, 164
CDP	2	404	404, 324	404	404, 324, 164
CMP	1	324	324, 164	324	324, 164
3′,5′-cCMP	1	306	306	306, 324	306, 324, 164
3′,5′-cCDP	2	386	386	386, 404	386, 404, 324
2′-O-P-3′,5′-cCMP	2	386	386	386, 404	386, 404, 306, 324
2′-O-asp-3′,5′-cCMP	1	421	421, 306	421, 324	306, 324, 164, 134
2′-O-glu-3′,5′-cCMP	1	435	435, 306	435, 324	306, 324, 164, 148

result of any aqueous incubation of CTP; thus it was the identification of the remaining four products that was crucial. The putative cyclic CMP was provisionally identified on the basis of its chromatographic behavior relative to chemically synthesized standards and a ³²P/¹⁴C ratio confirming a single phosphate was present. Its identity was unambiguously proven by FAB/MIKES analysis, with the mass spectrum containing intense peaks at m/z 306, 328, 350, 398, and 420 corresponding to the protonated molecule of cyclic CMP and its glycerol and sodium adducts, and the CID/MIKES spectrum containing the diagnostic [BH₂]⁺, [BH₂+28]⁺, and [BH₂+42]⁺ ions of 3′,5′-cyclic CMP arising from the S₁ and S₂ cleavages, at m/z 112, 154, and 140, respectively. The unlabeled compound was unchanged in respect of its chromatography and FAB/MIKES spectra after exposure to nucleotidase activity, (capable of hydrolyzing a phosphomonoester bond), but displayed changed chromatography and spectra consistent with 5′-CMP after incubation with phosphodiesterase (capable of hydrolyzing a phosphodiester bond), and with 5′- and 3′-CMP after acid hydrolysis (Table 4.1). The same strategy and rationale led to the identification of two of the remaining compounds as cytidine 3′,5′-cyclic pyrophosphate and 2′-phosphocytidine 3′,5′-cyclic monophosphate, while the third was determined to be in fact two compounds, 2′-O-aspartylcytidine 3′,5′-cyclic monophosphate and 2′-O-glutamylcytidine 3′,5′-cyclic monophosphate, which initially cochromatographed (Figure 4.7).[28] The unequivocal identification of the 3′,5′-cyclic CMP, plus the further demonstration that the four novel cyclic CMP derivatives generated during the reaction cross reacted in a cyclic CMP immunoassay, providing the evidence that the enzyme was indeed a cytidylyl cyclase and allowed the development of a specific assay for its estimation and the impetus for further study.[29]

4.4 MASS SPECTROMETRIC ANALYSIS OF SYNTHETIC CYCLIC NUCLEOTIDE ANALOGS

In addition to structural elucidation of naturally occurring substituted cyclic nucleotides, FAB/MIKES has proved invaluable in examining the products of the synthesis of cyclic nucleotide analogs and in optimizing the reaction conditions for their production. Cyclic nucleotides do not readily pass through membranes, so direct application to cells or tissues has little effect. The simplest, commonly used cell-permeating cyclic nucleotide derivatives are the dibutyryl derivatives, N^6,2′-O-dibutyryladenosine 3′,5′-cyclic monophosphate, N^2,2′-O-dibutyrylguanosine 3′,5′-cyclic monophosphate, and N^4,2′-O-dibutyrylcytidine 3′,5′-cyclic monophosphate.[30] In many instances the identification of the synthetic product is based solely on its chromatographic behavior, with the assumption that there is a sequential synthesis, with the ribose ring substitution proceeding much

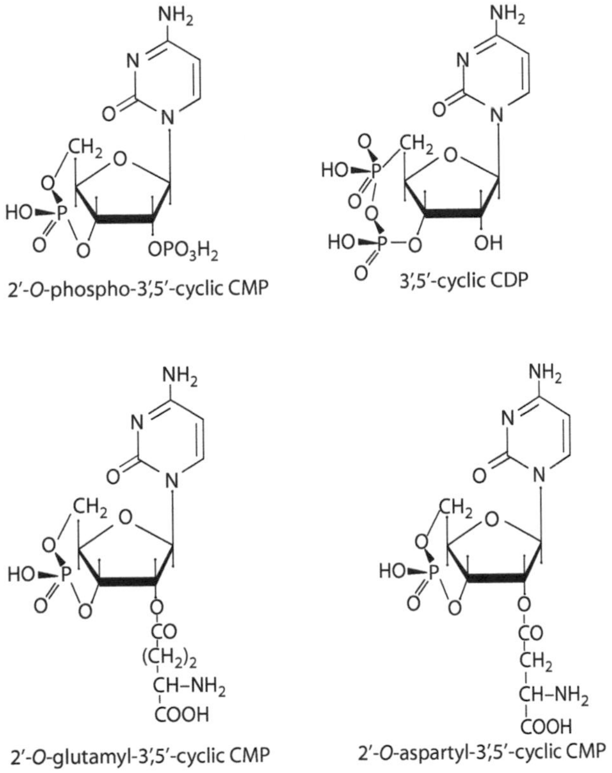

FIGURE 4.7 Side products produced by mammalian cytidylyl cyclase in addition to the major reaction product, cytidine 3′,5′-cyclic monophosphate.

more rapidly than the heterocyclic substitution. However with the synthesis, for example, of $N^4,2'$-O-dibutyrylcytidine 3′,5′-cyclic monophosphate under different reaction conditions and times it is possible to also erroneously generate a dibutyryl noncyclic mononucleotide, if one of the phosphodiester bonds has been hydrolyzed; a monoester derivative if only one of the butyryl substitutions has taken place; a dibutyryl derivative, but with substitution at positions other than the intended N^4 and 2′-O; and even a dibutyryl 2′,3′-cyclic nucleotide after hydrolysis and recyclization of the phosphodiester moiety. The retention of the 3′,5′-cyclic phosphate group and substitution of one butyryl residue at the 2′-O and the other at the heterocyclic amine are necessary for pharmacological experimentation, thus any misidentification of any of the side products or their presence as contaminants in any preparation administered to experimental animals would seriously compromise any subsequent data. In the case of $N^4,2'$-O-dibutyrylcytidine 3′,5′-cyclic monophosphate, reaction parameters were able to be optimized by examination of the mass spectrum obtained in different experiments with varied conditions. Under the final optimized conditions FAB-MS scans were produced containing intense peaks at m/z 446 ([M+H]⁺), m/z 468 ([M+Na]⁺), and m/z 490 ([M–H+2Na]⁺), but with no significant peaks relating to monobutyryl derivatives at m/z 376 ([M+H]⁺), m/z 398 ([M+Na]⁺), and m/z 420 ([M–H+2Na]⁺), nor to dibutyryl noncyclic nucleotide derivatives, that is, dibutyryl cytidine 3′- or 5′-monophosphate, at m/z 464 ([MH]⁺), m/z 486 ([MNa]⁺), and m/z 508 ([MNa₂]⁺) were present. The position of the two butyryl substitutions was shown by the ions at m/z 265 ([butyrylribose-cyclic PO₄]⁺), 287 ([Na-butyrylribose-cyclic PO₄]⁺), and 309 ([Na₂-butyrylribose-cyclic PO₄]⁺), which indicated a substitution of the ribose ring, while ions at m/z 182 ([butyrylcytosine]⁺), m/z 204 ([butyrylcytosine-Na]⁺), and m/z 206 ([butyrylcytosine-Na₂]⁺) confirmed the presence of the butyrylated cytosine moiety (Figure 4.8). MIKE spectra of the protonated molecule of the putative $N^4,2'$-O-dibutyrylcytidine 3′,5′-cyclic monophosphate

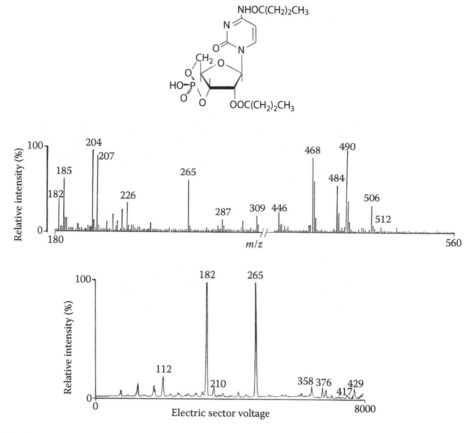

FIGURE 4.8   Mass spectrometric analysis of cell-permeating cyclic nucleotide derivatives. FAB mass spectrum and CID/MIKES spectrum of [MH]+ of N^4,$2'$-O-dibutyrylcytidine $3'$,$5'$-cyclic monophosphate. (Modified from Newton, R.P. et al., *Organic Mass Spectrom.* 24, 679, 1989.)

contained intense peaks at *m/z* 182 ([butyrylcytosine]+), *m/z* 204 ([butyrylcytosine-Na]+), and *m/z* 206 ([butyrylcytosine-Na$_2$]+), with further fragmentations consistent with butyryl substitution at the N^4-position. No ions at *m/z* 224 or 246, which would arise from an S_1 fragmentation, were apparent, but peaks corresponding to an S_2 cleavage occurred at *m/z* 232 and 210. The MIKE spectra obtained from the sodium adduct of butyrylribose cyclic monophosphate (*m/z* 287 in the precursor ion mass spectrum) had the major fragment ion at *m/z* 216, corresponding to a loss of CH3CH2CH2COOH, and other fragment ions consistent with the substitution of the butyryl group on the ribose ring at the $2'$-O position (Figure 4.8).[30]

A similar strategy has been employed to optimize derivatives required for cyclic nucleotide radioimmunoassay.[31] Such derivatization is necessary because cyclic nucleotides are not themselves immunogenic, being of too low a molecular size, thus a cyclic nucleotide–protein conjugate is synthesized for immunization to generate the requisite antibodies. A cyclic nucleotide derivative that will readily undergo radioiodination also has to be produced, allowing the synthesis of a high specific activity radiolabeled cyclic nucleotide as a labeled antigen in the competitive immunoassay. The first stage in both processes is the synthesis of the $2'$-O-succinyl cyclic nucleotide; this is then linked to a protein such as thyroglobulin, serum albumin, or hemocyanin to generate the cyclic nucleotide–protein conjugate for immunization, or alternatively to tyrosine methyl ester prior to radiolabeling with Na[^{125}I]. In both cases the retention of the cyclic phosphodiester and the site of succinylation at the $2'$-O position are essential. In the development of a radioimmunoassay for cyclic

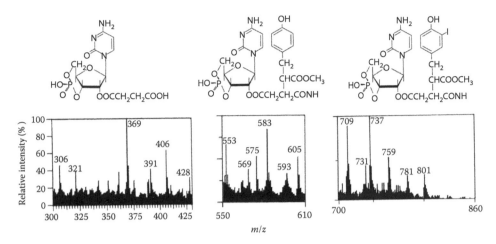

FIGURE 4.9 Mass spectrometric analysis of cyclic nucleotide derivatives synthesized for use in radioimmunoassay. Positive-ion FAB mass spectra of 2'-*O*-succinyl 3',5'-cCMP, 2'-*O*-succinyltyrosinyl methyl ester 3',5'-cyclic CMP, and 2'-*O*-succinyliodotyrosinyl methyl ester 3',5'-cyclic CMP. (Modified from Newton, R.P. et al., *Organic Mass Spectrom.* 28, 899, 1993.)

CMP, the putative 2'-*O*-succinyl 3',5'-cyclic CMP had a FAB mass spectrum with $[M+H]^+$ at m/z 404, and the absence of any significant peak at m/z 422 indicated that the phosphodiester moiety had not been hydrolyzed (Figure 4.9), while the CID/MIKE spectrum from the protonated molecule contained $[BH_2]^+$ at m/z 112 and $[BH_2+28]^+$ at m/z 140 from the S_2 cleavage, while the S_1 cleavage product normally at m/z 154 ($[BH_2+42]^+$) is displaced to m/z 254, confirming that the succinylation has occurred at the 2'-*O* position.[31] For the succinyltyrosinyl cyclic CMP methyl ester the FAB mass spectrum had $[M+H]^+$ at m/z 583, with a prominent peak at m/z 569 reflecting the loss of the methyl group; the absence of any significant peak at m/z 601 indicated that the phosphodiester moiety had not been hydrolyzed (Figure 4.9). Using unlabeled NaI, the conditions for the iodination of the succinyltyrosinyl cyclic CMP methyl ester were also optimized by monitoring the presence of $[M+H]^+$ at m/z 707 and the absence of a significant m/z 725, the latter indicative of hydrolysis of the cyclic phosphodiester, and the absence of a peak at m/z 835, indicative of a di-iodo derivative, which with a radiolabeled derivative would be unstable (Figure 4.9).[31]

In addition to the dibutyryl, succinyl, and succinyltyrosinyl methyl esters described above, a series of other cyclic nucleotide derivatives have had their molecular structure similarly confirmed by FAB/MIKES analysis, including a number of cyclic nucleotide-dependent protein kinase agonists and antagonists.[32–37] Cyclic nucleotide analogs are valuable tools in establishing mechanisms of kinase activation and effect in APK- and GPK-mediated signal transduction pathways, so that specific analogs can bind selectively at the regulatory sites of a particular cyclic nucleotide-dependent protein kinase, and act as agonists or antagonists of the natural effectors. Different cyclic nucleotide-binding sites on kinase isoforms differ in the rate of association of their natural cyclic nucleotide and also differ in their affinities for cyclic nucleotide analogs, thus unlike the natural agonists these analogs can be used to differentiate between the regulatory binding sites and selectively activate a single isoform. Such analogs include adenosine 3',5'-cyclic monophosphothioate (cAMPS) and 8-chloroadenosine-3',5'-cyclic monophosphate (8-ClcAMP); 8-(6-aminohexyl)-aminoadenosine-3',5'-cyclic monophosphate (8-AHAcAMP), N^6-benzyladenosine-3',5'-cyclic monophosphate (6-Bn cAMP), and 8-piperadenosine-3',5'-cyclic monophosphate (8-PIP cAMP); 1, N^6-ethenoadenosine-3',5'-cyclic monophosphate (ε-cAMP), 2'-aza-1, N^6-ethenoadenosine-3',5'-cyclic monophosphate (2'-aza-ε-cAMP), *N*-phenyl-adenosine-3',5'-cyclic monophosphate (6-Phe-cAMP), 6-chloropurineriboside-3',5'-cyclic monophosphate (6-Cl-PuMP), 5,6-dichloro-1-β-D-ribofuranosylbenzimidazole 3',5'-cyclic monophosphate (5,6-diCl-BIMP), and 8-(4-chlorophenylthio)-adenosine-3',5'-cyclic monophosphate

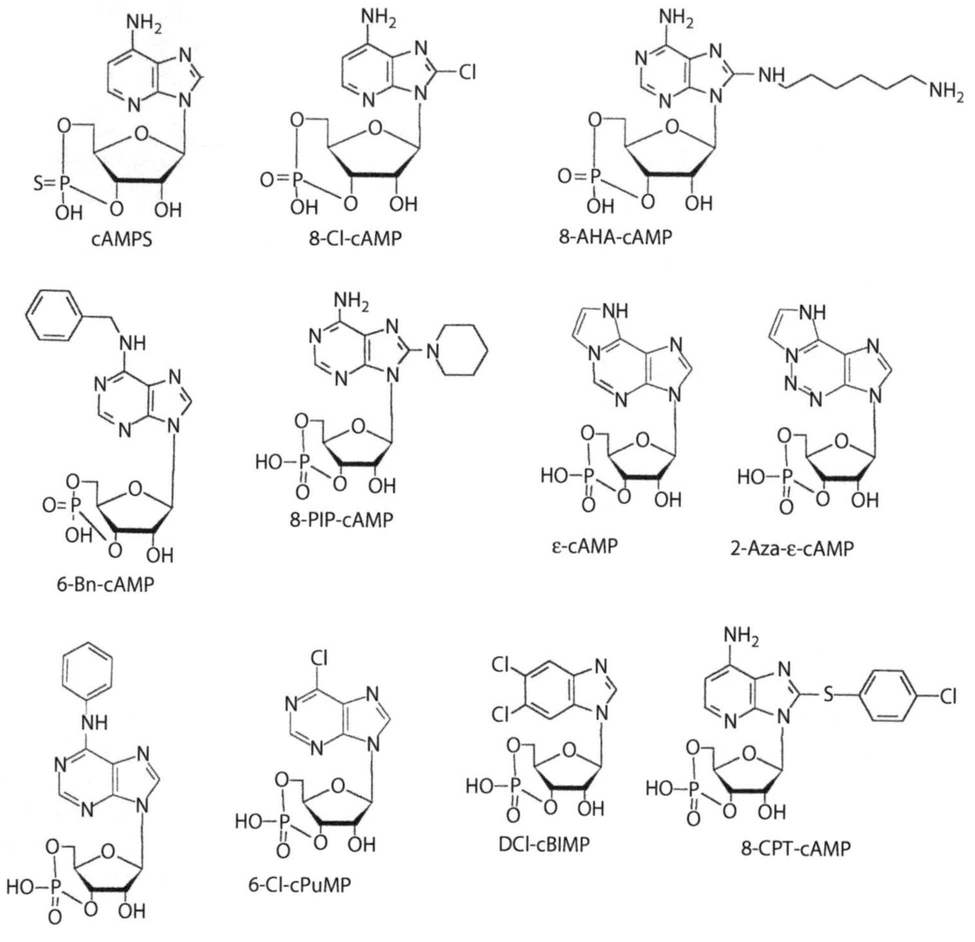

FIGURE 4.10   cAMP analogs used as site-selective probes of cyclic nucleotide-dependent protein kinases.

(8-Cl-PT-cAMP) [32–37] (Figure 4.10). As an example, the fragmentation of 8-piperadenosine-3′,5′-cyclic monophosphate (8-PIP cAMP) is shown in Figure 4.11.[36,37]

A further use of synthetic analogs in the cyclic nucleotide context is the use of cyclase substrate analogs. While the natural substrate of the nucleotidyl cyclases is a simple nucleoside triphosphate, with adenylyl cyclase in particular this natural substrate, ATP, is hydrolyzed by another enzyme ATPase, present in the cell and membrane preparations at much higher concentration and activity than the cyclase analyte. While this problem of substrate sequestration can be circumvented by the inclusion of ATP-regenerating systems in the adenylyl cyclase assay, another approach is to use alternative adenylyl cyclase substrates that are not susceptible to ATPase activity, such as adenylyl-(β,γ-methylene)-diphosphate (AMP-PCP) and adenylyl-imididiphosphate (AMP-PNP); again MS analysis of these analogs has been invaluable.[38]

4.5   QUANTITATION OF THE ACTIVITY OF CYCLIC NUCLEOTIDE-RELATED ENZYMES

While the majority of MS applications, particularly FAB-MS, in cyclic nucleotide research have been qualitative, quantitative analyses of the related enzyme systems (the nucleotidyl cyclases,

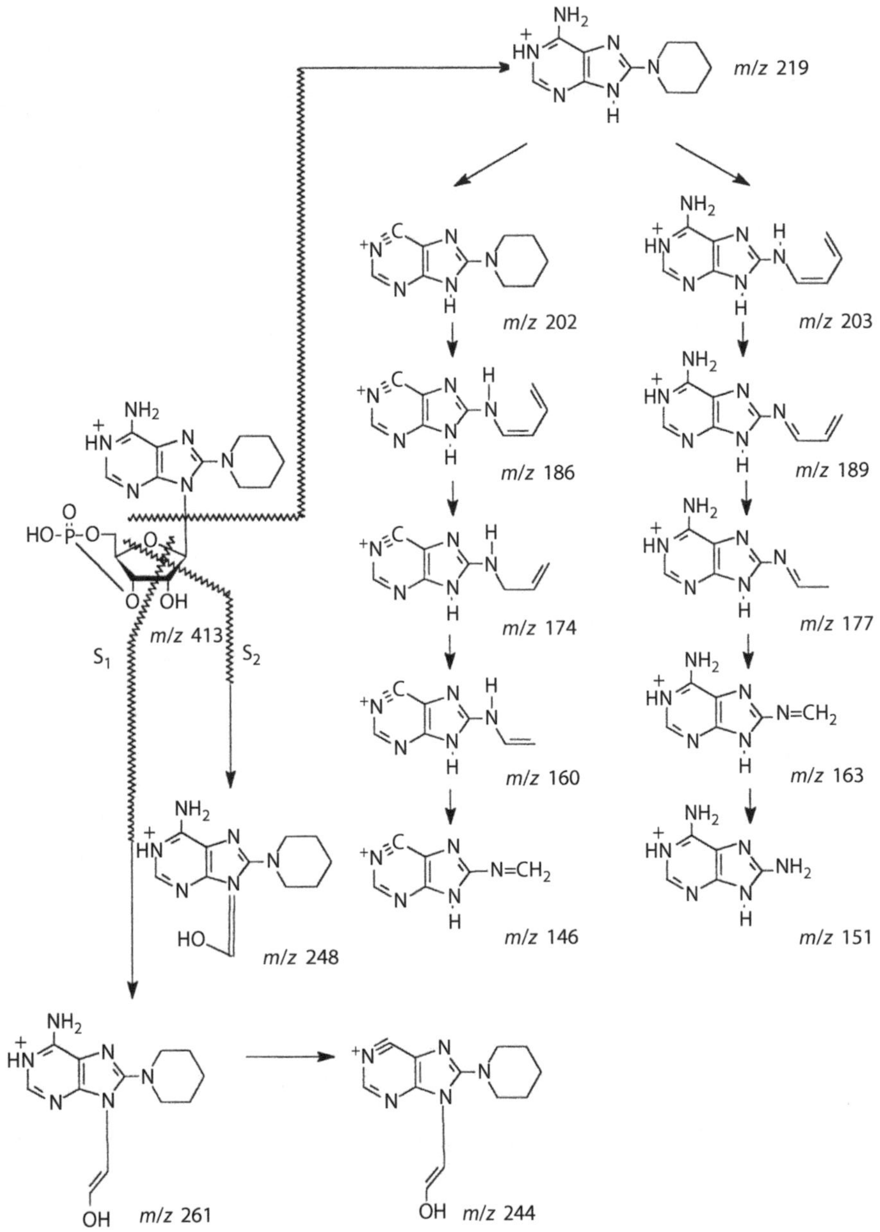

FIGURE 4.11 Fragmentation of 8-PIP-cAMP. (Modified from Pereira, M.L.M., *Analysis of Pharmaceutical Products and Nucleotides by LCMS and Tandem MS*, PhD thesis, Swansea University, 1998; Pereira, M.L.M., Games, D.E.G., and Newton, R.P., unpublished observations.)

cyclic nucleotide phosphodiesterases, and cyclic nucleotide-responsive protein kinases) have also been utilized beneficially. The majority of the classical non-MS-based assays of these enzymes use a radiolabeled substrate, ion exchange or other separation of labeled product from substrate, and only produce one datum point per incubation. The potential advantage of MS-based assays, aside from avoiding the use of radiolabeled material, lies in the potential for multiple component monitoring, that is, if several compounds have characteristic peaks that are not isobaric in the mass spectrum then these compounds may all be estimated in the same sample incubate.

Several approaches have been successful. The simplest has been continuous flow FAB, which with a highly purified phosphodiesterase provided data of adequate quality for useful kinetic analysis.[39–41] However, the assay proved very time consuming due to frequent blockages in the flow, and was deemed ineffective except for specific purposes typically with a homogeneous preparation of an enzyme.

A second approach was to obtain the MIKE spectra of $[M+H]^+$ ions from the FAB mass spectra of both the substrate and the product in an enzyme incubation sample, then to obtain a second spectrum after a duplicate set of samples were spiked with known quantities of either substrate or product as an identical internal standard.[42–45] Activity is calculated by

$$\frac{[P]_E}{[S]_E} = \frac{\text{sum of relative intensities of product diagnostic ions}}{\text{sum of relative intensities of substrate diagnostic ions}}$$

and

$$\text{rate} = \frac{[S]_0}{t} \times \frac{[P]_E}{[S]_E},$$

where $[S]_0$ is the initial concentration, $[P]_E$ and $[S]_E$ are the product and substrate concentrations at the completion of incubation time, t, and relative intensity is the peak intensity at m/z n in the MIKE spectrum divided by the change in peak intensity of m/z n after spiking with spikes of equal magnitude of standard substrate and product. The diagnostic peaks monitored for both cyclic and noncyclic nucleotides are $[BH_2]^+$, $[MH–17]^+$, $[BH_2–17]^+$, $[BH_2+28]^+$, and $[BH_2+42]^+$. This method has been used to good effect to assay cyclic UMP, IMP, and dTMP phosphodiesterase activities for which radioactively labeled substrates are not available for the conventional assays, and the accuracy of this strategy has been validated by the good correlation of data when used in parallel with radiometric assays for cAMP- and cyclic CMP-phosphodiesterases.[42–45] The method is not of wide applicability because of the demands of time, skill of the experimenter, and quantity of the substrate and the product required. However, it has been of inestimable value in assays in which it is advantageous to identify and determine the relative proportions of isomeric products.

One such example relates to the examination of a cyclic CMP-specific phosphodiesterase and a multifunctional phosphodiesterase, two enzymes capable of hydrolyzing cyclic CMP. Just as the cyclic nucleotide isomers can be readily differentiated by the CID/MIKE spectra of their protonated molecules, the mononucleotide products can be similarly distinguished.[46–48] The prospective products of phosphodiesterase hydrolysis of a 3′,5′-cyclic CMP substrate, 3′-CMP, and 5′-CMP can be differentiated by the relative intensities at m/z cyclic 213, 226, 232, 263, and 268 (Figure 4.12); examination of the enzymic reaction products showed that the cyclic CMP-specific phosphodiesterase produced only 5′-CMP whereas the multifunctional phosphodiesterase produced both 3′- and 5′-isomers. Analysis of the products of a series of incubations with preparations of these two mammalian enzymes facilitated elucidation of the differences between their mechanisms of action.[48] A similar strategy was applied with a chloroplast phosphodiesterase preparation to ascertain if the presence or absence of specific cations alters the nature of the phosphodiesterase product. In the absence of Li^+, Na^+, K^+, Mg^{2+}, Mn^{2+}, and Fe^{3+}, the 3′-ester linkage of the 3′,5′-cyclic substrates was hydrolyzed exclusively, while the addition of monovalent and divalent cations results in hydrolysis of both 3′- and 5′-ester linkages, with for example Li^+, Na^+, and K^+ producing approximately equimolar mixtures of 3′ products while Mg^{2+} and Mn^{2+} produce both isomers but with 5′-mononucleotides as the major product.[49] In concert with molecular modeling studies, these data have yielded valuable insight into the molecular mechanisms taking place at the catalytic site.

While the use of CID/MIKE spectra has been productive in the analysis of some phosphodiesterase preparations, kinetic analysis of enzyme activity using quantitative FAB-MS has been of

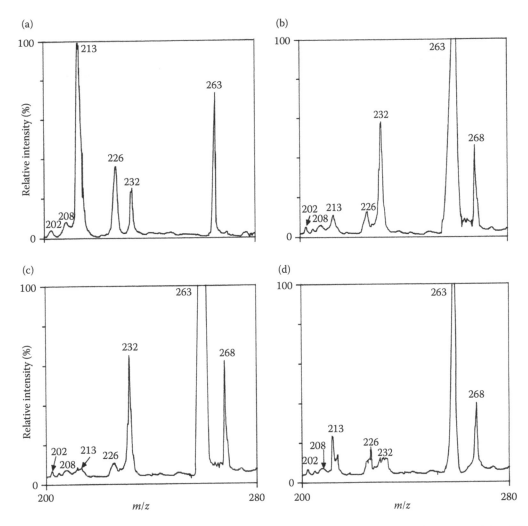

FIGURE 4.12 Identification of mononucleotide products of cyclic nucleotide phosphodiesterase activity. CID/MIKE spectra of (a) 3′-CMP, (b) 5′-CMP, (c) the product of 3′,5′-cyclic CMP hydrolysis by cyclic CMP-specific phosphodiesterase, and (d) the products of 3′,5′-cyclic CMP hydrolysis by multifunctional phosphodiesterase. (Reproduced from Newton, R.P. et al., *Rapid Commun. Mass Spectrom.* 13, 574–585, 1999. With permission.)

more general applicability to cyclic nucleotide-related enzymes.[40–45,48,50–58] In this method, the peak heights of the protonated molecules, their adducts, and their characteristic fragments are determined together with major peaks emanating from the glycerol matrix. Quantitation is then achieved by a proportionation of the sums of the relative peak intensities. Since the mass spectra from enzyme incubates are very complex, due to the presence of several different molecular species, it is necessary to carry out parallel series of incubations containing active and inactive enzyme preparations to obtain viable kinetic data. Change in the concentration of analyte X in the enzyme-catalyzed reaction is then calculated from the diagnostic peak intensities by

$$\Delta[X] = \frac{(\varphi InX/\varphi In\mathrm{matrix})_{\mathrm{expt}} - (\varphi InX/\varphi In\mathrm{matrix})_{\mathrm{control}}}{(\varphi InX/In\mathrm{matrix})_{\mathrm{control}}},$$

where φInX and $\varphi Inmatrix$ are the sum of the relative peak intensities of the characteristic ions of analyte X and the glycerol matrix, respectively. For cyclic nucleotides the peak intensities monitored are those of the protonated molecule, its cation, glycerol and buffer adducts, plus any adducts formed with any other incubation component, and each of these ions minus 17 and 18 mass units, corresponding to the loss of NH_3 and H_2O, respectively.

This quantitative MS was initially applied successfully to kinetic studies of cyclic CMP- and cAMP-responsive protein kinases, by determining the relative conversion of the phosphate donor ATP to ADP in the presence of the cyclic nucleotide agonist and an appropriate protein phosphorylation substrate when compared to controls in which the agonist, substrate, or active kinase is absent.[52,54,59] Data obtained by such MS quantitation show good correlation with data obtained by the conventional radiometric assay data, but provide the additional facility to measure the binding and activating potency of competing cyclic nucleotides, thereby providing information on specificity unavailable from the routinely used protein kinase assay protocol. This led to the first demonstration of cyclic CMP-specific protein kinase activity, together with its integral ATPase and phosphodiesterase activity, and to the discovery that the cAMP-dependent protein kinase (previously reported to have absolute specificity for cAMP) could in fact be activated by other cyclic nucleotides.[52,54]

Similar approaches have been successfully applied to quantitative MS studies of cyclic nucleotide phosphodiesterases and nucleotidyl cyclases, again providing data from multiple component monitoring unavailable from the conventional non-MS-based assays.[48,49,56,58,60,61] As described above, the CID/MIKES analyses of their products enabled differentiation between the 2′-, 3′-, and 5′-mononucleotide product isomers and insight into characteristic differences between a cyclic CMP-specific phosphodiesterase and a multifunctional enzyme in respect of major products released. An extensive kinetic analysis of these two enzymes by quantitative FAB-MS illustrated the extra facility of the MS-based assay to monitor several components simultaneously to study the concurrent hydrolysis of alternate cyclic nucleotide substrates and provide kinetic parameters of significance in interpreting substrate–enzyme interactions.[48] Simple visual inspection of the spectra (Figure 4.13a–d) indicates that, from comparison between spectra from the denatured control and the active cyclic CMP-specific enzyme incubated with just 3′,5′-cyclic CMP as substrate, there is a loss of peak intensities at m/z 306, 328, 350, and 398. This corresponds to the cyclic CMP substrate protonated molecule, and the sodium, disodium, and glycerol adducts, respectively. The increase in peak intensities at m/z 324, 346, and 368, corresponds to the CMP product protonated molecule, and the sodium and disodium adducts, respectively (Figure 4.13a and b), reflecting the hydrolysis of cyclic CMP by the active enzyme. When both cyclic CMP and cAMP are added (Figure 4.13c and d), the same changes in the characteristic cyclic CMP and CMP peaks listed above are observed, but there is no change in the peaks at m/z 330, 352, and 374, corresponding to the protonated molecule, and the sodium and disodium adducts of cAMP, respectively, indicating that while cyclic CMP is again hydrolyzed no cAMP hydrolysis has taken place. With the multifunctional enzyme, a similar pair of spectra to those with the cyclic CMP phosphodiesterase are obtained with cyclic CMP alone (Figure 4.13e and f), but with both cyclic CMP and cAMP as substrates (Figure 4.13g and h) the active enzyme shows a diminution in the characteristic peaks of the substrates and an increase in those of the mononucleotide products, indicating that both cyclic nucleotides are hydrolyzed by this enzyme.[48] Quantitation of the activities by the above formula showed good correlation between MS-derived data and the those from the routine radiometric assay, as exemplified in the Lineweaver–Burk plots obtained by each method (Figure 4.14a or b); V_{max} and K_m parameters were determined for 10 cyclic nucleotide substrates for each of the two enzymes, with the MS quantitation enabling the determination of turnover of several substrates simultaneously. Thus, for example, in addition to the hydrolysis of cyclic CMP being monitored against variation in concentration of cAMP, the same spectra could be used to monitor the reciprocal effect, that is, hydrolysis of cAMP in the presence of varied concentration of cyclic CMP (Figure 4.14c and d); thus in addition to K_m, K_i values for each are determined. With the cyclic CMP-specific phosphodiesterase, the K_m and K_i values for all

10 substrates in paired incubations were the same for each, indicative of binding at a single site, but in contrast with the multifunctional phosphodiesterase, there were differences between the K_m and K_i values of some of the cyclic nucleotide pairs, indicating that with this enzyme some of the inhibitory effects between the two substrates are just competitive and others are by other than simple direct competition, information which taken together with molecular modeling data gives insight into the differences in the binding of substrates with both enzymes.[48]

Similar kinetic studies of nucleotidyl cyclases have additionally led to the first unequivocal demonstration of adenylyl and guanylyl cyclase activities in isolated chloroplasts, showing an association between the activities but indicating that two distinct active sites were involved, to a novel method for comparison of the efficiency of ATP-regenerating systems and of phosphodiesterase inhibitors plus assessment of alternative substrates with mammalian particulate adenylyl cyclase, and to the simultaneous monitoring of the production of cyclic CMP and the side products cytidine 3',5'-cyclic pyrophosphate, 2'-phosphocytidine 3',5'-cyclic monophosphate, 2'-O-aspartylcytidine 3',5'-cyclic monophosphate, and 2'-O-glutamylcytidine 3',5'-cyclic monophosphate by cytidylyl cyclase under varied conditions.[56,60,61]

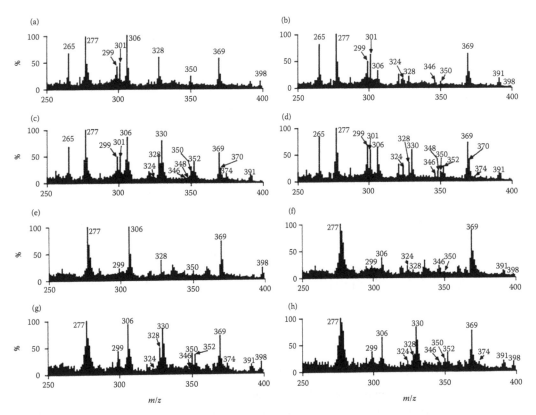

FIGURE 4.13 Quantitation of cyclic nucleotide phosphodiesterase activity by FAB-MS. Positive-ion FAB mass spectra of incubates containing (a) inactive cyclic CMP-specific phosphodiesterase with 3',5'-cyclic CMP substrate, (b) active cyclic CMP-specific phosphodiesterase with 3',5'-cyclic CMP substrate, (c) inactive cyclic CMP-specific phosphodiesterase with 3',5'-cyclic CMP and 3',5'-cAMP substrates, (d) active cyclic CMP-specific phosphodiesterase with 3',5'-cyclic CMP and 3',5'-cAMP substrates, (e) inactive multifunctional phosphodiesterase with 3',5'-cyclic CMP substrate, (f) active multifunctional phosphodiesterase with 3',5'-cyclic CMP substrate, (g) inactive multifunctional phosphodiesterase with 3',5'-cyclic CMP and 3',5'-cAMP substrates, and (h) active multifunctional phosphodiesterase with 3',5'-cyclic CMP and 3',5'-cAMP substrates.

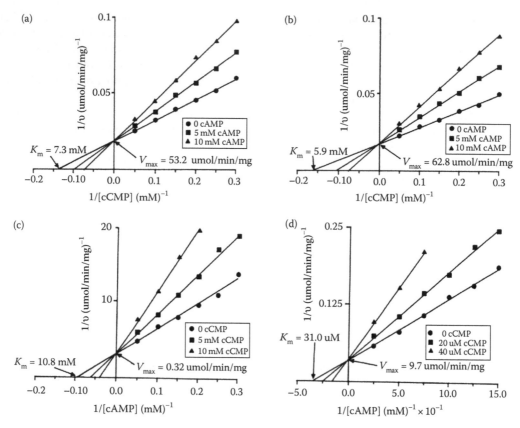

FIGURE 4.14 Lineweaver–Burk plots for kinetic analysis of cyclic nucleotide hydrolysis: (a) hydrolysis of 3′,5′-cyclic CMP, in the presence and absence of 3′,5′-cAMP, by cyclic CMP-specific phosphodiesterase determined by radiometric assay, (b) hydrolysis of 3′,5′-cyclic CMP, in the presence and absence of 3′,5′-cAMP, by cyclic CMP-specific phosphodiesterase determined by FAB-MS quantitation, (c) hydrolysis of 3′,5′-cAMP, in the presence and absence of 3′,5′-cyclic CMP, by cyclic CMP-specific phosphodiesterase determined by FAB-MS quantitation, and (d) hydrolysis of 3′,5′-cAMP, in the presence and absence of 3′,5′-cyclic CMP, by multifunctional phosphodiesterase determined by FAB-MS quantitation. (Reproduced from Newton, R.P. et al., *Rapid Commun. Mass Spectrom.* 13, 574–585, 1999. With permission.)

4.6 ELECTROSPRAY MS OF CYCLIC NUCLEOTIDES

Following the successful application of FAB-MS and CID/MIKES analyses to cyclic nucleotides and related enzymes, other ionization methods commonly used with comparable polar molecules, such as nucleotides and oligonucleotides, offered apparent additional advantages for cyclic nucleotide MS. While matrix-assisted laser desorption (MALDI) has been little utilized in the structural studies of cyclic nucleotides and their derivatives, it has been of value in monitoring the macromolecular consequences of cyclic nucleotide action, as will be discussed later. Electrospray ionization (ESI) has been used increasingly in preference to FAB-MS for cyclic nucleotide studies because of its improved performance of increased sensitivity and its ability to be coupled to a separation technology such as high-performance liquid chromatography (HPLC), while on a technical perspective the need for a glycerol matrix with its consequent plethora of matrix peaks in the spectra and the requirement for frequent FAB probe cleaning sessions is avoided. However, as will be discussed below, FAB-MS does retain at least two advantages over ESI-MS in respect of cyclic nucleotide analyses.

The first report of techniques other than FAB-MS to identify cyclic nucleotides was the use of ion-pair reverse-phase HPLC to determine 13 nucleotides, including cAMP and cGMP, by UV spectrophotometry, with thermospray MS used to confirm their structures in *Octopus* retina extracts.[62] Coupling of HPLC to ESI-MS provides the additional benefit of an on-line separation when several cyclic nucleotides (and analogs) are to be identified and estimated in a single sample, although this can cause additional problems related to buffer composition and flow rate of the mobile phase. However, successful protocols for LC-MS and LC-MS/MS of cyclic nucleotides in biological tissue extracts have been developed and provide reproducible quantitation in the femto-mole region, a level of sensitivity comparable to the radioimmunoassay conventionally used. In the first such report negative-ion ESI was preferred; since in positive-ion ESI the total ion content was spread over a number of cation species such as $[M+H]^+$, $[M+Na]^+$, $[M+K]^+$, and $[M+NH_4]^+$, a problem has not incurred with negative-ion ES-MS/MS.[63] For quantitation purposes, it was found that the pH used for ion-suppression reverse-phase HPLC, using acetic acid/methanol with a 5 uM C-18 column, was also optimal for the ESI process, and that this gave a superior performance relative to ion-pair conditions. Later, ion-pair chromatography was linked to ES using tetrabutylammonium bromide and the loss of sensitivity from ion-pair chromatography was remedied by using a 2-propanol coaxial sheath flow.[64] Diagnostic transitions were used for multiple reaction monitoring (MRM) for 3',5'-cAMP; 2',3'-cAMP; 3',5'-cyclic IMP; 3',5'-cGMP; and 3',5'-cyclic CMP (Figure 4.15), with the $[M-H]^-$ to $[B]^-$ transition being used in each case (Table 4.2). Excellent linearity was obtained for integrated peak area against sample quantity over a range of 100 fmol to 20 pmol. Further reduction of the liquid chromatography (LC) column dimensions of the setup has enhanced this sensitivity exponentially,[64,65] with the introduction of a capillary column switching method now allowing routine detection of cyclic nucleotides as low as 25 fmol.[64,65–68] By virtue of the increased sensitivity, it has been possible to apply this LC-ESI-MS/MS method to demonstrate cAMP formation by enzymic synthesis in plasma membrane preparations from *Phaseolus* and chloroplasts from *Spincacea oleracea* and *Nicotiana tabacum*, and to identify and quantitate 3',5'-cAMP; 3',5'-cGMP; and 3',5'-cyclic CMP in microextracts from tobacco BY2 cell cultures and a number of mouse cell lines.[69–73] The determination of cyclic nucleotides at such low levels has enabled their estimation in cell lines being subjected to a variety of pharmacological challenges with the concomitant monitoring of other biochemical and metabolic parameters, greatly facilitating elucidation of the molecular mechanisms of action.

The fragmentation patterns observed with ESI-MS/MS relative to FAB-MS/MIKES appear less consistent and more dependent on experimental conditions, thus, while invaluable when run in combination with authenticated standards, *de novo* interpretation is more difficult with ESI. In the first such report negative ion ESI was preferred, since in the positive-ion ESI, the total ion content was spread over a number of cation species such as $[MH]^+$, $[MNa]^+$, $[MK]^+$, and $[MNH_4]^+$, a problem not incurred with negative-ion ESI-MS/MS.[63]

The product ion spectra, however, can be used to identify isomeric cyclic nucleotides as they exhibit different fragmentation patterns in ESI-MS/MS. In the product ion spectra of the protonated molecules of 2',3'- and 3',5'-cAMP, the protonated adenine base appears in both spectra at m/z 136, although it is only the base peak for 2',3'-cAMP (Figure 4.16). Product ions at m/z 313, 250, and 232 formed by the loss of NH_3, HPO_3, and H_3PO_4, respectively, are present in both spectra, although they are in higher abundance when formed from 3',5'-cAMP. The most distinguishing feature in the spectrum of the 3',5'- isomer is that m/z 313 is the most abundant ion rather than the protonated base. The remaining features that also distinguish the spectra include the presence of m/z 214, 204, and 190 only in the 3',5'-cAMP spectrum and that of m/z 202 only in the spectrum from 2',3'-cAMP. These diagnostic ions at m/z 214, 204, and 202 arise from m/z 232 from the loss of H_2O, C_2H_4, and CH_2O, respectively.[36,37]

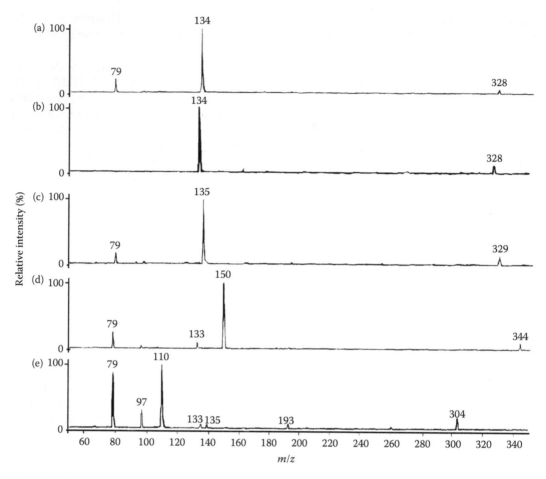

FIGURE 4.15 Quantitation of cyclic nucleotides by negative-ion electrospray mass spectrometry. Product ion spectra from [M–H]⁻ of (a) 3′,5′-cAMP, (b) 2′,3′-cAMP, (c) 3′,5′-cyclic IMP, (d) 3′,5′-cGMP, and (e) 3′,5′-cyclic CMP. (Modified from Newton, R.P. et al., *Organic Mass Spectrom.* 28, 899, 1993.)

In the product ion spectra of the protonated molecules of 2′,3′- and 3′,5′-cGMP, the protonated guanine base at *m/z* 152 is the most abundant product ion in both spectra; however, the peak at *m/z* 135, resulting from the loss of NH_3 from the base, only occurs in the spectrum of 3′,5′-cGMP. Further ions at *m/z* 328, 230, 202, and 110 in this 3′,5′-cGMP spectrum, formed by the loss of H_2O, H_3PO_4, and C_2H_4, respectively, are not observed in that of 2′,3′-cGMP.

TABLE 4.2
Diagnostic Ion Transitions Used in MRM for Quantitation of Cyclic Nucleotides by Negative LC/ES-MS/MS

Cyclic Nucleotide *m/z*	[M–H]⁻	Transition
3′,5′-cAMP	328	328 → 134
2′,3′-cAMP	328	328 → 134
3′,5′-cIMP	329	329 → 135
3′,5′-cGMP	344	344 → 150
3′,5′-cCMP	304	304 → 110

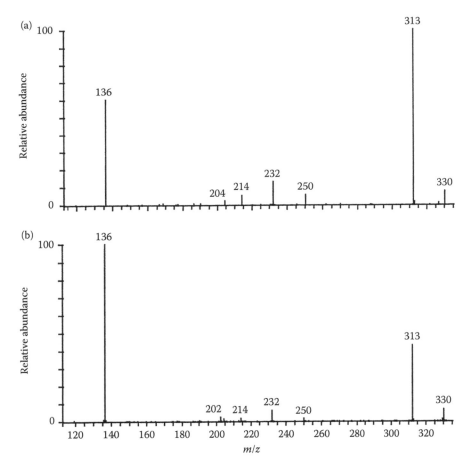

FIGURE 4.16 Positive-ion electrospray tandem mass spectrometry of purine cyclic nucleotides. Product ion spectra from [MH]+ of (a) 3′,5′-cAMP and (b) 2′,3′-cAMP.[37]

In the product ion spectra of the protonated molecules of 2′,3′- and 3′,5′-cyclic CMP, the most abundant fragment corresponds to the protonated cytidine base $[BH_2]^+$ at m/z 112, which subsequently loses NH_3 to form m/z 95 (Figure 4.17). Other product ions common to both spectra include m/z 289, 226, 208, and 190 formed by the loss of NH_3, HPO_3, H_3PO_4, and $H_3PO_4+H_2O$, respectively. The intensity of all four of these product ions is higher in the 3′,5′-isomer of cyclic CMP. These differences along with the formation of m/z 178 by the loss of CH_2O from m/z 208, observed only in 2′,3′-cyclic CMP spectrum, allow the isomers to be distinguished.

While negative ESI-MS is more productive for quantitation via on-line LC-MS, less fragmentation information of diagnostic value is obtained from negative ESI relative to that described above for positive ESI-MS. In the negative mode, fragmentation of the deprotonated molecule of 3′,5′-cAMP (m/z 328), for example, produces a major peak at m/z 134 ([B–H]−) and a minor peak at m/z 79, emanating from the pyrophosphate ion. In comparison, MS/MS of 2′,3′-cAMP showed only the adenine-derived peak at m/z 134. With 3′,5′-cGMP, the analogous guanine base (m/z 150) and pyrophosphate (m/z 79) fragments are again evident in negative ESI-MS/MS, together with a peak at m/z 133, identified as [guanine–NH_3]−. With 3′,5′-cyclic CMP, more extensive fragmentation is observed with the major ion at m/z 110 ([B–H]−), plus the pyrophosphate ion at m/z 79, and a second phosphate species, the dihydrogen phosphate at m/z 97. Minor fragment ions were observed at m/z 261, corresponding to a retro-Diels–Alder product of cytosine after the loss of isocyanate from the base, m/z 193, N^3-formyl cytosine, and m/z 139, ribose cyclic phosphate.[36,37,63] A generic fragmentation profile for 2′,3′- and 3′,5′-cyclic nucleotides is shown in Figure 4.18.

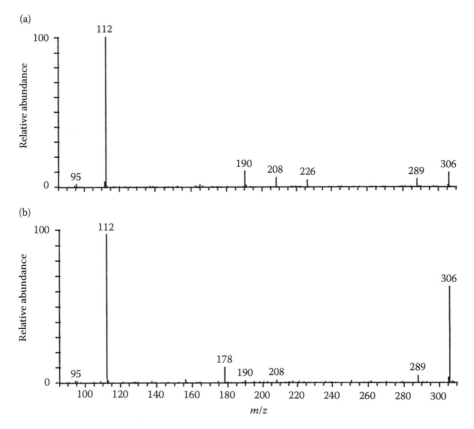

FIGURE 4.17 Positive-ion electrospray tandem mass spectrometry of pyrimidine cyclic nucleotides. Product ion spectra from [MH]$^+$ of (a) 3′,5′-cyclic CMP and (b) 2′,3′-cyclic CMP.[37]

As alluded to above, ESI-MS/MS fragmentations of cyclic nucleotides are more complex and more condition dependent than are the fragmentations in the FAB/MIKES spectra, and the complexity of nucleobase and nucleotide fragmentations are discussed elsewhere in the chapter of Dudley and in references [74–76]. Nevertheless, ESI-MS/MS is of value in the structural examination of cyclic nucleotide analogs, and is currently been applied in our laboratories in the examination of halogenated derivatives.[77]

When used for cyclic nucleotide-related enzyme quantitation, ESI-MS when compared to FAB-MS has the advantages of greater sensitivity, by at least one to two orders of magnitude, of a less complex spectrum, due to the absence of glycerol matrix-derived peaks, and the facility to be hyphenated to a chromatographic separation. The advantage that FAB-MS has is that of greater dynamic range. Thus with guanylyl cyclase incubates, the comparison of inactive and active enzymes by ESI-MS (Figure 4.19c and d) readily shows the appearance of cGMP at m/z 346, 368, and 390, corresponding to the protonated molecule and its mono- and disodium adducts, while the analogous GTP substrate peaks at m/z 524, 546, and 568 are evident in the inactive, but not in the active incubation spectrum. In the FAB mass spectra (Figure 4.19a and b), albeit with a more concentrated set of samples, the peaks described above relating to both the substrate and the product are visible in the active spectrum allowing monitoring not only of product (cGMP) appearance but also of substrate (GTP) disappearance, together with the appearance of additional GTP hydrolysis products GDP (m/z 444, 466, and 488) and GMP (m/z 364, 386, and 408). This multiple component monitoring, with the inclusion of effector turnover and binding parameters cannot readily be achieved by direct injection ESI-MS. However, such kinetic data can be obtained by on-line LC with ESI-MS, if it is accepted that the time of acquisition

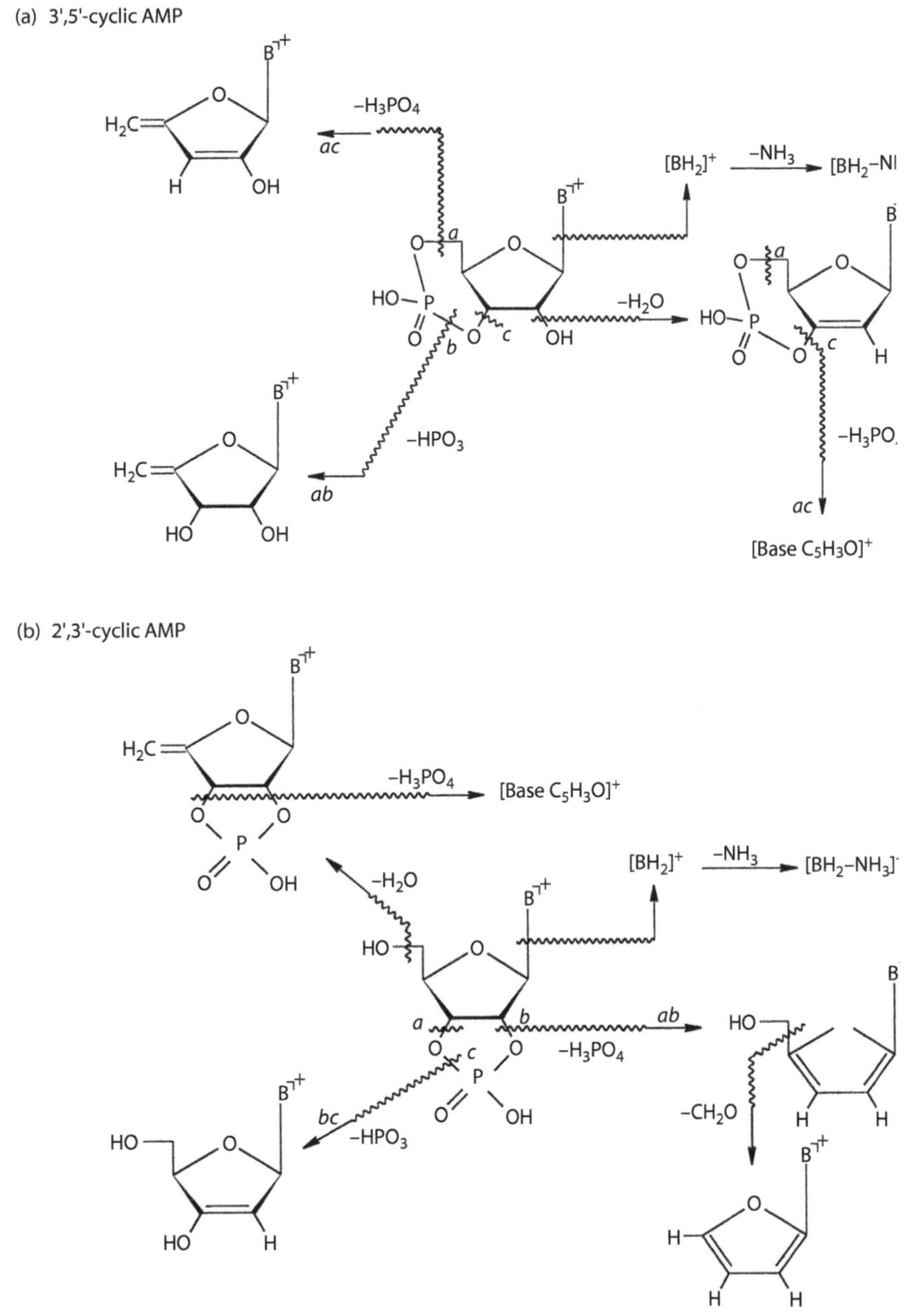

(a) 3',5'-cyclic AMP

(b) 2',3'-cyclic AMP

FIGURE 4.18 Generic fragmentation profiles from [MH]$^+$ of cyclic nucleotides in positive-ion electrospray tandem mass spectrometry. (a) 3',5'-Cyclic nucleotides and (b) 2',3'-cyclic nucleotides.[37]

is longer: This has been achieved with guanylyl cyclase, with the V_{max} and K_m parameters in the presence of varied concentrations of an agonist, NPS242, obtained by the conventional radioimmunoassay and by LC-ESI-MS quantitation showing reasonable correlation (Table 4.3); the FAB-MS method however has the advantage of allowing the additional monitoring of GTP disappearance, GDP and GMP appearance, and of NPS242 binding and metabolism (Figure 4.20).[78]

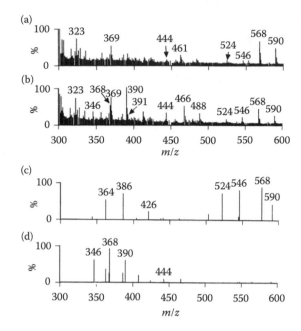

FIGURE 4.19 Quantitation of guanylyl cyclase activity by mass spectrometry. Positive-ion FAB mass spectra of incubates containing (a) inactive guanylyl cyclase and GTP substrate, (b) active guanylyl cyclase and GTP substrate; positive-ion electrospray mass spectra of incubates containing (c) inactive guanylyl cyclase and GTP substrate, and (d) active guanylyl cyclase and GTP substrate.[78]

Theoretically the use of LC-ESI-MS should provide the ideal basis for quantitation, utilizing MRM for multiple component monitoring. In practice the situation is more complex: Figure 4.21 presents a set of Lineweaver–Burk plots for a cAMP phosphodiesterase preparation, the plots being obtained by the traditional radiometric assay, by FAB-MS quantitation, and by LC-ESI-MS quantitation by SRM for substrate (cAMP) disappearance and for product (AMP) appearance.[78] The data obtained by the first three methods show good linearity and agreement: the data obtained by monitoring appearance of product do not. One plausible explanation of the latter lack of correlation is the recently discovered "IMAC-like" effect of the stainless steel probe and other instrument components in retaining AMP.[79]

TABLE 4.3
Comparison of Kinetic Data for Guanylyl Cyclase Obtained by Radiometric and Electrospray Mass Spectrometric Quantitation

	RIA		ES-MS	
	V_{max} (nmol/min/mg)	K_m (uM)	V_{max} (nmol/min/mg)	K_m (uM)
0 ng NPS242	360	126	400	141
4 ng NPS242	706	98	708	120
8 ng NPS242	1142	108	1236	127
16 ng NPS242	1560	96	1326	77
32 ng NPS242	1200	48	1554	55
64 ng NPS242	1600	43	1360	41

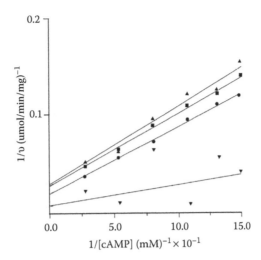

FIGURE 4.20 Lineweaver–Burk plots for kinetic analysis of cyclic AMP hydrolysis by phosphodiesterase. Kinetic data obtained by (●) radiometric assay, (■) FAB-MS quantitation, (▲) ESI-MS quantitation of cAMP depletion, and (▼) ESI-MS quantitation of AMP production.[78]

4.7 MACROMOLECULAR STUDIES

Direct use of MALDI in cyclic nucleotide studies has been restricted in its application. It has, although, proved useful in the examination of the cyclic nucleotide–protein conjugates used as immunogens to produce antisera. 2′-O-Succinyl-3′,5′-cyclic nucleotides are coupled to carrier proteins, such as thyroglobulin, bovine serum albumen, and so on, to render them immunogenic. This conjugation is not aimed at a 1:1 stoichiometry; rather the greater the number of cyclic nucleotide molecules bound on the surface of the protein carrier, the greater is the potency of the immunogen. MALDI provides a useful means of monitoring the effectiveness of the conjugate synthesis: In

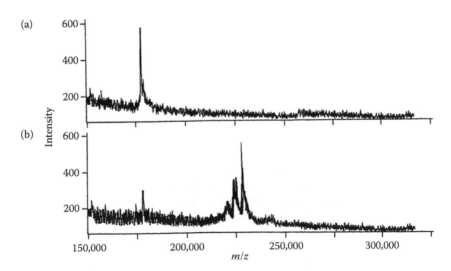

FIGURE 4.21 Mass spectrometric analysis of macromolecular cyclic nucleotide–protein conjugates. MALDI-TOF spectra of (a) poly-L-lysine polymer and (b) succinyl cyclic CMP-poly-L-lysine conjugate.[80]

Figure 4.21a, the MALDI mass spectrum of a synthetic protein carrier, poly-L-lysine is shown with peaks evident at m/z 178,000, while in Figure 4.21b, the major peaks of the purified conjugate are at m/z 222,000, 224,000, and 228,000, indicating that 110–130 molecules of succinyl cyclic CMP are conjugated to a single molecule of poly-L-lysine.[80]

In addition to the direct examination by MS of the cyclic nucleotides analogs, derivatives, and precursors described above, the advent of proteomic analysis by MS has also made immense contribution to our understanding of the mechanisms of action of cyclic nucleotide systems. A rapid and potentially useful means of differentiation of mononucleotides and cyclic nucleotides is that there is a hydrogen and deuterium exchange with mononucleotides in the gas phase but not with cyclic nucleotides.[81–83] While such exchanges remain at present only of potential direct use with cyclic nucleotides, they are already applied in examination of cyclic nucleotide-related proteins. For example, MALDI was used to identify peptic fragments from protein complexes that retained amide deuterium under hydrogen exchange conditions due to decreased solvent accessibility at the interface of the complex. The protein–protein interface was identified by the retention of more deuterons by peptides in the complex compared to controls in which only one peptide was present, enabling the identification of the kinase inhibitor and ATP-binding sites in the cAMP-dependent protein kinase.[84] The changes in backbone hydrogen/deuterium exchange in the regulatory subunit of cAMP-dependent protein kinase were probed by MALDI-TOF-MS in the presence of the catalytic subunit or cAMP; the data obtained indicated that the mutually exclusive binding of either cAMP or the catalytic subunit is controlled by binding at one site transmitting long-distance changes to the other site.[84]

Contributions to our understanding of the enzymes of cyclic nucleotide synthesis by MS have included the determination of the minimal stable and active catalytic C1 domain of 219 amino acids in adenylyl cyclase by limited proteolysis and ESI-MS/MS.[85] Similar applications into examination of cyclic nucleotide hydrolysis include a cAMP-specific phosphodiesterase in yeast that had its sites of phosphorylation identified as two serine residues, and that the sites were contained in the recognition sites for mitogen-activated protein kinase, suggesting that this enzyme could contribute to cross-talk between the cAMP and MAP kinase pathways.[86] Identification by ESI-MS/MS of a specific serine (Ser138) residue in calmodulin-stimulated phosphodiesterase phosphorylated by cAMP-dependent protein kinase that reduces the binding affinity of the phosphodiesterase for calmodulin, has demonstrated a switch-off mechanism for the latter enzyme by cAMP as a means of avoiding a futile biochemical cycle.[87] Other phosphodiesterase contributions include demonstration of the integration of the cAMP- and insulin-signaling pathways by mass spectrometric identification of the sites of the phosphorylations on a phosphodiesterase subjected to regulation by both systems and of the existence of ectophosphodiesterase activity, extracellular in the proximal renal tubules, and elucidation of its pharmacological characteristics by microdialysis coupled with ion-trap MS.[88–90]

In nonmammalian systems, MS has made a number of highly significant contributions to the knowledge of cyclic nucleotide-mediated pathways. In tobacco BY2 cells, proteins shown to possess cAMP-binding activity were identified as glyceraldehydes 3-phosphate kinase and two nucleoside diphosphate kinases by MS after combination with surface plasmon resonance and conventional affinity chromatography.[91] Furthermore, a peak in cAMP content during S and G_1 phases of the cell cycle in synchronously grown tobacco BY2 cells was monitored and identified by ESI-MS/MS; observation of this peak during a number of effector-based manipulations led to the proposal of a prostaglandin-dependent adenylyl cyclase activity, analogous to mammalian cyclases.[92] In slime molds, ESI-MS was used in the identification of platelet-activating factor, 1-O-alkyl-Δ^2-O-sn-glycero-3-phosphocholine, as a transient pulse in response to cAMP elevation in slime molds, and MS was also used to identify a novel 45-kDa protein, EppA, which regulates cAMP-induced relay and chemotaxis of the slime mold *Dictyostelium*.[93,94] In trematode worms, cAMP has now been demonstrated by ESI-MS/MS, while proteome study by MS has provided the first evidence of cAMP-mediated gene regulation in *Mycobacterium tuberculosis*, the etiological agent of tuberculosis, and locusta-diuretic peptide was first identified by ESI-MS as a potent stimulant of cAMP production and of fluid secretion in locust Malpighian tubules.[95-98] Further invertebrate examples include crustaceans; regulation of the molting cycle in decapod crustaceans involves a Y-organ in the cephalothorax, in which ecdysteriodogenesis is inhibited by two neuropeptides MIH and CHH. This inhibition is suppressed by cAMP and cGMP, and MS proteome analysis identified an increase in 170 proteins and decrease in 89, their identification indicating that phosphorylation of nitric oxide synthase is involved in neuropeptides signaling the pathway containing a nitric oxide-sensitive guanylyl cyclase.[99] A significant phosphorylation has also been examined in sea urchins. When sea urchin sperms contact the egg jelly, an exocytotic acrosome reaction is necessary for the fertilization to occur. This includes increases in adenylyl cyclase, cAMP, and cAMP-dependent protein kinase, and ESI-MS/MS has been used to identify six proteins that are phosphorylated during this process.[100]

A major result of the development of proteomics/MS technology has been its application in the study of cyclic nucleotide-dependent kinases and cyclic nucleotide-binding proteins. MS was used to identify proteins that bind to cAMP and cGMP by immobilizing the two cyclic nucleotides, selectively eluting cobinding proteins with a greater affinity for noncyclic nucleotides, oligonucleotides, and nucleic acids, and then identifying these proteins by MS. This approach was utilized to identify *inter alia* sphingosine kinase type 1-interacting protein as a protein binding to cAMP-dependent protein kinase in mammalian heart tissue.[101] Chemical proteomics, with immobilized cGMP together with MALDI-TOF-MS, were used to identify cGMP-binding proteins, with the surprising finding that mitogen-activated protein kinase 1 was one so identified. To extend the sensitivity of the process, surface plasmon resonance has now been included as a means of identifying low-abundance cyclic nucleotide-binding proteins.[102,103] IMAC-ESI-MS/MS has been used to identify proteins phosphorylated during the capacitation of mammalian sperms, a process requiring a cAMP-dependent increase in tyrosine phosphorylation. More than 60 phosphorylated targets were mapped including valosin-containing protein and two members of the A-kinase-anchoring protein family (AKAP).[104]

In addition to identifying proteins that are substrates of the cyclic nucleotide-dependent kinase, MS has enabled identification of the specific phosphorylation sites involved. The regulation of surface localization of the small conductance Ca^{2+}-activated K channel, Sk2, was identified by combination of mutagenesis and MS to involve four serine phosphorylation sites that are targets for cAMP-dependent protein kinase while quadrupole time-of-flight (Q-TOF) was used to identify autophosphorylation sites in the type II cGMP-dependent protein kinase, generating knowledge of the switch-off mechanism involving autoinhibition.[105,106] A good example of the value of HPLC-MS neutral loss scanning for large-scale identification and quantitation of protein phosphorylation in cyclic nucleotide-mediated signaling pathways is a study of vasopressin signaling in the renal collecting duct, in which protein phosphorylation plays a key role. Phosphoproteomic analysis was used to identify 714 phosphorylation sites found on 223 phosphoproteins that were located in the inner medullary collecting duct. This analysis involved immobilized metal affinity chromatographic enrichment and LC-MS neutral loss scanning. A number of phosphorylation sites were significantly changed in response to vasopressin.[107] A cAMP-dependent protein kinase, found in ram sperms, has several unusual properties; during investigation by MS, the catalytic unit was found to be 890 Da smaller than the C unit of the predominant form in somatic sources. ESI-MS/MS revealed that the amino terminal myristate and 14 amino acids in the somatic C are replaced by amino terminal acetate and six different amino acids; this altered the structure explaining the different localization process in the sperm flagellum.[108] ESI-MS/MS was also used to demonstrate that a regulator of microtubule dynamics, Oncoprotein 18, has two specific sites that, when phosphorylated by cAMP-dependent protein kinase, switch off Oncoprotein 18 in intact cells.[109]

The phosphorylation of the oncoprotein is one of many such examples that directly impinge on current applications in a medical context. Others include the finding that renal cyst enlargement was increased by cAMP. In polycystic kidney disease patients, a stable lipophilic molecule was identified by its HPLC-ESI-MS behavior as forskolin, which is a known activator of adenylyl cyclase that is produced in high quantities in *Coleus* roots, and sold for weight management and as a cardiovascular tonic, prompting recent warnings on its use.[110] The cAMP system was also implicated in alterations of neuronal functions in brains of chronic cocaine users by identifying proteomic alterations by peptide mass fingerprinting.[111] Similarly, MS has been used to identify the five proteins that are decreased with long-term morphine treatment. These five proteins include four α inhibitory and β G-protein subunits, providing a plausible explanation of opiate-induced sensitization of adenylyl cyclase.[112] Kruppel-like factor 6 is a zinc-finger transcription factor and a tumor suppressor that is inactivated in a number of human cancers. The activation of Kruppel-like factor 6 occurs via a process involving the cAMP-response element-binding protein, which resulted in a specific lysine acetylation. This lysine acetylation was first identified by MS and the loss of this acetylyl group appeared to contribute to the failure to suppress tumor growth.[113] Endosulfine is the natural agonist of the high-affinity binding sites that are coupled to K^+-ATP channels by which the antidiabetic drugs, sulfonylureas, act. By using MS, HPLC, and Edman degradation, the porcine endosulfine A has been identified as cAMP-regulated phosphoprotein-19, a previously known substrate of cAMP-dependent protein kinase in bovine brain.[114] In Guadeloupe, atypical Parkinson's disease is linked with use of fruit and teas from *Annonaceae* plants; annonacin, from these plants, was detected in the brain parenchyma by MALDI-TOF-MS and was associated with significant depletion in dopaminergic neurons in the *substantia nigra* and in cAMP-regulated phosphoprotein GABAergic neurons, lending credence to the hypothesis that some forms of Parkinson's disease might be induced by environmental toxins.[115] Final medical representative examples are the demonstrations that serine phosphorylation of nitric oxide synthase3 occurs in human umbilical vein endothelial cells in response to stimulation of beta$_2$ adrenoreceptors, activating nitric oxide production; it is regulated by both cAMP-dependent protein kinase and phosphatidylinositol 3-kinase, and occurs at a specific site (Ser1177) identified by ESI-MS/MS, and that D-mannose was identified by ESI-MS as the carbohydrate that stimulates mammalian retinal ganglion cells to regenerate their axons in the presence of elevated intracellular cAMP levels.[116,117]

4.8 CONCLUDING REMARKS

MS has made significant contributions to the exciting developments in cyclic nucleotide research and their subsequent application, and further contributions are in prospect. The importance of the mass spectrometric contribution can be exemplified by consideration of just one cyclic nucleotide, cyclic CMP. MS data were crucial to demonstrating the natural occurrence of the compound in mammals, plants, and bacteria, and in unequivocally demonstrating it as a product of cytidylyl cyclase activity and identifying side products, thereby resolving debate over the latter enzyme's existence. MS was the first means of demonstrating cyclic CMP-responsive protein kinase activity, it was also used to optimize the production of cyclic CMP derivatives for radioimmunoassay development and synthesis of cell-permeating derivatives, and as a method for quantitation of the cyclic nucleotide itself and of the enzymes involved in its metabolism and functioning. More recently the ability to examine macromolecules by MALDI- and ES-based proteomic analysis has enabled our laboratory to identify Rab23 as a protein selectively phosphorylated in murine brain in response to elevated cyclic CMP levels, providing a vital step along the route to determining the mechanism of action of cyclic CMP.[118]

Future contributions, as our mass spectrometric techniques are further developed and refined, with respect to sensitivity, resolution, and capacity, will hopefully aid greater understanding of the mechanisms of action of the naturally occurring six cyclic nucleotides. It will also enable examination of naturally occurring analogs, such as the deoxy- and aminoacyl-cyclic nucleotides, and help determining whether they are merely metabolic errors or also possess a specific function.

REFERENCES

1. Rall, T.W., Sutherland, E.W., and Berthet, J., The relation of epinephrine and glucagon to liver phosphorylase, *J. Biol. Chem.* 224, 1957, 1987.
2. Robison, G.A., Butcher, R.W., and Sutherland, E.W., *Cyclic AMP*, Academic Press, New York, 1971.
3. Strader, C.D., Fong, T.M., and Toat, M.R., Structure and function of G-protein coupled receptors, *Ann. Rev. Biochem.* 63, 101, 1994.
4. Gilman, A.G., G-proteins: Transducers of receptor-generated signals, *Ann. Rev. Biochem.* 56, 615, 1987.
5. Francis, S.H. and Corbin, J.D., Structure and function of cyclic nucleotide-dependent protein kinases, *Ann. Rev. Physiol.* 56, 237, 1994.
6. Beavo, J. and Houslay, M., *Cyclic Nucleotide Phosphodiesterases; Structure, Regulation and Drug Activity*, Wiley, New York, 1990.
7. Cesare, D. De, Fimiaa, G.M., and Sassone-Corsi, P., Signalling routes to CREM and CREB, *Trends Biochem. Sci.* 24, 281, 1999.
8. Bradley, J., Heterotrimeric olfactory cyclic nucleotide-gated channels; a subunit that confers increased sensitivity to cAMP, *Proc. Natl. Acad. Sci. USA* 91, 8890, 1994.
9. De Rooij, J., Zwartkruis, E.J.T., and Verheijen, M.H.G., Epac is a Rap 1 guanine nucleotide exchange factor directly activated by cyclic AMP, *Nature* 396, 474, 1998.
10. Lawson, A.M., Stillwell, R.N., Tacker, M.M., et al., Mass spectrometry of nucleic acid components-trimethylsilyl derivatives of nucleotides, *J. Am. Chem. Soc.* 93, 1014, 1971.
11. Newton, R.P., Gibbs, N., Moyse, C.D., et al., Mass spectrometric identification of adenosine 3′,5′-cyclic monophosphate isolated from a higher plant tissue, *Phytochemistry* 19, 1909, 1980.
12. Newton, R.P. and Brown, E.G., The biochemistry and physiology of cyclic AMP in higher plants, in *Receptors in Plants and Cellular Slime Moulds*, C.M. Garrod and D.R. Chadwick (eds), p. 115, Cambridge University Press, Cambridge, 1986.
13. Newton, R.P., Roef, L., Witters, E., et al., Cyclic nucleotides in higher plants: The enduring paradox: Tansley review, *New Phytologist* 143, 427, 1999.
14. Newton, R.P. and Smith, C.J., Cyclic nucleotides, *Phytochemistry* 65, 2423, 2004.
15. Barber, M., Bordoli, R.S., Sedgwick, R.D., et al., Fast atom bombardment of solids (FAB) a new ion-source for mass spectrometry, *J. Chem. Soc. Chem. Commun.* 7, 325 1981.
16. Kingston, E.E., Beynon, J.H., and Newton, R.P., The identification of cyclic nucleotides in living systems using collision-induced dissociation of ions generated by fast atom bombardment, *Biomed. Mass Spectrom.* 11, 367, 1984.
17. Kingston, E.E., Beynon, J.H., Newton, R.P., et al., The differentiation of isomeric biological compounds using collision-induced dissociation of ions generated by fast atom bombardment, *Biomed. Mass Spectrom.* 12, 525, 1985.
18. Newton, R.P., Kingston, E.E., and Overton, A., Identification of novel nucleotides found in the red seaweed, *Porphyra umbilicalis*, *Rapid Commun. Mass Spectrom.* 9, 305, 1995.
19. Newton, R.P., Kingston, E.E., Evans, D.E., et al., Occurrence of guanosine 3′,5′-cyclic monophosphate (cyclic GMP) and associated enzyme systems in *Phaseolus vulgaris* L., *Phytochemistry* 23, 1367, 1984.
20. Newton, R.P., Mass spectrometric analysis of cyclic nucleotides and related enzymes, in *Applications of Modern Mass Spectrometry in Plant Science Research*, R.P. Newton and T.J. Walton (eds), p. 159, Oxford University Press, Oxford, 1996.
21. Pacini, B., Petrigliano, A., Diffley, P., et al., Adenylyl cyclase activity in roots of *Pisum sativum*, *Phytochemistry* 34, 899, 1993.
22. Cooke, C.J., Smith, C.J., Walton, T.J., et al., Evidence that cyclic AMP is involved in the hypersensitive response of *Medicago sativa* to a fungal elicitor, *Phytochemistry* 35, 899, 1994.
23. Newton, R.P., Salih, S.G., Salvage, B.J., et al., Extraction, purification and identification of cytidine 3′,5′-cyclic monophosphate from rat tissues, *Biochem. J.* 221, 665, 1984.
24. Newton, R.P., Kingston, E.E., Hakeem, N.A., et al., Extraction, purification, identification and metabolism of 3′,5′-cyclic UMP, 3′,5′-cyclic IMP and 3′,5′-cyclic dTMP from rat tissues, *Biochem. J.* 236, 431, 1986.
25. Newton, R.P., Chiatante, D., Ghosh, D., et al., Identification of cyclic nucleotide constituents of meristematic and non-meristematic tissues of *Pisum sativum* roots, *Phytochemistry* 28, 2243, 1989.
26. Newton, R.P., Kingston, E.E., and Overton, A., Mass spectrometric identification of cyclic nucleotides released by the bacterium *Corynebacterium murisepticum* into the culture medium, *Rapid Commun. Mass Spectrom.* 12, 729, 1998.

27. Newton, R.P., Cytidine 3′,5′-cyclic monophosphate: A third cyclic nucleotide secondary messenger? *Nucleos. Nucleot.* 14, 743, 1995.

28. Newton, R.P., Hakeem, N.A., Salvage, B.J., et al., Cytidylate cyclase activity: Identification of cytidine 3′,5′-cyclic monophosphate and four novel cytidine cyclic phosphates as biosynthetic products from cytidine triphosphate, *Rapid Commun. Mass Spectrom.* 2, 118, 1988.

29. Newton, R.P., Evans, A.M., van Geyschem, J., et al., Radioimmunoassay of cytidine 3′,5′-cyclic mono-phosphate: Unambiguous assay by means of an optimized protocol incorporating a trilayer column sepa-ration to obviate cross-reactivity problems, *J. Immunoassay* 15, 317, 1994.

30. Newton, R.P., Walton, T.J., Basaif, S.A., et al., Identification of butyryl derivatives of cyclic nucleotides by positive ion fast atom bombardment mass spectrometry and mass-analysed ion kinetic energy spec-trometry, *Organic Mass Spectrom.* 24, 679, 1989.

31. Newton, R.P., Evans, A.M., Hassan, H.G., et al., Identification of cyclic nucleotide derivatives synthe-sized for radioimmunoassay development by fast-atom bombardment with collision-induced dissociation and mass-analysed ion kinetic energy spectroscopy, *Organic Mass Spectrom.* 28, 899, 1993.

32. Newton, R.P., Bayliss, M.A., Wilkins, A.C.R., et al., Fast-atom bombardment mass spectrometric analy-sis of cyclic-nucleotide analogues used in studies of cyclic nucleotide-dependent protein kinases, *J. Mass Spectrom. Rapid Commun Mass Spectrom.* S107, 1995.

33. Bayliss, M., Pereira, L., Games, D.E., et al., Fast-atom bombardment mass spectrometry of site-specific effectors of cyclic nucleotide-dependent protein kinases, *Atti MS Pharmaday* 2, 67, 1996.

34. Pereira, M.L., Walton, T.J., Newton, R.P., et al., Mass spectrometry of site-selective effectors of cyclic nucleotide-dependent protein kinases, *Proc. IIIrd Portugese Mass Spectrometry National Meeting*, p. 273, 1997.

35. Walton, T.J., Bayliss, M.A., Pereira, M.L., et al., Fast-atom bombardment tandem mass spectrometry of cyclic nucleotide analogues used as site-selective activators of cyclic nucleotide-dependent protein kinases, *Rapid Commun. Mass Spectrom.* 12, 449, 1998.

36. Pereira, M.L.M., *Analysis of Pharmaceutical Products and Nucleotides by LCMS and Tandem MS*, PhD thesis, Swansea University, 1998.

37. Pereira, M.L.M., Games, D.E.G., and Newton, R.P., unpublished observations.

38. Langridge, J.I., Evans, A.M., Walton, T.J., et al., Fast-atom bombardment mass spectrometric analysis of cyclic nucleotide and nucleoside triphosphate analogues used in studies of cyclic nucleotide-related enzymes, *Rapid Commun. Mass Spectrom.* 7, 725, 1993.

39. Langridge, J.I., Evans, A.M., Ghosh, D., et al., Application of continuous-flow fast atom bombardment mass spectrometry to cyclic nucleotide biochemistry, *Anal. Chem. Acta* 247, 177, 1991.

40. Langridge, J.I., Brenton, A.G., Walton, T.J., et al., Analysis of cyclic nucleotide-related enzymes by continuous-flow fast-atom bombardment mass spectrometry, *Rapid Commun. Mass Spectrom.* 7, 293, 1993.

41. Langridge, J.I., Brenton, A.G., Walton, T.J., et al., Qualitative and quantitative analysis of cyclic nucle-otides and related enzymes by static and dynamic fast-atom bombardment mass spectrometry, *Nucleos. Nucleot.* 14, 739, 1995.

42. Newton, R.P., Brenton, A.G., Walton, T.J., et al., Assay of cyclic nucleotide phosphodiesterase activity by FAB mass spectrometry with MIKES scanning, *Adv. Mass Spectrom.* 11, 1234, 1988.

43. Newton, R.P., Walton, T.J., Brenton, A.G., et al., Quantitation of fast atom bombardment/mass-analysed ion kinetic energy spectrometry; kinetic analysis of cyclic nucleotide phosphodiesterase activity, *Rapid Commun. Mass Spectrom.* 3, 178, 1989.

44. Newton, R.P., Brenton, A.G., Ghosh, D., et al., Quantitative and qualitative mass spectrometric analysis of cyclic nucleotides and related enzymes, *Anal. Chem. Acta* 247, 161, 1991.

45. Newton, R.P., Khan, J.A., Ghosh, D., et al., Determination of enzyme kinetic parameters of cyclic CMP-specific phosphodiesterase by quantitative fast atom bombardment mass spectrometry, *Organic Mass Spectrom.* 26, 447, 1991.

46. Walton, T.J., Ghosh, D.E., Newton, R.P., et al., Differentiation of isomeric purine and pyrimidine mono-nucleotides by fast atom bombardment tandem mass spectrometry, *Nucleos. Nucleot.* 9, 967, 1991.

47. Ghosh, D., Newton, R.P., Brenton, A.G., et al., Fast atom bombardment tandem mass spectrometry in the identification of isomeric ribomononucleotides, *Anal. Chem. Acta* 247, 187, 1991.

48. Newton, R.P., Bayliss, M.A., Khan, J.A., et al., Kinetic analysis of cyclic CMP-specific and multifunc-tional phosphodiesterases by quantitative positive-ion mass spectrometry, *Rapid Commun. Mass Spectrom.* 13, 574, 1999.

49. Smith, C.J., Roef, L., Walton, T.J., et al., Variation in isomeric products of phosphodiesterase from the chloroplasts of *Phaseolus vulgaris* in response to cations, *Plant Biosystems* 135, 143, 2001.

50. Walton, T.J., Langridge, J.I., Khan, J.A., et al., Analysis of kinetic parameters of cyclic nucleotide phosphodiesterase systems by fast atom bombardment tandem mass spectrometry, *Biochem. Soc. Trans.* 19, 397S, 1991.

51. Newton, R.P., Brenton, A.G., Harris, F.M., et al., Application of modern mass spectrometric methods to biochemical studies of second messengers, *Chimica Oggi* 10(5), 21, 1992.

52. Newton, R.P., Khan, J.A., Brenton, A.G., et al., Quantitation by fast atom bombardment mass spectrometry: Assay of cytidine 3′,5′-cyclic monophosphate-responsive protein kinase, *Rapid Commun. Mass Spectrom.* 6, 601, 1992.

53. Newton, R.P., Contributions of fast-atom bombardment mass spectrometry to studies of cyclic nucleotide biochemistry, *Rapid Commun. Mass Spectrom.* 7, 528, 1993.

54. Newton, R.P., Evans, A.M., Langridge, J.I., et al., Assay of cyclic AMP-dependent protein kinase by quantitative fast-atom bombardment mass spectrometry, *Anal. Biochem.* 224, 32, 1995.

55. Newton, R.P., Mass spectrometric analysis of cyclic nucleotides and cyclic nucleotide-related enzymes, *Atti MS Pharmaday* 2, 30, 1996.

56. Newton, R.P., Groot, N., van Geyschem, J., et al., Estimation of cytidylyl cyclase activity and monitoring of side-product formation by fast-atom bombardment mass spectrometry, *Rapid Commun. Mass Spectrom.* 11, 189, 1997.

57. Newton R.P., Kinetic analyses of nucleotidyl cyclases, cyclic nucleotide phosphodiesterases, and cyclic nucleotide-responsive protein kinases by quantitative mass spectrometry, *Atti MS-Pharmaday* 3, 48, 1999.

58. Newton, R.P., Quantitation of nucleotidyl cyclase and cyclic nucleotide-sensitive protein kinase activities by fast-atom bombardment mass spectrometry: A paradigm for multiple component monitoring by quantitative MS, in *Peptide and Protein Analysis; Advances in the Use of Mass Spectrometry*, J.R. Chapman (ed.), Methods in Molecular Biology Series Publications, p. 369, Humana Press, Totowa, NJ, 2000.

59. Evans, A.M., Harris, F.M., Walton, T.J., et al., Determination of cyclic nucleotide-responsive protein kinase activity by quantitative fast atom bombardment mass spectrometry, *Biochem. Soc. Trans.* 20, 1525, 1992.

60. Newton, R.P., Bayliss, M.A., van Geyschem J., et al., Kinetic analysis and multiple component monitoring of effectors of adenylyl cyclase activity by quantitative fast-atom bombardment mass spectrometry, *Rapid Commun. Mass Spectrom.* 11, 1060, 1997.

61. Newton, R.P., Bayliss, M.A., Langridge, J.A., et al., Product identification and kinetic studies of nucleotidyl cyclase activity in isolated chloroplasts by quantitative fast-atom bombardment mass spectrometry, *Rapid Commun. Mass Spectrom.* 13, 979, 1999.

62. Fathi, M., Tsacopoulos, M., Raverdino, V., et al., Separation of nucleotides in homogenates of Octopus retina by ion-pair reversed-phase liquid chromatography and identification by mass spectrometry, *J. Chromatogr. Biomed. Appl.* 563, 365, 1991.

63. Witters, E., Roef, L., Newton, R.P., et al., Quantitation of cyclic nucleotides in biological samples by negative electrospray tandem mass spectrometry coupled to ion suppression liquid chromatography, *Rapid. Commun. Mass Spectrom.* 10, 225, 1996.

64. Witters, E., VanDongen, W., Esmans, E.L., et al., Ion-pair liquid chromatography electrospray mass spectrometry for the analysis of cyclic nucleotides, *J. Chromatogr. B* 694, 55, 1997.

65. Vanhoutte, K., Van Dongen, W., Hoes, I., et al., Development of a nanoscale liquid chromatography-electrospray mass spectrometry method for the detection and identification of DNA adducts, *Anal. Chem.* 69, 3161, 1997.

66. Witters, E., Vanhoutte, K., Van Dongen, W., et al., Qualitative analysis of cyclic nucleotides and cytokinins using capillary column switching ES-LC-MSMS, *Proc. 45th ASMS Conference on Mass Spectrometry and Allied Topics*, Palm Springs California. p. 163, June 1–5, 1997.

67. Witters, E., Vanhoutte, K., Dewitte, W., et al., Analysis of cyclic nucleotides and cytokinins in minute plant samples using phase-system switching capillary electrospray-LC-MSMS, *Phytochem. Anal.* 10, 143, 1999.

68. Ehsan, H., Reichheld, J.-P., Roef, L., Witters, E., Lardon, F., Van Bockstaele, D., Van Montagu, M., Inzé, D., and Van Onckelen, H., Effect of indomethacin on cell cycle dependent cyclic AMP fluxes in tobacco BY-2 cells, *FEBS Lett.* 442, 165–169, 1998.

69. Roef, L., Witters, E., Gadeyne, J., et al., Analysis of 3′,5′ cAMP and adenylyl cyclase activity in higher plants using polyclonal chicken egg yolk antibodies, *Anal. Biochem.* 233, 188, 1996.

70. Roef, L., 1997. *Het 3′,5′-cAMP Metabolisme in Hogere Planten: Bijdrage tot de Karakterisatie van Adenylyl Cyclase en cAMP-afhankelijk Proteïne Kinase*. PhD thesis, University of Antwerp (UIA), Belgium.

71. Witters, E., Quanten, L., Bloemen, J., et al., Product identification and adenylyl cyclase activity in chloroplasts of *Nicotiana tabacum*, *Rapid Commun. Mass Spectrom.* 18, 499, 2004.

72. Richards, H., Das, S., Smith, C.J., et al., Cyclic nucleotide content of tobacco BY-2 cells, *Phytochemistry* 61, 531, 2002.

73. Geisbrecht, A., van Geyschem, Newton, R.P., unpublished observations.

74. Banoub, J.H., Newton, R.P., Esmans, E., et al., Recent developments in mass spectrometry for the characterization of nucleosides, nucleotides, oligonucleotides and nucleic acids, *Am. Chem. Soc. Rev.* 105, 1869, 2005.

75. Tuytten, R., Lemiere, F., Van Dongen, W., et al., Intriguing mass spectrometric behaviour of guanosine under low energy collision-induced dissociation: H_2O-adduct formation and gas phase reactions in the collision cell, *J. Am. Soc. Mass Spectrom.* 16, 1291, 2005.

76. Tuytten, R., Herrebout, W., Lemiere, F., et al., In-source CID of guanosine: Gas phase ion-molecule reactions, *J. Am. Soc. Mass Spectrom.* 17, 1050, 2006.

77. Bond, A., Dudley, E., Newton, R.P., et al., unpublished observations.

78. Oliver, M., Geisbrecht, A., Newton, R.P., et al., unpublished observations.

79. Tuytten, R., Lemiere, F., Witters E., et al., The stainless steel electrospray probe: A dead end for phosphorylated organic compounds, *J. Chromatogr. A* 1104: 209–221, 2006.

80. Geisbrecht, A., van Cleef, J., Newton, R.P., et al., unpublished observations.

81. Robinson, J.M., Greig, M.J., Griffey, R.H., et al., Hydrogen/deuterium exchange of nucleotides in the gas phase, *Anal. Chem.* 70, 3566, 1998.

82. Mandell, J.G., Falick, A.M., and Komives, E.A., Identification of protein–protein interfaces by decreased amide proton solvent accessibility, *Proc. Natl. Acad. Sci. USA* 95, 14705, 1998.

83. Mandell, J.G., Falick, A.M., and Komives, E.A., Measurement of amide hydrogen exchange by MALDI-TOF mass spectrometry, *Anal. Chem.* 70, 3987, 1998.

84. Anand, G.S., Hughes, C.A., Jones, J.M., et al., Amide H/H-2 exchange reveals communication between the cAMP and catalytic subunit-binding sites in the R-I alpha subunit of protein kinase A, *J. Mol. Biol.* 323, 377, 2002.

85. Zhang, G.Y., Liu, Y., Qin, J., et al., Characterization and crystallization of a minimal catalytic core domain from mammalian type II adenylyl cyclase, *Protein Sci.* 6, 903, 1997.

86. Lenhard, J.M., Kassel, D.B., Rocque, W.J., et al., Phosphorylation of a cyclic AMP-specific phosphodiesterase by mitogen-activated protein kinase, *Biochem. J.* 316, 751, 1996.

87. Florio, V.A., Sonnenburg, W.K., Johnson, R., et al., Phosphorylation of the 61-K-Da calmodulin-stimulated cyclic nucleotide phosphodiesterase at serine-120 reduces its affinity for calmodulin, *Biochemistry* 33, 8948, 1994.

88. Lind, R., Ahmad, F., Resjo, S., et al., Multisite phosphorylation of adipocyte and hepatocyte phosphodiesterase 3B, *Biochim. Biophys. Acta. Mol. Cell Res.* 1773, 584, 2007.

89. Jackson, E.K., Zacharia, L.C., Zhang, M., et al., cAMP-adenosine pathway in the proximal tubule, *J. Pharmacol. Exp. Therap.* 317, 1219, 2006.

90. Jackson, E.K., Ren, J., Zacharia, L.C., et al., Characterization of renal ecto-phosphodiesterase, *J. Pharmacol. Exp. Therap.* 321, 810, 2007.

91. Laukens, K., Roef, L., Witters, E., et al., Cyclic AMP affinity purification and ESI-QTOF MS-MS identification of cytosolic glyceraldehydes 3-phosphate dehydrogenase and two nucleoside diphosphate kinase isoforms from tobacco BY-2 cells, *FEBS Lett.* 508, 75, 2001.

92. Ehsan, H., Reichheld, J.P., Roef, L., et al., Effect of indomethacin on cell cycle dependent cyclic AMP fluxes, *FEBS Lett.* 422, 165, 1998.

93. Bussolino, F., Sordano, C., Benfenati, E., et al., Dictyostelium cells produce platelet-activating-factor in response to cyclic AMP, *Eur. J. Biochem.* 196, 609, 1991.

94. Chen, S.Y., Segall, J.E., EppA, a putative substrate of DdERK2, regulates cyclic AMP relay and chemotaxis in *Dictyostelium discoideum*, *Eukaryotic Cells* 5, 1136, 2006.

95. Ehsan, H., Rashid, K.A., Van Dongen, W., et al., Cyclic adenosine monophophshate in *Gastrothylax crumenifer* and *Explanatum explanatum*, *J. Helminthol.* 71, 147, 1997.

96. Gazdik, M.A., and McDonough, K.A., Identification of cyclic AMP-regulated genes in Mycobacterium tuberculosis complex bacteria under low oxygen conditions, *J. Bacteriol.* 187, 2681, 2005.

97. Patel, M., Chung, J.S., Kay, I., et al., Localization of Locusta-DP in locust CNS and haemolymph satisfies initial hormonal criteria, *Peptides* 15, 591, 1994.

98. Coast, G.M., Rayne, R.C., Hayes, T.K., et al., A comparison of the effects of 2 putative diuretic hormones from *Locusta migratoria* on isolated locust Malpghian tubules, *J. Exp. Biol.* 175, 1, 1993.

99. Lee, S.G. and Mykles, D.L., Proteomics and signal transduction in the crustacean molting gland, *Integr. Comp. Biol.* 46, 965, 2006.

100. Su, Y.H., Chen, S.H., Zhou, H.L., et al., Tandem mass spectrometry identifies proteins phosphorylated by cyclic AMP-dependent protein kinase when sea urchin sperm undergo the acrosome reaction, *Dev. Biol.* 285, 116, 2005.
101. Scholten, A., Poh, M.K., van Veen, T.A.B., et al., Analysis of the cGMP/cAMP interactome using a chemical proteomics approach in mammalian heart tissue validates sphingosine kinase type1-interacting protein as a genuine and highly abundant AKAP, *J. Proteome Res.* 5, 1435, 2006.
102. Kim, E. and Park, J.M., Identification of novel target proteins of cyclic GMP signaling pathways using chemical proteomics, *J. Biol. Mol. Biol.* 36, 299, 2003.
103. Visser, N.F.C., Scholten, A., van den Heuvel, R.H.H., et al., Surface plasmon based chemical proteomics; efficient specific extraction and semiquantitative identification of cyclic nucleotide-binding proteins from cellular lysates by using a combination of surface plasmon resonance, sequential elution and liquid chromatograph-tandem mass spectrometry, *Chembiochemistry* 8, 298, 2007.
104. Ficarro, S., Chertihin, O., Westbrook, V.A., et al., Phosphoproteome analysis of capacitated human sperm-Evidence of tyrosine phosphorylation of a kinase-anchoring protein 3 and valosin-containing protein/p97 during capacitation, *J. Biol. Chem.* 278, 11579, 2003.
105. Ren, Y.J., Barnwell, L.F., Alexander, J.C., et al., Regulation of surface localization of the small conductance Ca^{2+}-activated potassium channel, Sk2, through direct phosphorylation by cAMP-dependent protein kinase, *J. Biolog. Chem.* 281, 11769, 2006.
106. Vaandrager, A.B., Hogema, B.M., Edixhoven, M., et al., Autophosphorylation of cGMP-dependent protein kinase type 11, *J. Biol. Chem.* 278, 28651, 2003.
107. Hoffert, J.D., Pisitkun, T., Wang, G.H., et al. Quantitative phosphoproteomics of vasopressin-sensitive renal cells: Regulation of aquaporin-2 phosphorylation at two sites, *Proc. Natl. Acad. Sci. USA* 103, 7159, 2006.
108. San Agustin, J.T., Leszyk, J.D., Nuwaysir, L.M., et al., The catalytic subunit of the cAMP-dependent protein kinase ovine sperm flagella has a unique amino-terminal sequence, *J. Biol. Chem.* 273, 24874, 1998.
109. Gradin, H.M., Larsson, N., Marklund, U., et al., Regulation of microtubule dynamics by extracellular signals: Cyclic AMP-dependent protein kinase switches off the activity of oncoprotein 18 in intact cells, *J. Cell Biol.* 140, 131, 1998.
110. Putnam, W.C., Swenson, S.M., Reif, G.A., et al., Identification of a forskolin-like molecule in human renal cysts, *J. Am. Soc. Nephrol.* 18, 934, 2007.
111. Tannu, N., Mash, D.C., and Hemby, S.E., Cytosolic proteomic alterations in the nucleus accumbens of cocaine overdose victims, *Mol. Psychiatr.* 12, 55, 2007.
112. Mouledous, L., Neasta, J., Uttenweiler-Joseph, S., et al., Long-term morphine treatment enhances proteasome-dependent degradation of G beta in human neuroblastoma SH-SY5Y cells: Correlation with onset of adenylate cyclase sensitization, *Mol. Pharm.* 68, 467, 2005.
113. Li, D., Yea, S., Dolios, G., et al., Regulation of Kruppel-like factor 6 tumor suppressor activity by acetylation, *Cancer Res.* 65, 9216, 2005.
114. Virsolvyvergine, A., Salazar, G., Sillard, R., et al., Endosulphine, endogenous ligand from the sulphonylurea receptor: Isolation from porcine brain and partial structural determination of the alpha from, *Diabetologia* 39, 135, 1996.
115. Champy, P., Hoglinger, G.U., Feger, J., et al., Annnonacin, a lipophilic inhibitor of mitochondrial complex 1, induces nigral and striatal neurodegeneration in rats: Possible relevance for atypical parkinsonism in Guadeloupe, *J. Neurochem.* 88, 63, 2004.
116. Queen, L.R., Ji, Y., Xu, B., et al., Mechanisms underlying beta(2)-adrenoceptor-mediated nitric oxide generation by human umbilical vein endothelial cells, *J. Physiol.* 576, 585, 2006.
117. Li, Y.M., Irwin, N., Yin, Y.Q., et al., Axon regeneration in goldfish and rat retinal ganglion cells: Differential responsiveness to carbohydrates and cAMP, *J. Neurosci.* 23, 7830, 2003.
118. Bond, A.E., Dudley, E., and Tuytten, R., Mass spectrometric identification of Rab23 phosphorylation as a response to challenge by cytidine 3′,5′-cyclic monophosphate in mouse brain, *Rapid Commun. Mass Spectrom.* 21, 1, 2007.

5 Analysis of Urinary Modified Nucleosides by Mass Spectrometry

Ed Dudley

CONTENTS

5.1 INTRODUCTION

Nucleic acids are polymers of nucleotides, composed of a nucleobase attached to the sugar moiety at the anomeric C-1′ position and a phosphate group attached at the 5′ position. Nucleosides are nucleotide monomers without the phosphate group on the 5′ position of the sugar moiety. Unmodified (or "normal") nucleosides are considered to be the majority of nucleic acid macromolecules. These nucleosides are adenosine, cytosine, guanosine, and uridine in ribonucleic acid (RNA) molecules. The deoxynulceosides (DNA) have the same nucleobases as in the RNA molecules, except that thymidine replaces uridine in the DNA structures. Nucleoside modifications are based both on the nucleobases and on the available sugars. The modified nucleosides represent these same compounds with either additional moieties added to the basic nucleoside structures or an isomeric change in the molecule. Methylation is the most common form of modification detected in all nucleic acids, with up to three positions on the nucleoside molecule being methylated. These methyl groups are more commonly found on the base of the nucleoside but may also be present on the sugar moiety. Other modifications include acetylation, formylation, and isomeric changes in which the nucleobase side is attached to the C-1′ anomeric carbon atom of the sugar via a carbon atom of the base rather than the usual glycosidic (N–C) bond.

An overview of some of the modified nucleosides is shown in Figure 5.1. Internet databases of the known modified nucleosides and their biosynthetic pathways are available.[1,2] These nucleic acid

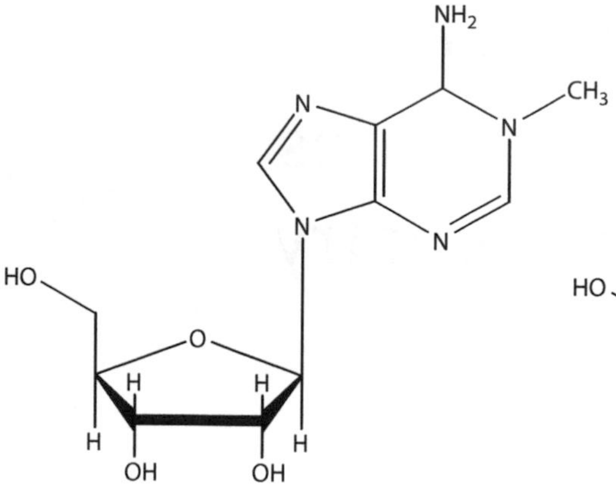

1-Methyladenosine

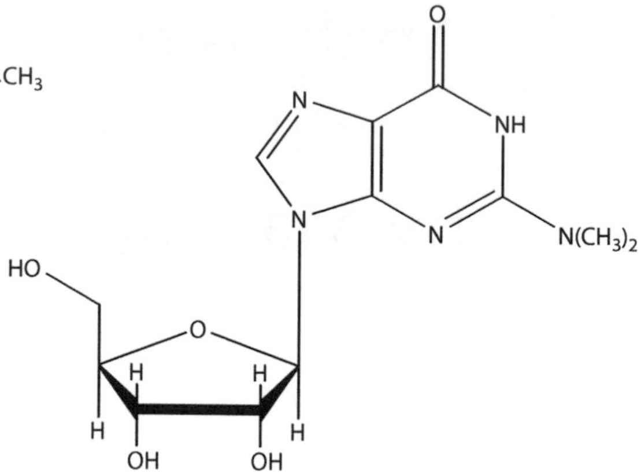

N^2, N^2-Dimethylguanosine

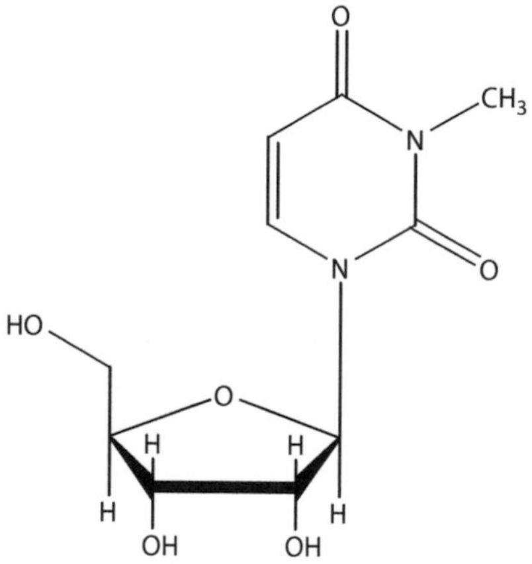

3-Methyluridine

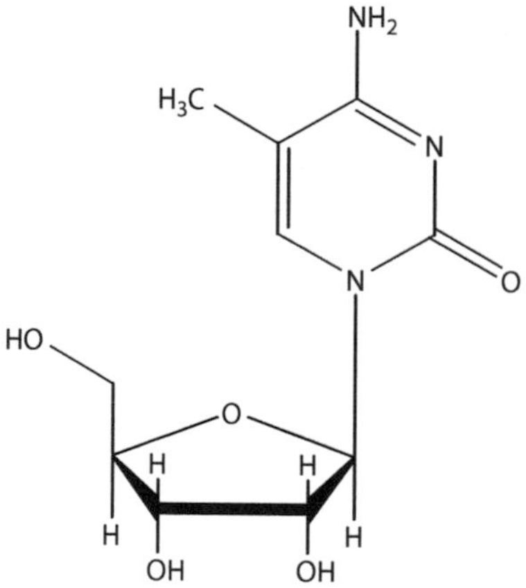

5-Methylcytidine

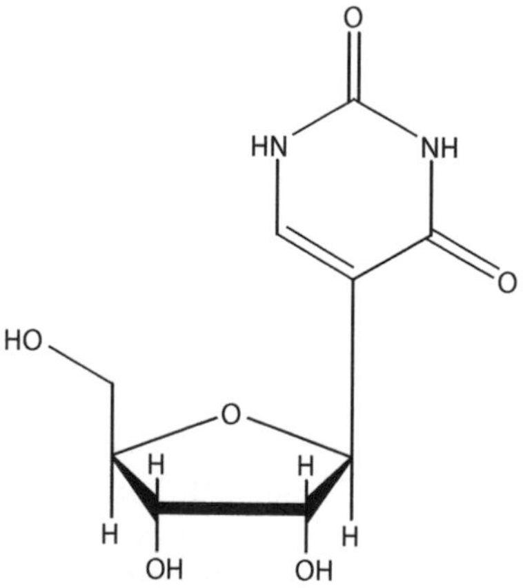

Pseudouridine

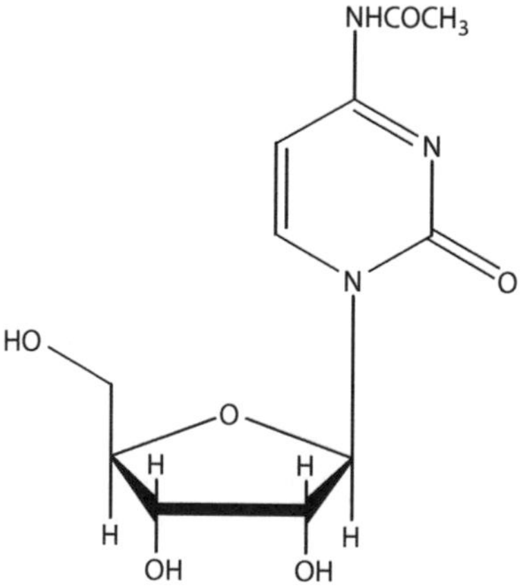

N^4-Acetylcytidine

FIGURE 5.1 Examples of some of the modified nucleosides.

modifications are mostly post-transcriptional and are known to occur naturally. These modifications are also highly conserved with respect to the type of modification; the base that is modified; and the respective location within the nucleic acid that is modified. However, the role that the modification plays is thought to be diverse and is still being investigated. A number of reports have identified nucleic acid modification as important in the adaptation of thermophilic species to their specific environments and also the role of DNA methylation in the development of some species such as zebra fish.[3,4]

There are other modifications, many of which are considered to be genotoxic, that occur via oxidative stress or by reactions with other compounds. However, the focus of this chapter is on the naturally occurring modifications that are found in the urinary excreted nucleoside profiles.

The modified nucleosides are normally excreted in urine, due to their aqueous solubility and because they cannot be salvaged for reutilization by the cells. While there are some unmodified nucleosides also present in the urine, it has been shown that, because of their relatively higher concentration, the modified nucleosides are more readily detected. Cell extracts that produce marginal amounts of nucleosides are not a good source of modified nucleosides.

The major interest in the modified nucleosides excreted in the urine relates to their potential use as biomarkers for cancer diagnosis and prognosis.[5–9] It has also been suggested that the analysis of the modified urinary nucleosides could be useful in the study of rheumatoid arthritis inflammation and in the study of nucleic acid metabolism in returning astronauts for the determination of the effects of space travel.[10–12] A large number of studies have been carried out monitoring the urinary nucleoside levels using HPLC with UV detection (generally at 254–260 nm), capillary electrophoresis (CE), and radioimmunoassay, the latter for only a specific subset of the urinary nucleosides.[5–9,13,14] However, the availability of mass spectrometric techniques, which are more easily hyphenated to chromatographic systems as the detection method for urinary nucleosides, has meant that this is becoming a more common method for both quantitation and structural elucidation of urinary nucleosides.

5.2 MASS SPECTROMETRIC BEHAVIOR OF MODIFIED AND UNMODIFIED NUCLEOSIDES

The study of the mass spectrometric behavior of the nucleosides and their collision-induced dissociation (CID) analyses are of great value in interpreting the data generated when analyzing these compounds from such complex biological matrices as urine. The electron impact (EI) spectra of a large number of nucleosides have been previously recorded and cataloged. Unknown ribosides can then be identified by comparing their fragmentation patterns with the cataloged patterns of known ribosides.[15] In addition, the fragmentation processes of the examined nucleosides may permit the differentiation between isobaric nucleosides and hence their specific and individual quantitation may be achieved. Numerous mass spectrometric techniques have been utilized to qualitatively identify and, in many cases, quantitate the levels of both unmodified and modified nucleosides. Gas chromatography-mass spectrometry with an electron ionization interface (GC-EI-MS), of the trimethylsilanyl derivatives of nucleosides generates a number of ions that allow more structural information to be obtained from the full-scan MS.[16] These ions can be split into three categories: the intact nucleoside or M series, the free base or B series, and the free sugar moiety or S series, and all have been well characterized (Table 5.1), giving much information concerning the nucleoside.[16,17]

MS studies of nucleosides are generally carried out in the positive ionization modes. The basic nucleobase can accept the proton addition more easily than the proton loss that would be required for the formation of negative ions. Note that the sugar moiety offers little ionization potential.

The sensitivity in positive and negative modes of ionization has been compared, in both fast-atom bombardment-mass spectrometry (FAB-MS) and electrospray ionization-mass spectrometry (ESI-MS) experiments. It was shown that the positive ionization mode was substantially more sensitive than the negative ionization mode.[18,19] The increases in sensitivity range from threefold to

TABLE 5.1

Ions Generated by Nucleosides During GC-MS Analysis

Molecular Ion Series Ions		Base Ion Series Ions		Sugar Ion Series Ions[a]	
M–15	M–CH$_3$	B–14	B–CH$_2$	S	S
M–90	M–C$_3$H$_{10}$OSi	B	B	S–1	S–H
M–103	M–C$_4$H$_{11}$OSi	B+1	B+H	S–90	S–CH$_3$OH
M–105	M–C$_4$H$_{13}$OSi	B+2	B+H$_2$	S–91	S–(CH$_3$)$_3$SiOH
M–118	M–C$_4$H$_{10}$O$_2$Si	B+13	B+CH	S–104	S–H–(CH$_3$)$_3$SiOH
M–131	M–C$_5$H$_{11}$O$_2$Si	B+30	B+CH$_2$O	S–106	S–CH$_4$–(CH$_3$)$_3$SiOH
M–180	M–C$_6$H$_{20}$O$_2$Si$_2$	B+41	B+C$_2$HO	m/z 230	C$_4$H$_4$O$_2$ (Me$_3$Si)$_2$
M–195	M–C$_7$H$_{23}$O$_2$Si$_2$	B+58	B+C$_2$H$_6$Si	m/z 217	C$_3$H$_3$O$_2$ (Me$_3$Si)$_2$
		B+74	B+C$_3$H$_{10}$Si	S–180	S–2(CH$_3$)$_3$SiOH
		B+100	B+C$_4$H$_8$OSi	m/z 103	CH$_2$O Me$_3$Si
		B+102	B+C$_4$H$_{10}$OSi		
		B+116	B+C$_5$H$_{12}$OSi		
		B+128	B+C$_6$H$_{12}$OSi		
		B+132	B+C$_5$H$_{12}$O$_2$Si		
		B+188	B+C$_8$H$_{20}$OSi$_2$		
		B+204	B+C$_8$H$_{20}$O$_2$Si$_2$		

[a] Based on the sugar being a ribose sugar.

195-fold in the FAB-MS analyses and 1.5–33-fold in the comparative ESI-MS experiments. The data are summarized as normalized intensity in Table 5.2.

The conventional full-scan spectra of nucleosides in the ESI-MS analysis are generally dominated by two nucleoside-related ions, and glycerol adducts are also prominently detected in the comparative FAB-MS spectra, while the sodium and potassium adducts occur more often in ESI data.

The major ions of interest, some of which are generated by in-source fragmentation of the nucleoside, in the full-scan mass spectra of nucleosides, are the protonated molecule [M+H]$^+$, and the protonated base moiety [B+H]$^+$ which occur by release of the sugar moiety (Figure 5.2).[19]

In some ESI experiments, sodium and potassium adducts and mobile-phase component adducts (ammonium from ammonium acetate buffers) are observed, as well as some dimer formation. The relative intensities of these alternative ions were investigated and are summarized in Table 5.3.

An investigation into the influence of the pK_a and the gas-phase basicity of the nucleosides in ESI was undertaken, and it was concluded that solution-phase chemistry is important for all nucleosides, whereas gas-phase chemistry is only important in the ionization of compounds with a pK_a of less than 3, when ammonium salt addition can improve ionization.[20]

A "soft" ionization technique, such as electrospray, usually causes less in-source fragmentation of the molecule than does GC-MS analysis, with only some techniques producing the protonated base as discussed. This results in the fact that tandem mass spectrometry (MS/MS) with CID analysis has been utilized to gain more structural information. The ESI-MS of the nucleosides and their CID-MS/MS pathways have been investigated widely and have been shown to exhibit primarily the cleavage of the glycosidic bond between the sugar and base moiety, leaving the protonated base as the major product ion (Figure 5.3). However, occasionally, an ion derived from the sugar moiety can be seen in relatively low abundance in the fragmentation spectra.[19] Occasionally, other low-abundance ions, representing the loss of small neutral molecules, such as the loss of a water molecule from the sugar moiety of the nucleoside, are detected. These ions, however, are minor product ions, with

TABLE 5.2
Comparative Sensitivity of FAB-MS and ESI-MS for Some
Nucleosides by Positive and Negative Ionization

Nucleoside	FAB-MS Data Normalized Intensity		ESI-MS Data Normalized Intensity	
	Positive	Negative	Positive	Negative
Adenosine	n.d.	n.d.	1	0.04
N^2,N^2-dimethylguanosine	n.d.	n.d.	1	0.2
Pseudouridine	n.d.	n.d.	1	0.21
Cytidine	n.d.	n.d.	1	0.11
Uridine	1	1	1	0.6
Tubercidin	1	0.33	1	0.02
Inosine	n.d.	n.d.	1	0.35
Guanosine	1	0.083	1	0.17
Xanthosine	n.d.	n.d.	1	0.25
N^1-methylguanosine	1	0.14	n.d.	n.d.
2-O'-methylguanosine	1	0.125	n.d.	n.d.
Thymidine	1	0.11	n.d.	n.d.
Dihydrouridine	1	0.05	n.d.	n.d.
2-O'-methyluridine	1	0.14	n.d.	n.d.

n.d. = not determined.

relative intensities of between 2% and 5% of that of the protonated base.[19] There has been further investigation of the primary (nucleobase) ion fragmentation by MS^3 or higher.

Rice and Dudek, Nelson and McCloskey, and Gregson and McCloskey studied the fragmentation pathways of various protonated nucleobases extensively.[21–23] This seminal work included the study of nucleobases and analysis of the mass shifts in the product ions formed when the comparative MS/MS data of the modified nucleobases were obtained. The labeling of the base

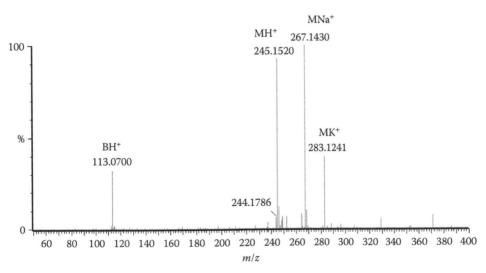

FIGURE 5.2 Full-scan mass spectrum of the nucleoside uridine.

TABLE 5.3
**Comparative Abundances of the Protonated Molecule and Ions
Related to the Nucleoside in Their Full-Scan Mass Spectrum**

Nucleoside	Normalized Relative Abundance (%)				
	M+H$^+$	M+Na$^+$	M+K$^+$	M$_2$+H$^+$	B+H$^+$
Pseudouridine	100	0	0	0	0
Uridine	95	100	40	0	32
Cytidine	100	80	13	0	41
Inosine	100	19	0	4	42
Tubercidin	100	6	1.5	5	2.5
Guanosine	100	21	0	0	21
Xanthosine	100	25	0	0	14
Adenosine	100	7	0	0	5
N^2,N^2-dimethylguanosine	100	77	10	0	9

M = Nucleoside molecule.
B = Nucleobase molecule.

with N^{15} and O^{18} at specific positions was also undertaken and comparative MS/MS spectra were acquired. They concluded that fragmentation of the protonated base occurred by the opening of the base ring moiety and the loss of low-molecular-weight neutral components that were dependent on the base structure. Common losses detected were that of NH_3, CO, NHCO, and HCN from the base structure.

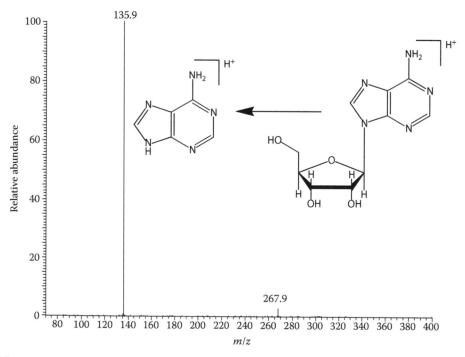

FIGURE 5.3   MS/MS spectrum of adenosine and the fragmentation pathway observed.

The study of guanine fragmentation has been the emphasis of a number of publications since 1967. The first study utilized guanine and guanine methylated at differing positions to study the fragmentation behavior of this nucleobase and this study established that the loss of cyanamide was the predominant pathway of dissociation.[21] A 1997 report utilized guanine molecules, labeled with stable isotopes at individual nitrogen and oxygen atoms, to study the fragmentation in more depth.[23] The comparative fragmentation spectra allow clear differences to be seen between differently labeled guanines (Figure 5.4). This study noted two major (initial) pathways of fragmentation proceeding via the loss of a molecule of ammonia and the loss of a molecule of cyanamide. By comparing the relative intensities of the product ions formed, it was determined that the pathway contributions are roughly equal as they generate product ions in approximately equal relative abundance.

When the fragmentation patterns of the nitrogen-labeled guanines were compared, it was noted that the loss of ammonia from the unlabeled guanine occurred (m/z 152 → m/z 135), whereas the N^3-labeled guanine was found to retain the labeled nitrogen as m/z 136 was detected. The N^1 and N^2 labeled guanine showed ions of equal intensity at m/z values 17 and 18 Da less than the precursor ions at m/z 136 and 135. This is explained by a "Dimroth-like" ring opening occurring and the randomization of the N^1 and N^2 nitrogen (Figure 5.5) and their equal loss. The loss of the cyanamide group from the unlabeled guanine (m/z 152 → m/z 110) results in the fully randomized loss of both the N^1 and N^2 as seen in the comparative spectra. A further loss of cyanamide from the product ion is detected with partial randomization being exhibited between N^3 and N^9.

Further fragmentation analysis starting from guanosine, combined with the generation of accurate mass data for the fragment ions, identified further pathways (Figure 5.6). When the MS^3 fragmentation pattern of guanosine was analyzed, some, initially unanticipated, mass losses were noted. The first of these was the observation that in the CID-MS^3 scan of guanine recorded with an ion trap, there was a reproducible low relative abundance ion, 24 Da lower than the protonated molecule that occurred (Figure 5.7).[24] This mass loss is difficult to explain in terms of simple neutral molecular losses; however, the observation was reproduced in both a triple quadrupole and a quadrupole time-of-flight mass spectrometry (Q-ToF-MS) instrument and similar losses were also observed in the methylated guanosines. Further experiments, including spiking the collision gas supply with deuterium-labeled water, provided the explanation of this occurrence as gas-phase reactions of the nucleobase fragment ions in the collision cell. The 24 Da difference was shown to arise from the formation of a water adduct of the fragment ion at m/z 110, which corresponds to the loss of HNCNH from the protonated guanine (Figure 5.6). It was also noted that the mass of this unusual product ion changed when deuterated water was added into the collision gas (Figure 5.8).[24] Further experiments have also demonstrated other adduct ions and also conversion of guanine to xanthine in the gas phase by the replacement of an oxygen atom with an amino species.[25] The data obtained suggest that, while these observations are not limited to the study of nucleosides, care should be taken when interpreting new ions arising during CID experiments.

The past investigations of adenine suggest that the three successive losses of HCN occur from the protonated nucleobase as a major fragmentation pathway.[21] Partial labeling of positions in the molecule with deuterium indicated that multiple pathways contribute to these consecutive losses. The loss of ammonia was indicated to be a minor pathway unless the amino group was substituted by a methyl in which case the loss of the alkylamino group was detected.

The fragmentation of uracil was also studied by the utilization of specifically labeled forms of the nucleobase.[22] The study demonstrated three major pathways of fragmentation with possible product ions indicated in Figure 5.9. These represent the initial loss of ammonia, water, and HNCO from uracil. Studying the loss of ammonia with stable isotope-labeled compounds indicated that 93% of the losses occurred via the loss of N^3 with the remaining 7% resulting from the loss of N^1, whereas the loss of water involves oxygen atoms at positions 2 and 4 equally, and the loss of HNCO occurs through three separate pathways with the major pathway resulting in the loss of N^3, C^2, and O^2 (87% total intensity).

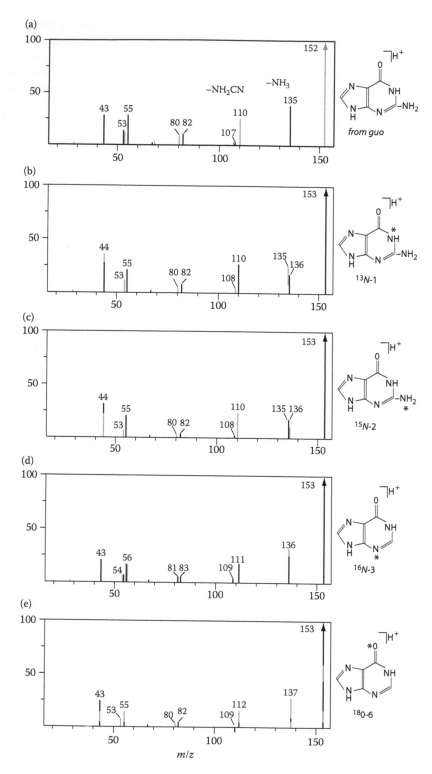

FIGURE 5.4 Fragmentation of the guanine nucleobase labeled with stable isotopes at differing positions of the molecule. (From Gregson, J.M. and McCloskey, J.A., *Int. J. Mass Spectrom. Ion Processes* 165/166, 475, 1997. With permission.)

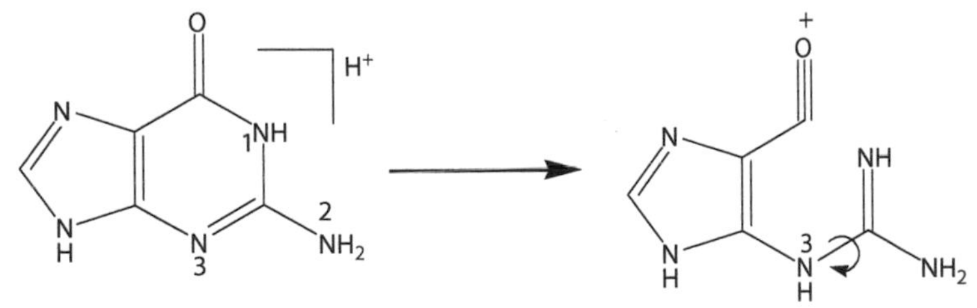

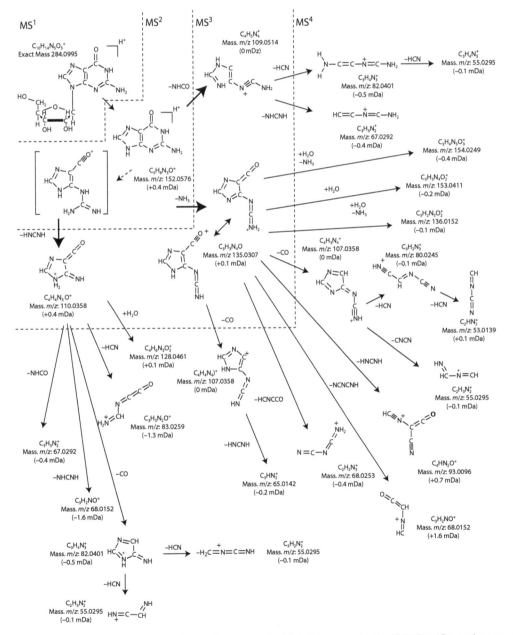

FIGURE 5.5 Randomization of N^1 and N^2 during ring opening of guanine.

FIGURE 5.6 Guanosine fragmentation pathways as elucidated by combined MS/MS, MS^n, and accurate mass analysis. (From Tuytten, R. et al., *J. Am. Soc. Mass Spectrom.* 16, 1291, 2005. With permission.)

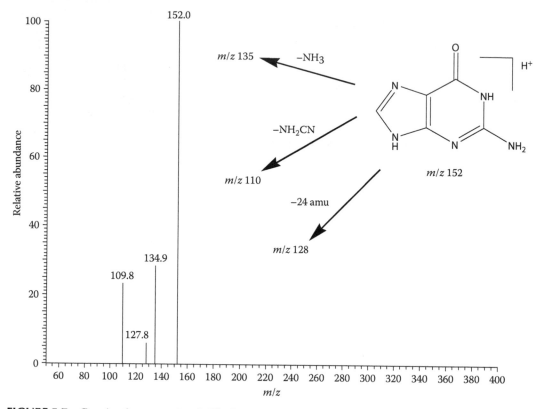

FIGURE 5.7   Guanine fragmentation (MS³ of guanosine) resulting in an unusual loss of 24Th.

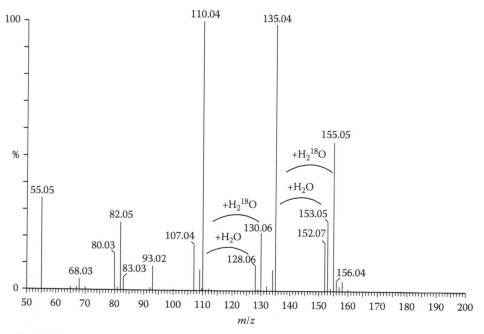

FIGURE 5.8   Effect of addition of deuterated water to the collision gas during MS/MS of guanine; formation of deuterated water adduct ions in the collision cell. (From Tuytten, R. et al., *J. Am. Soc. Mass Spectrom.* 16, 1291, 2005. With permission.)

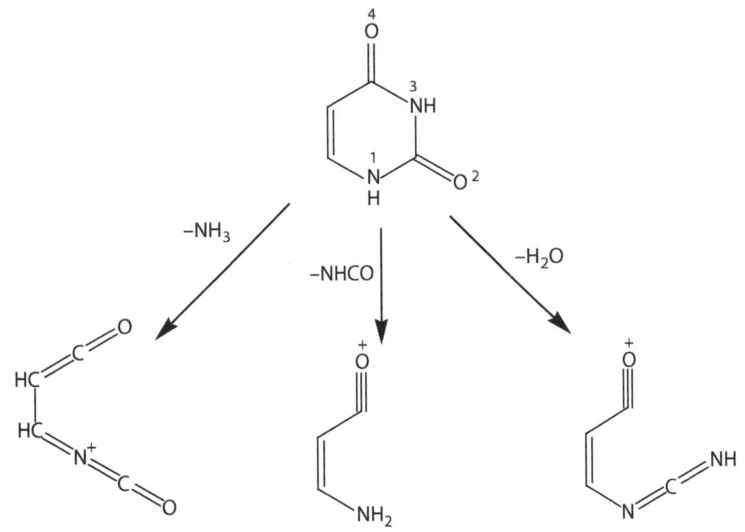

FIGURE 5.9   Fragmentation pathways of uracil.

The fragmentation of cytidine was also studied in more detail using stable isotope labeling at specific positions of the nucleobase.[26] After the loss of the sugar, further MS^n experiments gave rise to two major fragment ions at m/z 95 and 69, resulting from the loss of ammonia and HNCO, respectively. A minor product ion, due to the loss of a water molecule (m/z 94), was also detected. Study of the spectra generated from the labeled cytidine compounds indicated that the loss of ammonia equally involves the loss of N^3 and N^4 (unlike the ammonia loss from uracil) and that the loss of HNCO involves mainly N^3 with some contribution from N^4 as seen in the comparative fragmentation spectra (Figure 5.10).

Atmospheric pressure photoionization mass spectrometry (APPI-MS) has been investigated for the analysis of nucleosides, in which the behavior of the nucleosides, under both direct and dopant-assisted conditions were studied.[27] It was reported that direct analysis produced spectra similar to those obtained by thermospray; however the spectra generated with the use of a dopant were different. For example, uridine recorded with a dopant generated very little protonated nucleoside and produced a strong sodium adduct ion, a disodiated adduct ion, and a protonated nucleobase ion. When toluene was used as the dopant, the protonated base was the most abundant ion in the spectra and the amount of the sodium adduct was reduced. When acetone was used as the dopant, two additional ions appeared in the spectrum, one related to the loss of water from protonated uridine and the second at m/z 155 was due to the S$_1$ fragmentation of the sugar. Although APPI generates more ions, the sensitivity is reduced since the ion current for each nucleoside is spread over multiple ions and therefore little additional structural information is acquired.

The type of modification that is present on the nucleoside may have an effect on the CID-MS/MS fragmentation spectra obtained. Firstly, in the positive ionization mode, the major ion detected is the protonated nucleobase that resulted from the loss of the sugar with its attached methyl groups. Any further MS/MS fragmentation of this product ion generally finds that the modifier is maintained in its original position and the fragmentation pathway is analogous to that of the nucleobase.

Differentiation of methylated nucleoside isomers is based on the differences that are caused by the position of the methyl group on the nucleobase. Methylated derivatives of uracil occur at N^3 and N^5 and the two isomers are easily distinguished by MS/MS fragmentation.[22] Methyl addition at the N^3 position is indicated by a further mass loss of 14 Da, as the methyl modification is lost from the product ions that were generated by the initial loss of ammonia and NHCO; this is not the case for the N^5-modified derivative.

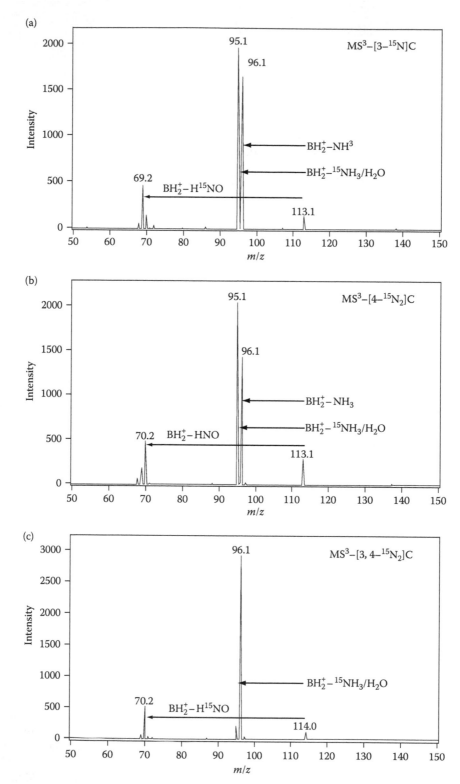

FIGURE 5.10 Fragmentation spectra of cytidine labeled at positions: (a) N^3, (b) N^4, and (c) both N^3 and N^4. (From Jensen S.S. et al. *J. Mass Spectrom.* 6, 49, 2006. With permission.)

Cytidine can be methylated at the 3, 4, and 5 positions of the nucleobase and comparison of the MS^3 spectra of the isomers suggest that there is a diagnostic ion for each.[26] The 3-methylcytidine is the only isomer that does not exhibit water loss from the protonated nucleobase, as the methylation prevents the two hydrogen atoms from being available for H_2O formation. In a similar analysis, 4-methylcytidine generates a product ion due to the loss of CH_3NH_2, while 5-methylcytidine does not generate this product ion but generates its unique product ion from the loss of H_2NCHO (Figure 5.11). Interestingly, in the case of acetylcytidine it was demonstrated that after the initial loss of the ribose sugar the acetyl group was lost producing the protonated "normal" cytosine base. In addition, it was noted that a minor ion was generated by the loss of the ribose moiety; and that the acetyl group from the normal cytosine base was also detectable (Figure 5.12a and b).[28]

The study of the isomers of the singly methylated guanosine nucleoside also indicates that the methyl group is located on the nucleobase at either the N^1, the N^2, or the N^7 position. It has been reported that EI allows the differentiation of these three isomers.[15] However, this finding is inconsistent with the reported isotopic labeling CID experiment in which it was shown that after "Dimroth"-like ring opening of the guanine base, randomization of the N^1 and N^2 positions was exhibited.[21] Differentiation between the N^1 or N^2 isomer and the N^7 isomer can be achieved by MS^3 analysis. MS/MS analysis does not differentiate between the isomers as only the ribose sugar is lost and the nucleobase-derived product ions are isobaric. MS^3 fragmentation indicated that there was retention of the N^7 atom and its associated modification, and there was a randomized loss of N^1 and N^2 due to ring opening. The observation during ammonia loss is that alkylation at either N^1 or N^2 appears to stabilize it, and therefore the loss of the other nonalkylated nitrogen was favored. The loss of NH_2CN from N^7-methylguanosine resulted in both N^1 and N^2 being lost, thus differentiation of the isomers is not possible (Figure 5.13). It has been reported that ion-trap fragmentation of these isomers allowed their differentiation, but this finding could not be explained due to the randomization around the N^1 and N^2 positions and it did not concur with previously published ion-trap data.[29]

Reliable differentiation of N^1 and N^2 methylguanosine is dependent of their retention properties on reverse-phase high-performance liquid chromatography linked to mass spectrometry (RP-HPLC-MS).

Differentiation between nucleosides modified on the base and on the sugar is easily achieved due to the complete loss of the uncharged sugar during the ESI-MS/MS analysis (the + ion mode) with the base-modified nucleosides losing the ribose (132 Da), as in the case of RNA modifications.

The divergence from the initial MS/MS analysis of the protonated molecule into the product ion results from the constant neutral loss transition representing the nucleoside sugar moiety modification. In the case of the methylated ribose, this results in the loss of methylribose (148 Da) from the parent ion.

There has been an ongoing problem in determining the position of the modifier in nucleoside isomers that have sugar moiety modifications present. This is because the base generally retains the charge during the MS/MS analysis in the positive ionization mode. There were some early EI-MS studies that suggested that diagnostic ions could differentiate between 2-O'-methylated and 3-O'-methylated nucleosides.[30] However, usually a combination of separation methods, such as GC and HPLC, was used to distinguish between the possible isomers that exhibited different retention times. There has also been an investigation of sugar-modified nucleosides using EI-MS in the negative ionization mode.[31] This study also concluded that 2-O'- and 3-O'-methylated ribonucleosides could provide ions that were diagnostic for their separate identification. The cytidine, guanosine, and adenosine with 2-O'-methylation exhibited loss of 90 Da while their 3-O'-derivatives did not; 3-O'-methyluridine and cytidine exhibited a loss of 60 Da that is not detected in the spectra for the 2-O'-isomers (Figure 5.14). Pseudouridine is an RNA-derived modification of uridine. In pseudouridine, the uracil base has a C^5 attachment to the ribose sugar as compared to the N^1 base attachment that is seen in uridine. These modified nucleosides are referred to as C-nucleosides and due to differences in the bond energies between the sugar and base moiety (62 kcal mol^{-1} for nucleosides, 80 kcal mol^{-1} for C-nucleosides), there are significant differences in their fragmentation. This is illustrated in the fragmentation scheme for uridine and pseudouridine shown in Figure 5.15.[32]

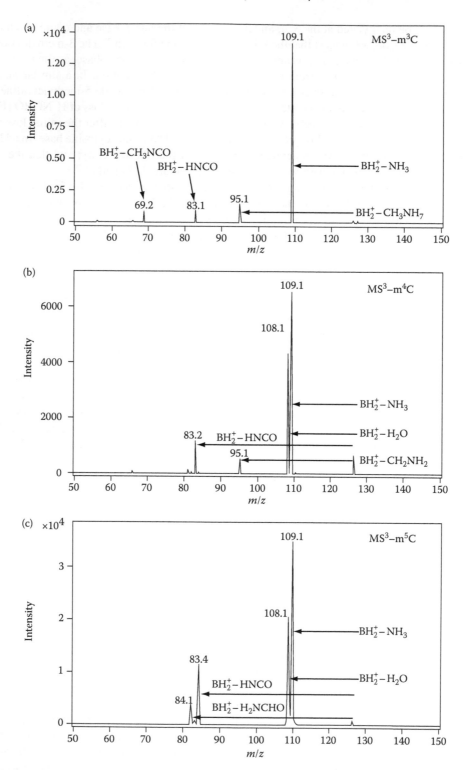

FIGURE 5.11 Comparative fragmentation spectra of (a) 3-methylcytidine, (b) 4-methylcytidine, and (c) 5-methylcytidine. (From Jensen S.S. et al. *J. Mass Spectrom.* 6, 49, 2006. With permission.)

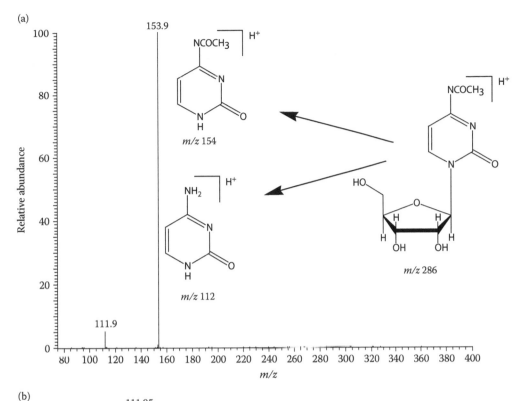

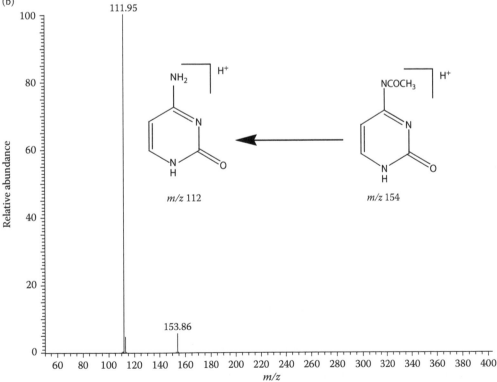

FIGURE 5.12 (a) MS/MS spectrum of acetylcytidine and (b) MS³ spectrum of acetylcytidine (via acetyl-cytosine from MS/MS analysis), both indicating the formation of the protonated unmodified cytosine base.

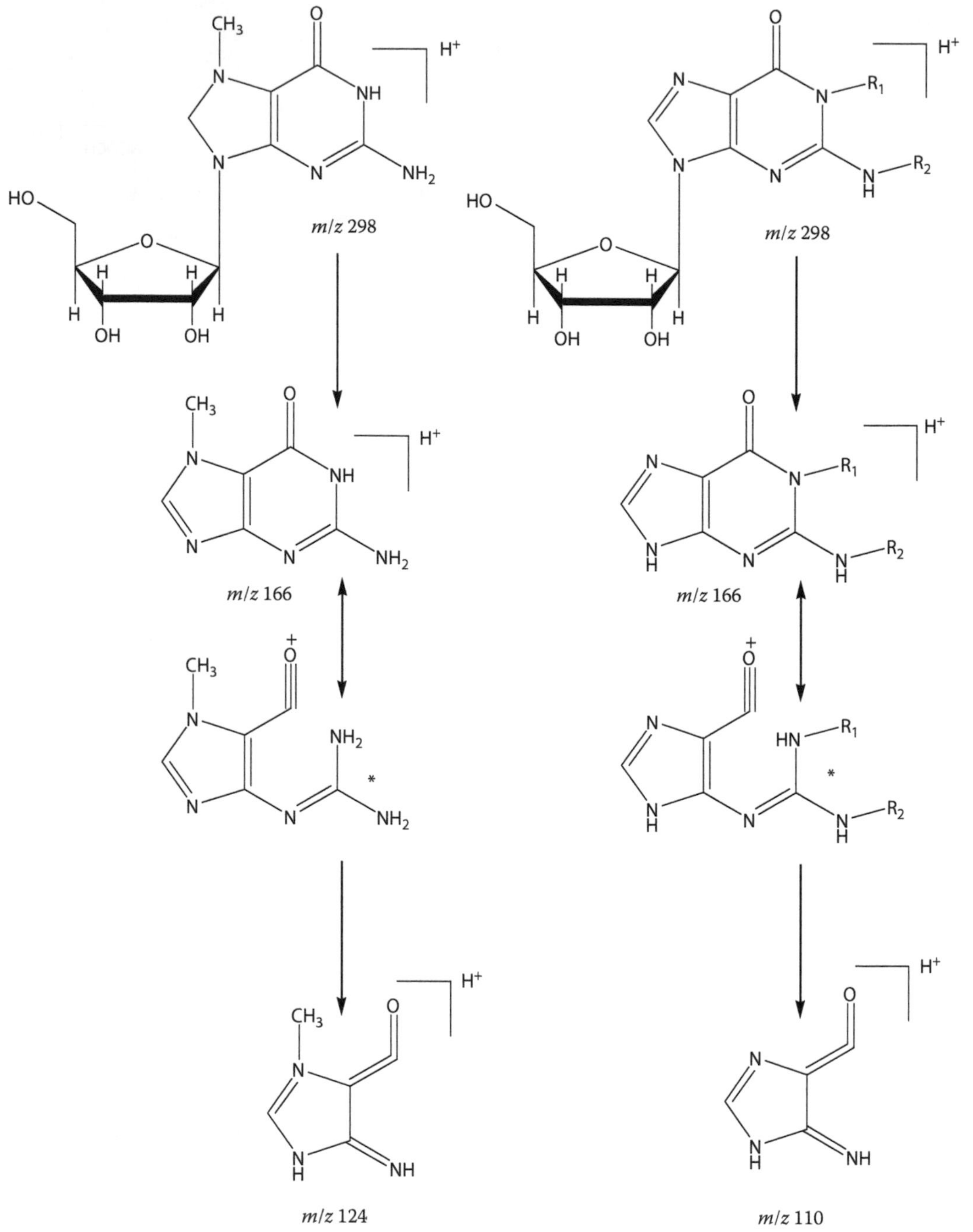

N^1-methylguanosine : R_1 = Methyl R_2 = H

N^2-methylguanosine : R_1 = H R_2 = Methyl

* = randomization around carbon atom after ring opening

FIGURE 5.13 Formation of a diagnostic fragment ion that allows distinction of N^7-methylguanosine and N^1 and/or N^2-methylguanosine (via MS^3).

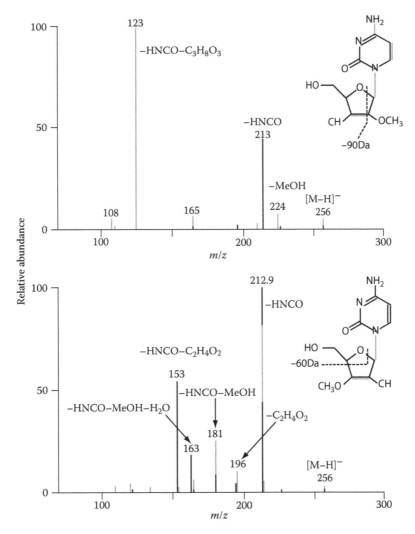

FIGURE 5.14 Negative ionization fragmentation of 2-*O'*- and 3-*O'*-methylated ribonucleosides. (From Zhang, Q. and Wang, Y., *J. Am. Soc. Mass Spectrom.* 17, 1096, 2006. With permission.)

The fragmentation scheme of uridine depicted in Figure 5.15 also shows the presence of a protonated sugar molecule at *m/z* 133.

The MS^n fragmentation of pseudouridine has been investigated by a Q-ToF-MS instrument and these data were compared with the results obtained from an ion-trap instrument.[33,34] The data obtained with the conventional ion trap compared favorably with the data from the Q-ToF. The Q-ToF-MS/MS analyses used in-source CID of the protonated nucleoside to generate product ions in the first quadrupole, followed by multiple MS/MS analyses with stringent data interpretation, to distinguish the nucleoside ions from the background signals. The data combined all the analyses and demonstrated that pseudouridine had three main MS fragmentation pathways.

The first of these proceeds through the consecutive loss of three molecules of water from the ribose sugar, generating product ions at *m/z* 227, 209, and 191. This is followed by losses from the base, mimicking those observed in uracil. A doubly protonated carbon was also prominent at *m/z* 125. The second pathway proceeded via the combined loss of two water molecules and CH$_2$O from the sugar moiety, giving rise to a product ion at *m/z* 179. Further fragmentation of this ion resulted in a stepwise loss of small neutral groups from the sugar, followed by fragmentation of the base.

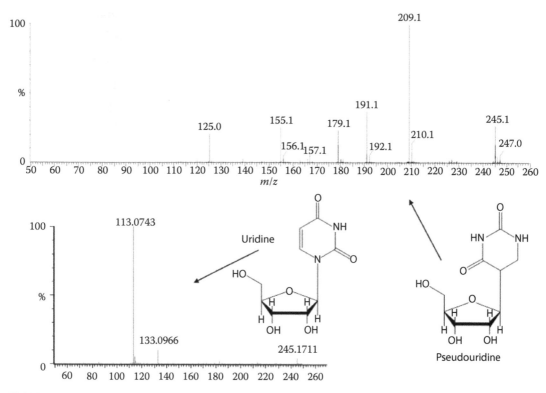

FIGURE 5.15 Comparative fragmentation of the isobaric nucleosides uridine and pseudouridine.

The third pathway occurred with the cleavage of a large fragment of the sugar, with only the CH_2CHO of the sugar being retained by the protonated uracil base and giving rise to a product ion at m/z 155. This has been referred to as the S_1 fragmentation pathway. This fragment ion can then undergo a ring-opening process with the subsequent loss of small neutral molecules. The C-nucleoside isomers can easily be distinguished from the other nucleoside isomers by their distinctive fragmentation patterns.

5.3 ANALYSIS OF FREE MODIFIED NUCLEOSIDES IN BIOLOGICAL SAMPLES

5.3.1 INTRODUCTION

There has been a great deal of research involved in the identification and quantitation of modified nucleosides in biological fluids, primarily urine and serum. These are thought to represent the turnover of RNA and constitute a potential marker of many diseases, especially various cancers.[5–9] The modified nucleosides do not have any specific biochemical salvage pathways for their reincorporation into nucleic acids as the "normal" nucleosides do. The modified nucleosides, which can become a toxic hazard if allowed to accumulate in the cell, are excreted from the cell into the blood and eliminated from the body via the urine. Urinary nucleoside levels are commonly normalized against urinary creatinine, as daily creatinine excretion is known to be constant and therefore can be used to compensate for individual variations in the volume of produced urine.[35] The traditional methods for urinary creatinine quantitation involve either the reaction of creatinine with picric acid to form a colorimetrically distinguishable product (Jaffe method) or by the use of clinical instruments that are specific for creatinine measurement. More recently, an MS/MS method has been utilized for the creatinine analysis.[36,37]

5.3.2 SINGLE NUCLEOSIDE ANALYSIS

It is usually the entire profile of the urinary nucleosides that is looked at when urinary levels of the compounds are studied; however there are cases of individual nucleosides being investigated. Urinary adenosine has been studied utilizing a column-switching technique for high-throughput HPLC-UV analysis.[38] Pseudouridine has been identified in nucleoside mixtures by mass spectrometric methods.[39] The study utilized the chemical reaction of pesudouridine with methyl vinyl sulfone (MVS) in order to study pseudouridine specifically in a nucleoside mixture as occurs in urine (Figure 5.16).

Pseudouridine, in a full-scan MS, has the same mass and empirical formula as uridine and thus is challenging for MS analysis. The early MS identification of pseudouridine involved chemical modifications to it before the MS analysis.[39] In order to study pseudouridine, it was reacted with MVS (Figure 5.16). The reaction was shown to also derivatize up to 10% of the uridine; however, HPLC-MS analysis of the derivatized uridine and derivatized pseudouridine was shown to have significantly different retention times that allowed the specific quantitation of pseudouridine.

HPLC-MS comparison of the sensitivity of the underivatized and derivatized modified nucleoside using selective ion monitoring (SIM) of the $[M+Na]^+$ and $[M+H]^+-H_2O$ ions indicated the advantage of the derivatization.

The derivatization process was shown to improve the overall sensitivity for the detection of pseudouridine. To date, most of the methods developed for the detection of a specific nucleoside have been applied to the analysis of the selected nucleoside in an RNA sample that was either partially or totally hydrolyzed. These techniques can be applied to a comparative study of urinary nucleoside levels but would only provide quantitative information regarding the single modified nucleoside. The detection of an individual nucleoside is, by its nature, more sensitive than the analysis of a whole class of compounds; however, the total urinary nucleoside profile is of more value and more reliable as a diagnostic biomarker for the development of cancer than is the level of excretion for any individual nucleoside.

5.3.3 NUCLEOSIDE PURIFICATION

As mentioned in the previous section, urinary modified nucleosides are generally studied as a whole set of compounds rather than in specific groups or as specific individual nucleosides. A brief

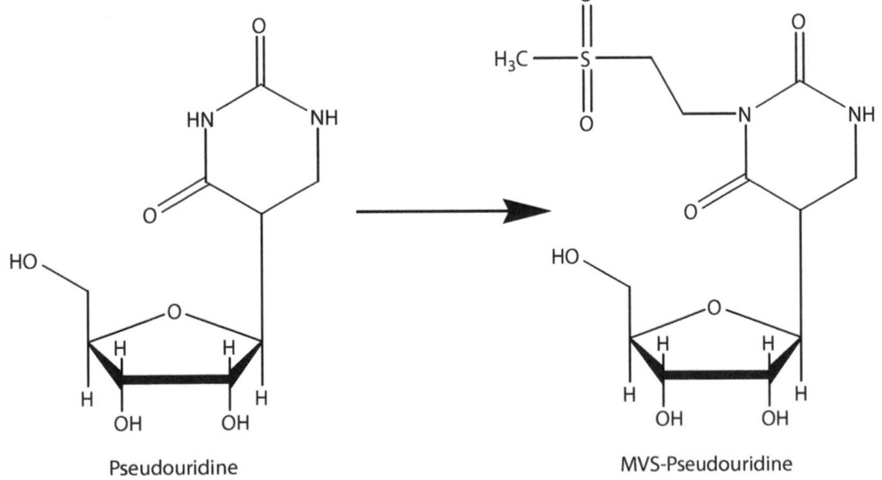

Pseudouridine   MVS-Pseudouridine

FIGURE 5.16 Derivatization of pseudouridine.

discussion of the specific purification of this class of compounds is therefore noteworthy before the mass spectrometric analysis of the urinary nucleosides is discussed. This is especially relevant considering that urine represents a biologically complex matrix from which the analysis of a particular species of comparatively low abundance compared to other urinary components can be problematic. The nucleosides are almost always purified from the urine prior to their analysis (with few exceptions, as discussed later in this chapter) and this purification has followed the same extraction principle for the past 30 years. The selective enrichment of the nucleosides from the other more highly abundant species present in urine takes advantage of the *cis*-diol group found on the ribose sugar of many of the nucleosides derived from RNA. DNA-derived nucleosides contain a deoxyribose sugar and are therefore not purified in the same manner. The *cis*-diol group can be selectively retained using boronate containing stationary phases (usually phenylboronate), which form a specific complex with the *cis*-diol group at high pH (8.8) and release the retained moieties when the pH is reduced, usually by the application of 0.1 M formic acid (Figure 5.17). This complexation was first recognized by Böeseken.[40] Other moieties within urine are still copurified using this approach (urinary sugars and catecholamines for example) and further separation may be required to remove these more abundant *cis*-diol-containing compounds. This has been accomplished by utilizing a combination of both cation and anion exchange columns after the initial affinity purification.[41] This step has the advantage of removing compounds that, due to their elevated urinary

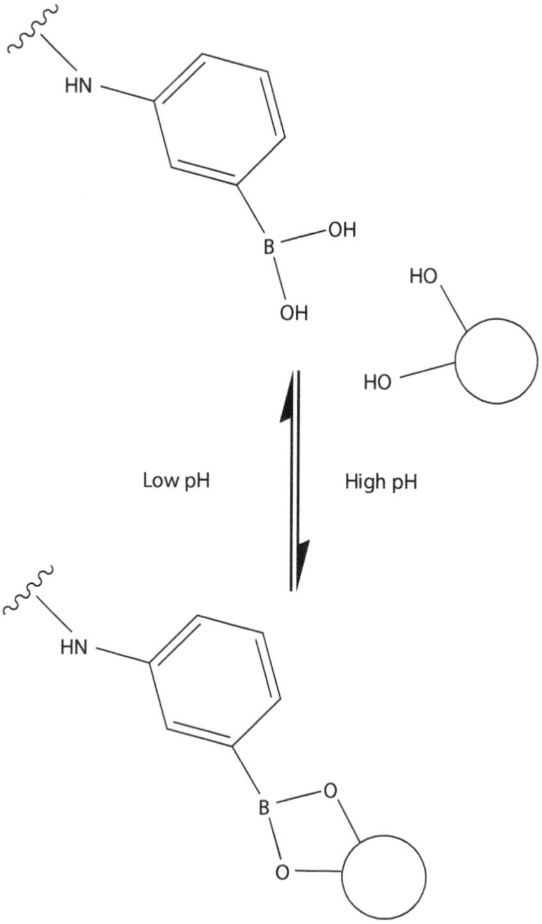

FIGURE 5.17 Boronate affinity chromatography.

levels compared to the comparatively minor levels of nucleosides, may affect the accurate quantitation of urinary nucleosides. It should be noted that this combination of ionic exchange columns also had the effect of separating different classes of the modified (and unmodified) nucleosides. The majority of the nucleosides were retained by the first cation exchange column whereas the sugars were eluted in the flow through. Uridine and its modified derivatives eluted with the sugar fraction (presumably due to their lower basicity and therefore lesser positive charge under the column conditions) and so required the further utilization of an anion exchange column in order to separate them from the higher abundance, more neutral urinary sugars. However, with improvements in the selectivity of the mass spectrometric detector toward nucleosides (as discussed later), these additional steps may have become redundant. One major disadvantage of the phenylboronate affinity purification protocol is that any 2′ or 3′ modified nucleosides are lost during purification, as are any modified nucleobases formed by the natural hydrolysis of their comparative nucleosides, as both groups lack the required *cis*-diol group. Despite this fact, the specificity of the affinity chromatography toward the majority of the modified nucleoside species has meant that the technique has not to date been surpassed and is still commonly utilized more than 30 years after its first application to the field. The purification is usually performed under low-pressure conditions with sample application and elution performed under gravity using primarily Affi-Gel 601 phenylboronate affinity gel from BioRad; however HPLC compatible boronate columns and HPLC packing material have become available.[42]

Interestingly, an initial study investigating the application of this HPLC compatible material to the on-line cleanup of urine samples indicated that specific binding did not occur and that elution occurred under loading conditions with different nucleosides exhibiting differing retention characteristics.[43] This suggested that retention was not, at least exclusively, based on the presence of a *cis*-diol moiety. Further experiments indicated that the Lewis-base properties of available nitrogens played an important role in the retention and the use of the material as an on-line trap column (as used in many miniaturized column switching setups) was only possibly in aprotic conditions (100% acetonitrile being the mobile phase used in the final setup). A pilot study of this more different trapping behavior was conducted and an on-line sample cleanup—the HPLC-MS/MS method—was developed as discussed later.

5.3.4 Urinary Nucleoside Profile Determination by MS

Before the use of MS detection of the levels of urinary modified nucleosides, they were primarily measured using low-pressure affinity purification protocol followed by HPLC separation and UV detection (254–260 nm).[5–9] While urinary modified nucleosides may be useful as potential biomarkers, given the large number of urinary nucleosides present and the relatively short HPLC analysis time, the actual identity of any given UV peak is difficult to substantiate. A potential drawback of this approach is the coelution of many of the nucleosides studied and these different compounds cannot be identified solely by UV. The application of MS detection in this analysis allows the separation and identification of coeluting nucleosides. The initial studies were based on the low-pressure phenylboronate-based affinity purification of the nucleosides and used microbore HPLC-MS via a direct liquid injection interface. It was demonstrated that the combination of techniques was viable and that it was feasible to identify nucleosides (such as 5,6-dihydrouridine) that were almost undetectable by HPLC-UV because they had low UV absorbance characteristics.[44] When thermospray was applied to the analysis of nucleoside mixtures containing coeluting species, a greater nucleoside selectivity, as compared to HPLC-UV analysis, was achieved.[45] Three main ions were detected for each nucleoside: the protonated molecule, the protonated base (derived from in-source fragmentation of the nucleoside), and the mobile-phase-derived ammonium adduct of the sugar. The MS scans obtained were comparatively more complex than the scans from the direct liquid injection approach. The next step in the study of these compounds was however involved in the use of the GC-MS as a tool for the study of urinary nucleosides. Langridge et al. reported that after phenyl-boronate

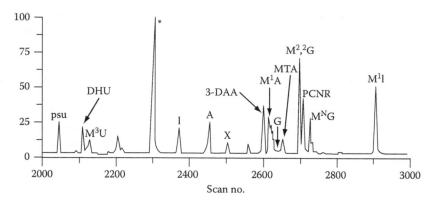

FIGURE 5.18   GC-MS separation and analysis of urinary nucleosides. (From Langridge, J.I. et al., *Rapid. Commun. Mass Spectrom.* 7, 427, 1993. With permission.)

affinity purification and derivitization with trimethylsilane (TMS), the modified urinary nucleoside levels could be studied by GC-MS and that a good separation of the purified nucleoside fraction was obtained (Figure 5.18).[46] HPLC-MS was reinvestigated as a technique for the separation and identification of urinary nucleosides because of the development of ESI as well as the wide availability of ESI-MS instrumentation along with the improvement in the ionization and transfer of the protonated nucleosides from the liquid phase to the gas phase. The optimization of the HPLC separation and the EI parameters for urinary nucleoside analysis resulted in an optimized LC-MS protocol.[19] The separation of a standard mixture of nucleosides is shown in Figure 5.19. This optimized LC-MS protocol was later used in a comparative study of standard GC-MS methods and miniaturized HPLC-MS methods.[47]

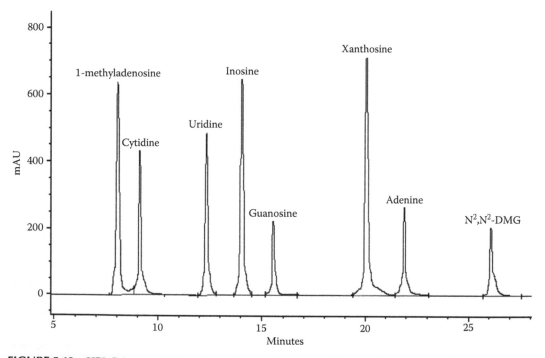

FIGURE 5.19   HPLC-ion-trap-MS analysis of nucleoside and modified nucleoside standards.

Another approach used capillary liquid chromatography linked to frit-FAB-MS to study modified nucleosides from the urine and serum in both uremic patients and healthy individuals.[48] They found that the results of these experiments were comparable with the results obtained by electrospray and thermospray studies of nucleosides. However, this approach is not commonly used due to the difficulty in hyphenating HPLC and FAB-MS and due to the formation of glycerol adducts that complicated the spectra.

Instead of HPLC, Liebich et al. used CE as the separation technique for the analysis of urinary nucleosides purified by *cis*-diol affinity. This group was also successful in combining both CE and HPLC separation with off-line matrix-assisted laser desorption ionization-time-of-flight mass spectrometry (MALDI-TOF-MS).[49,50] In this approach, UV absorbance was used for the quantitation of the nucleosides and MS was used for the identification of the nucleosides present in each individual UV peak. A major limitation to MALDI-TOF-MS for the analysis of the nucleosides is the overabundance of matrix-derived ions in the mass range that may coelute and compete for ionization, thereby complicating the spectra. HPLC coupled to quadrupole ion-trap MS (HPLC-QIT-MS) was used to accurately detect and quantify many urinary nucleosides, indicating the presence of a possible biomarker specific for head and neck cancer. This was achieved using HPLC with a low-pressure affinity chromatography cleanup.[51] Since these protocols were first developed, there have been many improvements made in the speed of separation and with complexity of mixtures that can be separated. Ultraperformance liquid chromatography (UPLC), for example, used in combination with a ToF-MS analyzer is three times faster than comparable HPLC runs and detects more nucleosides (94 versus 79) with improved sensitivity and resolution.[52]

Further developments of high-throughput techniques for the study of the urinary modified nucleosides by MS have taken two paths. The first of these examines the combination of differing separation and cleanup techniques and/or various MS data-dependent analyses for selectively studying the nucleoside. The second approach uses the selectivity of MS/MS detection to identify and quantitate specific modified nucleosides directly from urine samples. Liebich et al., as an example of the first approach, used HPLC-QIT-MS to develop an automated MS³ method to generate large amounts of MS/MS and MS³ data.[29] These data were then searched for the diagnostic fragment of 132Da, representing the loss of ribose from the nucleoside, so the most relevant MS³ data were selected and used to differentiate between the detected nucleosides.

Figure 5.20 shows the comparative full-scan MS data, the LC-UV data, and the constant neutral loss of 132 Da total ion chromatogram from the analysis (split into six segments with differing optimal mass spectrometric conditions).

The data-dependent acquisition (DDA) analysis allowed the extraction of nucleoside-specific data from the samples. The DDA experiment is outlined as one of three such potential methods in Figure 5.21. A second development by the same group used Affi-Gel purified urinary nucleoside samples followed by HPLC-ESI-TOF-MS for nucleoside detection. The generated automatically acquired data were then processed and the ions exhibiting the characteristic loss of 132 Da were selected and their accurate mass obtained.[53] The accurate mass and isotope distribution were then utilized in order to tentatively identify the corresponding nucleosides empirical formula and their identity (Figure 5.21).

Many of the traditional sample cleanup protocols used Affi-Gel resin, but due to its low-pressure resistance and swelling properties under equilibrium–elution conditions, it is not adaptable to on-line, high-pressure, purification methods. As an alternative to Affi-Gel resin, boronate affinity materials have been investigated for their suitability in automated, high-throughput method for the study of urinary nucleosides by HPLC-MS.[43] After the extensive study of the retention properties of the stationary phases, it was determined that ion trapping was achieved only under aprotic conditions with the addition of an aqueous mobile phase being necessary for elution. This trapping behavior, however, was not specific for *cis*-diol-containing compounds and retention relied more on

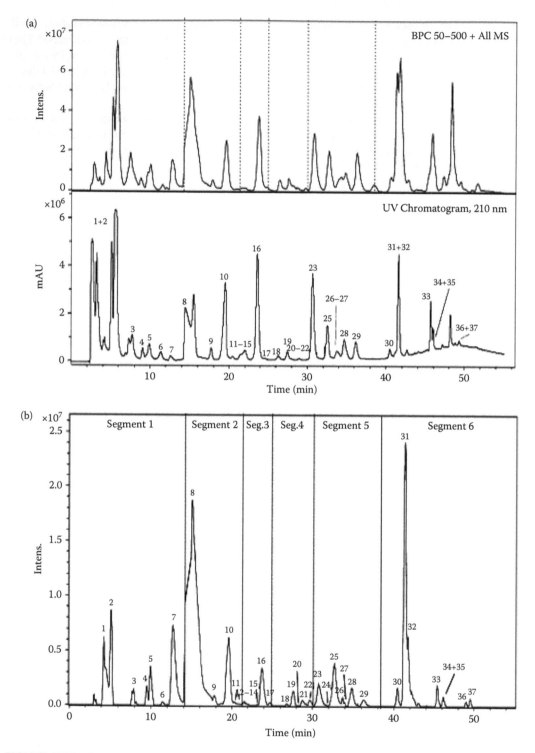

FIGURE 5.20 Data-dependent HPLC-ion-trap-MS analysis indicating mass spectrometric base peak ((a) upper trace), HPLC-UV data ((a) lower trace) and (b) constant neutral loss (−132 Da) peaks representing suspected nucleosides. (From Kammerer, B. et al., *Anal. Bioanal. Chem.* 382, 1017, 2005. With permission.)

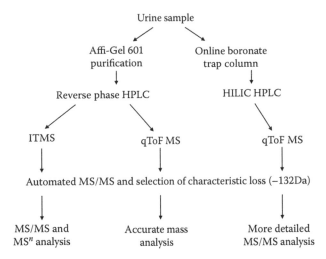

FIGURE 5.21 Outline of three possible advanced analyses of urinary nucleosides by HPLC-MS utilizing recent improvements in separation and/or selective detection.

the Lewis base properties of the examined compounds. Despite this lack of selectivity, the boronate trap was utilized in an on-line purification method combined with hydrophilic interaction chromatography (HILIC) and information-dependent MS data acquisition utilizing the characteristic transition of the nucleoside fragmentation due to the cleavage of the glycosidic bond (Figure 5.21). This single experiment method was used to both identify and quantify urinary nucleosides and creatinine, which is necessary for urinary nucleoside level normalization.

This experiment and the studies of Liebich et al. offer the possibility of DDAs, with the additional advantage of on-line urine sample cleanup by the boronate trap method, and the utilization of modern bioinformatics for the analyses of urinary nucleosides in a manner similar to the automated procedures for proteomic analyses.

The alternative approach has been to side step the purification and inject the urine directly into the HPLC-MS, making available for analysis a subset of nucleosides and other compounds that would have been lost in the purification process. To obtain the required sensitivity and selectivity of this system-selective reaction monitoring (SRM) is used in order to quantify certain nucleosides (Figure 5.22).[54] Interestingly, of the 13 compounds chosen for the study, two were not detected in any of the samples analyzed and some of the nucleosides from the normal population exhibited standard deviations that exceeded the mean values that were determined after normalization with urinary creatinine and the elimination of postcolumn infusion matrix effects by injecting "blank" urine.

The method provides quick analyses with a minimum of error introduced in sample preparation, but only a set number of compounds can be studied at one time, which may be disadvantageous, especially as urinary nucleoside "patterns" are considered to offer more potential as biomarkers than the quantification of specific nucleosides. The technique may also have a loss of specificity or sensitivity due to variations in the levels of other urinary components that are present in more abundance.

Another important advancement in urinary nucleoside analyses is the use of bioinformatics for the interpretation of the increasingly complex results that are obtained by MS or other high-throughput analyses. Recent articles have used learning vector quantization and artificial neural network analysis in order to evaluate the usefulness of the urinary nucleoside levels obtained as biomarkers for disease.[55,56]

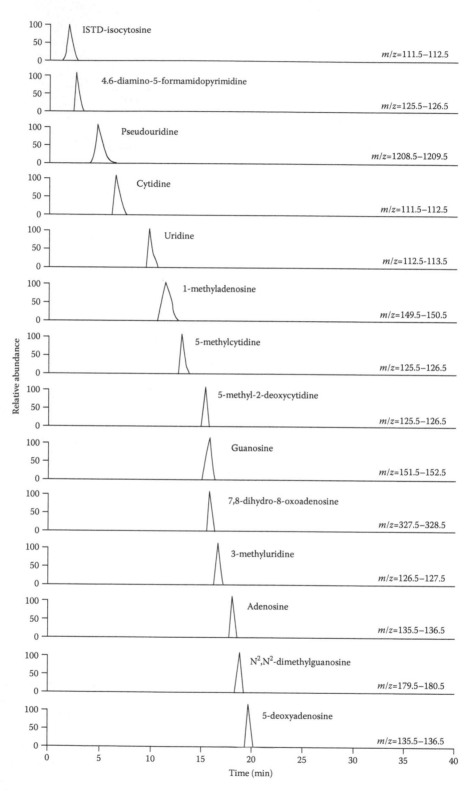

FIGURE 5.22 Data obtained from spiked urine samples for 13 nucleosides and related compounds obtained by direct injection of urine and SRM-MS. (From Lee, S.H. et al., *Rapid. Comm. Mass Spectrom.* 18, 973, 2004. With permission.)

5.4 MASS SPECTROMETRIC STUDIES OF THE LEVELS OF MODIFIED NUCLEOSIDES AS POTENTIAL CANCER BIOMARKERS

GC-MS was used in order to study the urinary modified nucleoside profile, normalized to urinary creatinine levels, from both cancer patients and the healthy (control) population.[46] Monitoring the levels of the modified nucleoside N^2,N^2-dimethylguanosine in both the control and cancer patient populations showed that the levels altered with various stages of disease progression. A slight elevation was detected between the healthy population samples and Stage 2 cancer patients with more significant increases detected with further tumor progression. The total urinary nucleoside pattern of the combined levels of all of the excreted compounds was also shown to mimic the pattern exhibited. A pilot study utilizing HPLC-QIT-MS also studied the effect of disease progression on the levels of many modified nucleosides excreted in the urine. The effect of disease progression on the urinary levels of 2-methylguanosine is shown in Figure 5.23 and shows a general increase in the observed levels with the increase in disease stage; however, it should be noted that the standard deviations observed were sufficiently large so as to make discrimination between stages difficult. Similar analyses by another group studying 1-methylinosine in healthy individuals versus breast cancer patients (undefined stage) again indicated an elevated level of the excreted compound in cancer patients; however, as seen in previous studies, the standard deviations exhibited overlap between the two sample sets.[29] Despite this overlap, the levels were shown to be statistically valid as a biomarker response and more sensitive to disease progression when a combination of nucleosides was studied. The data obtained to date from the MS analysis of these compounds as possible cancer biomarkers suggest that, as expected from previous reports, the study of individual nucleosides is not sufficient for the identification of disease. However, it can be seen that elevations in the excreted levels of modified nucleosides are detected more commonly in cancer patients compared to the normal population. The study of a number of modified nucleosides in a single analysis seems to offer a more promising benefit in terms of disease identification. Therefore further development of both the technology for accurate identification and quantitation of the nucleosides and bioinformatic interpretation of the obtained data is required in order to fully explore the potential of these compounds as biomarkers of cancer and other aberrations of metabolic control. A number of recent publications have utilized C_{18}-octadecyl solid-phase extraction (SPE) cartridges, which the authors claim to have *cis*-diol binding functionality, to purify both the nucleosides and creatinine.[56,57] This is followed by HPLC-UV analysis and determination of the nucleoside levels in relation to the creatinine levels simultaneously. However it seems unlikely that the SPE cartridge has *cis*-diol affinity

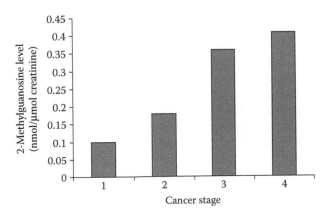

FIGURE 5.23 Change in the levels of N²-methylguanosine with differing stage of cancer development as monitored by HPLC-ion-trap-MS.

and if it did bind nucleosides in this way, it should not bind creatinine that lacks a *cis*-diol group; thus the data obtained regarding the levels of specific nucleosides (based on UV absorbance and retention time) should be treated with caution. While this analysis generates data that can be used for the diagnosis of cancer, further work is required to substantiate the identification of the nucleoside(s) in the UV peaks.

The discussion of whether a biomarker is required to be fully identified by MS or whether the change in a UV peak of unconfirmed identity is sufficient will guide the future direction of the research and should become a matter of further debate. It would seem to this author, however, that for further validation of the biomarker the unambiguous identification of the molecule is essential for it to be reliable in a clinical setting and also to gain an understanding of the biochemical processes that cause the biomarkers prevalence.

5.5   IDENTIFICATION OF NOVEL MODIFIED URINARY NUCLEOSIDES

Since urine contains relatively more of the modified nucleosides, due to the inability of the body to reutilize them, many novel nucleosides have been discovered in the urine instead of being identified from the "parent" or actual nucleic acid molecule. Much of the work involving the identification of novel modified nucleosides has relied upon the organic synthesis of the suspected modified nucleoside and comparison of the naturally occurring and synthesized molecule by UV absorbance, chromatographic behavior in a number of solvent systems, and MS data. Such techniques have in the past been utilized to identify novel modified nucleosides such as N^6-succinyladenosine from human urine.[58] More recently, the combination of MS/MS, MS^n, and accurate-mass (high-resolution) MS analysis has been used to, at least tentatively, identify novel nucleosides from urinary samples.[28,59] This identification relies to a great extent on the knowledge of nucleoside fragmentation schemes as outlined previously with the ability to perform further fragmentation of MS/MS product ions being essential, given the lack of structural information gained by initial fragmentation analyses of the nucleosides. Accurate mass instruments capable of obtaining mass accuracy of less than 5 ppm in both MS and MS/MS mode have added to the identification of new nucleosides. McCloskey elaborated some key factors to be considered when utilizing accurate mass data of nucleosides,[60] namely the total number of rings and double bonds is highly unlikely to fall outside the range 4–12; the nitrogen rule should always be followed; nitrogen content must be greater than or equal to 2; and for ribonucleosides, the oxygen atom content must be greater than 4.

Accurate mass analysis assisted to a great extent, together with the MS^n data obtained, in the identification of a number of unusual pyrimidine nucleosides isolated from urine.[28] A series of related nucleosides were found, which from the MS^n analysis due to the production of the cytosine base were thought to originate from a modification of cytidine (acetylcytidine, *m/z* 286). The nucleosides differed by 14 and 28 Da from acetylcytidine (*m/z* 300 and 314) and hence were suspected of being methylated and ethylated derivatives of this modified nucleoside. The MS/MS and MS^n patterns of all three compounds were shown to mimic each other exactly and accurate mass confirmed that both protonated molecules and fragment ions were in agreement with the modification suspected. The comparative data from cytidine, acetylcytidine, and the suspected modified acetylcytidines are shown in Tables 5.4a and 5.4b. Furthermore, the study of the constant neutral loss of 132 Da allowed the reconstructed ion chromatogram for this loss from the parent ion *m/z* 314 and 286 to be extracted. While the peak for the acetylcytidine (*m/z* 286) was clear in the latter, two other peaks were shown to occur which coeluted with the compounds responsible for the ions at *m/z* 300 and 314 (Figure 5.24). It was therefore concluded that these ions are capable of generating a fragment by in-source fragmentation of *m/z* 286, which due to the characteristic loss of 132 Da, was deduced to be acetylcytidine. This added further weight to the identification of these compounds as further modified versions of acetylcytidine. A second important factor in determining the structure of a novel nucleoside was shown by the identification of an ion at *m/z* 228, which when fragmented produced a fragment ion identical to the protonated cytosine base.[51] The ion was

TABLE 5.4a
MS/MS and MS^n Analysis Data of Cytidine, Acetylcytidine, and Suspected Methyl- and Ethyl-Acetylcytidine

Identification	Ion Seen (Th) [Identity]	MS/MS (Th) [Identity]	MS^3 (Th) [Identity] (Parent)	MS^4 (Th) [Identity] (Parent)
Cytidine	244 [MH$^+$]	112 [BH$^+$]	95 [BH$^+$–NH$_3$]	n.d.
N-acetylcytidine	286 [MH$^+$]	154 [BH$^+$], 112 [cytosineH$^+$]	112 [cytosineH$^+$], 95 [cytosineH$^+$–NH$_3$]	n.d.
Methyl-acetylcytidine	300 [MH$^+$]	168 [BH$^+$], 112 [cytosineH$^+$]	150 [BH$^+$–H$_2$O], 112 [cytosineH$^+$], 95 [cytosineH$^+$–NH$_3$] (168)	95 [cytosineH$^+$–NH$_3$] (150)
Ethyl-acetylcytidine	314 [MH$^+$]	184 [BH$^+$], 112 [cytosineH$^+$]	164 150 [BH$^+$–H$_2$O], 112 [cytosineH$^+$], 95 [cytosineH$^+$–NH$_3$] (184)	95 [cytosineH$^+$–NH$_3$] (164)

TABLE 5.4b
Accurate Mass Data of Cytidine, Acetylcytidine, and Suspected Methyl- and Ethyl-Acetylcytidine

Identification	Empirical Formula of Ion Seen (Scan Type) [Identity]	Theoretical m/z	Experimental m/z	Mass Difference (ppm)
Cytidine	C$_9$H$_{14}$N$_3$O$_5$ (full scan) [MH$^+$]	244.0933	244.0940	2.7
	C$_4$H$_6$N$_3$O (MS/MS) [BH$^+$]	112.0511	112.0515	3.7
N-acetylcytidine	C$_{11}$H$_{16}$N$_3$O$_6$ (full scan) [MH$^+$]	286.1039	286.1035	–1.5
	C$_6$H$_8$N$_3$O$_2$ (MS/MS) [BH$^+$]	154.0617	154.0640	15.0
Methyl-acetylcytidine	C$_{12}$H$_{18}$N$_3$O$_6$ (full scan) [MH$^+$]	300.1196	300.1185	–3.7
	C$_7$H$_{10}$N$_3$O$_2$ (MS/MS) [BH$^+$]	168.0773	168.0784	6.7
Ethyl-acetylcytidine	C$_{13}$H$_{20}$N$_3$O$_6$ (full scan) [MH$^+$]	314.1352	314.1323	–9.2
	C$_8$H$_{12}$N$_3$O$_2$ (MS/MS) [BH$^+$]	182.0930	182.0929	–0.3

therefore classified as a deoxyribose cytidine; however, in order for the compound to be purified (again by the *cis*-diol affinity methodology), the 2′ and 3′ positions of the sugar must not be modified. The compound was therefore considered to be a 5′-*O*-methylcytidine with the deoxy group situated on the 5′ position of the ribose sugar. This modification had not been detected before, and interestingly, was found *only* in patients with neck and head cancer. Given that the retention time of the compound was identical to 2′-deoxycytidine, the possibility that interaction of 2′ deoxycytidine with other nucleosides containing *cis*-diol groups, or its retention by other mechanisms, was tested and shown not to occur. This allows the investigator to acquire more information concerning unknown compounds, because when using low-pressure Affi-Gel chromatography the *cis*-diol group is essential for retention, and so the 2′ and 3′ positions of the sugars are eliminated as possible positions of modification. This also assisted in the tentative elucidation of a 5′-*O*-formyl cytidine nucleoside.[28]

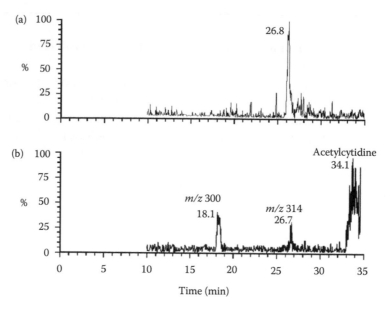

FIGURE 5.24   Reconstructed ion chromatogram of CNL of 132Th from (a) *m/z* 314 and (b) *m/z* 286.

5.6   CONCLUSION

The application of MS to the study of the modified nucleosides as metabolic by-products and as potentially valuable biomarkers of cancer has benefited to a great extent from early mass spectrometric study of nucleobases, as these are the most abundant product ions generated from MS/MS analysis of nucleosides. Although a great deal of positive data were generated by the early HPLC-UV studies of the potential application of urinary modified nucleosides as biomarkers for diseases such as cancer, mass spectrometric analysis has provided the ability to distinguish these compounds more successfully. This allows vital differences in the levels to be studied and the identification of nucleosides that, from preliminary data, at least, seem to be unique to specific cancers. Recent advances in separation science, data-dependent mass spectrometric acquisition techniques, and bioinformatic interpretation of the data generated offer scope for more high-throughput and automated analysis of urinary nucleosides. However, a finalized method is yet to be reported that covers all three required aspects. Such a method now seems achievable with modern techniques and software and would allow the analysis of a wide range of these compounds to be studied in large patient cohorts in order to further validate urinary nucleosides as metabolic biomarkers. In addition to urinary nucleoside profile application in clinical diagnosis and prognosis, MS is also vital in the successful identification of previously unknown modified nucleosides found in the urine, for which the fundamental biochemical role is yet to be defined.

REFERENCES

1. Rozenski, J., Crain, P.F., and McCloskey, J.A. *Nucleic Acids Res.* 27, 196–197, 1999, http://library.med.utah.edu/RNAmods/
2. Dunin-Horkawicz, S., Czerwoniec, A., Gajda, M.J., Feder, M., Grosjean, H., and Bujnicki, J.M., MODOMICS: A database of RNA modification pathways, *Nucleic Acids Res.* 1(34), D145–9, 2006. http://modomics.genesilico.pl/
3. Edmonds, C.G. et al., Posttranslational modification of tRNA in thermophilic *Archaea* (Archaebacteria), *J. Bacteriol.* 173, 3138, 1991.

4. Rai, K. et al., Zebra fish Dnmt1 and Suv39h1 regulate organ-specific terminal differentiation during development, *Mol. Cell. Biol.* 26, 7077, 2006.
5. Borek, E. et al., High turnover rate of transfer RNA in tumor tissue, *Can. Res.* 37, 3362, 1977.
6. Gehrke, C.W. et al., Patterns of urinary excretion of modified nucleosides, *Can. Res.* 39, 1150, 1979.
7. Borek, E., Modified nucleosides and cancer, *Can. Res.* 42, 2099, 1982.
8. Koshida, K. et al., Urinary modified nucleosides as tumor markers in cancer of the urinary organs or female genital tract, *Urol. Res.* 13, 213, 1985.
9. Nakano, K. et al., Urinary excretion of modified nucleosides as biological marker of RNA turnover in patients with cancer and AIDS, *Clin. Chim. Acta* 218, 169, 1993.
10. Tebib, J.G. et al., Relationship between urinary excretion of modified nucleosides and rheumatoid arthritis process, *Br. J. Rheum.* 36, 990, 1997.
11. Kawai, Y. et al., Endogenous formation of novel halogenated 2′-deoxycytidine. Hypohalous acid-mediated DNA modification at the site of inflammation, *J. Biol. Chem.* 279, 51241, 2004.
12. Szabo, L.D. et al., Study of nucleic acid metabolism in two astronauts, *Adv. Space Res.* 4, 15, 1984.
13. Liebich, H.M. et al., Analysis of normal and modified nucleosides in urine by capillary electrophoresis, *Chromatographia* 24, 396, 1997.
14. Sasco, A.J. et al., Breast cancer prognostic significance of some modified urinary nucleosides, *Can. Lett.* 108, 157, 1996.
15. Hecht, S.M., Gupta, A.S., and Leonard, N.J., Mass spectra of nucleoside components of tRNA, *Anal. Biochem.* 30, 249, 1969.
16. McCloskey, J.A., *Methods in Enzymology*, Chapter 45, p. 193, Academic Press, New York, 1990.
17. Pang, H. et al., Mass spectrometry of nucleic acid constituents. Trimethylsilyl derivatives of nucleosides, *J. Org. Chem.* 47, 3924, 1982.
18. Crow, F.W. et al., Fast atom bombardment combined with tandem mass spectrometry for the determination of nucleosides, *Anal. Biochem.* 139, 243, 1984.
19. Dudley, E. et al., Analysis of urinary nucleosides I. Optimisation of high performance liquid chromatograghy/electrospray mass spectrometry, *Rapid Commun. Mass Spectrom.* 14, 1200, 2000.
20. Yen-Yang, Y., Charles, M.J., and Voyksner, R.D., Processes that affect electrospray ionization—mass spectrometry of nucleobases and nucleosides, *J. Am. Soc. Mass Spectrom.* 7, 1106, 1996.
21. Rice, J.M. and Dudek, G.O., Mass spectra of nucleic acid derivatives. II. Guanine, adenine and related compounds, *J. Am. Chem. Soc.* 89, 2719, 1967.
22. Nelson, C.C. and McCloskey, J.A., Collision-induced dissociation of uracil and its derivatives, *J. Am. Soc. Mass Spectrom.* 5, 339, 1994.
23. Gregson, J.M. and McCloskey, J.A., Collision-induced dissociation of protonated guanine, *Int. J. Mass Spectrom. Ion Processes* 165/166, 475, 1997.
24. Tuytten, R. et al., Intriguing mass spectrometric behaviour of guanosine under low energy collision induced dissociation: H₂O adduct formation and gas phase reactions in the collision cell, *J. Am. Soc. Mass Spectrom.* 16, 1291, 2005.
25. Tuytten, R. et al., In-source CID of guanosine: Gas phase ion–molecule reactions, *J. Am. Soc. Mass Spectrom.* 17, 1050, 2006.
26. Jensen, S.S. et al., Collision induced dissociation of cytidine and its' derivatives, *J. Mass Spectrom.* 6, 49, 2006.
27. Bagag, A., Giuliana, A., and Laprevote, O., Atmospheric pressure photoionization mass spectrometry of nucleic bases, ribonucleosides and ribonucleotides, *Int. J. Mass Spectrom.* 264, 1, 2007.
28. Bond, A.E. et al., Analysis of urinary nucleosides V. Identification of urinary pyrimidine nucleosides by liquid chromatography/electrospray mass spectrometry, *Rapid Commun. Mass Spectrom.* 20, 137, 2006.
29. Kammerer, B. et al., Mass spectrometric identification of modified urinary nucleosides used as potential biomedical markers by LC-ITMS coupling, *Anal. Bioanal. Chem.* 382, 1017, 2005.
30. Howlett, H.A. et al., Mass spectral analysis of modified ribonucleosides obtained by degradation of 14 alkali-stable dinucleotides isolated from yeast ribonucleic acid, *Anal. Biochem.* 39, 429, 1971.
31. Zhang, Q. and Wang, Y., Differentiation of 2′-O- and 3′-O-methylated ribonucleosides by tandem mass spectrometry, *J. Am. Soc. Mass Spectrom.* 17, 1096, 2006.
32. Rice, J.M. and Dudek, G.O., Mass spectra of uridine and pseudouridine: Fragmentation patterns characteristic of a carbon–carbon nucleosidic bond, *Biochem. Biophys. Commun.* 35, 383, 1969.
33. Felden, B. et al., Presence and location of modified nucleotides in *Escherichia coli* tmRNA: Structural mimicry with tRNA acceptor branches, *EMBO J.* 17, 3188, 1998.
34. Dudley, E. et al., Study of the mass spectrometric fragmentation of pseudouridine: Comparison of fragmentation data obtained by matrix-assisted laser desorption/ionisation post-source decay, electrospray ion

trap multistage mass spectrometry, and by a method utilising electrospray quadrupole time-of-flight tandem mass spectrometry and in-source fragmentation, *Rapid Commun. Mass Spectrom.* 19, 3075, 2005.

35. Honda, I. et al., Creatinine at the evaluation of urinary 1-methyladenosine and pseudouridine excretion, *Tohuku J. Exp. Med.* 188, 133, 1999.

36. Narayanan, S. and Appleton, H.D., Creatinine: A review, *Clin. Chem.* 26, 1119, 1980.

37. Huskova, R. et al., Determination of creatinine in urine by tandem mass spectrometry, *Clin. Chim. Acta* 350, 99, 2004.

38. Taniai, H. et al., A simple quantitative assay for urinary adenosine using column switching high performance liquid chromatography, *Tohuku J. Exp. Med.* 208, 57, 2006.

39. Emmerechts, G., Herdewijn, P., and Rozenski, J., Pseudouridine detection improvement by derivatisation with methyl vinyl sulfone and capillary HPLC-mass spectrometry, *J. Chromatogr. B* 825, 233, 2005.

40. Böeseken, J., The use of boric acid for the determination of the configuration of carbohydrates, *Adv. Carbohydr. Chem. Biochem.* 4, 189, 1949.

41. Dudley, E. et al., Development of a purification procedure for the isolation of nucleosides from urine prior to mass spectrometric analysis, *Nucleot. Nucleos. Nucleic Acids.* 19, 545, 2000.

42. Sclimme, E. et al., Dual column HPLC analysis of modified ribonucleosides as urinary pathobiochemical markers in clinical research, *Nucleos. Nucleot.* 9, 407, 1993.

43. Tuytten, R. et al., The role of nitrogen Lewis basicity in boronate affinity chromatography of nucleosides, *Anal. Chem.*, 79, 6662, 2007.

44. Esmans, E.L. et al., Direct liquid introduction LC/MS for the analysis of nucleoside material present in human urine, *Biomed. Mass Spectrom.* 12, 241, 1985.

45. Edmonds, C.G., Vestal, M.L., and McCloskey, J.A., Thermospray liquid chromatography-mass spectrometry of nucleosides and of enzymatic hydrolysates of nucleic acids, *Nucleic Acids Res.* 13, 8197, 1985.

46. Langridge, J.I. et al., Gas chromatography/mass spectrometric analysis of urinary nucleosides in cancer patients; Potential of modified nucleosides as tumour markers, *Rapid Commun. Mass Spectrom.* 7, 427, 1993.

47. Dudley, E. et al., Analysis of urinary nucleosides II. Comparison of mass spectrometric methods for the analysis of urinary nucleosides, *Rapid Commun. Mass Spectrom.* 15, 1701, 2001.

48. Takeda, N., Yoshizumi, H., and Toshimitsu, N., Detection and characterisation of modified nucleosides in serum and urine of uremic patients using capillary liquid chromatography-frit fast atom bombardment mass spectrometry, *J. Chromatogr. B* 746, 51, 2000.

49. Kammerer, B. et al., MALDI-TOF MS analysis of urinary nucleosides, *J. Am. Soc. Mass Spectrom.*, 16, 940, 2005.

50. Liebich, H.M. et al., Chromatographic, capillary electrophoretic and matrix assisted laser desorption ionization time-of-flight mass spectrometry analysis of urinary modified nucleosides as tumor markers, *J. Chromatogr. A* 1071, 271, 2005.

51. Dudley, E. et al., Analysis of urinary nucleosides. III. Identification of 5′-deoxycytidine in urine of a patient with head and neck cancer, *Rapid Commun. Mass Spectrom.* 17, 1132, 2003.

52. Zhao, X. et al., Urinary profiling investigation of metabolites with *cis*-diol structure from cancer patients based on UPLC-MS and HPLC-MS as well as multivariate statistical analysis, *J. Sep. Sci.* 29, 2444, 2006.

53. Bullinger, D. et al., Identification of urinary nucleosides by ESI-TOF-MS, *Bruker Appl. Book*, 16, 2005.

54. Lee, S.H. et al., A rapid and sensitive method for quantitation of nucleosides in human urine using liquid chromatography/mass spectrometry with direct urine injection, *Rapid Commun. Mass Spectrom.* 18, 973, 2004.

55. Dieterle, F. et al., Urinary nucleosides as potential tumor markers evaluated by learning vector quantization, *Art. Intell. Med.* 28, 265, 2003.

56. Seidel, P., Seidel, A., and Herbarth, O., Multilayer perceptron tumour diagnosis based on chromatography analysis of urinary nucleosides, *Neur. Netw*, 20, 646, 2007.

57. Sediel, A. et al., Modified nucleosides: An accurate tumour marker for clinical diagnosis of cancer, early detection and therapy control, *Br. J. Cancer* 94, 1726, 2006.

58. Chheda, G.B., Isolation and characterisation of N^6-succinyladenosine from human urine, *Nucleic Acids Res.* 4, 739, 1977.

59. Dudley, E. et al., Analysis of urinary nucleosides IV. Identification of urinary purine nucleosides by liquid chromatography/electrospray mass spectrometry, *Rapid Commun. Mass Spectrom.* 18, 2730, 2004.

60. McCloskey, J.A., *Methods in Enzymology*, Chapter 41, p. 193, Academic Press, New York, 1990.

6 Mass Spectrometric Determination of DNA Adducts in Human Carcinogenesis

Filip Lemière

CONTENTS

6.1   INTRODUCTION

What are DNA adducts? What do they have to do with cancer? What are the sources of DNA adducts? And what role does mass spectrometry (MS) have in this?

6.2   HUMAN CARCINOGENESIS—QUESTIONS!

Tumor incidence and mortality are growing yearly, independent of the aging of the population. In some industrialized countries, cancer is liable for almost 30% of the total mortality. The real contribution of lifestyle, personal habits, and diet to this increase cannot be underestimated, but they are overshadowed by the influence of environmental changes caused by industry. Since World War II industrial expansion has been increasing ever faster, releasing thousands of natural and man-made agents into the environment. It is evident that the environment is not confined solely to the industrial production site but the air, soil, water, buildings, drinking water, food, drugs, and so on may also be contaminated.

The number of known carcinogens is growing in parallel with the number of carcinogenicity studies. Several industrial megacompounds, which have been produced and used for a long time, have lately been discovered to be carcinogenic. This is worrisome if one considers that the carcinogenicity of the majority of industrial chemicals has not yet been investigated. This seems to be a problem of the modern industrialized era. But exactly how modern are occupational diseases, especially cancer? Is there a relationship between the different carcinogens with respect to chemical class, structure, solubility, and metabolism? What is the malignant interaction between a carcinogen and a cell? How can we detect and describe this interaction? What can liquid chromatography (LC)-MS/MS do in this area?

6.3   DNA ADDUCTS: A HISTORICAL INTRODUCTION

6.3.1   OCCUPATIONAL CARCINOGENESIS

In 1531, the physician Paracelsus described a lung disease, *Malla metallorum*, which killed many miners at an early age. In his 12 volumes counting *De Re Metallica*, Agricola dealt with the (current in 1556) practices of mining and refining metals and the accidents and lung diseases commonly encountered by the workers. Some of these lung diseases were almost certainly lung cancer, tuberculosis, and silicosis. Although not recognized as such at the time, these were probably the first accounts describing occupational (lung) cancer.[1]

The father of occupational medicine, Bernardino Ramazzini (1633–1714) (Figure 6.1), made detailed observations of working conditions in a great range of occupations from cleaning out cesspits and other dirty trades to singers, nuns, and notaries.[2] He described the ailments suffered by many as a result of their callings, recommended measures for their prevention, and advised doctors, when visiting workers in their homes to ask what their occupation was. The first clear recognition of an occupational cancer was by Percivall Pott (1775), a surgeon at London's St. Bartholomew's Hospital, who described the common occurrence of cancer of the scrotum in chimney sweeps.[3]

FIGURE 6.1   Ramazzini, engraving by J.G. Seiller. *Archiv fur Kunst und Geschichte*, West Berlin.

From these common histories, Pott concluded: the occupation of these men as young boys was directly and causally related to their malignant disease and the soot to which they were excessively exposed in their work was the causative agent of the cancer.

More than a century later (1892), Henry T. Butlin reported the relative rarity of scrotal cancer in chimney sweeps in the European continent compared with those in England.[4] Apparently, frequent bathing and the protective clothing worn by the continental workers lowered the incidence of the disease. In the same period (1895), occupational cancer of the urinary bladder was recognized in a small group of German dyestuff workers.[5] The commercial production of aromatic amines started in the middle of the nineteenth century for the production of aniline-based dyes, for example, rosaniline and mauveine (Figure 6.2). Leichtenstern suggested that there could be a causative relationship between the exposure to these aromatic amines and urinary bladder cancer.[6] Epidemiological studies made it incontrovertible; the aromatic amines 2-naphtylamine, benzidine (Figure 6.3), and 4-aminobiphenyl caused bladder cancer in humans.

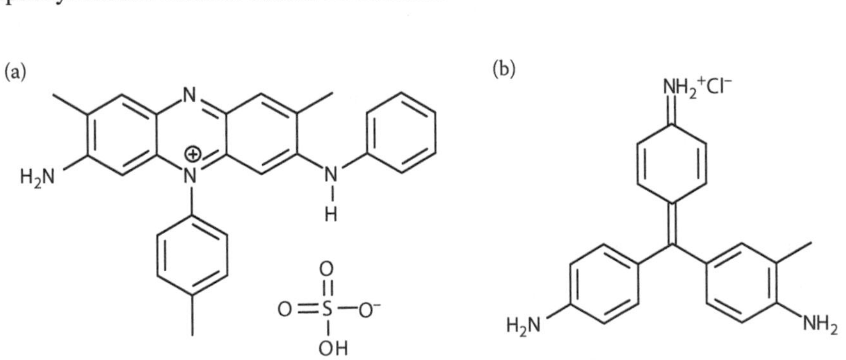

(a)   (b)

FIGURE 6.2   Structures of the two aniline dyes: (a) mauveine or aniline purple; (b) rosaniline or basic fuchsin.

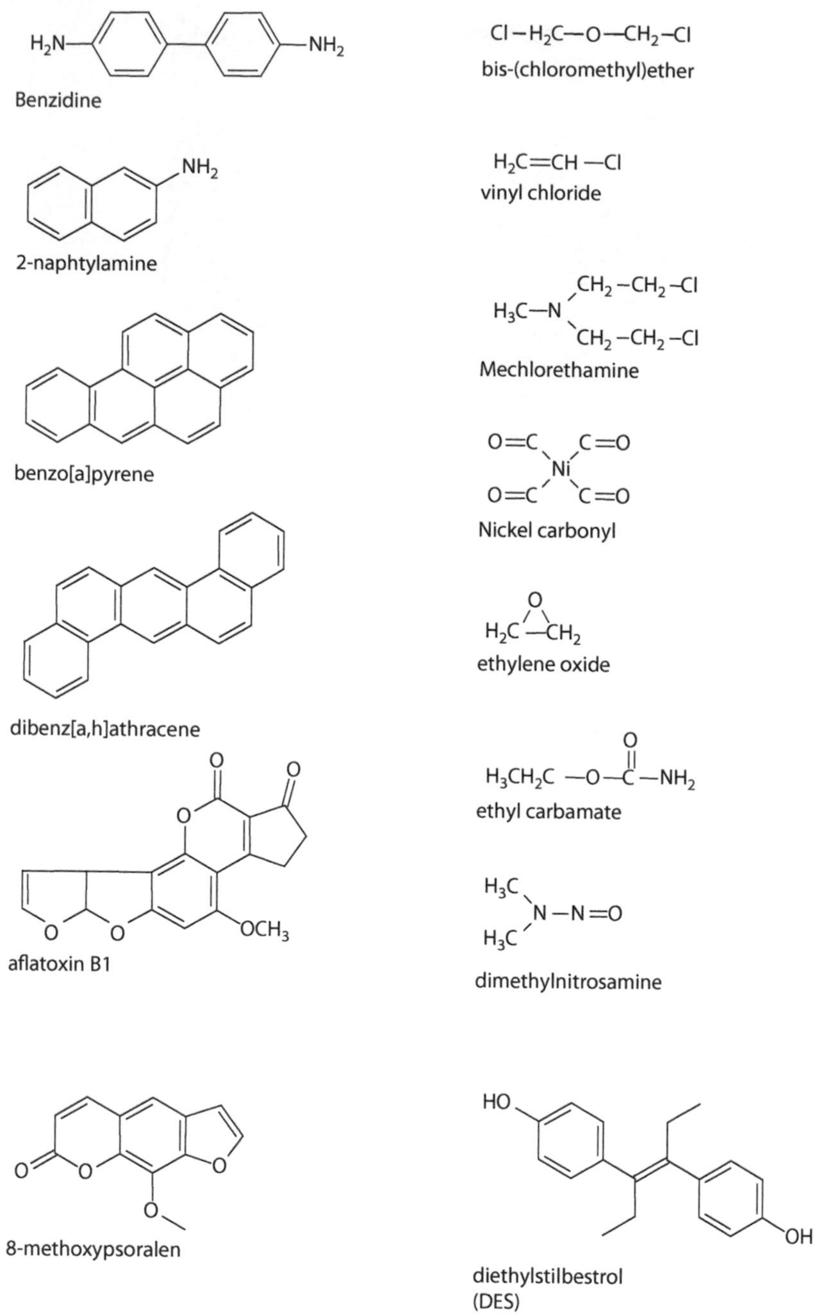

FIGURE 6.3 Selection of some known carcinogenic chemicals.

6.3.2 Chemical Carcinogenesis

While lifestyle factors such as diet, sexual mores, and reproductive history may not be considered under the heading of chemical carcinogenesis in the usual sense, chemicals or mixtures of chemicals are the predominant components that cause human cancer. Potential exceptions in this group are those neoplasms developing as a result of sexual promiscuity, where the human papillomavirus may play a major role in genital cancers. Although there is increasing data on viral induction of cancer, maybe even surpassing the data on chemical carcinogenesis, our understanding of the mechanisms of chemical carcinogenesis preceded the recognition of viral cancer induction.[7]

6.3.3   Chemical Carcinogens

Studies on experimental chemical carcinogenesis started in 1915 when K. Yamagawa and K. Ichikawa demonstrated the chemical induction of skin carcinomas on the ears of rabbits after repeated topical applications of coal tar over an extended period.[8] These experimental studies, coupled with the epidemiological studies in the nineteenth century, more than justified the proposals Dr. Pott had made almost 150 years earlier. The experimental chemical induction of cancer by coal tar, and the observation of the "aniline-dye–cancer" relationship that implied that relatively pure aromatic amines could be carcinogenic, triggered the hunt for specific carcinogenic compounds in mixtures. Cook et al. isolated the highly carcinogenic benzo[a]pyrene (Figure 6.3) from coal tar (1933).[9] A few years earlier (1930), a related synthetic polyaromatic compound, dibenz[a,h]anthracene, was demonstrated by Kennaway and Hieger to be carcinogenic.[10] The aromatic amines, suspected from the aniline dyes, were demonstrated to be carcinogenic in their pure form by feeding dogs 2-naphtylamine, which induced cancer of the bladder.[11] Since those early days, a large number of synthetic and naturally occurring chemicals and mixtures have been found to be carcinogenic in experimental animals, and by epidemiological investigations to be cancerous in humans (Table 6.1).[12] There are several other agents, such as physical (radiation, asbestos), microorganisms (*Helicobacter pylori*, hepatitis B/C virus), as well as chemical agents, that can induce cancer.

   Table 6.1 and the structures in Figure 6.3 show the extremely diverse nature of the presented chemicals, although there are structural similarities between compounds from the same chemical class. Many carcinogens contain aromatic residues and are essentially insoluble in aqueous media, but the carcinogenic *nitroso* compounds do not have this property. Although many carcinogens are organic nitrogen compounds, there are others (such as the hydrocarbons) which show that nitrogen is not *required* for carcinogenic activity. Since all these chemicals lead to the same overall biological response of cancer, it might be expected that some common causative property would be present in these various carcinogenic structures. Discovering this common property, however, proved to be a perplexing challenge that took many years to resolve.

6.3.4   Mechanism of Chemical Carcinogenesis

This disparity in the characteristics of carcinogens presented a serious problem to the experimental oncologist in the 1950s. They were uncertain if there was one mechanism for causing cancer common to all the carcinogen classes or if there were numerous mechanisms, unique for each chemical or class of chemicals. The apparent lack of common properties between the classes of carcinogens initially resulted in structure–activity analyses being confined to an individual class of carcinogens.

   A major breakthrough in the understanding of the mechanisms involved in chemical carcinogenesis was the Miller and Miller demonstration that reactive metabolites of carcinogenic chemicals were critical intermediates in the induction of neoplasia.[13] Their work indicated that the responsibility for initiating the carcinogenic process frequently rested not with the carcinogen itself but with one of its metabolites, termed the ultimate carcinogen. For example, N-2-fluorenylacetamide is metabolized to its N-hydroxy derivative and esterified. This ester was found, in the rat, to be a more potent carcinogen than the parent compound, suggesting that carcinogenic activity can be expressed only after metabolic activation of the carcinogen.

   Further evidence for the involvement of metabolism in the action of carcinogens was given by reports on covalent bonding of carcinogens to biomolecules in tissue. Aminoazo dyes were found to be bound to protein in rat liver and the intense fluorescence of benzo[a]pyrene was used to show its binding to mouse skin proteins.[14,15] Using radiochemical labeling techniques, Brookes and Lawley

TABLE 6.1
List of All Agents, Mixtures, and Exposures Evaluated to Date as Evaluated in IARC Monographs Volumes 1–97

Group 1: Carcinogenic to Humans (102)

Agents and Groups of Agents
- 4-Aminobiphenyl [92-67-1] (Vol. 1, Suppl. 7; 1987)
- Arsenic [7440-38-2] and arsenic compounds (Vol. 23, Suppl. 7; 1987)

(NB: This evaluation applies to the group of compounds as a whole and not necessarily to all individual compounds within the group)

- Asbestos [1332-21-4] (Vol. 14, Suppl. 7; 1987)
- Azathioprine [446-86-6] (Vol. 26, Suppl. 7; 1987)
- Benzene [71-43-2] (Vol. 29, Suppl. 7; 1987)
- Benzidine [92-87-5] (Vol. 29, Suppl. 7; 1987)
- Benzo[*a*]pyrene [50-32-8] (Vol. 32, Suppl. 7, Vol. 92; in preparation)

(NB: Overall evaluation upgraded from 2B to 1 based on mechanistic and other relevant data)

- Beryllium [7440-41-7] and beryllium compounds (Vol. 58; 1993)
- *N,N*-Bis(2-chloroethyl)-2-naphthylamine (Chlornaphazine) [494-03-1] (Vol. 4, Suppl. 7; 1987)
- Bis(chloromethyl)ether [542-88-1] and chloromethyl methyl ether [107-30-2] (technical-grade) (Vol. 4, Suppl. 7; 1987)
- 1,3-Butadiene [106-99-0] (Vol. 71, Vol. 97; in preparation)
- 1,4-Butanediol dimethanesulfonate (Busulphan; Myleran) [55-98-1] (Vol. 4, Suppl. 7; 1987)
- Cadmium [7440-43-9] and cadmium compounds (Vol. 58; 1993)
- Chlorambucil [305-03-3] (Vol. 26, Suppl. 7; 1987)
- 1-(2-Chloroethyl)-3-(4-methylcyclohexyl)-1-nitrosourea (Methyl-CCNU; Semustine) [13909-09-6] (Suppl. 7; 1987)
- Chromium[VI] (Vol. 49; 1990)
- Ciclosporin [79217-60-0] (Vol. 50; 1990)
- Cyclophosphamide [50-18-0] [6055-19-2] (Vol. 26, Suppl. 7; 1987)
- Diethylstilboestrol [56-53-1] (Vol. 21, Suppl. 7; 1987)
- Epstein-Barr virus (Vol. 70; 1997)
- Erionite [66733-21-9] (Vol. 42, Suppl. 7; 1987)
- Estrogen–progestogen menopausal therapy (combined) (Vol. 72, Vol. 91; in preparation)
- Estrogen–progestogen oral contraceptives (combined) (Vol. 72, Vol. 91; in preparation)

(NB: There is also convincing evidence in humans that these agents confer a protective effect against cancer in the endometrium and ovary)

- Estrogens, nonsteroidal (Suppl. 7; 1987)

(NB: This evaluation applies to the group of compounds as a whole and not necessarily to all individual compounds within the group)

- Estrogens, steroidal (Suppl. 7; 1987)

(NB: This evaluation applies to the group of compounds as a whole and not necessarily to all individual compounds within the group)

- Estrogen therapy, postmenopausal (Vol. 72; 1999)
- Ethanol [64-17-5] in alcoholic beverages (Vol. 96; 2007)
- Ethylene oxide [75-21-8] (Vol. 60, Vol. 97; in preparation)

(NB: Overall evaluation upgraded from 2A to 1 based on mechanistic and other relevant data)

- Etoposide [33419-42-0] in combination with cisplatin and bleomycin (Vol. 76; 2000)
- Formaldehyde [50-00-0] (Vol. 88; 2006)
- Gallium arsenide [1303-00-0] (Vol. 86; 2006)
- [Gamma Radiation: see X- and Gamma (g)-Radiation]

continued

TABLE 6.1   (continued)
List of All Agents, Mixtures, and Exposures Evaluated to Date As Evaluated in IARC Monographs Volumes 1–97

- Helicobacter pylori (infection with) (Vol. 61; 1994)
- Hepatitis B virus (chronic infection with) (Vol. 59; 1994)
- Hepatitis C virus (chronic infection with) (Vol. 59; 1994)
- Human immunodeficiency virus type 1 (infection with) (Vol. 67; 1996)
- Human papillomavirus types 16, 18, 31, 33, 35, 39, 45, 51, 52, 56, 58, 59, and 66 (Vol. 64, Vol. 90; in preparation)

(NB: The HPV types that have been classified as carcinogenic to humans can differ by an order of magnitude in risk for cervical cancer)

- Human T-cell lymphotropic virus type I (Vol. 67; 1996)
- Melphalan [148-82-3] (Vol. 9, Suppl. 7; 1987)
- 8-Methoxypsoralen (Methoxsalen) [298-81-7] plus ultraviolet A radiation (Vol. 24, Suppl. 7; 1987)
- MOPP and other combined chemotherapy including alkylating agents (Suppl. 7; 1987)
- Mustard gas (Sulfur mustard) [505-60-2] (Vol. 9, Suppl. 7; 1987)
- 2-Naphthylamine [91-59-8] (Vol. 4, Suppl. 7; 1987)
- Neutrons (Vol. 75; 2000)

(NB: Overall evaluation upgraded from 2B to 1 with supporting evidence from other relevant data)

- Nickel compounds (Vol. 49; 1990)
- N'-Nitrosonornicotine (NNN) [16543-55-8] and 4-(N-Nitrosomethylamino)-1-(3-pyridyl)-1-butanone (NNK) [64091-91-4] (Vol. 37, Suppl. 7, Vol. 89; in preparation)

(NB: Overall evaluation upgraded from 2B to 1 based on mechanistic and other relevant data)

- [Oestrogen: see Estrogen]
- Opisthorchis viverrini (infection with) (Vol. 61; 1994)
- [Oral contraceptives, combined estrogen–progestogen: see Estrogen–progestogen oral contraceptives (combined)]
- Oral contraceptives, sequential (Suppl. 7; 1987)
- Phosphorus-32, as phosphate (Vol. 78; 2001)
- Plutonium-239 and its decay products (may contain plutonium-240 and other isotopes), as aerosols (Vol. 78; 2001)
- Radioiodines, short-lived isotopes, including iodine-131, from atomic reactor accidents and nuclear weapons detonation (exposure during childhood) (Vol. 78; 2001)
- Radionuclides, α-particle-emitting, internally deposited (Vol. 78; 2001)

(NB: Specific radionuclides for which there is sufficient evidence for carcinogenicity to humans are also listed individually as Group 1 agents)

- Radionuclides, β-particle-emitting, internally deposited (Vol. 78; 2001)

(NB: Specific radionuclides for which there is sufficient evidence for carcinogenicity to humans are also listed individually as Group 1 agents)

- Radium-224 and its decay products (Vol. 78; 2001)
- Radium-226 and its decay products (Vol. 78; 2001)
- Radium-228 and its decay products (Vol. 78; 2001)
- Radon-222 [10043-92-2] and its decay products (Vol. 43, Vol. 78; 2001)
- Schistosoma haematobium (infection with) (Vol. 61; 1994)
- Silica [14808-60-7], crystalline (inhaled in the form of quartz or cristobalite from occupational sources) (Vol. 68; 1997)
- Solar radiation (Vol. 55; 1992)
- Talc containing asbestiform fibers (Vol. 42, Suppl. 7; 1987)
- Tamoxifen [10540-29-1] (Vol. 66; 1996)

(NB: There is also conclusive evidence that tamoxifen reduces the risk of contralateral breast cancer)

- 2,3,7,8-Tetrachlorodibenzo-para-dioxin [1746-01-6] (Vol. 69; 1997)

(NB: Overall evaluation upgraded from 2A to 1 with supporting evidence from other relevant data)

continued

TABLE 6.1   (continued)
List of All Agents, Mixtures, and Exposures Evaluated to Date As Evaluated in IARC Monographs Volumes 1–97

- Thiotepa [52-24-4] (Vol. 50; 1990)
- Thorium-232 and its decay products, administered intravenously as a colloidal dispersion of thorium-232 dioxide (Vol. 78; 2001)
- Treosulfan [299-75-2] (Vol. 26, Suppl. 7; 1987)
- Vinyl chloride [75-01-4] (Vol. 19, Suppl. 7, Vol. 97; in preparation)
- X- and Gamma (g)-Radiation (Vol. 75; 2000)

Mixtures

- Aflatoxins (naturally occurring mixtures of) [1402-68-2] (Vol. 56, Vol. 82; 2002)
- Alcoholic beverages (Vol. 44, Vol. 96; 2007)
- Areca nut (Vol. 85; 2004)

(NB: Overall evaluation based on human data, animal data, and mechanistic and other relevant data)

- Betel quid with tobacco (Vol. 85; 2004)
- Betel quid without tobacco (Vol. 85; 2004)
- Coal-tar pitches [65996-93-2] (Vol. 35, Suppl. 7; 1987)
- Coal-tars [8007-45-2] (Vol. 35, Suppl. 7; 1987)
- Herbal remedies containing plant species of the genus Aristolochia (Vol. 82; 2002)
- Household combustion of coal, indoor emissions from (Vol. 95; in preparation)
- Mineral oils, untreated and mildly treated (Vol. 33, Suppl. 7; 1987)
- Phenacetin, analgesic mixtures containing (Suppl. 7; 1987)
- Salted fish (Chinese-style) (Vol. 56; 1993)
- Shale-oils [68308-34-9] (Vol. 35, Suppl. 7; 1987)
- Soots (Vol. 35, Suppl. 7; 1987)
- Tobacco, smokeless (Vol. 37, Suppl. 7, Vol. 89; in preparation)
- Wood dust (Vol. 62; 1995)

Exposure circumstances

- Aluminium production (Vol. 34, Suppl. 7; 1987)
- Arsenic in drinking-water (Vol. 84; 2004)
- Auramine, manufacture of (Suppl. 7; 1987)
- Boot and shoe manufacture and repair (Vol. 25, Suppl. 7; 1987)
- Chimney sweeping (Vol. 92; in preparation)
- Coal gasification (Vol. 34, Suppl. 7, Vol. 92; in preparation)
- Coal-tar distillation (Vol. 92; in preparation)
- Coke production (Vol. 34, Suppl. 7, Vol. 92; in preparation)
- Furniture and cabinet making (Vol. 25, Suppl. 7; 1987)
- Haematite mining (underground) with exposure to radon (Vol. 1, Suppl. 7; 1987)
- Involuntary smoking (exposure to secondhand or "environmental" tobacco smoke) (Vol. 83; 2004)
- Iron and steel founding (Vol. 34, Suppl. 7; 1987)
- Isopropyl alcohol manufacture (strong-acid process) (Suppl. 7; 1987)
- Magenta, manufacture of (Vol. 57; 1993)
- Painter (occupational exposure as a) (Vol. 47; 1989)
- Paving and roofing with coal-tar pitch (Vol. 92; in preparation)
- Rubber industry (Vol. 28, Suppl. 7; 1987)
- Strong-inorganic-acid mists containing sulfuric acid (occupational exposure to) (Vol. 54; 1992)
- Tobacco smoking and tobacco smoke (Vol. 83; 2004)

This list contains all hazards evaluated to date, according to the type of hazard posed and to the type of exposure. Where appropriate, chemical abstract numbers are given [in square brackets]. For details of the evaluation, the relevant monograph should be consulted (volume number given in round brackets, followed by year of publication of latest evaluation).

were able to show that in mouse skin, carcinogenic potency was correlated with a series of aromatic hydrocarbons binding to DNA, but not to protein.[16] These findings made it evident that some kind of *in vivo* metabolism was involved, since neither azo dyes nor hydrocarbons form covalent bonds with these macromolecules.

It eventually became evident that, in most cases, it was not the "carcinogen" itself but the production of reactive metabolites from it that triggered the destructive process called cancer. This explains why, when initially looking at structure–function relationships, the common factor among chemical carcinogens was obscured, since all of the attention was focused on the parent compounds and not on their reactive metabolites. Once the role of the metabolites was recognized, most of the known chemical carcinogens were thought of as "procarcinogens," which were initially metabolized to "proximate carcinogens" and finally to "ultimate carcinogens (Figure 6.4)." The reactivity of these metabolically produced ultimate carcinogens, with their ability to form covalent bonds (adducts) with biologically important macromolecules such as proteins and DNA, are responsible for the carcinogenicity of the parent compound. The Millers proposed that the majority of carcinogens are (or have metabolites that are) electrophilic reactants and these reactants can covalently bond to cellular macromolecules, of which DNA is the most critical target.

Although the critical significance of electrophilic metabolites was demonstrated, for a number of carcinogens, the "ultimate carcinogenic form" was yet to be determined. Already in 1950 the ultimate carcinogen form of polycyclic hydrocarbons was suggested to be the epoxides.[17] The existence of these metabolites in biological systems was detected 20 years later by Jerina et al.[18] A number of investigators searched for the ultimate carcinogen form of the first isolated coal tar carcinogen, benz[*a*]pyrene. Finally, the structure of (+)-*anti*-benzo[*a*]pyrene-7,8-dihydrodiol-9,10-epoxide was put forward.[19,20]

6.3.5   DNA Adducts, Mutations, and Cancer

"What is the relationship between DNA adducts and cancer? It is a chronic messing up of our DNA," K. Hemminki[21]

The development of a cancer is a multistep process in which chemical carcinogens may play a critical role. Three major stages are involved in the development of cancer: initiation, promotion, and propagation. It is generally accepted that neoplastic growth is initiated for most chemical carcinogens by their interaction with DNA. This DNA damage leads to the alteration of the activity of certain critical genes, by inducing mutations that result in abnormal regulation of gene expression. Multiple mutations may be required to produce neoplastic growth, and the rate and frequency of progression toward neoplasia are greatly influenced by the extent of exposure to tumor promoters; there is good evidence that the formation and persistence of DNA damage correlates with the potency of a number of chemical carcinogens. This was already stated above, in the experiments of Brookes and Lawley who showed that in mouse skin, for a series of aromatic hydrocarbons that binding to DNA (but not to protein) correlated with carcinogenic potency.[16] This covalent binding between a chemical and DNA or one of its constituents (heterocyclic bases, nucleosides, nucleotides) is known as adduct formation and the resulting covalently bound end product as a DNA adduct.

There are various sites of the DNA chain that due to their nucleophilic character are susceptible to electrophile attack and, therefore, are prone to adduct formation. When alkylating DNA, several sites for alkylation are known to occur, including nitrogen, oxygen, and carbon (C8 of the pyrimidine bases) at the base moiety and the phosphate of the DNA backbone (Figure 6.5).[22–25] The nature of the electrophile can influence the selectivity for adduct formation on different nucleophilic sites of the nucleobase.[26]

It is evident, from the structure and function of DNA, that modifications on the base moiety, especially at the hydrogen-bonding sites, may disturb the genetic code causing mutations, and thereby affecting the proper functioning of the cell or organism and this can result in cancer.[27–29]

Several reviews have suggested that O^6 of guanine- and O^4 of thymine-oxygen adducts are the most probable causes of mutations that can initiate carcinogenesis; however, other adducts cannot be

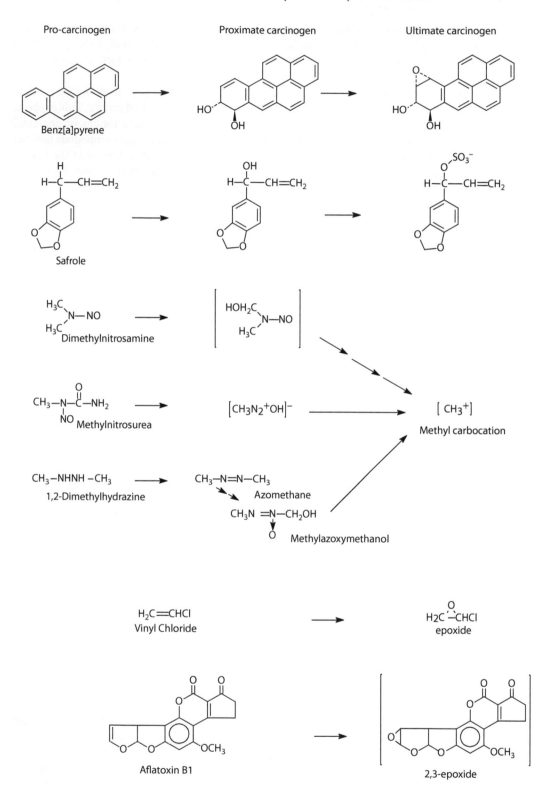

FIGURE 6.4 From procarcinogen to the ultimate carcinogenic metabolite.

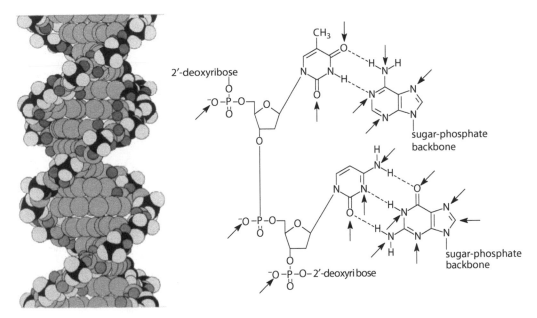

FIGURE 6.5   Structure of DNA, base pairing and nucleophilic sites indicated with arrows.

ruled out.[28–35] For the O[6] alkylation of guanine, where the carbonyl on C-6 is fixed in its *enol* tautomer, convincing evidence exists that this adduct induces G.C → A.T transitions.[36,30,37] Loveless proposed that this fixed *enol* tautomer facilitates the base pairing of the O[6]-guanine adduct with thymine (Figure 6.6), inducing a point mutation when the DNA code is duplicated.[30] Similarly, the presence of O[4] alkylthymine leads to the misincorporation of guanine producing an A.T → G.C transition.

The majority of and most prominent DNA adducts are the result of alkylation on one of the nitrogen atoms present in the base moiety. One of the main sites of alkylation in DNA is the N-7 atom of guanine (the most nucleophilic site), although examples can be given for alkylation on any of the nitrogen atoms in heterocyclic DNA bases.[22] The main product from alkylation with the potent carcinogen, benzo[*a*]pyrene diol epoxide (*vide supra*), is the *N*[2]-Gua adduct.[38,39] The *N*-acetyl-aminofluorene reacts mainly at the C-8 of guanine. All of the tested epoxides reacted with deoxyguanosine at N-7. The most reactive epoxides also formed 1,7-dialkylation products. N-1 of dAdo and N-3 of dCyd were found to be common alkylation sites with epoxides.[22] The effect of alkylation varies from base-pairing difficulties in guanine O[6], N-1, N[2]; cytosine O[2], N-3, N[4]; adenine N-1, N[6]; thymine O[4], N-3 (Figures 6.5 and 6.6). It can also cause conformational distortions in guanine C-8 of N-acetyl-aminofluorene (AAF), interstrand cross-linking, depurination and apurinic sites, hydrolysis of the imidazole ring, and hydrolytic deamination (β-hydroxy on N-3 of cytidine).[40]

Although not immediately linked to carcinogenesis, alkylation of the phosphate moiety neutralizes the negatively charged phosphodiester. This neutral phosphotriester (PTE) might alter the interaction of the DNA with other molecules (e.g., proteins). The phosphate adducts formed by simple alkylating agents tend to be stable, unless the alkyl group has an O, N, or S in the β-position (e.g., adducts of epoxides), and are either not repaired or are slowly repaired. This makes them possible candidates to monitor exposure to carcinogens.[41–44]

Once the importance of metabolic activation of a chemical carcinogen was recognized, adduct formation became generally accepted as the initiator of chemical carcinogenesis.[45] Unless repaired, DNA adducts lead to cell death, mutation, or cancer (Scheme 6.1).[46–48] The effect of the adducts on the malignant process and the techniques for their detection are still the subject of many investigations.

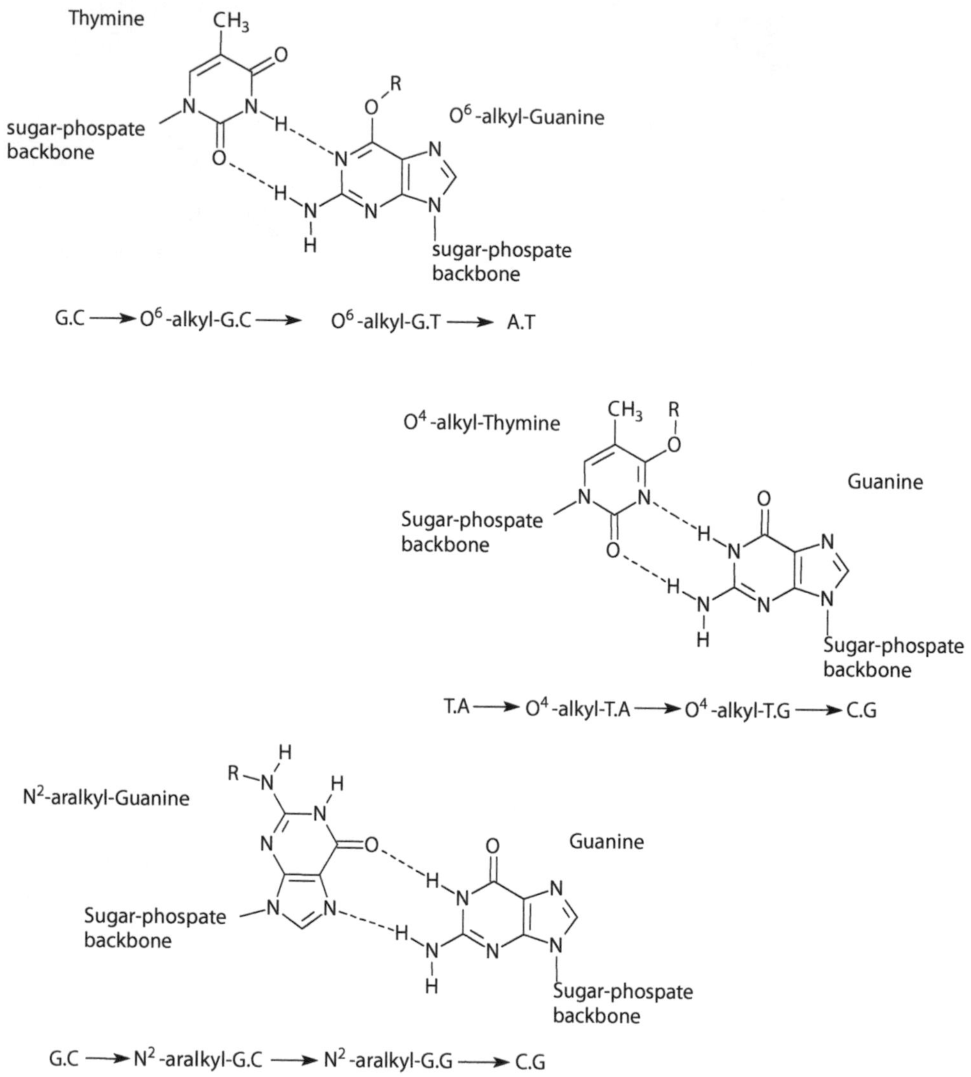

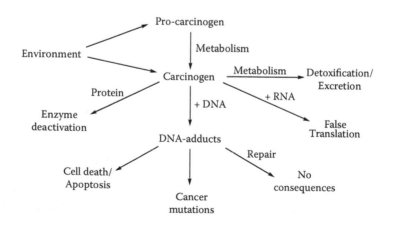

Thymine

sugar-phospate backbone

O⁶-alkyl-Guanine

sugar-phospate backbone

G.C ⟶ O⁶-alkyl-G.C ⟶ O⁶-alkyl-G.T ⟶ A.T

O⁴-alkyl-Thymine

Sugar-phospate backbone

Guanine

Sugar-phospate backbone

T.A ⟶ O⁴-alkyl-T.A ⟶ O⁴-alkyl-T.G ⟶ C.G

N²-aralkyl-Guanine

Sugar-phospate backbone

Guanine

Sugar-phospate backbone

G.C ⟶ N²-aralkyl-G.C ⟶ N²-aralkyl-G.G ⟶ C.G

FIGURE 6.6 Miscoding of DNA bases and their induced mutation.

Pro-carcinogen

Environment

Metabolism

Carcinogen

Metabolism ⟶ Detoxification/ Excretion

Protein

Enzyme deactivation

+ DNA

+ RNA

False Translation

DNA-adducts

Cell death/ Apoptosis

Repair

No consequences

Cancer mutations

SCHEME 6.1 Schematic overview of the formation of DNA adducts and the initiation of cancer.

6.4 SOURCES OF DNA-ALKYLATING COMPOUNDS

There are several sources of DNA-alkylating compounds that we come into contact with daily. We can either avoid exposure to some of these sources of mutagens and carcinogens (e.g., tobacco smoke, heating of food) or limit our exposure to them (e.g., aflatoxin). Accidental contact can sometimes yield very high-dose exposures (e.g., industrial accidents) and there are some cases where contact with known carcinogens is required (e.g., therapeutic drugs).

6.4.1 ENVIRONMENT/POLLUTION/LIFESTYLE

DNA adducts that resulted from pollution, occupational or accidental exposure, were the first environmental carcinogens that were investigated. Polyaromatic hydrocarbons (PAHs) adducts that originated from industrial/domestic coal heating, smoking, traffic exhaust, and wood fire were the predominant class initially investigated.[49–53] There are other examples in this adduct category that are related to smoking (4-aminobiphenyl-, etheno-), industrial chemicals (styrene, ethylene oxide (EO), 1,3-butadiene (BD), acrylamide, vinylchloride), and hair dyes.[54–66]

6.4.2 FOOD

Food, indispensable for life, should not be a source of potentially life threatening carcinogenic molecules; however, contamination with mycitoxins (aflatoxin) due to the presence of mold on food or animal feed is a well-documented source of DNA-alkylating compounds.[67,68] Carcinogenic *N*-nitrosamines can be present in food as a result of the curing process (meat, fish) or can be formed upon uptake of nitrite added to food in combination with amines already present. Even if no carcinogens are present in the fresh food, the cooking process can produce different toxic, and in some cases, highly carcinogenic compounds. Fried food is a potential source of acrylamide, grilled meat is a notorious source of heterocyclic amines; and PAHs are readily produced by barbecuing meat.[52,53,68–72]

6.4.3 DRUGS

The recognition of DNA adducts as an important factor in the process of chemical carcinogenesis makes it surprising that some of the most important drugs used in cancer therapy (Figure 6.7) are, in fact, themselves alkylating compounds that are carcinogenic. However, knowing the DNA replication mechanism that is essential for cell division, the link between DNA adducts and potential therapeutic application is readily made. Therefore, both carcinogens and chemotherapeutic agents are found in DNA-adduct studies (mitomycin-C, melphalan, ellipicine, misplatin; see also the Section *Chemotherapeutica*).[73–79]

There are also other drugs known to form DNA adducts. Synthetic derivatives of endogenous compounds (ethinylestradiol), or even the endogenous estradiol, which are widely used estrogen hormones for contraception and hormone replacement therapy are shown to produce DNA adducts.[80–82]

6.4.4 RADIATION

Radiation, although not a chemical, can cause modification of DNA. Otherwise, harmless and abundantly available molecules produce highly reactive radicals (e.g., OH·) as a result of the absorption of energy from electromagnetic radiation, which results in the formation of typical DNA adducts.[83–85] These radicals and other reactive oxygen species (ROS) can also be produced during metabolism of food substances and from endogenous compounds (hormones) resulting in the same oxidative damage to DNA. Even the DNA bases, upon UV irradiation, react with other DNA bases yielding a range of photoproducts such as cylobutane pyrimidine dimers (Figure 6.8); both the monomeric and dimeric DNA photoproducts are studied using LC-MS.[86]

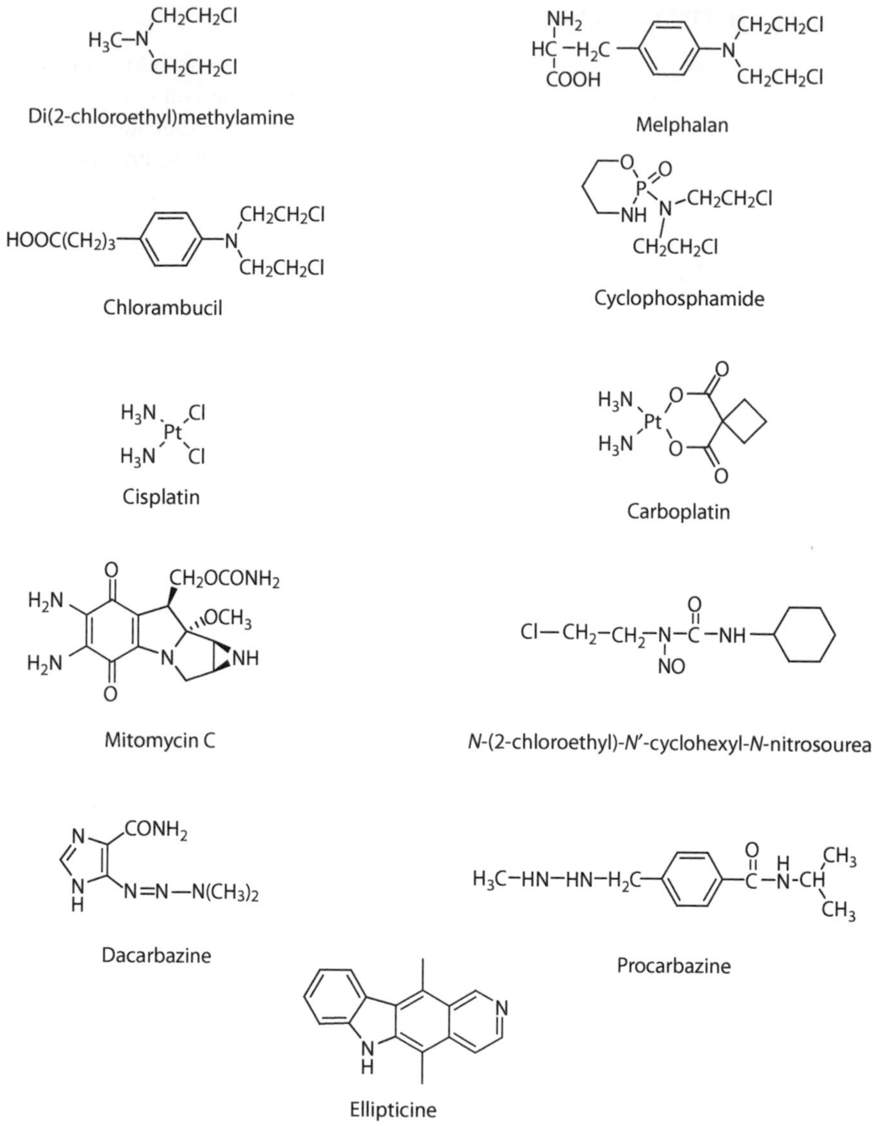

FIGURE 6.7   Selection of DNA alkylating chemotherapeutica.

6.4.5   ENDOGENOUS COMPOUNDS

Exposure to the DNA-alkylating compounds described above can be controlled by avoiding the source (smoking, food contamination, by-products of the cooking or preserving methods of food, drugs).[87 and references therein] However, one cannot limit the exposure to endogenous compounds, such as the hormone estradiol, and normal food products that produce DNA-alkylating compounds during metabolism, such as the etheno adducts from polyunsaturated fatty acids or the metabolites of the lipid peroxidation of polyunsaturated fatty acids of the cell membrane.[80–93] Some metabolic pathways produce ROS resulting in oxidative damage of DNA, and the metabolism of estradiols results in both estradiol adducts and oxidative damage to DNA (Figure 6.9).[94,95] Since cell metabolism is an important factor in the formation of "endogenous carcinogens," it can be expected that the pathological status (inflammation, infection) of cell metabolism can also alter the production of DNA

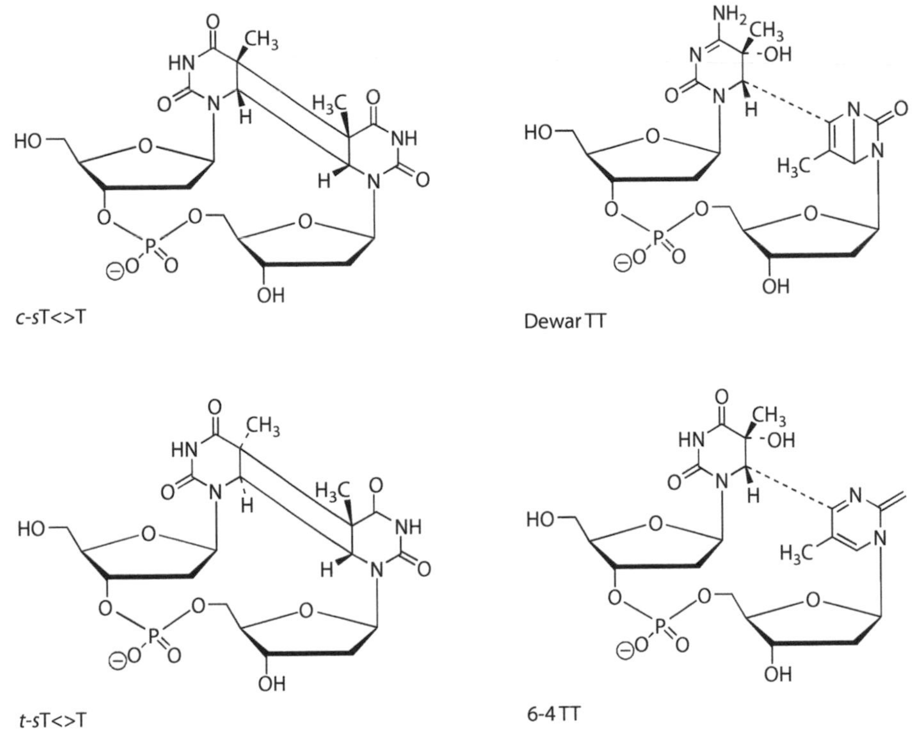

FIGURE 6.8 UV-induced thymine photoproducts. (From Douki, T. et al., *J. Biol. Chem.*, 275, 11678, 2000. With permission.)

adducts.[92] This is exemplified by the presence of 4-hydroxycstrone-1-N[3]adenine in the urine of men diagnosed with prostate cancer and the increased level of the same adduct in breast tissue of women with breast cancer.[88] Controls showed a significantly lower level of adduct in healthy breast tissue (of women), and the urine of healthy male volunteers was virtually blank.

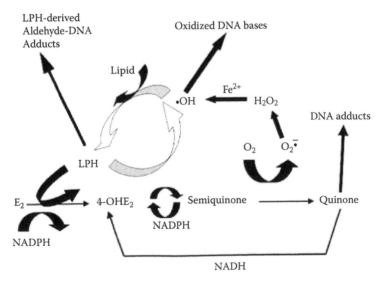

FIGURE 6.9 Estradiol-induced formation of lipid hydroperoxides (LPH) and DNA damage. (From Cavalieri, E. et al., *J. Natl. Cancer Inst. Monogr.*, 27, 75, 2000. With permission.)

6.4.6 Oxidative Damage

Oxidative damage to DNA is not the result of a specific chemical that modifies the DNA, but is the result of the interaction of an ROS or reactive nitrogen species (RNS) with the DNA. While the origin of the ROS and RNS is very diverse (see also the sections above), the reactive species is always the same, and therefore, modifications to the DNA are independent of the source of the ROS/RNS (Figure 6.10). ROS is a collective term, which includes oxygen radicals [superoxide $(O_2^{\bullet})^-$, hydroxyl (OH$^\bullet$), peroxyl (RO$_2^{\bullet}$), and alkoxyl (RO$^\bullet$)] and certain nonradicals that are either oxidizing agents or can be easily converted into radicals, such as HOCl, ozone (O_3), peroxynitrite (ONOO$^-$), singlet oxygen ($^1O_2$), and H_2O_2. Reactive is a term that is relative as H_2O_2 and $(O_2^{\bullet})^-$ react at diffusion-controlled rates with very few molecules, whereas (OH$^\bullet$) reacts at diffusion-controlled rates with almost anything, including RNA and DNA. (RO$_2^{\bullet}$), (RO$^\bullet$), (HOCl), (ONOO$^-$), and (O_3) have

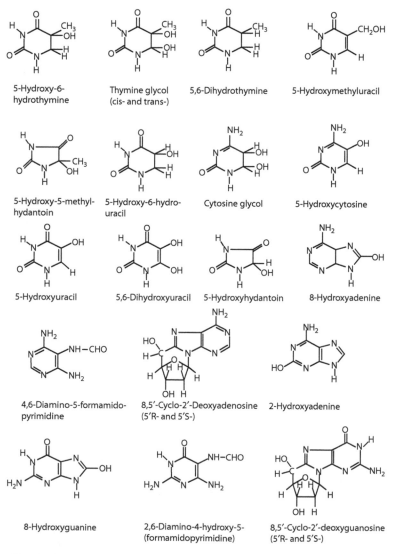

FIGURE 6.10 Chemical structures of some of the oxidation products of DNA. These modified bases (except 5,6-dihydrothymine, a product of attack on thymine by H? or hydrated electrons) can all be formed when hydroxyl radicals (OH$^\bullet$) attack DNA. (From Guetens, G. et al., *Crit. Rev. Clin. Lab. Sci.* 39, 331, 2002. With permission.)

intermediate reactivity. Hydroxyl radicals may generate further ROS and organic radicals by inter-action with biological macromolecules.[96,97]

6.5 DNA ADDUCTS IN DIFFERENT MATRICES

The detection, identification, or quantification of DNA adducts in the cell is the ultimate goal, but a multitude of other sources of DNA adducts have been analyzed. The source of DNA adducts can origi-nate in cells of a tissue sample, either human or animal, as well as from cell cultures. The extracted and hydrolyzed DNA can be treated under controlled conditions *in vitro*, with alkylating agents to produce DNA adducts in amounts higher than what would be available from an *in vivo* source. Similarly, nucleo-sides, nucleotides, or oligonucleotides can be reacted *in vitro*, also resulting in adducts without the need to hydrolyze the DNA. Naturally occurring adducts, already excised from the DNA, can be found in urine samples, with the detection of 7-alkylguanine adducts in urine being very common since *N*-7-alkylated guanines have a very strong tendency to depurinate, releasing the alkylated guanine from the DNA strand. All these different matrices have their place in DNA-adduct research protocols.

6.5.1 *IN VITRO*

The driving force behind the research and development of new methods for the structural and quan-titative analysis of DNA adducts is often biological and/or medical. The adducts themselves are often synthesized *in vitro*, since the *in vivo* levels of DNA-adduct formation are very low (1 adduct in 10^6–10^{10} nucleotides) and vary with the electrophile and the amount of exposure to it. In many studies, the *in vitro* preparation of the DNA adducts starts with DNA, (oligo)nucleotides, or nucleo-sides, and this permits the optimization of the analytical pathway from DNA hydrolysis to sample enrichment (using chromatography) to structural elucidation (using MS/MS and NMR).[60,98–105]

The MS fragmentation pattern is used to optimize the experimental conditions for the quantita-tive detection of DNA adducts from both *in vitro* and *in vivo* sources.[100,106,107] The DNA adducts produced *in vitro* permit detailed MS studies on the stereochemistry, the isomers, and the stability of the DNA adducts.[103,104,108,109]

The electrophile used determines the approach for the *in vitro* adduct formation. In the direct approach, the electrophile reacts immediately with the DNA, the nucleotide, or the nucleoside. However, the active electrophile is usually a carcinogen metabolite, for example, PAHs have to be oxidized before they can bind to DNA; NADH-cytochrome c reductase acts on mitomycin C to acti-vate it; cytochrome P450 activates cyclophosphamide; the use of living organisms or cells mimics the *in vivo* formation of DNA adducts; the use of organelles yields *in vitro* formation of the metabolites that react to form adducts.[10,16,98,110–112] Chemical activation of the carcinogen is also possible; examples of this are estrogen oxidation by MnO_2 or by Fremy's salt, poylaromatic compounds, and mitomycin C reduction by H_2/PtO_2 or $Na_2S_2O_4$.[101,102,104,110,111] Organic synthesis is another approach to the produc-tion of DNA adducts. The reactions and conditions used depend on the adduct that is required. The starting materials (e.g., 6-chloropurine) and the mechanism for the formation of the covalent bond can be different from the *in vivo* adduct formation pathway.[60,113,114] Although this approach sometimes requires multiple synthetic steps, the inclusion of protective groups and deprotection results in a spe-cific DNA adduct in a good yield. This approach also allows the production of isotope-labeled adducts that can be used in MS as an internal standard for quantitative analyses or in fragmentation mecha-nism studies that are often utilized in the identification of isomers.[104,108,115] Most *in vitro* studies start with the formation of adducts. These adducts may be formed with synthesized free nucleosides or nucleotides or with nucleosides or nucleotides that result from hydrolyzed DNA; in either case even-tually modified nucleobases are released. This approach has merits in terms of the detection and characterization of adducts, but in most cases has the disadvantage of the loss of all the DNA sequence information. There are also several MS studies published using the *in vitro* produced adducts of oligonucleotides (for some selected examples, see the section on *Adducts on oligonucleotides*).

6.5.2 *In Vivo*

The natural source for DNA is from living cells; however, the amount of extracted DNA available from a cell is very limited. The amount of cells or tissue available may also be limited. Therefore, the maximum amount of DNA available for analysis from such a source may be less than a few hundred micrograms. Adducts levels are often in the range of 1 adduct in 10^6–10^{10} nucleotides; this means that there are high requirements on the sensitivity of the analytic techniques employed for *in vivo* DNA adduct analysis. On the other hand, studying *in vitro* DNA adducts is meaningless without the possibility of comparing the (theoretical) *in vitro* results with the actual *in vivo* data. Numerous research groups have reported their *in vivo* DNA adducts results, and selected papers are reviewed here.

6.5.2.1 Cell Cultures

There is an indistinct line between the *in vitro* and *in vivo* study of DNA adducts. The *in vitro* culture of living cells is used as a model for both *in vitro* solutions of synthesized free DNA molecules and *in vivo* human cancer tissue samples. Tretyakova et al. used TK6 human lymphoblastoid cell cultures in between calf thymus DNA and DNA extracted from mice and rats to study the adduct formation of BD.[116] The use of cultured human cells replicates the human situation, as close as is possible, without actually using animals or patients. Cell lines are chosen for their specific properties. An example of human cells are the Jurkat cell lines as applied by Hoes and Van den Driessche in their study on the anticancer drug melpahalan.[99,117] Martins et al. used V79 Chinese hamster cells, a mammalian cell line devoid of cytochrome P450 (CYP) activity, to study the adduct formation of acrylamide that is epoxydized to glycidamide by cytochrome P450.[118]

6.5.2.2 Animal Models

The use of whole animals allows the study of DNA adduct formation in a situation that even more closely resembles the human situation. Metabolic pathways may differ from species to species; therefore the choice of the animal model limits the conclusions that can be extrapolated to the human studies. While the use of primates is not very convenient, using them, as Gangl et al. did with cynomolgus monkeys (*Macaca fuscicularis*), did permit to test the feasibility of the capillary LC-ESI-MSMS and ^{32}P postlabeling techniques they developed for the detection of *in vivo* formed 2-amino-3-methylimidazo[4,5-f]quinoline adducts.[119] Their results established a good basis for future applications to human samples. Huang et al. used fish in their investigation of 1,2-dihaloethanes from pollution in aquatic environments, with the idea that this adduct could be used as a biological marker indicating fish exposure to environmental carcinogens.[120]

Although there are other species used for specific purposes, most of the studies are conducted on rodents. Apart from reasons mentioned already, animal studies also allow the study of adduct formation in different cell types and tissues within one specimen. Examples of this approach is presented by Vodicka et al. who studied the level of styrene-DNA adducts in lung and liver of mice as a model for human inhalation exposure at the workplace.[56] Similarly, Van den Driessche et al. analyzed melphalan adducts in liver, bone marrow, blood, and kidney of the same rat; Le Pla et al. compared the adduct levels of oxaliplatin-DNA intrastrand cross-links in mouse liver, kidney, and lung DNA.[75,115,121]

6.5.2.3 Human Samples

A better understanding of mutagenesis and carcinogenesis can be had with the *in vitro* study of DNA adducts, and can lead to improved therapies and monitoring of carcinogen exposure, but the ultimate goal is to have the ability to measure these adducts in humans. There are ethical considerations that prohibit the intentional exposure of human subjects to potentially fatal compounds for experimental purposes; likewise, the availability of samples that come from actual patients is also obviously limited. Therefore, researchers rely on *in vitro* samples, cell culture, and animal studies to first optimize instrumentation conditions and to standardize methodologies in order to maximize the probability of detecting any analytes from (the precious) human samples.

Based on their rat experiments Chaudhary et al. successfully detected malondialdehyde (MDA) adducts from human liver samples.[122a] Li reported the detection of N-(deoxyguanosin-8-yl)-4-aminobiphenyl in human pancreas.[122b]

Optimal choice of the analytical technique is important when comparing different analytical methods (GC-MS, LC-ESI-MS/MS, and LC fluorescence) as did Chen et al. on human placenta.[123]

Embrechts et al. demonstrated that the use of specific scanning methods allowed the detection of several different adducts from the same sample. They were able to detect with a series of single reaction monitoring (SRM) transitions the adducts of estrone, estradiol, equilenin, ethynylestradiol, and benzo[a]pyrene in human breast tissue samples (Table 6.2).

TABLE 6.2
DNA Adduct as Detected in Human Breast Tissue Samples

Patient and Kind of Tissue[a]	Injection Volume (μL)[b]	Estrone–dCyd–adduct	Estradiol–dCyd–adduct	Equilenin–dCyd–adduct	Estrone–Thy–adduct	Estradiol–Thy–adduct	Estrone–dAdo–adduct	Estradiol–dAdo–adduct	Ethynyl-estradiol–dCyd–adduct	Equilenin–dAdo–adduct	Estrone–dGuo–adduct	Ethynyl-estradiol–Thy–adduct	Estradiol–dGuo–adduct	Ethynyl-estradiol–dAdo–adduct	Equilenin–dGuo–adduct	Benzo-[a]-pyrene–dGuo–adduct	Ethynyl-estradiol–dGuo–adduct	Known Background of Patient
1. Tumor	100	.	.	.	.	+	.	.	.	+	.	+	.	.	+	.	.	Unknown
1. Normal	100	.	.	.	.	.	.	.	+	+	.	+	.	.	+	.	.	Unknown
2. Tumor	100	.	.	.	.	.	.	.	.	.	.	+	+	.	+	.	.	Estradiol derivatives
2. Normal	100	.	.	.	.	+	.	.	.	+	.	+	.	.	.	.	.	Estradiol derivatives
3. Tumor	100	.	+	+	.	.	.	.	.	.	.	+	+	.	+	+	.	Unknown
3. Normal	100	.	.	.	.	.	.	.	.	.	.	+	.	.	.	.	.	Unknown
4. Tumor	20	.	.	.	.	.	.	.	.	+	+	+	.	.	O[c]	.	.	Premarin
4. Normal	20	.	.	.	.	.	.	.	.	.	.	.	.	.	O	.	.	Premarin
5. Tumor	20	.	.	.	.	.	.	.	.	.	.	.	.	.	O	.	.	Premarin
5. Normal	20	.	.	.	.	.	.	.	.	.	.	.	.	.	O	.	.	Premarin
6. Tumor	100	.	.	+	.	.	.	.	.	.	.	+	+	.	.	.	.	No HST[d]
7. Tumor	100	.	.	.	.	+	.	.	.	+	+	+	+	.	.	.	.	Premarin
8. Tumor	100	.	.	.	.	+	.	.	.	+	+	+	+	.	+	.	.	Premarin
9. Tumor	100	.	.	.	.	+	.	.	.	+	.	+	+	.	.	.	.	Probably no HST
10. Tumor	100	.	.	.	.	+	.	.	.	+	.	+	+	.	.	.	.	No HST
11. Tumor	100	.	.	.	.	+	+	.	.	+	.	+	+	.	.	.	+	No HST
12. Tumor	100	.	.	.	.	.	.	.	.	+	+	.	.	.	.	.	+	No HST
13. Tumor	100	.	.	.	.	+	.	.	.	+	+	.	.	+	.	.	.	Premarin

Source: From Embrechts, J. et al. *J. Am. Soc. Mass Spectrom.* 14, 482, 2003. With permission.

[a] Numbering refers to the origin of the sample, that is, patient.

[b] 100 μL injection volume it 38 μg DNA, 20 μL is 7.6 μg.

[c] O = Not tested on the compound.

[d] HST = Hormone substitute therapy.

6.5.2.4   Urine

Some adducts that are formed *in vivo* and are excreted in the urine can be used as an indicator of adduct formation in the body even though they are no longer present in the DNA. Human tissue samples from, target organs can be very difficult or impossible to obtain, urine, however, is easily accessible, but it is a complex matrix requiring extensive purification prior to the analysis of adducts.

Most of the DNA adducts in urine are modified nucleobases; for example, 4-hydroxyestrone-1-*N*-3adenine in the urine of men diagnosed with prostate cancer, and 4-hydroxyestrone-1-*N*-3adenine, 7-alkylguanine in the urine of mice exposed to styrene.[56,88] Modified nucleosides can also be detected in urine. Ribonucleosides originating from RNA turnover can be detected in urine as can DNA-derived 2′-deoxynucleoside adducts.[124–128] Hillestrom et al. developed a method (Table 6.3) for the analysis of 1,*N*-6-etheno-2′-deoxyadenosine in urine, an adduct that may be an indicator of human colon cancer.[129] Cigarette smoke is a known cause of lung cancer in humans and contains high levels of oxidants and other carcinogenic compounds. Adducts resulting from cigarette smoke, in combination with gender effect, are studied by analyzing the nucleobase adducts found in urine.[55,130]

6.5.3   DNA Extraction and Hydrolysis

When DNA adducts are formed *in vivo*, the DNA needs to be extracted from the tissue or cells prior to hydrolysis and analysis. There are several known protocols for the extraction of DNA, with specific variations in procedure depending on the starting tissue and the quality that is required for

TABLE 6.3
Gradient Elution Program and Schedule of Timed Events as Used for the Analysis of 1,N-6-Etheno-2′-Deoxyadenosine in Urine

Step	Time (min)	Flow (μL/min)	Mobil Phase[a] (% A)	(% B)	Valve	Comments
0	0	800	98	2	Switch (position 1)	Sample load on column 1 and column 2 re-equilibration
1	10	800	90	10		
2	13	600	89	11	Switch (position 2)	Heart-cut transfer at time 16.1–17.8 min
3	18	600	87.5	12.5		
4	20	1250	0	100	Switch (position 3)	Column 1 cleanup
5	21	700	92	8	Switch (position 4)	Separate, detect analyte of interest on column 2 and column 1 re-equilibration
6	43	700	31	69		
7	45	700	95	5		
8	49	700	95	5	Switch (position 1)	System preparing for next injection
9	49	800	98	2		
10	54	800	98	2		

Source: Hillestrom, P.R. et al., *Free Radical Biol. Med.* 36, 1383, 2004. With permission.

Note: The components for the two-dimensional separation consisted of column 1, a Luna HPLC column C18(2) (75 × 4.6 mm, 5 μm) protected with a C18 (ODS) guard column (4.0 × 3.0 mm), and column 2, a Synergi Polar-RP analytical column (150 × 4.6 mm, 4 μm).

[a] Mobile-phase content Solvent A: 5 mM ammonium acetate (pH 5), solvent B: 100% methanol.

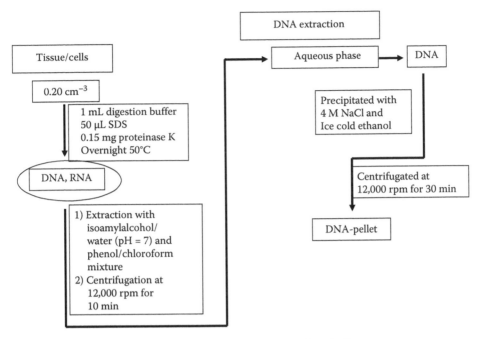

SCHEME 6.2 Typical procedure for the extraction of DNA from tissue/cells.

the extracted DNA. A typical procedure is given in Scheme 6.2 and there are standard commercial DNA extraction kits available.

In order to study the DNA adducts with MS and independent of the source the DNA, it needs to be hydrolyzed to either nucleotides or nucleosides, depending on the study. A typical DNA hydrolysis protocol is given in Scheme 6.3. Different procedures are available in the literature, but they should be chosen with care since the hydrolysis procedure can affect the presence and nature of the adducts. Van den Driessche compared three methods for the hydrolysis of DNA that was treated

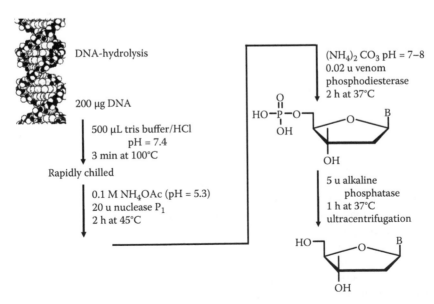

SCHEME 6.3 Typical procedure for the enzymatic hydrolysis of DNA to nucleosides.

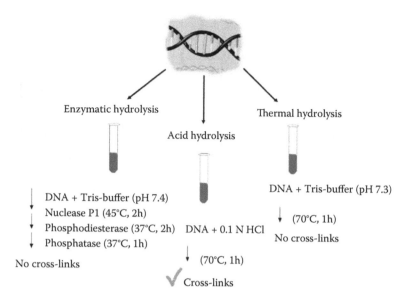

FIGURE 6.11   Overview of the hydrolytic procedures used in the sample preparation of melphalan-treated calf thymus DNA: Enzymatic versus acidic and thermal hydrolysis. (From Van den Driessche, B. et al., *Rapid Commun. Mass Spectrom.* 19, 449, 2005. With permission.)

with melphalan, a bifunctional nitrogen mustard.[131] The results of this study show that the way in which DNA samples are processed, prior to DNA adduct analysis, may have a dramatic impact on the analytical results. A comparison among enzymatic, acidic, and thermal hydrolysis of nitrogen mustard modified calf thymus DNA proved that acidic hydrolysis gave rise to cross-linked adducts, while the samples produced by thermal or enzymatic hydrolysis did not (Figure 6.11).

The enzymatic hydrolysis of DNA to nucleotides or nucleosides is in most cases intended to hydrolyze the DNA to the mononucleotide level; it is, however, known that the cleavage of the phosphodiester adjacent to an alkylated base moiety is often incomplete; as a consequence of this, dinucleotide adducts can be present in the hydrolyzate. We encountered examples of this in the hydrolyzates of DNA treated with phenylglycidyl ether (PGE) (Figure 6.12) and with adducts of melphalan.[132,133,99] The alkylation of phosphodiesters leads to PTEs that are resistant to enzymatic hydrolysis with the nucleases that are routinely used in DNA hydrolysis. These adducts can be analyzed as dinucleotides (see also subsection 6.6.2.2.1.13).[44]

6.6   DETECTION OF ADDUCTS

The study of the interaction of carcinogens or their electrophilic metabolites with DNA implies the characterization, detection, and quantitation of the corresponding DNA adducts and DNA constituents in general. Further, there is a need for an accurate means of assessing the risk of human exposure to environmental carcinogens. The presence of DNA adducts in tissues or cells from individuals suspected (or known) of being exposed to genotoxic agents is one of several ways by which the exposure can be monitored and the risk assessed.[134]

There are several analytical techniques and approaches being used for DNA adducts analysis, which are determined by the source of the DNA adduct and how complex the mixtures are.[48,135]

There is often a separation and detection step included, such as high-performance liquid chromatography (HPLC) in combination with fluorescence, UV, or UV-diode array detection.[136,137] Randerath et al. developed the ^{32}P postlabeling method that combines a planar separation technique with the high sensitivity of autoradiography.[138,139] This technique is used by many groups, often in combination with other techniques, when there is a lack of structural information available.[140,141]

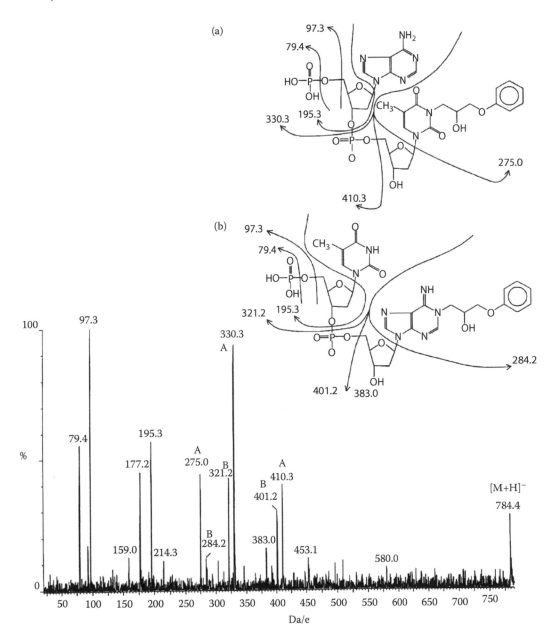

FIGURE 6.12 Low-energy CAD product ion spectrum of the [M−H]⁻ ion at *m/z* 784. Identified as a mixture of two coeluting adducts of pApT(PGE), structure (a) the specific fragments for molecule A are denoted in the spectrum with A and pTpA(PGE); structure (b), with specific fragments (collision energy was 33 eV) present in the DNA hydrolysate after sample clean-up. (From Deforce, D.D. et al., *Carcinogenesis* 19, 1077, 1998. With permission.)

The ^{32}P-postlabeling method is very sensitive, but it requires the use of relatively large amounts of radioactive material and it is time consuming. Several immunoassays and radioimmunoassays have also been developed for the selective isolation and detection of specific adducts.[142,143] Gas chromatography (GC) and LC, coupled to MS combine the possibilities of an efficient separation technology with the sensitivity and structural information of MS/MS rendering these techniques increasingly popular in DNA adduct analysis.[144–148 and references therein]

6.6.1 CHROMATOGRAPHY

As mentioned above, studies on DNA adducts involve samples that have varying degrees of complexity in their composition. Even in the simple case where only one nucleoside is reacted *in vitro* with one alkylating compound, you can have a mixture of unmodified nucleoside, one or more (isomeric) nucleoside adducts, depurination products, and degradation products of the reagents (Figure 6.13). Multiple chromatographic steps can be used, each with its specific goal: SPE to desalt the samples; a precolumn to concentrate the sample; and an analytical separation of the compounds of interest. Several modes of separation can be selected for the analysis of adducts, with selection depending on several parameters. The first parameter is the method of detection. The very sensitive but time demanding autoradiographic detection used in [23]P postlabeling requires (2D)-planar

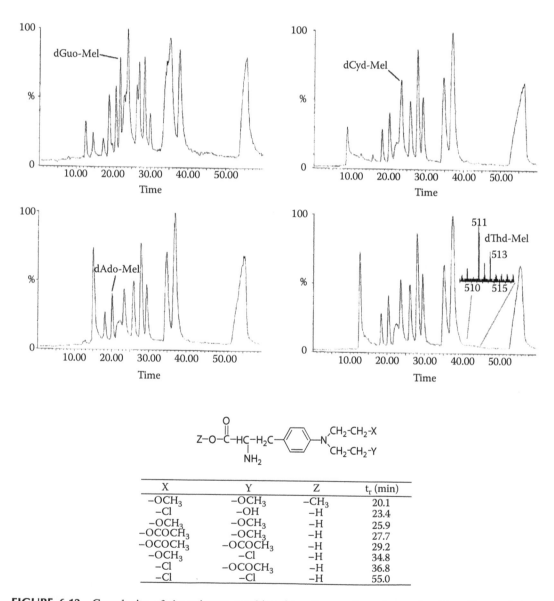

X	Y	Z	t_r (min)
–OCH$_3$	–OCH$_3$	–CH$_3$	20.1
–Cl	–OH	–H	23.4
–OCH$_3$	–OCH$_3$	–H	25.9
–OCOCH$_3$	–OCH$_3$	–H	27.7
–OCOCH$_3$	–OCOCH$_3$	–H	29.2
–OCH$_3$	–Cl	–H	34.8
–Cl	–OCOCH$_3$	–H	36.8
–Cl	–Cl	–H	55.0

FIGURE 6.13 Complexity of the mixture resulting from the reaction of a nucleoside with melphalan. (From Van den Driessche, B. et al., *J. Chromatogr. B Anal. Technol. Biomed. Life Sci.* 785, 21, 2003. With permission.)

chromatography.[147,148,52] Fluorescence detection requires a liquid-phase separation. MS is readily coupled with GC, but coupling with LC and capillary electrophoresis (CE) became really accessible only after the development of atmospheric pressure ionization methods such as electrospray ionization (ESI) and atmospheric pressure chemical ionization (APCI).

A second parameter determining the choice of the separation mode is the nature of the analytes. For example, intact DNA and oligonucleotides can be separated using gel electrophoresis. Capillary zone electrophoresis can be used to separate nucleotides; nucleosides are readily separated using RP-HPLC and nucleobases are separated by GC often after derivatization. Here we limit ourselves to those separation methods that are routinely coupled to MS.

6.6.2 MASS SPECTROMETRY

The use of MS in the study of DNA adducts should not be surprising as it has the potential of sensitive limits of detection providing and can also provide molecular weight and structural information.

This structural information, apart from being part of the identity of the DNA adducts, might provide insight into the metabolism of the carcinogen prior to adduct formation and the mechanisms behind the carcinogenic potential of specific adducts (e.g., N-7 versus O^6 alkylation of guanine). Techniques that provide direct structural information, such as x-ray and NMR, require more material than is available in most cases, whereas the more sensitive methods of ^{32}P postlabeling yield no structural information.

Prior to the 1980s, the ionization methods, such as EI and CI, that were available for MS studies required that the molecules be in the gas phase before they could be converted to ions, and therefore the analytes had to be either volatile or converted to a volatile derivative.

These conditions limited the application of MS to relatively apolar compounds and thermally stable molecules. The introduction of each new ionization method—fast atom bombardment (FAB), matrix-assisted laser desorption (MALDI) often in combination with coupling of the MS with LC; direct liquid introduction; dynamic FAB; thermospray; APCI; and ESI—allowed larger, more polar, and thermally unstable molecules such as nucleosides and nucleotides, and even aggregates of molecules to be analyzed.[48,73,91,102,149–156] This coupling of LC with MS is a rapidly evolving domain and DNA-adduct researchers follow the advances closely.[146,155,156] Currently, ESI is the dominant ionization technique used in MS of DNA adducts.

6.6.2.1 Gas Chromatography–Mass Spectrometry

The efficient separations that can be obtained with GC require that the analytes be volatilized. Often this requires derivatization of the polar DNA adducts, which introduces uncertainties about derivatization yield, completeness of the reaction, and possible artifact production. It is documented that the workup, hydrolysis, and final derivatization for GC-MS analysis of the DNA yields artifacts that distort the quantification especially of the oxidation products.[157–159] The combination of GC with either electron capture detection (GC-ECD) or negative ion chemical ionization mass spectrometry (GC-NCI-MS) offers the prospect of highly sensitive methods of adduct detection, even given that a derivatization step is required. Using these techniques with electrophoric labeling (e.g., introduction of pentafluorobenzoyl groups) of the nucleobases yields detection limits in the femtomole range, as little as 0.04 fmol O^4-ethylthymidine has been detected by GC-ECD.[160] Since the sample still has to be volatilized the less polar compounds are more amenable to GC analysis. Since larger molecules have generally a higher boiling point, it can be noticed that small and/or apolar modification to the bases are more easily analyzed using GC. Therefore, GC is mainly used for the analysis of DNA adducts, at the nucleobase level; either the modified nucleobases are available in the samples (urine samples) or the DNA is hydrolyzed to the nucleobases.[55,88,130,161,162] Most adducts analyzed using GC-MS are small groups such as etheno-adducts or oxidation products attached to the nucleobase.[55,91,96 and references therein,163] A selection of GC-MS applications is given in Table 6.4[54,96,122,123,157,158,164–176] and the analyzed adducts are shown in Figure 6.14.

TABLE 6.4
Selected Applications of GC-MS to DNA Adducts

Modified Base/Method	Results
Pyrimido[1,2-α]purine-10(3H)-one (M₁G), GC-EC-NICI MS	M₁G was reduced and derivatized. The resulting depurinated PFB-derivative was analyzed. Method developed for analysis of rat liver samples,[122] and applied to human samples (liver[164] and leukocytes[165]). Later M₁G was confirmed with LC-MS[166]
Cis-thymine glycol, GC-MS-SIM	Method for quantification of this species in human placental DNA has been developed[167]
5-(hydroxymethyl)uracil, GC-MS	Extant GC-MS methods for the analysis of 5-(hydroxymethyl)uracil in acid hydrolysates of DNA were found to be inadequate, and a modification to the standard protocol was proposed[168]
5-hydroxy-5,6-dihydrothymine, HPLC with GC-MS	Excision of this base from γ-irradiated DNA by endonuclease III has been monitored[169]
N^2,3-ethenoguanine, 7-(2-hydroxyethyl)guanine, GC-EC-NICI-HRMS	High-resolution MS allows the detection of N^2,3-ethenoguanine and 7-(2-hydroxyethyl)guanine below the femtomole level. This sensitivity allows analysis at endogenous concentrations in human tissue.[170,171] The GC-MS analysis of 7-(2-hydroxyethyl)guanine was compared with a ^{32}P-labeling protocol using rat DNA. Good agreement was found between the two methods[172]
8-hydroxyguanine and other oxidized bases, GC-MS	Detailed study of modified bases in calf thymus DNA addresses the extent of artifactual formation during derivatization for GC-MS. This study reveals that, contrary to previous claims, only 8-hydroxyguanine presents a problem that is circumvented by derivatization at room temperature without the use of TFA[157]
1,N-6-ethenoadenine, GC-EC-NICI MS	1,N-6-ethenoadenine was detected in human placenta DNA by three independent assays (GC/MS, LC/MS, and HPLC/ fluorescence) and a level of 2.3 adducts per 106 Ade bases was found in commercial and in fresh placental DNA. These levels were confirmed using LC-MS[123]
1,N^2-ethenoguanine, GC-EC-NICI-HRMS	Use of an immunoaffinity column to trap 7-(2-hydroxyethyl) guanine improved the selectivity of the analysis. Different hydrolysis procedures were compared.[173] Later the assay was improved to permit simultaneous analysis of 7-(2-hydroxyethyl)guanine and its isomer 1,N^2-ethenoguanine[174]
α-nitrosamino-aldehydes, GC-MS-SIM	Quantitative analysis of *in vitro* deamination of bases in calf thymus DNA by nitrosation shows that guanine is more reactive than adenine. Reaction rates were measured[175]
3,N^4-ethenocytosine, GC-NICI-MS	An ultrasensitive assay was developed, which detects this etheno adduct at the femtomole level[176]. Compound 3,N^4-ethenocytosine was detected in the urine of smokers
DNA-adducted 4-amino-biphenyl, GC-NICI-MS	Using tissue from bladder tumors, a significant association was found, in high-grade tumors, between adducted 4-aminobiphenyl and a smoking habit in the patient[54]
Products of oxidative damage to DNA, GC-MS (structures see Figure 6.11)	Elaborated reviews on oxidative damage to DNA. mechanisms, biological significance, and analytical methods of oxidative damage are discussed. Merits and problems of the GC-MS technique are presented[158,96]

Source: Adapted from Banoub, J.H. et al., *Chem. Rev.* 105, 1869, 2005. With permission.

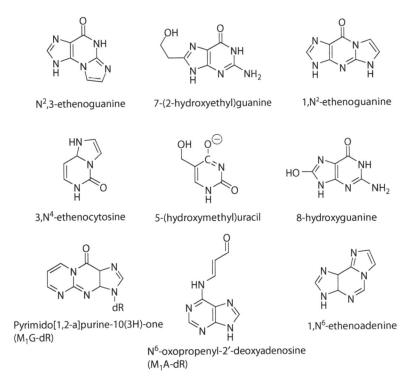

N^2,3-ethenoguanine   7-(2-hydroxyethyl)guanine   1,N^2-ethenoguanine

3,N^4-ethenocytosine   5-(hydroxymethyl)uracil   8-hydroxyguanine

Pyrimido[1,2-a]purine-10(3H)-one
(M$_1$G-dR)

N^6-oxopropenyl-2'-deoxyadenosine
(M$_1$A-dR)

1,N^6-ethenoadenine

FIGURE 6.14   Selected structures of the DNA adducts analyses by GC-MS (cfr. Table 6.2) and LC-MS.

6.6.2.2   Liquid Chromatography–Mass Spectrometry

LC is a separation method that is good for polar compounds such as nucleosides and nucleotides.[155] This is a technique that introduces less energy into the analytes, yielding more molecular weight data but less structural information data due to the reduced fragmentation profile in the MS scan. This problem can be addressed by the input of extra energy to the ions in the MS source; for instance, in-source low-energy collision-activated dissociation (CAD)-ESI of oligonucleotides, or the use of tandem mass spectrometers (MS/MS) where ion(s) selected by the first analyzer are fragmented and monitored by the second analyzer.[177] CAD-ESI-MS/MS provides the needed structural information about the exact position of the alkyl group on the pyrimidine or purine moiety or phosphate.[103,104,178–180] Once tandem mass spectrometers were introduced, several types of LC-MS/MS experiments could be done and applied to the detection, characterization and quantification of DNA adducts (see subsection below).

The ESI also allows miniaturizing the LC methods without the loss of the mass spectrometric sensitivity. As demonstrated by several research groups, ESI-MS and ESI-MS/MS hyphenated to capillary-, nano-HPLC, or capillary zone electrophoresis (CZE) offers a unique analytical tool capable of analyzing DNA adducts at the nucleoside or nucleotide level.[178–181]

6.6.2.2.1   Selected Applications of LC-MS to Analysis of DNA Adducts

There are numerous aspects to consider when using LC-MS to investigate DNA adducts. These include qualitative (adduct of what to which nucleobase, distinguish isomers, etc.) and quantitative questions. What techniques to be used? Where do the samples come from (*in vitro*, animal tissue, human)? In this section, a selection of applications and examples are given, showing how different research groups have used LC-MS to study DNA adducts, highlighting the differences in the studies.

The ESI interface, while common to both LC-MS and CE-MS methods, has adaptations that are specific for the LC or the CE. Since CE is technically different from LC separations specific problems (e.g., limited loadability) and specific solutions (e.g., sample stacking) may be encountered. The use of CE is of particular interest when the compounds being studied are ions in solution, or may be ionized by alteration of the pH. This is obviously an attractive method for the study of nucleotides and DNA fragments as can be seen in the examples in this section.

6.6.2.2.1.1 Oxidative Damage   Although originally investigated by GC-MS analysis, the products of oxidative damage to DNA are increasingly being analyzed by LC-MS methods. The group of Cadet developed LC-MS methods using stable isotope dilution MS for the analysis of DNA-(photo)oxidation products.[182–184] Characteristic for the LC approach is that the adducts needed no derivatization and that the products could be analyzed as nucleosides, whereas the GC-MS method analyzes *N,O*-bis(trimethylsilyl)trifluoroacetamide derivatives of the nucleobases. This includes the possibility in LC-MS/MS to detect the loss of the 2′-deoxyribose moiety and use the characteristic transition of the protonated molecule to the protonated base in an multiple reaction monitoring (MRM) method enhancing the sensitivity and selectivity of the method. This typical neutral loss of 116 Da of the 2′-deoxyribose moiety is used in most 2′-deoxynucleoside adduct studies (see also the Section A selection of MS/MS techniques applied to DNA adducts). The method used isotope labeled standards to maximize the accuracy of the technique. However, it was noted that using the HPLC-MS/MS method, a good correlation coefficient was also obtained without the use of the labeled internal standard (external calibration). This is not the case when the calibration curve was established by GC-MS.[183]

6.6.2.2.1.2 Pyrimido[1,2-α]purine-10(3H)-one (M₁G-dR) and N⁶-oxopropenyl-2′-deoxy-adenosine (M₁A-dR)   Malondialdehyde (MDA) is a product of lipid peroxidation that causes mutations in bacterial and mammalian cells and cancer in rats. In the earliest studies, M_1G, from both rat and human liver, was analyzed by GC-MS and confirmed by LC-MS.[122,164,185] The M_1G-dR adduct that was detected was also shown to be present in MDA modified calf thymus DNA using an LC-MS approach. The same LC-MS method allowed the detection of M_1A-dR. This study exemplifies the use of different MS/MS techniques for different purposes. The constant neutral loss (CNL) analysis shows the presence of several (modified) 2′-deoxynucleosides and the product ion spectra give structural information on the adducts, whereas the SRM analysis enhances the signal-to-noise (S/N) ratio.

6.6.2.2.1.3 7-(2-Hydroxyethyl)guanine   Ethylene oxide (EO), a strong electrophile that does not need prior metabolic activation to form covalent bonds to DNA, is a widely used sterilant and a chemical intermediate in the polymer industry. It is known to be genotoxic and carcinogenic in rodents and probably in humans. Reaction at the most nucleophilic N-7-position guanine yields the depurination product 7-(2-hydroxyethyl)guanine (7HEG). Using a single quadrupole LC-ESI-MS method, Leclercq et al. developed a quantitative method to detect 7HEG after calf thymus DNA and human blood (10 ml blood uptakes) exposure to various doses of EO.[186] For both sample types a clear dose–response relationship was obtained. The detection limit is *approximately* three modified bases on 10^8 intact nucleotides in blood samples using an single ion recording (SIR) method for monitoring the presence of the protonated molecule.

6.6.2.2.1.4 S-[2-(N-7-guanyl)ethyl]-glutathione   1,2-Dichloroethane (EDC) and 1,2-dibromoethane (EDB) are produced and used in multiple industrial processes, such as additives in gasoline, intermediates of vinyl chloride, vinyl bromide production, and other halogenated organics. They are also found as components of grain or soil fumigants, and as solvents for cleaners and other industrial products. These compounds have been found to be mutagenic and/or carcinogenic in experimental animals. After bioactivation with glutathione, EDC and EDB react with guanine residues in DNA resulting in the formation of *S-[2-(N-7-guanyl)ethyl]-glutathione*. Sensitive and specific methods were developed to quantify the level of this guanine adduct in the DNA of animals exposed to

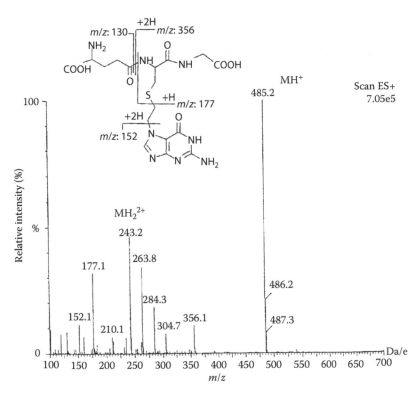

FIGURE 6.15   Full-scan spectrum of S[2-(N7-guanyl)ethyl]glutathione and proposed structures of fragment ions. Note the presence of the doubly charged ion at m/z 243.2, typical for peptides. (From Huang, H. et al., *Anal. Biochem.* 265, 139, 1998. With permission.)

1,2-dihaloethanes.[120] The product ions of the doubly charged molecular ion were monitored using an isotope-labeled internal standard and the MRM mode. This methodology was applied to determine the adduct levels in rat and fish samples. In the full scan MS spectrum, a protonated molecule [M+H]+ is observed at m/z 485.2, as well as the doubly charged molecule [M+2H]2+ at m/z 243.2 (Figure 6.15). The presence of a doubly or multiply protonated molecule, typical for peptides, was the most sensitive way to detect this adduct at lower sample concentrations. It is important to understand the chemical and mass spectrometric behavior of the DNA adducts. For example, the study of nucleoside adducts would be uninformative since N-7-alkyated guanosines are prone to internal depurination. Therefore, chemical release of the adducts is affected by thermal hydrolysis of the DNA followed by removal of the partial apurinic nucleic acid by precipitation and centrifugation. The mass spectrometric behavior of this specific adduct is dominated by ions formed from the ethylglutathion part, and doubly charged molecule [M+2H]2+ results from the presence of the peptide portion. Unmodified nucleobases under similar conditions are detected as [M + H]+ singly protonated molecules and the product ions are related to fragmentation of the ethyl-glutathion part.

6.6.2.2.1.5   Adducts of 1,3-Butadiene   Butadiene (BD) is a high-volume industrial chemical and probable human carcinogen. There are several structurally diverse adducts that have been detected, both *in vitro* (calf thymus DNA and TK6 cell structures) and *in vivo*, after animals were exposed to BD or to its epoxy derivatives.[116]

Based on earlier studies of N-7-alkylguanines (e.g., N-7-(2-hydroxyethyl)guanine and N-7-(2-hydroxypropyl)guanine), the initial approach to the analysis of butadiene adducts was using GC-ECNI-HRMS.[64,187] It was found that derivatization obtained with pentafluorobenzyl bromide (PFBBr) was inefficient, probably due to steric hindrance created by the first PFB group that entered the molecule.

It was also noted that the GC-MS approach was difficult because of the total mass of the obtained PFB-derivatized adduct. The direct analysis (no derivatization) of some of the BD adducts by LC-ESI-MS, was possible; however, sensitivity of this technique was lower than the GC-ECNI-HRMS method.

6.6.2.2.1.6 1,N-6-ethenoadenine   There have been reports of exocyclic DNA adducts being derived from various exogenous and endogenous sources, and therefore, these adducts can be detected even in untreated DNA. 1,*N*-6-ethenoadenine (εAde) was detected in human placenta DNA by three independent assays (GC-MS, LC-MS, and HPLC/fluorescence), and a level of 2.3 adducts per 10^6 Ade bases was found in both commercial and fresh placental DNA. The more sensitive detection of these adducts was achieved by using GC-MS, but the presence and identity of εAde were further confirmed by using LC-ESI-MS/MS.[123]

6.6.2.2.1.7 Adducts of Styrene Oxide   Styrene is a very important chemical with widespread applications. Polystyrene, or styrene in combination with other monomers, is used in the production of plastics, foams, polymeric resins, in both industrial and domestic applications. Since the styrene monomer can be liberated during the processing of the polymers, the occupational exposure of workers to styrene can occur. After inhalation, the most common route of exposure, styrene is metabolized to styrene oxide. The research group of Linscheid studied the adduct formation with styrene oxide *in vitro* using different MS approaches. Both LC and CZE approaches were used for the detection of oligonucleotide styrene adducts.[106,107,188,189] Using 2D plots with the *m/z* versus scan number, the presence of several oligonucleotides and oligonucleotide adducts in a partially hydrolyzed styrene-treated DNA sample is presented in a very clear way (Figure 6.16). An electrophoresis analysis of intact oligonucleotide adducts, as compared to nucleoside adducts, is facilitated by the presence of the negatively charged phosphate group on the analyte. CE allows high-resolution separations, although loadability (absolute amount on column) of the capillary is a general limitation

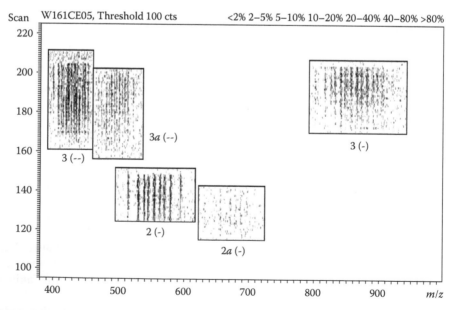

FIGURE 6.16   Two-dimensional separation of alkylated and digested DNA using MS 1. Scan range, *m/z* 300–1000; sheath flow, 10 μL/min propanol-2; ESI voltage, −3 kV; separation voltage, 23 kV; and buffer, NH₄HCO₃, pH 6.7. The number indicates the chain length; the number of charges is given and adducts are indicated by *a*. (From Schrader, W. et al., *Arch. Toxicol.* 71, 588, 1997. With permission.)

of this technique and sample stacking might be necessary in order to increase the sample concentration. Although the application of sample stacking reduces electrophoretic resolution, it increases the detection power by about a factor of 10 and this gain allows the detection of di-adducts, which would otherwise be undetected due to their low concentration.[106] The presence of the phosphate group, common to all oligonucleotides, permits the detection of styrene adducts by ICP-MS. Since the phosphorus is detected as an elemental species (phosphorus at m/z 30.97), it was used for the development of a quantitative method that used one internal standard for all nucleotides. The use of one internal standard also eliminated the influence of changing solvent composition during the separation gradient after a correction function was constructed.

ICP-MS (phosphorus signal) was used for the quantitative determination of nucleotides, with a detection limit of three adducts in 10^7 native nucleotides, from DNA modified *in vitro* by styrene oxide.[189] Using LC-ESI-MS with single ion monitoring (SIM), the detection limit for styrene oxide adducts of nucleotides was determined to be 14 modified residues in 10^8 unmodified nucleotides, similar to the results obtained with ^{32}P postlabeling.[188] The same approach was applied to adducts of melphalan, a mustard drug used in cancer chemotherapy. The reported absolute detection limit of 45 fmol corresponds to the detection of three modified nucleotides among 10^7 native nucleotides from 50 μg DNA.[76]

Styrene adducts, at least in mice, are excreted via urine. Vodicka et al. used an LC-ESI-MS/MS approach (SRM m/z 272 → 152, loss of the styrene part from the guanine adduct) to quantify the urinary excretion of two isomeric forms of styrene-modified guanine (N-7αG and N-7βG).[56] Mice were exposed to styrene levels that were comparable with the levels of human workplace exposure, and using the ^{32}P approach they studied, in parallel, the level of DNA adducts in the lungs and livers of these mice,. The styrene level in the blood was also measured. The correlation between styrene dose and the measured values was used to probe rates of metabolism, adduct formation, and DNA repair.

6.6.2.2.1.8 Adducts of Heterocyclic Aromatic Amines (HAAs)   2-Amino-3-methyl-imidazo [4,5-f]quinoline (IQ) is a known hepatocarcinogen that is derived from food. IQ is a member of the wider class of HAAs that can be found in both cigarette smoke and cooked food. Several types of HAAs are commonly found in cooked meat, fish, poultry, and pan residues (Figure 6.17) and the

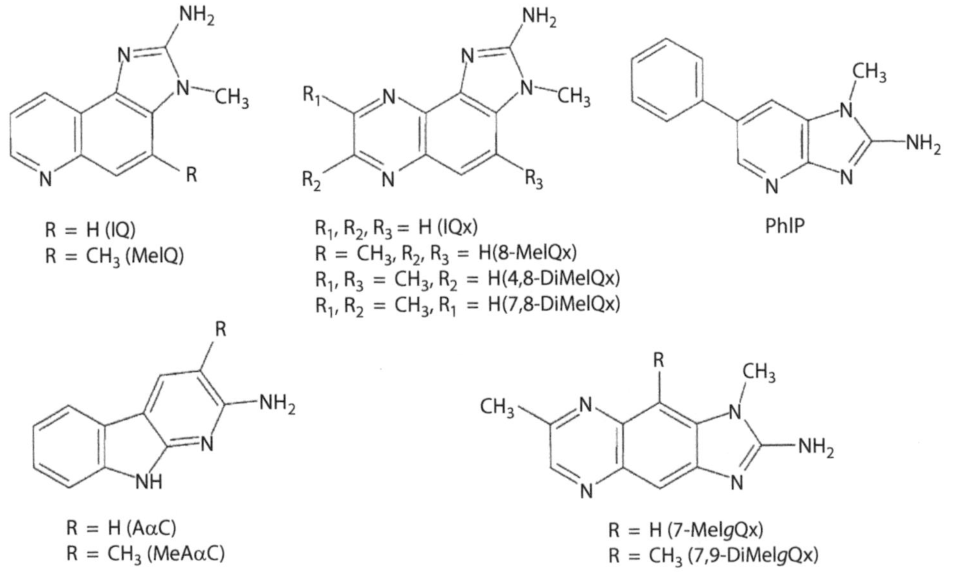

R = H (IQ)
R = CH$_3$ (MeIQ)

R$_1$, R$_2$, R$_3$ = H (IQx)
R = CH$_3$, R$_2$, R$_3$ = H(8-MeIQx)
R$_1$, R$_3$ = CH$_3$, R$_2$ = H(4,8-DiMeIQx)
R$_1$, R$_2$ = CH$_3$, R$_1$ = H(7,8-DiMeIQx)

PhIP

R = H (AαC)
R = CH$_3$ (MeAαC)

R = H (7-MeIgQx)
R = CH$_3$ (7,9-DiMeIgQx)

FIGURE 6.17 Chemical structures of representative HAAs formed in cooked meats. More than 20 HAAs have been identified in cooked meats. (From Turesky, R.J., *Toxicol. Lett.* 168, 219, 2007. With permission.)

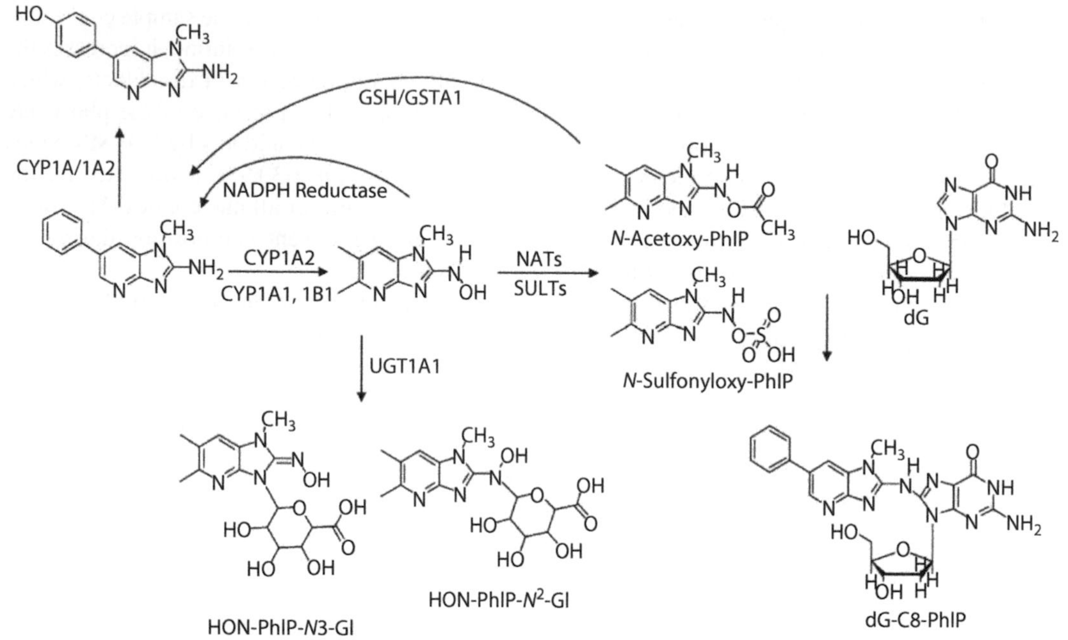

Major pathways of metabolism of PhIP in experimental animals and humans. Bioactivation occurs by cytochrome P450 CYP1A2 in liver or by CYP1A1 and CYP1B1 in extrahepatic tissues. The HONH-PhIP undergoes esterification reactions with acetyltransferase (NATs) or sulfotransferase (SULTs) to produce reactive intermediates that form adducts with dG in DNA to produce the dG-C8-PhIP. Detoxication occurs by 4_-hydroxylation of PhIP or by glucuronidation of HNOH-PhIP with UGT1A1, or by reduction of reactive N-acetoxy-PhIP to PhIP, presumably through a GSH-conjugate, which is catalyzed by Glutathione-S-transferase (GSTA1). Direct glucuronidation of PhIP is another prominent pathway of detoxication.

FIGURE 6.18 Example of the metabolism and DNA adduct formation of by a heterocyclic aromatic amine (2-amino-1-methyl-6-phenylimidazo[4,5-b]pyridine, PhIP). (From Turesky, R.J., *Toxicol. Lett.*, 168, 219, 2007. With permission.)

chronic exposure to these chemicals poses a serious health threat to the public. These aromatic amines are metabolized to *N*-hydroxides and then converted to esters (Figure 6.18) that react as electrophiles with DNA.[190] The group of Vouros and Turesky developed a highly optimized method for the analysis of HAA–DNA adducts with the resulting data being consistent with the ^{32}P post-labeling data.[119,191,192] All steps of their methodology were optimized: solid-phase extraction of the adducts maximized sample recovery and enrichment; in-house packed columns (10 cm × 75 μm, 5 μm nucleosil C18 100A) allowed large-volume injections; a custom-built electrospray interface and a gradient afforded efficient chromatography; optimizing the LC setup. The conditions for MS analysis were optimized by first determining which of the different scanning methods (product ion, CNL, and SRM) would provide the optimal conditions for the quantitative studies. The product ion spectra allowed the differentiation of isomeric adducts (Figure 6.19).[119,193] The final LC-MS/MS analysis used SRM as a sensitive scanning mode. Two adducts of 2-deoxyguanosine with this food-derived heterocyclic amine were detected in monkey pancreas tissue with detection limits of one adduct in 10^9 unmodified bases.[119,191] Using this system, a dose–response study was used to measure the adducts formed in rats upon treatment with 2-amino-3-methylimidazo[4,5-*f*]quinoline.[193] The results of this study clearly demonstrate the effectiveness of capillary LC-μESI-MS for DNA-adduct analysis in *in vivo* systems using microgram quantities of DNA and at levels approaching human exposure.

The release of the adduct from the DNA is another approach used to study adduct formation. This was applied to GC-MS analysis of adducts formed by the HAA 2-amino-1-methyl-6-phenylimidazo[4,5-*b*]

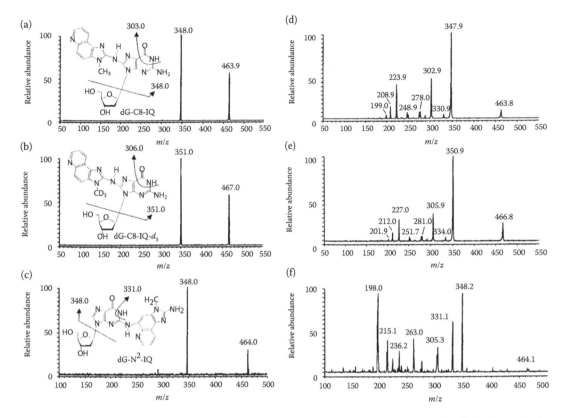

FIGURE 6.19 Flow injection analysis (FIA) of dG-C8-IQ (a), dG-C8-IQ-d3 (b), and dG-N^2-IQ (c) standards. (a–c) represent product ion spectra of the [M+H]$^+$ ion of each respective standard utilizing a collision offset voltage of −20 V. (d–f) represent product ion spectra of the [M+H]$^+$ion of each standard, only the collision offset voltage was increased to −50 V. Prior to FIA, the collision cell pressure was adjusted to 0.9 m Torr. (From Soglia, J.R. et al., *Anal. Chem.* 73, 2819, 2001. With permission.)

pyridine (PhIP). PhIP was released from adducted DNA by alkaline hydrolysis and analyzed as the di(3,5-bistrifluoromethylbenzyl) derivative; the nucleoside N-(29-deoxyguanosin-8-yl)PhIP was generated by enzymic digestion of DNA and analyzed with LC-MS (MRM, LOQ of 200 pg/500 mg DNA). The studied adducts were produced in cell lines exposed to PhIP. The LC-MS assay for N-(2′-deoxyguanosin-8-yl)PhIP in enzymic digests of DNA has a similar LOQ to the GC-MS assay, with the advantage that a specific adduct rather than a hydrolysis product is being measured.[194]

6.6.2.2.1.9 Adducts of Aflatoxin Aflatoxin (Figure 6.3) is a known hepatocarcinogen. DNA adducts can be formed upon metabolic activation by epoxidizing the A ring (Figure 6.4). Aflatoxin N-7-guanine and its imidazole ring-opened derivative were detected and quantified in the urine from rats that were dosed with aflatoxin allowing assessment of the efficacy of oltipraz, a chemopreventive agent. Other metabolites derived from the conjugation and/or oxidation of aflatoxin B$_1$ were also measured in the same study.[195] The imidazole ring-opened product is a typical by-product of alkylation at the N-7 position of guanine. When first interpreting the MS data obtained for any "unknown" DNA adducts, one should be aware of secondary reactions, such as the addition of H$_2$O to the purine base moiety, which could potentially complicate the mass spectrum interpretation.

6.6.2.2.1.10 Adducts of Estrogens Administration of estrogens to rodents leads to tumors in estrogen-responsive tissues. Epidemiological studies show that there is a strong association between estrogens and breast and prostate cancer risk.[80,88] The relationship between estrogens and breast

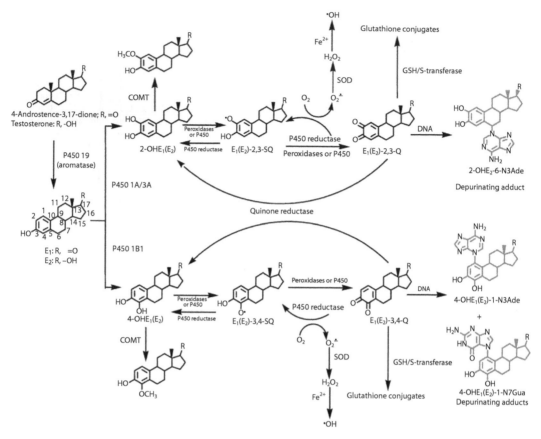

FIGURE 6.20   Formation, metabolism and DNA adducts of estrogens. (From Cavalieri, E. et al., *Biochim. Biophys. Acta Rev. Cancer* 1766, 63, 2006. With permission.)

cancer development appears to be complex (Figures 6.9 and 6.20), and includes the direct binding of estrogen–quinone metabolites to DNA as well as a variety of DNA modifications resulting from oxidative damage. Radicals responsible for this damage are produced in the metabolism of estrogens. Estrogen treatment also increases lipid peroxidation, and DNA exposure to the products of this process results in MDA adducts. Several groups are investigating estrogen–DNA adduct formation because estrogens are both endogenous and exogenous. Estrogens are in widespread use in contraceptives and hormone replacement therapies, and there is increasing evidence of their carcinogenic activity. The presence and types of adducts in different samples such as *in vitro* nucleoside/nucleotides/oligonucleotides and *in vivo* tissues/urine provide information on mechanisms of adduct formation. It also affords rationalization of the metabolism of protection from the adverse effects of estrogen exposure and statistical data on the correlation between estrogen and adduct levels and disease states.[82,104,108,196–199]

Characterization of the adduct by MS analysis is always the first step in the application of MS to the study of any DNA adduct. Van Aerden et al. produced *in vitro* adducts of estradiol with dG and dC and investigated their chromatographic separation and the MS fragmentation patterns.[196] They were able to identify dG isomeric adducts and depurination products. The isomeric forms could be separated and are clearly distinguished in the recorded product ions spectra (Figure 6.21); however, these spectra did not allow a precise structural identification for each adduct formed in the various reactions. Similar differences were found by Embrechts et al. for the adducts formed by equilenin with dG, dA, and dC.[104] The different diastereomers could be distinguished from their characteristic product ion spectra and different fragmentation pathways could be explained related to the

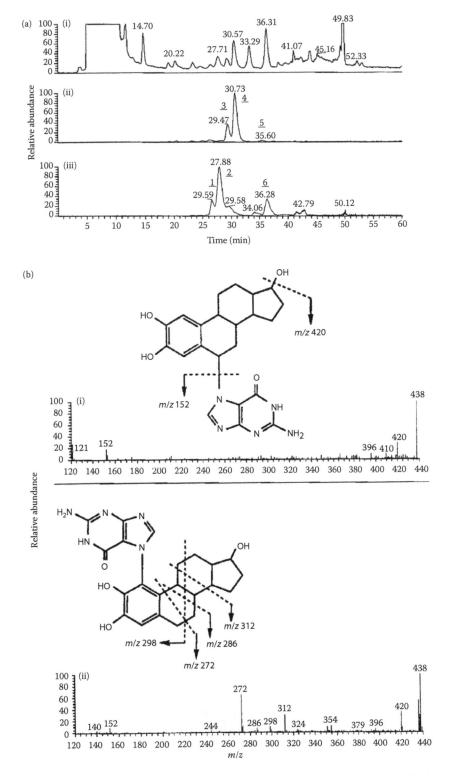

FIGURE 6.21   (a) LC-UV-ESI-MS analysis of an E-2,3-Q-dG crude reaction mixture. i) UV chromatogram ($\lambda = 280$ nm) and reconstructed selected ion chromatograms for ii) m/z 554 (E-2,3-Q-dG) and iii), m/z 438. (E-2,3-Q-Gua). (b) CID mass spectra obtained from MH$^+$ ions (m/z 438) of, i) peak 2 and ii) peak 6 from Figure 6.19a. (From Van Aerden, C. et al., *Analyst* 123, 2677, 1998. With permission.)

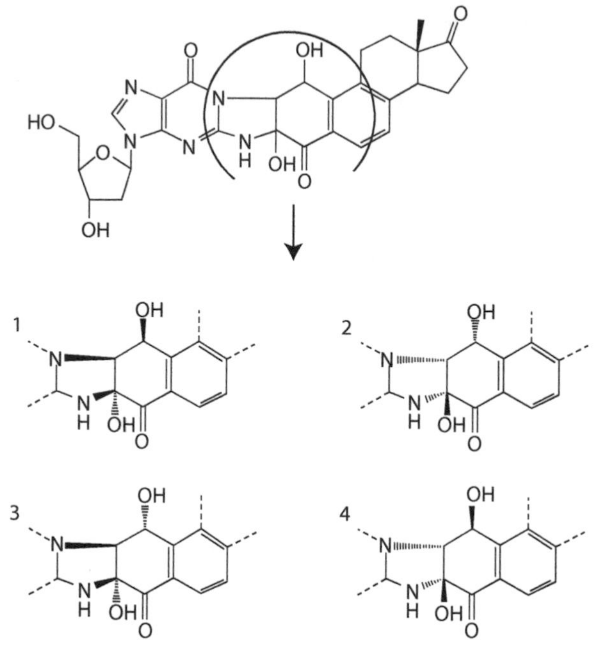

FIGURE 6.22   Stereochemistry of the four diastereomeric *cis*-annellated 4OHEN–dGuo adducts. (From Embrechts, J. et al., *J. Mass Spectrom.* 36, 317, 2001. With permission.)

stereochemistry of the DNA adducts (Figure 6.22). The mechanisms that resulted in product ion differences for the various isomers were supported by the use of deuterium labeled analogues. This showed that the deuterium transfer to the base moiety from the equilenin portion was dependent on the stereochemistry of the adduct (Figure 6.23).

Using a miniaturized column-switching setup, Embrechts performed an initial study on the presence of estrogen DNA adducts in breast tumor tissue.[81] Endogenous and exogenous estrogen adducts as well as benzo-[*a*]-pyrene–dGuo–adduct could be detected in both breast tumor tissue and adjacent normal breast tissue. No quantitative data were shown.

6.6.2.2.1.11   Adducts of Chemotherapeutica   As DNA adducts of carcinogens were being investigated using LC-MS techniques, the adducts formed by chemotherapeutic agents (Figure 6.7) were also being investigated. For example, the adducts formed by mitomycin-C, porfiromycin, and thiotepa were studied by Musser et al. using thermospray as the interface between the RP-LC separation and the MS.[73]

Cis-diaminedichloroplatinum(II) or cisplatin has antineoplastic activity due to its ability to coordinate with guanine in the DNA strand. Although this drug has been used for about 30 years, the details of its mode of action are still under investigation. The question of what the contribution to the antitumor activity is of each of the cisplatin adducts is unanswered. CZE-ESI-MS methods allow to separate and characterize the adducts formed by cisplatin and different nucleotides.[78] This method was applied to cisplatin adducts with dGMP and some related nucleosides and nucleotides, providing evidence for the possibility of O^6–N-7 chelation as a coordination mode of platinum with oxopurine nucleotides.[79] Studies like these provide information that increases our understanding of antitumor activity and can be used to improve existing drugs and to design new drugs. The fact that cisplatin adducts (and adducts with other related drugs) contain platinum permits the use of specific detection methods such as ICP-MS, which can focus the platinum signal. Examples of this can be found in the study of monoclonal antibody against cisplatin-modified DNA as presented by Meczes et al. and a study by Sar et al. on the speciation of cisplatin adducts.[200,201]

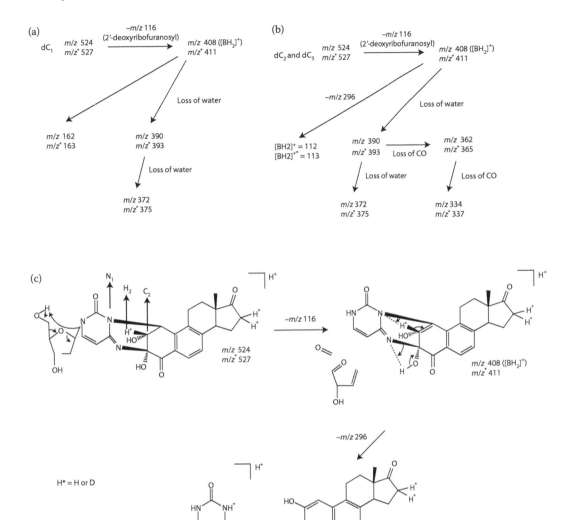

FIGURE 6.23   (a) Fragmentation pattern for dC_1 (the $m/z*$ values are the m/z values found for the adducts formed with the deuterated compound 4OHEN (2, 4, 16, 16-d4). (b) Fragmentation pattern for dC2 and dC3 (the $m/z*$ values are the m/z values found for the adducts formed with the deuterated compound 4OHEN (2, 4, 16, 16-d4). (c) Fragmentation pattern for dC2 and dC3 to m/z 112 (the $m/z*$ values are the m/z values when the H* were replaced by deuterium atoms; adducts made with 4OHEN (2, 16, 16-d3). (From Embrechts, J. et al., *J. Mass Spectrom.* 36, 317, 2001. With permission.)

Tamoxifen (Z)-1-[4-[2-(dimethylamino)ethoxy]phenyl]-1,2-diphenyl-1-butene is used in breast cancer therapy. Although it was approved by the FDA, an increased incidence of endometrial cancers was observed in both healthy women enrolled in a chemopreventive trial and in tamoxifen-treated patients. Umemoto et al. used LC-ESI-MS to detect nine adducts in liver samples of mice that had been exposed to tamoxifen.[202] Da Costa et al. developed a quantitative method for the detection of tamoxifen adducts.[203] The limit of detection for the HPLC-ES-MS/MS method was approximately five adducts in 10^9 nucleotides. The method was tested on rat liver samples where 496 ± 16 adducts/10^8 nucleotides for (E)-alpha-(deoxyguanosin-N-2-yl)tamoxifen and 626 ± 18 adducts/10^8 nucleotides for (E)-alpha-(deoxyguanosin-N-2-yl)-N-desmethyltamoxifen were detected. The data indicate that the HPLC-ES-MS/MS methodology has sufficient sensitivity and precision to be useful in the analysis of tamoxifen DNA adducts formed *in vivo* in experimental models and may be able to detect tamoxifen DNA adduct formation in human tissue samples. Endometrial samples from five women who had taken tamoxifen were analyzed using LC-MS/MS. The results indicated that there was no evidence of tamoxifen–DNA adduct formation in the endometrial samples from these women, although previous ^{32}P labeling studies claimed that tamoxifen adducts could be found in this kind of sample.[204] Since the sensitivity of the method applied was sufficient to detect the adduct at the earlier reported levels and the high chemical specificity of the MRM method using deuterium labeled internal standard, it was concluded that no tamoxifen adducts were present. Using the high sensitivity of accelerator MS, coupled with HPLC separation of enzymatically digested DNA, authentic dG-*N*-2-tamoxifen adduct was detected, at levels ranging from 1 to 7 adducts/10^9 nucleotides, in DNA samples extracted from the colon tissue of three patients who had standard therapeutic or chemopreventive dose of tamoxifen.[205] Melphalan or L-phenylalanine mustard, like other nitrogen mustards (chlorambucil, mechlorethamine, phosphormamide) (Figure 6.7), is a valuable chemotherapeutic agent for the treatment of several types of cancers. Although these mustards are among the oldest anticancer agents known, they are still extensively used in the chemotherapy of malignant diseases. Due to the bifunctional alkylating properties, it has been stated that inter and intrastrand DNA cross-links, and cross-links between two protein molecules or even between a protein and a DNA molecule, could be formed. These cross-linked adducts are believed to be a critical factor in the general antitumor activity of the alkylating agent. However, patients treated with this drug display bone marrow suppression, hypersensitivity reactions, gastrointestinal toxicity, and pulmonary toxicity. Long-term toxicities are related to infertility and secondary cancers. The latter might also originate from the formation of covalent bonds between the DNA and the chemotherapeutic agent.

Esmans' group developed LC-MS methods for the analysis of melphalan nucleotide adducts using column-switching capillary- and nano-LC-MS methods. The method they developed was for the analysis of melphalan adducts that were produced in Jurkat cell lines and this method can be used for the analysis of adducts in patients' blood samples.[117] The separation of the adducts proved to be better when they were analyzed as nucleosides.[99] In this study, the DNA obtained from calf thymus DNA and treated Jurkat cells was enzymatically hydrolyzed to the nucleoside level; in spite of this procedure, a dAMP adduct was found. Most likely the enzyme activity of the alkaline phosphatase is hampered by the presence of the melphalan on the phosphate ester. The fragmentation behavior of the obtained adducts was studied allowing characterization of adducts by their product ion spectra.

Based on this experiment, a qualitative and quantitative study of melphalan adducts in treated rats was carried out.[75,115] Liver, bone marrow, blood (only in the qualitative study), and kidney were analyzed, using an ^{15}N-labeled dGuo-melphalan standard, for the presence of melphalan adducts using SRM (Figure 6.24). The presence of this adduct could also be quantified (Table 6.5).

In the course of this study, the importance of the method used for the hydrolysis of DNA was highlighted. The presence of cross-linked adducts in the samples appears to depend on the hydrolytic method (see also *DNA extraction and hydrolysis*).[131]

The same ICP-MS approach used by Siethoff and Edler for the detection of styrene adducted to nucleotides was applied by Edler to the quantification of melphalan adducts.[76,188,189] This technique allows for the quantification of the nucleotide adduct regardless of the chemical nature of the modification. The absolute detection limit of 45 fmol corresponds to the detection of three modified nucleotides in 10^7 native nucleotides (based on 50 μg DNA). The melphalan adducts at the phosphate groups of the four nucleotides can be separated by reversed-phase chromatography using a three-step gradient elution with methanol as an organic modifier. A series of base-alkylated adducts is detected but only partially resolved chromatographically (Figure 6.25). The detailed structure of the detected adducts is provided by a parallel experiment using ESI-MS/MS.

6.6.2.2.1.12. Adducts on Oligonucleotides   Banoub et al. used an ESI-MS approach to identify and locate benzo[*a*]pyrene and acetylaminofluorene adducts in oligomers up to 18 nucleotides

(a)

Adduct	SRM Function	Cone Voltage (V)	Collision Energy (eV)	Dwell Time (s)
dGuo–mel	536 → 420	27	20	0.1
dAdo–mel	520 → 404	27	20	0.1
dCyd–mel	496 → 380	27	20	0.1
Gua–mel	420 → 269	27	20	0.1
Ade–mel	404 → 269	27	20	0.1
Cyt–mel	380 → 269	27	20	0.1

(b)

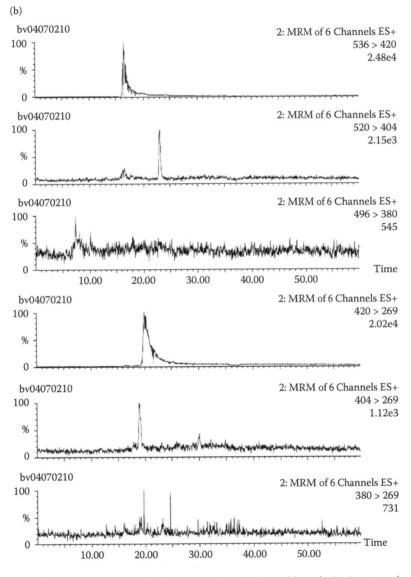

FIGURE 6.24 (a) SRM-functions used for the search of melphalan adducts in *in vivo* treated rats. (b) SRM signals of the nucleoside adducts, depurinated and depyrimidinated base adducts, measured in the hydrolyzed liver DNA of melphalan-treated rats. 20-Deoxyguanosine– and 20-deoxyadenosine–melphalan adducts can be detected for SRM functions. (From Van den Driessche, B. et al., *J. Mass Spectrom.* 39, 29, 2004. With permission.)

TABLE 6.5
Levels of dGuo-mel Adducts Relative to the Number of
Unmodified Nucleosides from DNA From Bone Marrow,
Liver and Kidney, after 24 h Melphalan Incubation in the Rat

	Amount DNA Isolated (μg)	1 Adduct per No. of Nucleosides
Bone marrow		
24 h	422	$4.7' \times 10^6$
Kidney		
24 h	489	10^7
Liver		
24 h	245	$2.7' \times 10^8$

Source: From Van den Driessche, B. et al., *Rapid Commun. Mass Spectrom.* 19, 1999, 2005. With permission.

long.[177] These oligonucleotide adducts, containing multiple phosphate groups, are detected as a series of multiply deprotonated molecules (charges in the range 5⁻ to 9⁻). Fragmentation of the molecular anions was achieved by in-source fragmentation by increasing the cone voltage. It was observed that this method of inducing fragments was more efficient than using the collision cell of the triple quadrupole instrument for CID product ion spectra. The obtained product ions confirmed the presence of the alkylated guanine and allowed for the sequencing of the oligonucleotides, including the positioning of the modified base.

Adduct formation of benzo[*a*]pyrene diol epoxide (Figure 6.4), the reactive metabolite of benzo[*a*]pyrene, appears to be dependent on the stereochemistry at positions 7, 8, 9, and 10 of the

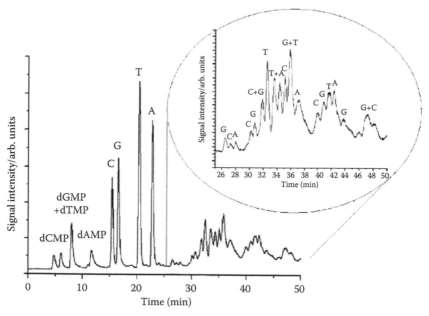

FIGURE 6.25 LC-ICP-MS (31P) chromatogram of melphalan adducts in the nucleotide mixture; the remaining nucleotides (first group), the phosphate adducts (second group, major signals) and the mixture of base adducts (blow up) are assigned. (From Edler, M. et al., *J. Mass Spectrom.* 41, 507, 2006. With permission.)

diol epoxide. Tretyakova et al. used synthetic DNA strands in which N-15-labeled dG was placed at defined positions within DNA duplexes containing 5-methylcytosine at all physiologically methylated sites. These oligonucleotides were reacted with (+/ −)-anti-BPDE. After enzymatic hydrolysis, capillary HPLC-ESI-MS/MS was used to establish the amounts of (–)-trans-N^2-BPDE-dG, (+)-cis-N^2-BPDE-dG, (–)-cis-N^2-BPDE-dG, and (+)-trans-N^2-BPDE-dG originating from the ^{15}N-labeled bases. These analyses allowed them to conclude that the (+)-trans-$N2$-BPDE-dG isomer contributed most (70.8% and 92.9%) to the adduct formation, and that alkylation preferably occurred at guanines with a methylcytosine adjacent to it, probably as a result of favorable hydrophobic interactions between BPDE and 5-methylcytosine.

Xiong et al. applied ion-pair reversed-phase nano-HPLC to the separation of oligonucleotides modified with benzo[*a*]pyrene diol epoxide.[206] Two oligonucleotide strands, 5′-PO4–ACCCGCGTCCGCGC-3′ (primary strand) and 5′-GCGCGGGCGCGGGT-3′ (complementary strand) were reacted with the diol epoxide. The 14-mer oligonucleotide selected for this study represents the major mutational hotspot codons 157 and 158 in human lung cancers and the benzo[*a*]pyrene preferential site in the *p53* tumor suppressor gene. This nano-LC method could separate and characterize nine isomeric covalent BPDE oligonucleotide adducts. Isomeric differentiation was achieved using MS/MS experiments.

Debrauwer et al. investigated the adducts formed by estradiol, extending their studies on the reactivity of estradiol-2,3-quinone to T-rich oligonucleotides containing one or two reactive bases (not T).[101] A reversed-phase LC method was developed using mixtures of ammonium triethylacetate (5 mM)–CH_3CN in the solvent gradient using an ion trap MS. To distinguish the isomeric forms of the G containing oligomers, energy-resolved product ion spectra were recorded. Where individual product ion spectra provided important information, the isomeric forms of the TTTTTG*TTTTTT estradiol adducts were readily distinguished from their breakdown curves (Figure 6.26).

6.6.2.2.1.13 Phosphate Alkylation and Nucleotide Adducts DNA–phosphate adducts are known to be formed by a variety of alkylating agents. Due to little or no repair of adducts, they may offer increased possibilities for both identification and quantification of DNA adducts. The formation of DNA–phosphate adducts leads to a complete esterification of the phosphate group giving rise to a phosphotriester (PTE) configuration.

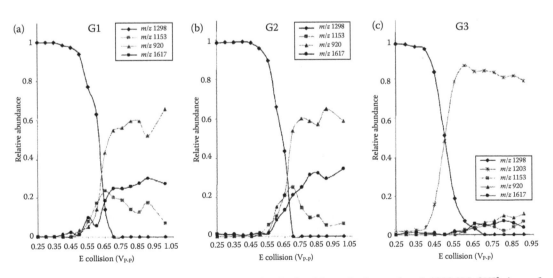

FIGURE 6.26 Energy-resolved breakdown graphs obtained from the isomeric *m/z* 1298 [M−3H]$^{3-}$ ions of (a) adduct G1, (b) adduct G2, and (c) adduct G3 from modified TTTTTGTTTTTT. (From Debrauwer, L. et al., *J. Chromatogr. A*. 976, 123, 2002. With permission.)

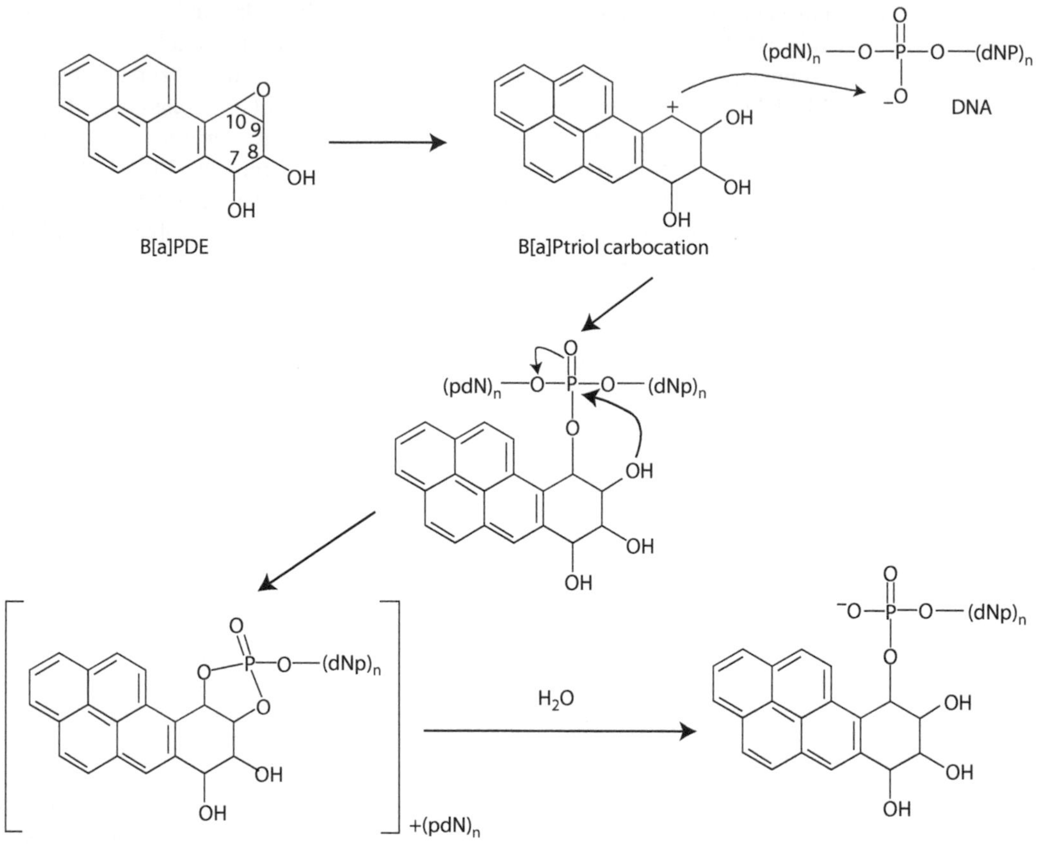

FIGURE 6.27 Reaction of B[*a*]PDE with the sugar phosphate backbone of DNA and postulated mechanism for strand scission. (From Gaskell, M. et al., *Nucleic Acids Res.* 35, 5014, 2007. With permission.)

PTEs are chemically stable under physiological conditions, unless the alkyl group contains an oxygen, a sulfur, or a nitrogen in the β-position with regard to the ester linkage.[41] The biological consequences of alkyl PTE adducts is not fully understood. Potentially, the altered properties of the phosphate group induces changes in the binding or function of proteins such as DNA repair or replication enzymes.[207] PTEs in DNA are also resistant to hydrolysis by certain nucleases; the enzymes that normally hydrolyze diester bonds are not capable of cleaving the internucleotide bonds adjacent to a completely esterified phosphate group. This leads to dinucleotide formations that are usually overlooked, as their characteristics are different from the known characteristics of the expected base-modified mononucleotides. To date, the compounds investigated for the formation of phosphodiester or PTE adducts include alkylating agents, such as dialkylsulfates, alkyl methanesulfonates and *N*-nitroso compounds, cyanoethylene oxide, cyclophosphamide, melphalan, phenyl glycidyl ether, aziridinylbenzoquinones, and benzo[*a*]pyrene-7,8-dihydrodiol-9,10-epoxide.[42,43,99,112,132,133,178,180,208–212]

Strand breaks in the adducted DNA is mentioned in several reports as an indication of phosphate alkylation.[132,212] The basis for these strand breaks is found in the presence of a group on the β position relative to the phosphate, catalyzing the hydrolysis of the PTE.[41] This configuration is also found when adducts are formed by epoxides, such as benzo[*a*]pyrene-7,8-dihydrodiol-9,10-epoxide and PGE. The mechanism is shown in Figure 6.27.

In contrast, PTEs of simple alkyl groups are stable and are subject to little or no repair, which results in their accumulation and thereby making them good candidates for exposure monitoring. Haglund et al. developed a method using an extremely strong nucleophile [cob(I)alamin] that transfers the electrophile from the phosphostriester to the cobalamine nucleophile.[209] The alkylcobalamine can be quantified using capillary column switching with an LC-ESI-MS/MS method. In the quest for increasing sensitivity, it would be advantageous, if all of the potential different PTE dinucle-otides (10 for every alkyl group that forms a PTE) formed by the transalkylation method are converted to alkyl-cobalamin (1 analyte). Furthermore, every new compound to be studied requires 10 different products to be characterized and quantified using isotope-substituted standards. However, analysis of the individual PTEs by LC-ESI-MS/MS yields qualitative information such as preferences for specific nucleotide combinations where in the transalkylation procedure, this structural information would be lost. Both procedures, direct analysis of the PTEs and indirect analysis of the cobalamin transalkylation product, are available. The approaches are complementary and useful for different purposes.[44]

6.6.3 CHROMATOGRAPHIC WAYS TO IMPROVE DNA-ADDUCT ANALYSIS

6.6.3.1 Miniaturization

The available samples for analysis of DNA adducts are often small or in low concentration, and the analytical methods have to be adapted to this. As can be seen in the different applications, the techniques adopted by different research groups often rely on miniaturized separation systems. The LC columns had a standard diameter of 4.6 mm for many years, but today this internal diameter is ever decreasing (1 mm ID micro columns, 300 μm capillary columns, and 75 μm nano-columns).

Basic chromatographic theory, as described by the Van Deemter equation, dictates that the loadability of a column is given by

$$M_s = \frac{C_m \pi d^2 L \varepsilon (1 + k')}{2\sqrt{N}}.$$

So, if all other parameters are kept equal (resolution, column length, separation), the amount of material that can be injected on a 75 μm ID column compared to a 4.6 mm ID column is given by

$$\frac{M_{s(75\mu m)}}{M_{s(4.6mm)}} = \frac{d^2_{75\mu m}}{d^2_{4.6mm}} = \frac{(75\mu m)^2}{(4.6mm)^2} \quad \text{or} \quad M_{s(75\mu m)} = \frac{1}{3761} M_{s(4.6mm)}.$$

This tells us that the same chromatographic result can be obtained on a 75 μm ID column as on a 4.6 mm ID column, but we need more than 3000 times less material.

The chromatographic peak volume is given by

$$V_p = \frac{\pi d^2 L (1 + k')}{\sqrt{N}}.$$

This gives the same downscaling factor for the chromatographic peak volume:

$$V_{p(75\mu m)} = \frac{1}{3761} V_{p(4.6mm)}.$$

It should be noted that the ratio of the *amount of analyte* over *peak volume* is equal for both the 4.6 mm and 75 μm columns. This means that the concentration of the analyte eluting from both columns is the same, although on the miniaturized column an amount 3761 times smaller is injected. Detection systems that depend on the concentration of the analyte in the detector will give, in both cases, the same readout. Within certain limits, ESI behaves like a concentration-dependent detector. As we have only minute amounts of DNA adduct material available, this is very promising.

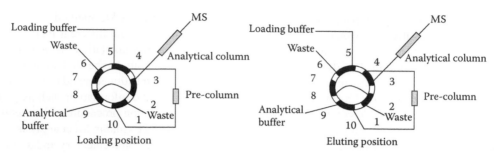

FIGURE 6.28 Schematic diagram of the column switch setup. (From Van den Driessche, B. et al., *J. Mass Spectrom*. 39, 29, 2004. With permission.)

Vanhoutte[179] demonstrated that this gain was achievable in practice as well. Using adenosine as a test compound, the detection limit of his SIR-LC-MS setup using a 4.6 mm column was 400 fmol. The same solution injected on a capillary 300 μm column resulted in a detection limit of 2.4 fmol, increased by a factor of 177 (theoretically $4600^2/300^2 = 235$).

In a second study by the same author, a nano-column of 75 μm ID was used.[181] Using this further miniaturized system, a detection limit of 30 fg (120 amol) for adenosine was obtained. This is an increase in mass sensitivity by a factor of 3300 compared to the 4.6 mm conventional LC-MS system.

6.6.3.2 Column Switching

There are some practical issues to deal with when a miniaturized LC is used; for example, the flow rate needs to be reduced by the same downscaling factor, posing serious demand on the LC pumps. We need flow rates of 200–300 nL min^{-1}. If we keep to the theoretical requirements, the injection volume also needs to be reduced resulting in injection volumes of a few nanoliters.

In order to handle the samples, we need a certain liquid volume which is a limiting factor on the concentration of the samples; although nanoliter injection volumes are feasible, they are not very practical. These problems can be overcome by using a column-switching setup (Figure 6.28).

Column switching adds two advantages to the analytical method.

1. *Sample fractionation*: Since it adds a second column to the setup, it also adds a second dimension of separation to the method. This allows for the fractionation of the sample prior to its analytical separation in the second column. Salt and unmodified nucleosides can be eliminated and only adducts need to be separated on the analytical column (Figure 6.29). This extra separation step is also advantageous in classic 4.6 mm ID methods. In order to handle the complexity of the urine samples and to preconcentrate the analytes, Hillestrom et al. developed a methodology consisting of a simple prepurification step with SPE followed by column-switching HPLC and quantification by isotope dilution MS/MS (Table 6.3).[129] They validated the method with respect to linearity, precision, and accuracy; so the analysis of real samples was demonstrated and this assay may be useful for large-scale studies. Sample volume was not an issue in this study with only 3 mL urine being used.

2. *Sample concentration*: The precolumn also allows concentration of the sample prior to the analytical separation. It was shown that by using a solvent with low elution strength the high polarity analytes and salts will elute from a reversed-phase precolumn (Figure 6.29), whereas compounds with low polarity will stick to the stationary phase. This effectively focuses these analytes on the top of this column in a narrow band. This band can be back-flushed to the analytical column using the analytical gradient, finally eluting the analytes in a chromatographic peak with a volume depending on the (miniaturized) columns dimensions, thereby effectively concentrating the sample.

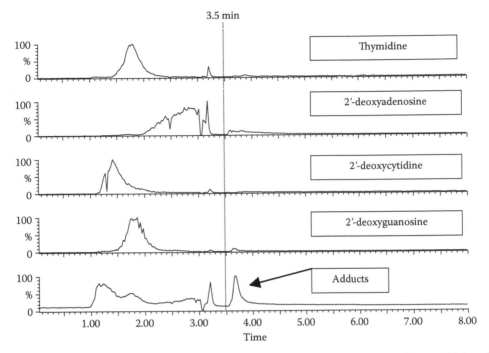

FIGURE 6.29 Column-switching after 3.5 min in order to improve the concentration sensitivity of the LC-MS system and to separate the unmodified nucleosides from the adducts prior to HPLC analysis. (From Van den Driessche, B. et al., *J. Chromatogr. B Anal. Technol. Biomed. Life Sci.* 785, 21, 2003. With permission.)

This is also demonstrated by Vanhoutte et al. using a column-switching setup with a 75 μm ID analytical column.[181] Increasing the injection volume from 3 nl to 1000 nl, using split injection and a column-switching method, should give an increase in signal in both UV- and ESI-MS by a factor of 333 (Table 6.6). The experimental value of 278 for UV detection and 211 for ESI-MS detection was somewhat lower than the theoretical value.

This column-switching technique in combination with miniaturized chromatography is successfully applied in our research for the analysis of DNA adducts and related subjects. [44,81,104,213]

TABLE 6.6
Comparison of the Three Injection Methods Used in Nanoflow ES LC/MS by Injection of the dGMP/BPADGE *in Vitro* Reaction[a]

Injection Type	Injected Volume (nL)	UV Response	Full Scan MS Response	Theoretical Gain
Split	3	1	1	1
Large volume	20	3	8	7
Large volume	100	32	64	33
Large volume	500	161	167	166
Column switching	1000	278	211	333

Source: From Vanhoutte, K. et al., *Anal. Chem.* 69, 3161, 1997. With permission.

[a] Adduct A is the phosphate-alkylated dGMP/BPADGE adduct. Comparison was made by normalizing the relative area of the peak of adduct A in UV and RIC ([M–H]$^-$ in the Nanoflow ES LC/MS system with split injection to 1.

6.6.4   (Tandem) Mass Spectrometric Techniques Applied to DNA Adducts

6.6.4.1   Product Ion Spectra

Once it is recognized that there is a DNA adduct present in a DNA sample, identification of its structure becomes the fundamental question. Although not always easy to interpret, the product ions formed from an initial (de)protonated nucleobase/nucleoside/nucleotide can reveal structural information. Product ion spectra are produced using an MS/MS instrument, either separated in space (triple quadrupole instruments) or in time (ion trap instruments). An ion produced in the source is selected by the first mass analyzer, fragmented in a collision cell and the product ions are analyzed in the final mass analyzer. For thermolabile analytes, such as nucleosides and nucleotides, using soft ionization techniques such as ESI, the protonated molecules are first produced and then fragmented. Nucleobases, analyzed with GC-MS/MS, produce radical cations (EI) or protonated (CI) molecules depending on the ionization technique applied. In order to understand the mass spectra obtained from modified DNA, one needs first to understand the fragmentation behavior of the naturally occurring nucleotides, nucleosides, and nucleobases, followed by the fragmentation behavior of the adducts.

The basic fragmentation behavior of nucleosides and nucleotides is well documented in literature. The first bond to be broken is the anomeric bond connecting the heterocyclic base moiety to the (deoxy)ribose (Figure 6.19). Since the masses of the unmodified bases are known, any deviation of these masses is a first indication of the nature of the alkyl part of the adduct.

How the protonated base ions break up further depends on the nature of the base and the covalent modifications or adducts (Figure 6.30). The protonated base, often, as in Figure 6.30, though certainly not always, as in Figure 6.19, can be detected in the product ion spectra. This again gives information on the nature of base moiety and the alkyl group attached to it. The typical characteristics of the "alkyl" moiety have a profound influence on the spectrum. For example, due to the glutathione part, the full scan (Figure 6.15) and product ion (not shown) ESI spectra of S-[2-(N7-guanyl)ethyl]glutathione look like a peptide spectrum, with the characteristic doubly protonated molecule and the fragmentation by breaking the peptide bond. Even closely related structures (adduct isomers) can yield product ion spectra with distinct differences (Figure 6.21b). In the negative ion mode, the basic fragmentation is the same but, due to its nature, ions from the base moiety are less readily formed resulting in weaker spectra. However, in some cases, very distinctive spectra are found, allowing one to distinguish isomeric forms of DNA adducts based on their negative product ion spectra (Figure 6.31).

Similar to the 2D plots (*m/z versus* scan number) in full scan MS mode (Figure 6.16) revealing the presence of groups of similar structures (here oligonucleotide adducts), the 2D plots of product ion spectra chromatograms reveal differences in fragmentation behavior of similar (isomeric) adducts (Figure 6.32).

Although the fragmentation of a nucleotide adduct is similar to that of a nucleoside adduct, two major differences can be observed. At first, the presence of the phosphate group on the nucleotide makes the detection of negatively charged ions much easier compared to nucleosides and nucleobases. Secondly, the phosphate group is also a possible alkylation site. This means that product ions can be found derived from the phosphate group with the alkyl group attached to it (Figure 6.33).

6.6.4.2   Energy-Resolved Product Ion Spectra

Energy-resolved product ion spectra are a series of product ion spectra recorded from the same precursor ion using different collision energies to produce product ions. Ions producing otherwise identical product ions might do so at different rates and/or relative abundances under different activation conditions. Typical examples are the two isomeric forms of a DNA adduct (Figure 6.26 and Lemière et al.) where the same nucleobase is alkylated with the same electrophile and the adducts differ only in the position of the alkylation on the nucleobase.[180] Interpretation of these sorts of differences is a possible key toward identification of the isomers, but it requires knowledge

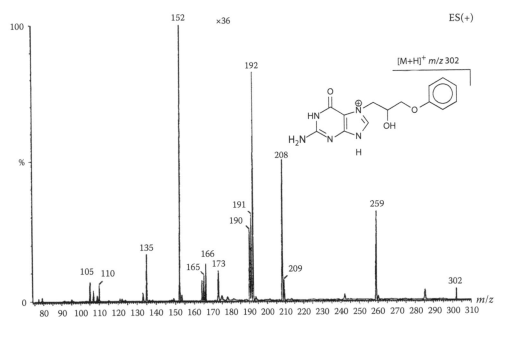

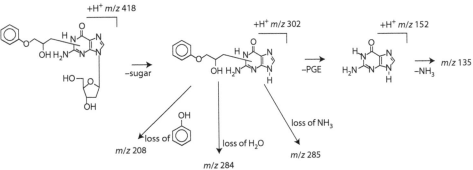

FIGURE 6.30 Low-energy CAD product ion spectrum of the in-source CAD-generated [BH2]C of the PGE dGuo isomer 1 (*m/z* 302) and suggested fragmentation pattern explaining the product ions of N7-PGE dGuo. (From Lemière, R. et al., *J. Mass Spectrom.* 34, 820, 1999. With permission.)

and interpretation of the energies of the different isomeric ions and the transition states connecting the consecutive product ions. The relative abundances of the ions as a function of collision energy can also be used to choose optimal conditions for later SRM or MRM analyses (see below).

6.6.4.3   SIR and SRM/MRM

Once the DNA adducts are identified using full product ion spectra, the question arises as to the amount of adducts that are present in a given sample. Most, if not all, MS based studies quantifying DNA adduct rely on SIR and/or SRM/MRM for the detection and subsequent quantification. Selecting one ion to monitor (single ion recording/monitoring, SIR/SIM) instead of scanning a wide mass range, drastically increases the sensitivity of the detection. Monitoring one ion using one mass filter leaves the possibility that isobaric ions are present in the analyzed material giving rise to a signal not originating from the adducts and giving rise to false positive results and so-called "chemical noise." Producing product ions and selecting one (SRM) or more (MRM) product ions of the selected parent ion makes it virtually impossible for an arbitrary isobaric ion to produce a product ion that also passes the second MS filter. This chance is even further reduced if multiple product

(a)

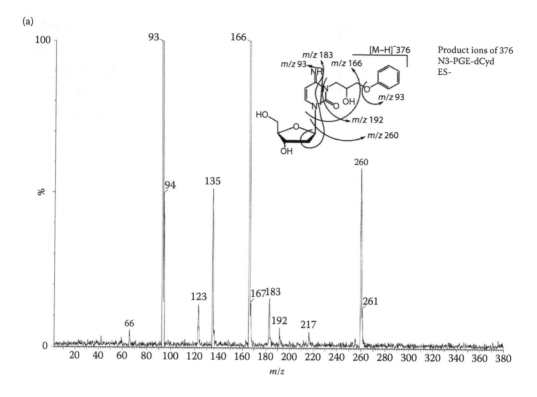

(b)

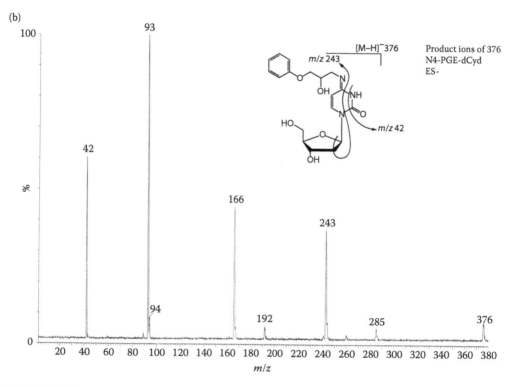

FIGURE 6.31 Negative ion mode (ESI) low-energy CAD product ion spectra of two isomeric PGE/dCyd adducts showing unique product ions allowing for the unambiguous location of the alkyl chain on the nucleobase. (From Lemière, F. et al., *J. Am. Soc. Mass Spectrom.* 7, 682, 1996. With permission.)

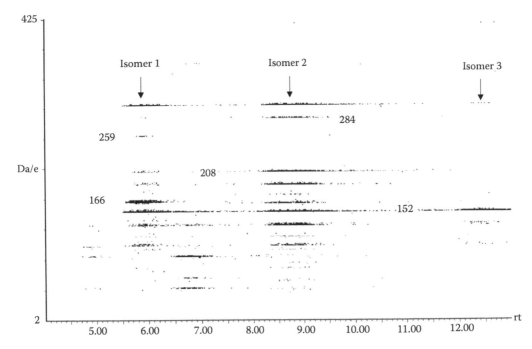

FIGURE 6.32   2D map of an analysis of the product ions of PGE dGuo ([MH]⁺ m/z 418) showing the presence of three isomers. (From Lemière, F. et al., *J. Mass Spectrom.* 34, 820, 1999. With permission.)

ions are monitored. Due to the high selectivity of SRM/MRM, the noise in the analysis is strongly reduced. Using an extra fragmentation step and subsequent MS filtering also reduces the MS signal but, in most cases, to a lesser extent, resulting in an increased S/N ratio and more sensitive detection. A prerequisite for this is that the product ion monitored in the second MS stage has a strong abundance. Therefore the fragmentation parameters (collision gas pressure, collision energy) need to be chosen with care. Recording the product ion spectra at varying energies (energy-resolved spectra) is explicit or implicit, and is often included in the method development. In most studies, SIR and SRM/MRM are combined with the application of isotope-labeled internal standards. For example, applying SIR and SRM/MRM (Table 6.4 and subsection 6.6.2.2.1).

6.6.4.4   Constant Neutral Loss/Precursor Ion Spectra

A study on DNA adducts, demonstrating different MS/MS techniques on both *in vitro* and *in vivo* samples, was presented by Gangl et al.[119] CNL scanning showed the presence of adducts in a study on the food-derived mutagen 2-amino-3-methylimidazo[4,5-*f*]quinoline. A detection of one adduct in 10⁴ unmodified bases was achieved. A lower detection limit was obtained using the SRM approach, with one adduct in 10⁷ unmodified bases using a 300 μg of DNA. This capillary LC/microelectrospray MS/MS setup was used to detect N-(deoxyguanosin-8-yl)-2-amino-3-methylimidazo[4,5-*f*]quinoline and 5-(deoxyguanosin-*N*-2-yl)-2-amino-3-methylimidazo[4,5-*f*]quinoline in kidney tissues of chronically treated cynomolgus monkeys.

CNL scanning is a powerful tool for the rapid screening of mixtures for a well-defined class of compounds. The loss of 116 Da is indicative of the occurrence of 2′-deoxynucleosides with an unmodified 2′-deoxyribofuranosyl moiety. Owing to the specificity of this scanning technique, interferences are eliminated, yielding neat chromatograms and spectra revealing only the 2′-deoxynucleoside derivatives of interest. In our own research, CNL was applied to detect adducts formed by PGE.[180] This showed the presence of two bis-PGE-dGuo adducts (Figure 6.34), which would have escaped our attention without this CNL data. Plotting the same data retention

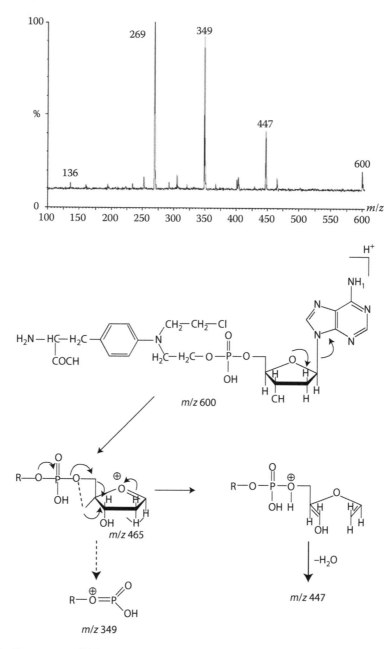

FIGURE 6.33 Low-energy CAD product ion spectra of a phosphate-alkylated dAMP found in melphalan-treated calf thymus DNA hydrolysates. Diagnostic ions, proving phosphate alkylation, are *m/z* 447 and 348; and (bottom) tentative fragmentation scheme of the phosphate-alkylated dAMP adduct. (From Van den Driessche, B. et al., *J Chromatogr B Anal. Technol. Biomed. Life Sci.* 785, 21, 2003. With permission.)

time *versus m/z* ratio yields a 2D map showing even more information. Besides the two bis-PGE adducts, a guanine–imidazole ring-opened bis-PGE-dGuo was also present, characterized by a mass of 18 Da (addition of H_2O) above the normal guanine adducts. The sample that was supposed to contain only 2′-deoxyguanosine and its PGE adducts was clearly shown to also contain 2′-deoxyadenosine (and adducts thereof) in the 2D map (Figure 6.35) (PGE/dAdo *m/z* 402 and 2PGE/dAdo *m/z* 586).

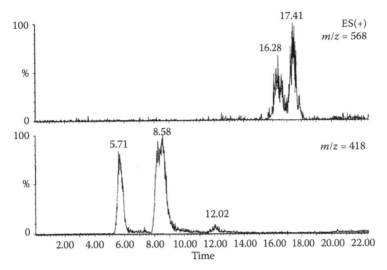

FIGURE 6.34 CNL scan (116u) of a column switching LC/MS experiment showing the presence of three isomers of PGE dGuo and two isomers of bis-PGE dGuo. (From Lemière, F. et al., *J. Mass Spectrom.* 34, 820, 1999. With permission.)

The general applicability of CNL of 116 Da to 2′-deoxynucleosides can be used to detect unknown DNA adducts. This opens the door for the analysis of all deoxynucleoside adducts present at a given moment in a given DNA sample: the adductome.[214]

6.7 THE FUTURE?

Nobody knows what the future will bring. But looking at the past, it is certain that every new mass spectrometric and separation technique will be explored for its possibilities for the detection, characterization, and quantification of DNA adducts.

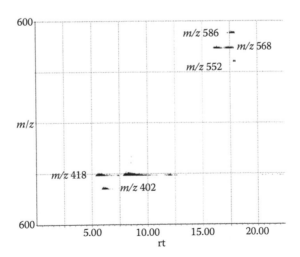

FIGURE 6.35 2D plot of a CNL of 116u, column switch experiment showing the presence of several 2′-deoxynucleosides, including PGE/dGuo (three isomers, *m/z* 418) and 2PGE/dGuo (two isomers, *m/z* 568), imidazole ring opened-2PGE/dGuo (*m/z* 586), PGE/dAdo (*m/z* 402) and 2PGE/dAdo (*m/z* 552). (Adapted from F. Lemière, Unpublished data PhD dissertation, University of Antwerp, 1998.)

REFERENCES

1. Searle, C.E. and Teale, O.J., Occupational Carcinogenesis, in *Chemical Carcinogenesis and Mutagenesis I*, Springer, Berlin, 1990.
2. Ramazzini, B., De morbis diatriba, 1713, Translated by W.C. Wright, in *History of Medicine*, Vol. 23, New York Academy of Medicine, New York, 1964.
3. Pott, P., Chirurgical observations relative to the cataract, the polypus of the nose, the cancer of the scrotum, the different kinds of ruptures, and the mortification of the toes and feet. Hawes, Clarke and Collins, London, 1775, *Natl. Cancer Inst. Monogr.* 10, 7, 1963.
4. Butlin, H.T., Cancer of the scrotum in chimney-sweeps and others. II. Why foreign sweeps do not suffer from scrotal cancer, *Br. Med. J.* 21, 1892.
5. Rehn, L., lasengeschwulste bei Fuchsin-Arbeitern, *Arch. Klin. Chir.* 50, 588, 1895.
6. Leichtenstern, O., Uber Harnblasenentzündung and Harnblasengeschwulste bei Arbeitern in Farbfabriken, *Deut. Med. Wochenschr.* 24, 709, 1898.
7. Pitot, H.C., Mechanisms of chemical carcinogenesis: Theoretical and experimental bases, in *Chemical Carcinogenesis and Mutagenesis I*, Springer, Berlin, 1990.
8. Yamagawa, K. and Ichikawa, K., Experimentelle Studie über die Pathogenese der Epithelialgeschwulste, *Mitteilungen Med. Facultat Kaiserl. Univ. Tokyo.* 15, 295, 1915.
9. Cook, J.W., Hewett, C.L., and Hieger, I., The isolation of a cancer-producing hydrocarbon from coal tar. Parts I, II, and III, *J. Chem. Soc.* 395, 1933.
10. Kennaway, K.H. and Hieger, I., Carcinogenic substances and their fluorescence spectra, *Br. Med. J.* 1, 1044, 1930.
11. Hueper, W.C., Wiley, F.H., and Wolfe, H.D., Experimental production of bladder tumors in dogs by administration of beta-naphtylamine, *J. Ind. Hyg. Toxicol.* 20, 46, 1938.
12. Overall Evaluations of Carcinogenicity to Humans. List of all agents, mixtures and exposures evaluated to date, http://monographs.iarc.fr/ENG/Classification/crthall.php
13. Miller, J.A. and Miller, EC., The metabolic activation of carcinogenic aromatic amines and amides, *Prog. Exp. Tumor Res.* 11, 273, 1969.
14. Miller, E.C. and Miller, J.A., The presence and significance of bound aminoazo dyes in the livers of rats fed p-dimethyl- aminoazobenzene, *Cancer Res.* 7, 468, 1947.
15. Miller, E.C., Studies on the formation of protein-bound derivatives of 3,4-benzpyrene in the epidermal fraction of mouse skin, *Cancer Res.* 11, 100, 1951.
16. Brookes, P. and Lawley, P.D., Evidence for the binding of polynuclear aromatic hydrocarbons to the nucleic acids of mouse skin: Relation between carcinogenic power of hydrocarbons and their binding to deoxyribonucleic acid, *Nature* 202, 781, 1964.
17. Boyland, E., The biological significance of metabolism of polycyclic compounds, *Biochem. Soc. Symp.* 5, 40, 1950.
18. Jerina, D.M., Dali, J.W., Witkop, B., Zaltzman-Niremberg, P., and Udenfriend, S., 1,2-Naphtaleneoxide as an intermediate in the microsomal hydroxylation of naphtalene, *Biochemistry* 9, 147, 1970.
19. Yang, S.K., Enzymatic conversion of benzo[a]pyrene leading predominantly to thediol-epoxide r-7,t-8-dihydroxy-t-9,10-oxy-7,8,9,10-tetrahydrobenzo[a]pyrene through a single enantiomer of r-7,t-8-dihydroxy-7,8-dihy, *Proc. Natl. Acad. Sci. USA* 73, 2594, 1976.
20. Harvey, R.G., Activated metabolites of cacinogenic hydrocarbons, *Acc. Chem. Res.* 14, 218, 1981.
21. Hemminki, K., DNA adducts, mutation and cancer, *Carcinogenesis* 14, 2007, 1993.
22. Hemminki, K., Nucleic acid adducts of chemical carcinogens and mutagens, *Arch. Toxicol.* 52, 249, 1983.
23. Beranek, D.T., Weis, C.C., and Swenson, D.H., A comprehensive quantitative-analysis of methylated and ethylated DNA using high-pressure liquid-chromatography, *Carcinogenesis* 1, 595, 1980.
24. Andrews, P.J., Quilliam, M.A., McCarry, B.E., Bryant, D.W., and McCalla, D.R., Identification of the DNA adduct formed by metabolism of 1,8-dinitropyrene in Salmonella typhimurium, *Carcinogenesis* 7, 105, 1986.
25. Singer, B. and Essigmann, J.M., Site-directed mutagenesis using DNA adducts: Historical perspective and future directions, *Carcinogenesis* 12, 949, 1991.
26. Dipple, A., DNA-adducts of chemical carcinogens, *Carcinogenesis* 16, 437, 1995.
27. Lawley, P.D., Carcinogenesis by alkylating agents, in ACS Monograph 173, Searle, C.E., ed., American Chemical Society, Washington, DC, 1976.
28. Singer, B., Alkylation of the O^6 of guanine is only one of many chemical events that may initiate carcinogenesis, *Cancer Invest.* 2, 233, 1984.

29. Basu, A.K., and Essigmann, J.M., Site-specifically modified oligonucleotides as probes for the structural and biological effects of DNA-damaging agents, *Chem. Res. Toxicol.* 1, 1, 1988.

30. Loveless, A., Possible relevance of O-6 alkylation of deoxyguanosine to the mutagenicity and carcinogenicity of nitrosamines and nitrosamides, *Nature* 223, 206, 1969.

31. Singer, B. and Grunberger, D., *Molecular Biology of Mutagenesis and Carcinogenesis*, Plenum Press, 1983.

32. Dosanjh, M.K., Singer, B., and Essigmann, J.M., Comparative mutagenesis of O6-methylguanine and O4-ethylthymine in *Escherichia Coli*, *Biochemistry* 30, 7027, 1991.

33. Swenberg, J.A., Richardson, F.C., Boucheron, J.A., and Dyroff, M.C., Relationships between DNA adduct formation and carcinogenesis, *Environ. Health Perspect.* 62, 177, 1985.

34. Essigmann, J.M. and Wood, M.L., The relationship between the chemical structures and mutagenic specificities of the DNA lesions formed by chemical and physical mutagens, *Toxicol. Lett.* 67, 29, 1993.

35. Lutz, W.K., *In vivo* covalent binding of organic chemicals to DNA as a quantitative indicator in the process of chemical carcinogenesis, *Mutat. Res.* 65, 289, 1979.

36. Frei, J.V., Swenson, D.H., Warren, W., and Lawley, P.D., Alkylation of deoxyribonucleic acid *in vivo* in various organs of C57BL mice by the carcinogens *N*-methyl-*N*-nitrosourea, *N*-ethyl-*N*-nitrosourea and ethyl methanesulphonate in relation to induction of thy, *Biochem. J.* 174, 1031, 1978.

37. Warren, J.J., Forsberg, L.J., and Beese, L.S., The structural basis for the mutagenicity of O-6-methyl-guanine lesions, *Proc. Natl. Academy Sci. USA* 103, 19701, 2006.

38. Straub, K.M., Meehan, T., Burlingame, A.L., and Calvin, M., Identification of major adducts formed by reaction of benzo[*a*]pyrene diol epoxide with dna invitro, *Proc. Natl. Academy Sci. USA* 74, 5285, 1977.

39. Meehan, T., Straub, K., and Calvin, M., Benzo[*A*]pyrene diol epoxide covalently binds to deoxyguanosine and deoxyadenosine in DNA, *Nature* 269, 725, 1977.

40. Fuchs, R.P.P. and Bintz, R., Activity of carcinogens that bind to the C8 position of guanine residues in an assay specific for the detection of −2 frameshift mutations in a defined hot spot, *Environ. Health Perspect.* 88, 83, 1990.

41. Walles, S. and Ehrenberg, L., Effects of b-hydroxyethylation and b-methoxyethylation on DNA *in vitro*, *Acta Chem. Scand.* 22, 2727, 1968.

42. Deforce, D.L.D., Lemière, F., Esmans, E.L., De Leenheer, A., and Van den Eeckhout, E.G., Analysis of the DNA damage induced by phenyl glycidyl ether using capillary zone electrophoresis-electrospray mass spectrometry, *Anal. Biochem.* 258, 331, 1998.

43. Singh, R., Sweetman, G.M.A., Farmer, P.B., Shuker, D.E.G., and Rich, K.J., Detection and characterization of two major ethylated deoxyguanosine adducts by high perfomance liquid chromatography, electrospray mass spectrometry, and 32P-postlabeling, Development of an approach for detection of phosphotriesters, *Chem. Res. Toxicol.* 10, 70, 1997.

44. Haglund, J., Van Dongen, W., Lemière, F., and Esmans, E.L., Analysis of DNA-phosphate adducts *in vitro* using miniaturized LC-ESI-MS/MS and column switching: Phosphotriesters and alkyl cobalamins, *J. Am. Soc. Mass Spectrom.* 15, 593, 2004.

45. Otteneder, M. and Lutz, W.K., Correlation of DNA adduct levels with tumor incidence: Carcinogenic potency of DNA adducts, *Mutat. Res.-Fundam. Mol. Mech. Mutagen.* 424, 237, 1999.

46. Sedgwick, B., Bates, P.A., Paik, J., Jacobs, S.C., and Lindahl, T., Repair of alkylated DNA: Recent advances, *DNA Repair* 6, 429, 2007.

47. Farmer, P.B., DNA and protein adducts as markers of genotoxicity, *Toxicol. Lett.* 149, 3, 2004.

48. Singh, R. and Farmer, P.B., Liquid chromatography-electrospray ionization-mass spectrometry: The future of DNA adduct detection, *Carcinogenesis* 27, 178, 2006.

49. RamaKrishna, N.V.S., Gao, F., Padmavathi, N.S., Cavalieri, E.L., Rogan, E.G., Cerny, R.L., and Gross, M.L., Model adducts of benzo[*a*]pyrene and nucleosides formed from its radical cation and diol epoxide, *Chem. Res. Toxicol.* 5, 293, 1992.

50. Wellemans, J., George, M., Cerny, R.L., and Gross, M.L., Improving the sensitivity of FAB tandem mass spectrometry for PAH-DNA adducts, *Polycyclic, Aromat. Compd.* 6, 103, 1994.

51. Wolf, S.M. and Vouros, P., Incorporation of sample stacking techniques into the capillary electrophoresis of CF-FAB mass spectrometric analysis of DNA adducts, *Anal. Chem.* 67, 891, 1995.

52. Schoket, B., DNA damage in humans exposed to environmental and dietary polycyclic aromatic hydrocarbons, *Mutat. Res.-Fundam. Mol. Mech. Mutagen.* 424, 143, 1999.

53. Rothman, N., Poirier, M.C., Baser, M.E., Hansen, J.A., Gentile, C., Bowman, E.D., and Strickland, P.T., Formation of polycyclic aromatic hydrocarbon DNA adducts in peripheral white blood-cells during consumption of charcoal-broiled beef, *Carcinogenesis* 11, 1241, 1990.

54. Airoldi, L., Orsi, F., Magagnotti, C., Coda, R., Randone, D., Casetta, G., Peluso, M., Hautefeuille, A., Malaveille, C., and Vineis, P., Determinants of 4-aminobiphenyl-DNA adducts in bladder cancer biopsies, *Carcinogenesis* 23, 861, 2002.

55. Chen, H.J.C., Wu, C.F., and Huang, J.L., Measurement of urinary excretion of 5-hydroxymethyluracil in human by GC/NICI/MS: Correlation with cigarette smoking, urinary TBARS and etheno DNA adduct, *Toxicol. Lett.* 155, 403, 2005.

56. Vodicka, P.E., Linhart, I., Novak, J., Koskinen, M., Vodickova, L., and Hemminki, K., 7-Alkylguanine adduct levels in urine, lungs and liver of mice exposed to styrene by inhalation, *Toxicol. Appl. Pharm.* 210, 1, 2006.

57. Teixeira, J.P., Gaspar, J., Roma-Torres, J., Silva, S., Costa, C., Roach, J., Mayan, O., Rueff, J., and Farmer, P.B., Styrene-oxide N-terminal valine haemoglobin adducts in reinforced plastic workers: Possible influence of genetic polymorphism of drug-metabolising enzymes, *Toxicology* 237, 58, 2007.

58. Thier, R. and Bolt, H.F., Carcinogenicity and genotoxicity of ethylene oxide: New aspects and recent advances, *Crit. Rev. Toxicol.* 30, 595, 2000.

59. Yong, L.C., Schulte, P.A., Kao, C.Y., Giese, R.W., Boeniger, M.F., Strauss, G.H.S., Petersen, M.R., and Wiencke, J.K., DNA adducts in granulocytes of hospital workers exposed to ethylene oxide, *Am. J. Ind. Med.* 50, 293, 2007.

60. Dorr, D.Q., Murphy, K., and Tretyakova, N., Synthesis of DNA oligodeoxynucleotides containing structurally defined N-6-(2-hydroxy-3-buten-1-yl)-adenine adducts of 3,4-epoxy-1-butene, *Chemico-Biol. Interact.* 166, 104, 2007.

61. Besaratinia, A. and Pfeifer, G.P., DNA adduction and mutagenic properties of acrylamide, *Mutat. Res.-Genetic Toxicol. Environm. Mutagen.* 580, 31, 2005.

62. Granath, F. and Tornqvist, M., Who knows whether acrylamide in food is hazardous to humans? *J. Natl. Cancer Inst.* 95, 842, 2003.

63. Muller, M., Belas, F.J., Blair, I.A., and Guengerich, F.P., Analysis of 1,N2-ethenoguanine and 5,6,7,9-tetrahydro-7-hydroxy-9-oxoimidazo[1,2-a]purine in DNA treated with 2-chlorooxirane by high performance liquid chromatography/electrospray mass spectrometry a, *Chem. Res. Toxicol.* 10, 242, 1997.

64. Fedtke, N., Boucheron, J.A., Turner, M.J.J., and Swenberg, J.A., Vinyl chloride-induced DNA adducts. I: Quantitative determination of Ný,3-ethenoguanine based on electrophore labeling, *Carcinogenesis* 11, 1279, 1990.

65. Ambrosone, C.B., Abrams, S.M., Gorlewska-Roberts, K., and Kadlubar, F.F., Hair dye use, meat intake, and tobacco exposure and presence of carcinogen-DNA adducts in exfoliated breast ductal epithelial cells, *Arch. Biochem. Biophys.* 464, 169, 2007.

66. Turesky, R.J., Freeman, J.P., Holland, R.D., Nestorick, D.M., Miller, D.W., Ratnasinghe, D.L., and Kadlubar, F.F., Identification of aminobiphenyl derivatives in commercial hair dyes, *Chem. Res. Toxicol.* 16, 1162, 2003.

67. Wang, J.S. and Groopman, J.D., DNA damage by mycotoxins, *Mutat. Res.-Fundam. Mol. Mech. Mutagen.* 424, 167, 1999.

68. Goldman, R. and Shields, P.G., Food mutagens, *J. Nutr.* 133, 965S, 2003.

69. Tareke, E., Rydberg, P., Karlsson, P., Eriksson, S., and Tornqvist, M., Analysis of acrylamide, a carcinogen formed in heated foodstuffs, *J. Agric. Food Chem.* 50, 4998, 2002.

70. Tareke, E., Rydberg, P., Karlsson, P., Eriksson, S., and Tornqvist, M., Acrylamide: A cooking carcinogen? *Chem. Res. Toxicol.* 13, 517, 2000.

71. Wakabayashi, K., Nagao, M., Esumi, H., and Sugimura, T., Food-derived mutagens and carcinogens, *Cancer Res.* 52, S2092, 1992.

72. Sugimura, T., Wakabayashi, K., Nakagama, H., and Nagao, M., Heterocyclic amines: Mutagens/carcinogens produced during cooking of meat and fish, *Cancer Sci.* 95, 290, 2004.

73. Musser, S.M., Pan, S.S., and Callery, P.S., Liquid-chromatography thermospray mass-spectrometry of DNA adducts formed with mitomycin-C., porfiromycin and thiotepa, *J. Chromatogr.* 474, 197, 1989.

74. Warren, A.J., Mustra, D.J., and Hamilton, J.W., Detection of mitomycin C-DNA adducts in human breast cancer cells grown in culture, as xenografted tumors in nude mice, and in biopsies of human breast cancer patient tumors as determined by P-32-postlabeling, *Clin. Cancer Res.* 7, 1033, 2001.

75. Van den Driessche, B., Lemière, F., Van Dongen, W., Van der Linden, A., and Esmans, E.L., Qualitative study of *in vivo* melphalan adduct formation in the rat by miniaturized column-switching liquid chromatography coupled with electrospray mass spectrometry, *J. Mass Spectrom.* 39, 29, 2004.

76. Edler, M., Jakubowski, N., and Linscheid, M., Quantitative determination of melphalan DNA adducts using HPLC-inductively coupled mass spectrometry, *J. Mass Spectrom.* 41, 507, 2006.

77. Poljakova, J., Frei, E., Gomez, J.E., Aimova, D., Eckschlager, T., Hrabeta, J., and Stiborova, M., DNA adduct formation by the anticancer drug ellipticine in human leukemia HL-60 and CCRF-CEM cells, *Cancer Lett.* 252, 270, 2007.
78. Warnke, U., Gysler, J., Hofte, B., Tjaden, U.R., van der Greef, J., Kloft, C., Schunack, W., and Jaehde, U., Separation and identification of platinum adducts with DNA nucleotides by capillary zone electrophoresis and capillary zone electrophoresis coupled to mass spectrometry., *Electrophoresis* 22, 97, 2001.
79. Warnke, U., Rappel, C., Meier, H., Kloft, C., Galanski, M., Hartinger, C.G., Keppler, B.K., and Jaehde, U., Analysis of platinum adducts with DNA nucleotides and nucleosides by capillary electrophoresis coupled to ESI-MS: Indications of guanosine 5'-monophosphate O-6-N7 chelation, *Chembiochemistry* 5, 1543, 2004.
80. Roy, D. and Liehr, J.G., Estrogen, DNA damage and mutations., *Mutat. Res.-Fundam. Mol. Mech. Mutagen.* 424, 107, 1999.
81. Embrechts, J., Lemière, F., Van Dongen, W., Esmans, E.L., Buytaert, P., Van Marck, E., Kockx, M., and Makar, A., Detection of estrogen DNA-adducts in human breast tumor tissue and healthy tissue by combined nano LC-nano ES tandem mass spectrometry, *J. Am. Soc. Mass Spectrom.* 14, 482, 2003.
82. Cavalieri, E. and Rogan, E., Catechol quinones of estrogens in the initiation of breast, prostate, and other human cancers—keynote lecture, *Estrogens Human Dis.* 1089, 286, 2006.
83. Cadet, J., Delatour, T., Douki, T., Gasparutto, D., Pouget, J.P., Ravanat, J.L., and Sauvaigo, S., Hydroxyl radicals and DNA base damage, *Mutat. Res.-Fundam. Mol. Mech. Mutagen.* 424, 9, 1999.
84. Cadet, J., Sage, E., and Douki, T., Ultraviolet radiation-mediated damage to cellular DNA, *Mutat. Res.-Fundam. Mol. Mech. Mutagen.* 571, 3, 2005.
85. Cadet, J., Douki, T., Gasparutto, D., and Ravanat, J.L., Radiation-induced damage to cellular DNA: Measurement and biological role, *Radiat. Phys. Chem.* 72, 293, 2005.
86. Douki, T., Court, M., Sauvaigo, S., Odin, F., and Cadet, J., Formation of the main UV-induced thymine dimeric lesions within isolated and cellular DNA as measured by high performance liquid chromatography-tandem mass spectrometry, *J. Biol. Chem.* 275, 11678, 2000.
87. Gupta, R.C. and Lutz, W.K., Background DNA damage from endogenous and unavoidable exogenous carcinogens: A basis for spontaneous cancer incidence? *Mutat. Res.-Fundam. Mol. Mech. Mutagen.* 424, 1, 1999.
88. Cavalieri, E., Chakravarti, D., Guttenplan, J., Hart, E., Ingle, J., Jankowiak, R., Muti, P., Rogan, E., Russo, J., Santen, R., and Sutter, T., Catechol estrogen quinones as initiators of breast and other human cancers: Implications for biomarkers of susceptibility and cancer prevention, *Biochim. Biophys. Acta-Rev. Cancer* 1766, 63, 2006.
89. Nair, J., Barbin, A., Velic, I., and Bartsch, H., Etheno DNA-base adducts from endogenous reactive species, *Mutat. Res.-Fundam. Mol. Mech. Mutagen.* 424, 59, 1999.
90. Luczaj, W. and Skrzydlewska, E., DNA damage caused by lipid peroxidation products, *Cellular Mol. Biol. Lett.* 8, 391, 2003.
91. Hillestrom, P.R., Covas, M.I., and Poulsen, H.E., Effect of dietary virgin olive oil on urinary excretion of etheno-DNA adducts, *Free Radical Biol. Med.* 41, 1133, 2006.
92. Bartsch, H. and Nair, J., Ultrasensitive and specific detection methods for exocyclic DNA adduct: Markers for lipid peroxidation and oxidative stress, *Toxicology* 153, 105, 2000.
93. Nair, U., Bartsch, H., and Nair, J., Lipid peroxidation-induced DNA damage in cancer-prone inflammatory diseases: A review of published adduct types and levels in humans, *Free Radical Biol. Med.* 43, 1109, 2007.
94. Epe, B., Role of endogenous oxidative DNA damage in carcinogenesis: What can we learn from repair-deficient mice? *Biol. Chem.* 383, 467.
95. Cavalieri, E., Frenkel, K., Liehr, J.G., Rogan, E., and Roy, D., Estrogens as endogenous genotoxic agents—DNA adducts and mutations, *J. Natl. Cancer Inst. Monogr.* 27, 75, 2000.
96. Guetens, G., De Boeck, G., Highley, M., van Oosterom, A.T., and de Bruijn, E.A., Oxidative DNA damage: Biological significance and methods of analysis, *Crit. Rev. Clin. Lab. Sci.* 39, 331, 2002.
97. Poulsen, H.E., Oxidative DNA modifications, *Exp. Toxicol. Pathol.* 57, 161, 2005.
98. Fang, A.H., Smith, W.A., Vouros, P., and Gupta, R.C., Identification and characterization of a novel benzo[a] pyrene-derived DNA adduct. *Biochem. Biophys. Res. Commun.* 281, 383, 2001.
99. Van den Driessche, B., Lemière, F., Van Dongen, W., and Esmans, E.L., Alkylation of DNA by melphalan: Investigation of capillary liquid chromatography-electrospray ionization tandem mass spectrometry in the study of the adducts at the nucleoside level, *J. Chromatogr. B Anal. Technol. Biomed. Life Sci.* 785, 21, 2003.
100. Van den Driessche, B., Lemière, F., Van Dongen, W., and Esmans, E.L., Structural characterization of melphalan modified 2'-oligodeoxynucleotides by miniaturized LC-ES MS/MS, *J. Am. Soc. Mass Spectrom.* 15, 568, 2004.

101. Debrauwer, L., Rathahao, E., Couve, C., Poulain, S., Pouyet, C., Jouanin, J., and Paris, A., Oligonucleotide covalent modifications by estrogen quinones evidenced by use of liquid chromatography coupled to negative electrospray ionization tandem mass spectrometry, *J. Chromatogr. A* 976, 123, 2002.

102. Hanson, A.A., Rogan, E.G., and Cavalieri, E.L., Synthesis of adducts formed by iodine oxidation of aromatic hydrocarbons in the presence of deoxyribonucleosides and nucleobases, *Chem. Res. Toxicol.* 11, 1201, 1998.

103. Lemière, F., Joos, P., Vanhoutte, K., Esmans, E.L., De Groot, A., Claeys, M., and Van den Eeckhout, E., Phenylglycidyl ether adducts of 2′-deoxycytidine and 2′-deoxyadenosine: Stability in solution and structure analysis by electrospray tandem mass spectrometry, *J. Am. Soc. Mass Spectrom.* 7, 682, 1996.

104. Embrechts, J., Lemière, F., Van Dongen, W., and Esmans, E.L., Equilenin-2′-deoxynucleoside adducts: Analysis with nano-liquid chromatography coupled with nano-electrospray tandem mass spectrometry, *J. Mass Spectrom.* 36, 317, 2001.

105. Andrews, C.L., Harsch, A., and Vouros, P., Analysis of the *in vitro* digestion of modified DNA to oligonucleotides by LC-MS and LC-MS/MS, *Int. J. Mass Spectrom.* 231, 169, 2004.

106. Schrader, W. and Linscheid, M., Determination of styrene oxide adducts in DNA and DNA components, *J. Chromatogr. A* 717, 117, 1995.

107. Schrader, W. and Linscheid, M., Styrene oxide adducts: *In vitro* reaction and sensitive detection of modified oligonucleotides using capillary zone electrophoresis interfaced to electrospray mass spectrometry, *Arch. Toxicol.* 71, 588, 1997.

108. Debrauwer, L., Rathahao, E., Jouanin, I., Paris, A., Clodic, G., Molines, H., Convert, O., Fournier, F., and Tabet, J.C., Investigation of the regio- and stereo-selectivity of deoxyguanosine linkage to deuterated 2-hydroxyestradiol by using liquid chromatography/ESI-ion trap mass spectrometry, *J. Am. Soc. Mass Spectrom.* 14, 364, 2003.

109. Nourse, B.D., Hettich, R.L., and Buchanan, M.V., Methyl guanine isomer distinction by hydrogen/deuterium exchange using a Fourier transform mass spectrometer, *J. Am. Soc. Mass Spectrom.* 4, 296, 1992.

110. Paz, M.M., Das, A., and Tomasz, M., Mitomycin C linked to DNA minor groove binding agents: Synthesis, reductive activation, DNA binding and cross-linking properties and *in vitro* antitumor activity, *Bioorganic Med. Chem.* 7, 2713, 1999.

111. Palom, Y., Belcourt, M.F., Musser, S.M., Sartorelli, A.C., Rockwell, S., and Tomasz, M., Structure of adduct X, the last unknown of the six major DNA adducts of mitomycin C formed in EMT6 mouse mammary tumor cells, *Chem. Res. Toxicol.* 13, 479, 2000.

112. Maccubbin, A.E., Caballes, L., Riordan, J.M., Huang, D.H., and Gurtoo, H.L., A cyclophosphamide/DNA phosphoester adduct formed *in vitro* and *in vivo*, *Cancer Res.* 51, 886, 1991.

113. Takamura-Enya, T., Ishikawa, S., Mochizuki, M., and Wakabayashi, K., Chemical synthesis of 2′-deoxyguanosine – C8 adducts with heterocyclic amines: An application to synthesis of oligonucleotides site-specifically adducted with 2-amino-1-methyl-6-phenylimidazo[4,5-b] pyridine, *Chem. Res. Toxicol.* 19, 770, 2006.

114. Takamura-Enya, T., Enomoto, S., and Wakabayashi, K., Palladium-catalyzed direct N-arylation of nucleosides, nucleotides, and oligonucleotides for efficient preparation of dG-N-2 adducts with carcinogenic amino-/nitroarenes, *J. Organic Chem.* 21-7-71, 5599, 2006.

115. Van den Driessche, B., Esmans, E.L., Van der Linden, A., Van Dongen, W., Schaerlaken, E., Lemière, F., Witters, E., and Berneman, Z., First results of a quantitative study of DNA adducts of melphalan in the rat by isotope dilution mass spectrometry using capillary liquid chromatography coupled to electrospray tandem mass spectrometry, *Rapid Commun. Mass Spectrom.* 19, 1999, 2005.

116. Tretyakova, N.Y., Chiang, S.Y., Walker, V.E., and Swenberg, J.A., Quantitative analysis of 1,3-butadiene-induced DNA adducts *in vivo* and *in vitro* using liquid chromatography electrospray ionization tandem mass spectrometry, *J. Mass Spectrom.* 33, 363, 1998.

117. Hoes, I., Van Dongen, W., Lemière, F., Esmans, E.L., Van Bockstaele, D., and Berneman, Z., Comparison between capillary and nano liquid chromatography-electrospray mass spectrometry for the analysis of minor DNA-melphalan adducts, *J. Chromatogr. B* 748, 197, 2000.

118. Martins, C., Oliveira, N.G., Pingarilho, M., da Costa, G.G., Martins, V., Marques, M.M., Beland, F.A., Churchwell, M.I., Doerge, D.R., Rueff, J., and Gaspar, J.F., Cytogenetic damage induced by acrylamide and glycidamide in mammalian cells: Correlation with specific glycidamide-DNA adducts, *Toxicol. Sci.* 95, 383, 2007.

119. Gangl, E.T., Turesky, R.J., and Vouros, P., Determination of *in vitro*- and *in vivo*-formed DNA adducts of 2-amino-3-methylimidazo[4,5-f]quinoline by capillary liquid chromatography/microelectrospray mass spectrometry, *Chem. Res. Toxicol.* 12, 1019, 1999.

120. (a) Huang, H., Jemal, A., David, C., Barker, S.A., Swenson, D.H., and Means, J.C., Analysis of DNA adduct, S-[2-(N-7-guanyl)ethyl]glutathione, by liquid chromatography mass spectrometry and liquid chromatography tandem mass spectrometry, *Anal. Biochemi.* 265, 139 1998; (b) Donghui Li., Andertson M.D. in *Solid Tumor Series; Panceratic Cancer*, B. Douglas Evans, W.T. Peter Pisters, and James L. Abbruzze, eds, Part 1; Springer, New York, pp. 3–13, 2006.

121. Le Pla, R.C., Ritchie, K.J., Henderson, C.J., Wolf, C.R., Harrington, C.F., and Farmer, P.B., Development of a liquid chromatography-electrospray ionization tandem mass spectrometry method for detecting oxaliplatin-DNA intrastrand cross-links in biological samples, *Chem. Res. Toxicol.* 20, 1177, 2007.

122. Chaudhary, A.K., Nokubo, M., Marnett, L.J., and Blair, I.A., Analysis of the malondialdehyde-2′-deoxyguanosine adduct in rat liver DNA by gas chromatography/electron capture negative chemical ionization mass spectrometry, *Biol. Mass Spectrom.* 23, 457, 1994.

123. Chen, H.J.C., Chiang, L.C., Tseng, M.C., Zhang, L.L., Ni, J.S., and Chung, F.L., Detection and quantification of 1,N-6-ethenoadenine in human placental DNA by mass spectrometry, *Chem. Res. Toxicol.* 12, 1119, 1999.

124. Bond, A., Dudley, E., Lemière, F., Tuytten, R., El Sharkawi, S., Brenton, A.G., Esmans, E.L., and Newton, R.P., Analysis of urinary nucleosides. V. Identification of urinary pyrimidine nucleosides by liquid chromatography/electrospray mass spectrometry, *Rapid Commun. Mass Spectrom.* 20, 137, 2006.

125. Dudley, E., El Sharkawi, S., Games, D.E., and Newton, R.P., Analysis of urinary nucleosides. 1. Optimisation of high performance liquid chromatography/electrospray mass spectrometry, *Rapid. Commun. Mass. Spectrom.* 14, 1200, 2000.

126. Dudley, E., Lemière, F., Van Dongen, W., Langridge, J.I., El Sherkawi, S., Games, D.E., Esmans, E.L., and Newton, R.P., Analysis of urinary nucleosides.II. Comparison of mass spectromtric methods for the analysis of urinary nucleosides, *Rapid Commun. Mass Spectrom.* 15, 1701, 2001.

127. Dudley, E., Lemière, F., Van Dongen, W., Langridge, J.I., El Sharkawi, S., Games, D.E., Esmans, E.L., and Newton, R.P., Analysis of urinary nucleosides. III. Identification of 5′-deoxycytidine in urine of a patient with head and neck cancer, *Rapid Commun. Mass Spectrom.* 17, 1132, 2003.

128. Dudley, E., Lemière, F., Van Dongen, W., Tuytten, R., El Sharkawi, S., Brenton, A.G., Esmans, E.L., and Newton, R.P., Analysis of urinary nucleosides. IV. Identification of urinary purine nucleosides by liquid chromatography/electrospray mass spectrometry, *Rapid Commun. Mass Spectrom.* 18, 2730, 2004.

129. Hillestrom, P.R., Hoberg, A.M., Weimann, A., and Poulsen, H.E., Quantification of 1,N-6-etheno-2′-deoxyadenosine in human urine by column-switching LC/APCI-MS/MS, *Free Radical Biol. Med.* 36, 1383, 2004.

130. Chen, H.J.C. and Kao, C.F., Effect of gender and cigarette smoking on urinary excretion of etheno DNA adducts in humans measured by isotope dilution gas chromatography/mass spectrometry, *Toxicol. Lett.* 169, 72, 2007.

131. Van den Driessche, B., Lemière, F., Witters, E., Van Dongen, W., and Esmans, E.L., Implications of enzymatic, acidic and thermal hydrolysis of DNA on the occurrence of cross-linked melphalan DNA adducts, *Rapid Commun. Mass Spectrom.* 19, 449, 2005.

132. Deforce, D.L.D., Lemière, F., Hoes, I., Millecamps, R.E.M., Esmans, E.L., De Leenheer, A., and Van den Eeckhout, E.G., Analysis of the DNA adducts of phenyl glycidyl ether in Calf Thymus DNA hydrolysate by capillary zone electrophoresis-electrospray mass spectrometry: Evidence for phosphate alkylation, *Carcinogenesis* 19, 1077, 1998.

133. Hoes, I., Lemiere, F., Van Dongen, W., Vanhoutte, K., Esmans, E.L., Van Bockstaele, D., Berneman, Z., Deforce, D., and Van den Eeckhout, E.G., Analysis of melphalan adducts of 2′-deoxynucleotides in calf thymus DNA hydrolysates by capillary high-performance liquid chromatography-electrospray tandem mass spectrometry, *J. Chromatogr. B.* 736, 43, 1999.

134. Sharma, R.A. and Farmer, P.B., Biological relevance of adduct detection to the chemoprevention of cancer, *Clin. Cancer Res.* 10, 4901, 2004.

135. Sturla, S.J., DNA adduct profiles: Chemical approaches to addressing the biological impact of DNA damage from small molecules, *Curr. Opin. Chem. Biol.* 11, 293, 2007.

136. Penmetsa, K.V., Shea, D., Leidy, R.B., and Bond, J.A., Analysis of benzo[a]pyrene-DNA adducts by capillary electrophoresis with laser-induced fluorescence detection, *J. High Resolut. Chromatogr.* 18, 719, 1995.

137. Weston, A., Bowman, E.D., Shields, P.G., Trivers, G.E., Poirier, M.C., Santella, R.M., and Manchester, D.K., Detection of polycyclic aromatic hydrocarbon-DNA adducts in human lung, *Environm. Health Perspect.* 99, 257, 1993.

138. Randerath, K. and Randerath, E., 32P-postlabeling methods for DNA adducts detection: Overview and critical evaluation, *Drug Metab. Rev.* 26, 67, 1994.

139. Randerath, K., Reddy, M.V., and Gupta, R.C., 32P-Labeling test for DNA damage, *Proc. Natl. Acad. Sci. USA* 78, 6126, 1981.

140. Singh, R., Sweetman, G.M.A., Farmer, P.B., Shuker, D.E.G., and Rich, K.J., Detection and characterization of two major ethylated deoxyguanosine adducts by high perfomance liquid chromatography, electrospray mass spectrometry, and 32P-postlabeling. Development of an approach, *Chem. Res. Toxicol.* 10, 70, 1997.

141. Talaska, G., Roh, J.H., and Getek, T., 32P-Post labelling and mass spectrometric methods for analysis of bulky, polyaromatic carcinogen-DNA adducts in humans, *J. Chrom. Biomed. Appl.* 580, 293, 1992.

142. Muller, R. and Rajewsky, M.F., Antibodies specific for DNA components structurally modified by chemical carcinogens, *J. Cancer Res. Clin. Oncol.* 102, 99, 1981.

143. Hsu, I.C., Poirier, M.C., Yuspa, S.H., Grunberger, D., Weinstein, I.B., Yolken, R.H., and Harris, C.C., Measurement of benzo(A)pyrene-DNA adducts by enzyme immunoassays and radioimmunoassay, *Cancer Res.* 41, 1091, 1981.

144. Saha, M., Abushamaa, A., and Giese, R.W., General method for determining ethylene oxide and related N-7-guanine DNA adducts by gas chromatography-electron capture mass spectrometry, *J. Chromatogr. A* 712, 345, 1995.

145. Langridge, J.I., McClure, T.D., El Shakawi, S., Fielding, A., Schram, K.H., and Newton, R.P., Gas chromatography/mass spectrometric analysis of urinary nucleosides in cancer patients—potential of modified nucleosides as tumour markers, *Rapid Commun. Mass. Spectrom.* 7, 427, 1993.

146. Esmans, E.L., Broes, D., Hoes, I., Lemière, F., and Vanhoutte, K., Liquid chromatography-mass spectrometry in nucleoside, nucleotide an modified nucleotide characterization, *J. Chromatogr. A* 794, 109, 1998.

147. Reddy, M.V. and Randerath, K., Nuclease P1-mediated enhancement of sensitivity of 32P-postlabeling test for structurally diverse DNA adducts, *Carcinogenesis* 7, 1543, 1986.

148. Gupta, R.C., Enhanced sensitivity of p-32-postlabeling analysis of aromatic carcinogen-DNA adducts, *Cancer Res.* 45, 5656, 1985.

149. Mitchum, R.K., Evans, F.E., Freeman, J.P., and Roach, D., Fast atom bombardment mass spectrometry of nucleoside and nucleotide adducts of chemical carcinogens, *Int. J. Mass Spectrom. Ion Process.* 46, 383, 1983.

150. Stemmler, E.A., Buchanan, M.V., Hurst, G.B., and Hettich, R.L., Structural characterization of polycyclic aromatic hydrocarbon dihydrodiol epoxide DNA adducts using matrix-assisted laser desorption/ionization Fourier transform mass spectrometry, *Anal. Chem.* 66, 1274, 1994.

151. Esmans, E.L., Geboes, P., Luyten, Y., and Alderweireldt, F.C., Direct liquid introduction LC/MS microbore experiments for the analysis of nucleoside material present in human urine, *Biomed. Mass Spectrom.* 12, 241, 1985.

152. Wolf, S.M., Annan, R.S., Vouros, P., and Giese, R.W., Characterization of amino polyaromatic hydrocarbon-DNA adducts using continuous-flow fast atom bombardement and collision-induced dissociation: Positive and negative ion spectra, *Biol. Mass Spectrom.* 21, 647, 1992.

153. Lemière, F., Esmans, E.L., Van Dongen, W., Van den Eeckhout, E., and Van Onckelen, H., Evaluation of liquid chromatography-thermospray mass spectrometry in the determination of some phenylglycidyl ether-2′-deoxynucleoside adducts, *J. Chromatrogr.* 647, 211, 1993.

154. Jajoo, H.K., Burcham, P.C., Goda, Y., Blair, I.A., and Marnett, L.J., A thermospray liquid chromatography-mass spectrometry method for analysis of human urine for the major malondialdehyd-guanine adduct, *Chem. Res. Toxicol.* 5, 870, 1992.

155. Andrews, C.L., Vouros, P., and Harsch, A., Analysis of DNA-adducts using high-performance separation techniques coupled to electrospray ionization mass spectrometry, *J. Chromatogr. A* 856, 515, 1999.

156. Banoub, J.H., Newton, R.P., Esmans, E., Ewing, D.F., and MacKenzie, G., Recent developments in mass spectrometry for the characterization of nucleosides, nucleotides, oligonucleotides, and nucleic acids, *Chem. Rev.* 105, 1869, 2005.

157. Senturker, S. and Dizdaroglu, M., The effect of experimental conditions on the levels of oxidatively modified bases in DNA as measured by gas chromatography-mass spectrometry: How many modified bases are involved? Prepurification or not? *Free Radical Biol. Med.* 27, 370, 1999.

158. Dizdaroglu, M., Jaruga, P., Birincioglu, M., and Rodriguez, H., Free radical-induced damage to DNA: Mechanisms and measurement, *Free Radical Biol. Med.* 32, 1102, 2002.

159. Cadet, J., Douki, T., and Ravanat, J.L., Artifacts associated with the measurement of oxidized DNA bases, *Environm. Health Perspect.* 105, 1034, 1997.

160. Koch, S.A.M., Turner, M.J., and Swenberg, J.A., Quantitation of O-4-ethyldeoxythymidine using electrophore post-labeling, *Proc.Am. Assoc. Cancer Res.* 28, 93, 1987.

161. Yen, T.Y., Holt, S., Sangaiah, R., Gold, A., and Swenberg, J.A., Quantitation of 1,*N*-6-ethenoadenine in rat urine by immunoaffinity extraction combined with liquid chromatography electrospray ionization mass spectrometry, *Chem. Res. Toxicol.* 11, 810, 1998.

162. Yen, T.Y., ChristovaGueoguieva, N.I., Scheller, N., Holt, S., Swenberg, J.A., and Charles, M.J., Quantitative analysis of the DNA adduct *N*-2,3-ethenoguanine using liquid chromatography/electrospray ionization mass spectrometry, *J. Mass Spectrom.* 31, 1271, 1996.

163. Gonzalez-Reche, L.M., Koch, H.M., Weiss, T., Muller, J., Drexler, H., and Angerer, J., Analysis of ethenoguanine adducts in human urine using high performance liquid chromatography-tandem mass spectrometry, *Toxicol. Lett.* 134, 71, 2002.

164. Chaudhary, A.K., Nokubo, M., Reddy, G.R., Yeola, S.N., Morrow, J.D.B., Blair, I.A., and Marnett, L.J., Detection of endogenous malondialdehyde-deoxyguanosine adducts in human liver, *Science* 265, 1580, 1994.

165. Rouzer, C.A., Chaudhary, A.K., Nokubo, M., Ferguson, D.M., Reddy, G.R., Blair, I.A., and Marnett, L.J., Analysis of the malondialdehyde-2′-deoxyguanosine adduct pyrimidopurinone in human leukocyte DNA gas chromatography/electron capture/negative chemical ionization/mass spectrometry, *Chem. Res. Toxicol.* 10, 181, 1997.

166. Chaudhary, A.K., Nokubo, M., Oglesby, T.D., Marnett, L.J., and Blair, I.A., Characterization of endogenous DNA adducts by liquid chromatography/electrospray ionization tandem mass spectrometry, *J. Mass Spectrom.* 30, 1157, 1995.

167. Farooq, S., Bailey, E., Farmer, P.B., Jukes, R., Lamb, J.H., Hernandez, H., Sram, R., and Topinka, J., Determination of cis-thymine glycol in DNA by gas chromatography mass spectrometry with selected ion recording and multiple reaction monitoring, *J. Chromatogr. B* 702, 49, 1997.

168. LaFrancois, C.J., Yu, K., and Sowers, L.C., Quantification of 5-(hydroxymethyl)uracil in DNA by gas chromatography/mass spectrometry: Problems and solutions, *Chem. Res. Toxicol.* 11, 786, 1998.

169. D'Ham, C., Ravanat, J.L., and Cadet, J., Gas chromatography mass spectrometry with high-performance liquid chromatography prepurification for monitoring the endonuclease III mediated excision of 5-hydroxy-5,6-dihydrothymine and 5,6-dihydrothymine from gamma-irradiated DNA, *J. Chromatogr. B* 710, 67, 1998.

170. Ranasinghe, A., Scheller, N., Wu, K.Y., Upton, P.B., and Swenberg, J.A., Application of gas chromatography electron capture negative chemical ionization high-resolution mass spectrometry for analysis of DNA and protein adducts, *Chem. Res. Toxicol.* 11, 520, 1998.

171. Wu, K.Y., Scheller, N., Ranasinghe, A., Yen, T.Y., Sangaiah, R., Giese, R., and Swenberg, J.A., A gas chromatography/electron capture/negative chemical ionization high-resolution mass spectrometry method for analysis of endogenous and exogenous N7-(2-hydroxyethyl)guanine in rodents and its potential for human biological monitoring, *Chem. Res. Toxicol.* 12, 722, 1999.

172. Eide, I., Zhao, C.Y., Kumar, R., Hemminki, K., Wu, K.Y., and Swenberg, J.A., Comparison of P-32-postlabeling and high-resolution GC/MS in quantifying N7-(2-hydroxyethyl)guanine adducts, *Chem. Res. Toxicol.* 12, 979, 1999.

173. Ham, A.J.L., Ranasinghe, A., Morinello, E.J., Nakamura, J., Upton, P.B., Johnson, F., and Swenberg, J.A., Immunoaffinity/gas chromatography/high-resolution mass spectrometry method for the detection of *N*-2,3-ethenoguanine, *Chem. Res. Toxicol.* 12, 1240, 1999.

174. Morinello, E.J., Ham, A.J.L., Ranasinghe, A., Sangaiah, R., and Swenberg, J.A., Simultaneous quantitation of *N*-2,3-ethenoguanine and 1,*N*-2-ethenoguanine with an immunoaffinity/gas chromatography/high-resolution mass spectrometry assay, *Chem. Res. Toxicol.* 14, 327, 2001.

175. Park, M. and Loeppky, R.N., *In vitro* DNA deamination by alpha-nitrosaminoaldehydes determined by GC/MS-SIM quantitation, *Chem. Res. Toxicol.* 13, 72, 2000.

176. Chen, H.J.C., Lin, T.C., Hong, C.L., and Chiang, L.C., Analysis of 3,*N*-4-ethenocytosine in DNA and in human urine by isotope dilution gas chromatography/negative ion chemical ionization/mass spectrometry, *Chem. Res. Toxicol.* 14, 1612, 2001.

177. Banoub, J., Combden, S., Miller-Banoub, J., Sheppard, G., and Hodder, H., Structural characterization of intact covalently linked DNA adducts by electrospray mass spectrometry, *Nucleos. Nucleot.* 18, 2751, 1999.

178. Deforce, D.L., Ryniers, F.P.K., Van den Eeckhout, E.G., and Lemière, F., Esmans, E.L., Analysis of DNA adducts in DNA hydrolysates by capillary zone electrophoresis and capillary zone electrophoresis electrospray mass spectrometry, *Anal. Chem.* 68, 3575, 1996.

179. Vanhoutte, K., Joos, P., Lemière, F., Van Dongen, W., and Esmans, E.L., Comparision of conventional capillary electrospray liquid chromatography/mass spectrometry and liquid chromatography mass spectrometry for the detection of 2′-deoxynucleoside-bisphenol a diglycidyl, *J. Mass Spectrom. Rapid Commun. Mass Spectrom.* Special, 143, 1995.
180. Lemière, F., Vanhoutte, K., Jonckers, T., Marek, R., Esmans, E.L., Claeys, M., Van den Eeckhout, E., and Van Onckelen, H., Differentiation between isomeric phenylglycidyl ether adducts of 2′-deoxyguanosine and 2′-deoxyguanosine-5′-phosphate using liquid chromatography/electrospray tandem mass spectrometry, *J. Mass Spectrom.* 34, 820, 1999.
181. Vanhoutte, K., Van Dongen, W., Hoes, I., Lemière, F., Esmans, E.L., Van Onckelen, H., Van den Eeckhout, E., van Soest, R.E.J., and Hudson, A.J., development of a nanoscale liquid chromatography/electrospray mass spectrometry methodology for the detection and identification of DNA adducts, *Anal. Chem.* 69, 3161, 1997.
182. Ravanat, J.L., Duretz, B., Guiller, A., Douki, T., and Cadet, J., Isotope dilution high-performance liquid chromatography–electrospray tandem mass spectrometry assay for the measurement of 8-oxo-7,8-dihydro-2′-deoxyguanosine in biological samples, *J. Chromatogr. B-Anal. Technol. Biomed. Life Sci.* 715, 349, 1998.
183. Ravanat, J.L., Remaud, G., and Cadet, J., Measurement of the main photooxidation products of 2′-deoxyguanosine using chromatographic methods coupled to mass spectrometry, *Arch. Biochem. Biophys.* 374, 118, 2000.
184. Frelon, S., Douki, T., Ravanat, J.L., Pouget, J.P., Tornabene, C., and Cadet, J., High-performance liquid chromatography-tandem mass spectrometry measurement of radiation-induced base damage to isolated and cellular DNA, *Chem. Res. Toxicol.* 13, 1002, 2000.
185. Chaudhary, A.K., Reddy, G.R., Blair, I.A., and Marnett, L.J., Characterization of an N6-oxopropenyl-2′-deoxyadenosine adduct in malondiadehyde-modified DNA using liquid chromatography/electrospray tandem mass spectrometry, *Carcinogenesis* 17, 1167, 1996.
186. Leclercq, L., Laurent, C., and De Pauw, E., High-performance liquid chromatogaphy/electrospray mass spectrometry for the analysis of modified bases in DNA: 7-(2-hydroxyethyl)guanine, the major ethylene oxide-DNA adduct, *Anal. Chem.* 69, 1952, 1997.
187. RiosBlanco, M.N., Plna, K., Faller, T., Kessler, W., Hakansson, K., Kreuzer, P.E., Ranasinghe, A., Filser, J.G., Segerback, D., and Swenberg, J.A., Propylene oxide: Mutagenesis, carcinogenesis and molecular dose, *Mutat. Res.-Fundam. Mol. Mech. Mutagen.* 380, 179, 1997.
188. Siethoff, C., Feldmann, I., Jakubowski, N., and Linscheid, M., Quantitative determination of DNA adducts using liquid chromatography/electrospray ionization massspectrometry and liquid chrmomatography/high resolution inductively coupled plasma mass spectrometry, *J. Mass Spectrom.* 34, 412, 1999.
189. Edler, M., Jakubowski, N., and Linscheid, M., Styrene oxide DNA adducts: Quantitative determination using P-31 monitoring, *Anal. Bioanal. Chem.* 381, 205, 2005.
190. Turesky, R.J., Formation and biochemistry of carcinogenic heterocyclic aromatic amines in cooked meats, *Toxicol. Lett.* 168, 219, 2007.
191. Gangl, E.T., Turesky, R.J., and Vouros, P., Detection of *in vivo* formed DNA adducts at the part-per-billion level by capillary liquid chromatography/microelectrospray mass spectrometry, *Anal. Chem.* 73, 2397, 2001.
192. Turesky, R.J. and Vouros, P., Formation and analysis of heterocyclic aromatic amine-DNA adducts *in vitro* and *in vivo*, *J. Chromatogr. B-Anal. Technol. Biomed. Life Sci.* 802, 155, 2004.
193. Soglia, J.R., Turesky, R.J., Pachler, A., and Vouros, P., Quantification of the heterocyclic aromatic amine DNA adduct N-(deoxyguanosin-8-yl)-2-amino-3-methylimidazo[4,5-f]quinoline in livers of rats using capillary liquid chromatography/microelectrospray ma, *Anal. Chem.* 73, 2819, 2001.
194. Crosbie, S.J., Murray, S., Boobis, A.R., and Gooderham, N.J., Mass spectrometric detection and measurement of N-2-(2′-deoxyguanosin-8-yl)PhIP adducts in DNA, *J. Chromatogr. B* 744, 55, 2000.
195. Walton, M., Egner, P., Scholl, P.F., Walker, J., Kensler, T.W., and Groopman, J.D., Liquid chromatography electrospray-mass spectrometry of urinary aflatoxin biomarkers: Characterization and application to dosimetry and chemoprevention in rats, *Chem. Res. Toxicol.* 14, 919, 2001.
196. Van Aerden, C., Debrauwer, L., Tabet, J.C., and Paris, A., Analysis of nucleoside-estrogen adducts by LC-ESI-MS-MS, *Analyst* 123, 2677, 1998.
197. Convert, O., Van Aerden, C., Debrauwer, L., Rathahao, E., Molines, H., Fournier, F., Tabet, J.C., and Paris, A., Reactions of estradiol-2,3-quinone with deoxyribonucleosides: Possible insights in the reactivity of estrogen quinones with DNA, *Chem. Res. Toxicol.* 15, 754, 2002.

198. Lu, F., Zahid, M., Saeed, M., Cavalieri, E.L., and Rogan, E.G., Estrogen metabolism and formation of estrogen-DNA adducts in estradiol-treated MCF-10F cells: The effects of 2,3,7,8-tetrachlorodibenzo-p-dioxin induction and catechol-O-methyltransferase inhibition, *J. Steroid Biochem. Mol. Biol.* 105, 150, 2007.

199. Markushin, Y., Zhong, W., Cavalieri, E.L., Rogan, E.G., and Small, G.J., Yeung, E.S., Jankowiak, R., Spectral characterization of catechol estrogen quinone (CEQ)-derived DNA adducts and their identification in human breast tissue extract, *Chem. Res. Toxicol.* 16, 1107, 2003.

200. Meczes, E.L., Azim-Araghi, A., Ottley, C.J., Pearson, D.G., and Tilby, M.J., Specific adducts recognised by a monoclonal antibody against cisplatin-modified DNA, *Biochem. Pharm.* 70, 1717, 2005.

201. Sar, D.G., Montes-Bayon, M., Gonzalez, E.B., and Sanz-Medel, A., Speciation studies of cis-platin adducts with DNA nucleotides via elemental specific detection (P and Pt) using liquid chromatography-inductively coupled plasma-mass spectrometry and structural characterization by electrospray mass spectrometry, *J. Anal. Atomic Spectrom.* 21, 861, 2006.

202. Umemoto, A., Monden, Y., Suwa, M., Kanno, Y., Suzuki, M., Lin, C.X., Ueyama, Y., Momen, M.A., Ravindernath, A., Shibutani, S., and Komaki, K., Identification of hepatic tamoxifen-DNA adducts in mice: Alpha-(*N*-2-deoxyguanosinyl)tamoxifen and alpha-(*N*-2-deoxyguanosinyl)tamoxifen *N*-oxide, *Carcinogenesis* 21, 1737, 2000.

203. da Costa, G.G., Marques, M.M., Beland, F.A., Freeman, J.P., Churchwell, M.I., and Doerge, D.R., Quantification of tamoxifen DNA adducts using on-line sample preparation and HPLC-electrospray ionization tandem mass spectrometry, *Chem. Res. Toxicol.* 16, 357, 2003.

204. Beland, F.A., Churchwell, M.I., Doerge, D.R., Parkin, D.R., Malejka-Giganti, D., Hewer, A., Phillips, D.H., Carmichael, P.L., da Costa, G.G., and Marques, M.M., Electrospray ionization-tandem mass spectrometry and P-32-postlabeling analyses of tamoxifen-DNA adducts in humans, *J. Natl. Cancer Inst.* 96, 1099, 2004.

205. Brown, K., Tompkins, E.M., Boocock, D.J., Martin, E.A., Farmer, P.B., Turteltaub, K.W., Ubick, E., Hemingway, D., Horner-Glister, E., and White, I.N.H., Tamoxifen forms DNA adducts in human colon after administration of a single [C-14]-labeled therapeutic dose, *Cancer Res.* 67, 6995, 2007.

206. Xiong, W.N., Glick, J., Lin, Y.Q., and Vouros, P., Separation and sequencing of isomeric oligonucleotide adducts using monolithic columns by ion-pair reversed-phase nano-HPLC coupled to ion trap mass spectrometry, *Anal. Chem.* 79, 5312, 2007.

207. Tsujikawa, L., Weinfield, M., and Reha-Krantz, L.J., Differences in replication of a DNA template containing an ethyl phosphotriester by T4 DNA polymerase and Escherichia coli DNA polymerase I, *Nucleic Acids Res.* 31, 4965, 2003.

208. Lawley, P.D., Reaction of *N*-methyl-*N*-nitrosourea (Mnua) with P-32-labeled DNA—evidence for formation of phosphotriesters, *Chemico-Biol. Interactions* 7, 127, 1973.

209. Haglund, J., Ehrenberg, L., and Tornqvist, M., Studies of transalkylation of phosphotriesters in DNA: Reaction conditions and requirements on nucleophiles for determination of DNA adducts, *Chemico-Biol. Interact.* 1997 108, 119.

210. Yates, J.M., Fennell, T.R., Turner, M.J., Recio, L., and Sumner, S.C.J., Characterization of phosphodiester adducts produced by the reaction of cyanoethylene oxide with nucleotides, *Carcinogenesis* 15, 277, 1994.

211. Skibo, E.B. and Schulz, W.G., Pyrrolo[1,2-A]benzimidazole-based aziridinyl quinones—a new class of DNA-cleaving agent exhibiting G-base and A-base specificity, *J. Med. Chem.* 36, 3050, 1993.

212. Gaskell, M., Kaur, B., Farmer, P.B., and Singh, R., Detection of phosphodiester adducts formed by the reaction of benzo[*a*]pyrene diol epoxide with 2'-deoxynucleotides using collision-induced dissociation electrospray ionization tandem mass spectrometry, *Nucleic Acids Res.* 35, 5014, 2007.

213. Haglund, J., Silvari, V., Esmans, E., and Tornqvist, M., Cobalamin as an analytical tool for analysis of oxirane metabolites of 1,3-butadiene: Development and validation of the method, *J. Chromatogr A* 1119, 246, 2006.

214. Kanaly, R.A., Hanaoka, T., Sugimura, H., Toda, H., Matsui, S., and Matsuda, T., Development of the adductome approach to detect DNA damage in humans, *Antioxidants Redox Signaling* 8, 993, 2006.

7 Sequence Distribution of Nucleobase Adducts Studied by Isotope Labeling of DNA–Mass Spectrometry

Natalia Y. Tretyakova

CONTENTS

7.1 INTRODUCTION

The formation of covalent nucleobase lesions (DNA adducts) is the first critical step of chemical carcinogenesis.[1] If not repaired, DNA adducts can cause mispairing during DNA replication and be converted to heritable mutations. Accumulation of genetic changes in tumor suppressor genes and protooncogenes is a hallmark of cancer.[2]

The ability of DNA adducts to induce mutagenesis and carcinogenesis is dependent on their chemical structure, stability, and their ability to be recognized by specialized DNA repair proteins.[3,4] Local DNA sequence context plays a major role in mediating the rate of nucleobase adduct repair and their mispairing potency.[5] Furthermore, the exact location of a DNA adduct within the gene determines whether the resulting mutation affects the structure and function of the gene product.

Previous investigations have largely relied on alkali- or repair endonuclease-induced DNA strand cleavage to map the formation of DNA lesions along gene sequences following chemical exposure.[6,7] These indirect approaches, for example, ligation mediated polymerase chain reaction (LMPCR),[8] do not analyze DNA adducts *per se*, but rather detect DNA nicks produced at the sites of nucleobase modifications by the action of DNA repair enzymes or hot piperidine. The adduction sites are identified from the size of radioactively end-labeled DNA fragments analyzed on a denaturing polyacrylamide gel electrophoresis (PAGE) gel. Although simple and convenient, these approaches require the use of radioisotopes and suffer from several important limitations. Firstly, gel electrophoresis-based methods cannot determine the stereochemistry and the structural identity of nucleobase lesions giving rise to DNA strand breaks. Therefore, the usefulness of LMPCR and related techniques for mapping specific types of DNA damage is limited. Secondly, these approaches are limited to alkali-labile and endonuclease-cleavable lesions. For example, some of the most mutagenic DNA modifications, for example, O^6-alkyl-G, cannot be converted to single-strand breaks and therefore cannot be detected by gel electrophoresis. Finally, these indirect methods are prone to errors, since DNA sequence context may affect the rates of adduct repair[5] and/or base-catalyzed hydrolysis required for strand cleavage.

Since the beginning of the field of chemical carcinogenesis, mass spectrometry (MS) has played a leading role in identifying and quantifying chemical modifications of DNA by carcinogens and their metabolites (DNA adducts). A combination of high sensitivity, selectivity, and rich structural information provided by MS experiments provides a powerful tool for quantitative and qualitative analyses of DNA modifications.[9] Notably, most of the previous studies have focused on determining the structures and total DNA adduct numbers in a given sample, rather than analyzing their distribution along DNA sequences. Because a typical MS experiment involves digesting DNA to the corresponding deoxynucleosides or free bases, any sequence information is lost. However, the exact locations of DNA lesions within a gene sequence can dictate their biological fate by influencing the repair rates, efficiency of DNA polymerase bypass, and their ability to induce mutations at critical sites within the genome. Tandem mass spectrometry (MS/MS) can be used to determine the sequence of DNA oligomers and identify the positions of structurally modified bases.[9] However, this methodology is limited to relatively short sequences (<15), requires large amounts of pure material, and is not suitable for analyzing mixtures of differentially modified oligonucleotides resulting from exposure to a DNA-modifying agent. Similar limitations apply to an exonuclease ladder sequencing approach, in which DNA ladders produced by partial exonuclease digestion are analyzed by matrix assisted laser desorption ionization (MALDI) or electrospray ionization (ESI)-MS.[10,11]

To overcome the limitations of these earlier methodologies, we have developed an approach based on stable isotope labeling of DNA, carcinogen exposure, enzymatic or chemical digestion, and MS analysis of the resulting isotopomeric DNA nucleosides. This methodology, which has been termed ILD-MS (isotope labeling of DNA-mass spectrometry), provides accurate information about the distributions of specific nucleobase adducts along a DNA sequence.

7.2 STABLE ISOTOPE LABELING OF DNA: ISOTOPE RATIO MASS SPECTROMETRY (ILD-MS) APPROACH

The ILD-MS (mass tagging) approach involves placing ^{15}N- or ^{13}C-labeled nucleobases at specific positions within DNA oligodeoxynucleotides representing gene sequences of interest, followed by carcinogen exposure, enzymatic or acid hydrolysis of alkylated DNA, and mass spectral analysis of the resulting nucleoside adducts (Scheme 7.1). DNA adducts formed at the ^{15}N- or ^{13}C-labeled base can be distinguished from lesions originating at other sites by their molecular weight, which is increased due to the presence of ^{15}N and/or ^{13}C atoms in the molecule. The extent of reaction at the labeled nucleobase is directly calculated from the peak areas (A) in the extracted ion chromatograms corresponding to unlabeled and stable isotope-labeled adducts (Scheme 7.1): percent reaction at $X = A_{15_{N-X}}/(A_X + A_{15_{N-X}})$. By preparing a series of oligomers of the same sequence, but with different label positions (e.g., Table 7.1), the extent of the adduct formation at each site within the sequence

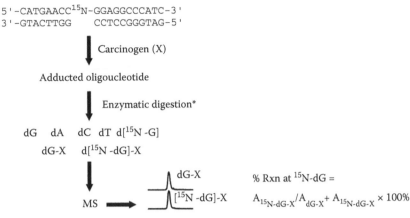

5'-CATGAACC^{15}N-GGAGGCCCATC-3'
3'-GTACTTGG   CCTCCGGGTAG-5'

Carcinogen (X)

Adducted oligoucleotide

Enzymatic digestion*

dG dA dC dT d[^{15}N -G]

dG-X   d[^{15}N -dG]-X

MS

dG-X
[^{15}N -dG]-X

% Rxn at ^{15}N-dG =

$A_{^{15}N\text{-}dG\text{-}X}/A_{dG\text{-}X} + A_{^{15}N\text{-}dG\text{-}X} \times 100\%$

* chemical digestion to free bases can be used

SCHEME 7.1 Strategy for quantitation of dG adducts at specific sites within DNA by stable ILD-MS approach.

can be quantified. Since the stable isotope-labeled nucleobase is chemically identical to the unlabeled base, its reactivity will represent the extent of reaction in unlabeled DNA. Because of the limited duplex stability at the ends of the DNA molecule, isotope tags are positioned at least three nucleotides away from the ends of the duplex.

The presence of naturally occurring isotopes (e.g., ^{13}C) can interfere with mass spectral analysis and requires that the isotopically tagged nucleobase differs from the normal base by at least three mass units.[12] In our previous studies of guanine adducts, we have employed nucleobase labels containing between three and five labels, for example, $^{15}N_3$; $^{15}N_3,^{13}C_1$; and $^{15}N_5$.[10,12–15] Purine nucleobases contain five nitrogen atoms, all of which can be replaced with ^{15}N. Pyrimidine nucleobases contain only 2–3 nitrogen atoms, and ^{13}C labeling may be required to achieve the required mass shift. One must also take into account the potential loss of nitrogen labels during adduct formation, for example, via deamination or oxidative degradation.[16]

The molecular weight of the nucleoside adducts produced at the isotopically labeled nucleoside are increased by a value corresponding to the number of ^{13}C and ^{15}N atoms in the molecule. The same mass shift is observed for the isotopically labeled nucleobases. For example, high performance liquid

TABLE 7.1

DNA Sequences Used to Investigate the Distribution of Guanine Adducts Along a DNA Duplex Representing _K-ras_ Codons 10–15 Using the ILD-MS Approach

ID	Sequence	Calculated Molecular Weight	Observed Molecular Weight
[$^{15}N_3,^{13}C_1$]-_K-ras_-G3	GGA[$^{15}N_3,^{13}C_1$G]CTGGTGGCGTAGGC	5640.7	5639.7
[$^{15}N_3,^{13}C_1$]-_K-ras_-G4	GGAGCT[$^{15}N_3,^{13}C_1$–G]GTGGCGTAGGC	5640.7	5640.0
[$^{15}N_3,^{13}C_1$]-_K-ras_-G5	GGAGCTG[$^{15}N_3,^{13}C_1$–G]TGGCGTAGGC	5640.7	5639.9
[$^{15}N_3,^{13}C_1$]-_K-ras_-G6	GGAGCTGGT[$^{15}N_3,^{13}C_1$–G]GCGTAGGC	5640.7	5640.0
[$^{15}N_3,^{13}C_1$]-_K-ras_-G7	GGAGCTGGTG[$^{15}N_3,^{13}C_1$–G]CGTAGGC	5640.7	5640.0
[$^{15}N_3,^{13}C_1$]-_K-ras_-G8	GGAGCTGGTGGC[$^{15}N_3,^{13}C_1$–G]TAGGC	5640.7	5640.0
[$^{15}N_3,^{13}C_1$]-_K-ras_-G9	GGAGCTGGTGGCGTA[$^{15}N_3,^{13}C_1$–G]GC	5640.7	5640.0
(−)-_K-ras_	GCCTACGCCACCAGCTCC	5365.5	5365.0

chromatography (HPLC)-ESI-MS/MS quantitation of the benzo[a]pyrene diolepoxide-derived N^2-BPDE-dG adducts is based on the molecular ions of the adducted nucleoside ($m/z = 570.2$, M + H) undergoing a neutral loss of deoxyribose ($M = 116$) under collision-induced dissociation conditions, leading to a predominant fragment at $m/z = 453.8$ (Figure 7.1b). $^{15}N_3$-labeled N^2-BPDE-dG gives rise to analogous molecular and fragment ions whose m/z values are increased by 3, which corresponds to the number of ^{15}N atoms in the molecule (Figure 7.1c). The extent of adduct formation at the isotopically labeled nucleobase is calculated directly from the ratios of the areas under the HPLC-ESI-MS/MS peaks corresponding to $^{15}N_3$-labeled and unlabeled adducts, respectively (Figure 7.1).

The ILD-MS approach requires that MS detection sensitivity is sufficient to accurately quantify DNA adducts formed at a specific nucleobase within the selected oligonucleotide sequence. For example, to study the formation of N^2-BPDE-dG lesions at a single G within a DNA sequence containing 20 guanines, the sensitivity must be 20-fold greater than for a "normal" HPLC-ESI-MS/MS assay measuring the sum of all N^2-BPDE-dG adducts. We typically use capillary HPLC-ESI-MS/MS at flow rates of 8–15 µL/min. The use of capillary HPLC enables at least a 10-fold increase in sensitivity relative to typical 1 mm columns operating at flow rates of 60–70 µL/min,[17] resulting in detection limits in the low fmol region (20 fmol for N^2-BPDE-dG at S/N = 3). Other laboratories have employed GC-MS following acid hydrolysis of DNA and silylation of the resulting free bases,[16] which achieves similar sensitivity to that of capillary HPLC-ESI-MS/MS. An even greater sensitivity (on attomolar levels) can be achieved with nanoflow HPLC-nanospray MS.

Oligonucleotide purity is an important requirement for ILD-MS analyses. Any contaminating DNA fragments can generate nucleobase adducts upon carcinogen treatment, interfering with accurate quantification. Our typical oligonucleotide isolation strategy is to perform reversed-phase HPLC purification, followed by a purity check on a different HPLC column (e.g., anion exchange) and, if necessary, further repurification using a different buffer system. The molecular weight and purity of HPLC-purified DNA is established by capillary HPLC-ESI-MS or MALDI-TOF-MS using 3-hydroxypicolinic acid as the MALDI matrix.

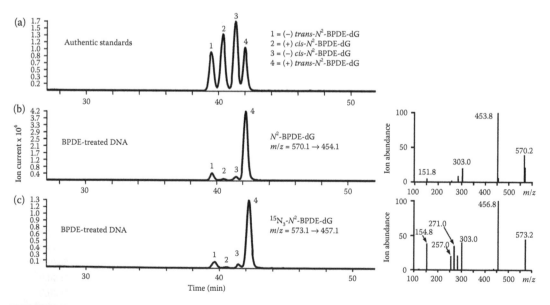

FIGURE 7.1 Capillary HPLC-ESI-MS/MS analysis of N^2-BPDE-dG diastereomers originating from unlabeled (b) and [$^{15}N_3$]-labeled guanines (c) in comparison with authentic standards (a). Inset: MS/MS spectra of unlabeled (b) and [$^{15}N_3$]-labeled N^2-BPDE-dG (c). (From Matter, B. et al., *Chem. Res. Toxicol.* 17, 735, 2004. Copyright 2004 American Chemical Society. With permission.)

In the beginning of each investigation, concentration dependence curves are obtained with unlabeled DNA to identify exposure conditions that yield a sufficient number of adducts to enable the detection of the lesions originating from the labeled nucleobase.[15,18,19] The overall level of modification should be less than 5% to maintain sequence specificity. Under these conditions, no more than one adduct per DNA molecule is formed ("single-hit" conditions), and the sequence selectivity is maintained.

Quantitative enzymatic digestion of carcinogen-modified DNA to 2'-deoxyribonucleosides is necessary to enable the quantitative analysis of DNA adducts formed at specific bases within DNA sequences. Some DNA adducts, for example, N^2-BPDE-dG and O^6-POB-dG, can block phosphodiesterase enzymes, which may lead to incomplete digestion of modified DNA and possible errors in ILD-MS analysis.[10,20,21] However, incubation times, enzyme amounts, and buffer pH/composition can be optimized to achieve quantitative hydrolysis of modified oligomers. We found that using Nuclease P1 allows for a rapid and efficient digestion of most structurally modified oligomers, without an "end bias" that is sometimes observed for exonuclease enzymes.

7.2.1 Applications of Stable Isotope Labeling: HPLC-ESI-MS/MS (ILD-MS) Approach

The ILD-MS approach is universal, for example, it can be applied to analyze the distribution of nucleobase adducts of any type or structure. As described below, our laboratory has employed a stable isotope labeling—HPLC-ESI-MS/MS approach to map the formation of DNA adducts induced by a variety of electrophilic species, including acetaldehyde, tobacco-specific nitrosamines, benzo[a]pyrene diolepoxide, reactive oxygen species, and butadiene diepoxide. More recently, other groups have employed a similar approach to examine hypochlorous acid-induced damage to cytosine residues in CG dinucleotides.[16]

7.2.1.1 DNA Damage by Benzo[a]pyrene diol Epoxide

Polycyclic aromatic hydrocarbons (PAHs) are produced by the incomplete combustion of organic materials during cigarette burning. The best studied carcinogenic PAH in cigarette smoke, benzo[a] pyrene (B[a]P, Scheme 7.2), is a potent systemic and local carcinogen known to induce skin, lung, and stomach tumors in laboratory animals.[22] B[a]P is a suspect human carcinogen based on its potent tumorigenic effects and the similarity of B[a]P-DNA damage in animal studies and in human cells.[22] B[a]P requires metabolic activation to its ultimate carcinogenic compound, *anti-7r,8t*-dihydroxy-*c*9,10-epoxy-7,8,9,10-tetrahydrobenzo[a]pyrene [(+) *anti*-BPDE, Scheme 7.2].[23] Following the intercalation between DNA base pairs, the epoxide moiety of *anti*-BPDE attacks the exocyclic amino group of guanine to produce diastereomeric N^2-guanine lesions, (+) *trans*, (−) *trans*, (+) *cis*, and (−) *cis-anti*-7R,8S,9S-trihydroxy-10S-(N^2-deoxyguanosyl)-7,8,9,10-tetrahydrobenzo[a]pyrene.[24] The predominant nucleoside adduct, (+)-*trans*-N^2-BPDE-dG (Scheme 7.2),[24] is a promutagenic lesion that accumulates in target tissues of B[a]P-treated animals[25] and is found in tissues of smokers and in occupationally exposed humans.[26–28]

Bulky N^2-BPDE-dG lesions represent strong blocks for major replicative DNA polymerases, but can be bypassed by specialized lesion bypass polymerases.[29,30] Depending on the adduct stereochemistry, sequence context, and polymerase identity, translesion synthesis can be error free or can lead to G → T transversions and G → A transition mutations.[31–34]

We have analyzed the formation of N^2-BPDE-dG lesions within double-stranded oligodeoxynucleotides representing frequently mutated regions of the *K-ras* protooncogene and the *p53* tumor suppressor gene:

    I. 5'-G$_1$G$_2$AG$_3$CTG$_4$G$_5$TG$_6$G$_7$C G$_8$TAG$_9$G$_{10}$C-3' (**G$_4$G$_5$T** = *K-ras* codon 12).

    II. 5'CC^{Me}CG$_1$G$_2$CACC^{Me}CG$_3$^{Me}CG$_4$TC^{Me} CG$_5$^{Me}CG$_6$-3' (**G$_4$TC** = *p53* codon 157, ^{Me}**CG$_5$**^{Me}**C** = codon 158).

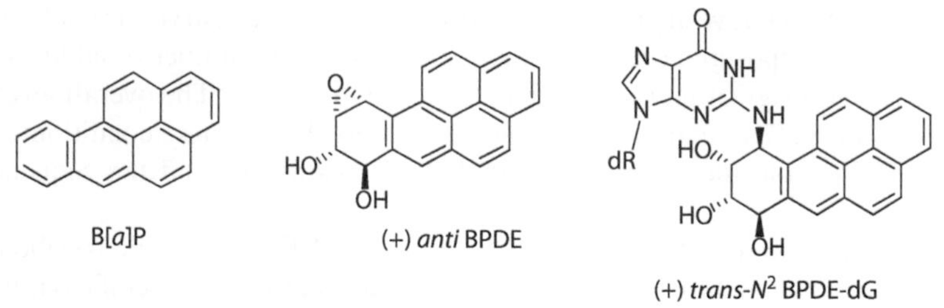

B[a]P   (+) anti BPDE

(+) trans-N^2 BPDE-dG

SCHEME 7.2 Chemical structures of benzo[*a*]pyrene (B[*a*]P), *anti*-7*r*,8*t*-dihydroxy-*c*9,10-epoxy-7,8,9,10-tetrahydrobenzo[*a*]pyrene ((+) *anti*-BPDE), and its dominant guanine adduct (+)-*trans*-N^2-BPDE-dG.

III. 5′-ATG₁G₂G₃ᴹᵉCG₄G₅CATG₆AACᴹᵉCG₇G₈AG₉G₁₀CCCA(G₄GC = *p53*codon245,ᴹᵉCG₇G₈ = codon 248).

IV. 5′-G₁CTTTG₂AG₃G₄TG₅ᴹᵉC G₆TG₇ T TT G₈TG₉-3′ (ᴹᵉCG₆T = *p53* codon 273).

For each sequence, a series of strands was prepared containing a [1,7,NH₂–¹⁵N₃] or [1,7,NH₂–¹⁵N₃,2-¹³C₁] isotope label at one of the highlighted guanines. This label served as an isotope "tag" to enable the quantification of guanine lesions originating from that specific position (Scheme 7.1). 5-Methylcytosine (ᴹᵉC) was incorporated at all physiologically methylated sites. Each isotopically labeled oligomer was carefully purified, structurally characterized, and annealed to the complementary unlabeled strand. Following the treatment with (±) *anti*-BPDE (10 μM), the adducted DNA was enzymatically hydrolyzed to 2′-deoxynucleosides in the presence of DNAse I, phosphodiesterase I, and alkaline phosphatase. Capillary HPLC-ESI⁺-MS/MS was used to establish the relative amounts of N^2-BPDE-dG lesions originating from the stable isotope labeled dG and from unlabeled dGs elsewhere in the sequence. For example, quantitative analysis of N^2-BPDE-dG and ¹⁵N₃-N^2-BPDE-dG was based on selected reaction monitoring (SRM) of mass transitions *m/z* 570.2 → 454.1 and *m/z* 573.2 → 457.1, respectively (Figure 7.1), and percent reaction at the labeled site was established from HPLC-ESI⁺-MS/MS areas as shown in Scheme 7.1. A triple stage quadriplole (TSQ) Quantum mass spectrometer (ThermoQuest, Palo Alto, CA) was interfaced with an Agilent 1100 capillary HPLC operated at a flow rate of 15 μL/min. All four N^2-BPDE-dG diastereomers, (+) *trans*, (–) *trans*, (+) *cis*, and (–)-*cis*-N^2-BPDE-dG, were resolved and quantified separately (Figure 7.1).[13]

Because *K-ras*-derived duplex **I** contains a total of 13 guanines, a strictly random reaction with (±) *anti*-BPDE would result in 100%/13 = 7.7% of N^2-BPDE-dG adducts originating from each G. Our stable isotope labeling experiments have shown that the extent on N^2-BPDE-dG adduct formation was not uniform across the different guanine nucleobases. The greatest alkylation efficiency (16.9 %) was observed at the first guanine of *K-ras* codon 12 (**G**₄GT), a major hot spot for activating G → T transversions in smoking-induced lung adeno-carcinoma.[35,36] A smaller number of adducts (8–10%) was produced at **G**₅ and **G**₆, and only 3.5% of the lesions originated at **G**₃ (Figure 7.2). These results indicate that, in double-stranded DNA, N^2-BPDE-dG adduct yields are strongly affected by the local sequence context, leading to the preferential modification of the first G in *K-ras* codon 12.[12]

We next investigated the formation of N^2-BPDE-dG within DNA duplexes derived from the *p53* gene. We found that BPDE adducts were formed preferentially at the endogenously methylated CG dinucleotides, including those within *p53* codons 157, 158, 245, 248, and 273 (Figure 7.3).[13] Many of these sites coincide with known hot spots for G → T transversions detected in smoking-induced lung cancer.[37] Targeted formation of N^2-BPDE-dG at ᴹᵉCG dinucleotides within the *p53* gene (Figure 7.3) has been previously detected by endonuclease incision–ligation-mediated PCR assay[7,38] and is consistent with the prevalence of G → T transversions at these sites.

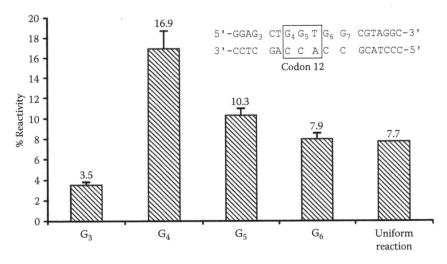

FIGURE 7.2   Distribution of N^2-BPDE-dG adducts along DNA duplex representing *K-ras* codons 10–15. (From Tretyakova, N. et al., *Biochemistry* 41, 9541, 2002. Copyright 2002 American Chemical Society. With permission.)

7.2.1.2   DNA Adducts of Tobacco-Specific Nitrosamine (NNK)

Tobacco-specific *N*-nitrosamines, for example, 4-(methylnitrosamino)-1-(3-pyridyl)-1-butanone (NNK, Scheme 7.3), are generated during the curing of tobacco from corresponding alkaloids.[39] NNK is a potent pulmonary carcinogen that induces lung tumors in laboratory animals, independent of the route of administration.[40] Metabolic activation of NNK yields methyl- and pyridyloxobutyldiazonium ions that can alkylate DNA to produce *N*7-methyldeoxyguanosine (*N*7-Me-dG), O^6-methyldeoxyguanosine (O^6-Me-dG), O^6-pyridyloxobutyl-deoxyguanosine (O^6-POB-dG), and other pyridyloxobutylated lesions (Scheme 7.3).[41–47] *N*7-Me-dG accounts for 70–90% of all DNA adducts following exposure to methylating agents, giving rise to potentially promutagenic abasic sites via spontaneous depurination ($t_{1/2} \approx 155$ h in *ds* DNA).[1,48–50] The O^6-Me-dG lesions, although minor (6% of total methylation), are considered critical for NNK mutagenesis based on their strong mispairing properties[5,51,52] and a correlation between the accumulation of O^6-Me-dG and the incidence of lung tumors in NNK-treated mice.[42]

The ILD-MS approach was employed to analyze the formation of *N*7 and O^6-guanine lesions induced by NNK metabolites, within synthetic DNA duplexes representing *K-ras* codons 10–15 and *p53* codons 153–159, 243–250, and 269–275. 5-Methylcytosine (^{Me}C) was incorporated in both strands of the physiologically methylated CG dinucleotides.[53] The DNA-reactive species of NNK, methyldiazohydroxide and pyridyloxobutyl diazohydroxide, were generated by the esterase-catalyzed hydrolysis of (acetoxylmethyl)methylnitrosamine (AMMN) and 4-(acetoxymethylnitrosamino)-1-(3-pyridyl)-1-butanone (NNKOAc), respectively. ILD-MS[19] was used to determine the relative yields of NNK adducts at each guanine base of interest.

[1,7,NH$_2$-^{15}N$_3$]-dG or [1,7,NH$_2$-^{15}N$_3$,2-^{13}C$_1$]-dG containing DNA duplexes were treated with 2 mM AMMN or 10 mM NNKOAc, followed by DNA precipitation with cold ethanol. The alkylated DNA was heated to release *N*7-pyridyloxobutylguanine (POBG) and *N*7-MeG, and partially depurinated DNA backbone was precipitated with ethanol and enzymatically digested to deoxynucleosides. Methylated DNA was hydrolyzed as described above for N^2-BPDE-dG adducts, while phosphodiesterase II was used for pyridyloxobutylated DNA.[54] This modification of the digestion method was necessary because O^6-POB-dG adducts were found to block phosphodiesterase I.[20] Quantitative analyses of O^6-Me-dG, O^6-POB-dG, and their isotopomers were performed using a Finnigan Quantum Discovery TSQ-MS interfaced with an Agilent 1100 capillary HPLC. We monitored the neutral loss of deoxyribose (116 mass units) from protonated molecules of each nucleoside

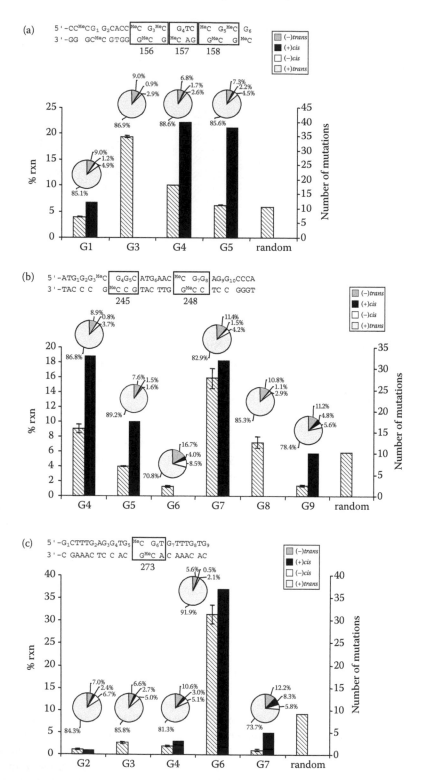

FIGURE 7.3 Relative formation of N^2-BPDE-dG adducts (striped bars) and the frequency of G to T transversions in human lung cancer (black bars) along DNA duplexes derived from *p53* exon 5 (a), *p53* exon 7 (b), and *p53* exon 8 (c). Bar graphs: Diastereomeric composition of N^2-BPDE-dG adducts generated at each site. (Adapted from Matter, B. et al., *Chem. Res. Toxicol.* 17, 731, 2004. Copyright 2004 American Chemical Society. With permission.)

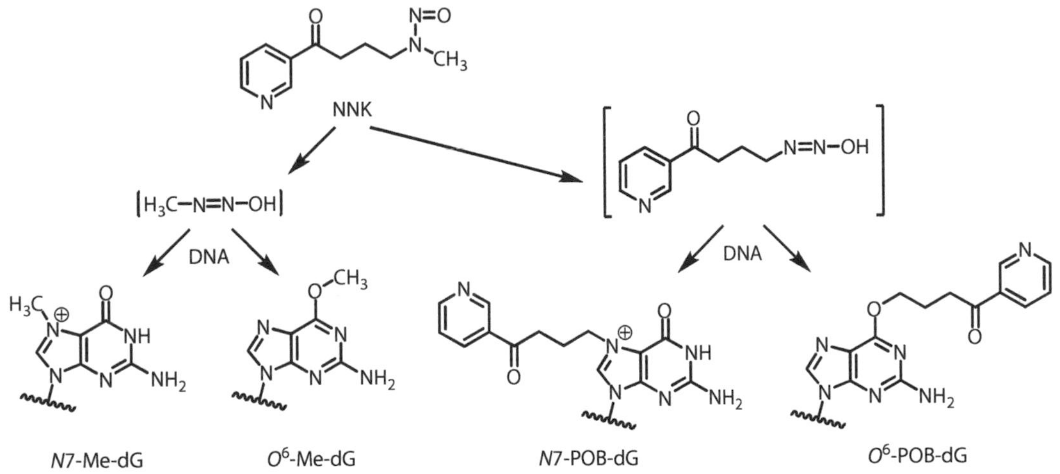

SCHEME 7.3   Formation of guanine adducts of the tobacco-specific nitrosamine, NNK.

adduct [m/z 282 → 166 and m/z 285 → 169 for O^6-Me-dG and its isotopomer (Figure 7.4a); m/z 415.4 → 148, 152 and m/z 419.4 → 148, 156 for O^6-POB-dG and its isotopomer, respectively (Figure 7.4b)].[19,54] N7-POBG and $^{15}N_3,^{13}C_1$ isotope analog were quantified by using the transitions m/z 299 → 148, 152, and 303.1 and m/z 299 → 148, 156, respectively. N7-MeG and $^{15}N_3,^{13}C_1$-N7-MeG were quantified on an Agilent Technologies ion-trap mass spectrometer operated in the full-scan mode. Quantitative analysis was performed from reconstructed ion chromatograms of m/z 166 and m/z 170 for N7-MeG and $^{15}N_3,^{13}C_1$-N7-MeG, respectively.[54]

Our experiments revealed that both O^6-Me-dG and O^6-POB-dG adducts formed preferentially at the second position of *K-ras* codon 12 (**GG**T) (Figure 7.5), supporting the involvement of NNK-induced O^6-guanine lesions in *K-ras* gene mutagenesis.[19] Both N7-Me-G and N7-POB-G lesions were overproduced at the 3′-guanine bases within polypurine runs, while the formation of O^6-Me-dG and O^6-POB-dG was specifically preferred at the 3′-guanine base of 5′-GG and 5′-GGG sequences (Figures 7.6 and 7.7), consistent with previous studies that observed a stimulating effect of 5′-flanking purine bases on methylation of guanine bases.[55] These results demonstrate that mapping alkali-labile N7-alkyl-G adducts by gel electrophoresis cannot be used to predict the distribution of promutagenic O^6-guanine lesions. In contrast with our results for N^2-BPDE-dG adducts (Figure 7.3), low O^6-Me-dG and O^6-POB-dG yields were detected at endogenously methylated CG dinucleotides, including the sites of increased mutational frequency (*p53* codons 157, 158, 245, 248, and 273) (Figures 7.6 and 7.7).[54]

7.2.1.3   DNA Damage by Acetaldehyde

Acetaldehyde (AA, Scheme 7.4) is among the most abundant carcinogens in tobacco smoke with concentrations in cigarette smoke 1000-fold greater than those of PAHs and tobacco-specific nitro-samines.[56] Like BPDE, AA modifies the N^2 position of guanine in DNA.[57] The resulting N^2-ethylidene-dG adducts can be reduced to irreversible N^2-ethyl-dG lesions (Scheme 7.4).[58] Vitamin C, glutathione, and other antioxidants may act as reducing agents *in vivo*, while NaBH$_3$CN is commonly used in the laboratory.[58] N^2-ethylidene-dG adducts are the predominant AA–DNA lesions detected in humans.[59–61] While the biological effects of N^2-ethylidene-dG have not been examined because of their limited stability, N^2-ethyl-dG lesions are capable of inducing G to C transversions, although human DNA polymerase η bypasses the adduct in an error-free manner.[62,63]

Because N^2-ethylidene-dG and N^2-ethyl-dG lesions are not recognized by the *uvr*ABC nuclease and are not cleavable by hot piperidine, their sequence locations cannot be readily determined by gel electrophoresis methods. We investigated the DNA sequence preferences for the formation of

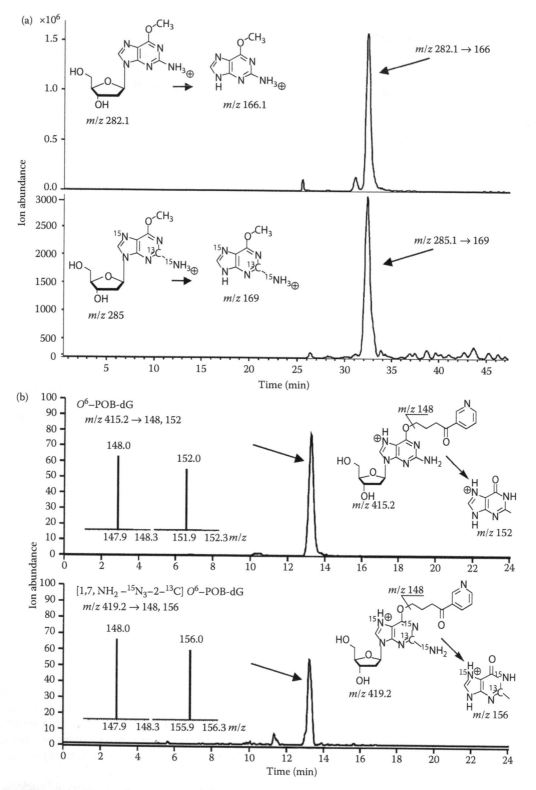

FIGURE 7.4 Capillary HPLC-ESI$^+$-MS/MS analysis of O^6-Me-dG (a) and O^6-POB-dG adducts (b) originating from unlabeled (top) and [^{15}N$_3$,^{13}C$_1$]-labeled guanines (bottom) following exposure of isotopically labeled duplexes to AMMN (A) or NNKOAc (B). (From Ziegel, R. et al., *Chem. Res. Toxicol.* 16, 545, 2003. Copyright 2004 American Chemical Society. With permission.)

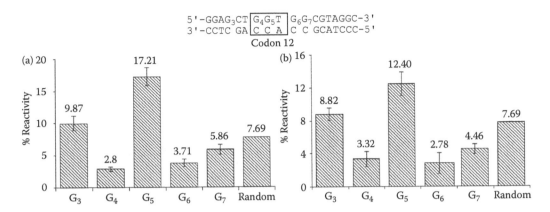

FIGURE 7.5 Relative formation of O^6-Me-dG (a) and O^6-POB-dG (b) at guanine residues within DNA duplex representing *K-ras* codons 10–15. Codon 12 = G_4G_5T, where G_5 is a mutational hot spot for G → A transitions. (Adapted from Ziegel, R., *Chem. Res. Toxicol.* 16, 545, 2004 and from Rajesh, M. et al., *Biochemistry* 44, 2203, 2005. Copyright American Chemical Society. With permission.)

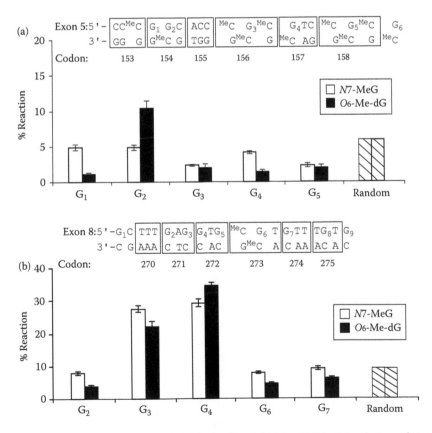

FIGURE 7.6 Relative formation of *N*7-MeG (white bars) and O^6-Me-dG (black bars) at guanine nucleobases within a DNA duplex derived from *p53* exon 5 (a) and *p53* exon 8 (b). Random reactivity values were calculated from the number of Gs in a duplex. (From Rajesh, M. et al., *Biochemistry* 44, 2202, 2005. Copyright 2003 American Chemical Society. With permission.)

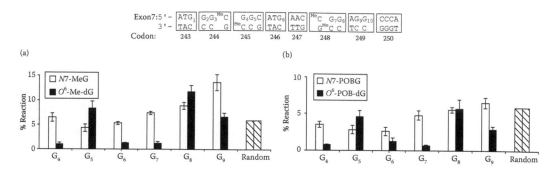

FIGURE 7.7 Relative formation of methylated (a) and pyridyloxobutylated dG adducts (b) along *p53* exon 7-derived DNA duplex. Random reactivity value was calculated from the total number of guanines in both strands. (From Rajesh, M. et al., *Biochemistry* 44, 2203, 2005. Copyright 2003 American Chemical Society. With permission.)

acetaldehyde–nucleobase adducts using the ILD-MS approach (Scheme 7.1). DNA duplexes representing major *K-ras* and *p53* lung cancer mutational hot spots and their surrounding sequences were employed. In each duplex, $^{15}N_3, ^{13}C_1$ stable isotope tags were sequentially introduced at one of the highlighted Gs:

I. 5'-$G_1G_2AG_3CTG_4G_5TG_6G_7C$ $G_8TAG_9G_{10}C$-3' (G_4G_5T = *K-ras* codon 12).

II. 5'CC$^{Me}CG_1G_2CACC^{Me}CG_3^{Me}CG_4TC^{Me}CG_5^{Me}CG_6$-3' ($G_4TC$ = *p53* codon 157, $^{Me}CG_5^{Me}C$ = codon 158).

$^{15}N_3, ^{13}C_1$-dG-containing DNA duplexes were treated with 4 mM acetaldehyde for 96 h, and the resulting N^2-ethylidene-dG lesions were stabilized by their reduction to N^2-ethyl-dG in the presence of NaBH$_3$CN,[58] followed by DNA precipitation with cold ethanol. Under these conditions, the total number of N^2-ethyl-dG adducts was 4 pmol/nmol DNA. N^2-ethyl-dG containing DNA was digested to free deoxynucleosides, and N^2-ethyl-dG and its $^{15}N_3, ^{13}C_1$ isotopomer were isolated by solid-phase extraction.[15]

Quantitative analysis of N^2-ethyl-dG and $^{15}N_3, ^{13}C_1$-N^2-ethyl-dG in DNA hydrolyzates was performed by capillary HPLC-ESI$^+$-MS/MS (Figure 7.8). A Thermo Scientific TSQ Quantum Ultramass spectrometer interfaced with an Agilent Technologies 1100 capillary HPLC was operated in the SRM mode. N^2-ethyl-dG and $^{15}N_3, ^{13}C_1$-N^2-ethyl-dG were quantified by monitoring the transitions m/z 296.0 → m/z 180.0 and m/z 300.0 → m/z 184.0, respectively, corresponding to a neutral loss of ribose from protonated molecules of the adducts (Figure 7.8).[15]

We found that the distribution of N^2-ethyl-dG adducts along duplexes **I** and **II** was only weakly effected by DNA sequence context (Figure 7.9).[15] This is in contrast with the sequence-dependent formation of N^2-BPDE-dG, NNK, and oxidative lesions along the same DNA duplexes (e.g., Figures 7.2 and 7.5). This apparent lack of sequence specificity for AA-mediated alkylation of DNA may be

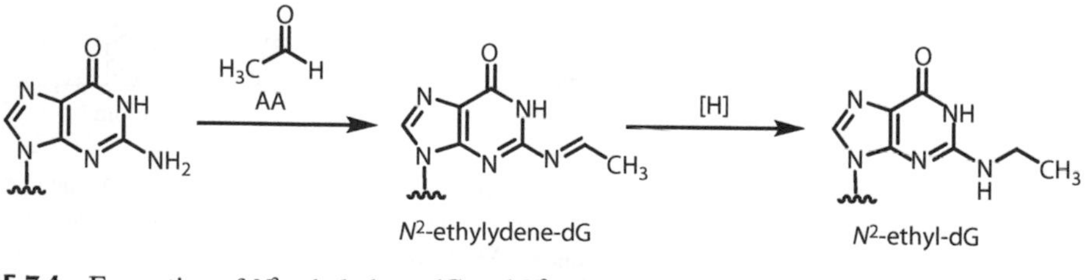

SCHEME 7.4 Formation of N^2-ethylydene-dG and N^2-ethyl-dG from guanine reactions with acetaldehyde.

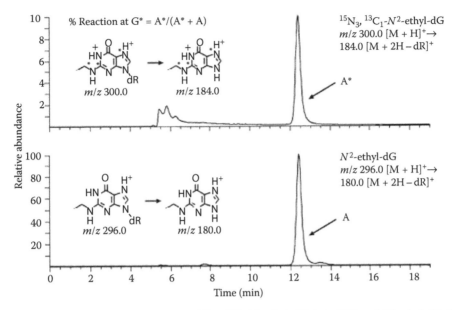

FIGURE 7.8 HPLC-ESI⁺-MS/MS analysis of $^{15}N_3$,$^{13}C_1$-N^2-ethyl-dG (*m/z* 300) and N^2-ethyl-dG (*m/z* 296) following acetaldehyde treatment and NaBH₃CN reduction of the isotopically tagged duplex. (From Matter, B. et al., *Chem. Res. Toxicol.* 20, 1383, 2007. Copyright 2004 American Chemical Society. With permission.)

a result of the reversible nature of Shiff base adducts. It is possible that the initially formed N^2-ethylidene-dG adducts redistribute between various DNA sites before they are reductively stabilized as N^2-ethyl-dG.

7.2.1.4   DNA Damage by Reactive Oxygen Species

Oxidative DNA damage is hypothesized to play an important role in aging, cancer, and neurodegenerative disease. In particular, it has been implicated in the development of lung cancer. Tobacco

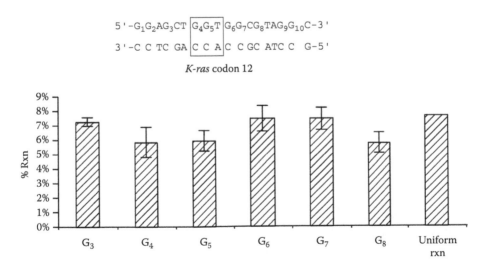

FIGURE 7.9 Distribution of acetaldehyde-induced N^2-ethyl-dG adducts along *K-ras*-derived DNA duplex. Random reactivity (uniform reaction) value was calculated from the total number of guanines in both strands. (From Matter, B. et al., *Chem. Res. Toxicol.* 20, 1383, 2007. Copyright 2004 American Chemical Society. With permission.)

smoke contains high concentrations of nitric oxide (NO), which can be oxidized to the nitrating and oxidizing agent, nitrogen dioxide.[64,65] Furthermore, phenolic and polyphenolic species present in cigarette tar and formed during carcinogen metabolism can undergo redox cycling yielding superoxide anion radicals, O_2^-.[66–68] For example, PAH-derived *o*-quinones have been recently shown to induce oxidative DNA damage under redox cycling conditions.[69] Superoxide combines with NO at a diffusion-controlled rate to yield a strong oxidant, peroxynitrite.[70] O_2^- is also subject to spontaneous or enzymatic dismutation to hydrogen peroxide (H_2O_2), which can undergo the Fenton reaction with Fe^{2+} to produce highly reactive hydroxyl radicals (HO•).[71]

Because of their low redox potential as compared with other DNA bases, guanine nucleobases are the primary targets of reactive oxygen species.[72] Oxidation of guanine gives rise to a complex mixture of oxidative products, including 8-oxo-7,8-dihydro-2′-deoxyguanosine (8-oxo-dG); spiroiminodihydantoin (Sp); guanidinohydantoin (Gh); 2,2-diamino-4-[(2-deoxy-β-D-*erythro*-pentofuranosyl)amino]-4*H*-imidazol-4-one (imidazolone); and its hydrolysis product 2,2-diamino-4-[(2-deoxy-β-D-*erythro*-pentofuranosyl)amino]-2,5-dihydrooxazol-5-one (oxazolone).[73–76] Peroxynitrite-mediated nitration of guanine produces 8-nitro-G.[77]

Oxidative DNA lesions induced by smoking are likely to contribute to lung cancer initiation. For example, knockout mice deficient at repairing oxidative lesions are predisposed to the development of lung adenocarcinoma.[78]

We employed the ILD-MS approach (Scheme 7.1) to analyze the distribution of 8-oxo-dG, oxazolone, and 8-nitro-G (Scheme 7.5) along *K-ras* and *p53* gene-derived DNA sequences following riboflavin-mediated photooxidation, oxidation in the presence of H_2O_2, and treatment with peroxynitrite (Matter and Tretyakova, unpublished data). Following the oxidation of $^{15}N_3,^{13}C_1$-dG containing DNA duplexes in the presence of 50 µM riboflavin/light; 1 mM peroxynitrite/25 mM sodium bicarbonate; or 3.5 mM H_2O_2/12 mM ascorbic acid, DNA was enzymatically digested to deoxynucleosides in the presence of nuclease P1 and alkaline phosphatase. The formation of oxazolone, 8-oxo-dG, and 8-nitroguanine lesions at each site following the oxidative treatment was followed by capillary HPLC-ESI-MS/MS (Figure 7.10). We found that in double-stranded DNA, oxidized and nitrated lesions were generated with a high degree of specificity. The distribution patterns for each adduct were dependent on the identity or reactive oxygen species. For example, photooxidation-mediated DNA oxidation was concentrated at the 5′ guanines in 5′-GG-3′ dinucleotides, including known mutational "hot spots," for example, the first position of *K-ras* codon 12 (Figure 7.11). This reactivity can be explained by electronic factors: according to computational studies, over 70% of the highest occupied molecular orbital is localized at the 5′-G of 5′-GG-3′ dinucleotides, increasing the nucleophilicity of 5′-G as compared with 3′-G.[79] In contrast, peroxynitrite-mediated oxidation had a different distribution, and single-stranded DNA was oxidized in a random manner (Matter and Tretyakova, unpublished data). Unlike gel electrophoresis-based methods, this approach provides important structural information about oxidized adducts formed at each nucleotide of interest.

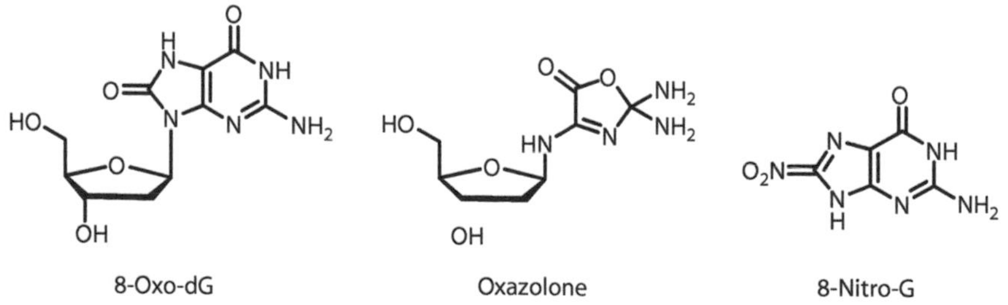

8-Oxo-dG   Oxazolone   8-Nitro-G

SCHEME 7.5 Chemical structures of oxidative DNA lesions, 8-oxo-dG, oxazolone, and 8-nitro-dG.

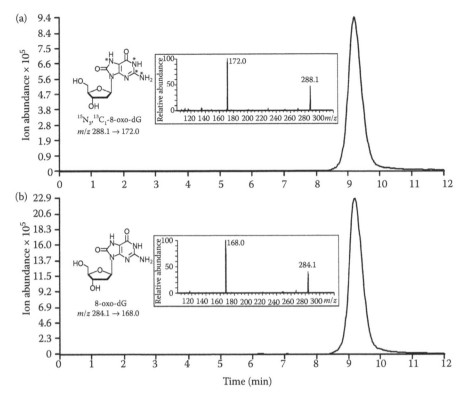

FIGURE 7.10 HPLC-ESI⁺-MS/MS analysis of $^{15}N_3,^{13}C_1$-8-oxo-dG (m/z 288, a) and 8-oxo-dG (m/z 284, B) following photooxidation of isotopically tagged duplex. Inset: MS/MS spectra of $^{15}N_3,^{13}C_1$-8-oxo-dG and 8-oxo-dG.

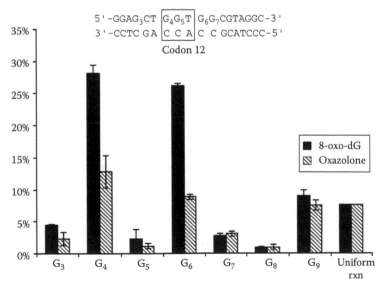

FIGURE 7.11 Distribution of 8-oxo-dG adducts along *K-ras*-derived DNA duplex following photooxidation in the presence of riboflavin.

7.3 EFFECTS OF ENDOGENOUS CYTOSINE METHYLATION ON REACTIVITY OF NEIGHBORING GUANINE BASES

In mammalian cells, all CG dinucleotides along exons 5–8 of the *p53* tumor suppressor gene contain 5-methylcytosine (MeC).[53] Although the biological role of this endogenous methylation is unknown, the majority of lung cancer mutational hot spots are observed at MeCG sites, for example, codons 157, 158, 245, 248, and 273.[80] Our adduct distribution studies have revealed that methylated CG dinucleotides of the *p53* gene (e.g., codons 245, 247, and 273) are the preferred binding sites for BPDE, but not NNK (Figures 7.3, 7.6, and 7.7). However, questions remain regarding the mechanism of this effect, especially since CG dinucleotides contain two MeC nucleobases (one in each DNA strand), both of which may influence the reactivity of the targeted G. To examine how 5'-neighboring MeC and the base-paired MeC influence guanine adduct formation, we placed methylated cytosine in either one or in both DNA strands of a sequence containing CG dinucleotides (G = isotopically labeled dG).[13,81] Following the treatment with (±) *anti*-BPDE, acetaldehyde, or acetylated precursors to NNK diazohydroxides, the extent of adduct formation at the isotopically tagged G was determined by ILD-MS and correlated to cytosine methylation status. We found that while the introduction of a MeC group within CG dinucleotides of the same DNA strand did not alter the extent of N^2-BPDE-dG formation, a large increase in adduct yield was observed in the presence of MeC base paired with the target G (Figure 7.12a).[12] On the contrary, the presence of 5'-neighboring MeC strongly inhibited the formation of O^6-Me-dG and O^6-POB-dG at the neighboring G, with the greatest decrease observed in fully methylated dinucleotides and at G's preceded by MeC (Figure 7.12b).[81] In addition, the O^6-Me-dG/N7-Me-G molar ratios were decreased in the presence of the 5'-neighboring MeC, suggesting that the regioselectivity of guanine alkylation by NNK metabolites is changed in endogenously methylated CG dinucleotides.[81] Acetaldehyde-dG adduct formation was similar in the presence and absence of a neighboring MeC (Figure 7.12c), consistent with the random reactivity of AA observed for nonmethylated sequences.[15]

Taken together, these results suggest that the endogenous methylation of cytosine has opposing effects on the extent of guanine adduct formation by BPDE and NNK, leading to characteristic adduct distribution patterns within a *p53* gene sequence. The preferential formation of N^2-BPDE-dG yields at guanines base paired with MeC is likely a result of precovalent interactions between BPDE and MeC. On the other hand, the protective effect of MeC against O^6-Me-dG adduct formation may be caused by local helical disruptions in the presence of MeC, leading to a decreased accessibility and/or altered charge density in the vicinity of the O^6 position of guanine. Furthermore, electronic effects of the methyl group transmitted through the MeC:G base pair may alter the regiospecificity of methylation, reducing O^6-alkylguanine adduct yields.[81]

7.4 DNA–DNA CROSS-LINKING BY 1,2,3,4-DIEPOXYBUTANE

1,2,3,4-Diepoxybutane (DEB, Scheme 7.6) is a genotoxic *bis*-electrophile produced on the metabolic activation of 1,3-butadiene (BD), a high-volume industrial chemical also found in cigarette smoke and in automobile exhaust. Epidemiological studies suggest that workers occupationally exposed to BD are at an increased risk of leukemia and other hematopoietic cancers, and BD has been recently identified as one of the major carcinogens in tobacco smoke.[82] Available experimental evidence suggests that DEB may be responsible for many of the adverse biological effects of BD. DEB is 50–100-fold more genotoxic and mutagenic than other metabolites of BD, and induces pronounced genetic changes, including point mutations at AT and GC base pairs and large deletions.[83,84] While all three possible stereoisomers of DEB (*S,S*; *R,R*; and *meso*) are produced on metabolic activation, the *S,S* isomer exhibits the most potent genotoxicity.[85–87]

DEB is an S_N2-type alkylating agent that targets many nucleophilic sites in DNA, including the N7 position of guanine, N3 of adenine, and N1 of adenine.[88,89] Sequential alkylation of two nucleophilic sites within DNA by DEB can produce bifunctional DNA lesions (DNA–DNA cross-links),

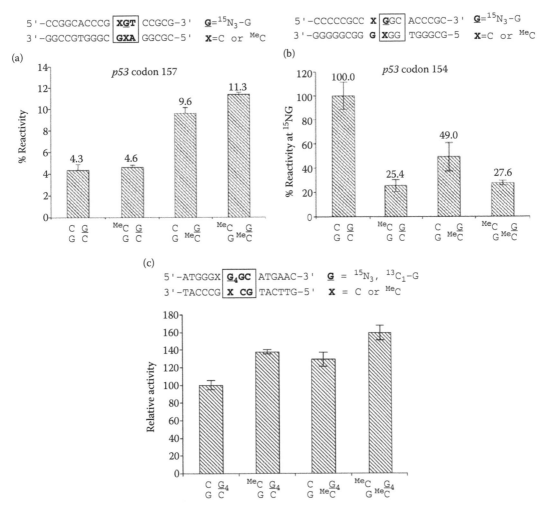

FIGURE 7.12 Effects of neighboring ^{Me}C on the formation of N^2-BPDE-dG (a), O^6-Me-dG (b), and N^2-ethyl-dG (c) within CG dinucleotides. (From Tretyakova, N.Y. et al., *Biochemistry* 41, 9541, 2002; Ziegel, R. et al., *Biochemistry* 43, 546, 2004; and Matter, B. et al., *Chem. Res. Toxicol.* 20, 1385, 2007. Copyright American Chemical Society. With permission.)

for example, 1,4-*bis*-(guan-7-yl)-2,3-butanediol (*bis*-N7G-BD, Scheme 7.6).[90] Depending on their type (inter- or intrastrand), such cross-linked adducts can have a different biological effect. Interstrand DNA–DNA adducts are known to prevent DNA strand separation, blocking DNA replication, transcription, and repair and leading to cell death. In contrast, intrastrand cross-links can be bypassed by DNA polymerases, potentially leading to mispairing and mutagenesis.[91]

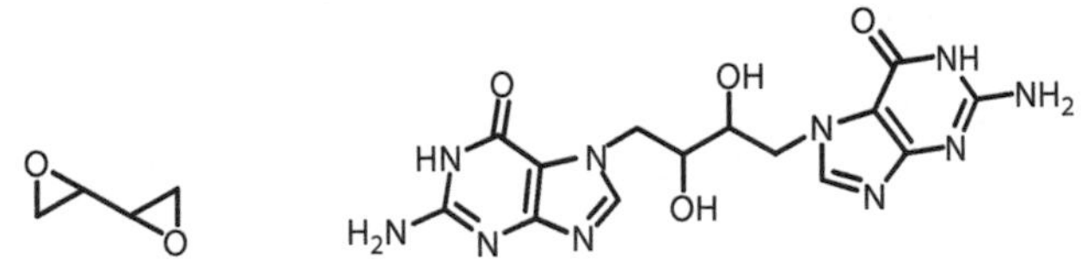

Diepoxybutane (DEB)   *bis*-N7G-BD

SCHEME 7.6 Chemical structures of DEB and its major DNA–DNA cross-link, 1,4-*bis*-(guan-7-yl)-2,3-butanediol (*bis*-N7G-BD).

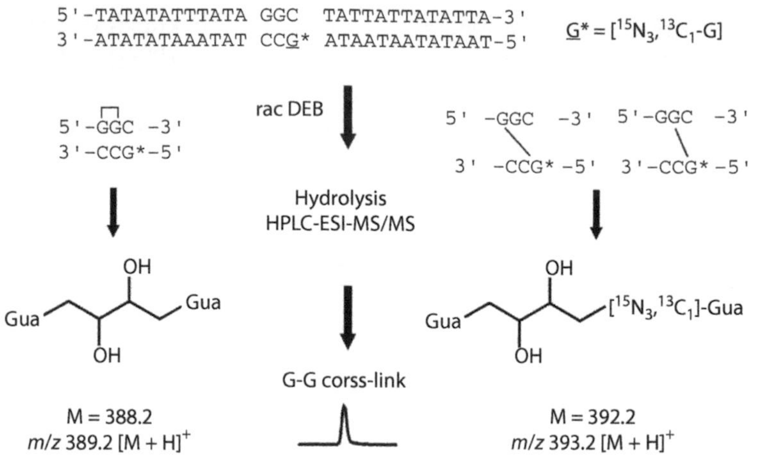

SCHEME 7.7 ILD-MS approach to distinguish between intrastrand and interstrand *bis-N*7G-BD cross-link formation by DEB in 5'-GGC-3' sequences. (From Park, S. et al., *J. Am. Chem. Soc.* 127, 14361, 2005. Copyright 2005 American Chemical Society. With permission.)

The ILD-MS methodology was used to establish the cross-linking specificity of DEB stereoisomers.[90] DNA duplexes containing a single 5'-GGC-3' trinucleotide flanked by two AT-rich sequences were prepared. Only native bases were incorporated into the (+) strand, while the (−) strand contained a single $[^{15}N_{3,}{}^{13}C_1]$-guanine opposite GGC (Scheme 7.7):

$$5'\text{-GG}\cdots\cdots\cdots\text{C-}3'$$
$$3'\text{-CC}[^{15}N_{3,}{}^{13}C_1\text{-G}]\text{-}5'$$

Since the only guanine present in the (−) strand had the $^{15}N_{3,}{}^{13}C_1$ isotope tag, any interstrand G–G conjugates originating from this duplex should contain the $^{15}N_{3,}{}^{13}C_1$ label, with a corresponding (+4) mass shift detectable by MS. In contrast, intrastrand G–G DEB lesions would originate exclusively from the unlabeled (+) strand and thus would not contain the $^{15}N_{3,}{}^{13}C_1$ label. HPLC-ESI⁺-MS/MS analysis of DNA hydrolyzates was employed to distinguish between $[^{15}N_{3,}{}^{13}C_1]$-labeled and unlabeled *bis-N*7G-BD conjugates [*m/z* 393.2 and 389.2 (M + H)⁺, respectively] (Figure 7.13).

We found that the relative contribution of intrastrand G–G cross-linking to the total number of *bis-N*7G-BD adducts was strongly dependent on diepoxide stereochemistry, accounting for 4% of total *bis-N*7G-BD lesions for *S,S*-DEB, 19% for *R,R*-DEB, and 51% for *meso*-DEB (Table 7.2).[92] Molecular models suggest that the observed differences between the ability of DEB isomers to induce DNA–DNA lesions may be caused by different orientations of functional groups in stereoisomeric *N*7-(2-hydroxy-3,4-epoxybut-1-yl)-guanine (*N*7-HEBG) intermediates.[92]

7.5 HYPOCHLOROUS ACID-INDUCED DAMAGE TO CYTOSINE RESIDUES IN DNA

Kang and Sowers[16] recently reported the use of isotope tagging to investigate the formation of 5-chlorocytosine and 5-chlorouracil at CpG sites in the presence of hypochlorous acid. Synthetic DNA duplexes were constructed containing one CpG dinucleotide. All cytosine residues outside the CpG dinucleotide were labeled with ^{15}N. Following HOCl treatment, DNA was subjected to acid hydrolysis to release free bases. 5-Chlorocytosine and 5-chlorouracil were quantified by GC-MS as silyl derivatives following derivatization with *N-tert*-butyldimethylsilyl-*N*-methyltrifluoroacetamide (MTBSTFA) (Figures 7.14 and 7.15). A series 6890 gas chromatograph was interfaced with a 5973

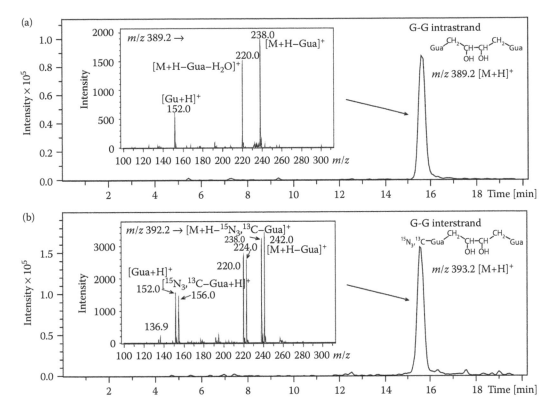

FIGURE 7.13 HPLC-ESI+-MS/MS analysis of 1,4-*bis*(guan-7-yl)-2,3-butanediol (*bis-N*7G-BD) adducts in DNA hydrolyzates derived from *meso*-DEB-treated duplex (5′-TAT ATA TTT ATA GGC TAT TAT TAT ATT A) (+ strand) and (5′-TAA TAT AAT AAT A[1,7,NH₂-¹⁵N₃-2-¹³C-G]C CTA TAA ATA TAT A) (− strand). Extracted ion chromatograms corresponding to intrastrand (unlabeled) and interstrand (¹⁵N₃, ¹³C₁-labeled) *bis-N*7G-BD cross-links were obtained by monitoring the transitions m/z 389.2 [M + H]+ → 152.0 [Gua + H]+ and m/z 393.2 [M + H]+ → 155.0 [¹⁵N₃¹³C₁ − Gua + H]+, 152.0 [Gua + H]+, respectively. Inset: MS/MS spectra corresponding to unlabeled (a) and ¹⁵N₃, ¹³C₁-labeled *bis-N*7G-BD (b). (From Park, S. et al., *J. Am. Chem. Soc.* 127, 14362, 2005. Copyright 2005 American Chemical Society. With permission.)

TABLE 7.2
Quantitative Analysis of 1,3-Interstrand; 1,2-Interstrand; and 1,2-Intrastrand 1,4-*bis*(Guan-7-yl)-2,3-Butanediol Cross-Links Generated in 5′-GGC-3′/3′-CCG-5′ Sequence Context

DEB Isomer	Total Adduct Number, pmol/nmol DNA[a]	% 1,3 Interstrand 5′–GGC–3′ 3′–CCG–5′	% 1,2 Interstrand 5′–GGC 3′–CCG	% 1,2 Intrastrand ⌐ 5′–GGC–3′ 3′–CCG–5′	Molar Ratio Interstrand/ Intrastrand
S,S	3.3 ± 0.4	96	0	4	24
R,R	1.2 ± 0.3	68	13	19	4.2
Racemic	2.4 ± 0.2	90	3	7	13
Meso	0.8 ± 0.2	49	0	51	0.96

Source: From Park, S. et al., *J. Am. Chem. Soc.* 127, 14362, 2005. Copyright 2005 American Chemical Society. With permission.

[a] Average ± SD of three measurements.

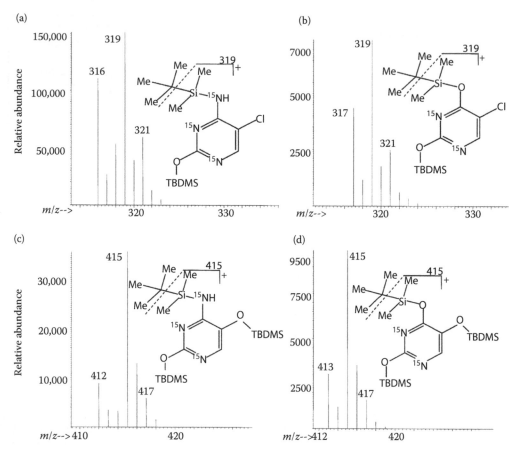

FIGURE 7.14 EI mass spectra of 5-chlorocytosine (a), 5-chlorouracil (b), 5-hydroxycytosine (c), and 5-hydroxyuracil (d) from isotopically tagged oligodeoxynucleotides reacted with 500 μM HOCl for 1 h at 37°C. (From Kang, J.I., Jr. and Sowers, L.C., *Chem. Res. Toxicol.* 21, 1211–1218, 2008. Copyright 2008 American Chemical Society. With permission.)

mass-selective detector (Agilent Technologies). The authors found that the overall reactivity of hypochlorous acid toward all cytosine residues was similar, while the extent of chlorination relative to oxidation damage was twofold greater at the CpG site compared to other cytosine residues in the sequence.[16]

7.6 CURRENT LIMITATIONS AND FUTURE DIRECTIONS

The most serious limitation of the ILD-MS approach is that the experiments must be conducted *in vitro* using synthetically prepared DNA duplexes. For most applications, the use of synthetic duplexes as mimics of nuclear DNA is justified as chromatin structure appears to have limited influence on sequence preferences of DNA-reactive electrophiles.[38,93] However, the length of DNA strands that can currently be used in ILD-MS experiments (typically 18–25 nucleotides long) is limited by the current sensitivity of the mass spectrometric detection. The use of more sensitive methodologies, for example, nanoflow HPLC-nanospray MS/MS or CE-MS, will make it possible to expand the length of the DNA fragments that can be employed in ILD-MS studies.

Although ILD-MS studies performed to date have employed a single-agent treatment, it should be applicable to analyzing the distributions of DNA adducts resulting from exposure to mixtures of carcinogens. Sequence preferences of various electrophiles are likely to be affected in the presence

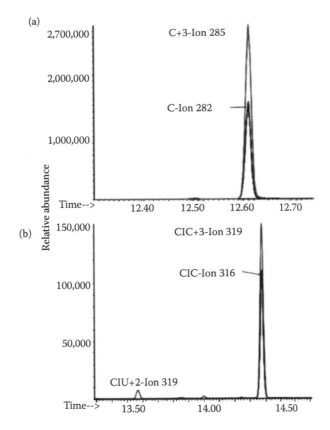

FIGURE 7.15 Extracted ion chromatograms of (a) Ions 282 and 285 for cytosine from isotopically enriched oligonucleotide incubated in buffer for 1 h at 37°C (negative control) and (b) Ions 316 and 319 for 5-chloro-dC from ^{15}N-labeled oligodeoxynucleotides reacted with 500 μM HOCl for 1 h at 37°C. (From Kang, J.I., Jr. and Sowers, L.C., *Chem. Res. Toxicol.* 21, 1211–1218, 2008. Copyright 2008 American Chemical Society. With permission.)

of competing reactions. Such experiments would more closely reflect biologically relevant conditions, for example, exposure of cigarette smokers to multiple components of cigarette smoke. Furthermore, the use of flexible scan modes (e.g., neutral loss of deoxyribose from DNA nucleosides) in HPLC-ESI-MS/MS experiments will allow mapping the formation of previously unidentified DNA lesions.

7.7 CONCLUSION

The ILD-MS approach is a logical extension of stable isotope labeling and isotope ratio analyses commonly used in MS of chemically modified DNA nucleosides. For example, widely accepted isotope dilution methods employ isotopically labeled nucleoside or nucleobase adduct analogs that are added to samples as internal standards for quantitation.[94–96] In contrast, in ILD-MS, isotope tags are introduced into DNA strands prior to carcinogen treatment and enzymatic or chemical digestion to monomers, making it possible to distinguish between adducts originating from different sites within a DNA duplex. This methodology has already provided a wealth of information about sequence preferences of various DNA modifying agents and the sequence distribution of specific DNA adducts produced on chemical exposure.[10,13,15,19,54,81] Future developments in MS instrumentation will allow further expansion of this approach to new exciting areas or research, generating insights into the mode of action and sequence preferences of DNA modifying carcinogens and drugs.

ACKNOWLEDGMENTS

All isotopically labeled dG phosphoramidites for these studies were provided by Professor Roger Jones (Rutgers University). This research was supported by a grant from the National Cancer Institute (CA-100670).11.

REFERENCES

1. Singer, B. and Grunberger, D., *Molecular Biology of Mutagens and Carcinogens*, Plenum Press, New York and London, 1983.
2. Malejka-Giganti, D. and Tretyakova, N., Molecular mechanisms of carcinogenesis, in *Carcinogenic and Anticarcinogenic Food Components*, W. Baer-Dubowska, A. Bartoszek, and D. Malejka-Giganti, eds, CRC Press, Boca Raton, FL, pp. 13–36, 2006.
3. Delaney, J.C. and Essigmann, J.M., Context-dependent mutagenesis by DNA lesions, *Chem. Biol.* 6, 743, 1999.
4. Loechler, E.L. et al., The role of carcinogen DNA adduct structure in the induction of mutations, *Prog. Clin. Biol. Res.* 340A, 51, 1990.
5. Delaney, J.C. and Essigmann, J.M., Effect of sequence context on O^6-methylguanine repair and replication *in vivo*, *Biochemistry* 40, 14968, 2001.
6. Burrows, C.J. and Muller, J.G., Oxidative nucleobase modifications leading to strand scission, *Chem. Rev.* 98, 1109, 1998.
7. Denissenko, M.F. et al., Preferential formation of benzo[a]pyrene adducts at lung cancer mutational hotspots in *p53*, *Science* 274, 430, 1996.
8. Pfeifer, G.P., Denissenko, M.F., and Tang, M.S., PCR-based approaches to adduct analysis, *Toxicol. Lett.* 102–103, 447, 1998.
9. Jacobson, K.B. et al., Applications of mass spectrometry to DNA sequencing, *Genet. Anal. Tech. Appl.* 8, 223, 1991.
10. Tretyakova, N. et al., Locating nucleobase lesions within DNA sequences by MALDI-TOF mass spectral analysis of exonuclease ladders, *Chem. Res. Toxicol.* 14, 1058, 2001.
11. Brown, K. et al., Structural characterization of carcinogen-modified oligodeoxynucleotide adducts using matrix-assisted laser desorption/ionization mass spectrometry, *J. Mass Spectrom.* 38, 68, 2003.
12. Tretyakova, N. et al., Formation of benzo[a]pyrene diol epoxide-DNA adducts at specific guanines within *K-ras* and *p53* gene sequences: Stable isotope-labeling mass spectrometry approach, *Biochemistry* 41, 9535, 2002.
13. Matter, B. et al., Formation of diastereomeric benzo[a]pyrene diol epoxide-guanine adducts in *p53* gene-derived DNA sequences, *Chem. Res. Toxicol.* 17, 731, 2004.
14. Matter, B. et al., Quantitative analysis of the oxidative DNA lesion, 2,2-diamino-4-(2-deoxy-β-D-*erythro*-pentofuranosyl)amino]-5(2*H*)-oxazolone (oxazolone), *in vitro* and *in vivo* by isotope dilution-capillary HPLC-ESI-MS/MS, *Nucleic Acids Res.* 34, 5449, 2006.
15. Matter, B. et al., Sequence distribution of acetaldehyde-derived N^2-ethyl-dG adducts along duplex DNA, *Chem. Res. Toxicol.* 20, 1379, 2007.
16. Kang, J.I., Jr. and Sowers, L.C., Examination of hypochlorous acid-induced damage to cytosine residues in a cpg dinucleotide in DNA, *Chem. Res. Toxicol.* 21, 1211–1218. 2008.
17. Gangl, E.T., Turesky, R.J., and Vouros, P., Detection of *in vivo* formed DNA adducts at the part-per-billion level by capillary liquid chromatography/microelectrospray mass spectrometry, *Anal. Chem.* 73, 2397, 2001.
18. Tretyakova, N.Y. et al., Formation of benzo[a]pyrene diol epoxide-DNA adducts at specific guanines within *K-ras* and *p53* gene sequences: Stable isotope-labeling mass spectrometry approach, *Biochemistry* 41, 9535, 2002.
19. Ziegel, R. et al., *K-ras* gene sequence effects on the formation of 4-(methylnitrosamino)-1-(3-pyridyl)-1-butanone (NNK)-DNA adducts, *Chem. Res. Toxicol.* 16, 541, 2003.
20. Park, S. et al., 3'-Exonuclease resistance of DNA oligodeoxynucleotides containing O^6-[4-oxo-4-(3-pyridyl)butyl]guanine, *Nucleic Acids Res.* 31, 1984, 2003.
21. Mao, B. et al., Opposite stereoselective resistance to digestion by phosphodiesterases I and II of benzo[a]pyrene diol epoxide-modified oligonucleotide adducts, *Biochemistry* 32, 11785, 1993.
22. Intenational Agency for Research on Cancer (IARC), *IARC Monographs on the Evaluation of the Carcinogenic Risk of Chemicals to Humans: Polynuclear Aromatic Compounds, Part I, Chemical,*

Environmental, and Experimental Data, Vol. 32, Intenational Agency for Research on Cancer, Lyon, France, 1983.

23. Yang, S.K., Roller, P.P., and Gelboin, H.V., Enzymatic mechanism of benzo[*a*]pyrene conversion to phenols and diols and an improved high-pressure liquid chromatographic separation of benzo[*a*]pyrene derivatives, *Biochemistry* 16, 3680, 1977.

24. Osborne, M.R. et al., The reaction of (+/ −)-7a, 8b-dihydroxy-9b, 10b-epoxy-7,8,9,10-tetrahydrobenzo[*a*] pyrene with DNA, *Int. J. Cancer* 18, 362, 1976.

25. Talaska, G. et al., Chronic, topical exposure to benzo[*a*]pyrene induces relatively high steady-state levels of DNA adducts in target tissues and alters kinetics of adduct loss, *Proc. Natl. Acad. Sci. USA* 93, 7789, 1996.

26. Kriek, E. et al., Polycyclic aromatic hydrocarbon-DNA adducts in humans: Relevance as biomarkers for exposure and cancer risk, *Mutat. Res.* 400, 215, 1998.

27. Phillips, D.H., DNA adducts in human tissues: Biomarkers of exposure to carcinogens in tobacco smoke, *Environ. Health Perspect.* 104(Suppl 3), 453, 1996.

28. Randerath, E. et al., Covalent DNA damage in tissues of cigarette smokers as determined by ^{32}P-postlabeling assay, *J. Natl. Cancer Inst.* 81, 341, 1989.

29. Johnson, R.E. et al., Bridging the gap: A family of novel DNA polymerases that replicate faulty DNA, *Proc. Natl. Acad. Sci. USA* 96, 12224, 1999.

30. Guengerich, F.P., Interactions of carcinogen-bound DNA with individual DNA polymerases, *Chem. Rev.* 106, 420, 2006.

31. Alekseyev, Y.O. and Romano, L.J., *In vitro* replication of primer-templates containing benzo[*a*]pyrene adducts by exonuclease-deficient *Escherichia coli* DNA polymerase I (Klenow fragment): Effect of sequence context on lesion bypass, *Biochemistry* 39, 10431, 2000.

32. Zhang, Y. et al., Error-prone lesion bypass by human DNA polymerase h, *Nucleic Acids Res.* 28, 4717, 2000.

33. Huang, X. et al., Effects of base sequence context on translesion synthesis past a bulky (+)-*trans-anti*-B[*a*]P-N^2-dG lesion catalyzed by the Y-family polymerase pol k, *Biochemistry* 42, 2456, 2003.

34. Zhao, B. et al., Polh, Polz and Rev1 together are required for G to T transversion mutations induced by the (+)- and (−)-*trans-anti*-BPDE-N^2-dG DNA adducts in yeast cells, *Nucleic Acids Res.*, 34, 417, 2006.

35. Rodenhuis, S. and Slebos, R.J., Clinical significance of *ras* oncogene activation in human lung cancer, *Cancer Res.* 52, 2665s, 1992.

36. Westra, W.H. et al., *K-ras* oncogene activation in atypical alveolar hyperplasias of the human lung, *Cancer Res.* 56, 2224, 1996.

37. Harris, C.C., Structure and function of the *p53* tumor suppressor gene: Clues for rational cancer therapeutic strategies, *J. Natl. Cancer Inst.* 88, 1442, 1996.

38. Denissenko, M.F. et al., Cytosine methylation determines hot spots of DNA damage in the human *p53* gene, *Proc. Natl. Acad. Sci. USA* 94, 3893, 1997.

39. Hoffmann, D. and Hecht, S.S., Nicotine-derived *N*-nitrosamines and tobacco-related cancer: Current status and future directions, *Cancer Res.* 45, 935, 1985.

40. Hecht, S.S. et al., Induction of respiratory tract tumors in Syrian golden hamsters by a single dose of 4-(methylnitrosamino)-1-(3-pyridyl)-1-butanone (NNK) and the effect of smoke inhalation, *Carcinogenesis* 4, 1287, 1983.

41. Hecht, S.S. et al., Comparative tumorigenicity and DNA methylation in F344 rats by 4-(methylnitrosamino)-1-(3-pyridyl)-1-butanone and *N*-nitrosodimethylamine, *Cancer Res.* 46, 498, 1986.

42. Peterson, L.A. and Hecht, S.S., O^6-methylguanine is a critical determinant of 4-(methylnitrosamino)-1-(3-pyridyl)-1-butanone tumorigenesis in A/J mouse lung, *Cancer Res.* 51, 5557, 1991.

43. Wang, L. et al., Pyridyloxobutyl adduct O^6-[4-oxo-4-(3-pyridyl)butyl]guanine is present in 4-(acetoxymethylnitrosamino)-1-(3-pyridyl)-1-butanone-treated DNA and is a substrate for O^6-alkylguanine-DNA alkyltransferase, *Chem. Res. Toxicol.* 10, 562, 1997.

44. Wang, M. et al., Identification of adducts formed by pyridyloxobutylation of deoxyguanosine and DNA by 4-(acetoxymethylnitrosamino)-1-(3-pyridyl)-1-butanone, a chemically activated form of tobacco specific carcinogens, *Chem. Res. Toxicol.* 16, 616, 2003.

45. Upadhyaya, P. et al., Identification of adducts produced by the reaction of 4-(acetoxymethylnitrosamino)-1-(3-pyridyl)-1-butanol with deoxyguanosine and DNA, *Chem. Res. Toxicol.* 16, 180, 2003.

46. Hecht, S.S. et al., Identification of O^2-substituted pyrimidine adducts formed in reactions of 4-(acetoxymethylnitrosamino)-1-(3-pyridyl)-1-butanone and 4-(acetoxymethylnitros-amino)-1-(3-pyridyl)-1-butanol with DNA, *Chem. Res. Toxicol.* 17, 588, 2004.

47. Sturla, S.J. et al., Mass spectrometric analysis of relative levels of pyridyloxobutylation adducts formed in the reaction of DNA with a chemically activated form of the tobacco-specific carcinogen 4-(methylnitrosamino)-1-(3-pyridyl)-1-butanone, *Chem. Res. Toxicol.* 18, 1048, 2005.

48. Gentil, A. et al., Mutagenic properties of a unique abasic site in mammalian cells, *Biochem. Biophys. Res. Commun.* 173, 704, 1990.

49. Takeshita, M. and Eisenberg, W., Mechanism of mutation on DNA templates containing synthetic abasic sites: Study with a double strand vector, *Nucleic Acids Res.* 22, 1897, 1994.

50. Ide, H. et al., On the mechanism of preferential incorporation of damp at abasic sites in translesional DNA synthesis. Role of proofreading activity of DNA polymerase and thermodynamic characterization of model template-primers containing an abasic site, *Nucleic Acids Res.* 23, 123, 1995.

51. Essigmann, J.M., Loechler, E.L. and Green, C.L., Genetic toxicology of O^6-methylguanine, *Prog. Clin. Biol. Res.* 209A, 433, 1986.

52. Basu, A.K. and Essigmann, J.M., Site-specifically alkylated oligodeoxynucleotides: Probes for mutagenesis, DNA repair and the structural effects of DNA damage, *Mutat. Res.* 233, 189, 1990.

53. Tornaletti, S. and Pfeifer, G.P., Complete and tissue-independent methylation of cpg sites in the *p53* gene: Implications for mutations in human cancers, *Oncogene* 10, 1493, 1995.

54. Rajesh, M. et al., Stable isotope labeling-mass spectrometry analysis of methyl- and pyridyloxobutyl-guanine adducts of 4-(methylnitrosamino)-1-(3-pyridyl)-1-butanone in *p53*-derived DNA sequences, *Biochemistry* 44, 2197, 2005.

55. Sendowski, K. and Rajewsky, M.F., DNA sequence dependence of guanine-O^6 alkylation by the *N*-nitroso carcinogens *N*-methyl- and *N*-ethyl-*N*-nitrosourea, *Mutat. Res.* 250, 153, 1991.

56. Hoffmann, D. and Hoffmann, I., The changing cigarette, 1950–1995, *J. Toxicol. Environ. Health* 50, 307, 1997.

57. Vaca, C.E., Fang, J.L., and Schweda, E.K., Studies of the reaction of acetaldehyde with deoxynucleosides, *Chem. Biol. Interact.* 98, 51, 1995.

58. Wang, M. et al., Identification of DNA adducts of acetaldehyde, *Chem. Res. Toxicol.* 13, 1149, 2000.

59. Fang, J.L. and Vaca, C.E., Detection of DNA adducts of acetaldehyde in peripheral white blood cells of alcohol abusers, *Carcinogenesis* 18, 627, 1997.

60. Wang, M. et al., Identification of an acetaldehyde adduct in human liver DNA and quantitation as N^2-ethyldeoxyguanosine, *Chem. Res. Toxicol.* 19, 319, 2006.

61. Zhang, S. et al., Analysis of crotonaldehyde- and acetaldehyde-derived 1, N^2-propanodeoxyguanosine adducts in DNA from human tissues using liquid chromatography electrospray ionization tandem mass spectrometry, *Chem. Res. Toxicol.* 19, 1386, 2006.

62. Terashima, I. et al., Miscoding potential of the N^2-ethyl-2'-deoxyguanosine DNA adduct by the exonuclease-free Klenow fragment of *Escherichia coli* DNA polymerase I, *Biochemistry* 40, 4106, 2001.

63. Perrino, F.W. et al., The N^2-ethylguanine and the O^6-ethyl- and O^6-methylguanine lesions in DNA: Contrasting responses from the "bypass" DNA polymerase h and the replicative DNA polymerase a, *Chem. Res. Toxicol.* 16, 1616, 2003.

64. Pryor, W.A., Prier, D.G., and Church, D.F., Electron-spin resonance study of mainstream and sidestream cigarette smoke: Nature of the free radicals in gas-phase smoke and in cigarette tar, *Environ. Health Perspect.* 47, 345, 1983.

65. Shafirovich, V. et al., Nitrogen dioxide as an oxidizing agent of 8-oxo-7,8-dihydro-2'-deoxyguanosine but not of 2'-deoxyguanosine, *Chem. Res. Toxicol.* 14, 233, 2001.

66. Kim, S.G. and Novak, R.F., Role of P450IIE1 in the metabolism of 3-hydroxypyridine, a constituent of tobacco smoke: Redox cycling and DNA strand scission by the metabolite 2,5-dihydroxypyridine, *Cancer Res.* 50, 5333, 1990.

67. Yu, D. et al., Reactive oxygen species generated by PAH *o*-quinones cause change-in-function mutations in *p53*, *Chem. Res. Toxicol.* 15, 832, 2002.

68. Park, J.H. et al., Formation of 8-oxo-7,8-dihydro-2'-deoxyguanosine (8-oxo-dguo) by PAH o-quinones: Involvement of reactive oxygen species and copper(II)/copper(I) redox cycling, *Chem. Res. Toxicol.* 18, 1026, 2005.

69. Park, J.H. et al., Polycyclic aromatic hydrocarbon (PAH) *o*-quinones produced by the aldo-keto-reductases (akrs) generate abasic sites, oxidized pyrimidines, and 8-oxo-dguo via reactive oxygen species, *Chem. Res. Toxicol.* 19, 719, 2006.

70. Alvarez, M.N., Trujillo, M., and Radi, R., Peroxynitrite formation from biochemical and cellular fluxes of nitric oxide and superoxide, *Methods Enzymol.* 359, 353, 2002.

71. Halliwell, B., Gutteridge, J.M., and Cross, C.E., Free radicals, antioxidants, and human disease: Where are we now?, *J. Lab. Clin. Med.* 119, 598, 1992.

72. Prat, F., Houk, K.N., and Foote, C.S., Effect of guanine stacking on the oxidation of 8-oxoguanine in B-DNA, *J. Am. Chem. Soc.* 120, 845, 1998.

73. Cadet, J. and Treoule, R., Comparative study of oxidation of nucleic acid components by hydroxyl radicals, singlet oxygen and superoxide anion radicals, *Photochem. Photobiol.* 28, 661, 1978.

74. Douki, T. and Cadet, J., Peroxynitrite mediated oxidation of purine bases of nucleosides and isolated DNA, *Free Radical Res.* 24, 369, 1996.

75. Luo, W. et al., Characterization of spiroiminodihydantoin as a product of one-electron oxidation of 8-oxo-7,8-dihydroguanosine, *Org. Lett.* 2, 613, 2000.

76. Luo, W. et al., Characterization of hydantoin products from one-electron oxidation of 8-oxo-7,8-dihydroguanosine in a nucleoside model, *Chem. Res. Toxicol,* 14, 927, 2001.

77. Yermilov, V. et al., Formation of 8-nitroguanine by the reaction of guanine with peroxynitrite *in vitro*, *Carcinogenesis* 16, 2045, 1995.

78. Sakumi, K. et al., *Ogg1* knockout-associated lung tumorigenesis and its suppression by *Mth1* gene disruption, *Cancer Res.* 63, 902, 2003.

79. Saito, I. et al., The most electron-donating sites in duplex DNA: Guanine-guanine stacking rule, *Nucl. Acids Symp. Ser.* 191, 1995.

80. Bennett, W.P. et al., Molecular epidemiology of human cancer risk: Gene-environment interactions and *p53* mutation spectrum in human lung cancer, *J. Pathol.* 187, 8, 1999.

81. Ziegel, R. et al., Endogenous 5-methylcytosine protects neighboring guanines from $N7$ and O^6-methylation and O^6-pyridyloxobutylation by the tobacco carcinogen 4-(methylnitrosamino)-1-(3-pyridyl)-1-butanone, *Biochemistry* 43, 540, 2004.

82. Fowles, J. and Dybing, E., Application of toxicological risk assessment principles to the chemical constituents of cigarette smoke, *Tob. Control* 12, 424, 2003.

83. Cochrane, J.E. and Skopek, T.R., Mutagenicity of butadiene and its epoxide metabolites: I. Mutagenic potential of 1,2-epoxybutene, 1,2,3,4-diepoxybutane and 3,4-epoxy-1,2-butanediol in cultured human lymphoblasts, *Carcinogenesis* 15, 713, 1994.

84. Recio, L. et al., Mutational spectrum of 1,3-butadiene and metabolites 1,2-epoxybutene and 1,2,3,4-diepoxybutane to assess mutagenic mechanisms, *Chem. Biol. Interact.* 135–136, 325, 2001.

85. Verly, W.G., Brakier, L., and Feit, P.W., Inactivation of the T7 coliphage by the diepoxybutane stereoisomers, *Biochim. Biophys. Acta* 228, 400, 1971.

86. Matagne, R., Induction of chromosomal aberrations and mutations with isomeric forms of L-threitol-I,4-bismethanesulfonate in plant materials, *Mutat. Res.* 7, 241, 1969.

87. Bianchi, A. and Contin, M., Mutagenic activity of isomeric forms of diepoxybutane in maize, *J. Heredity* 53, 277, 1962.

88. Tretyakova, N. et al., Adenine adducts with diepoxybutane: Isolation and analysis in exposed calf thymus DNA, *Chem. Res. Toxicol.* 10, 1171, 1997.

89. Tretyakova, N.Y., et al., Synthesis, characterization, and *in vitro* quantitation of N-7-guanine adducts of diepoxybutane, *Chem. Res. Toxicol.* 10, 779, 1997.

90. Park, S. and Tretyakova, N., Structural characterization of the major DNA-DNA cross-link of 1,2,3,4-diepoxybutane, *Chem. Res. Toxicol.* 17, 129, 2004.

91. Rajski, S.R. and Williams, R.M., DNA cross-linking agents as antitumor drugs, *Chem. Rev.* 98, 2723, 1998.

92. Park, S. et al., Interstrand and intrastrand DNA-DNA cross-linking by 1,2,3,4-diepoxybutane: Role of stereochemistry, *J. Am. Chem. Soc.* 127, 14355, 2005.

93. Boger, D.L. et al., Bifunctional alkylating agents derived from duocarmycin SA: Potent antitumor activity with altered sequence selectivity, *Bioorg. Med. Chem. Lett.* 10, 495, 2000.

94. Frelon, S. et al., High-performance liquid chromatography–tandem mass spectrometry measurement of radiation-induced base damage to isolated and cellular DNA, *Chem. Res. Toxicol.* 13, 1002, 2000.

95. Tretyakova, N.Y. et al., Quantitative analysis of 1,3-butadiene-induced DNA adducts *in vivo* and *in vitro* using liquid chromatography electrospray ionization tandem mass spectrometry, *J. Mass Spectrom.* 33, 363, 1998.

96. Knize, M.G. et al., Liquid chromatography-tandem mass spectrometry method of urine analysis for determining human variation in carcinogen metabolism, *J. Chromatogr. A* 914, 95, 2001.

8 Electrospray Mass Spectrometry of Noncovalent Complexes between Small Molecule Ligands and Nucleic Acids

Valérie Gabelica

CONTENTS

8.1 INTRODUCTION

8.1.1 ESI-MS of Noncovalent Complexes

Shortly after the development of electrospray ionization mass spectrometry (ESI-MS),[1–3] it was reported that this method could be used to detect intact noncovalent complexes.[1–4] Since this seminal report in 1991, the literature concerning supramolecular complex analysis by ESI-MS has been constantly growing. The complexes studied to date include from synthetic assemblies (cation–macrocycle, supramolecular assemblies, etc.) to complexes of biochemical interest (protein–protein, protein–ligand, protein–DNA, protein–RNA associations, etc.), and this list is not exhaustive.[5–9] The ESI-MS of noncovalent complexes has now found important applications as a screening tool in drug discovery,[10–14] including for DNA- or RNA-targeting drugs.[10–14]

This chapter focuses on the uses of ESI-MS for the study of noncovalent complexes between nucleic acids and small molecule ligands. Two reviews appeared on the subject in 2001, and another in 2008; useful information on drug–nucleic acid interaction studies can also be found in more general reviews.[12,15–20] The ionization sources, other than electrospray, that have been successfully used to detect DNA duplexes and higher-order structures, or nucleic acid–ligand noncovalent complexes, are matrix-assisted laser desorption/ionization (MALDI), cold-spray ionization (CSI), laser spray ionization (LSI), and laser-induced liquid beam ion desorption (LILBID).[21–27] These methods and their results will not be discussed in this chapter, but most of the general principles described here also hold for these other ionization methods.

8.1.2 Nucleic Acid Targeting by Small Molecules

Two major nucleic acid targets are RNA structures (often hairpins with bulges), and double-stranded DNA (dsDNA). As prokaryotic and eukaryotic RNA sequences significantly differ, RNA is a validated target for antibacterial therapy. RNA constitutes the genetic material of viruses, and several RNA substructures such as HIV-1 TAR RNA are key targets for antiviral therapies. The chemistry of small molecules that target RNA has been reviewed.[28,29] The most widely studied class of RNA-targeting ligands are aminoglycosides.[30,31] The driving force for aminoglycoside binding is electrostatic interactions between the positively charged ligand and the negative nucleic acid and displacement of tightly bound ions.[32] Furthermore, RNA is intrinsically flexible and ligand-induced fit of the binding site is sometimes significant. This has led to the development of a novel class of RNA molecules called aptamers, which are selected *in vitro* for binding particular targets such as small molecules or proteins.[33] In that case, the RNA becomes the ligand and the small molecule the target. An example structure is shown in Figure 8.1a with the neomycin B aptamer bound to its aminoglycoside ligand.[34]

dsDNA is another common target for nucleic acid ligands. dsDNA is the major constituent of our chromosomes, and targeting dsDNA causes major problems in DNA replication and cell division. Cancer cells, which divide at anomalously high rates, are therefore more affected by dsDNA ligands than normal cells. This is why dsDNA binders have been used, and are still used today, in cancer chemotherapies.[35,36] For example, daunomycin and mitoxantrone bind dsDNA by intercalation between base pairs. Intercalators usually have 2–5 conjugated aromatic rings and they insert their planar aromatic chromophore between base pairs.[37–39] Intercalators can have a very simple structure like ellipticine (Figure 8.1b), or can bear substituents interacting with the minor or the major groove which enhance the intermolecular interactions. During the intercalation process, the double helix must unwind and elongate to form an intercalation site.[37] Molecular interactions, such as stacking interactions between the heteroaromatic planes of the intercalator and the base pairs, hydrogen bonding, and van der Waals contacts, stabilize the complex.[40,41] Pure intercalators are not sequence specific and have, at the most, a moderate preference for GC-rich sequences. Some sequence specificity can be conferred by side chains interacting with the grooves of the DNA.

Another class of small molecules interacting with dsDNA is the minor groove binders. The most studied compounds are fluorescent dyes such as DAPI, Hoechst 33258, or Hoechst 33342, and polyamides such as netropsin and distamycin A. These five molecules bind to AT-rich sequences in the minor groove of DNA. This preferred interaction is driven by the narrowed minor groove in A/T-rich region in comparison with the minor groove in G/C-rich regions. Additionally, the electrostatic potential is more negative in the AT region of minor groove, favoring the electrostatic interactions with positively charged drugs. The crescent shape of minor groove binders allows fitting in the groove with little distortion of the double helix (Figure 8.1c).[42–44] The edges of the base pairs have

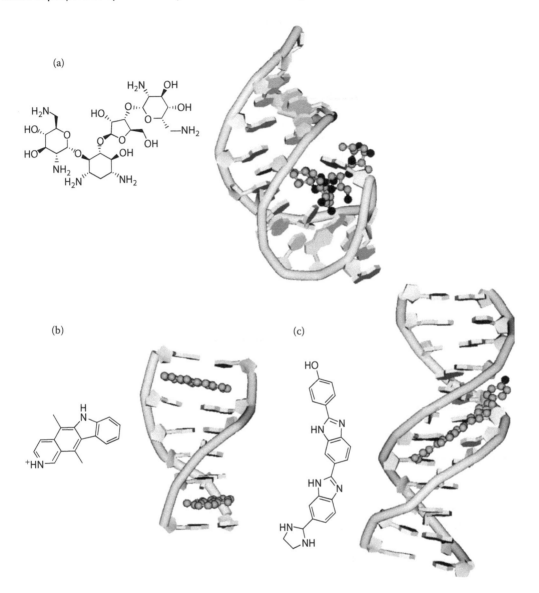

FIGURE 8.1 (a) NMR structure of the complex between ligand neomycin B and the RNA aptamer sequence GGACUGGGCGAGA AGUUUAGUCC [protein data bank (PDB) structure 1NEM].[34] (b) X-ray diffraction structure of intercalation complex between dsDNA (dCGATCG)$_2$ and two ellipticin ligands (PDB structure 1Z3F).[37] (c) X-ray diffraction structure of minor groove binding complex between the ligand 5-(2-imidazolinyl)-2-[2-(4-hydroxyphenyl)-5-benzimidazolyl]benzimidazole and dsDNA (dCGCGAATTCGCG)$_2$ (PDB structure 109D).[42]

hydrogen bond donor and acceptor groups. Minor groove binders can form either 1:1 or 2:1 complexes in their binding site. Dervan and coworkers have pioneered the development of "pairing rules" for recognition of the DNA minor groove by pyrrole (Py)–imidazole (Im) polyamides.[45–47] Im/Py is specific for G–C and Hp (*N*-methyl-3-hydroxypyrrole)/Py for T–A. This model is at the basis of the development of sequence-selective ligands of dsDNA. Minor groove binders of almost any sequence selectivity can now be synthesized at will. More recently, there has also been a revival of interest in DNA structures other than dsDNA, such as triplexes, DNA junctions, i-motif, and G-quadruplex DNA.[48–55] The rationale is to reduce the drug toxicity associated with nonspecific binding to multiple sites in dsDNA and increase drug activity toward particular structures associ-ated with particular cellular mechanisms.

8.2 ELECTROSPRAY MASS SPECTROMETRY OF NUCLEIC ACID NONCOVALENT COMPLEXES

The major strength of MS is its ability to resolve complex mixtures. As opposed to most other spec-troscopic techniques, MS gives one signal for each species that differs by mass. This is especially valuable for the study of noncovalent complexes because it allows documenting the stoichiometries of complexes with unprecedented detail. The second aspect is that relative intensities in the mass spectra are related to the relative abundances of the corresponding species in solution. It is explained below that, in some conditions, ESI-MS can even permit the determination of equilibrium binding constants in a fast and accurate manner.

8.2.1 STOICHIOMETRIES: NUMBER OF STRANDS AND DETECTION OF NUCLEIC ACID HIGHER-ORDER STRUCTURES

The first observations of dsDNA (also called duplex DNA) and of quadruple-stranded DNA (qua-druplex DNA) by ESI-MS date back to 1993.[56–58] The specificity of the dsDNA observed in ESI-MS was first discussed in 1995 by Ding and Anderegg.[59] Since then, several studies by MS/MS, ion mobility spectrometry, and molecular modeling have led to the conclusion that the Watson–Crick base pairing is retained in the gas phase, and that the helical structure is distorted for highly charged ions but conserved for low-charged ions.[60–67] Accurate mass measurement of a dsDNA of 500 base pairs (309 kDa) can be achieved after proper sample preparation with an Fourier transform ion cyclotron resonance (FTICR)-MS operated in the negative ion mode.[68]

Structures of a higher-order DNA than dsDNA can also be detected by ESI-MS. For example, triplexes formed by interaction of a third strand with the major groove of a Watson–Crick base pair have been detected, with either DNA or PNA (peptide nucleic acid) as the third strand.[69–72] DNA quadruplexes have a different structure, based on the formation of guanine quartets; that is, four guanines in a square planar configuration (G-quartets).[73] G-quartets can be formed from solutions containing only guanine or guanine derivatives, and ESI- or CSI-MS scans of these kinds of sam-ples usually show guanine assemblies with fourfold symmetry.[74–80] Nucleic acid strands containing several consecutive guanines can form the so-called G-quadruplexes, which are stabilized by G-quartets and cations between these quartets, and ESI-MS also has been successfully used to detect these G-quadruplex architectures, their complexes with noncovalently bound ligands, and higher-order assemblies based on G-quadruplex units.[23,58,69,81–93]

8.2.2 STOICHIOMETRIES: NUMBER OF BOUND LIGANDS

In 1994, the observation of noncovalent complex between actinomycin D and a single-stranded DNA (ssDNA) was reported by Hsieh et al., and the observation of dsDNA–drug noncovalent com-plexes by ESI-MS was reported by Gale et al.[94,95] The observed 2:1 stoichiometries for the complex between dsDNA and distamycin were consistent with solution data obtained using NMR spectroscopy.[96–98] Initial results on minor groove binders suggested that ESI-MS could be an effec-tive analytical technique for the detection of specific drug–dsDNA noncovalent complexes and that the complexes observed by ESI-MS are reflecting the stoichiometries in solution. However, this was less clear in the case of intercalators, because these molecules are less sequence specific and can bind to DNA at multiple sites. Therefore, based on the stoichiometry alone, the fact that multiple stoichiometries are observed could be interpreted as either due to nonspecific complex formation during electrospray or due to nonspecific (but nevertheless real) binding that was present in solution.[97] However, Sheil and coworkers reported the observation of daunomycin and nogalamycin complexes with dsDNA, with stoichiometries consistent with the neighbor exclusion principle, which states that intercalation between a base pair will prevent intercalation of a molecule in the neighboring site.[99] Later, MS/MS experiments aimed at probing the binding energetics of intercalators bound to dsDNA proved that the complexes detected with intercalators are indeed specific, in the sense that they are preserved from the solution to the gas phase, and not formed during the electrospray process.[100] Our group also made several test experiments on duplex–ligand systems where no binding was expected, and indeed, no binding was detected using ESI-MS.[100] This is also an excellent indication that the complexes detected by ESI-MS are representative of the species present in solution. Biophysicists are using ESI-MS more and more to assess ligand binding with nucleic acids.[101–111]

8.2.3 Role of Cations in Nucleic Acid Structure and Ligand Binding

Cations play a key role in solution because nucleic acids are negatively charged under physiological conditions. The most abundant cations in the intracellular medium are K^+ (~140 mM), Na^+ (~10 mM), and Mg^{2+} (~10 mM). Monovalent cations form strong electrostatic interactions with the phosphate groups in solution, thereby blocking the negative charges. Higher-order structures of nucleic acids can be formed only if the electrostatic repulsion between phosphates is screened by appropriate ionic force. Also, in order to be relevant to *in vivo* conditions, ligand binding to nucleic acid structures should be tested in solutions of ionic force corresponding to ~150 mM monovalent cation. A big limitation to the use of ESI-MS is that it is incompatible with the use of 150 mM sodium or potassium. Ammonium acetate is compatible with ESI-MS because the ammonium cations can give back a proton to the nucleic acid and depart as neutral NH_3, and concentrations up to 150 mM can be used without problem.

In addition, some nucleic acids require binding of specific cations to adopt a particular structure. For example, G-quadruplexes are stabilized by cations bound between two consecutive guanine tetrads. Fortunately, G-quadruplex forming sequences usually adopt a structure in ammonium acetate that is similar to its structure in potassium, with ammonium cations being coordinated between tetrads; but there are particular cases where the structure in potassium is different than the structure in ammonium. An example of this exception is the telomeric sequence GGGTTAGGGTTAGGGTTAGGG that adopts a mostly parallel structure in potassium, and an antiparallel structure in ammonium and in sodium.[84,86,112,113] A challenge is therefore to find experimental conditions that mimic the native structure while remaining compatible with ESI-MS.

Another example is RNA structures, like some aptamers, that require Mg^{2+} for proper folding. In that case, addition of a few equivalents of Mg^{2+} to the nucleic acid solution remains compatible with ESI-MS; however, using 10 mM $MgCl_2$ to mimic physiological conditions is problematic.[114] Finally, minute amounts of divalent cations (Co^{2+}, Ni^{2+}, Cu^{2+}, and Zn^{2+}) may be required for the ligand–DNA interaction itself. The cations make a complex with the ligand and confer a positive charge on it; the cation-bound ligand interacts with DNA with much higher affinity than the cation-free analog. In that case, adding a few equivalents of metal cation to the mixture allows recording the ESI-MS spectra in good conditions and detecting the ternary complexes [DNA + x ligands + y cations]$^{z-}$.[105,115]

8.2.4 Determination of Equilibrium Binding Constants

As mentioned above, ESI-MS allows detection of intact noncovalent complexes and permits the counting of the number of DNA strands, the number of specifically bound cations, and the number of bound ligands. Now we will describe the use of ESI-MS to determine equilibrium binding constants in the most accurate manner. The three steps are (1) the preparation of the sample, (2) the recording of the ESI-MS spectra using appropriate instrumental conditions, and (3) data processing.

Sample preparation is crucial for two reasons. First, the nucleic acid needs to be properly folded. Second, the determination of the binding constant requires knowing the exact total concentrations of each component of the solution. Nucleic acid is first dissolved in water and its concentration is determined by UV spectrophotometry, using tabulated extinction coefficients (e.g., http://www.idtdna.com/analyzer/Applications/OligoAnalyzer/Default.aspx). Then, folding of the nucleic acid is ensured by proper preparation of the nucleic acid stock solution: suitable ionic force (150 mM ammonium acetate mimics physiological ionic force and is compatible with ESI-MS) and sufficient amount of time to allow folding. An annealing procedure (heating of the solution followed by slow cooling) may help to accelerate folding into the thermodynamically stable structure. Ligand concentration in its stock solution must also be determined as accurately as possible. The complex is prepared by mixing the nucleic acid stock solution with the ligand stock solution, adding cofactors (e.g., minute amounts of metal cations) if needed, and diluting to the desired concentration. In case ligand binding is slow, equilibration time is required before injection.[116] In that case, ESI-MS can be used to monitor ligand binding kinetics. Slow ligand binding is usually an indication that the ligand-binding mechanism involves conformational rearrangements of the nucleic acid.[117]

Most mass spectrometers allow recording ESI-MS spectra from purely aqueous solution in the negative ion mode, but usually the signal is much higher when adding some methanol to the solution prior to injection. We generally use up to 20% methanol in the samples, and methanol addition was not found to induce major changes in the relative intensities. Nevertheless, as a precaution we add methanol at the last moment, just before ESI infusion, and DNA solutions were never stored for a long time in methanol.

The next step is recording the ESI-MS spectra. The instrumental parameters must allow proper evaporation of the droplets and desolvation of the ions to obtain reasonably large ion signals, and minimize extra internal energy uptake by the ions to avoid disruption of the noncovalent interactions between the nucleic acid and the ligand. Crucial parameters are therefore source or capillary temperatures (kept as low as possible), and all acceleration voltages (cone, skimmer, and lens voltages, including before the collision cell, must be kept low). When using ammonium acetate solutions, the issue is to disrupt noncovalent interactions between phosphate and ammonia although keeping intact the noncovalent interactions between the nucleic acid and the ligand. A good indication of the softness of source conditions is therefore the detection of a few ammonia adducts on the nucleic acid anions.[118]

Figure 8.2 shows ESI-MS spectra recorded for a mixture of 12-mer dsDNA and the ligand monomeric quinacridine (MMQ1), at three different radio frequency (RF) lens voltages (acceleration lens in the transfer optics just after the ESI source).[119] The relative intensity of adducts decreases as the RF Lens1 voltage increases from 60 to 100 V. Increasing the source pressure can help in achieving better desolvation although still keeping the noncovalent interactions intact.[120] The reason is that higher source pressures result in narrower internal energy distribution,[1] and the fragmentation of some ions while others are not yet fully desolvated is avoided.[121,122]

Data processing includes determination of the peak areas, calculation of the concentration of each species at equilibrium, and calculation of the equilibrium binding constants. The procedure for the determination of peak areas depends on the presence or absence of cation adducts on the free DNA and on the complexes, and whether the adduct peaks must be summed up with the nonadducted peak. In the example shown in Figure 8.2, the adduct peaks must be summed up to account for the total amount of each species. Next, the relative concentrations of free nucleic acid and each complex are calculated from the total nucleic acid concentration and the peak areas using the following equations (NA stands for nucleic acid, which can be DNA, RNA, and the like; 1:1 means one ligand bound to one NA, 2:1 means two ligands bound to one NA, etc.):

$$[NA] = [NA]_{total} \times \frac{A(NA)}{A(NA) + A(1:1) + A(2:1) + A(3:1)}, \tag{8.1}$$

$$[1:1] = [NA]_{total} \times \frac{A(1:1)}{A(NA) + A(1:1) + A(2:1) + A(3:1)} \tag{8.2}$$

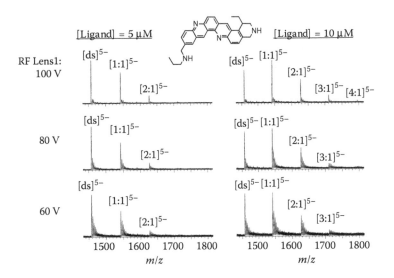

FIGURE 8.2 Full-scan ESI-MS spectra of mixtures of 5 μM dsDNA (dCGCGAATTCGCG)$_2$ and 5 μM (left) or 10 μM (right) ligand MMQ1 (structure shown) in 80% water containing 150 mM NH$_4$OAc and 20% methanol added just before injection. Spectra were recorded on a Q-TOF (quadrupole time-of-flight) Ultima Global (Micromass, Manchester, UK) mass spectrometer in the negative ion mode. Capillary voltage = −2.2 kV, cone voltage = −100 V, collision energy = 10 V, source pressure readback = 4.0 mbar, and RF lens 1 V is varied as indicated.

$$[2:1] = [NA]_{total} \times \frac{A(2:1)}{A(NA) + A(1:1) + A(2:1) + A(3:1)} \tag{8.3}$$

$$[3:1] = [NA]_{total} \times \frac{A(3:1)}{A(NA) + A(1:1) + A(2:1) + A(3:1)} \tag{8.4}$$

The total concentration of bound ligand is then calculated from the concentration of each complex (Equation 8.5), and the concentration of free ligand is equal to the total ligand concentration minus the concentration of bound ligand (Equation 8.6):

$$[Ligand]_{bound} = [1:1] + 2 \times [2:1] + 3 \times [3:1], \tag{8.5}$$

$$[Ligand]_{free} = [Ligand]_{total} - [Ligand]_{bound} \tag{8.6}$$

The concentrations of all species at equilibrium therefore allow the calculation of the equilibrium binding constants. The stepwise binding constants are defined in Equations 8.7 through 8.9:

$$K_1 = \frac{[1:1]}{[NA] \times [Ligand]_{free}} \tag{8.7}$$

$$K_2 = \frac{[2:1]}{[1:1]} \times [Ligand]_{free} \tag{8.8}$$

$$K_3 = \frac{[3:1]}{[2:1]} \times [Ligand]_{free} \tag{8.9}$$

The binding constants can therefore be determined from a single mass spectrum. However, it is highly recommended to verify the binding constants by performing a second measurement with a

TABLE 8.1

Equilibrium Association Constants Determined from the ESI-MS Spectra of Figure 8.2, and Average Equilibrium Association Constants

RF Lens1	$[DNA]_0$ (µM)	$[Ligand]_0$ (µM)	$[DNA]$ (µM)	$[1:1]$ (µM)	$[2:1]$ (µM)	$[3:1]$ (µM)	$[Ligand]_{bound}$ (µM)	$[Ligand]_{free}$ (µM)	$\log K_1$	$\log K_2$	$\log K_3$
60 V	5	5	2.83	1.77	0.40	0.00	2.57	2.43	5.41	4.97	—
80 V	5	5	2.68	1.83	0.49	0.00	2.80	2.20	5.49	5.08	—
100 V	5	5	2.62	1.88	0.51	0.00	2.89	2.11	5.53	5.11	—
60 V	5	10	1.70	2.10	1.00	0.20	4.69	5.31	5.37	4.95	4.57
80 V	5	10	1.54	2.06	1.11	0.29	5.15	4.85	5.44	5.05	4.73
100 V	5	10	1.44	2.02	1.18	0.36	5.46	4.54	5.49	5.11	4.83
Average									5.45 ± 0.06	5.04 ± 0.07	4.71 ± 0.13

Notes: Raw data were smoothed (mean function, 2 × 80 channels) and background subtracted, and peak areas are calculated. Concentrations of each species at equilibrium were determined from the peak areas using Equations 8.1 through 8.6, and binding constants were determined using Equations 8.7 through 8.19.

---

different concentration of ligand. Equilibrium binding constants are the same whatever the concentration, if the solution is ideal. Table 8.1 summarizes the concentration of all species at equilibrium determined from the ESI-MS spectra of Figure 8.2, from two different concentration ratios, and for all three voltages. We see that the constants are the same for both concentration ratios.

8.2.5 Are the Relative Intensities Proportional to the Abundances in Solution?

All equations described above are based on the assumption that the intensity ratios determined in the ESI-MS spectra are equal to the concentration ratios in solution. It is therefore assumed that free and complexed nucleic acid ions have the same response factors. What about the validity of this assumption?

Response factors are affected by all the parameters affecting ionization efficiency, transmission efficiency, and detection efficiency in the mass spectrometer. Parameters such as the mass spectrometer's transmission and detection efficiencies depend on the instrument, not on the system under study. Usually, species with similar m/z transmit equally well, and species with the same charge z are detected with the same efficiency on MCP detectors and in FT-ICR-MS. When investigating complexes between nucleic acids and small molecules, the peaks of the free nucleic acid and its complexes at a given charge state are therefore not subjected to large differential response due to the mass spectrometer.

Another factor that plays a role when analyzing noncovalent complexes is the possible disruption of complexes on their way from the source to the mass analyzer. If the complex is more fragile than the free nucleic acid, as in the case for loosely bound ligands, then the binding constants would be underestimated. If however the free nucleic acid is more fragile than the complex, this can happen, for example, when the nucleic acid is itself a noncovalent complex like dsDNA; then the binding constants would be overestimated. It is usually good practice to determine the binding constants by using different source parameters to determine how collisional activation in the source influences the relative intensities. For example, in Table 8.1, it can be seen that the relative amount of free dsDNA decreases slightly when increasing the RF Lens1 voltage. In any case, the binding constants recorded at low voltages should always be preferred.

The most unpredictable factor is however the electrospray response factor, that is, the efficiency of production of the ions from the species in the charged droplets. Ideally, the ionization efficiencies should be the same for all species used for quantification. Mechanistic studies of the electrospray process established that the electrospray response depends mainly on the analyte partitioning between the core of the droplet and its surface.[123] More hydrophobic analytes tend to move to the droplet surface while hydrophilic analytes tend to stay in the bulk droplet.[124,125] When the analyte concentrations are low compared to the amount of charges on the droplet surface (low analyte concentrations and low flow rates), then all analytes can efficiently compete with the droplet surface and can become ionized, and there is no marked difference of response factors between analytes.[126,127] However, when analyte concentrations are higher compared to the available charges on the surface, competition for ionization is biased toward the most hydrophobic analytes.

In our own experience, when performing ESI-MS determination of equilibrium binding constants at 4 µL min^{-1} injection flow rates from solutions containing maximum 10 µM nucleic acid, with duplex minor groove binders, good agreement is obtained between ESI-MS binding constants and those determined by other methods, and the response factors differed by a factor of maximum 2.[128,129] The case of minor groove binders is particularly favorable because no distortion of the duplex is associated with ligand binding. In contrast, studies of ligand bound to RNA aptamers that undergo conformational rearrangement upon binding showed significant discrepancies between abundances in ESI-MS and binding constants in solution.[114] Fortunately, even if the absolute values of binding constants might be taken with caution, the relative affinities determined by ESI-MS usually match closely the ranking obtained by other methods, thereby validating ESI-MS as an approach for screening of new ligands or of ligand selectivity.[87,89,107,130,131]

8.3 CHARACTERIZING NONCOVALENT LIGAND BINDING BY MS/MS OF THE COMPLEXES

8.3.1 OVERVIEW OF DISSOCIATION PATHWAYS

In this section, we discuss the dissociation pathways that can be observed when performing MS/MS experiments on negatively charged complexes between duplex DNA (dsDNA, hereafter denoted as DS for conciseness) and noncovalently bound ligands: $[DS + L]^{n-}$. In Section 8.3.2, we will discuss how the influence of experimental conditions on the pathways that can be observed. Then, in the following sections, we will discuss how these pathways and the collision energy at which they appear can be used to obtain information on noncovalent ligand binding.

Four possible reaction pathways are possible for a $[DS + L]^{n-}$ complex, as shown in Scheme 8.1. The first pathway is the rupture of all noncovalent interactions between the duplex and the ligand, resulting in loss of neutral ligand (Ia) or of negatively charged ligand (Ib). Pathway II is the rupture of all noncovalent interactions between the two single strands (noted "ss") of the duplex. The other two pathways involve rupture of covalent bonds in the oligonucleotides: either loss of neutral base (III) or loss of strand fragmentation (IV). According to the commonly accepted mechanism of DNA strand cleavage, base loss is a prerequisite for strand fragmentation to occur.[132] As the size of the

$$
\begin{align}
\text{(Ia)} \quad & [DS+L]^{n-} \longrightarrow [DS]^{n-} + L^0 \\
\text{(Ib)} \quad & [DS+L]^{n-} \longrightarrow [DS]^{(n-z)-} + L^{z-} \\
\text{(II)} \quad & [DS+L]^{n-} \longrightarrow [ss_1]^{(n-z)-} + [ss_2+L]^{z-} \\
\text{(III)} \quad & [DS+L]^{n-} \longrightarrow [DS+L-Base]^{n-} + Base^0 \\
\text{(IV)} \quad & [DS+L]^{n-} \longrightarrow [DS+L-frag(ss)]^{(n-z)-} + frag(ss)^{z-}
\end{align}
$$

SCHEME 8.1 Dissociation reaction pathways possible for negatively charged complexes between duplex DNA (DS) and noncovalently bound ligands (L).

nucleic acid increases, the probability of base loss increases. Also note that, in order to be detected, the loss of a neutral base or of a strand fragment from the complex must also involve the rupture of part of the noncovalent interactions between the strands.

8.3.2 How Observed Pathways Depend on Instrumental Parameters

Understanding the dissociation kinetics is a prerequisite to interpret MS/MS dissociation pathways. The observation of a given fragmentation pathway depends on whether the precursor ion received sufficient amount of internal energy to undergo fragmentation before the product ion spectrum is recorded. The two key instrumental parameters are the amount of internal energy given in the activation step, and the time allowed for fragmentation (in other words, the fragmentation time scale). In collision-induced dissociation (CID), the amount of internal energy is proportional to the collision energy. Typically, fragmentation in a triple quadrupole takes place in less than a millisecond, whereas in a quadrupole ion trap or an FT-ICR-MS/MS it takes from 30 ms to seconds. Note that the fragmentation time can be chosen by the user on most ion-trap instruments.

Example of interplay between internal energy, fragmentation time scale, and observed pathway is shown in Figure 8.3 for the dissociation of a 12-mer dsDNA.[133] The graph can be read as follows: Each curve represents the appearance of a particular dissociation pathway, and the shorter the fragmentation time scale, the more internal energy is needed for the ions to fragment within that time scale. The fact that base loss is a prerequisite for strand fragmentation is represented here by parallel curves, with the strand fragmentation curve shifted to higher internal energies than the base loss curve. However, when pathways are competitive instead of consecutive, the curves are not necessarily parallel. For example, the strand separation curve is steeper than the base loss curve, and the curves cross. This means that when using an instrument allowing long fragmentation time scales, the first reaction pathway observed when increasing the internal energy is base loss, then strand fragmentation, and strand separation is observed at even higher internal energy. Conversely, when using instruments allowing for short fragmentation time scale, the first reaction pathway observed is strand separation.

Parameters influencing the position of the curves in the case of oligonucleotide duplexes are the base composition and the charge of the duplex. Guanine loss is the most favored base loss, so duplexes containing many guanines lose neutral base at lower internal energy than duplexes containing few guanines.[134] Strand separation kinetics is also affected by the proportion of AT and GC base pairs, with GC-rich strands being more kinetically stable than AT-rich strands because of the higher degree of hydrogen bonding.[60–62,135] Finally, for a given duplex size, higher total charge leads to more favorable strand separation.[61,72,136]

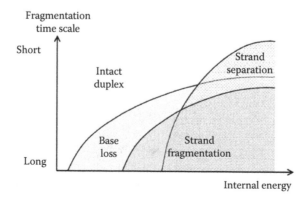

FIGURE 8.3 Kinetic diagram of reaction pathways encountered for a short duplex DNA as a function of the internal energy and fragmentation time scale.

8.3.3 HOW OBSERVED PATHWAYS DEPEND ON LIGAND BINDING

Let us now see how these curves are affected by noncovalent interactions between the duplex and a ligand. The three pathways above remain valid for the $[DS + L]^{n-}$ complex, and an additional pathway is the loss of ligand.

When the ligand is weakly bound, loss of ligand can be prevalent whatever the fragmentation time scale. This case is represented in Figure 8.4a. Curve (I) is left to all other curves, and when increasing the internal energy, the first pathway encountered is always pathway (I): intact complex → duplex. The position of curve (I) on the diagram depends on the following parameters. If the ligand departs with the neutral or negative charge it had in the isolated complex, the internal energy required to observe pathway (I) depends on the strength of noncovalent (electrostatic and van der Waals) interactions between the ligand and the duplex. However, if there was a proton transfer between the ligand and the duplex on the course of dissociation, the internal energy required to observe pathway (I) depends not only on the strength of noncovalent interactions, but also on the energy required for that proton transfer to occur (i.e., on the gas-phase acidity or basicity).[100,137,138]

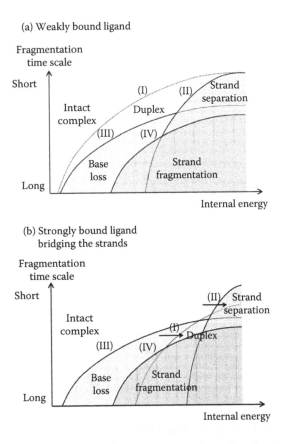

FIGURE 8.4 Kinetic diagram of reaction pathways encountered as a function of the collision energy and fragmentation time scale. The fragmentation time scale depends on the mass spectrometer used and on the instrumental settings. Pathways II, III, and IV can be observed in free duplex as well as in complexes of ligands with duplexes, while pathway I is proper to ligand–duplex complexes. (a) In the case of weakly bound ligands, loss of ligand occurs at lower collision energy than all other pathways. As the ligand binding energy increases, the curve corresponding to pathway I moves to the right (higher collision energies). Furthermore, if the ligand forms noncovalent bonds with both strands, the curve corresponding to pathway II (separation of the strands) is also displaced toward higher collision energies.

As nucleic acids are negatively charged biopolymers in solution, ligands having the highest affinities for nucleic acids are usually positively charged in solution. It is generally supposed that, in the $[DS + L]^{n-}$ complexes isolated in the gas phase, the ligand keeps its positive charge(s), and that accordingly some additional phosphates are deprotonated to give a total charge of n. Upon collisional activation, there are two possible outcomes: either proton transfer from the ligand to the DNA occurs, and the ligand can depart as a neutral, or no proton transfer occurs and the positively charged ligand sticks to the negatively charged DNA. In the latter case, represented in Figure 8.4b, curve (I) is shifted far right and pathway (I) cannot be observed. Then, base loss and subsequent strand fragmentation are observed when using long dissociation time scales, and strand separation without loss of ligand is observed when using short dissociation time scales.

Usually, another consequence of the binding of positively charged ligand that sticks to the strands is a shift in curve (II): provoking strand separation requires more internal energy if the ligand bridges the strands together by Coulomb and/or hydrogen bonding interactions. The consequence as curve (II) shifts to the right is that, for some fragmentation time scales, the preferred dissociation pathway becomes base loss and strand fragmentation in the presence of ligand instead of strand separation in the absence of ligand.

8.3.4 PROBING THE ENERGETICS OF LIGAND-DNA INTERACTIONS

Up to this point, the discussion was kept as general as possible. Let us now see how to use these concepts to derive useful and sensible information on ligand binding from MS/MS experiments. The first kind of information is the energetics of ligand–nucleic acid interactions. When the strength of noncovalent interactions between two partners increases, and provided that the dissociation mechanisms are the same (i.e., similar degree of rearrangement upon dissociation), the corresponding curve moves to the right.

For example, when pathway (Ia), Scheme 8.1, is the most favored for a series of ligands, the collision energy at which loss of ligand is observed depends on the strength of noncovalent interactions between the ligand and the nucleic acid in the gas phase. If the structure of the complex had been preserved from the solution to the gas phase, then the MS/MS experiments become a probe of the relative strength of noncovalent interactions between the ligand and the nucleic acid, a solvent-free contribution. Several studies have shown that, for neutral intercalators that complexes dissociate via pathway (Ia), the collision energy required to obtain 50% dissociation correlates very well with the dipole–dipole and hydrogen bonding interactions.[100,114,139] Another example in the same vein is the probe of ammonium loss from the central channel of DNA quadruplexes which correlates well with the ability of the bases to form a tetrad and coordinate the cation in solution.[85]

A second case where MS/MS experiments are useful is when pathway (II) is the most favored. In that case, the collision energy required to obtain 50% dissociation indicates the extent to which ligands bridge the two strands.[100,140] Pathway (II) is usually favored for positively charged minor groove binders that cannot fragment via pathway (I), and only in instruments characterized by a short fragmentation time scale. The ligand-induced stabilization of the duplex is correlated with the extent of hydrogen bonding and Coulomb interactions between the ligand and both strands.[100,140]

8.3.5 DETERMINING THE LIGAND BINDING MODE

Several papers reported the systematic characterization of the MS/MS dissociation pathways for a ligand of well-known mode in order to derive some general rules and hopefully use MS/MS to determine binding mode of unknown ligands. The first paper reporting such studies was published in 2000 by Gross and coworkers.[135] Using a quadrupole ion-trap mass spectrometer (QIT-MS), they found that complexes with the well-known minor groove binders Hoechst 33258, Hoechst 33342, berenil, DAPI, and distamycin A underwent dissociation by covalent strand breaks, whereas complexes with the intercalator actinomycin D lead to strand separation and loss of charged drug.

This is different from what our group reported on the same kind of complexes by using a Q-TOF-MS, either by quadrupole MS/MS[97] or by in-source CID.[97,140] This apparent discrepancy was explained later by the curves represented in Figures 8.4.[133] These two instruments have different fragmentation time scales (Q-TOF-MS/MS having shorter time scales than QIT-MS/MS), and because of the curve crossing, different pathways are favored on different instruments.

Furthermore, it is mentioned earlier that the preferred pathway is determined by the ligand charge, more than by the ligand binding mode. For example, complexes with ethidium (a positively charged intercalator) fragment via pathway (II) like most minor groove binders.[100] Ramos et al. also reported interesting results on porphyrin derivatives bearing from 1 to 4 positive charges, binding to short poly(dGdC) duplexes.[141] They found that the MS/MS dissociation pathways differ according to the charge rather than according to the binding mode, which was presumably the same for all compounds. Complexes with ligands having 1 or 2 positive charges did fragment in single strands; complexes with 4 positive charges did not separate in single strands and fragmented by covalent bond cleavage in the terminal parts of the strands; and complexes with 3 positive charges showed an intermediate behavior.

In conclusion, although encouraging results were obtained with some series of ligand, there are also notable exceptions, and these are mainly due to the prevalence of Coulombic interactions in the gas phase over contributions of other noncovalent interactions like nonionic hydrogen bonding and stacking. Therefore, using MS/MS dissociation pathways to deduce ligand binding mode should be done with extreme caution.

Another possibility to deduce the ligand binding mode is using energetics information instead of pathways, and to compare the energies at which a given pathway is observed with the interaction energies obtained by molecular modeling. Then the structural model for which the interaction energy ranking best matches the dissociation energy ranking is the most plausible. Examples have been published for the characterization of compounds binding 16S ribosomal RNA (rRNA), and the determination of intercalation geometry of ligands binding duplex and triplex DNA.[118,139]

8.3.6 Determining the Ligand Binding Site by MS/MS

Determination of the ligand binding site by MS/MS requires a fragmentation of the oligonucleotide by rupture of the covalent sugar–phosphate bonds while the ligand remains attached to its binding site by noncovalent interactions (pathway IV). In other words, the noncovalently bound ligand should behave like a covalent adduct of the oligonucleotide.

In rare cases, the ligand first binds to the DNA noncovalently in solution, and then covalently reacts with the DNA strand at the binding site. In those cases, the ligand binding site can be determined just like that of a covalently bound ligand.[142,143] For noncovalently bound ligands, based on what is said above, pathway (IV) is favored when the ligand bears a permanent positive charge. It is also favored compared to pathway II using long fragmentation time scales (slow activation methods). This kind of behavior has been observed in RNA complexes with the aminoglycoside neomycin B.[144] In that case, CID could be used to map the ligand binding site to HIV-1 ψ-RNA (a single strand forming a hairpin structure). Retention of the ligand has also been observed in the infrared multiphoton dissociation (IRMPD) of dsDNA–netropsin complexes.[145] Upon IRMPD of the duplex, separation into single strands that keep the bound drug is first observed. Then in an MS[3] IRMPD activation of this [ssDNA + drug] complex, the ligand binding site to the single strand can be determined.

When the ligand–oligonucleotide noncovalent bond is labile in the gas-phase complexes, a trick may consist in making the oligonucleotide covalent bonds even weaker. This trick has been used for determining the ligand binding site to 16S rRNA.[146] Three adenosine residues were mutated into deoxyadenosine. During CID of this modified RNA, the fragmentation occurs preferentially at these sites. Upon ligand binding to this modified site, a decrease in CAD fragmentation efficiency is observed. This influence on the CAD efficiency is only observed when the ligand binds close to the modified adenosine residues, not when it binds at distant sites. So this is a way of mapping the ligand binding site.

Finally, another way to determine ligand binding site is to use covalent chemical probes of the nucleic acid structure and use MS/MS to determine the location of these probes. In that case, MS/MS is not performed directly on the complex. Examples can be found in a recent study by Mazzitelli and Brodbelt, who used $KMnO_4$ oxidation of thymines to probe thymine accessibility in dsDNAs and their complexes with ligands, and in papers by Fabris and coworkers, who investigated RNA structures and RNA–protein complexes.[147–151]

8.4   CONCLUSION AND OUTLOOK

ESI-MS can now be used confidently to screen ligand for particular targets, to determine ligand selectivity among several possible targets, and even to determine equilibrium binding constants. The low sample consumption and possibilities of automation make ESI-MS a method of choice for these screening applications. However, the usefulness of MS/MS to obtain information on the ligand binding mode or binding site has not yet been universally demonstrated. MS/MS ligand binding strength in the gas phase (i.e., electrostatic and van der Waals contribution) for loosely bound ligands that were lost as neutrals upon collisional activation. The main problem with collisional activation is that ligand separation from the nucleic acid can be accompanied by nucleic acid denaturation and proton-transfer processes, which all perturb the complex initially produced from the solution.

Novel experimental approaches for the study of gas-phase ion structures can be applied to the study of ligand–nucleic acid binding. One such method is ion mobility mass spectrometry (IM-MS), which has already been applied to the study of ligand binding to G-quadruplexes.[86,152] Alternatively, methods keeping noncovalent interactions intact while fragmenting the DNA backbone could be used to map noncovalent ligand binding site on DNA. These methods might include electron detachment dissociation (EDD)[153] and electron photodetachment dissociation (EPD).[154] Finally, another class of methods that might reveal useful to investigate gas-phase ion structure is ion spectroscopy, either in the infrared spectroscopy to probe hydrogen bonding interactions or in the UV-visible spectroscopy to probe the environment of chromophore ligands.[155–162] Future studies with these methods may lead to novel approaches for investigating noncovalent DNA and RNA complexes in the gas phase.

REFERENCES

1. Yamashita, M. and Fenn, J.B., Electrospray ion source. Another variation on the free-jet theme, *J. Phys. Chem.* 88, 4451, 1984.
2. Fenn, J.B. et al., Electrospray ionization for mass spectrometry, *Science* 246, 64, 1989.
3. Fenn, J.B. et al., Electrospray ionization—principles and practice, *Mass Spectrom. Rev.* 9, 37, 1990.
4. Ganem, B., Li, Y.-T., and Henion, J.D., Detection of non-covalent receptor–ligand complexes by mass spectrometry, *J. Am. Chem. Soc.* 113, 6294, 1991.
5. Schalley, C.A., Molecular recognition and supramolecular chemistry in the gas phase, *Mass Spectrom. Rev.* 20, 253, 2001.
6. Baytekin, B., Baytekin, H.T., and Schalley, C.A., Mass spectrometric studies of non-covalent compounds: Why supramolecular chemistry in the gas phase?, *Org. Biomol. Chem.* 4(15), 2825, 2006.
7. Smith, R.D. and Light-Wahl, K.J., The observation of non-covalent interactions in solution by electrospray ionization mass spectrometry: Promise, pitfalls and prognosis, *Biol. Mass Spectrom.* 22, 493, 1993.
8. Smith, R.D. et al., New mass spectrometric methods for the study of non-covalent associations of biopolymers, *Chem. Soc. Rev.* 26, 191, 1997.
9. Breuker, K., The study of protein–ligand interactions by mass spectrometry—a personal view, *Int. J. Mass Spectrom.* 239(1), 33, 2004.
10. Siegel, M.M., Early discovery drug screening using mass spectrometry, *Curr. Top. Med. Chem.* 2, 13, 2002.

11. Glish, G.L. and Vachet, R.W., The basis of mass spectrometry in the twenty-first century, *Nat. Rev. Drug Discov.* 2, 140, 2003.
12. Hofstadler, S.A. and Sannes-Lowery, K.A., Applications of ESI-MS in drug discovery: *Interrogation* of noncovalent complexes, *Nat. Rev. Drug Discov.* 5, 585, 2006.
13. Zehender, H. and Mayr, L.M., Application of mass spectrometry technologies for the discovery of low-molecular weight modulators of enzymes and protein-protein interactions, *Curr. Opin. Chem. Biol.* 11(5), 511, 2007.
14. Annis, D.A. et al., Affinity selection-mass spectrometry screening techniques for small molecule drug discovery, *Curr. Opin. Chem. Biol.* 11(5), 518, 2007.
15. Hofstadler, S.A. and Griffey, R.H., Analysis of noncovalent complexes of DNA and RNA by mass spectrometry, *Chem. Rev.* 101, 377, 2001.
16. Beck, J. et al., Electrospray ionization mass spectrometry of oligonucleotide complexes with drugs, metals, and proteins, *Mass Spectrom. Rev.* 20, 61, 2001.
17. Rosu, F., De Pauw, E., and Gabelica, V., Electrospray mass spectrometry to study drug-nucleic acid interactions, *Biochimie* 90(7), 1074, 2008.
18. Hofstadler, S.A., Sannes-Lowery, K.A., and Hannis, J.C., Analysis of nucleic acids by FTICR MS, *Mass Spectrom. Rev.* 24(2), 265, 2005.
19. Sannes-Lowery, K.A. et al., High throughput drug discovery with ESI-FTICR, *Int. J. Mass Spectrom.* 238(2), 197, 2004.
20. Banoub, J.H. et al., Recent developments in mass spectrometry for the characterization of nucleosides, nucleotides, oligonucleotides, and nucleic acids, *Chem. Rev.* 105(5), 1869, 2005.
21. Kirpekar, F., Berkenkamp, S., and Hillenkamp, F., Detection of double-stranded DNA by IR- and UV-MALDI mass spectrometry, *Anal. Chem.* 71(13), 2334, 1999.
22. Yamaguchi, K., Cold-spray ionization mass spectrometry: Principle and applications, *J. Mass Spectrom.* 38, 473, 2003.
23. Sakamoto, S. and Yamaguchi, K., Hyperstranded DNA architectures observed by cold-spray ionization mass spectrometry, *Angew. Chem. Int. Ed.* 42, 905, 2003.
24. Shi, X. et al., Stability analysis for double-stranded DNA oligomers and their noncovalent complexes with drugs by laser spray, *J. Mass Spectrom.* 41(8), 1086, 2006.
25. Morgner, N. et al., Detecting specific ligand binding to nucleic acids: A test for ultrasoft laser mass spectrometry, *Z. Phys. Chem.* 221(5), 689, 2007.
26. Morgner, N., Barth, H.D., and Brutschy, B., A new way to detect noncovalently bonded complexes of biomolecules from liquid micro-droplets by laser mass spectrometry, *Aus. J. Chem.* 59(2), 109, 2006.
27. Morgner, N. et al., Binding sites of the viral RNA element TAR and of TAR mutants for various peptide ligands, probed with LILBID: A new laser mass spectrometry, *J. Am. Soc. Mass Spectrom.* 2008.
28. Xavier, K.A., Eder, P.S., and Giordano, T., RNA as a drug target: Methods for biophysical characterization and screening, *Trends Biotechnol.* 18, 349, 2000.
29. Gallego, J. and Varani, G., Targeting RNA with small-molecule drugs: Therapeutic promise and chemical challenges, *Acc. Chem. Res.* 34, 836, 2001.
30. Wallis, M.G. and Schroeder, R., The binding of antibiotics to RNA, *Prog. Biophys. Mol. Biol.* 67(2–3), 141, 1997.
31. Schroeder, R., Waldsich, C., and Wank, H., Modulation of RNA function by aminoglycoside antibiotics, *EMBO J.* 19(1), 1, 2000.
32. Hermann, T. and Westhof, E., RNA as a drug target: Chemical, modeling, and evolutionary tools, *Curr. Opin. Biotechnol.* 9(1), 66, 1998.
33. Famulok, M., Mayer, G., and Blind, M., Nucleic acid aptamers—from selection *in vitro* to applications in vivo, *Acc. Chem. Res.* (33), 591, 2000.
34. Jiang, L.C. et al., Saccharide-RNA recognition in a complex formed between neomycin B and an RNA aptamer, *Structure* 7(7), 817, 1999.
35. Nelson, S.M., Ferguson, L.R., and Denny, W.A., DNA and the chromosome—varied targets for chemotherapy, *Cell Chromosome* 3(1), 2, 2004.
36. Chabner, B.A. and Roberts, T.G., Timeline—chemotherapy and the war on cancer, *Nat. Rev. Cancer* 5(1), 65, 2005.
37. Canals, A. et al., The anticancer agent ellipticine unwinds DNA by intercalative binding in an orientation parallel to base pairs, *Acta Crystallogr. D Biol. Crystallogr.* 61, 1009, 2005.
38. Waring, M.J., Drugs which affect the structure and function of DNA, *Nature* 219, 1320, 1968.
39. Wang, A.H.J., Intercalative drug binding to DNA, *Curr. Opin. Struct. Biol.* 2, 361, 1992.

40. Reha, D. et al., Intercalators. 1. Nature of stacking interactions between intercalators (ethidium, dauno-mycin, ellipticine, and 4',6-diaminide-2-phenylindole) and DNA base pairs. Ab initio quantum chemical, density functional theory, and empirical potential study, *J. Am. Chem. Soc.* 124, 3366, 2002.

41. Ren, J., Jenkins, T.C., and Chaires, J.B., Energetics of DNA intercalation reactions, *Biochemistry* 39, 8439, 2000.

42. Wood, A.A. et al., Variability in DNA minor-groove width recognized by ligand-binding—the crystal structure of A bis-benzimidazole compound bound to the DNA duplex d(CGCGAATTCGCG)$_2$, *Nucleic Acids Res.* 23(18), 3678, 1995.

43. Pindur, J. and Fischer, G., DNA complexing minor groove-binding ligands: Perspective in antitumor and antimicrobial drug design, *Curr. Med. Chem.* 3, 379, 1996.

44. Geierstranger, B.H. and Wemmer, D.E., Complexes of the minor groove of DNA, *Annu. Rev. Biomol. Struct.* 24, 463, 1995.

45. Dervan, P.B., Design of sequence-specific DNA-binding molecules, *Science* 232, 64, 1986.

46. Dervan, P.B. and Bürli, R.W., Sequence-specific DNA recognition by polyamides, *Curr. Opin. Chem. Biol.* 3, 688, 1999.

47. Dervan, P.B., Molecular recognition of DNA by small molecules, *Bioorg. Med. Chem.* 9, 2215, 2002.

48. Belmont, P., Constant, J.-F., and Demeunynk, M., Nucleic acid conformation diversity: From structure to function and regulation, *Chem. Soc. Rev.* 30, 70, 2001.

49. Besch, R., Giovannangeli, C., and Degitz, K., Triplex-forming oligonucleotides—sequence-specific DNA ligands as tools for gene inhibition and for modulation of DNA-associated functions, *Curr. Drug Targets* 5(8), 691, 2004.

50. Chan, P.P. and Glazer, P.M., Triplex DNA: Fundamentals, advances, and potential applications for gene therapy. *J. Mol. Med.*, 1997, pp. 267–282.

51. Gueron, M. and Leroy, J.L., The i-motif in nucleic acids, *Curr. Opin. Struct. Biol.* 10(3), 326, 2000.

52. Phan, A.-T., Kuryavyi, V., and Patel, D.J., DNA architecture: From G to Z, *Curr. Opin. Struct. Biol.* 16, 288, 2006.

53. Burge, S. et al., Quadruplex DNA: Sequence, topology and structure, *Nucleic Acids Res* 34(19), 5402, 2006.

54. Han, H. and Hurley, L.H., G-quadruplex DNA: A potential target for anti-cancer drug design, *Trends Pharm. Sci.* 21, 136, 2000.

55. Jenkins, T.C., Targeting multi-stranded DNA structures, *Curr. Med. Chem.* 7, 99, 2000.

56. Ganem, B., Li, Y.-T., and Henion, J.D., Detection of oligonucleotide duplex forms by ionspray mass spectrometry, *Tetrahedron Lett.* 34, 1445, 1993.

57. Light-Wahl, K.J. et al., Observation of a small oligonucleotide duplex by electrospray ionization mass spectrometry, *J. Am. Chem. Soc.* 115, 803, 1993.

58. Goodlett, D.R. et al., Direct observation of a DNA quadruplex by electrospray ionization mass spectrometry, *Biol. Mass Spectrom.* 22, 181, 1993.

59. Ding, J. and Anderegg, R.J., Specific and non-specific dimer formation in the electrospray ionization mass spectrometry of oligonucleotides, *J. Am. Soc. Mass Spectrom.* 6, 159, 1995.

60. Schnier, P.D. et al., Activation energies for dissociation of double strand oligonucleotide anions: Evidence for Watson-Crick base pairing in vacuo, *J. Am. Chem. Soc.* 120, 9605, 1998.

61. Gabelica, V. and De Pauw, E., Collision-induced dissociation of 16-mer DNA duplexes with various sequences: Evidence for conservation of the double helix conformation in the gas phase, *Int. J. Mass Spectrom.* 219, 151, 2002.

62. Gabelica, V. and De Pauw, E., Comparison between solution-phase stability and gas-phase kinetic stability of oligodeoxynucleotide duplexes, *J. Mass Spectrom.* 36, 397, 2001.

63. Gidden, J. et al., Structural motifs of DNA complexes in the gas phase, *Int. J. Mass Spectrom.* 240, 183, 2004.

64. Gidden, J. et al., Duplex formation and the onset of helicity in poly d(CG)$_n$ oligonucleotides in a solvent-free environment, *J. Am. Chem. Soc.* 126, 15132, 2004.

65. Baker, E.S. and Bowers, M.T., B-DNA helix stability in a solvent-free environment, *J. Am. Soc. Mass Spectrom.* 18(7), 1188, 2007.

66. Perez, A. et al., Exploring the essential dynamics of B-DNA, *J. Chem. Theor. Comput.* 1(5), 790, 2005.

67. Rueda, M. et al., The structure and dynamics of DNA in the gas phase, *J. Am. Chem. Soc.* 125, 8007, 2003.

68. Muddiman, D.C., Null, A.P., and Hannis, J.C., Precise mass measurement of a double-stranded 500 base-pair (309 kDa) polymerase chain reaction product by negative ion electrospray ionization Fourier transform ion cyclotron resonance mass spectrometry, *Rapid Commun. Mass Spectrom.* 13, 1201, 1999.

69. Rosu, F. et al., Triplex and quadruplex DNA structures studied by electrospray mass spectrometry, *Rapid Commun. Mass Spectrom.* 16, 1729, 2002.

70. Mariappan, S.V.S. et al., Analysis of GAA/TTC DNA triplexes using nuclear magnetic resonance and electrospray ionization mass spectrometry, *Anal. Biochem.* 334(2), 216, 2004.

71. Baker, E.S. et al., PNA/dsDNA complexes: Site specific binding and dsDNA biosensor applications, *J. Am. Chem. Soc.* 128(26), 8484, 2006.

72. Delvolve, A. et al., Charge dependent behavior of PNA/DNA/PNA triplexes in the gas phase, *J. Mass Spectrom.* 41(11), 1498, 2006.

73. Davis, J.T., G-quartets 40 years later: From 5'-GMP to molecular biology and supramolecular chemistry, *Angew. Chem. Int. Ed.* 43(6), 668, 2004.

74. Fukushima, K. and Iwahashi, H., 1:1 complex of guanine quartet with alkali metal cations detected by electrospray ionization mass spectrometry, *Chem. Commun.* 895, 2000.

75. Manet, I. et al., An ESI-MS and NMR study of the self-assembly of guanosine derivates, *Helv. Chim. Acta* 84, 2096, 2001.

76. Koch, K.J. et al., Clustering of nucleobases with alkali metals studied by electrospray ionization tandem mass spectrometry: Implications for mechanisms of multistrand DNA stabilization, *J. Mass Spectrom.* 37, 676, 2002.

77. Aggerholm, T. et al., Clustering of nucleosides in the presence of alkali metals: Biologically relevant quartets of guanosine, deoxyguanosine and uridine observed by ESI-MS/MS, *J. Mass Spectrom.* 38(1), 87, 2003.

78. Sakamoto, S. et al., Formation and destruction of the guanine quartet in solution observed by cold-spray ionization mass spectrometry, *Chem. Commun.* 788, 2003.

79. Shi, X. et al., Lipophilic G-quadruplexes are self-assembled ion pair receptors, and the bound anion modulates the kinetic stability of these complexes, *J. Am. Chem. Soc.* 125, 10830, 2003.

80. Baker, E.S., Bernstein, S.L., and Bowers, M.T., Structural characterisation of G-quadruplexes in deoxyguanosine clusters using ion mobility mass spectrometry, *J. Am. Soc. Mass Spectrom.* 16, 989, 2005.

81. Krishnan-Ghosh, Y., Liu, D.S., and Balasubramanian, S., Formation of an interlocked quadruplex dimer by d(GGGT), *J. Am. Chem. Soc.* 126(35), 11009, 2004.

82. Krishnan-Ghosh, Y., Whitney, A.M., and Balasubramanian, S., Dynamic covalent chemistry on self-templating PNA oligomers: Formation of a bimolecular PNA quadruplex, *Chem. Commun.* 3068, 2005.

83. Datta, B. et al., Quadruplex formation by a guanine-rich PNA oligomer, *J. Am. Chem. Soc.* 127(12), 4199, 2005.

84. Baker, E.S. et al., G-quadruplexes in telomeric repeats are conserved in a solvent-free environment, *Int. J. Mass Spectrom.* 253(3), 225, 2006.

85. Gros, J. et al., Guanines are a quartet's best friend: Impact of base substitutions on the kinetics and stability of tetramolecular quadruplexes, *Nucleic Acids Res.* 35(9), 3064, 2007.

86. Gabelica, V. et al., Stabilization and structure of telomeric and c-myc region intramolecular G-quadruplexes: The role of central cations and small planar ligands, *J. Am. Chem. Soc.* 129, 895, 2007.

87. Carrasco, C. et al., Tight binding of the antitumor drug ditercalinium to quadruplex DNA, *ChemBioChem* 3, 100, 2002.

88. David, W.M. et al., Investigation of quadruplex oligonucleotide-drug interactions by electrospray ionization mass spectrometry, *Anal. Chem.* 74, 2029, 2002.

89. Rosu, F. et al., Selective interaction of ethidium derivatives with quadruplexes. An equilibrium dialysis and electrospray ionization mass spectrometry analysis., *Biochemistry* 42, 10361, 2003.

90. Li, W. et al., Interactions of daidzin with intramolecular G-quadruplex, *FEBS Lett.* 580(20), 4905, 2006.

91. Mazzitelli, C.L. et al., Evaluation of binding of perylene diimide and benzannulated perylene diimide ligands to DNA by electrospray ionization mass spectrometry, *J. Am. Soc. Mass Spectrom.* 17(4), 593, 2006.

92. Li, H., Yuan, G. and Du, D., Investigation of formation, recognition, stabilization, and conversion of dimeric G-quadruplexes of HIV-1 integrase inhibitors by electrospray ionization mass spectrometry, *J. Am. Soc. Mass Spectrom.* 19(4), 550, 2008.

93. Smargiasso, N. et al., G-quadruplex DNA assemblies: Loop length, cation identity, and multimer formation, *J. Am. Chem. Soc.* 130(31), 10208, 2008.

94. Hsieh, Y.L. et al., Studies of non-covalent interactions of actinomycin D with single stranded oligodeoxy-nucleotides by ion spray mass spectrometry and tandem mass spectrometry, *Biol. Mass Spectrom.* 23, 272, 1994.

95. Gale, D.C. et al., Observation of duplex DNA-drug non-covalent complexes by electrospray ionization mass spectrometry, *J. Am. Chem. Soc.* 116, 6027, 1994.

96. Fagan, P. and Wemmer, D.E., Cooperative binding of distamycin A to DNA in the 2:1 mode, *J. Am. Chem. Soc.* 114, 1080, 1992.

97. Gabelica, V., De Pauw, E., and Rosu, F., Interaction between antitumor drugs and double-stranded DNA studied by electrospray ionization mass spectrometry, *J. Mass Spectrom.* 34, 1328, 1999.

98. Gale, D.C. and Smith, R.D., Characterization of non-covalent complexes formed between minor groove binding molecules and duplex DNA by electrospray ionization mass spectrometry, *J. Am. Soc. Mass Spectrom.* 6, 1154, 1995.

99. Kapur, A., Beck, J.L., and Sheil, M.M., Observation of daunomycin and nogalamycin complexes with duplex DNA using electrospray ionization mass spectrometry, *Rapid Commun. Mass Spectrom.* 13, 2489, 1999.

100. Rosu, F. et al., Positive and negative ion mode ESI-MS and MS/MS for studying drug–DNA complexes, *Int. J. Mass Spectrom.* 253, 156, 2006.

101. Guo, X.H. et al., Structural features of the L-argininamide-binding DNA aptamer studied with ESI-FTMS, *Anal. Chem.* 78(20), 7259, 2006.

102. Zhou, J., Yuan, G., and Tang, F.L., Estimation of binding constants for complexes of polyamides and human telomeric DNA sequences by electrospray ionization mass spectrometry, *Rapid Commun. Mass Spectrom.* 20(15), 2365, 2006.

103. Ravikumar, M., Prabhakar, S., and Vairamani, M., Chiral discrimination of alpha-amino acids by the DNA triplet GCA, *Chem. Commun.* 392, 2007.

104. Smith, S.I. et al., Evaluation of relative DNA binding affinities of anthrapyrazoles by electrospray ionization mass spectrometry, *J. Mass Spectrom.* 42(5), 681, 2007.

105. Mazzitelli, C.L., Rodriguez, M., Kerwin, S.M., and Brodbelt, J.S., Evaluation of metal-mediated DNA binding of benzoxazole ligands by electrospray ionization mass spectrometry, *J. Am. Soc. Mass Spectrom.* 19, 209–218, 2007.

106. Evans, S.E. et al., End-stacking of copper cationic porphyrins on parallel-stranded guanine quadruplexes, *J. Biol. Inorg. Chem.* 12(8), 1235, 2007.

107. Monchaud, D. et al., Ligands playing musical chairs with G-quadruplex DNA: A rapid and simple displacement assay for identifying selective G-quadruplex binders, *Biochimie* 90(8), 1207, 2008.

108. Rosu, F. et al., Cooperative 2:1 binding of a bisphenothiazine to duplex DNA, *ChemBioChem*, accepted, 2008.

109. Talib, J. et al., A comparison of the binding of metal complexes to duplex and quadruplex DNA, *Dalton Trans.* (8), 1018, 2008.

110. Bahr, M. et al., Selective recognition of pyrimidine–pyrimidine DNA mismatches by distance-constrained macrocyclic bis-intercalators, *Nucleic Acids Res.* 36(15), 5000, 2008.

111. Wang, Z. et al., Evaluation of flavonoids binding to DNA duplexes by electrospray ionization mass spectrometry, *J. Am. Soc. Mass Spectrom.* 19(7), 914, 2008.

112. Ambrus, A. et al., Human telomeric sequence forms a hybrid-type intramolecular G-quadruplex structure with mixed parallel/antiparallel strands in potassium solution, *Nucleic Acids Res.* 34(9), 2723, 2006.

113. Xu, Y., Noguchi, Y., and Sugiyama, H., The new models of the human telomere d[AGGG(TTAGGG)$_3$] in K$^+$ solution, *Bioorg. Med. Chem.* 14(16), 5584, 2006.

114. Keller, K.M. et al., Electrospray ionization of nucleic acid aptamer/small molecule complexes for screening aptamer selectivity, *J. Mass Spectrom.* 40(10), 1327, 2005.

115. Reyzer, M. et al., Evaluation of complexation of metal-mediated DNA-binding drugs to oligonucleotides via electrospray ionization mass spectrometry, *Nucleic Acids Res.* 29(21), e103, 2001.

116. Greig, M.J. and Robinson, J.M., Detection of oligonucleotide–ligand complexes by ESI-MS (DOLCE-MS) as a component of high throughput screening, *J. Biomol. Screening* 5(6), 441, 2000.

117. Wilhelmsson, L.M., Lincoln, P., and Norden, B., Slow DNA binding, in *Sequence-Specific DNA Binding Agents*, M.J. Waring, ed., RSC Publishing, Cambridge, Chapter 4, p. 69, 2006.

118. Griffey, R.H. et al., Characterization of low-affinity complexes between RNA and small molecules using electrospray ionization mass spectrometry, *J. Am. Chem. Soc.* 122, 9933, 2000.

119. Teulade-Fichou, M.-P. et al., Selective recogniton of G-quadruplex telomeric DNA by a bis(quinacridine) macrocycle, *J. Am. Chem. Soc.* 125, 4732, 2003.

120. Schmidt, A., Bahr, U., and Karas, M., Influence of pressure in the first pumping stage on analyte desolvation and fragmentation in nano-ESI MS, *Anal. Chem.* 73, 6040, 2001.

121. Gabelica, V., De Pauw, E., and Karas, M., Influence of the capillary temperature and the source pressure on the internal energy distribution of electrosprayed ions, *Int. J. Mass Spectrom.* 231, 189, 2004.

122. Gabelica, V. and De Pauw, E., Internal energy and fragmentation of ions produced in electrospray sources, *Mass Spectrom. Rev.* 24, 566, 2005.

123. Enke, C.G., A predictive model for matrix and analyte effects in electrospray ionization of singly-charged ionic analytes, *Anal. Chem.* 69, 4885, 1997.
124. Cech, N.B. and Enke, C.G., Effect of affinity for droplet surfaces on the fraction of analyte molecules charged during electrospray droplet fission, *Anal. Chem.* 73(19), 4632, 2001.
125. Cech, N.B. and Enke, C.G., Relating electrospray ionization response to nonpolar character of small peptides, *Anal. Chem.* 72, 2717, 2000.
126. Schmidt, A., Karas, M., and Dulcks, T., Effect of different solution flow rates on analyte ion signals in nano-ESI MS, or: When does ESI turn into nano-ESI?, *J. Am. Soc. Mass Spectrom.* 14(5), 492, 2003.
127. Kuprowski, M.C. and Konermann, L., Signal response of coexisting protein conformers in electrospray mass spectrometry, *Anal. Chem.* 79(6), 2499, 2007.
128. Rosu, F. et al., Determination of affinity, stoichiometry and sequence selectivity of minor groove binder complexes with double-stranded oligodeoxynucleotides by electrospray ionization mass spectrometry, *Nucleic Acids Res.* 30(16), e82, 2002.
129. Gabelica, V. et al., Influence of response factors on determining equilibrium association constants of non-covalent complexes by electrospray ionization mass spectrometry, *J. Mass Spectrom.* 38, 491, 2003.
130. Guittat, L. et al., Interactions of cryptolepine and neocryptolepine with unusual DNA structures, *Biochimie* 85, 535, 2003.
131. Guittat, L. et al., Ascididemin and meridine stabilise G-quadruplexes and inhibit telomerase in vitro, *Biochim. Biophys. Acta* 1724(3), 375, 2005.
132. Wu, J. and McLuckey, S.A., Gas-phase fragmentation of oligonucleotide ions, *Int. J. Mass Spectrom.* 237(2–3), 197, 2004.
133. Gabelica, V. and De Pauw, E., Comparison of the collision-induced dissociation of duplex DNA at different collision regimes: Evidence for a multistep dissociation mechanism, *J. Am. Soc. Mass Spectrom.* 13, 91, 2002.
134. Klassen, J.S., Schnier, P.D., and Williams, E.R., Blackbody infrared radiative dissociation of oligonucleotide anions, *J. Am. Soc. Mass Spectrom.* 9, 1117, 1998.
135. Wan, K.X., Shibue, T., and Gross, M.L., Gas-phase stability of double-stranded oligodeoxynucleotide and their noncovalent complexes with DNA-binding drugs is revealed by collisional activation in an ion trap, *J. Am. Soc. Mass Spectrom.* 11, 450, 2000.
136. Keller, K.M. et al., Influence of initial charge state on fragmentation patterns for noncovalent drug/DNA duplex complexes, *J. Mass Spectrom.* 40(10), 1362, 2005.
137. Alves, S., Woods, A., and Tabet, J.C., Charge state effect on the zwitterion influence on stability of non-covalent interaction of single-stranded DNA with peptides, *J. Mass Spectrom.* 42(12), 1613, 2007.
138. Alves, S. et al., Influence of salt bridge interactions on the gas-phase stability of DNA/peptide complexes, *Int. J. Mass Spectrom.* 278, 122–128, 2008.
139. Rosu, F. et al., Ligand binding mode to duplex and triplex DNA assessed by combining electrospray tandem mass spectrometry and molecular modeling, *J. Am. Soc. Mass Spectrom.* 18(6), 1052, 2007.
140. Gabelica, V. et al., Gas phase thermal denaturation of an oligonucleotide duplex and its complexes with minor groove binders, *Rapid Commun. Mass Spectrom.* 14, 464, 2000.
141. Ramos, C.I.V. et al., Interactions of cationic porphyrins with double-stranded oligodeoxynucleotides: A study by electrospray ionisation mass spectrometry, *J. Mass Spectrom.* 40(11), 1439, 2005.
142. Pothukuchy, A. et al., Duplex and quadruplex DNA binding and photocleavage by trioxatriangulenium lone, *Biochemistry* 44(6), 2163, 2005.
143. David-Cordonnier, M.H. et al., Covalent binding of antitumor benzoacronycines to double-stranded DNA induces helix opening and the formation of single-stranded DNA: Unique consequences of a novel DNA-bonding mechanism, *Mol. Cancer Ther.* 4(1), 71, 2005.
144. Turner, K.B. et al., Mapping noncovalent ligand binding to stemloop domains of the HIV-1 packaging signal by tandem mass spectrometry, *J. Am. Soc. Mass Spectrom.* 17, 1401, 2006.
145. Wilson, J.J. and Brodbelt, J.S., Infrared multiphoton dissociation of duplex DNA/drug complexes in a quadrupole ion trap, *Anal. Chem.* 79(5), 2067, 2007.
146. Griffey, R.H. et al., Targeting site-specific gas-phase cleavage of oligoribonucleotides. Application in mass spetrometry-based identification of ligand binding sites, *J. Am. Chem. Soc.* 121, 474, 1999.
147. Mazzitelli, C.L. and Brodbelt, J.S., Probing ligand binding to duplex DNA using $KMnO_4$ reactions and electrospray ionization tandem mass spectrometry, *Anal. Chem.* 79(12), 4636, 2007.
148. Yu, E. and Fabris, D., Direct probing of RNA structures and RNA-protein interactions in the HIV-1 packaging signal by chemical modification and electrospray ionization Fourier transform mass spectrometry, *J. Mol. Biol.* 330(2), 211, 2003.

149. Yu, E. and Fabris, D., Toward multiplexing the application of solvent accessibility probes for the investigation of RNA three-dimensional structures by electrospray ionization-Fourier transform mass spectrometry, *Anal. Biochem.* 334(2), 356, 2004.

150. Yu, E.T., Zhang, Q.G., and Fabris, D., Untying the FIV frameshifting pseudoknot structure by MS3D, *J. Mol. Biol.* 345(1), 69, 2005.

151. Kellersberger, K.A. et al., Top-down characterization of nucleic acids modified by structural probes using high-resolution tandem mass spectrometry and automated data interpretation, *Anal. Chem.* 76(9), 2438, 2004.

152. Baker, E.S. et al., Cyclo[*n*]pyrroles: Size and site-specific binding to G-quadruplexes, *J. Am. Chem. Soc.* 128(8), 2641, 2006.

153. Mo, J.J. and Hakansson, K., Characterization of nucleic acid higher order structure by high-resolution tandem mass spectrometry, *Anal. Bioanal. Chem.* 386(3), 675, 2006.

154. Gabelica, V. et al., Electron photodetachment dissociation of DNA polyanions in a quadrupole ion trap mass spectrometer, *Anal. Chem.* 78(18), 6564, 2006.

155. Oh, H.B. et al., Infrared photodissociation spectroscopy of electrosprayed ions in a Fourier transform mass spectrometer, *J. Am. Chem. Soc.* 127(11), 4076, 2005.

156. Oomens, J. et al., Charge-state resolved mid-infrared spectroscopy of a gas-phase protein, *Phys. Chem. Chem. Phys.* 7(7), 1345, 2005.

157. Polfer, N.C. et al., Differentiation of isomers by wavelength-tunable infrared multiple-photon dissociation-mass spectrometry: Application to glucose-containing disaccharides, *Anal. Chem.* 78(3), 670, 2006.

158. Bush, M.F. et al., Infrared spectroscopy of cationized arginine in the gas phase: Direct evidence for the transition from nonzwitterionic to zwitterionic structure, *J. Am. Chem. Soc.* 129(6), 1612, 2007.

159. Gabelica, V. et al., Infrared signature of DNA G-quadruplexes in the gas phase, *J. Am. Chem. Soc.* 130, 1810, 2008.

160. Andersen, L.H. et al., Absorption of Schiff-base retinal chromophores in vacuo, *J. Am. Chem. Soc.* 127(35), 12347, 2005.

161. Gabelica, V. et al., Base-dependent electron photodetachment from negatively charged DNA strands upon 260-nm laser irradiation, *J. Am. Chem. Soc.* 129, 4706, 2007.

162. Gabelica, V. et al., Electron photodetachment dissociation of DNA anions with covalently or noncovalently bound chromophores, *J. Am. Soc. Mass Spectrom.* 18(11), 1990, 2007.

9 Electrospray Ionization–Mass Spectrometry for the Investigation of Protein–Nucleic Acid Interactions

Daniele Fabris, Kevin B. Turner, and Nathan A. Hagan

CONTENTS

Vital cellular processes including genome replication, transcription, translation, and repair rely on specific interactions between proteins and nucleic acids to complete diverse and complex tasks. Among the biophysical techniques employed to study the large biomolecular assemblies performing these tasks, mass spectrometry (MS) presents the greatest potential based on its intrinsic characteristics and accessible information. Ever since the initial reports describing the ability of electrospray ionization (ESI)[1,2] to desorb intact noncovalent complexes,[3–5] ESI-MS has been successfully applied to the characterization of a large variety of assemblies, as extensively reviewed in references.[6–9] In this chapter, we offer an account of our personal experience in employing this analytical platform for the functional investigation of selected protein–nucleic acid complexes of retroviral systems. The strategies followed in these types of studies, the practical issues faced in completing the experiments, and the information provided by the different approaches are discussed in the context of the recent advances in the mass spectrometric analysis of protein–nucleic acid complexes.

9.1   APPEAL OF ESI-MS ANALYSIS

The scope of the study of protein–nucleic acid assemblies can range from the identification of bound components to their complete structural elucidation, from determining the binding stoichiometry to evaluating the strength of the specific interaction, from recognizing the type of interaction to locating the position of the binding site, and more. Any technique employed to tackle these questions faces challenges posed by the intrinsic nature of these assemblies and their constitutive components. In general, charging properties, solubility, and stability are excellent predictors of the ability to successfully complete the MS analysis of a given sample. The polar character of proteins and nucleic acids is critical to the performance of ESI and determines the ability to achieve their direct desorption from predominantly aqueous solutions. With no need for specific chromophores or radioactive labels, the biomolecules involved in a certain complex are immediately recognized from their unique molecular masses that are narrowly defined by the chemical structures of their respective building blocks. By preserving the integrity of the noncovalent interactions, ESI-MS offers the rare opportunity to observe simultaneously all the species involved in a certain equilibrium, which can be unambiguously differentiated by the resolving power and mass accuracy afforded by the available analyzer. In this way, binding stoichiometries can be determined directly from observed masses, rather than from curve fitting of bulk data obtained from traditional spectroscopic and calorimetric techniques. The ability to tackle sample mixtures without separation procedures that may also induce complex dissociation offers the opportunity of acquiring direct information about binding affinities, inhibition constants, and other important parameters describing a system at equilibrium.

Whereas the assemblies' size may become an obstacle to the effective application of other structural techniques, the extensive charging associated with the ESI process allows for the immediate detection of very large analytes using mass analyzers of relatively limited range and cost. Furthermore, the ability to couple this ionization technique with virtually any type of analyzer has produced combinations that enhance the respective strengths of each component. For example, coupling ESI with Fourier transform ion cyclotron resonance (FTICR)[10,11] mass spectrometers enables the observation of multiply charged ions at the lower end of the mass range, where this type of analyzer affords its greatest resolving power. This very auspicious combination has led to the successful detection of nucleic acid samples with masses exceeding 1 MDa.[12,13] The virtually unlimited mass range afforded by time-of-flight (TOF)[14] analyzers enables their application to assemblies that manifest only limited charging. This situation is frequently encountered when the negative charges on the nucleic acid backbone are partially neutralized by positive charges on the cognate protein components. For this reason, a hybrid instrument using TOF technology was successfully employed to analyze ribonucleoprotein (RNP) complexes corresponding to intact and partially dissociated ribosomes with molecular masses approaching 900 kDa.[15,16]

In addition to features that are particularly appealing for the investigation of protein–nucleic acid complexes, this analytical platform offers all the typical benefits afforded by MS-based methods. Reaching routinely in the low femtomole range,[17,18] the amount of sample necessary to complete ESI-MS determinations compares very favorably with those reported for other biophysical techniques. Further, attomole-level sensitivities can be achieved in protein detection[19] and peptide sequencing[20] by coupling capillary electrophoresis or liquid chromatography through low flow electrospray interfaces (nanospray).[21] This figure of merit becomes particularly critical when the analyte of interest is not available in large quantities, for example, through recombinant techniques, but can be obtained only in limited amounts from actual cellular sources.

Finally, ESI-MS approaches offer the opportunity of performing tandem mass spectrometry (MS/MS)[22,23] to obtain structural information about the analyte of interest. In typical experiments, the target precursor ion is isolated in the analyzer from any other species present in the sample mixture, activated by different means in the gas phase, and then allowed to dissociate into fragments that reflect its composition and chemical structure. Because biopolymers produce characteristic ion series by cleavage of their respective backbones, MS/MS is routinely employed to determine the sequence of peptides and proteins,[24,25] as well as oligonucleotides[26–30] and nucleic acid adducts.[31–35] In contrast, its application to noncovalent complexes of nucleic acids has been limited mainly to assessing the stability of binding interactions in the gas phase.[36–39] The recent implementation of new forms of gas-phase activation is expected to pave the way for a more extensive utilization of MS/MS in the structural characterization of these important biomolecular assemblies.

9.2 DEMANDS OF SAMPLE PREPARATION

Sample composition and instrumental conditions play critical roles in ensuring that intact noncovalent assemblies are successfully detected by ESI-MS. This ionization process involves submitting sample solutions to a strong electric field to obtain fine droplets that undergo desolvation in the source interface and provide ions with characteristic charge-state distributions (Figure 9.1).[40,41] Multiple charging is generally facilitated by extreme conditions of pH, which induce protonation or deprotonation of basic/acidic functional groups on the target substrate. Therefore, typical solutions employed for analyses in the positive ion mode include small amounts of a weak acid, such as acetic or formic acid, whereas those performed in the negative ion mode call for a weak base, usually ammonia or triethylamine. Further, sample solutions often include sizeable percentages of an organic cosolvent (e.g., methanol or 2-propanol) to decrease the surface tension and enhance spray stability. Unfortunately, extreme pH and organic solvents are not compatible with the majority of biological assemblies and can induce denaturation and dissociation directly in the sample tube before the actual analysis.

In preparing solutions of noncovalent assemblies, a fine balance is often sought between conditions that enable optimal analytical performance and those necessary to preserve the position of the target equilibrium in solution (Figure 9.2a). Physiological values of pH and ionic strength are typically obtained by employing buffers and salts that do not hamper ESI-MS analysis. The presence of any alkaline cations (e.g., Na^+, K^+, etc.) in solution tends to induce the formation of stable nonspecific adducts with biomolecular analytes, which results in undesirable signal suppression and resolution degradation. In contrast, ammonium adducts are capable of dissociating into NH_3 and H^+ during the desolvation process, with the protons remaining attached to basic sites on the analyte structure.[42] The volatile nature of this species is at the basis of the utilization of small amounts of ammonium acetate, bicarbonate, or citrate as MS-friendly buffers.[43,44] Typical concentrations in the

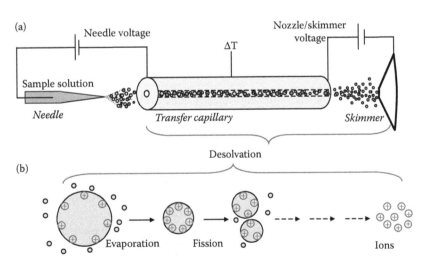

FIGURE 9.1 Processes involved in ESI. (a) A stable spray is achieved by applying an appropriate voltage between the electrospray needle and the transfer capillary to induce nebulization of the sample solution into small, charged droplets.[1,2] The droplets travel along the heated capillary, in which they evaporate and collide with solvent molecules and other droplets. Charged particles are accelerated by the difference of potential between the end of the capillary and the skimmer, which increases the energy of the collisional processes. (b) As each droplet shrinks due to solvent evaporation, the charges on its surface are forced close to each other, until Coulombic repulsion induces droplet fission. Reiteration of the evaporation–fission cycle culminates in complete solvent elimination and production of gas-phase ions.[58,59] Alternatively, ions are ejected directly from the droplet surface by a progressively stronger electrostatic field.[60]

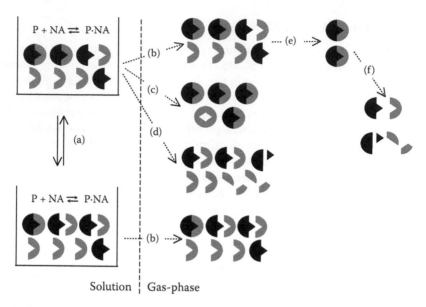

FIGURE 9.2 (a) The position of a typical binding equilibrium between a protein (P) and a nucleic acid substrate (NA) can be affected by solution conditions, such as pH and ionic strength, and by the presence of ligands and other additives in solution. Ideal solution conditions should avoid equilibrium perturbations before the analysis, while enabling the best possible analytical performance. (b) Ideal instrumental conditions should allow for the unbiased transfer of all analytes to the gas phase and should not affect the state of association of the target noncovalent complex, thus providing a faithful representation of the partitioning between free and bound species in solution. (c) Nonspecific aggregation between any of the species present in the initial sample may take place at high analyte concentrations. Control experiments are recommended to recognize the occurrence of this type of situation. (d) Harsh desolvation conditions may induce dissociation of the noncovalent complex, as well as fragmentation of the covalent backbones of its components. (e) MS/MS experiments are performed by isolating the precursor ion of interest from the other species present in the sample. (f) The precursor is then activated in the gas phase to obtain dissociation of the bound components, as well as fragmentation of their covalent backbones.

low mM range can offer sufficient buffering capacity at neutral pH to safeguard the stability of the majority of biological assemblies during sample preparation and analysis. In some cases, however, the reduced charging associated with mild pH conditions is expected to place increasing demands on the mass-to-charge range covered by the available analyzer.

Nucleic acid analytes are particularly prone to undesirable cation adducts due to the presence of negatively charged phosphates on their phosphodiester backbone. For this reason, a great deal of effort has been dedicated over the years to devise different methods for removing the source of such adducts from sample solutions. In addition to paying close attention to the quality of water and glassware employed in sample preparation, ion exchange,[45] reversed-phase high-performance liquid chromatography (RP-HPLC),[46] metal chelation,[47,48] ethanol precipitation,[43] ultrafiltration, and microdialysis[49–51] have been tested to achieve proper desalting of nucleic acid analytes for MS. While separation-based methods perform very well for initial nucleic acid substrates, their application to noncovalent complexes should be carefully scrutinized for the possibility that the differential removal of one of the components may unintentionally shift the binding equilibrium and induce dissociation. Whenever possible, it may be preferable to desalt separately the individual species involved in the equilibrium, determine their respective concentrations, and then assemble the complex of interest from the salt-free components under controlled conditions.

An alternative approach for controlling the extent of cation adduction takes advantage of ion–ion reactions in the gas phase.[52,53] Using a dual emitter setup that allows for different reactants to be sprayed simultaneously without direct mixing in solution, we have recently shown that nucleic acid adducts can be significantly reduced by reaction with chelating agents in the rf-only quadrupole at the front end of our FTICR mass spectrometer.[54] The beneficial effects of metal transfer are clearly evident when comparing spectra obtained from a 17-mer deoxy-oligonucleotide in the absence and presence of the metal chelator 1,2-diaminocyclohexanetetraacetic acid (Figure 9.3). Analyzed in the negative ion mode, a sample containing 5 μM analyte and 20 μM MgCl$_2$ in 150 mM ammonium acetate displayed a broad distribution of metal adducts, which depressed the observed signal-to-noise (S/N) ratio and overall signal intensity (panel a). When the sequestrating agent was introduced by activating the auxiliary emitter, the reaction between ions of like polarity induced dramatic reduction of the observed adducts, which resulted in decreased spectral complexity and improved S/N (panel b). The fact that reactants did not mix in solution, but shared the same ion path, ensured that the transfer process took place in the gas phase, after the desorption process was complete and equilibrium considerations were no longer a concern. For this reason, ion–ion reactions enabled us to control the extent of adduction exhibited by nucleic acid complexes with a variety of ligands, including proteins, without adversely affecting their association state.[54]

The large contribution of electrostatic interactions to the stability of protein–nucleic acid assemblies makes ionic strength a critical parameter. By shielding the charges carried by the species involved, ions can modulate the strength of the binding interactions and, thus, affect the position of the observed equilibrium.[55] In this respect, the majority of nucleic acid-binding proteins is basic in

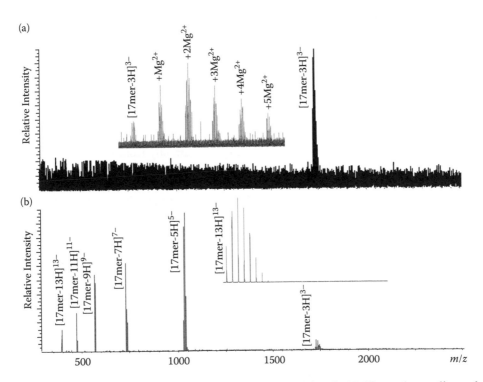

FIGURE 9.3 (a) Nanospray-FTICR analysis of a sample containing 5 μM 17-mer deoxy-oligonucleotide and 20 μM MgCl$_2$ in 150 mM ammonium acetate (pH 7.0) and 10% volume 2-propanol. In the absence of transfer reactions, the analyte provided a monoisotopic mass of 5134.903 Da (5134.909 Da calculated from sequence) and up to five magnesium adducts characterized by a 21.97-Da incremental mass (inset). In (b), performing transfer reactions with 250 mM 1,2-diaminocyclohexanetetraacetic acid in the auxiliary emitter and 90 s hexapole residence time resulted in the elimination of nonspecific magnesium adducts, as shown here by the highest charge state produced by the ion–ion reaction (inset).

nature and positively charged at neutral pH. The absence of charge shielding favors predominantly their electrostatic interactions with negatively charged nucleic acids, which are thus greatly strengthened by low salt concentrations. In contrast, nucleic acid–nucleic acid interactions are stabilized by high ionic strengths that reduce the Coulombic repulsion between like-charged phosphodiester backbones. In sample solutions for ESI-MS analysis, volatile salts can be judiciously employed to reach the desired salt content. In most cases, the buffer concentration itself can be appropriately adjusted to provide both buffering capacity and ionic strength. For example, samples containing up to 0.5 M concentrations of ammonium acetate have been successfully analyzed in our laboratory using nanospray ionization,[56] with concentrations up to 2 M being reported for the investigation of selected protein–DNA interactions.[18]

Finally, the widespread utilization of gas-assisted nebulizers[57] and nanoflow electrospray sources[21] can greatly reduce the need for organic cosolvents to improve spray stability. Alternative solutions for minimizing the impact on target complexes could use *ad hoc* hardware setups in which the organic component is introduced through a mixing-T located immediately before the ESI needle, or sprayed together with the analyte solution through a needle formed by coaxial layers. If direct mixing in the sample tube were still necessary, possible cosolvent effects on complex association should be carefully monitored, for example, by evaluating solutions that included decreasing amounts of the desired additive. For the vast majority of the biological systems considered in our laboratory, the addition of 2-propanol to a final concentration of 10% (v/v) has provided excellent spray stability in nanospray mode without detectable effects on the position of binding equilibria.

9.3 CRITICAL INSTRUMENTAL PARAMETERS

In ideal situations, ESI allows for the unbiased transfer of all analytes to the gas phase, thus producing a faithful snapshot of the actual partitioning between free and bound species in solution (Figure 9.2b). However, achieving this desirable outcome requires avoiding possible pitfalls associated with events that are integral parts of the ESI process. According to the accepted mechanisms (Figure 9.1), the initial droplets generated by a strong electric field travel through regions of differential vacuum, where they undergo solvent evaporation and collide with solvent molecules and other droplets. Collisions produce smaller droplets that keep evaporating and fissioning, until fully desolvated ions are left in the gas phase.[58,59] Alternatively, ions are ejected directly from solution by the increasing repulsive forces generated when the shrinking surface pushes charged species closer to one another.[60] It has been observed that, as the droplet size decreases, analyte concentrations can increase to the point where random association may take place between any of the solutes left in solution.[59,61] This effect could potentially induce an artificial increase of the signal provided by target complexes or lead to the detection of nonspecific aggregates that are devoid of any chemical/biological basis (Figure 9.2c).

The argument in favor of mass action as a possible factor inducing artifactual association does not take in account the time scale within which the solute concentrations may increase in the droplets during desolvation. In fact, theoretical calculations have shown that typical evaporation time frames fall in the ~μsec range and do not provide enough time for an equilibrium to shift substantially, even for reactants with association rate constants near the diffusion limit.[62,63] A different explanation invokes the possibility that raising concentrations could increase the probability of finding unbound components trapped in the same droplet after the final fission event has taken place. With no chance to segregate during the subsequent evaporation of residual solvent, the final result would consist of an aggregate that was not present in the initial solution. In this case, any statistical combination of the different species in solution could be potentially produced, thus accounting for the formation of nonspecific clusters as well as actual target complex. Consistent with a mechanism that is largely defined by the initial solute concentrations, only nonspecific aggregates have been obtained from very concentrated samples. In the case of nucleic acids, mixtures of complementary and noncomplementary oligonucleotides in concentrations exceeding 100 μM were shown to produce nonspecific complexes between strands that were not supposed to form duplexes.[64] These products were not observed after diluting the initial solution 10-fold, thus reinforcing the notion that analyte concentration should be kept as low as possible to avoid unexpected clustering effects.

In the ESI-MS analysis of noncovalent complexes, it is always important to know how to recognize the occurrence of nonspecific aggregation by designing *ad hoc* controls that require some knowledge of the system under investigation. When possible, chaotropic agents that disrupt productive interactions or mutant substrates devoid of binding capabilities should be employed to test whether the selected desolvation conditions may lead to unexpected association. We implemented a similar approach in the investigation of noncovalent assemblies formed by the nucleocapsid (NC) protein of human immunodeficiency virus type 1 (HIV-1) with RNA domains of the genome packaging signal (Ψ-RNA) (Figure 9.4).[56] As a control, a sample of apo-NC was obtained by denaturing its characteristic zinc-finger motifs to remove the two coordinated Zn(II) ions. Considering the crucial role played by zinc chelation in maintaining a three-dimensional (3D) fold that is conducive to specific RNA binding,[65] we mixed equimolar amounts of native and apo-NC with one of the target stem-loops. When the resulting solution was analyzed directly by ESI-FTICR, only the folded protein containing Zn^{2+} provided a stable RNA complex, whereas no interactions could be detected for the apo-form lacking the proper structure (Figure 9.5). On the one hand, this experiment demonstrated that the selected instrumental conditions did not induce artifactual aggregation. On the other, it confirmed the specific nature of the NC–stem-loop interactions, which requires the exquisite recognition between structural motifs, rather than nonspecific electrostatic contacts that may still take place after protein defolding.[56]

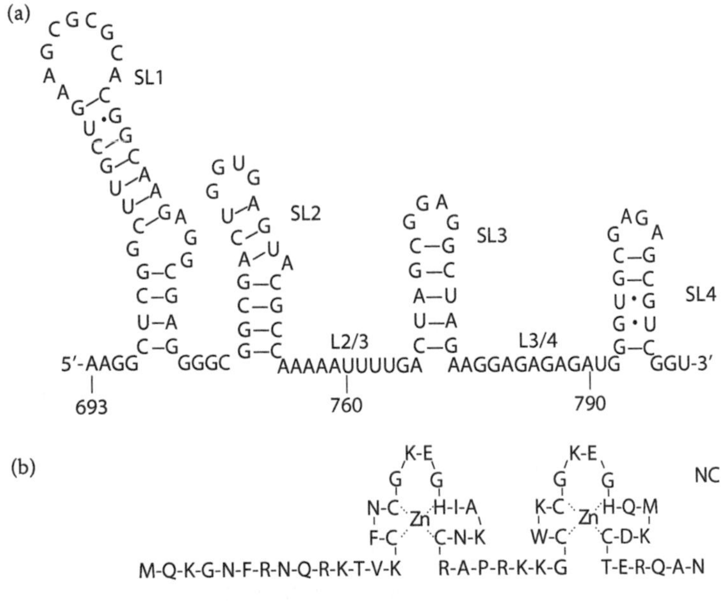

FIGURE 9.4 (a) Ψ-RNA constitutes the packaging signal of the HIV-1 genome. This highly conserved sequence folds into discrete stem-loop domains (named SL1 through 4) that are involved in different steps of the viral life cycle. Also labeled are the linker regions between SL2 and SL3 (L2/3) and between SL3 and SL4 (L3/4). The full-length construct corresponding to the subtype B sequence includes 115 nucleotides and presents a molecular mass of 37,984 Da. (b) The HIV-1 NC protein performs critical functions in the viral life cycle, which are mediated by its broad nucleic acid binding and chaperoning activities. Coordination of two Zn(II) ions is necessary to define the structure of two zinc-finger motifs of retroviral type (i.e., CCHC), which are responsible for NC's functions. This highly basic protein comprises 55 amino acid residues and presents a molecular mass of 6488.9 Da, including the coordinated Zn(II).

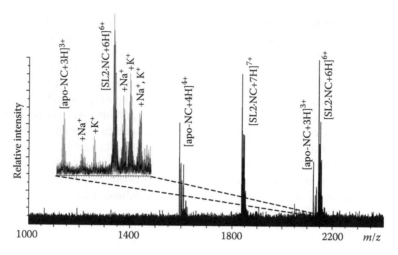

FIGURE 9.5 ESI-FTICR analysis of a sample obtained by adding SL2 to an equimolar mixture of apo- and holo-NC. Only the latter was proven capable of forming a stable 1:1 complex with the cognate RNA substrate (observed monoisotopic mass 12,618.0 ± 0.1 Da, 12,617.8 Da calculated from sequence). In the absence of Zn(II) coordination to stabilize the zinc-finger domains, no binding could be observed for the apo-form of NC. The inset expands the region containing the 3+ charge state of apo-NC and the 6+ ion of the noncovalent complex between holo-NC and SL2. Typical cation adducts are also observed.

A different effect produced by the collisional processes involved in desolvation consists of the possible activation of analyte ions in the source region of the mass spectrometer. Left unchecked, the energy and number of collisions between ionized species and neutral solvent can activate the dissociation of specific noncovalent interactions, as well as the cleavage of covalent bonds (Figure 9.2d). In general, complexes that are stabilized by significant hydrophobic contributions are significantly weakened by solvent removal and tend to dissociate more readily during the desolvation step.[66] In contrast, complexes involving at least some degree of electrostatic character, such as protein–nucleic acid assemblies, are less prone to dissociation during transfer to the gas phase.[6] While it is not normally possible or desirable to modify the nature of the assemblies of interest, selected instrumental parameters can be appropriately adjusted to minimize the destabilizing effects of the desolvation process.

One of the first parameters to consider is the temperature of the desolvation region, which has a significant impact on the outcome of ESI-MS analysis. In the case of nucleic acid substrates, complementary oligodeoxynucleotides have been effectively employed to test the effects of source temperature on the stability of pairing interactions.[67] Under the selected experimental conditions, it was shown that a double-stranded structure was readily detected below 180°C, whereas its single-stranded components were predominant above 200°C. Considering that this dissociation range is well above the melting point calculated for the duplex in solution, the results suggest that the actual droplet temperature may be significantly lower than expected during desolvation. A possible interpretation credits the process of solvent evaporation for dissipating most of the heat provided in the desolvation region, which would keep droplet temperatures low. A more plausible explanation is based on kinetic considerations indicating that droplet evaporation takes place in an estimated μsec scale, which is not sufficient for strand unzipping to reach completion within the analysis time frame.[67] Both arguments have merits and can be readily extended to protein–nucleic acid assemblies that are usually stabilized by a large number of electrostatic and H-bonding interactions. For this reason, the effects of source temperature on the association state should always be explored when tackling new systems and the results should be carefully examined to find the best possible instrumental settings.

Another important parameter is the so-called nozzle–skimmer voltage, or orifice voltage, depending on hardware design. These terms refer to the potential applied between two contiguous elements in the ESI source, which accelerate charged species through the source and toward the analyzer (Figure 9.1). Higher voltages increase the kinetic energy of the collisional processes involved in desolvation,[68,69] which may reach the threshold for activating the dissociation of target assemblies and the fragmentation of their covalent structures. This feature has been effectively utilized to assess the stability of nucleic acid complexes[15,69–71] and to obtain sequence information[67] in pseudo-MS/MS experiments lacking precursor ion selection. In similar fashion, the nozzle–skimmer voltage can provide an effective tool for revealing nonspecific interactions based on their apparent strength. For example, the voltage inducing complete dissociation of ammonium adducts from oligonucleotide ions can provide a practical boundary above which only the more stable and usually specific complexes are effectively detected (Figure 9.6). This criterion has been particularly helpful in the study of the binding modes of small molecule ligands and candidate inhibitors with selected nucleic acid substrates of biological interest.[72,73] Beyond specificity considerations, however, optimal voltages for this type of analysis should be capable of eliminating any adduct of water and cosolvent without destabilizing the actual target. In general, proper settings can be obtained by tuning the desolvation conditions directly on the assembly of interest, if sample availability is not an issue, while looking for signs of adduct formation or assembly dissociation. Alternatively, a complex of similar nature can be employed as a tuning standard to approach the desired conditions that can be subsequently refined using the actual analyte. For nucleic acid substrates, duplex constructs

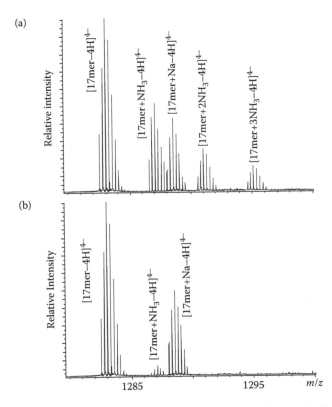

FIGURE 9.6 ESI-FTICR analysis of an unstructured 17-mer deoxy-oligonucleotide (17mer) in 10 mM ammonium acetate (pH adjusted to 7.0) and 10% (v/v) 2-propanol. Spectra (a) and (b) were obtained by using a nozzle–skimmer voltage of 115 and 165 V, respectively, while the other conditions were kept constant. The nearly complete elimination of ammonium adducts was achieved by raising the nozzle–skimmer voltage, whereas a similar outcome could not be obtained for the more stable sodium adduct.

of different length and base composition constitute very suitable standards for finding the proper desolvation conditions necessary to meet the demands of progressively weaker interactions.

9.4   WHAT IS IN A MOLECULAR MASS?

The unique relationship between molecular mass and chemical composition provides a wealth of information on the nature and structure of noncovalent complexes, which can translate into corresponding functional insights. In the case of nucleic acid assemblies, molecular mass determinations can lead to the identification of unknown ligands bound to the substrate of interest, and vice versa. When the possible binding partners are already known, the mass provides a direct reading of the stoichiometry of binding, which in turn can reveal the number of binding sites on a certain structure, or the arrangement of a multisubunit assembly. For these purposes, ESI-MS has been applied to characterize protein–nucleic acid complexes involving single-stranded oligodeoxynucleotides,[71] double-stranded oligodeoxynucleotides,[18,69,74–77] and a variety of RNA substrates.[56,70,78–80]

The ability to determine the stoichiometry of protein–RNA assemblies has enabled us to obtain valuable insights into the mechanisms of genome recognition, dimerization, and packaging mediated in HIV-1 by the NC protein. The ESI-FTICR analysis of samples obtained by adding increasing amounts of viral protein to selected oligodeoxynucleotides was carried out to determine the size of the binding site onto unstructured single-stranded nucleic acids. Based on the observed preference for exposed G-nucleotides, the titration of d(TG)n sequences of different length was followed not only by high-resolution MS, but also by surface plasmon resonance (SPR), tryptophan fluorescence quenching (TFQ), fluorescence anisotropy (FA), and isothermal calorimetry (ITC).[81,82] Whereas the data provided by these techniques required curve fitting to putative binding models, ESI-FTICR offered a direct and unambiguous determination of the binding stoichiometry in solution, which confirmed that a typical NC site spans at least five nucleotides. The ability to accommodate two protein units exhibited by the octamer d(TG)4, the second of which with lower affinity (vide infra), served to explain the lack of saturation observed in the binding curves obtained from spectroscopy experiments and the significant fitting discrepancies encountered during their data interpretation. In particular, the fact that a ternary complex was formed by a substrate that does not stretch the length of two full binding sites supports the hypothesis that NC could bridge across adjacent d(TG)4 strands to maximize its nucleotide contacts, thus accounting for the known ability of this viral protein to induce nucleic acid condensation.[81]

In the investigation of NC interactions with structured RNA substrates, ESI-FTICR was employed to identify the motifs that are capable of sustaining specific protein binding within the HIV-1 packaging signal (Figure 9.4). After preliminary data had shown that full-length Ψ-RNA was capable of binding a maximum of six protein units, a "divide-and-conquer" approach was implemented in which its 115-nucleotide sequence was subdivided into constructs corresponding to individual stem-loop and linker domains. After these species were mixed in equimolar amounts and titrated with increasing concentrations of NC, ESI-FTICR analysis showed that stem-loops 2–4 (SL2, SL3, and SL4) and the linker between SL3 and SL4 (L3/4) were capable of interacting with one NC each, whereas stem-loop 1 (SL1) bound up to two units (Figure 9.7).[83] These results were consistent with the known tropism of NC for unpaired G-nucleotides and for hairpin structures, in particular. The fact that the total number of binding events obtained from the separate constructs matched the maximum stoichiometry afforded by the full-length sequence ruled out the possibility that Ψ-RNA might fold into a compact structure that may hinder the access to some of the stem-loops, or create new binding sites with contributions from contiguous domains.

A concerted strategy using ad hoc mutants and ESI-FTICR analysis was implemented to investigate the multifaceted binding modes exhibited by SL1, which are complicated by its ability to initially form a metastable kissing-loop (KL) dimer that can subsequently isomerize into a more stable extended duplex (ED) form (Figure 9.8). For this reason, selected monomeric constructs had been previously shown to possess two specific sites by fluorescence quenching experiments[84,85] that

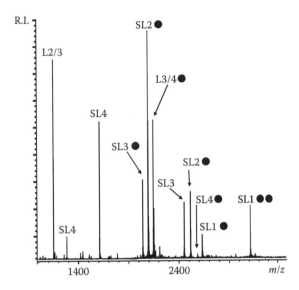

FIGURE 9.7 Competitive binding of NC (●) to isolated domains and linkers of Ψ-RNA. With the exception of L2/3, all RNA substrates were found capable of sustaining protein interactions under the selected experimental conditions. After considering the different charge states exhibited by the detected species (see text), the following scale of binding affinities was derived from the observed relative intensities: SL2 > SL3 > SL1′ ≈ SL1″ > L3/4 > SL4 ≫ L2/3. (Reproduced from Fabris, D. et al. *Eur. J. Mass Spectrom.* 13, 29–33, 2007. With permission.)

were not corroborated by native gel electrophoresis and NMR,[86] thus casting doubts on the number and location of the putative sites. In the absence of high-resolution 3D structures for any possible NC–SL1 complex, ESI-FTICR determinations provided unambiguous proof of the existence of two independent sites (as shown also in Figure 9.7) and led to the identification of their position on the stem-loop structure.[87] Following a systematic process of elimination, we evaluated the binding properties of mutants in which the palindromic loop or the stem-bulge were alternatively removed (SL1-BG1/SL1-BG2 and SL1-TR, Figure 9.8). The fact that each construct was still capable of binding at least one equivalent of NC proved beyond doubt that these salient motifs are responsible for sustaining the observed interactions (Figure 9.9). In similar fashion, conformer-specific mutants that are incapable of undergoing isomerization served to demonstrate that the stem-bulge site remains unaffected by dimerization, despite the fact that this motif is formed in the ED dimer by contributions from cognate strands. The binding capabilities of the loop site were instead severely inhibited by the participation of the palindrome sequence in interstrand contacts, regardless of the isomeric form (Figure 9.8).

In these examples, the nucleic acid component was mutated in systematic fashion and ESI-MS was employed to monitor the effects of nucleotide deletions or substitutions on the binding properties of the target assembly. Mirroring this approach, however, the cognate protein could also be mutated to highlight the contributions of selected amino acid residues to the binding mechanism, thus completing the spectrum of information about the specific roles played by the constitutive components of the complex. In addition, this analytical platform could be utilized to extend the scope of the investigation to include the possible effects of different third-party ligands that are not physiological components of the assembly. An example of this type of application consists of our study of archetypical nucleic acid ligands as possible inhibitors of the specific interactions between NC and selected Ψ-RNA domains.[73] Initially, we followed a multiplexed strategy in which groups of ligands were presented simultaneously to each individual RNA substrate, taking advantage of the

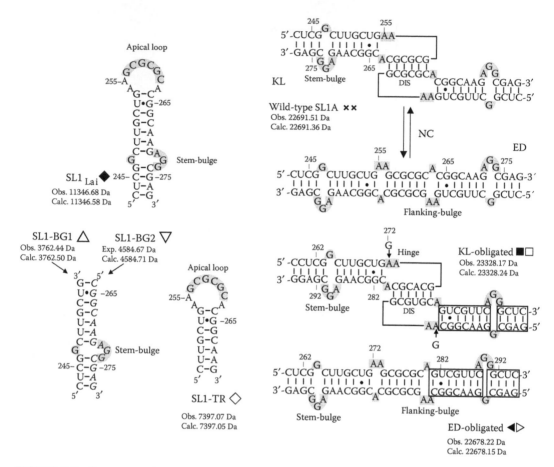

FIGURE 9.8 Sequence and secondary structure of constructs employed in the investigation of the process of dimerization and isomerization of HIV-1 SL1. After monomeric SL1 had shown the ability to bind two equivalents of NC (see Figure 9.7), the position of the specific binding sites was identified by using a construct devoid of apical loop, which was obtained by annealing the complementary strands SL1-BG1 and SL1-BG2, and a truncated mutant SL1-TR devoid of stem-bulge.[87] During dimerization, an initial KL complex is formed by the annealing of the palindrome sequence situated on the apical loop. The metastable KL dimer is converted by NC into a more stable ED structure. The mass spectrometric investigation of the isomerization mechanism was facilitated by designing mutants that could assume exclusively the KL or the ED conformation (KL- and ED-obligated constructs, respectively), with no possibility of interconversion.[97]

high resolution afforded by ESI-FTICR to correctly determine the composition of the various complexes formed in solution. Subsequently, individual ligands were added to solutions containing preformed NC–stem-loop complexes. The results were consistent with the known binding modes of the archetypical ligands and with the unique structural features displayed by each substrate. For example, the aminoglycosidic antibiotic neomycin B (NB) was found capable of displacing SL3 from its NC complex, whereas the mixed-mode binder mitoxantrone (MT) did not induce dissociation, but formed instead ternary assemblies with multiple stoichiometries (Figure 9.10). In similar fashion, the aminoglycoside caused dissociation of the analogous NC·SL4 complex, but bound instead to NC·SL2 in a stable ternary assembly. The different behaviors were interpreted on the basis of a direct competition between NB and NC for the same binding sites onto the SL3 and SL4 structures, as later confirmed by locating the aminoglycoside sites in the upper-stem regions of these stem-loops.[88] In contrast, the aminoglycoside site on SL2 presents only marginal overlap with that of NC, thus justifying the ability of this RNA substrate to bind both ligands simultaneously.[73]

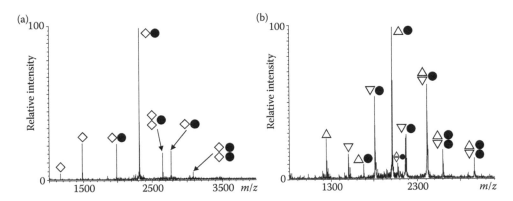

FIGURE 9.9 (a) ESI-FTICR analysis of a sample containing NC (●) and SL1-TR (◇) revealed the predominant formation of a 1:1 complex, thus indicating that elimination of the stem-bulge motif resulted in the loss of one binding site. Note that dimeric SL1-TR can still sustain weaker binding through specific interactions with the unpaired A nucleotides flanking the palindrome in the ED structure. (b) A 1:1 complex was also detected upon addition of NC to the SL1-BG1/SL1-BG2 construct (△▽), thus confirming the apical loop as the remaining binding site. In this case, the normal chaperoning activity of NC induced both partial and complete unzipping of the duplex structure, as evident from the detection of NC complexes with both single- and double-stranded species. (Adapted from Hagan, N.A. and Fabris, D., *J. Mol. Biol.* 365, 396–410, 2007. With permission.)

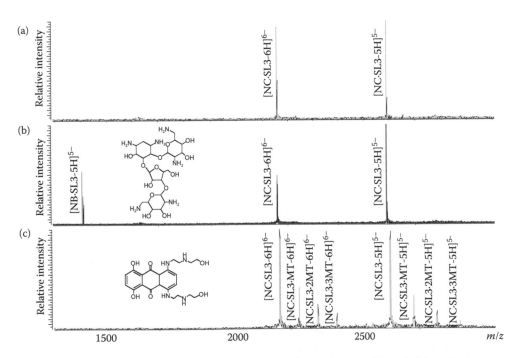

FIGURE 9.10 ESI-FTICR analysis of the NC·SL3 complex in the absence of ligands (a) and in the presence of either (b) the aminoglycosidic antibiotic NB, or (c) the mixed-mode anticancer drug MT. After the addition of NB, the protein–nucleic acid assembly was detected together with a new drug–nucleic acid complex, which is consistent with a direct competition for overlapping binding sites. The fact that addition of MT did not induce assembly dissociation, but produced stable ternary complexes, indicates that this ligand does not bind to the same site occupied by NC. (Reproduced from Turner, K.B., Hagan, N.A., and Fabris, D., *Nucleic Acids Res.* 34, 1305–1316, 2006. With permission.)

9.5   BEYOND MOLECULAR MASS DETERMINATIONS

The information provided by mass spectrometric determinations is not limited to the molecular mass of a certain species, which can be read on the mass-to-charge (m/z) ratio scale. The signal recorded for each analyte exhibits an intensity that is proportional to the number of ions detected, which is ultimately determined by the amount of analyte in the initial sample. For this reason, in addition to a positive identification of free and bound species in solution, ESI-MS is capable of providing quantitative information about their respective partitioning. In turn, this information provides the basis for calculating dissociation constants (K_d's), inhibition constants (K_i's), and other figures of merit concerning the equilibrium of interest. A great deal of attention, however, must be paid not only to the experimental approaches, but also to the procedures followed to translate the raw data into the desired quantity.[89] In typical titration and competition experiments, the selection of proper desolvation conditions is paramount to ensure that the observed spectrum reflects the actual situation in solution, as discussed earlier. Any situation affecting the position of the binding equilibrium during analysis constitutes a potential source of error, which should be eliminated or at least recognized by completing necessary control experiments. When the ideal situation is verified and there are no adverse effects on the state of association, the next step consists of finding the correct relationship between signal intensity and sample concentration. Due to intrinsic differences of ionization efficiency, such relationship may vary for the different species involved in the equilibrium. Although standard calibration curves could be built to obtain individual response factors,[90] this route may not be practical when serial dilutions of intact complex could induce its partial dissociation to satisfy the equilibrium concentrations of the corresponding unbound components. For this reason, alternative approaches have been devised to obtain accurate quantitative determinations of noncovalent complexes without direct knowledge of their response factors.[89]

As part of titration experiments aimed at determining the K_d of selected NC–stem-loop assemblies, we have employed the molar fractions of the species in equilibrium to obtain their actual concentrations.[56] This analysis involved adding increasing amounts of NC to a fixed concentration of each RNA substrate, using initial stock solutions of accurately known concentrations measured by UV spectroscopy. The partitioning between free and bound species was assessed by ESI-FTICR and the respective molar fractions were calculated directly from the observed relative intensities. Peak heights rather than peak areas were employed as initial raw data, following the guidelines suggested by Eyler and coworkers for quantitative determinations by FTICR.[91] Each peak height was divided by the respective charge state to correct for the fact that image currents detected in FTICR depend on the number of charges carried by the ion.[11] This quantity was calculated for all the charge states displayed by the same analyte and the values were summed together to obtain a fair and comprehensive measure of its contribution to the total ion current detected in the experiment.

This operation minimizes any bias generated by possible binding effects on the charging characteristics of the complex, which might result in the misrepresentation of its actual abundance in solution. Considering that the ESI-MS analysis of biomolecules relies on protonation/deprotonation of suitable functional groups, any changes in solvent exposure, H-bonding pattern, or near-neighbor effect may affect the pK_a of chargeable groups involved in binding. If the analyte were singly charged, this could result in ion neutralization and signal suppression. In multiply charged ions, however, binding tends to affect only a fraction of all available groups. The final result is a mere shift of the observed charge states distribution, rather than complete signal suppression. Therefore, adding all the signals detected for a given complex minimizes the impact of possible charging effects and provides a more faithful representation of its abundance in solution.

From the values representing the abundance of each species, the desired molar fractions were readily calculated and then multiplied by the initial concentrations of substrate or titrant to obtain the respective final concentrations at equilibrium. Repeated after each addition of titrant, this treatment provided concentrations that were in turn employed to calculate individual K_ds for each titration point, which were averaged together to reach the reported value.[56] Alternatively, the concentrations obtained from molar fractions could be employed to build canonical binding curves that could be fitted and analyzed to obtain the desired K_d's.[73] In the case of the NC–stem-loop assemblies, our MS-based approach produced K_d's matching with those obtained previously by ITC determinations,[92] while requiring far lower sample consumption and raising no ambiguities about binding stoichiometries.[56] In an analogous way, equilibrium concentrations obtained by ESI-FTICR were employed to evaluate the inhibitory effects induced by small ligands on selected NC–stem-loop assemblies, which were properly quantified by calculating the corresponding concentrations inducing 50% inhibition (IC_{50}'s).[73]

In many instances, the actual values of these figures of merit may not have immediate significance *per se*, but become critical to enabling meaningful correlations between related biological systems. Competitive binding experiments based on the ability of ESI-MS to resolve complex sample mixtures can provide direct comparisons with no need for rigorous quantitative treatments. An example is provided by our investigation of the binding sites of Ψ-RNA, in which equimolar amounts of the individual domains competed for NC binding.[83,87] After all the possible equilibria were established in solution, any free and bound species in the mixture were fully resolved by ESI-FTICR (Figure 9.7). A close inspection of the relative intensities revealed a distinctive partitioning pattern among the different substrates, as expected from their unique affinities toward NC. The simultaneous assessment of all binding events eliminated possible reproducibility errors that may be induced by spray instability, source contamination, decreased ion transmission, and any other cause of ion current fluctuations occurring between individual determinations. Accounting for the different charge states as discussed earlier, the competition provided the following scale of relative binding affinities: SL2 > SL3 > SL1′ ≈ SL1″ > L3/4 > SL4, whereas no binding was observed for L2/3 within the explored range of concentrations. In this study, establishing a firm ranking for the different domains provided helpful insights into their individual contributions to the process of genome recognition and packaging mediated by NC. In the investigation of archetypical nucleic acid ligands, a similar competition scheme was implemented to multiplex the application of possible inhibitors, thus expediting the selection of the tighter binders for further evaluation.[73] Based on the high resolution afforded by the FTICR platform, competitive binding experiments are expected to find increasing applications in drug discovery for screening combinatorial libraries and reducing the number of candidates to the point where actual biological testing becomes practical.

9.6 GAS-PHASE DISSOCIATION OF NONCOVALENT COMPLEXES

Approaches involving accurate determinations of sample concentrations are usually designed to evaluate the strength of a certain interaction and require that no perturbation of the binding equilibrium took place during analysis for the observed constant to be safely ascribable to a solution-phase process. However, additional information on the nature of the binding interaction can be acquired by purposely activating the gas-phase dissociation of the assembly of interest. This process can be completed either in the source region by increasing the energy involved in ion desolvation (Figure 9.2d) or in the mass analyzer by performing a full-fledged tandem experiment (Figure 9.2b, e, and f). In recent years, the former approach has been employed to assess the stability of protein complexes with single-[71] and double-stranded DNA,[69,74] and the organization of the protein components in ribosomal subunits.[15,70] The latter has been used to probe the interactions between the bacteriophage T4 regA protein and cognate RNA substrates.[93] In general, classic tandem experiments involve isolating the desired precursor ion from the sample mixture (Figure 9.2e), which enables the unambiguous determination of its dissociation pathway. This is not always the case for in-source dissociation methods, in which all possible precursor ions in the mixture are dissociated together in the same experiment (Figure 9.2d). Depending on the activation regime, typical outcomes can range from assembly dissociation into individual subunits,[15,69–71,74] which is entropy-driven and takes place under fast activation conditions, to partial fragmentation of the covalent backbones of bound components,[74,93] which is instead enthalpy-driven and is favored by slow activation.[94] Although a direct correlation between gas- and solution-phase behavior is sometimes difficult,[95] experiments of this type have produced an increasing body of evidence in favor of the gas-phase preservation of hydrogen bonds and stacking interactions in nucleic acid ions,[37,38,67,94,96] which has greatly increased the interest in the application of gas-phase approaches to study the higher order structure of these biomolecules and their noncovalent assemblies.

In the investigation of the mechanism of genome dimerization in HIV-1, we have employed MS/MS to interrogate the conformational state of the complexes formed by NC and the alternative dimeric forms of SL1 (Figure 9.8).[97] Initially, conformer-specific constructs that could only fold into either the KL or the ED structure were submitted to sustained off-resonance irradiation collision-induced dissociation (SORI-CID)[98] to test whether their unique base-paring patterns would produce distinctive gas-phase behaviors. This method involves isolating the precursor ion of interest using correlated rf sweeps (CHEF),[99] irradiating it with a slightly off-resonance pulse to increase its energy, and then colliding it with Ar to induce dissociation.[98] Under relatively mild activation conditions, the KL-obligated dimer was readily dissociated into individual hairpins, whereas the ED-obligated complex remained intact (Figure 9.11). These results are consistent with the fact that the process of strand dissociation in nucleic acid duplexes follows a multistep mechanism that is largely defined by

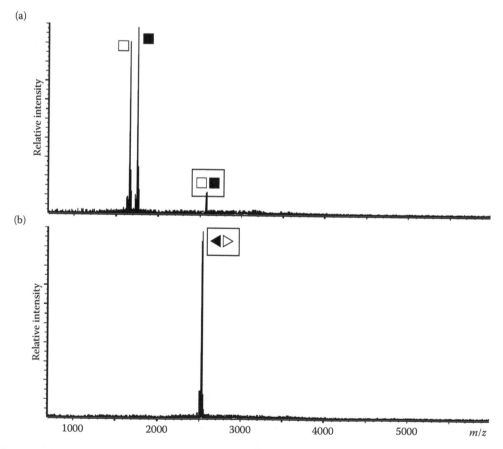

FIGURE 9.11 Product ion spectra obtained by submitting (a) the KL-obligated dimer (■□) and (b) the ED-obligated dimer (◄▷) to SORI-CID. Under the selected conditions, the KL structure readily dissociated into individual stem-loop components, whereas the ED conformer remained intact. This is consistent with the greater stabilization afforded by 28 interstrand base pairs in the latter versus only 6 in the former. (Reproduced from Turner, K.B., Hagan, N.A., and Fabris, D., *J. Mol. Biol.* 369, 812–828, 2007. With permission.)

kinetic considerations.[94] The presence of 28 interstrand base pairs in the ED dimer versus only 6 in the KL conformer accounts for the increased overall activation barrier and reaction time necessary to induce duplex unzipping. These distinctive structural features are also at the basis of the different kinetic stabilities exhibited by the NC complexes of dimeric SL1, which mirrored very closely those provided by the respective RNA conformers. Under the selected experimental conditions, assemblies formed by the KL-obligated construct underwent dissociation into individual NC–hairpin components with only minor loss of protein binding (Figure 9.12a). In contrast, activation of the ED-obligated complexes induced stepwise dissociation of NC units, but left the duplex core intact (Figure 9.12b).

The unique gas-phase behaviors observed for conformer-specific mutants offered the key to interpret the results obtained from wild-type SL1, which was capable of transitioning from KL to ED in the presence of NC (Figure 9.8). The fact that complexes of alternative conformations shared identical composition/stoichiometry did not allow for their form to be clearly recognized from their molecular masses alone (Figure 9.13a). However, completing MS/MS experiments for each assembly observed in solution enabled us to distinguish the two different populations based on their characteristic dissociation patterns (Figure 9.13b and c). Using a semiquantitative treatment analogous

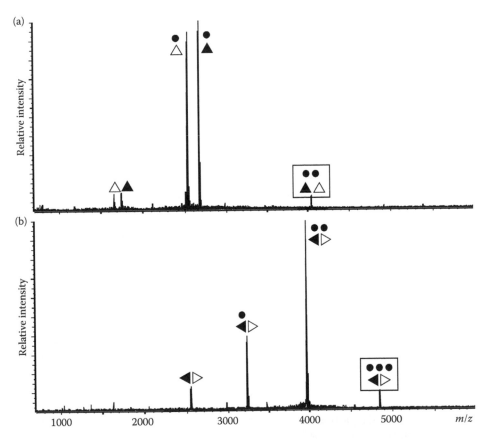

FIGURE 9.12 Product ion spectra obtained by submitting (a) 2:2 complex of NC (●) with KL-obligated dimer (▲△) and (b) its 3:2 complex with the ED-obligated dimer (◀▷) to SORI-CID. The ED assembly displayed stepwise losses of protein units, but no strand dissociation. In contrast, the KL complex underwent strand dissociation that provided individual stem-loop products bound to NC. (Adapted from Turner, K.B., Hagan, N.A., and Fabris, D., *J. Mol. Biol.* 369, 812–828, 2007. With permission.)

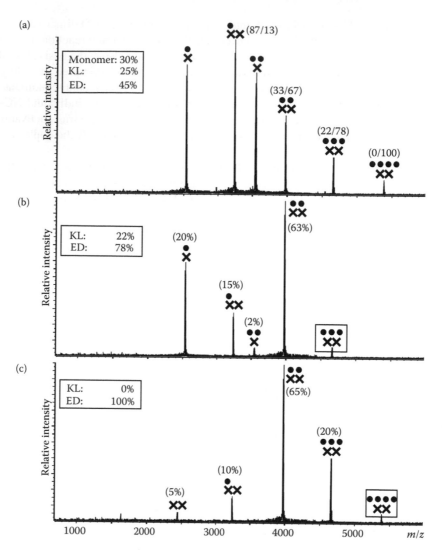

FIGURE 9.13 (a) ESI-FTICR mass spectrum of a sample containing NC (●) and wild-type SL1A (✗) in 150 mM ammonium acetate (pH 7.0) after 3 h incubation at 37°C. Panels b and c show the product ion spectra obtained by submitting the 3:2 and 4:2 complexes to SORI-CID, with boxed symbols identifying each precursor ion. In panels b and c, the percentages in parentheses indicate the normalized intensity of the corresponding signal compared to the total intensity of the product species, as described in the text. Dimeric products were assigned to the ED conformer, whereas monomeric products were assigned to KL. The percentages were then summed to determine the partitioning between the two conformers within each precursor ion population, as reported in the box. In panel a, the KL/ED proportions within each ion signal were used to obtain the overall proportions of monomeric, KL, and ED forms in solution. (Reproduced from Turner, K.B., Hagan, N.A., and Fabris, D., *J. Mol. Biol.* 369, 812–828, 2007. With permission.)

to that discussed earlier, the partitioning between KL and ED assemblies was estimated to obtain a measure of the ability of NC to mediate the isomerization process. In turn, this approach was applied to selected SL1 mutants to investigate the role of the different motifs in the mechanism of RNA dimerization and stabilization. The results showed that the broad binding and chaperoning activities of NC can produce very different outcomes that are defined by the structural context of the RNA domains involved in the specific interaction, rather than by the protein function itself.[97]

9.7   FUTURE DIRECTIONS

The ever-increasing availability of genetic information for a growing number of organisms has precipitated the need for correlating the product of each gene with its specific function and for elucidating its mechanism of action at the molecular level. The prominent roles played by protein–nucleic acid interactions in the cell life cycle and the potential for their complexes to constitute viable targets for therapeutic intervention justify the efforts aimed at developing new experimental approaches for their functional investigation. In this direction, their mass spectrometric analysis is expected to benefit greatly from the introduction of new techniques for desorbing progressively larger ions directly from cellular environments, from the refinement of alternative methods for the gas-phase activation of biomolecules, and from the increasing availability of instrumentation for ion mobility mass spectrometry (IMS).

The recent introduction of ambient ionization techniques, such as desorption electrospray ionization (DESI),[100] electrospray-assisted laser desorption/ionization (ELDI),[101] and matrix-assisted laser desorption electrospray ionization (MALDESI),[102] enables the examination of species submitted to analysis directly in their native environment without sample preparation. Applied to noncovalent complexes, these techniques are expected to minimize the risk of disrupting weak interactions, which is inherent in the established separation and purification procedures. These types of experiments could provide faithful snapshots of binding equilibria taken in the presence of the normal cellular components that are typically removed during sample processing procedures. In addition, the potential for utilizing these techniques to perform MS-imaging of tissue sections[103,104] could pave the way for mapping tissue-specific expression and composition of target assemblies, which could further the field of cell biology and help the diagnosis of pathological states.

The development of alternative methods for the activation of precursor ions in MS/MS will greatly expand the range of information that can be obtained from protein–nucleic acid assemblies. For example, it has been recently shown that submitting a drug–RNA complex to SORI-CID can induce facile fragmentation of the phosphodiester backbone, with the exception of the nucleotides in direct contact with the bound ligand.[88] In similar fashion, infrared multiphoton dissociation (IRMPD),[105,106] electron-capture dissociation (ECD),[107,108] and electron-transfer dissociation (ETD)[109] could be employed to induce selective fragmentation of the protein or nucleic acid components of target assemblies. In what amounts to a gas-phase footprinting experiment, the lack of fragmentation in specific regions would reveal protection effects induced by binding interactions and would help to map the interface between bound biomolecules. Direct information about the residues/nucleotides involved in intermolecular contacts would provide precious details about the actual mechanism followed by a certain assembly to complete its task.

Finally, novel approaches based on IMS[110,111] will spur an increasing number of applications aimed at investigating the global fold of large assemblies produced under controlled environmental conditions. The ability to obtain direct determinations of their cross sections in the gas phase provides valuable information about their structure in solution. Comparing the results obtained in the absence and presence of selected ligands will allow for a direct determination of conformational changes that could reveal important mechanistic details. This type of application is expected to facilitate the study of the effects of metabolites, cofactors, and messenger molecules on the spatial situation of large multisubunit receptors. In this way, it will be also possible to evaluate candidate inhibitors not only for their ability to cause assembly dissociation, but also for possible activity in stabilizing conformations that are not compatible with normal function.

In conclusion, ESI-MS affords a unique platform for investigating the biophysical properties of a vast array of biomolecular complexes, which has provided and will continue to provide a wealth of information regarding their structure, dynamics, and mechanism of activity. Bolstered by its modest sample requirements, broad applicability, and high-throughput capabilities, this technique has emerged as a very powerful tool for the investigation of vital cellular processes and for early-stage drug discovery. As new diseases come to the forefront and old infectious agents rear their head for the selection of resistant strains, MS-based technologies will continue to support the identification of target assemblies for the development of novel antiviral and antimicrobial therapies.

ACKNOWLEDGMENTS

This research was funded by the National Institutes of Health (R01-GM643208) and the National Science Foundation (CHE-0439067). N.A. Hagan was also supported by an NIH Chemistry Biology Interface Training Fellowship (T32-GM066706).

REFERENCES

1. Yamashita, M. and Fenn, J.B., Electrospray ion source. Another variation on the free-jet theme, *J. Phys. Chem.* 88, 4671–4675, 1984.
2. Aleksandrov, M.L., Gall, L.N., Krasnov, V.N., Nikolaev, V.I., Pavlenko, V.A., and Shkurov, V.A., Extraction of ions from solutions under atmospheric pressure: A method of mass spectrometric analysis of bioorganic compounds, *Doklady Akademii Nauk.* 277, 379–383, 1984.
3. Ganem, B., Li, Y.T., and Henion, J.D., Detection of non-covalent receptor–ligand complexes by mass spectrometry, *J. Am. Chem. Soc.* 113, 6294–6296, 1991.
4. Ganem, B., Li, Y.T., and Henion, J.D., Observation of noncovalent enzyme-substrate and enzyme-product complexes by ion-spray mass spectrometry, *J. Am. Chem. Soc.* 113, 7818–7819, 1991.
5. Ganguly, A.K., Pramanik, B.N., Tsarbopoulos, A., Covey, T.R., Huang, E., and Fuhrman, S.A., Mass spectrometric detection of the noncovalent GDP-bound conformational state of the human H-*ras* protein, *J. Am. Chem. Soc.* 114, 6559–6560, 1992.
6. Loo, J.A., Studying noncovalent protein complexes by electrospray ionization mass spectrometry, *Mass Spectrom. Rev.* 16, 1–23, 1997.
7. Pramanik, B.N., Bartner, P.L., Mirza, U.A., Liu, Y.H., and Ganguly, A.K., Electrospray ionization mass spectrometry for the study of non-covalent complexes: An emerging technology, *J. Mass Spectrom.* 33, 911–920, 1998.
8. Beck, J.L., Colgrave, M.L., Ralph, S.F., and Sheil, M.M., Electrospray ionization mass spectrometry of oligonucleotide complexes with drugs, metals, and proteins, *Mass Spectrom. Rev.* 20, 61–87, 2001.
9. Hofstadler, S.A., and Griffey, R.H., Analysis of noncovalent complexes of DNA and RNA by mass spectrometry, *Chem. Rev.* 101, 377–390, 2001.
10. Comisarow, M.B., and Marshall, A.G., Fourier transform ion cyclotron resonance, *Chem. Phys. Lett.* 25, 282–283, 1974.
11. Hendrickson, C.L., Emmett, M.R., and Marshall, A.G., Electrospray ionization Fourier transform ion cyclotron resonance mass spectrometry, *Annu. Rev. Phys. Chem.* 50, 517–536, 1999.
12. Chen, R., Cheng, X., Mitchell, D.W., Hofstadler, S.A., Wu, Q., Rockwood, A.L., Sherman, M.G., and Smith, R.D., Trapping, detection, and mass determination of coliphage T4 DNA ions of 10^8 Da by electrospray ionization Fourier transform ion cyclotron resonance mass spectrometry, *Anal. Chem.* 67, 1159–1168, 1995.
13. Cheng, X., Camp, D.G., 2nd, Wu, Q., Bakhtiar, R., Springer, D.L., Morris, B.J., Bruce, J.E., Anderson, G.A., Edmonds, C.G., and Smith, R.D., Molecular weight determination of plasmid DNA using electrospray ionization mass spectrometry, *Nucleic Acids Res.* 24, 2183–2189, 1996.
14. Cotter, R.J., *Time-of-Flight Mass Spectrometry. Instrumentation and Applications in Biological Research*, ACS Professional Reference Books, ACS, Washington, DC, 1997.
15. Rostom, A.A., Fucini, P., Benjamin, D.R., Juenemann, R., Nierhaus, K.H., Hartl, F.U., Dobson, C.M., and Robinson, C.V., Detection and selective dissociation of intact ribosomes in a mass spectrometer, *Proc. Nat. Acad. Sci. USA* 97, 5185–5190, 2000.
16. McKay, A.R., Ruotolo, B.T., Ilag, L.L., and Robinson, C.V., Mass measurements of increased accuracy resolve heterogeneous populations of intact ribosomes, *J. Am. Chem. Soc.* 128, 11433–11442, 2006.
17. Greig, M.J., Gaus, H.J., and Griffey, R.H., Negative ionization micro electrospray mass spectrometry of oligonucleotides and their complexes, *Rapid Commun. Mass Spectrom.* 10, 47–50, 1996.
18. Kapur, A., Beck, J.L., Brown, S.E., Dixon, N.E., and Sheil, M.M., Use of electrospray mass spectrometry to study the binding interactions between a replication terminator protein and DNA, *Protein Sci.* 11, 147–157, 2002.
19. Valaskovic, G.A., Kelleher, N.L., and McLafferty, F.W., Attomole protein characterization by capillary electrophoresis-mass spectrometry, *Science* 273, 1199–1202, 1996.

20. Martin, S.E., Shabanowitz, J., Hunt, D.F., and Marto, J.A., Subfemtomole MS and MS/MS peptide sequence analysis using nano-HPLC micro-ESI Fourier transform ion cyclotron resonance mass spectrometry, *Anal. Chem.* 72, 4266–4274, 2000.

21. Wilm, M., and Mann, M., Analytical properties of the nanoelectrospray ion source, *Anal. Chem.* 68, 1–8, 1996.

22. McLafferty, F.W., Tandem mass spectrometry, *Science* 214, 280–287, 1981.

23. Hunt, D.F., Shabanowitz, J., Yates, J.R. r., Zhu, N.Z., Russell, D.H., and Castro, M.E., Tandem quadrupole Fourier transform mass spectrometry of oligopeptides and small proteins, *Proc. Natl. Acad. Sci. USA* 84, 60–63, 1987.

24. Biemann, K., and Scoble, H., Characterization by tandem mass spectrometry of structural modifications in proteins, *Science* 237, 992–998.

25. Papayannopoulos, I.A., 1995, The interpretation of collision-induced dissociation tandem mass spectra of peptides, *Mass Spectrom. Rev.* 15, 49–73, 1987.

26. Cerny, R.L., Tomer, K.B., Gross, M.L., and Grotjahn, L., Fast atom bombardment combined with tandem mass spectrometry for determining structures of small oligonucleotides, *Anal. Biochem.* 165, 175–182, 1987.

27. McLuckey, S.A., and Habibi-Goudarzi, S., Decompositions of multiply charged oligonucleotide anions, *J. Am. Chem. Soc.* 115, 12085–12095, 1993.

28. Little, D.P., Chorush, R.A., Spier, J.P., Senko, M.W., Kelleher, N.L., and McLafferty, F.W., Rapid sequencing of oligonucleotides by high-resolution mass spectrometry, *J. Am. Chem. Soc.* 116, 4893–4897, 1994.

29. Limbach, P.A., Crain, P.F., and McCloskey, J.A., Characterization of oligonucleotides and nucleic acids by mass spectrometry, *Curr. Opin. Biotechnol.* 6, 96–102, 1995.

30. Nordhoff, E., Kirpekar, F., and Roepstorff, P., Mass spectrometry of nucleic acids, *Mass Spectrom. Rev.* 15, 67–138, 1996.

31. Kowalak, J.A., Pomerantz, S.C., Crain, P.F., and McCloskey, J.A., A novel method for the determination of post-transcriptional modification in RNA by mass spectrometry, *Nucleic Acids Res.* 21, 4577–4585, 1993.

32. Barry, J.P., Vouros, P., Van Schepdael, A., and Law, S.-J., Mass and sequence verification of modified oligonucleotides using electrospray tandem mass spectrometry, *J. Mass Spectrom.* 30, 993–1006, 1995.

33. Iannitti, P., Sheil, M.M., and Wickham, G., High sensitivity and fragmentation specificity in the analysis of drug-DNA adducts by electrospray tandem mass spectrometry, *J. Am. Chem. Soc.* 119, 1490–1491, 1997.

34. McCloskey, J.A., Graham, D.E., Zhou, S., Crain, P.F., Ibba, M., Konisky, J., Soll, D., and Olsen, G.J., Post-transcriptional modification in archaeal tRNAs: Identities and phylogenetic relations of nucleotides from mesophilic and hyperthermophilic Methanococcales, *Nucleic Acids Res.* 29, 4699–4706, 2001.

35. Kellersberger, K.A., Yu, E., Kruppa, G.H., Young, M.M., and Fabris, D., Top-down characterization of nucleic acids modified by structural probes using high-resolution tandem mass spectrometry and automated data interpretation, *Anal. Chem.* 76, 2438–2445, 2004.

36. Gale, D.C. and Smith, R.D., Characterization of noncovalent complexes formed between minor groove binding molecules and duplex DNA by electrospray ionization-mass spectrometry, *J. Am. Soc. Mass Spectrom.* 6, 1154–1164, 1995.

37. Gabelica, V., Rosu, F., Houssier, C., and De Pauw, E., Gas phase thermal denaturation of an oligonucleotide duplex and its complexes with minor groove binders, *Rapid Commun. Mass Spectrom.* 14, 464–467, 2000.

38. Wan, K.X., Gross, M.L., and Shibue, T., Gas-phase stability of double-stranded oligodeoxynucleotides and their noncovalent complexes with DNA-binding drugs as revealed by collisional activation in an ion trap, *J. Am. Soc. Mass Spectrom.* 11, 450–457, 2000.

39. David, W.M., Brodbelt, J.S., Kerwin, S.M., and Thomas, P.W., Investigation of quadruplex oligonucleotide-drug interaction by electrospray ionization mass spectrometry, *Anal. Chem.* 74, 2029–2033, 2002.

40. Fenn, J.B., Mann, M., Meng, C.K., Wong, S.F., and Whitehouse, C.M., Electrospray ionization for mass spectrometry of large biomolecules, *Science* 246, 64–71, 1989.

41. Cech, N.B. and Enke, C.G., Practical implications of some recent studies in electrospray ionization fundamentals, *Mass Spectrom. Rev.* 20, 362–387, 2001.

42. Amad, M.H., Cech, N.B., Jackson, G.S., and Enke, C.G., Importance of gas-phase proton affinities in determining the electrospray ionization response for analytes and solvents, *J. Mass Spectrom.* 35, 784–789, 2000.

43. Stults, J.T. and Marsters, J.C., Improved electrospray ionization of synthetic oligodeoxynucleotides, *Rapid Commun. Mass Spectrom.* 5, 359, 1991.

44. Pieles, U., Zurcher, W., Schar, M., and Moser, H.E., Matrix-assisted laser desorption ionization time-of-flight mass spectrometry: A powerful tool for the mass and sequence analysis of natural and modified oligonucleotides, *Nucleic Acids Res.* 21, 3191–3196, 1993.

45. Nordhoff, E., Ingendoh, A., Cramer, R., Overberg, A., Stahl, B., Karas, M., Hillenkamp, F., and Crain, P.F., Matrix-assisted laser desorption/ionization mass spectrometry of nucleic acids with wavelengths in the ultraviolet and infrared, *Rapid Commun. Mass Spectrom.* 6, 771–776, 1992.

46. Little, D.P., Thannhauser, T.W., and McLafferty, F.W., Verification of 50- to 100-mer DNA and RNA sequences with high-resolution mass spectrometry, *Proc. Natl. Acad. Sci. USA* 92, 2318–2322, 1995.

47. Limbach, P.A., Crain, P.F., and McCloskey, J.A., Molecular mass measurement of intact ribonucleic acids via electrospray ionization quadrupole mass spectrometry, *J. Am. Soc. Mass Spectrom.* 6, 27–39, 1995.

48. Muddiman, D.C., Cheng, X., Udseth, H.R., and Smith, R.D., Charge-state reduction with improved signal intensity of oligonucleotides in electrospray ionization mass spectrometry, *J. Am. Soc. Mass Spectrom.* 7, 697–706, 1996.

49. Liu, C., Wu, Q., Harms, A.C., and Smith, R.D., On-line microdialysis sample cleanup for electrospray ionization mass spectrometry of nucleic acid samples, *Anal. Chem.* 68, 3295–3299, 1996.

50. Xu, N., Lin, Y., Hofstadler, S.A., Matson, D., Call, C.J., and Smith, R.D., A microfabricated dialysis device for sample cleanup in electrospray ionization mass spectrometry, *Anal. Chem.* 70, 3553–3536, 1998.

51. Hannis, J.C. and Muddiman, D.C., Characterization of a microdialysis approach to prepare polymerase chain reaction products for electrospray ionization mass spectrometry using on-line ultraviolet absorbance measurements and inductively coupled plasma atomic emission spectroscopy, *Rapid Commun. Mass Spectrom.* 13, 323–330, 1999.

52. McLuckey, S.A. and Stephenson, J.L., Jr., Ion/ion chemistry of high-mass multiply charged ions, *Mass Spectrom. Rev.* 17, 369–407, 1998.

53. Newton, K.A., He, M., Amunugama, R., and McLuckey, S.A., Selective cation removal from gaseous polypeptide ions: Proton *vs.* sodium ion abstraction via ion/ion reaction, *Phys. Chem. Chem. Phys.* 6, 2710–2717, 2004.

54. Turner, K. B., Monti, S. and Fabris, D., Like polarity ion/ion reactions enable the investigation of specific metal interactions in nucleic acids and their non-covalent assemblies, *J. Am. Chem. Soc.* 30, 13353–13363, 2008.

55. McGhee, J.D. and von Hippel, P.H., Theoretical aspects of DNA-protein interactions: Co-operative and non-co-operative binding of large ligands to a one-dimensional homogeneous lattice, *J. Mol. Biol.* 86, 469–489, 1974.

56. Hagan, N. and Fabris, D., Direct mass spectrometric determination of the stoichiometry and binding affinity of the complexes between HIV-1 nucleocapsid protein and RNA stem-loops hairpins of the HIV-1 Ψ-recognition element, *Biochemistry* 42, 10736–10745, 2003.

57. Bruins, A.P., Covey, T.R., and Henion, J.D., Ion spray interface for combined liquid chromatography-atmospheric pressure ionization mass spectrometry, *Anal. Chem.* 59, 2642–2646, 1987.

58. Dole, M., Mack, L.L., Hines, R.C., Mobley, R.L., Ferguson, L.D., and Alice, M.B., Molecular Beams of Macroions, *J. Chem. Phys.* 49, 2240–2249, 1968.

59. Kebarle, P., A brief overview of the present status of the mechanisms involved in electrospray mass spectrometry, *J. Mass Spectrom.* 35, 804–817, 2000.

60. Iribarne, J.V. and Thomson, B.A., On the evaporation of small ions from charged droplets, *J. Chem. Phys.* 64, 2287–2294, 1976.

61. Mora, J.F., Van Berkel, G.J., Enke, C.G., Cole, R.B., Martinez-Sanchez, M., and Fenn, J.B., Electrochemical processes in electrospray ionization mass spectrometry, *J. Mass Spectrom.* 35, 939–952, 2000.

62. Verkerk, U.H., Peschke, M., and Kebarle, P., Effect of buffer cations and of H3O+ on the charge states of native proteins. Significance to determinations of stability constants of protein complexes, *J. Mass Spectrom.* 38, 618–631, 2003.

63. Peschke, M., Verkerk, U.H., and Kebarle, P., Features of the ESI mechanism that affect the observation of multiply charged noncovalent protein complexes and the determination of the association constants by the titration method, *J. Am. Soc. Mass Spectrom.* 15, 1424–1434, 2004.

64. Ding, J. and Anderegg, R.J., Specific and nonspecific dimer formation in the electrospray ionization mass spectrometry of oligonucleotides, *J. Am. Soc. Mass Spectrom.* 6, 159–164, 1994.

65. De Guzman, R.N., Wu, Z.R., Stalling, C.C., Pappalardo, L., Borer, P.N., and Summers, M.F., Structure of the HIV-1 nucleocapsid protein bound to the SL-3 Ψ-RNA recognition element, *Science* 279, 384–388, 1998.

66. Robinson, C.V., Chung, E.W., Kragelund, B.B., Knudsen, J., Aplin, R.T., Poulsen, F.M., and Dobson, C.M., Probing the nature of noncovalent interactions by mass spectrometry. A study of protein-CoA ligand binding and assembly, *J. Am. Chem. Soc.* 118, 8646–8653, 1996.

67. Gabelica, V. and De Pauw, E., Comparison between solution-phase stability and gas-phase kinetic stability of oligodeoxynucleotide duplexes, *J. Mass Spectrom.* 36, 397–402, 2001.

68. Schwartz, B.L., Bruce, J.F., Anderson, G.A., Hofstadler, S.A., Rockwood, A.L., Smith, R.D., Chilkoti, A., and Stayton, P.S., Dissociation of tetrameric ions of non-covalent streptavidin complexes by electrospray ionization, *J. Am. Soc. Mass Spectrom.* 6, 459–465, 1995.

69. Potier, N., Donald, L.J., Chernushevich, I., Ayed, A., Ens, W., Arrowsmith, C.H., Standing, K., and Duckworth, H.W., Study of a noncovalent trp repressor: DNA operator complex by electrospray ionization time-of-flight mass spectrometry, *Protein Sci.* 7, 1388–1395, 1998.

70. Benjamin, D.R., Robinson, C.V., Hendrick, J.P., Hartl, F.U., and Dobson, C.M., Mass spectrometry of ribosomes and ribosomal subunits, *Proc. Natl. Acad. Sci. USA* 95, 7391–7395, 1998.

71. Cheng, X., Harms, A.C., Goudreau, P.N., Terwilliger, T.C., and Smith, R.D., Direct measurement of oligonucleotide binding stoichiometry of gene V protein by mass spectrometry, *Proc. Natl. Acad. Sci. USA* 93, 7022–7027, 1996.

72. Griffey, R.H., Sannes-Lowery, K.A., Drader, J.J., Mohan, V., Swayze, E.E., and Hofstadler, S.A., Characterization of low-affinity complexes between RNA and small molecules using electrospray ionization mass spectrometry, *J. Am. Chem. Soc.* 122, 9933–9938, 2000.

73. Turner, K.B., Hagan, N.A., and Fabris, D., Inhibitory effects of archetypical nucleic acid ligands on the interactions of HIV-1 nucleocapsid protein with elements of Ψ-RNA, *Nucleic Acids Res.* 34, 1305–1316, 2006.

74. Cheng, X., Morin, P.E., Harms, A.C., Bruce, J.E., Ben-David, Y., and Smith, R.D., Mass spectrometric characterization of sequence-specific complexes of DNA and transcription factor PU.1 DNA binding domain, *Anal. Biochem.* 239, 35–40, 1996.

75. Veenstra, T.D., Johnson, K.L., Tomlinson, A.J., Craig, T.A., Kumar, R., and Naylor, S., Zinc-induced conformational changes in the DNA-binding domain of the vitamin D receptor determined by electrospray ionization mass spectrometry, *J. Am. Soc. Mass Spectrom.* 9, 8–14, 1998.

76. Craig, T.A., Benson, L.M., Tomlinson, A.J., Veenstra, T.D., Naylor, S., and Kumar, R., Analysis of transcription complexes and effects of ligands by microspray ionization mass spectrometry, *Nat. Biotechnol.* 17, 1214–1218, 1999.

77. Deterding, L.J., Kast, J., Przybylski, M., and Tomer, K.B., Molecular characterization of a tetramolecular complex between dsDNA and a DNA-binding leucine zipper peptide dimer by mass spectrometry, *Bioconj. Chem.* 11, 335–344, 2000.

78. Sannes-Lowery, K.A., Hu, P., Mack, D.P., Mei, H.Y., and Loo, J.A., HIV-1 Tat peptide binding to TAR RNA by electrospray ionization mass spectrometry, *Anal. Chem.* 69, 5130–5135, 1997.

79. Mei, H.Y., Mack, D., Galan, A.A., Halim, N.S., Heldsinger, A., Loo, J.A., Moreland, D.W., Sannes-Lowery, K.A., Sharmeen, L., Truong, H.N., and Czarnik, A.W., Discovery of selective, small-molecule inhibitors of RNA complexes-I. The Tat protein/TAR RNA complexes required for HIV-1 transcription, *Bioorg. Med. Chem.* 5, 1173–1184, 1997.

80. Loo, J.A., Holler, T.P., Foltin, S.K., McConnell, P., Banotal, C.A., Horne, N.M., Mueller, W.T., Stevenson, T.I., and Mack, D.P., Application of electrospray ionization mass spectrometry for studying human immunodeficiency virus protein complexes, *Proteins* 2(Suppl.), 28–37, 1998.

81. Fisher, R.J., Fivash, M.J., Stephen, A.G., Hagan, N.A., Shenoy, S.R., Medaglia, M.V., Smith, L.R., Worthy, K.M., Simpson, J.T., Shoemaker, R., McNitt, K.L., Johnson, D.J., Hixson, C.V., Gorelick, R.J., Fabris, D., Henderson, L.E., and Rein, A., Complex interactions of HIV-1 nucleocapsid protein with oligonucleotides, *Nucleic Acids Res.* 34, 472–484, 2006.

82. Stephen, A.G., Datta, S.A. K., Worthy, K.M., Bindu, L., Fivash, M.J., Turner, K.B., Fabris, D., Rein, A., and Fisher, R.J., Measuring the binding stoichiometry of HIV1 Gag to very low density oligonucleotide surfaces using surface plasmon resonance spectroscopy, *J. Biomol. Technol.* 18, 259–266, 2007.

83. Fabris, D., Chaudhari, P., Hagan, N.A., and Turner, K.B., Functional investigations of retroviral protein-ribonucleic acid complexes by nanospray Fourier transform ion cyclotron resonance mass spectrometry, *Eur. J. Mass Spectrom.* 13, 29–33, 2007.

84. Yuan, Y., Kerwood, D.J., Paoletti, A.C., Shubsda, M.F., and Borer, P.N., Stem of SL1 RNA in HIV-1: Structure and nucleocapsid protein binding for a 1 × 3 internal loop, *Biochemistry* 42, 5259–5269, 2003.

85. Shubsda, M.F., Paoletti, A.C., Hudson, B.S., and Borer, P.N., Affinities of packaging domain loops in HIV-1 RNA for the nucleocapsid protein, *Biochemistry* 41, 5276–5282, 2002.

86. Lawrence, D.C., Stover, C.C., Noznitsky, J., Wu, Z., and Summers, M.F., Structure of the intact stem and bulge of HIV-1 Psi-RNA stem-loop SL1, *J. Mol. Biol.* 326, 529–542, 2003.

87. Hagan, N.A. and Fabris, D., Dissecting the protein-RNA and RNA-RNA interactions in the nucleocapsid-mediated dimerization and isomerization of HIV-1 stemloop 1, *J. Mol. Biol.* 365, 396–410, 2007.

88. Turner, K.B., Hagan, N.A., Kohlway, A., and Fabris, D., Mapping noncovalent ligand binding to stem-loop domains of the HIV-1 packaging signal by tandem mass spectrometry, *J. Am. Soc. Mass Spectrom.* 17, 1401–1411, 2006.

89. Daniel, J.M., Friess, S.D., Rajagopalan, S., Wendt, S., and Zenobi, R., Quantitative determination of noncovalent binding interactions using soft ionization mass spectrometry, *Int. J. Mass Spectrom. Ion Processes* 216, 1–27, 2002.

90. Millard, B.J., *Quantitative Mass Spectrometry*, Heyden & Son Ltd., London, 1978.

91. Goodner, K.L., Milgram, K.E., Williams, K.R., Watson, C.H., and Eyler, J.R., Quantitation of ion abundances in Fourier-transform ion cyclotron resonance mass spectrometry, *J. Am. Soc. Mass Spectrom.* 9, 1204–1212, 1998.

92. Amarasinghe, G.K., De Guzman, R.N., Turner, R.B., Chancellor, K.J., Wu, Z.R., and Summers, M.F., NMR structure of the HIV-1 nucleocapsid protein bound to stem-loop SL2 of the Ψ-RNA packaging signal. Implications for genome recognition, *J. Mol. Biol.* 301, 491–511, 2000.

93. Liu, C., Pasa-Tolic, L., Hofstadler, S.A., Harms, A.C., Smith, R.D., Kang, C., and Sinha, N., Probing RegA/RNA interactions using electrospray ionization-Fourier transform ion cyclotron resonance-mass spectrometry, *Anal. Biochem.* 262, 67–76, 1998.

94. Gabelica, V. and De Pauw, E., Comparison of the collision-induced dissociation of duplex DNA at different collision regimes: Evidence for a multistep dissociation mechanism, *J. Am. Soc. Mass Spectrom.* 13, 91–98, 2002.

95. Heck, A.J. R. and van den Heuvel, R.H. H., Investigation of intact protein complexes by mass spectrometry, *Mass Spectrom. Rev.* 23, 368–389, 2004.

96. Schnier, P.D., Klassen, J.S., Strittmatter, E.F., and Williams, E.R., Activation energies for dissociation of double strand oligonucleotide anions: Evidence for Watson–Crick base pairing in vacuo, *J. Am. Chem. Soc.* 120, 9605–9613, 1998.

97. Turner, K.B., Hagan, N.A., and Fabris, D., Understanding the isomerization of the HIV-1 dimerization initiation domain by the nucleocapsid protein, *J. Mol. Biol.* 369, 812–828, 2007.

98. Gauthier, J.W., Trautman, T.R., and Jacobson, D.B., Sustained off-resonance irradiation for collision-activated dissociation involving Fourier transform mass spectrometry. Collision-activated dissociation technique that emulates infrared multiphoton dissociation, *Anal. Chim. Acta* 246, 211–225, 1991.

99. de Koning, L.J., Nibbering, N.M.M., van Orden, S.L., and Laukien, F.H., Mass selection of ions in a Fourier transform ion cyclotron resonance trap using correlated harmonic excitation fields (CHEF), *Int. J. Mass Spectrom. Ion Processes* 165/166, 209–219, 1997.

100. Takáts, Z., Wiseman, J.M., and Cooks, R.G., Ambient mass spectrometry using desorption electrospray ionization (DESI): Instrumentation, mechanisms and applications in forensics, chemistry, and biology, *J. Mass Spectrom.* 40, 1261–1275, 2005.

101. Shiea, J., Huang, M.-Z., Hsu, H.-J., Lee, C.-Y., Yuan, C.-H., Beech, I., and Sunner, J., Electrospray-assisted laser desorption/ionization mass spectrometry for direct ambient analysis of solids, *Rapid Commun. Mass Spectrom.* 19, 3701–3704, 2005.

102. Sampson, J.S., Hawkridge, A.M., and Muddiman, D.C., Generation and detection of multiply-charged peptides and proteins by matrix-assisted laser desorption electrospray ionization (MALDESI) Fourier transform ion cyclotron resonance mass spectrometry, *J. Am. Soc. Mass Spectrom.* 17, 1712–1716, 2006.

103. Stoeckli, M., Chaurand, P., Hallahan, D.E., and Caprioli, R.M., Imaging mass spectrometry: A new technology for the analysis of protein expression in mammalian tissues, *Nat. Med.* 7, 493–496, 2001.

104. McDonnell, L.A., and Heeren, R.M.A., Imaging mass spectrometry, *Mass Spectrom. Rev.* 26, 606–643, 2007.

105. Woodin, R.L., Bomse, D.S., and Beauchamp, J.L., Multiphoton dissociation of molecules with low power continuous wave infrared laser radiation, *J. Am. Chem. Soc.* 100, 3248–3250, 1978.

106. Little, D.P., Speir, J.P., Senko, M.W., O'Connor, P.B., and McLafferty, F.W., Infrared multiphoton dissociation of large multiply charged ions for biomolecule sequencing, *Anal. Chem.* 66, 2809–2815, 1994.

107. Zubarev, R.A., Kelleher, N.L., and McLafferty, F.W., Electron capture dissociation of multiply charged protein cations: A nonergodic process, *J. Am. Chem. Soc.* 120, 3265–3266, 1998.

108. Zubarev, R.A., Horn, D.M., Fridriksson, E.K., Kelleher, N.L., Kruger, N.A., Lewis, M.A., Carpenter, B.K., and McLafferty, F.W., Electron capture dissociation for structural characterization of multiply charged protein cations, *Anal. Chem.* 72, 563–573, 2000.

109. Syka, J.E. P., Coon, J.J., Schroeder, M.J., Shabanowitz, J., and Hunt, D.F., Peptide and protein sequence analysis by electron transfer dissociation mass spectrometry, *Proc. Natl. Acad. Sci. USA* 101, 9528–9533, 2000.

110. Kaneko, Y., Megill, L.R., and Hasted, J.B., Study of inelastic collisions by drifting ions, *J. Chem. Phys.* 45, 3741, 1966.

111. Clemmer, D.E. and Jarrold, M.F., Ion mobility measurements and their applications to cluster biomolecules, *J. Mass Spectrom.* 32, 577–592, 1967.

10 Characterization of Noncovalent Complexes of Nucleic Acids with Peptides and Proteins by Mass Spectrometry

William Buchmann

CONTENTS

10.1 INTRODUCTION

The study of nucleic acid–peptide/protein interactions is a highly active research field since such interactions are involved in many cellular processes, such as replication, regulation of gene expression, and DNA packaging. Nearly all the functions of nucleic acids are accomplished by interacting with proteins.[1] R. D. Kornberg received the Nobel Prize in Chemistry in 2006 for his studies on the molecular basis of eukaryotic transcription.[2] In 2006, A. Z. Fire Craig and C. Mello shared the Physiology/Medicine Nobel Prize for their discovery of RNA interference or gene silencing by double-stranded RNA.[2,3] Both transcription and gene silencing processes involve nucleic acid–protein interactions. It is both fundamentally important and topical to understand at the molecular level the basis of many cellular processes involving nucleic acids, which range from transcription–translation to DNA repair and DNA packaging in the cells.

Among DNA-binding proteins, regulatory proteins bind to specific sequences of duplex DNA in order to initiate or control the transcription; DNA cleavage proteins, such as restriction enzymes, can specifically cleave particular DNA sequences; repair proteins can identify a lesion and excise damaged DNA. There are some proteins, such as DNA topoisomerases, that are able to unwind DNA prior to replication and other proteins that play a structural role; some proteins, such as histones, are involved in DNA packaging. Finally, there are processing proteins that operate without sequence specificity: DNA and RNA polymerases bind to DNA without sequence specificity and use DNA as a template for nucleic acid synthesis.[4]

Specific RNA–protein interactions are involved in a variety of essential processes within the cell: gene regulation through transcriptional control, RNA processing, transport, and translational control. Pre-messenger RNA (pre-mRNA), binds to ribonucleoproteins (RNPs), forming the spliceosome during posttranscriptional regulation of gene expression. Transfer RNA (tRNA) binds specific aminoacyl–synthetases for gene code translation during protein synthesis. Ribosomal RNA (rRNA) along with many proteins composes the ribosome. Finally, understanding gene expression and its regulation requires understanding how both DNA and RNA interact with proteins.

Studies in sequencing and analysis of the human genome in 2001 estimated that ~13% of the whole human genome encoded nucleic acid-binding proteins (~4000 genes).[5,6] In April 2007, more than 1700 high-quality structures of nucleic acid–protein complexes were already known and available from protein data bank (PDB) (i.e., ~10 times the number of known structures in 1996).[7] These structures were distributed as 1207 DNA–protein versus 489 RNA–protein complexes arising from various eukaryotic and prokaryotic sources. X-ray crystallography and nuclear magnetic resonance (NMR) provided the majority of these high-resolution structures. These structures are extremely detailed but cannot always be obtained due to a lack of homogeneity, an insufficient amount of sample, or the impossibility of crystallization. Our knowledge of the interactions between DNA/RNA and proteins has therefore grown rapidly but still remains far from complete.

Crucial information can be obtained by mass spectrometry (MS) that plays an increasingly important role in this field.[8] MS can enable the direct observation of single-strand DNA (ssDNA), double-strand DNA (dsDNA), and DNA/RNA–peptide/protein noncovalent complexes, providing information about binding stoichiometry and ligand identity and giving the advantages of speed, sensitivity, and ease of use. The need for the presence of a specific cofactor, such as a small molecule, or a metal cation in a multimeric assembly, can be established by MS more easily and more quickly than by other methods. MS has the potential to identify a small chemical modification in one of the binding partners from a heterogeneous mixture. Moreover, this technique can also account for the relative binding affinities between the various binding partners.

In this chapter, the recent progress in the direct MS characterization of noncovalent nucleic acid–protein complexes is reviewed, including the basic principles of DNA/RNA–protein recognition and the description of MS techniques that are mainly applied in this field, with the emphasis on the two most widely used ionization techniques nowadays: matrix-assisted laser desorption/ionization (MALDI) and electrospray ionization (ESI). The utility of MS in analyzing multimeric and high-molecular-weight DNA/RNA–protein complexes will be illustrated by several examples.

The scope of this review has been limited to the direct characterization of noncovalent DNA/RNA–protein complexes by MALDI-MS and ESI-MS. The methods aiming for the identification of the partners involved in the complex, employing either the protein or the nucleic acid as a "hook" by affinity purifications, are outside the scope of this article and the interested reader is referred to another review on this topic.[9]

10.2 OVERVIEW OF DNA/RNA–PROTEIN COMPLEXES

10.2.1 MOLECULAR BASIS OF DNA–PROTEIN INTERACTIONS

The molecular basis of the DNA/RNA–protein interactions are here briefly reviewed with the aim of providing an introduction to the underlying principles governing nucleic acid–protein recognition.[4,10–12] This section should help MS specialists to understand what precisely can maintain together the different binding partners in solution before the ionization process, and what may still maintain together the partners after the transfer to the gas phase.

There is no simple code or general rule such as that one amino acid side chain binds one DNA base in order to describe, or predict, the DNA–protein recognition. While there is no simple code, some preferences of interactions between certain protein side chains and DNA bases do exist.[11]

DNA–protein complexes generally result from a complex network of contacts between both components and several different possible interactions between DNA and proteins can be involved: (a) hydrogen bonds between positively charged basic residues of amino acids and DNA phosphate groups (electrostatic interactions), (b) hydrogen bonds between polar amino acid residues and the functions exposed at the edges of the bases or with sugars, (c) Van der Waals (VdWs) contacts, and (d) water-mediated bonds.

Double-stranded DNA is a molecule with a double-helix shape consisting of a highly negatively charged sugar–phosphate backbone and stacked base pairs whose edges are exposed either in the major or in the minor groove. Figure 10.1 shows the available sites for hydrogen bonding in A–T and G–C base pairs. DNA–protein recognition essentially comes from specific interactions with the chemical surface resulting from the succession of each base pair exposed in either the major groove or the minor groove. This chemical surface acts as a chemical "signature" of each DNA sequence.[10] DNA–protein recognition can also depend on variations of the DNA flexibility as a function of its base sequence. The interactions between DNA and the protein backbone are not the most important factor for the recognition but can play a role in the stabilization of the complex. Lastly, it is worth

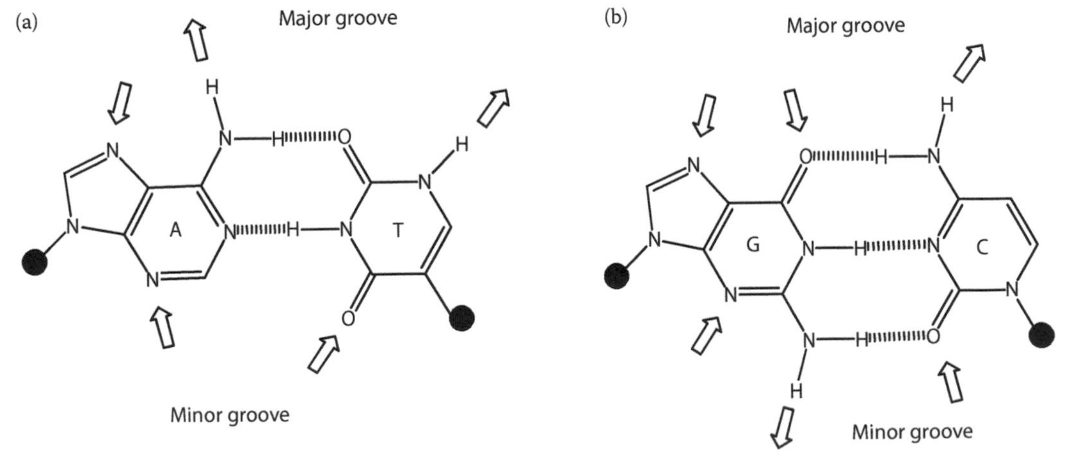

FIGURE 10.1 Available sites for hydrogen bonding in the Watson–Crick base pairs. The direction of the arrows indicates the acceptor or donor characteristics of the considered sites.

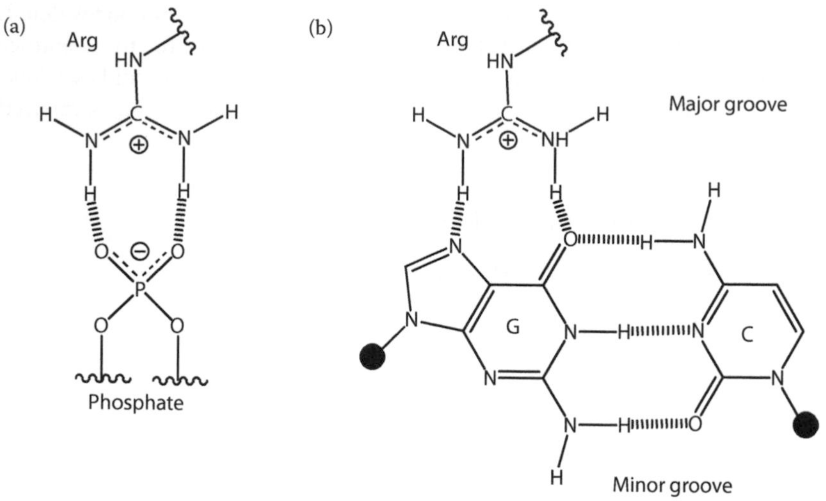

FIGURE 10.2 Possible interactions involving the basic residue of arginine and (a) a phosphate group and (b) a guanine base.

noting that ssDNA–protein interactions differ from dsDNA–protein because of the absence of base pairing.[13]

Luscombe et al. assessed from the detailed structural investigation of 129 dsDNA–protein complexes, the contribution of hydrogen bonds, VdWs contacts, and water-mediated bonds involved in the complexes according to the amino acid nature and the DNA base or backbone.[11] The dsDNA–protein complexes were solved by x-ray crystallography available from the Protein Data Bank and the Nucleic Acid Database.[7,14] The positively charged and polar residues contribute the most frequently to the hydrogen bonds between proteins and DNA (DNA bases + DNA backbone) according to the order: Arg (33%) > Lys > Ser > Thr > Asn > Gln. The most important interactions involve the phosphate groups of DNA backbone and the charged ends of basic residues of arginines or lysines. An example is given in Figure 10.2a with arginine. In fact, nearly two-third of hydrogen bonds correspond to bonds with phosphate groups; these interactions being independent on base sequence are not directly implicated in the specificity of recognition.

Among the various bases: thymine, cytosine, adenine, guanine, the latter shows the highest hydrogen-bonding potential (G > A > C > T). This is because guanine exposes on its edges the highest number of accessible sites for hydrogen bonding, then A and finally C/T (Figure 10.1). Each base pair can provide different possible interactions with the side chain of amino acids. In the 129 DNA–protein complexes that makes up the data set of Luscombe et al., guanine was involved in 50% of the hydrogen bonds with the amino acids side chains.[11] The arginine residue, which is involved in more than 40% of the amino acid–DNA base interactions (via H-bond), forms with guanine the most common H-bond (see Figure 10.2b, keeping in mind that other arrangements are possible), but arginine can also form favorable hydrogen bonds with thymine and adenine. The second most frequent H-bond corresponds to Lys–G. Amide residues of asparagine and glutamine display an affinity for adenine but to a lesser extent. Serine and threonine essentially interact with the phosphate backbone.

Surprisingly, VdW contacts account for two-third of all protein–DNA interactions. VdW interactions with the DNA backbone (phosphate + sugar) are by far the most prominent (>75%). The bases and amino acids involved in VdW interactions differ from those above described with H-bonds. The number of VdW interactions decreases in the following order according to the participating amino acid and DNA component: Lys > Arg > Thr > Phe > Asn > Gln and T > A > G > C. Between DNA bases and the amino acid side chains, threonine shows a particular affinity for thymine, phenylalanine and glutamine for adenine/thymine, and arginine for guanine.[11]

Water molecules can provide mediation between DNA and amino acid side chains. Water-mediated interactions, accounting for ~15% of all interactions, essentially occur between the phosphate group of the DNA backbone (~70%) and basic and polar amino acids (Arg, Lys, Asn, Ser and Thr, Glu, Gln and Asp).[11]

10.2.2 COMMON PROTEIN FOLDS IN DNA–PROTEIN COMPLEXES

A great number of x-ray and NMR structures of DNA–protein complexes are available from the Protein Data Bank and Nucleic Acid Database websites and the most common protein folds allowing DNA binding are now well identified.[7,14] Although proteins may bind to DNA through a variety of possible folds, they often present similar characteristic structural motifs for DNA recognition.[4,15] These amazing recognition motifs are here briefly described mainly from transcription factors.[10,12,15] The α-helix motif is the most common structural element encountered in DNA-binding proteins. It is generally embedded in more complex recognition structural motifs that are helix-turn-helix (HTH), zinc-finger, and basic region-leucine-zipper motifs. Generally, sequence-specific proteins bind to DNA in the major groove, rather than the minor groove, due to a greater number of possible interactions (Figure 10.1). In most cases, duplex DNA remains under its B-type conformation after protein binding, but some proteins can induce bending of DNA. The most detailed assemblies have been revealed from crystallographic data or in-solution NMR experiments.

The motif HTH has been identified in a large number of prokaryotic and eukaryotic proteins related to transcription regulation and enzymatic functions.[10,16,17] It is composed of approximately 20 amino acid residues; a first α-helix (5AA) is linked by a tight bend (5AA) to a second α-helix (15AA) which acts as a recognition element alone by snugly fitting into the major groove. This second helix, referred as the recognition helix, is the only one of the two helixes showing specific contacts with duplex DNA. Figure 10.3a shows, as an example, the engrailed homeodomain DNA complex consisting of three helices, two of them forming the HTH motif.[18] The interactions between DNA and the helix take place in the major groove. Bacterial HTH proteins generally bind to palindromic DNA sequences as homodimers whereas eukaryotic homeodomain proteins, including a HTH motif, bind to the target DNA as monomers. In the former case, bending of DNA is observed if the two protein domains bind on the same side of a DNA sequence, whereas no net bending is observed when the HTH motifs are located on opposite sides. In some cases, proteins such as homeodomain proteins, which only bind to DNA as monomers, use a sort of flexible peptide "arm" located at the N-terminal amino acids sequence, in order to bind to the minor groove in an analogous manner as the minor groove binding drugs such as netropsin. In addition to the interactions occurring in the DNA major groove with an α-helix, interactions between amino acid residues and

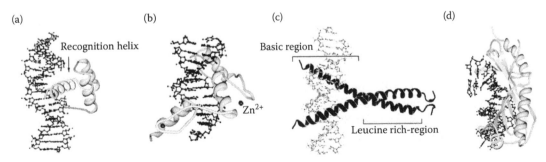

FIGURE 10.3 Representative examples of DNA-binding protein motifs. (a) HTH in engrailed homeodomain DNA complex[18] (3HDD). (b) Zinc finger in zif268–DNA complex[20] (1ZAA). (c) Basic region-leucine-zipper in GCN4–DNA complex[21] (1DGC). (d) β-Sheets in TATA-binding protein[22] (1YTB). PDB accession codes are shown in parentheses. Figures were created using coordinates deposited in the PDB.

the DNA minor groove can improve the overall complex stability. It is worth noting that an isolated HTH motif is not sufficient for DNA recognition, this process requires the whole protein and often its homo- or heterodimerization.[12]

Zif268 is a protein belonging to one of the largest eukaryotic DNA-binding motif families, which recognize diverse set of DNA sequences. The zinc-finger recognition motif is made of about 30 amino acids, and consists of two antiparallel β-sheets and an α-helix (ββα-domain), held together by a zinc ion coordination involving two cysteine and two histidine residues.[4,10,19] The presence of zinc is required to stabilize the protein folding as shown in Figure 10.3b with zif268–DNA complex.[20] The zinc-finger proteins generally contain several copies of the zinc-finger motif separated by short linkers of 7–8 amino acids. Each of these zinc fingers can interact with 3–4 consecutive DNA base pairs owing to the contact of an α-helix into the major groove. The inter-actions essentially involve Arg, His, and Glu amino acid residues of the zinc fingers and the most guanine-rich strand of DNA.

The basic region-leucine-zipper motifs (bZIP) can be encountered in eukaryotes transcription factors.[4,10,15] The yeast transcription factor CGN4[21] is shown in Figure 10.3c as a representative example. Two very long α-helices of about 60 residues interact together to form the so-called "leucine-zipper" motif. Two different domains can be distinguished in each helix, a C-terminal leucine-rich region (each helix possesses repeated leucines in every seven residues, all being ori-ented in the same direction, which favors a dimerization of the two helices), and a basic region of about 20 amino acids in direct contact with the major groove of DNA, the orientation of the leucine-zipper being perpendicular to the DNA helix.

The β-sheet is another structural motif allowing the binding of proteins with DNA. An example is the 5′-TATA box-binding protein (TBP, Figure 10.3d) involved in the initiation of transcription in eukaryotes cells.[22] The TBP protein exhibits a wide concave surface formed by 10-stranded anti-parallel β-sheets. Upon binding to the DNA minor groove by TBP, the DNA helix undergoes a dramatic bending (>90°) over 7–8 base pairs. The binding DNA region is highly distorted and the minor groove is both widened and flattened to permit an optimal binding to the β-sheets. Finally, other proteins are able to induce bending of DNA. High mobility group (HMG) proteins induce bending of DNA on binding in the minor groove. These proteins incorporate another DNA recogni-tion motif consisting of three helices arranged in an L-shape which is different from the HTH described above. The packaging of DNA in the chromosomes involves histone proteins and also leads to a smooth bending of DNA. Four histone proteins, H2A, H2B, H3, and H4, assemble them-selves in pairs to form an octameric structure around which 146 base pairs can wind. This assembly forms the repeating unit of the supercoiled DNA in the chromosomes.[23]

10.2.3 Molecular Basis of RNA–Protein Interactions

Although RNA–protein interactions are very important in many biological processes, the molecular basis and binding recognition motifs involved in RNA–protein interactions are still poorly understood.[13,24–28] Our knowledge concerning the recognition mechanisms is limited, in compari-son with that of DNA–protein interactions, due to a smaller number of structures that have been solved by x-ray crystallography and NMR. The reason is obviously related to difficulties encoun-tered during sample preparation before structural analysis (purification and crystallization) and also to the complexity of interactions between RNA and protein. While in DNA–protein complexes, DNA commonly exhibits a regular B-DNA double-helix structure, even short single-stranded RNAs in RNA–protein complexes can display a variety of complex secondary and tertiary structures.[1,4] The flexibility of single-stranded RNA enables the formation of short sequences of double-stranded RNA, hairpin (stem-loops), bulges, and multiple internal loops. Double-stranded RNA can be read-ily formed in an analogous manner to dsDNA by base pairing between A and U, C, and G. Because the resulting double-stranded RNA shares common features with A-DNA, it is known as A-RNA. Double-stranded RNA can contain defects, accommodate base mismatches, and exhibit unpaired bases outside RNA double helix in bulges and stem-loops. These secondary structure elements are often involved in RNA–protein interactions. In an early, detailed, computational analysis of 32 RNA–protein complexes, it was observed that the many unpaired bases in RNA structures could more readily make contact with amino acid residues than those of tightly paired dsDNA.[13] In dsDNA, since bases are less accessible to interact with proteins, interactions with the backbone are favored. To date, VdWs contacts have been found to be more numerous than hydrogen bonds. Proteins preferentially make contact with guanine instead of uracil, a feature shared with dsDNA–proteins complexes. It was also reported that RNA–protein interactions more frequently involved arginine, tyrosine, and phenylalanine than other amino acids.[13] In contrast to DNA–protein interactions, hydrogen bonds to bases and the phosphate backbone appeared in equal numbers. A comparison between RNA–protein and DNA–protein complexes revealed that RNA–protein complexes were less well packed, due to the tertiary structure that RNA can form.[13] From a data set of 89 RNA–protein complexes, it was confirmed in a recent study that VdW interactions predominate over hydrogen bonds.[29] The side chains of the proteins have a stronger preference to be involved in RNA interactions than the main chain, and contacts to the RNA sugar–phosphate backbone were found to be more frequent than base-specific contacts. In this way, the positively charged amino acid residues, Arg and Lys, displayed a clear propensity to bind to the phosphate groups of RNA. The most favored amino acid–nucleotide pairings observed were Lys–P (Phosphate), Tyr–U, Arg–P, Phe–A, and Trp–G. In contrast to the previous study, made from a data set of 32 complexes only, it was observed that proteins made preferential contact with guanine, then adenine instead of uracil.[13] Numerous other disparities can be read in the literature as a function of the data set size or definition of the various RNA–protein contacts as mentioned by Ellis et al.[29]

Interestingly, it appears that according to the required nucleobase sequence specificity, proteins which bind specifically to mRNA, tRNA, and viral RNA (vRNA) show more base-specific contacts and less backbone contacts than expected, whereas other rRNA–protein complexes display less base-specific contacts than expected.[29] Auweter et al.[30] conclude that stacking and electrostatic interactions provide affinity between binding partners, whereas hydrogen bonds contribute to sequence specificity. Although electrostatics enable the initial attraction that brings RNA and protein together, stacking and hydrogen bonds freeze the RNA in an appropriate orientation within the complex.[30]

10.2.4 RNA–Protein Structures

From the early crystallographic and NMR studies, RNA-binding proteins were roughly divided into two main classes: proteins that bind RNA helix grooves (RNA groove binders) and proteins that recognize single-stranded bases in a sequence-specific manner through a β-sheet (β-sheet binders).[25] RNA-groove binding proteins are able to fit the groove of an RNA helix with one of their secondary structure element such as an α-helix or an unstructured loop. In the other group, proteins use β-sheet surfaces to create pockets that recognize unpaired bases of single-stranded RNA. Recently, with a larger number of known structures, it appeared that recognition through β-sheets was used predominantly by RNA-binding proteins.[29,30] Two common RNA recognition motifs (RRMs) have been identified in RNA–protein complexes: the RNP domain, and the K-homology (KH) domain.[24,26,30–32] The RNP domain, which is also simply referred to as the RRM or the RNA-binding domain (RBD), is by far the most widespread and the best described. The RNP domain (~90 amino acids) has a βαββαβ topology and can fold into a four-stranded antiparallel β-sheet packed against two α-helices (Figure 10.4a). The surface formed from the four-stranded β-sheets and more particularly, the first and the third β-strands, play an essential role in RNA binding. The highly abundant KH domain (~70 amino acids) can be subdivided into two types as a function of topology (βααββα for type I and αβααββ for type II), containing in both cases, a three-stranded antiparallel β-sheet and three α-helices (Type I KH shown in Figure 10.4b). In these structures, the loops connecting the two consecutive α-helices (e.g., α1α2 for type I and α2α3 for type II) and between β2 and β3 in

(a)   (b)

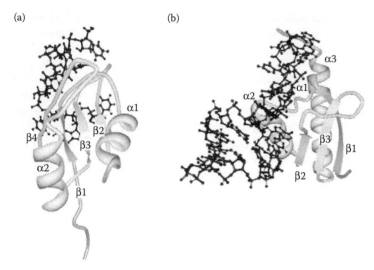

FIGURE 10.4   Representative examples of RNA-binding protein motifs. (a) RNP in Fox-1–RNA complex[31], (b) Type I KH domain in Nova–RNA complex[32]. Figures were created using coordinates deposited in the PDB.

type I, and β3-α2 in type II are strongly implicated in RNA binding. RNP and KH domains contain typically binding sites for four nucleotides.

Other recognition motifs have been identified:[26,30] the Pumilio homology domain, various zinc-binding domains and a double-stranded RNA-binding domain (dsRBD) which share similarities with the KH domain.[26,30] The dsRBD can binds dsRNA (but neither dsDNA nor DNA–RNA hybrids) selectively but without sequence specificity. dsRBD has an αβββα topology and folds into a three-stranded antiparallel β-sheet flanked on one side by two α-helices.[26] The first helix, the loops 2 (between β1 and β2) and 4 (between β3 and α2) has been shown to mediate decisive contacts for recognition. It is necessary to make two further points: in numerous cases, RNA-binding proteins can combine several RRM, dsRBD, and KH domains; and RNA recognition is ensured through cooperative contacts. Furthermore, conformational changes have been detected in proteins, RNAs, or both upon complex formation. These conformational rearrangements which seem necessary for RNA–protein recognition may also have a biological role.[27]

10.3   MS OF NONCOVALENT COMPLEXES

10.3.1   NONCOVALENT INTERACTIONS BY MS: GENERAL CONSIDERATIONS

Ganem and Henion were the first, followed by Katta and Chait, to describe, in 1991, the detection of noncovalent complexes by MS.[33–35] Since these early reports, the detection of a great variety of non-covalent complexes has been reported: protein–protein, protein–nucleic acid, protein–carbohydrate, protein–metal, and protein–ligand. The subject of protein complexes has been reviewed by several authors.[36–42] While the great potential of MS for direct characterization of protein–nucleic acid complexes was soon recognized, the number of papers dealing with noncovalent complexes involving nucleic acids, especially when associated with proteins, remained limited.[8,9,43–46]

All of the available ionization techniques do not provide the required soft transfer of the species from the solution phase into the gas phase, which is necessary to maintain many of the weak noncovalent interactions that preexist under physiological conditions. ESI and related low-flow techniques such as microelectrospray and nanospray have become the most widely used ionization techniques for non-covalent complex characterization owing to the gentleness of the process and their compatibility with nondenaturing conditions.[47,48] While most of the published MS studies on noncovalent studies concentrate on the use of ESI, MALDI has also been successfully used in this area.[49–51]

10.3.1.1 Matrix-Assisted Laser Desorption/Ionization Mass Spectrometry

The detection of a number of noncovalent complexes (protein/protein, protein/nucleic acid, peptide/metal, etc.) has been reported by MALDI-MS and this ionization technique appears as an alternative method to ESI in order to investigate noncovalent interactions.[51] The production of gaseous ions by MALDI results from the irradiation by a pulsed laser of a target composed of matrix crystals in which the analyte is embedded. Typically, the most widely used lasers emit in the UV with pulse durations of a few nanoseconds. The matrices used are small organic compounds (MW 100–250) absorbing at the laser wavelength. Sample preparation is of critical importance in MALDI-MS. The choice of the appropriate matrix is a key element in the success of the analysis. According to the most popular method ("dried droplet"), an aqueous solution of analyte is mixed with a highly concentrated matrix solution, and then this mixture is spotted on to the MALDI target and allowed to air dry. The most widely used matrices are acidic: 2,5-dihydroxy-benzoic acid (DHB), α-cyano-4 hydroxycinnamic acid (CHCA), 3,5-dimethoxy-4 hydroxycinnamic (sinapinic acid, SA), or 3-hydroxy picolinic acid (3-HPA). Obviously, the dried environment of these acidic organic matrices is not ideal for the conservation of noncovalent complexes under physiologically relevant conditions. Noncovalent complexes must yet survive within the matrix at the conclusion of target preparation, and then after laser irradiation and during the desorption–ionization process in order to be detected. For noncovalent complex detection, laser fluence is generally kept as low as possible, and in the case of nucleic acids–protein complexes, the best results have been obtained with nonacidic matrices: 6-aza-2-thiothymine (ATT), and 2,4,6-trihydroxyacetophenone (2,4,6-THAP) in the presence of ammonium acetate and/or ammonium citrate to reach a pH value compatible with complex stability. Avoiding organic solvents and drying under vacuum to accelerate the drying process may be preferable.

10.3.1.2 Electrospray Ionization Mass Spectrometry

The ESI process typically generates multiply charged ions. In the positive ion mode, molecular weights are deduced from $(M+zH)^{z+}$ ions where M corresponds to the molecular mass of the sample and z the number of charges (in the negative ion mode, $(M-zH)^{z-}$ ions are observed). ESI can provide a sufficient amount of energy for the efficient desolvation of the sample, but not too much to prevent the noncovalent bonds. The success of noncovalent complex detection is often the result of a fragile equilibrium of chosen source parameters (voltages, temperature, and pressures). In solution, noncovalent complex cohesion comes from a set of interactions: electrostatic interactions, hydrogen bonds, hydrophobic bonds, and VdWs forces (Sections 10.2.1 and 10.2.3). After the desolvation step, where ions are in the absence of water, electrostatic interactions are known to be reinforced while hydrophobic interactions are weakened.[52] Two important consequences ensue from this: the nature of forces keeping together the different components of a complex may be modified during the ionization process, and certain noncovalent complexes may not be transferable from the liquid phase to the gas phase, due to the important contribution of hydrophobic interactions in solution.

Under classical conditions, samples for ESI-MS analysis are dissolved in a mixture of water/methanol or water/acetonitrile in the presence of a few percent of an organic acid (acetic acid or formic acid). As a consequence, a sample such as a protein, which would be denatured under these conditions, appears in the ESI mass spectrum as a wide distribution of multiply charged ions in the low m/z range. In contrast, noncovalent complexes are prepared at near physiological conditions (buffered aqueous solutions and neutral pH) and their corresponding ions usually carry a lower number of charges due to a more compact conformation.

The difficulties related to the sample preparation required for DNA/RNA–protein complexes to make those suitable for MS detection may be one of the reasons for the small number of papers compared to the number of papers reporting protein–protein complexes. Indeed, biological samples, including noncovalent complexes, are often stored in nonvolatile salt/buffers and other solubilizing agents to ensure their solution stability. However, nonvolatile salt/buffers are highly undesirable because the ESI stability is compromised in their presence. As already stated above, ESI process stability requires the use of aqueous solutions containing volatile salts/buffers. Ammonium acetate and ammonium bicarbonate are very commonly used. Thus, samples must be desalted and buffer exchanged prior to MS analysis. Unfortunately, these sample handlings are time consuming and can lead to noncovalent complex disruption and loss of sample. The use of a small-sized exclusion cartridge on-line with ESI-MS has been proposed as a means of efficient and rapid desalting of DNA/RNA complexes before MS analysis.[53] Compared with an off-line approach, direct analysis enables minimal sample handling and maximal sample recovery.

The conventional ESI source operates with flow rates of about 2–5 µL min^{-1}. The nanospray ion source (NanoES) developed by Wilm et al. enables very low flow rates (20–40 nL min^{-1}) which leads to a lower sample consumption and gives much more time for analysis.[48] However, many other unique features make NanoES extremely attractive for noncovalent interaction studies. For example, a NanoES source is more tolerant toward salt contaminations and buffered solutions, which are important for noncovalent complex analysis. Moreover, a NanoES source employs capillaries with very small spraying orifices (~1–2 µm inner diameter) which enable very small droplets to be generated (100–1000 × smaller than those obtained by conventional ESI). The small size of the droplets strongly improves the desolvation process, and consequently, the overall sensitivity.[48] Zampronio et al. investigated the influence of the tip diameter and the coating of NanoES needles.[54] While the different types of needles tested (metallized or not) did not show any significant effect on mass spectra, it was found that tip diameter can affect both the signal itself and the signal reproducibility. Needles with diameters of 2 µm resulted in more reproducible spectra of noncovalent complexes than the larger tip diameters (6–20 µm).[54] NanoES preserves more faithfully the original solution composition or the equilibrium between free species and complexes in the case of noncovalent interactions than does conventional ESI.[55] From smaller droplets, the in-solution equilibrium before ionization would be less influenced by solvent evaporation. However, it must be kept in mind that several parameters such as capillary-to-cone distance and tip internal diameter can strongly affect the reproducibility of the detection.[55,56]

Because noncovalent complex ions usually carry a lower number of charges, m/z detection often requires analyzers with extended mass range. Time-of-flight (TOF) analyzers, which have theoretically unlimited mass range, are particularly appropriate mass analyzers for noncovalent assembly detection. Coupled with an ESI ion source, the continuous ion beam must be transformed into a pulsed one via an orthogonal injection system to be injected into the TOF section.[57] Quadrupole, Fourier transform ion cyclotron resonance (FT-ICR), and magnetic sectors have also been successfully used in conjunction with ESI. Q-TOF hybrid instruments are particularly well-suited instruments but MS/MS capabilities may be limited by the mass range restriction of the first quadrupole analyzer that routinely operates up to m/z 3000–4000 on commercial instruments. Sobott et al. have extended the mass range of their Q-TOF beyond m/z 90,000 (m/z 150,000 in theory) by reducing the frequency applied to the quadrupole.[58] It was also shown that it was possible to isolate a single ion at m/z 22,000, although the theoretical limit was claimed at m/z 32,000. In addition, the pressure in the different regions of the mass spectrometer, from the ion source and up to the mass analyzer, has to be carefully optimized for better ion transmission.[59] Typically, higher pressures than those used in normal conditions considerably improve ion transmission of high m/z value by collisional cooling and focusing in the RF-only multipole ion guides.[58,60] This parameter is sometimes so critical that there are no complex ions that can be detected using standard pressure settings.

10.3.2 Advantages and Drawbacks of MS and Comparison with Other Analytical Methods

Many different techniques, other than MS, can be applied for studying nucleic acids–protein interactions: gel mobility shift assay, filter binding, fluorescence spectroscopy, isothermal titration calorimetry, analytical centrifugation, surface plasmon resonance (SPR), x-ray crystallography, and NMR spectroscopy.[1,61] The advantages and drawbacks of each technique will be briefly discussed.

Equilibrium constants and stoichiometry can be obtained from filter-binding experiments and gel mobility shift assays, but both require radiolabeling of the DNA/RNA. Furthermore, artifacts can arise in filter-binding experiments from washing the filter and from using large DNA/RNAs. With gel mobility shift assays, problems can arise from the length of the experiment; where the separation can be too long for unstable species like RNA–protein complexes. Fluorescence spectroscopy needs the presence of a fluorophore (like tryptophan, which is often already present in the protein) but interferences with the intrinsic absorption of DNA/RNA nucleotides can occur. SPR can follow the association processes in real time, enabling the determination of the kinetics and thermodynamics of an interaction (kinetic rate constants and equilibrium constants). The major drawback is that because one of the binding partners must be immobilized in a gel or onto a surface before the SPR experiment, the immobilization can influence interactions. Moreover, limitations arise concerning the mass of the interactants. Isothermal titration calorimetry can provide equilibrium constants and stoichiometry but as filter-binding experiments and gel mobility shift assays, it does not allow the accurate characterization of the binding partners. In the same way, analytical centrifugation provides a lower accuracy in mass measurement than does MS. Finally, x-ray crystallography and NMR spectroscopy can provide magnificent high-resolution structures but are time consuming and require, in reality, a significant amount of sample. Moreover, some structures are too large to be directly determined by NMR and certain samples do not crystallize or a part of a structure appears too flexible to be solved by x-ray crystallography. Thus, it remains desirable to develop complementary approaches that can be applied easily and efficiently.

Although MS cannot provide the detailed structural information provided by x-ray crystallography or NMR and is a destructive method, it requires a very small amount of sample (<pmol) and can yield important information about binding stoichiometry and nature of the binding partners. The main advantages of MS are clear: speed and sensitivity. MS can provide a fast and efficient approach for noncovalent complex studies with low sample consumption. MS can also be an appropriate technique for high-throughput analyses. Partners of the complex do not need to be labeled in MS. The stoichiometry of the present species in the complex, and their identification, can simultaneously be deduced from accurate molecular weight measurements. MS and MS/MS experiments can provide more details about the nature of each partner within the complex (even in the case of small ligands), and address the question of the specificity and stability of the complexes. Gas-phase-binding energies and in-solution equilibrium between the different species can be examined by different methods.[52]

However, MS interaction studies require appropriate instrumentation, an ESI source for example, associated with an analyzer of sufficient mass range. As previously mentioned, noncovalent complexes are commonly observed through narrow charge distributions because the nondenaturing conditions used promote the formation of compact structures with low charge states. Detection of these noncovalent species requires high m/z mass range analyzers such as TOF analyzers. Mass range and resolution must be high enough to allow the analysis of the large molecular assemblies in which most of the DNA–protein complexes and multimeric species exist. Practical problems can arise from difficulties related to the solubilization of the complexes and from the limited choice of buffers that are compatible with MS. The peaks of complex ions can become weaker and broader with increasing mass due to adduct formation with metal, solvent, or buffer molecules. MALDI performance can be affected by shot-to-shot reproducibility, NanoES performance by tip-to-tip variability.

Several studies showed that a careful choice of experimental conditions allows fragile, noncovalent complexes to be maintained during the transition from the solution phase to the gas phase. A well-known question arises: Do the ions and their abundances really reflect the solution species? How can one be sure that the observed species do not simply come from artifactual nonspecific aggregation occurring during the ionization process? It is known that weak noncovalent bonds can be broken during the ionization process and the noncovalent complex can not be detected (false negatives). Conversely, solutions that are too highly concentrated can lead to the formation of nonspecific aggregates (false positives). Relevant information about stoichiometry and actual existence of complexes can be obtained only if great attention is paid to ensure that the chosen experimental parameters do not induce the formation of nonspecific aggregates. Rigorous control experiments and comparison with the results provided by other well-established techniques are unfortunately often required. As an example, a dsDNA detected by MS must correspond to the sum of the two complementary single-strands (ss1 + ss2). The presence of the species (2 × ss1) or (2 × ss2) would indicate nonspecific complex formation. If these species are detected, experimental conditions are not appropriate and they must be modified. In the following sections, many examples will illustrate the various possible strategies used to assess the relevance of the MS data.

Last but not least, it is very likely that the acceleration and desolvation steps of the ionization process may induce conformational modifications within the complexes that affect the strength of the interactions. Clearly, mass detection of a complex alone does not constitute a proof for solution structure conservation. The extent of the conformational changes upon the ionization process is not yet really known. Even if in several cases it has been shown that gas-phase ions conserve, at least, a major part of their solution structure; the evidence for conformation conservation is often indirectly obtained. In this context, ion mobility begins to provide useful information (see next section).

10.3.3   RECENT DEVELOPMENTS IN MS INSTRUMENTATION (POTENTIALLY) USEFUL FOR NONCOVALENT COMPLEX STUDIES

The past few years have witnessed considerable progress in MS instrumentation. In particular, the development of new ionization techniques represents a highly active research field in MS.[62] New ionization methods that have been recently developed include atmospheric pressure MALDI (AP-MALDI), desorption ionization on silicon (DIOS), atmospheric pressure photoionization (APPI), desorption–electrospray ionization (DESI), direct analysis in real time (DART), electrosonic spray ionization (ESSI), Cold-spray ionization (CSI), and Laser spray ionization (LSI).[63–72] Although AP-UV-MALDI has been recognized as a softer ionization technique than conventional UV-MALDI, no significant advantage has been reported until now from its use in noncovalent interaction studies. Similarly, AP-IR-MALDI, which can use water as a matrix, has not yet given the hoped-for results with high masses, but this technique is extremely promising and the mechanism of ion production from the aqueous solutions irradiated by a pulsed IR laser under atmospheric pressure is under investigation.[73,74] DIOS and DART remain limited to the analysis of low-molecular-weight compounds. Among these new methods cited above, the most promising for noncovalent complex detection seem likely to be DESI, ESSI, CSI, and LSI. The DESI source generates similar ions as a classical ESI (multiply charged ions) by electrospraying charged droplets onto a surface that supports the sample, but when compared to the ESI source, it is less sensitive and also displays some limitations for the ionization of high-molecular-weight species. However, when these technical difficulties will be overcome, the benefits of ambient ionization and the possible coupling with imaging on tissues might be of considerable interest.[75,76] In contrast, ESSI has already given impressive results. ESSI is a new variant of ESI that uses a supersonic nebulizing gas.[68,77] ESSI shows sensitivity that is comparable to NanoES but with considerably better reproducibility. The most useful part for noncovalent interaction studies is that ESSI leads to extremely narrow charge-state distributions, which suggests a better conservation of three-dimensional structures with the ionization process.[68] Another variant of ESI is CSI, which operates at a low temperature and provided interesting results with triplex, quadruplex DNA, and other hyperstranded DNA architectures of low melting temperature.[69] Laser spray, which has been recently applied to the study of DNA-protein complexes, will be described in more detail (Section 10.5.1).[71,72]

Concerning mass analyzers, the orbitrap system recently introduced by Makarov and commercialized by ThermoFinningan represents a real novelty in this field.[78] This analyzer provides high resolution (60,000–100,000) but the main drawback is the mass range limitation (m/z 4000). Finally, the most significant breakthrough might come from the gas-phase ion mobility measurements coupled with MS.[79,80] Complex ions formed by electrospray can be separated as a function of their size and shape owing to the measurement of their drift times in a cell filled with an inert gas. Successful results have recently been obtained with multiple protein and dsDNA complexes allowing a new insight on conservation of conformational structures upon gas-phase ionization.[81,82] With the introduction of commercial instruments, ion mobility will undoubtedly play an important complementary role in future noncovalent complex MS studies.

10.4   MALDI-MS OF NUCLEIC ACID–PEPTIDE/PROTEIN COMPLEXES

A list of nucleic acid–peptide/protein noncovalent complexes detected by MALDI-MS is given in Table 10.1. The heaviest noncovalent complex ever detected by MALDI-MS is an RNA–protein

TABLE 10.1
DNA and RNA Complexes with Peptides/Proteins Detected by MALDI-MS

MS Instrumentation	Studied Complexes	Upper Mass Recorded (Da)	Experimental Conditions for MS Analysis	Refs.
UV-MALDI (+) $\lambda = 337$ nm Linear TOF mode	ssDNA (dT$_{10}$ and dT$_{20}$) with basic polypeptides, histone H4 (calf thymus)	14,316 (for dT$_{10}$-H4; 1:1)	Matrix: SA, anthralinic acid and 3-amino-pyrazine-2-carboxylic acid	88
UV-MALDI (+) $\lambda = 337$ nm Linear TOF mode	ssDNA (dT$_{10->60}$) with basic dipeptides (His–His, Arg–Lys) and insuline, ubiquitin, cytochrome C, ribonuclease A, lysozyme	~18,000	Matrix: 2,5-dihydroxy-benzoic acid (DHB)	89
UV-MALDI (−) $\lambda = 337$ nm Linear TOF mode	ssDNA and dsDNA with basic peptides (ACTH fragment 11–24, a nonapeptide)	10,259 (for EcoRI adaptor dsDNA + ACTH 11–24)	Matrix: Saturated solution of ATT in 50% acetonitrile and 1% ammonium citrate	91
UV-MALDI (+) $\lambda = 337$ nm Linear TOF mode	Various ssDNA (5-mer) with a zinc-finger peptide (p55F1) in the presence of Zn^{2+}	3598 (for s(TTGTT)-Zn^{2+}-p55F1; 1:1:1)	Matrix: Aqueous solution of 2-amino-4-methyl-5-nitropyridine (AMNP)	92
UV-MALDI (+/−) $\lambda = 337$ nm Linear TOF mode	Various ssDNA (16-mer) with a peptide fragment of HTH HIV-1 reverse transcriptase	~7800	Matrix: Saturated solution of ATT in 50% acetonitrile and 1% ammonium citrate	93
UV-MALDI (−) $\lambda = 337$ nm Linear TOF mode	ssDNA and dsDNA (7-mer, 12-mer) with BD15 peptides (MW 1816)	7350 (for BD15-ssDNA; 2:1)	Matrix: Saturated solution of ATT in 50% acetonitrile + 8 mM NH$_4$OAc buffer	94
UV-MALDI (+) $\lambda = 355$ nm Linear TOF mode	Yeast seryl and tyrolsyl-tRNA synthetases with tRNASer, tRNATyr	163,444 (for seryl-tRNA synthetase-tRNASer, 2:2)	Matrix: ATT in 10 mM ammonium citrate or 12.5 mM ammonium acetate	83
UV-MALDI (−) $\lambda = 337$ nm Linear TOF mode	Various ssRNA (9-mer > 19-mer) with different polypeptides/proteins	8611 (for insulin-ssRNA 9-mer)	Matrix: 2,4,6-trihydroxyaceto-phenone (2,4,6-THAP)	97

complex with a molecular weight in excess of 150 kDa.[83] Various matrices have been used; aza-2-thiothymine in the presence of an ammonium salt provided the best results in terms of specificity. Both ion modes, negative and positive, have been employed successfully (Table 10.1). The TOF analyzer was always used in the linear mode. While dsDNA (or is it ssDNA) alone has been characterized by several groups, only a few studies involved dsDNA–protein complexes due to difficulties in finding the appropriate conditions for noncovalent detection.[84–87] From a practical point of view, ssDNA, dsDNA, or RNA are generally mixed with the peptide/protein in deionized aqueous solution and the matrix is added at the last moment just before the MALDI analysis. To form dsDNA, equimolar mixtures of complementary ssDNA sequences are mixed, heated to 90°C for a few minutes, and then slowly cooled to form specific duplexes. In the case of RNA, RNA is heated alone to achieve the proper folding. The nucleic acids studied to date are generally segments with lengths limited to a few dozen bases (Table 10.1).

10.4.1 MALDI-MS of DNA Complexes

Juhasz and Biemann reported early in 1994 the possible detection by UV-MALDI of noncovalent complexes formed between small ssDNAs (dT_{10} and dT_{20}) and basic polypeptides.[88] Noncovalent complexes were produced in the gas phase in the presence of various acidic matrices: SA, anthralinic acid, and 3-aminopyrazine-2-carboxylic acid. It was soon understood that complex formation mainly depended on the content of arginine in the polypeptides, histidine and lysine content appeared to be of a lesser importance. Histone H4 protein (11,236 Da), which contains 14 arginines and 11 lysines, was found to be the best complexing agent for oligonucleotides. It was formed with dT_{10} (2980 Da) to make the largest detected complex. Since it was already well known that metal cation attachment by Na^+, K^+, and so on was a common problem with nucleic acids analysis when the sample was not desalted (poly-sulfonated, -sulfated, and -phosphorylated biomolecules share the same behavior), the complex formation was introduced at that time as a useful analytical method to limit the peak broadening of highly acidic compounds due to metal cation attachment.[88]

The next year Vertes et al.[89] explored by means of MALDI-TOF-MS the noncovalent interactions between various polypeptides ranging from Arg–Lys, His–His basic dipeptides up to a selection of proteins (insulin, ubiquitin, cytochrome C, and lysozyme) and ssDNA homooligomers ($dT_{10->60}$, dA_9, dG_{10}, and dC_{10}). These authors investigated the effect of amino acids and nucleotide base composition, the role of the matrix and the effect of pH on the number and abundances of ssDNA/protein complexes. Among the four oligomers (dT_n, dA_n, dG_n, and dC_n), abundant complexes were detected in DHB for dT_n only. Arg–Lys and His–His gave similar results. The larger complexes were obtained from Arg–Lys/dT_{60} and ubiquitin/dT_{20} (~18 kDa). A lower sensitivity of ion detection was clearly observed in the higher masses. It was remarked that among three proteins of similar size (cytochrome C, ribonuclease A, and lysozyme), lysozyme formed the most abundant complex with T_{10} and also contained the largest number of Arg residues. The authors observed a strong correlation between the presence and the number of basic residues and the tendency to form complexes. Detection and intensity of protein–DNA complex ions are depended on the choice of matrix as well. Complexes were obtained with DHB and 3-HPA matrices, whereas there were no complexes detected in SA or CHCA. The particular role of thymine in the ssDNA–protein complexation was interpreted through an acid–base process, by predicting the protonation degree of the peptides and nucleotides as a function of the pH. Only T is expected to bear a negative net charge at pH 1.7 (DHB saturated solution pH), whereas the other nucleotides are under zwitterionic form (leading to an intramolecular ion pair), thus the presence of T was needed for complex formation. When the basic polypeptides were in large excess, multiple stoichiometries were observed. These authors also noted a remarkable improvement in the mass resolution of the signal in the presence of a basic peptide. The basic polypeptides were supposed to occupy the highly acidic sites on phosphates groups, displacing the alkali ions. The signal was consequently concentrated into a single peak rather than in an ensemble of cationized ions contributing to peak broadening.[89]

In early reports on the analysis of DNA alone by MALDI-MS, only ssDNA was detected, intact dsDNA was hard to detect. Sample preparation is obviously critical. The conditions (temperature, pH, ionic strength, and solvent must be appropriate) must be chosen to preserve the DNA duplex in the matrix. Even if the noncovalent complex survives this first step, it must also survive the ionization process. In order to detect specific intact dsDNA, Lecchi and Pannell introduced in 1995, the use of ATT as a matrix combined with ammonium citrate to avoid acidic pH during sample preparation.[84,90] In contrast, when 2′,4′,6′-THAP or 3-HPA was used, only individual DNA strands were detected. With the EcoR1 adaptor (which contains 12 complementary base pairs, ~8600 Da), these authors demonstrated that in addition to the duplex DNA signal, the detected single strands reflected species already present in solution since the addition of an enzyme (nuclease S1) that preferentially digests the ssDNA made the corresponding peaks rapidly disappear.[84] Later, several other authors detected by MALDI-MS stable dsDNA up to a 70-mer and dsDNA-peptide or protein complex studies became conceivable.[85–87]

Lin et al. studied the effects of pH, ionic strength, and matrix nature on the detection of noncovalent complexes between ss/dsDNA of various lengths (12-mer, 16-mer, 20-mer, and 25-mer) and different basic peptides.[91] Noncovalent complexes with dsDNA were observed only in the case of an ACTH fragment (**KPVGKKRRPVKVYP**) and a synthetic peptide (**VRKRTLRRL**) where at least five basic residues were in close proximity to each other and separated by aliphatic residues. The authors noted that this arrangement may be compared to the amino acid primary structure of H4 histone. dsDNA–peptide noncovalent complexes were detected only by using ATT as a matrix (pH 7); no complex was observed with the standard 3-HPA matrix. Typically, noncovalent complexes were detected using a saturated solution of ATT in 50% acetonitrile and 1% ammonium citrate. The DNA–peptide mixture in deionized water was first deposited on the MALDI plate, and then the ATT matrix solution was spotted onto the same place. Even if the DNA sequence did not seem to play a role in the formation of the DNA–peptide complexes, since the interactions between DNA and peptides were supposed to be mainly ionic, ATT clearly favored the AT-rich oligonucleotides' desorption–ionization when compared to the GC-rich oligonucleotides. The ionic nature of the interactions between the negatively charged phosphates of DNA and the positively charged residues of arginine and lysine was supported by the fact that when the pH was decreased to 2, or when the noncovalent complex was incubated in the presence of 50 mM ammonium citrate, the corresponding complexes disappeared from the mass spectra.[91]

MALDI-MS has not only the potential to rapidly and directly detect noncovalent DNA–peptide complexes, MALDI results can also, under carefully chosen conditions, reflect a solution-phase behavior as shown by Lehmann and Zenobi with the detection of a noncovalent zinc-finger peptide–ssDNA complex.[92] The 18-residue peptide sequence (named p55F1) corresponded to the first zinc-finger domain from the gag protein p55 of HIV-1, and the DNA was a short ssDNA fragment d(TTGTT). The MALDI spectrum obtained in positive polarity using 2-amino-4-methyl-5-nitropyridine (AMNP) as the MALDI matrix showed a specific complex at m/z 3598 corresponding to the expected complex d(TTGTT)-Zn^{2+}-p55F1 (1:1:1). Several control experiments were performed including variation of the pH, peptide competition experiments, replacement of Zn^{2+} by another metal ion, and replacement of d(TTGTT) by another ssDNA sequence. It was demonstrated that the MALDI mass spectra reflected a solution-phase behavior because decreasing the pH value from 6.5–7 to 5–5.5 led to a significant decrease of the specific complex d(TTGTT)-Zn^{2+}-p55F1 ion abundance. This result was consistent with circular dichroism spectroscopy (CD) data, which showed that the complex was stable only above pH 6. A similar decrease in the d(TTGTT)-Zn^{2+}-p55F1 ion abundance was noted when a peptide without the zinc-finger motif was used and when Zn^{2+} was replaced by Cu^{2+}. In the presence of Cu^{2+}, cysteine amino acids, which are essential for zinc-finger integrity, were oxidized and the resulting chemical modifications of the zinc-finger motif prevented complex formation. When complexation with other ssDNA sequences was investigated, the complex did not exhibit a dependence on the DNA sequence.

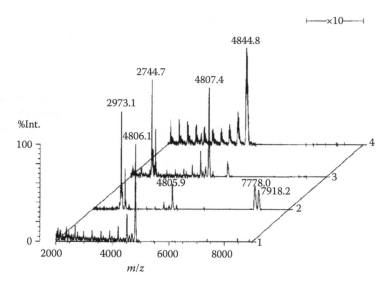

FIGURE 10.5 Positive ion linear MALDI-TOF mass spectra of noncovalent complexes of dT_{16} and the helix clamp. The matrix was a saturated solution of ATT in 50% acetonitrile and 1% ammonium citrate. Oligonucleotide [dT_{16}] concentration was 100 μM, and peptide concentration was 45 μM. Spectrum 1, dT_{16} alone led to [dT_{16}+H]$^+$ at m/z 4806.1; spectrum 2, dT_{16} was mixed with the helix clamp resulting in a non-covalent complex at m/z 7778 and 7918.2 (matrix adduct). m/z 2973.1 and 4805.9 corresponded to [M+H]$^+$ of dT_{16} and helix clamp, respectively; spectrum 3, dT_{16} was mixed with HTHK → A, noncovalent complex was not detected; spectrum 4, dT_{16} was mixed with ANP, noncovalent complex was not detected. (Reprinted from Lin et al. *Anal. Chem.* 72(11), 2635–2640, 2000. With permission. Copyright 2000 American Chemical Society.)

Additional evidence that MALDI results can clearly reflect an in-solution behavior was presented by Lin et al.[92] These authors studied interactions between a 26-residue peptide sequence ([259]KLVGKLNWASQIYAGIKVKQLCKLLR[284]) and various ssDNA (16-mers). This peptide called a "helix clamp" in their report is, in fact, the DNA-binding domain of an HTH motif of HIV-1 reverse transcriptase. The MALDI mass spectrum, recorded in the positive mode using ATT as the matrix in the presence of ammonium citrate, yielded the dT_{16} + helix clamp peptide complex at m/z 7778 with a matrix adduct being detected at m/z 7918 (Figure 10.5, spectrum 2). Spectrum 1 of this same figure corresponds to dT_{16} alone (m/z 4806). By replacing four lysines with four alanines in the "helix clamp" peptide sequence (the peptide sequence becomes [259]KLV-GALNWASQIYAGIAVAQLCALLR[284], noted HTHK → A), it was shown that the DNA-binding domain lost its ability to bind DNA. The signal corresponding to the complex disappeared in the MALDI spectrum (Figure 10.5, spectrum 3). It was also demonstrated that the presence of basic amino acids within the peptide sequence was not sufficient for DNA binding by replacing the "helix clamp" peptide by another peptide with a similar molecular weight (atrial natriuretic peptide, ANP), which also contained six basic amino acid residues in its sequence. As before, the complex was no longer detected in the MALDI spectrum (Figure 10.5, spectrum 4). Lastly, since a DNA–peptide complex was still detected by replacing the dT_{16} DNA sequence by another 16-mer DNA sequence (not shown), it was concluded that the helix clamp binding to DNA did not require a particular DNA sequence. All of these MALDI experiments were duplicated using SPR and the results were in perfect agreement.[93]

The previously described studies demonstrated the possible observation by MALDI of non-covalent complexes between single- or double-stranded DNA and peptides. Nevertheless, these interactions were not DNA sequence specific. The involved interactions were essentially ionic interactions occurring between the negative charges of the phosphate groups of the DNA backbone and the positively charged residues of the peptides. The first example of DNA sequence-specific interaction was recently reported by Luo et al.[94] These authors studied the attachment of the basic domain (BD15) of the c-Fos protein onto various different ss/dsDNA sequences. It was known from x-ray crystallographic data that the c-Fos protein binds specific DNA sequences containing the consensus sequence (5′-TGAGTCA-3′) under a dimeric form. As expected, the noncovalent DNA complexes involving BD15 were only detected under a dimeric form in the presence of DNA sequences containing the targeted DNA sequence. Moreover, MALDI spectra, accounting for this specific interaction, were recorded only when using ATT as the MALDI matrix. Using DHB as the matrix, the BD15 peptide binds to DNA under its monomeric form only, whereas no complexes were observed when HCCA was used as a matrix. It was thus concluded that ATT is a good matrix choice for studying DNA–peptide-specific noncovalent complexes. Unfortunately, it must be noted that a dsDNA + dimeric DB5 peptide complex was never detected. In the best case, either dsDNA was complexed by one BD15 peptide, or ssDNA was attached to two BD15 peptides.[94]

ATT as a matrix has been also employed by Ohara et al. in a recent study dealing with the interactions between highly basic sites of arginine derivatives (a set of 12 compounds, each of which include one guanidinium group) and single-stranded DNAs.[95] Strong interactions were expected owing to the complementary shapes and charges of the phosphate and guanidinium groups. Among the various MALDI matrices tested, 3-HPA, THAP, nicotinic acid (NA), and ATT, it was reported that ATT with ammonium citrate gave the best results. The strongest interactions between ssDNA and guanidinium compounds were found for compounds where the guanidinium group was connected to an aromatic structure enabling stabilization by π-stacking. The fact that no complex was detected with lysine was explained by an insufficient basicity of the latter, at least in the gas phase. The electrostatic nature of the interaction was demonstrated in several ways: when NaCl was added, the abundance of the complexes strongly decreased as expected, due to the competition of Na$^+$ with guanidinium to form ionic bonds with phosphate groups; no complex showed guanidinium adducts that exceeded the number of phosphate groups; and almost no complex was detected when a fully methylphosphonate ssDNA was used, which allowed neither electrostatic bonding nor H-bonding. Furthermore, a control experiment was performed to prove that the noncovalent complexes were first formed in solution and then transferred intact into the gas phase. Initially, ssDNA and the guanidinium compounds were prepared in their matrix solutions separately and then dried. The resulting powders were mixed without any solvent and deposited on the MALDI plate. Upon laser irradiation, only ssDNA was detected and no complex could be identified from the powder mixture. After the addition of an acetonitrile/water mixture to the deposit, all complexes previously seen under standard conditions could be detected again. Even if this work involved only arginine and related derivatives rather than peptides, it was demonstrated again that MALDI ionization can be efficiently used to probe noncovalent interactions involving DNA.[95]

10.4.2 MALDI-MS OF RNA COMPLEXES

A few studies have reported on the analysis of RNA-protein noncovalent complexes using MALDI-TOF-MS. Gruic-Sovulj et al. studied complexes formed between Yeast aminoacyl-tRNA synthetases (α) and tRNAs.[83] The formation of synthetase(α)–tRNAs complexes corresponds to an important step for the future-specific attachment of amino acids to their cognate tRNAs during gene translation. When seryl-tRNA synthetase (SerRS, 54 kDa) and tyrosyl-tRNA synthetase (TyrRS, 42 kDa) were first analyzed alone in the presence of ATT as the MALDI matrix, desorption of intact homodimeric species (2α) was observed, as expected, for native enzymes known to be dimeric species. It is worth noting that these homodimeric species were only observed for the very first laser shot. From the second shot, complex signal decreased, and then disappeared. Only the signal corresponding to the monomeric species survived. This phenomenon which is commonly described as the "first-shot phenomenon" (also reported by other authors) was not observed for the synthetase–tRNA complexes.[50,51]

A study by Horneffer et al. established that the first-shot phenomenon observation comes from intact complexes localized at the matrix crystal surface whereas complexes do not survive upon incorporation into the matrix crystals.[96] Horneffer et al. underlined that the first-shot phenomenon has been observed for protein–protein complexes only and not for nucleic acid–protein complexes.[96] In the presence of their cognate tRNA^{Tyr} (25 kDa) or tRNA^{Ser} (27 kDa), signals corresponding to synthetase–tRNA complexes were detected with various stoichiometries. As seen in Figure 10.6a, the mixture of SeRS with its cognate tRNA^{Ser} yielded α_1-tRNA^{Ser+} (m/z ~82,000), α_2-tRNA^{Ser+} (m/z ~136,000), and α_2-tRNA$_2^{Ser+}$ (m/z ~162,500). As only α_2-tRNA was expected and noncognate complexes ((TyrRS)$_2$-tRNA^{Ser} and (SerRS)$_2$-tRNA^{Tyr}) were also detected in crossed experiments (not shown), binding specificity between tRNAs and synthetases was investigated through control experiments. Layered deposits were prepared as described. As a first step, synthetase was spotted on the target and then allowed to dry. The tRNA was then added without mixing with synthetase. Complex formation between synthetase and tRNA was not supposed to occur under these conditions and no peaks corresponding to any protein–RNA aggregates were detected (Figure 10.6b).

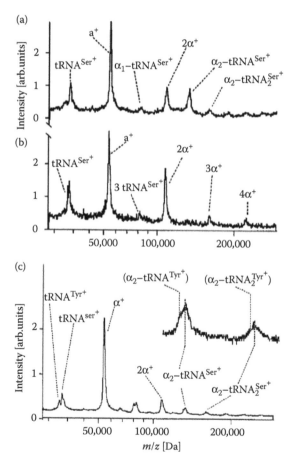

FIGURE 10.6 MALDI-mass spectra of SerRS and its complexes with cognate tRNA. The amounts of material were 0.86 pmol of SerRS, 3.2 pmol of tRNASer, 0.12 mmol of ATT, 0.4 mmol of glycerol, and 38 nmol of ammonium acetate (a). In (b) the amounts of material are the same as in (a), but the components were applied to the target separately (see text). In (c) MALDI-mass spectrum of a mixture of SerRS, tRNATyr, and tRNASer (tRNATyr and tRNASer in equimolar amount). The recordings (a), (b), and (c) show the sums of 16, 6, and 30 single-shot spectra, respectively. (Adapted from Gruic-Sovulj, I. et al., *J. Biol. Chem.* 272(51), 32084–32091, 1997. With permission. Copyright 1997 The American Society for Biochemistry and Molecular Biology.)

Specific recognition was demonstrated from equimolar mixtures of both cognate and noncognate tRNAs in competition for binding to one synthetase alone (Figure 10.6c). The predominant detection of cognate complexes in MALDI spectra clearly showed that artifacts did contribute to complex signals. In this study, it was shown that MALDI-MS enables the determination of RNA–protein complex stoichiometry with numerous advantages (direct deduction of the stoichiometry and speed) over gel electrophoresis. However, as artifact signals inherent to the MALDI process can be superimposed on specific complex signals, precautions must be taken and suitable control experiments are advised.[83]

Thiede et al. evaluated MALDI-TOF-MS as a fast screening method for studying RNA–peptide interactions.[97] Several complexes were formed between different model peptides containing 7–22 amino acids and various ssRNAs (from a 9-mer up to a 19-mer). RNA and peptide sequences were chosen to mimic known contact sites of the ribosome. Among the different MALDI matrices tested, DHB, 3-HPA, 2,3,4-THAP, and 2,4,6-THAP, 2,4,6-THAP was selected because it was the only matrix allowing the simultaneous detection of the peptide, RNA, and RNA–peptide complex in the same mass spectrum (in the negative mode). By mixing together one RNA with two different peptides followed by MALDI-MS analysis, and by repeating this experiment several times, taking into account the various possible combinations, an order of preference of the different peptides for RNA binding was established. The same order of preference of the peptides for complex formation was found for three different RNAs in contrast to the data of cross-linking experiments. It was clearly demonstrated that complex detection depended mainly upon the basic amino acid content in the peptides. As the abundance of arginines, lysines, and histidines increased in the peptides, the abundance of the detected RNA–peptide complex also increased with, arginine amino acids leading to the most significant effects. No significant effect due to RNA composition (A, U, G, C content) or RNA secondary structure (presence of hairpin loop, internal loop, etc.) was observed. This is in contrast to the results obtained from the MALDI-MS study of DNA–peptides by Tang et al.[89] Despite the fact that the necessary presence of basic amino acids within the peptide for complex formation tended to support the pertinence of the interactions observed by MALDI-MS, this point remains to be clarified. The authors point out that as peptides and RNAs are potential drugs and drug targets, MALDI-MS might be used within this context as a screening method, owing to its ability to perform fast and automated analyses.[97]

10.5 ESI-MS OF NUCLEIC ACID–PEPTIDE/PROTEIN COMPLEXES

10.5.1 ESI-MS OF DNA COMPLEXES

As previously stated (Section 10.4.1), the MALDI technique is not ideal for analyzing duplex DNA; in contrast, several ESI-MS studies have been published on the analysis of duplex DNA. Ganem et al. and Light-Wahl et al. demonstrated in 1993 that dsDNA was stable enough to survive the ESI process.[98,99] The conservation of Watson–Crick pairing in the gas phase was evidenced later by several authors.[100,101] Finally, ion mobility and molecular modeling showed that the double-helix conformation was maintained in the gas phase.[82]

To date, the largest dsDNA–protein noncovalent complex detected by ESI-MS is ~135 kDa and the largest RNA–protein complexes detected by ESI-MS are intact ribosomes with molecular weights in excess of MDa (2×10^6 Da).[102,103] Noncovalent complexes detected by ESI-MS are summarized in Table 10.2 for DNA–peptide/protein complexes and Table 10.3 for RNA complexes. ESI detection of nucleic acid–protein complexes has been carried out in both negative and positive polarity within an m/z 2000–5000 mass range. ESI and related low-flow techniques (microspray and nanospray) have been used in combination with various analyzers (quadrupole, EB/BE, TOF, Q-TOF, and FT-ICR). Control experiments with a great variety of approaches have been described in the literature to provide evidence that the complexes did not arise from nonspecific aggregation but rather reflected an in-solution behavior. In almost all cases, complexes were also characterized

TABLE 10.2
DNA–Peptide/Protein Complexes Detected by ESI-MS

MS Instrumentation	Studied Complexes	Upper Mass Recorded (Da)	Experimental Conditions for MS Analysis	Refs.
ESI (–) Quadrupole	ssDNA phosphorothioate (20-mer) with BSA protein	79,273 (for BSA-ssDNA; 1:2)	33 mM imidazole buffer-10 mM NH$_4$OAc buffer, pH 7	104
ESI (–) FT-ICR	ssDNA (MW ~4000–5500) with gene V protein (MW 96,88) from bacteriophage f1	43,650 (for gene V-16-mer DNA; 4:1)	10 mM NH$_4$OAc buffer, pH 7	105
ESI (–) FT-ICR	dsDNA (17-mer, MW 10,378) with transcription factor PU.1 Binding domain (MW 13,534)	23,913	10 mM NH$_4$OAc buffer, pH 7	106
ESI (–) EB	dsDNA (34-mer, MW 21,375) with vitamin D receptor-binding domain with Zn^{2+}(MW 12,944)	47,049 (for the 2:1 complex)	H$_2$O	107
ESI (+) TOF	Operator dsDNA (21-mer, MW 21,375) with *trp* repressor and tryptophan (MW 12,944)	37,675 (for TrpR dimer-dsDNA-tryptophans; 1:1:2)	5 mM NH$_4$OAc buffer, pH 6	108
μESI (–) FT-ICR	dsDNA (20-mer) with human XPA-binding domain (MW 14,767)	41,997 (for protein dimer-dsDNA)	10 mM NH$_4$OAc buffer, pH 6.7	109
μESI (+) BE	dsDNA (34-mer, MW 21,375) with vitamin D receptor (VDR, MW 48,660) and retinoid X receptor α (RXRα, MW 48,180) and 1α, 25-dihydroxyvitamin D$_3$ and 9-*cis*-retinoic acid	118,200 (for 4:1:1:1 Zn^{2+}/RXRα/VDR/dsDNA)	10 mM NH$_4$HCO$_3$ buffer, pH 8	110
ESI (+) Quadrupole	Various dsDNA (13-mer, 24-mer, 30-mer) with CGN4 (MW 7988)	~32,000	90/10 water/methanol	111
NanoESI (+) TOF	dsDNA (29 bp) with GclR and IclR-8 proteins (MW 29,109 and 28,933, respectively)	134,941 (for GclR tetramer–dsDNA)	10 mM NH$_4$HCO$_3$ buffer, 1 mM DTT	102
ESI (+) Q-TOF	Various dsDNA (21 bp) with Tus protein (MW 35,653) and mutants	49,613	10–2200 mM NH$_4$OAc buffer, pH 8	112
μESI (–) EB	dsDNA (30 bp) with AbrB protein (MW ~10,500)	~60,500 (for protein tetramer–dsDNA)	20 mM NH$_4$HCO$_3$ buffer, pH 8	53,113, 114
NanoESI (+) Q-TOF	dsDNA (23-mer) with BlaI protein and a mutant (MW ~15,400)	~45,500 (for protein dimer–dsDNA)	150 mM NH$_4$OAc buffer	56
ESI (+) Q-TOF	Hairpin DNA (40-mer) with C170 protein (MW 21,119)	33,405	10 mM NH$_4$OAc buffer	121
μESI (–) EB	Various dsDNA (30-mer and 43-mer) with AbrBN/AbrB proteins (MW ~6100/10,500)	~69,000 (for AbrB tetramer–43 bp dsDNA)	10 mM NH$_4$HCO$_3$ buffer, pH 8	115
Laser-ESI (+) o-TOF	dsDNA (22-mer) with c-Myb-binding domain protein (MW ~13,000)	~26,250	10 mM NH$_4$OAc buffer	71
Robotic chip-based NanoESI (+) TOF	ssDNA (17-mer) with E9 DNase (MW ~15,087) in the presence of Zn^{2+}/Ni^{2+}/Co^{2+}	20,375 (for 1:1:1; Ni^{2+}/E9 DNase/ssDNA)	50 mM NH$_4$OAc buffer, pH 6.7	120

continued

TABLE 10.2   (continued)
DNA–Peptide/Protein Complexes Detected by ESI-MS

MS Instrumentation	Studied Complexes	Upper Mass Recorded (Da)	Experimental Conditions for MS Analysis	Refs.
ESI (+) Q-TOF	Various dsDNA (16-mer, 22-mer) with c-Myb-binding domain protein (MW ~13,000)	~26,250	10–600 mM NH_4OAc buffer	123
Laser-ESI (+) o-TOF	Various dsDNA (16-mer, 22-mer) with c-Myb-binding domain protein (MW ~13,000)	~26,250	10–50 mM NH_4OAc buffer	122
NanoES (−) FT-ICR	ssDNA (($TG)_4$, MW 2470) with HIV-1 nucleocapsid protein (NC, MW 9688)	15,448 (for ssDNA-NC; 1:2)	150 mM NH_4OAc buffer, pH 7	147
NanoES (−) Q-TOF	ssDNA (12-mer) with penta-L-arginine (MW 799), penta-L-lysine (MW 659)	4454 (for ssDNA-R_5)	10 mM NH_4OAc buffer	119

by others techniques to enable comparisons. The main questions that MS can address include stoichiometry, complex stability, and the relative affinity between binding partners.

In 1995, Greig et al. detected, by ESI-MS in negative polarity, the first oligonucleotides/protein complexes by mixing bovine serum albumin protein (BSA, 66 kDa) with increasing concentrations of a 20-mer phosphothioate oligonucleotide (GCCCAAGCTGGCATCCGTCA, MW 6367) in the

TABLE 10.3
RNA–Peptide/Protein Complexes Detected by ESI-MS

MS Instrumentation	Studied Complexes	Upper Mass Recorded (Da)	Experimental Conditions for MS Analysis	Refs.
(Nano)ES (+/−) EBqQ	HIV1 TAR RNA (31-mer), various mutants with Tat protein and a Tat peptide fragment	~19,700	10 mM NH_4OAc buffer, pH 6.9, 10% methanol	129
μESI (+/−) FT-ICR	Four ssRNA (12-mer > 20-mer) with bacteriophage T4 regA protein	~21,000	50 mM NH_4OAc buffer, pH 7	130
μESI (+) EB	α-p50 RNA aptamer (31-mer) with NF-κB p50 homodimer	~83,350	20 mM NH_4HCO_3 buffer	53,131
ESI (−) FT-ICR	RNA hairpins of the Ψ-RNA recognition element with HIV-1 nucleocapsid protein p7 (NC)	12,963	10 mM NH_4OAc buffer, 10% (v/v) isopropanol, pH 7.5	133
ESI (+) FT-ICR	RNA hairpins of the Ψ-RNA recognition element with HIV-1 nucleocapsid protein	~12,600	10 mM NH_4OAc buffer, 10% (v/v) isopropanol	134
NanoES (+) Q-TOF	ssRNA (10-mer) with N-terminal domain of RNase E	~263,000	1 M NH_4OAc	132
NanoES (−) FT-ICR	RNA hairpins and mutant RNAs with HIV-1 nucleocapsid protein (NC)	~48,650 (for RNA–NC, 2:4)	10–150 mM NH_4OAc	135
NanoES (+) o-TOF	Intact 70S Ribosomes from *E. coli*	~2,300,000	1 M NH_4OAc, 5 mM Mg^{2+}	103
NanoES (+) Q-TOF	Intact 70S Ribosomes from *Thermus thermophilus*	~2,300,000	1 M NH_4OAc	136,137

presence of a imidazole and ammonium acetate buffer.[104] The K_D values were deduced from the sum of the integrated ion abundances of the various charge states, as it was assumed that the relative ion abundances reflected the in-solution proportions of free and associated species (1:1 and 1:2 as BSA:ssDNA complexes). The K_D values obtained by ESI-MS ($K_{D1} \sim 3.1 \pm 0.3$; $K_{D2} \sim 11.9 \pm 0.6$) were in good agreement with the values obtained using capillary electrophoresis ($K_{D1} \sim 2.8 \pm 0.3$; $K_{D2} \sim 10.0 \pm 0.2$). As the K_D value depended on the ionic strength and pH, it was suggested that electrostatic forces contributed significantly to the binding energy of the complex. It is important to note that no complex was detected when the oligonucleotide was mixed with a heated-denatured BSA.[104]

Cheng et al. studied the binding stoichiometries of gene V protein (MW 9688 Da) from bacteriophage f1 to several small single-stranded oligonucleotides (MW ~4000–5500) using ESI-MS.[105] The gene V protein alone in 10 mM ammonium acetate (pH 7) gives a dimeric species by ESI-MS, as expected, because the protein is known to exist as a dimer under physiological conditions. Upon addition of single-stranded oligonucleotides of various lengths, a 4:1 complex (protein:ssDNA) was mainly observed with an 18-mer and 16-mer, while 2:1 complexes were predominantly detected with shorter oligonucleotides (14-mer and 13-mer). Both 4:1 and 2:1 were simultaneously detected with a 15-mer. The gene V: 16-mer ssDNA stoichiometry was confirmed independently by size exclusion chromatography (SEC). Several control experiments were conducted to ensure that the data obtained by ESI-MS reflected the known in-solution behavior. For example, the complexes disappeared from mass spectra when acids or organic solvents were added; conversely, the results were not affected by modifying the capillary temperature. Moreover, a strong preference in binding to poly(dT) over poly(dA) was observed, and no complexes were detected with double-stranded DNA, as expected, since it was known that the gene V protein prefers to bind ssDNA over dsDNA.

The DNA sequence specificity of a protein–DNA complex was examined using ESI-MS through the noncovalent interactions of the DNA-binding domain of the PU.1 protein (PU1 DBD, MW 13,534 Da), a eukaryotic transcription factor, and two different double-stranded DNA sequences (17 and 19-bp).[106] One of the two dsDNA sequences contained the GGAA consensus binding sequence (the wild-type, 17-bp), whereas the other did not (the mutant, 19-bp). For the mutant, the GGAA sequence core was simply replaced by GCTA and a supplementary base was added at each end of the DNA strands. When PU1 DBD protein was incubated with an excess of the wild-type DNA sequence in 10 mM ammonium acetate, only the protein–dsDNA complex ions of 1:1 stoichiometry were detected. Even in the presence of a mixture containing a 20-fold excess of the mutant 19-bp dsDNA relative to the wild-type 17-bp, only the 1:1 specific complex with the wild-type 17-bp dsDNA was detected. These MS results, consistent with gel electrophoresis mobility assay results, clearly showed that the PU1 DBD protein specifically recognized the GGAA consensus sequence as expected.[106]

The necessity of the presence of Zn^{2+} (or Cd^{2+}) for the formation of a noncovalent complex between a DNA-binding domain of vitamin D receptor (VDR DBD) and dsDNA was characterized by Veenstra et al.[107] Whereas the VDR DBD protein is supposed to contain two high-affinity Zn^{2+}-binding sites (two putative Zn^{2+} finger regions), both the monomer and dimer protein/dsDNA complexes containing two Zn^{2+} or two Cd^{2+} per protein were detected by ESI-MS in the negative mode. No complex was observed in the absence of Zn^{2+} or Cd^{2+}, or with a randomized dsDNA sequence. It was also shown that the formation of the protein/dsDNA complexes depended on the metal concentration range. An excessive increase in Zn^{2+} or Cd^{2+} concentration disrupted the complex.

Potier et al. chose the *Escherichia coli* trp repressor–operator system as a model to study a DNA-binding protein whose affinity for DNA is regulated by a cofactor.[108] The noncovalent interactions between the trp apo-repressor (TrpR), a DNA-binding protein (MW 12,224 Da) involved in the regulation of tryptophan biosynthesis, its corepressor tryptophan and three different operator dsDNA sequences were investigated by ESI-MS. The authors reported that no protein-DNA complex was observed in 5 mM ammonium acetate because TrpR protein was partially denatured under these conditions (this fact was confirmed by circular dichroism). When the complex was dialyzed first in 90 mM NaCl, 20 mM NaH_2PO_4 (pH 6), then in 10 mM NH_4OAc (pH 6), and finally diluted in 5 mM NH$_4$OAc, a [1:1] [TrpR dimer–dsDNA] complex (MW 37,260) was detected as expected. Moreover, the specific complex was observed only when dsDNA possessed the two CTAG consensus sequences separated by four base pairs. When the TrpR protein was mixed with dsDNA sequences with a different spacing between the two CTAG consensus sequences (i.e., 2 or 6 bp), no complexes were detected. Even if the trp corepressor was not needed for complex formation with DNA, the specific addition of two L-tryptophan molecules to the protein–DNA complex was noted in contrast to D-tryptophan. This work demonstrated the potential of ESI-MS to assess the sequence specificity of protein–DNA complexes.[108]

The study of interaction of the human xeroderma pigmentosum A (XPA) minimal-binding domain (WPA-MDB, MW 14,767) with a damaged dsDNA (20-mer) was reported using ESI-FT-ICR by Xu et al.[111] The human XPA protein is supposed to be involved in the early step of the repair process of damaged DNA. Cisplatin was used as a damaging agent leading to a cross-linking between two adjacent guanines on the same DNA strand. Both damaged and undamaged dsDNA 20-mer formed 1:1 complexes with XPA-MDB; only a slight preferential binding with the damaged dsDNA was noted. In contrast, a 2:1 protein–DNA complex was only observed with damaged dsDNA. A control experiment was performed with cytochrome C instead of XPA–MDB, and no protein–DNA complex was detected in this case. The authors underlined the benefits of MS over gel mobility shift assays. MS appeared more appropriate due to the relatively weak binding of XPA–MDB with dsDNA, and was sensitive to the small mass difference between damaged dsDNA and undamaged dsDNA (only 229 Da).[109]

The usefulness of μESI-MS in assessing the interactions of the various components of transcriptional complexes depending on the presence of ligands was demonstrated by Craig et al.[110] The vitamin D receptor (VDR) and retinoid X receptor α (RXRα) are ligand-dependent transcription factors whose activity in the presence of Zn^{2+} can be modulated by at least two known ligands: 1α,25-dihydroxyvitamin D$_3$ and 9-cis-retinoic acid. The use of ESI-MS permitted the rapid detection of specific complexes with the osteopontin vitamin D response element (OPVDR, a 34-mer dsDNA). In the absence of DNA, both VDR and RXRα proteins are essentially detected alone under their respective monomeric form. When each protein was mixed with the OPVDR dsDNA sequence, VDR did not form any noncovalent complexes, whereas RXRα formed a homodimeric complex (4:2:1 Zn^{2+}/RXRα/dsDNA; MW 117,750). Upon the addition of the equimolar VDR + RXRα protein mixture to the OPVDR dsDNA, a protein–DNA complex (4:1:1:1 Zn^{2+}/RXRα/VDR/dsDNA; MW 118,200) was recorded, which is one of the largest protein–DNA complexes detected to date. A control experiment revealed that no complex was detected with random dsDNA. The results were in relatively good agreement with those obtained from gel shift analyses. The addition of 9-cis-retinoic acid led to an increase of the homodimeric complex (4:2:1 Zn^{2+}/RXRα/dsDNA; MW 117,750 Da) whereas the addition of 1α,25-dihydroxyvitamin D$_3$ resulted in a mass increase, which has been attributed to the binding of 1α,25-dihydroxyvitamin D$_3$ to VDR within the Zn^{2+}/RXRα/VDR/dsDNA complex. The results were consistent with previously published data obtained using gel shift or SPR.[110]

At the time of writing, the heaviest DNA–protein complex ever transferred intact from solution phase into the gas phase by ESI-MS was that reported in 2001 by Donald et al.[102] This complex, with a molecular weight of 134,941 Da (mass spectrum shown in Figure 10.7), consisted of one dsDNA (29 bp) plus one tetrameric form of the GclR protein (glycoxylate carboligase repressor). This detection record was achieved by combining a nanospray source and a TOF analyzer (constructed at the University of Manitoba). The sample was prepared in 10 mM NH$_4$HCO$_3$ and 1 mM dithiothreitol (DTT). Despite the fact that nondenaturing conditions seemed to produce broad peaks with low resolution, probably due to a partial desolvation, the ions corresponding to the complex bearing 22–28 positive charges are readily identifiable in the m/z 5000–6000 range (Figure 10.7). This example clearly illustrates how the need to have an extended mass range can be crucial. The deconvoluted spectrum shows two additional minor species (134,941–314 and 134,941–633 Da) that have been attributed to the successive loss of two 5′ adenine nucleotides from the DNA ends.[102]

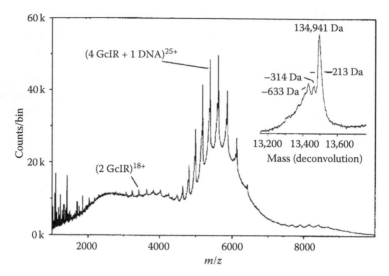

FIGURE 10.7   Nanospray spectrum of 13 M GclR protein with 6 µM consensus DNA in 10 mM NH_4HCO_3, 1 mM DTT. This spectrum was acquired continuously over 40 min. Only one major ion series is found, with the 22–28$^+$ of a complex of mass 134,939 Da, which can be explained as a tetrameric protein with one dsDNA. Addition of excess DNA made the peaks broader (increase in the FWHM value from 213 Da), but did not produce any different ions. The deconvolution of the major ion series (inset) shows three species, the most abundant representing a complete dsDNA and four protein subunits. The minor species probably represent failure synthesis of the DNA. The ions at m/z 3000–4000, are the 14–20$^+$ ions from protein dimer. (Reprinted from Donald, L.J. et al., *Protein Sci.* 10(7), 1370–1380, 2001. With permission.)

DNA recognition by CGN4 proteins, which are involved in the regulation of gene transcription in yeast, has been investigated by Deterding et al. using ESI-MS.[111] The CGN4 proteins contain a bZIP structural motif (Section 10.2.2) that consists of a basic DNA-binding domain and a leucine zipper dimeric domain. In this study, two kinds of CGN4 proteins were used: CGN4-bR, a peptide model containing the basic DNA-binding domain only and a disulfide bond taking place as a linker instead of the leucine-zipper; and GCN4-61, a fragment of CGN4 that contains the entire bZIP structural motif and has to dimerize itself to bind DNA. Alone, CGN4-bR and GCN4-61 were detected as monomeric species only. In the presence of a dsDNA (13, 24, or 30 bp) containing the binding element ATGA(C/G)TCAT, GCN4-61 protein led specifically to a 1:2 DNA–protein complex, whereas CGN4-bR gave only a 1:1 complex as expected. In the case of GCN4-61, no ion corresponding to the GCN4-61 protein dimer alone or the GCN4-61 monomer attached to the dsDNA was detected. As a confirmation of the biological relevance of the results, no complex was observed with a dsDNA (24 bp) that did not contain the specific recognition sequence. Finally, the authors combined differential limited proteolysis of the CGN4 protein–DNA complexes with subsequent MS identification of the products to probe the interaction between the CGN4 amino acids and the DNA. By comparing the digestion products of CGN4-bR bound and unbound to DNA with different endoproteases at different reaction times, the authors showed that certain cleavages were strongly reduced upon DNA complexation, thus leading to the determination of the amino acid residues involved in DNA binding. The results agreed well with the structure previously obtained from x-ray.[111]

Kapur et al. used ESI-MS to study the interactions between *E. coli* Tus protein (MW 35,653) and its DNA recognition sequence, TerB (21 bp).[112] Tus protein is known to be able to halt the replication upon binding as a monomer to specific DNA sequences. Under optimized conditions (10 mM ammonium acetate, pH 8), the authors reported positive polarity detection of the Tus–TerB complex with a 1:1 stoichiometry as expected. Under the same conditions, a mixture of Tus with a nonspecific DNA sequence gave predominantly free Tus protein, suggesting that the 1:1 Tus–TerB complex observed in the gas phase was not the result of nonspecific aggregation but rather was representative of in-solution species. This work described the first comparison of the relative binding affinities of native and mutant proteins with specific DNA sequences by using ESI-MS. No distinction could be made between bindings of Tus protein with different DNA sequences upon substitution of only one base pair within the DNA sequence. Only very stable complex ions were observed; no ions from free ligands were detected. Experiments with mutants of Tus protein led to the same conclusion. It was impossible to obtain the relative orders of binding affinities. The authors proposed, therefore, to weaken the noncovalent binding by increasing the range of ammonium acetate concentration in the spray solvent from 10 to 2200 mM at pH 8.0. Under these conditions, it was possible to determine the relative order of binding affinities by comparing the ammonium acetate concentration at which each complex was 50% dissociated. The stronger complexes were those disrupting at the higher ammonium concentrations. It was reported that the obtained data were in reasonable agreement with the SPR data and with the results of multiple competition experiments conducted in 800 mM ammonium acetate with various 1:1:1 mixtures (mutant 1:mutant 2:dsDNA).[112]

The preparation of biological complexes requires high concentrations of nonvolatile salts that have to be removed before MS analysis. Cavanagh et al. described the use of a cell allowing the mixing and incubation of the complex reactants before analysis owing to the coupling of SEC with ESI-MS, the cell being directly connected to the SEC-MS coupling.[113] The interest of such a coupling was multiple: reduction of sample losses upon handling, removal of nonvolatile salt and buffers and replacement with volatile salts, subsequent SEC separation of the free ligands and complex depending on the hydrodynamic volumes, and then unambiguous identification of the complex with the exact stoichiometry by MS on the basis of the molecular weight measurement. The power of this coupling was illustrated by the analysis of the *Bacillus subtilis* transition-state regulator AbrB (MW ~10,500) attached with its *SinIR* target dsDNA sequence (30 bp) in positive polarity. The results clearly showed that the AbrB protein binds to DNA as a tetramer.

Another report provided a more accurate picture of the AbrB–*SinR* dsDNA interactions. A complex of the N-terminal DNA-binding domain of AbrB (AbrBN53, residues 1–53) with dsDNA was detected by ESI-MS when it was in its dimeric form; hence it was naturally concluded that recognition elements were located in the N-terminal dimer domain of AbrB.[114] Furthermore, because the DNA–protein complex (1:4; MW 60,529) eluted after the AbrB protein homotetramer (MW 41,972) in SEC-MS coupling, it was concluded that binding of AbrB to dsDNA resulted in a significant reduction in hydrodynamic volume of the DNA–protein complex compared to the tetrameric protein alone. These authors underlined that some limitations inherent to microdialysis membranes are overcome by the use of their SEC-MS coupling.[114]

The next year, the same authors reported the use of a micro-on-line SEC cartridge in order to improve the desalting process and reduce analysis time.[53] Because AbrB protein regulates the expression of more than 60 different genes and binds DNA sites that do not share any apparent consensus sequence, a complementary study was carried out by the same group to further investigate DNA characteristics that may contribute to the binding specificity of AbrB protein.[115] By choosing DNA targets with similar ionization efficiencies, it was shown that the relative binding affinities of AbrB protein against various dsDNA sequences could be rapidly deduced from ESI-MS spectra of the complexes. The resulting binding affinity hierarchy for AbrB–DNA interactions was then compared with the obtained results from three different and independent spectroscopic methods (circular dichroïsm, tyrosine fluorescence emission, and UV spectroscopy), and their agreement validated the ESI-MS findings.[115]

Most of the ESI-MS studies on DNA–protein noncovalent complexes have been carried out using a classical electrospray source but the device has sometimes been modified to allow delivery of flow rates <1 µL min^{-1} to the spray needle.[106,107,109,110] Reducing the flow rate is of great interest since the ionization process consumes less of the sample and results in the strong improvement of both ionization efficiency and sensitivity. Consequently, the introduction of the commercially available NanoES allowing flow rates <100 nL min^{-1} was a major advance.[48] However, the lack of reproducibility of this source, especially when the nanospray tips have to be cut manually, has been pointed out by Gabelica et al.[56] In the study of a noncovalent complex between the repressor BlaI protein dimer (wild type and GM2 mutant) and its dsDNA operator (23 bp), the relative intensities of the observed species in the NanoESI mass spectra strongly depended on the shape of the capillary tips and on the capillary-to-cone distances. Evidently, these parameters were not possible to control because the nanospray tips have to be cut manually just before the MS experiment as mentioned above. In contrast, MS/MS experiments were highly reproducible and interestingly, the MS/MS experiment on the [dsDNA–protein dimer]$^{14+}$ led to the loss of a highly charged protein monomer [M]$^{9+}$ accounting for one-third of the initial complex mass but carrying two-third of the initial charge. That is to say that the small ion keeps the most important number of charges. As mentioned by the authors, this result recalls a behavior already observed within the dissociation of protein–protein complexes and recently discussed.[116–118] The loss of a highly charged subunit during the dissociation process would be induced by the unfolding of this subunit within the noncovalent complex.[116] Consequently, the charges would be redistributed in the most coulombic favorable manner according to the relative surface areas of the unfolded subunit and the remaining complex. The charge reparation that appears asymmetric with respect to the mass of the ions would be in fact symmetric with respect to the surface area of the ions as recently explained by Benesch et al.[117] This charge partitioning can depend on the ion activation method used. Jones et al. showed that collision-induced dissociation (CID) activation gives an asymmetric repartition of charges whereas surface-induced dissociation (SID) activation does not.[118] This result was explained by the fact that CID provokes a slow increase of internal energy of the complex through multiple collisions in comparison with the sudden activation provided by SID.

Another recent study by Terrier et al. focused on CID fragmentation mechanisms of ssDNA–peptides complexes produced by NanoES.[119] CID spectra of (12-mer ssDNA + penta-L-lysine) and (12-mer ssDNA + penta-L-arginine) complexes were recorded at several translational energies from 5-, 4-, and 3-charged ions. Different dissociation pathways were observed according to the nature of the basic polypeptide and to the initial charge state of the complex. The (12-mer ssDNA + penta-L-arginine) complex underwent fragmentations of covalent bonds independently of its initial charge state. In contrast, the dissociation pathway of the (12-mer ssDNA + penta-L-lysine) complex has been shown to be clearly charge-state dependent. Noncovalent dissociation, the separation of the two binding partners driven by coulombic repulsion, occurred for the higher-charged complexes, whereas fragmentation of covalent bonds was the main pathway for the lower-charged complexes. Multiple proton transfers from peptides to ssDNA have been observed upon noncovalent dissociation, and it has been proposed that possible proton transfers before the CID step could make the nature of the interactions in solution different than in the gas phase.[119]

The lack of reproducibility of the traditional nanospray source can be avoided by using an automated chip-based nanospray source. The "nanomate" source commercialized by Advion Biosciences is a chip consisting of 10×10 nozzles in a silicon wafer, which can improve reproducibility and increase analysis throughput. Van der Heuvel et al. described the use of this source to monitor, in real-time, the enzymatic hydrolysis of an ssDNA (17-mer) by E. coli colicin E9 DNase (15 kDa), a metallo-dependent endonuclease belonging to the widespread HNH family.[120] The reactivity of Mg^{2+}, Zn^{2+}, Ni^{2+}, and Co^{2+} was also tested. In the presence of Mg^{2+}, only the binary complex (ssDNA:E9 DNase) was observed. In the presence of Zn^{2+}, the ternary complex (ssDNA:E9 DNase:Zn^{2+}) was identified but no digest fragment was detected even after extended incubation. In contrast, in the presence of Ni^{2+} or Co^{2+}, a ternary complex (ssDNA:E9 DNase:metal^{2+}) appeared. The intensity of this complex decreased over time, while complex fragments appeared signifying enzymatic hydrolysis. NanoES-MS was used to follow the disappearance of the noncovalent complex (ssDNA-E9 DNase-Ni^{2+}) as a function of time to obtain more information about hydrolysis kinetics. Interpretation of the complex fragments enabled the authors to uncover precisely the location of the cleavage sites. Van der Heuvel et al. clearly showed that NanoES-MS was a fast and very powerful technique to study the reactivity, selectivity, and detailed kinetics of an enzymatic DNA hydrolysis.[120] Compared to gel shift assays, DNA footprinting, and electrochemical methods, this method is faster and does not require any chemical modification.

It was already known that the folding and unfolding of proteins could be readily monitored by ESI-MS from the analysis of the charge-states distributions. Indeed, modification of pH, addition of organic solvent, or temperature increase can induce the denaturation of a protein, which is generally followed by a shift of the observed charge states in MS spectra. By studying the DNA binding by the catalytic domain of bacteriophage λ integrase (C170, MW 21,119), Kamadurai et al. reported a change of the charge-state distribution of C170 protein ions when the protein was in the presence of the cognate DNA.[121] In the absence of DNA, three distinct species were identified in solution corresponding to a folded, a partially folded, and a dimeric species. Upon addition of a cognate DNA sequence, the ions coming from the unfolded protein disappeared and a 1:1 DNA–protein complex (MW 33,405) appeared. The signals coming from the free C170 protein exhibited only the narrow distribution typical of a compact conformation. The folding of the C170 protein upon DNA addition was interpreted as the result of a stabilization of the folded protein C170 owing to its interaction with the DNA. The authors suggested that the DNA could serve as a sort of "chaperone."[121]

CID experiments on DNA–ligand noncovalent complexes can lead to the cleavage of covalent bonds rather than selective dissociation of the noncovalent interactions. While such behavior depends on the strength of the interactions between the various partners of the complex, it can also depend on the charge state of the complex.[119] The more stable the noncovalent complex is, the more the cleavage of covalent bonds over noncovalent dissociation is expected. These covalent fragmentations are welcome when more information is required on the position of the ligand along the DNA helix, but are rather undesirable when the study deals with the affinity of the protein for the DNA, or with the stability of the complex. In this case, specific cleavage of the noncovalent bonds is highly preferred. Hiraoka et al. developed a promising technique for the characterization of DNA–protein noncovalent complexes.[72] The principle consists of the irradiation of an electrosprayed solution containing the sample by a 10.6 μM infrared laser. The laser is focused at the end of the electrospray tip and the laser power can be adjusted from 0 W (pure electrospray) to an optimized value (not more than 1.6 W) to control the degree of fragmentation. This device was used to probe a dsDNA–protein complex between the DNA-binding domain of a transcription factor (c-Myb DBD) and various dsDNAs, ranging from 16- to 22-mer.[71,122] The results were compared to those obtained by classical ESI-MS.[123] In one report, the classical ESI-MS investigation of the dsDNA–protein complexes showed that these could not be readily dissociated into their free components in the gas phase by CID of the multiply protonated species when the sample was prepared in 10 mM ammonium buffer, even if high cone voltage values were applied.[123] The peak broadening that was observed at the higher voltages was assigned to depurination of DNA. Interestingly, the authors remarked, as previously mentioned by Kapur et al., that noncovalent bonds appeared less stable than covalent bonds when the ammonium concentration was increased.[112] High salt concentrations weaken the noncovalent interactions and destabilize the complexes. To enable the evaluation by ESI-MS of the relative stability of the various dsDNA–c-Myb complexes over a wide range of K_D values (10^{-7} and 10^{-9} M), the ammonium acetate concentration was then optimized and kept at 500 mM to permit the progressive dissociation of the complexes by increasing the cone voltage. The cone voltage value at which 50% of the complex was dissociated ($V_{50\%}$) was compared to the relative binding free energy ($\Delta\Delta G$) determined separately by a filter-binding assay. A positive correlation between gas-phase and solution-phase stabilities was then reported, but it was suggested that a more rigorous quantification of the binding affinities would require avoiding covalent bond cleavage during the dissociation process.[123] The laser spray experiments on the stability of the same DNA–protein complexes (c-Myb DBD with 16-mer and 22-mer dsDNA) showed that the complexes could be dissociated into their subunits without any covalent fragmentation (Figure 10.8). Typically, the complex with the dsDNA of the highest K_D value yielded a narrow distribution of ions by ESI-MS

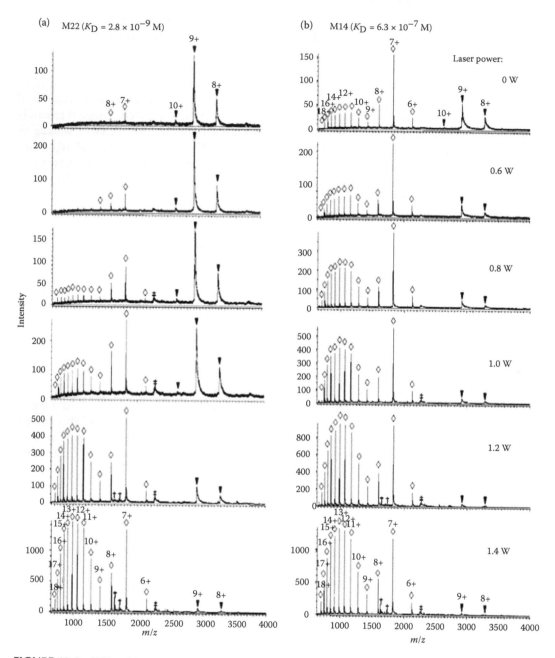

FIGURE 10.8 ESI and laser spray mass spectra of complexes of (a) the c-Myb DBD with m22 dsDNA and (b) the c-Myb DBD with m14 dsDNA in 50 mM ammonium acetate obtained at a laser powers of 0, 0.6, 0.8, 1.0, 1.2, and 1.4 W. Multiply charged peaks of the c-Myb DBD and protein–dsDNA complex are marked with open diamond and filled inverted triangle, respectively. The single- and double-dagger symbols represent 4+ charged peaks of two ssDNA and 6+ charged peaks of dsDNA, respectively. (Reprinted from Hiraoka, K., *J. Mass Spectrom.* 39(4), 341–350, 2004. Copyright 2006 from Elsevier.)

(laser OFF, 0 W). Free c-Myb DBD was detected as a narrow distribution only upon increasing the cone voltage, but this was accompanied by peak broadening of the complex (depurination of DNA, data not shown). In contrast, when the laser was firing, the progressive dissociation of the complex was observed without any cleavage of covalent bonds upon increasing the laser power; as the narrow distribution of ions corresponding to the complex disappeared progressively, a wide distribution appeared, accounting for free c-Myb DBD over several highly charged ions, suggesting a denaturation of the protein (Figure 10.8a). By progressively increasing the laser power, even if the sample was prepared with a low ammonium acetate concentration (10 mM), it was possible to determine the laser power at which 50% of the complex was dissociated ($E_{50\%}$). Another example of a weaker interaction is given in Figure 10.8b. Here, the correlation coefficient between $E_{50\%}$ (reflecting the gas-phase stability) and relative binding free energy ($\Delta\Delta G$) was reported to be much better than that obtained with ESI (0.9808 versus 0.859). The authors indicate that observed differences with ESI reside in the fact that the laser spray device permits a prompt heating of the sample during the ionization process enabling the specific cleavage of the hydrogen bonds. A difference in the ESI process is that noncovalent dissociations would occur in solution rather than the gas phase. This feature of laser spray (prompt heating) leads to stronger ion signals than electrospray, especially in negative polarity with aqueous solutions.[72,122]

Peptide nucleic acids (PNA) are DNA analogues introduced in 1991 by Nielsen et al., in which the sugar–phosphate backbone has been replaced by an achiral peptide backbone.[124,125] PNAs can bind DNA and RNA with high specificity and lead to PNA–DNA, (PNA)$_2$–DNA, and PNA–RNA hybrids, which display an improved stability over the corresponding nucleic acid complexes. Their biostability and resistance against nucleases and proteases makes PNAs particularly attractive in molecular biology. Only a limited number of papers deal with the MS detection of these PNA–DNA complexes.[126–128] Greig et al. reported the negative polarity ESI-MS detection of a (14-mer) PNA–DNA complex from a 25 mM ammonium acetate solution in a water/isopropanol mixture (50/50).[126] Griffith et al. have determined the stoichiometry of binding of both single and bis-PNAs (two single PNAs linked by a neutral or a positively charged spacer) to ssDNA (16-mer) in 50 mM ammonium acetate.[127] They found that bis-PNAs mainly formed 1:1 complexes with ssDNA whereas a single PNA led to a 2:1 complex as expected. More recently, Sforza et al. used ESI-MS to confirm that modified PNAs (chiral D-lysine-based PNAs bearing three adjacent chiral monomers in the middle of the strand) were able to bind DNA with a high direction control (i.e., in binding to antiparallel versus parallel DNA).[128] The ESI-MS spectra recorded in negative polarity (ammonium formate 10 mM, pH 7) of both an achiral PNA–antiparallel ssDNA duplex and an achiral PNA–parallel ssDNA duplex showed two stable duplexes (10 bp, MW 5750) with no particular difference, which suggested a lack of direction specificity in DNA binding by a standard achiral PNA. In contrast, a chiral PNA–antiparallel ssDNA duplex (10 bp, MW 5964) gave rise to a strong signal with no detectable ssDNA signal, whereas a chiral PNA–parallel ssDNA duplex gave only an abundant free ssDNA signal with no detectable duplex.[128]

10.5.2 ESI-MS OF RNA COMPLEXES

Sannes-Lowery et al. reported the first ESI-MS study dealing with RNA–protein noncovalent interactions, which concluded that specific interactions could be maintained in the gas phase then observed by MS.[129] They investigated the Tat protein, which is essential for HIV viral replication by binding to a specific RNA target sequence (named TAR). The formation of noncovalent complexes between HIV1 Tat protein (9.8 kDa) and TAR RNA (31-mer, 9.9 kDa) were investigated using ESI-MS. Tat40, a peptide fragment of Tat protein keeping the basic region of Tat protein, yielded a 1:1 peptide–RNA complex in both negative and positive ion modes. Although 2:1 and 3:1 stoichiometries were observed in the presence of an excess of Tat peptide, the fact that similar stoichiometries were obtained in both modes led the authors to conclude that the MS results reflected solution interactions and that the 1:1 stoichiometry was the only one related to specific interactions.[129]

A 1:1 RNA–protein noncovalent complex with the full-length Tat protein (~19.7 kDa) was detected in positive polarity from a 10 mM ammonium acetate, 10% (v/v) methanol solution, in the presence of a reducing agent (1 mM DTT). DTT was needed to maintain the Tat protein in the disulfide reduced state to form a complex with Tar RNA. Control experiments were performed to check whether ESI-MS data reflected the actual Tat peptide and RNA interactions. Because solution-phase methods have shown that a three nucleotide (UCU) bulge of Tar RNA is essential for Tat binding, complexes with three RNA mutants were studied. In one mutant, the bulge was simply removed; it was replaced in a second by a poly(ethylene glycol) (PEG) linker, and the final mutant carried a loop substituted by a PEG. As expected, only the mutations in the bulge region reduced the affinity of Tat peptide to Tar RNA.[129]

Liu et al. investigated the interactions of bacteriophage T4 regA protein, a translational regulator, with various RNAs using ESI-FT-ICR MS.[130] In all cases, a 1:1 RNA–protein complex was detected. Mass accuracy and resolving power of FT-ICR were essential to readily determine the charge state, and thus the molecular weight, from the isotopic spacing. Indeed, complex peaks were observed in most cases under a single charge state only, which can be interpreted as a strong indication of conformational preservation during transfer from solution to the gas phase. As the unintentional degradation of one of the studied RNAs led to an unexpected competition experiment, it was noted that a simple loss of one or two nucleotides could strongly modify the binding affinity of the RNA to regA protein. Competition experiments through a more classical approach gave a similar relative binding affinity order to that obtained by *in vitro* repression experiments. Lastly, CID experiments by sustained off-resonance (SORI-CID) revealed the strong potential of dissociation experiments for probing RNA–protein complexes. Assuming that protein portions not involved in contact with RNA should be more predisposed to fragmentation, the results suggested that the regA N-terminal part was less likely to be involved in binding with RNA.[130]

Two complementary methods, analytical centrifugation and μESI-MS, were used by Cassiday et al. to confirm the stoichiometry of an RNA–protein complex consisting of an RNA aptamer (α-p50 RNA, 31-mer) and a homodimer form (p50)$_2$ of the human transcription factor NF-κB.[131] Previous studies showed that α-p50 RNA aptamer competes with DNA for binding to (p50)$_2$. The access of the transcription factor to its binding site on DNA can consequently be blocked and the transcriptional activity inhibited. In the absence of α-p50 RNA, p50 protein was detected under its monomeric (MW 36,736) and dimeric form (MW 73,473) by ESI-MS. The presence of the p50 monomer was explained to be the result of partial decomposition of the dimer occurring during the ESI process. Then, in the presence of α-p50 RNA, p50 protein was detected under its dimeric form only, complexed by RNA aptamer (α-p50 RNA)/(p50)$_2$ (MW 83,372). Strikingly, p50 monomer alone was no longer observed. Furthermore, no (α-p50 RNA)$_2$/(p50)$_2$ complex was detected even in excess of α-p50 RNA. These results led the authors to propose that RNA binding stabilizes the p50 dimer. ESI-MS results were consistent with the data obtained by ultracentrifugation. The specificity of the interactions between (α-p50 RNA) and (p50)$_2$ was demonstrated by replacing α-p50 RNA by a nonspecific RNA. As a consequence, (p50)$_2$ was detected again and only a low-abundant (nonspecific RNA)$_2$/(p50)$_2$ complex (MW 89,510) was observed, which could be interpreted as the result of weak nonspecific binding. Finally, the data obtained by ESI-MS were consistent with the hypothesis that the RNA aptamer specifically recognizes the DNA-binding groove formed by the junction of the two p50 monomers of NF-κB transcription factor.[131]

Another example that demonstrates whether MS can provide a direct method for stoichiometry determination has been given by Callaghan with a complex RNase E-ssRNA. RNase E is an endoribonuclease involved within the degradosome in the processing and degradation of mRNA in *E. coli*.[132] Using nanospray MS in conjunction with other biophysical techniques (SEC, analytical ultracentrifugation, SDS-page gel, crystallographic analyses), it was clearly shown that the catalytic domain (N-terminal) of RNase E forms a tetramer (MW 247,537). The MS detection of the tetrameric species was achieved using a modified Q-TOF with an extended mass

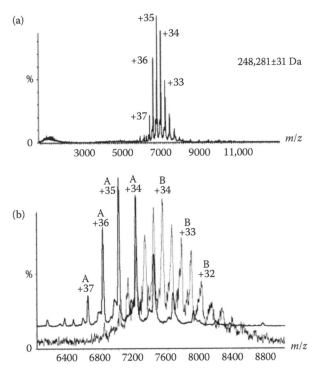

FIGURE 10.9 Nanoflow-ES mass spectra of the N-terminal domain of RNase E. (a) ESI mass spectrum of the N-terminal domain under nondissociating conditions. Individual charged species of the tetrameric form are resolved by the mass-to-charge ratio (*m/z*), and five abundant charged states are labeled. Ordinate axis is the percent abundance relative to the maximum peak height (+35 species). (b) The top spectrum (black line) is shown vertically offset from zero for clarity and corresponds to an expansion from *m/z* 6300–9000 of the spectrum in panel A. The charge states are labeled as series A (+37 to +34). The bottom spectrum (gray line) was recorded for a solution of the RNase E N-terminal domain after addition of RNA. This spectrum has equivalent charge states at higher masses because of the RNA bound to the tetramer (series B, +34 to +32 labeled). (Reprinted from Callaghan, A.J. et al., *Biochemistry* 42(47), 13848–13855, 2003. With permission. Copyright 2003 American Chemical Society.)

range. By carefully adjusting pressures and acceleration voltages on the ion path to preserve noncovalent interactions, ions carrying about 32–37 positive charges were found centered in the area around *m/z* 7200 (Figure 10.9a) corresponding to a slightly higher molecular weight (MW ~248,281) than expected. The overestimation of the molecular weight of the tetrameric assembly has been ascribed to a partial desolvation. Water and buffer molecules remained attached to the complex, which is considered a natural consequence of the soft ESI process as reported elsewhere by Robinson's group.[59] In the presence of a slight stoichiometric excess of RNA (10-mer), mass shifts on the mass spectrum (Figure 10.9b) corroborated by simulations, enable up to four RNA molecules (MW 3933) to be identified as bound to the tetrameric protein corresponding to a molecular weight of about 263 kDa. This result, consistent with the binding of one RNA molecule per assembly subunit, also suggests that longer RNA substrates might be bound simultaneously. These findings contributed to a better understanding of RNase E organization within the degradosome.[132]

Fabris and coworkers are the authors of a series of papers focusing on the characterization by MS of HIV RNA–protein complexes[133–135] and their work is described in detail in Chapter 9 of this book.

10.6 RIBOSOMES AND VIRUSES: MS DETECTION OF MASSIVE NONCOVALENT COMPLEXES

The detection by MS of very large noncovalent assemblies such as ribosomes and viruses is among the most impressive advances in the MS field of the last decade. The molecular weights of the species studied here (>MDa) are significantly higher than those previously reported, and these assemblies cannot be detected with commercial instruments under normal operation conditions.

Ribosomes are molecular machines enabling the translation of the genetic code by achieving protein biosynthesis in the cell. In bacteria, ribosomes consist of two noncovalently bound subunits (30S and 50S). The small (30S) and large subunits (50S) are associated to form the entire (70S) ribosome (2.5 MDa), which contains 54 different proteins and three large RNA molecules. The detection of intact ribosomes by MS has been described in a series of papers by Robinson's group.[103,136–140] The difficulties in obtaining mass spectra of such assemblies from solutions of ribosomes not only come from their size, but are also related to the presence of RNAs that can bind various cations and lead to numerous adducts. Furthermore, several populations of ribosomes can coexist in the same sample. Certain proteins can be missing in the ribosome; others which are present can show various posttranslational modifications. The first mass spectra obtained from intact ribosomes (70S from *E. coli*), were recorded using a NanoES source in the positive mode coupled to a quadrupole analyzer in 1998.[138] At that time, ions from the intact assembly could not be detected due to mass range limitations, but individual proteins coming from dissociation of the intact ribosome were identified and it was thought that their presence could, at least in part, reflect the interactions between components within ribosomes. The most readily released proteins were those interacting with RNA to a lesser extent.[138] The first mass spectra showing ions of intact 70s ribosomes from *E. coli* and their intact 50S and 30S subunits were published in 2000 by Rostom et al. using NanoES combined with TOF analysis allowing detection of ions beyond *m/z* 20,000.[103] In particular, transmission of high mass ions from the NanoES source to the TOF analyzer was improved by optimizing the pressure gradients to allow collisional cooling. Furthermore, to maintain the interaction between ribosome subunits in an MS compatible solution, it was also necessary to prepare the ribosomes in an aqueous volatile buffer (ammonium acetate, neutral pH) in the presence of high concentration of Mg^{2+} (5 mM). Although the recording of a mass spectrum required averaging over several successive acquisitions (up to 1 h), it was difficult to determine the mass and charges of the ions because of broadening of the signals, except for the signal of the 30S subunit. The ions from the 70S intact ribosomes were detected around *m/z* 22,000 with an average charge state of +105.[103] Decreasing Mg^{2+} concentration led to 70S disruption into 30S and 50S subunits that were detected around *m/z* 13,000 (+65) and *m/z* 15,900 (+90), respectively. Sufficient resolution of the charge states in the case of the 30S subunits enabled the molecular mass to be estimated at MW 85,2187, a value within 0.6% the calculated value from the masses of 21 small subunit proteins and the 16S rRNA. This value was higher than expected and was attributed to the trapping of small molecules or counterions, which is common in noncovalent MS studies.[103]

In more recent studies, considerable improvements in resolution have been obtained using a Q-TOF modified for high mass transmission. Figure 10.10a shows a mass spectrum recorded by Ilag et al. from an aqueous solution of 70S *Thermus thermophilus* ribosomes using a NanoES source.[138] Charge-state series are assigned to the 2.3 MDa intact 70S ribosomes and both the 30S and 50S subunits. Figure 10.10b displays a mass spectrum of the 50S subunit alone. The molecular weights of the three species have been determined as $810,208 \pm 963$ Da, $1,516,052 \pm 1986$ Da, and $2,325,463 \pm 2003$ Da for 30S, 50S, and 70S, respectively. These molecular weight measurements were complemented by MS/MS experiments (not shown), possible both on the 30S and 50S subunits despite their high molecular weights (0.8 and 1.5 MDa, respectively). In particular, MS/MS revealed unexpected stoichiometries of associated proteins and the presence of phosphorylated proteins within a particularly not well-defined region of the ribosome, named the "stalk."[136] Large protein

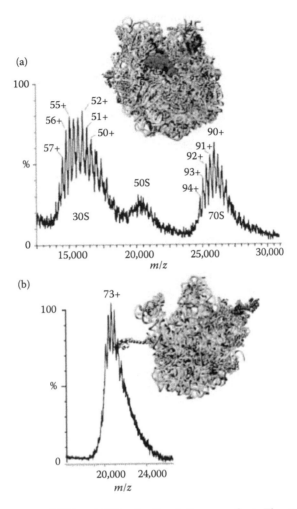

FIGURE 10.10 Mass spectra of 70S and 50S subunits of ribosomes from *Thermus thermophilus*. (a) The MS of 70S were recorded with a 100-V offset on the collision cell and reveals well-resolved charge states for the 30S and 70S subunits. The 50S subunit was separated chromatographically, and a spectrum of the isolated subunit (b) was recorded. The structures of the 70S and 50S particles were produced from the Protein Data Bank coordinates 1GIX and 1GIY. (Reprinted from Ilag, L.L. et al., *Proc. Natl. Acad. Sci. USA* 102(23), 8192–8197, 2005. With permission. Copyright 2005 from National Academy of Sciences, USA.)

assemblies such as ribosomes tend to form adducts with salts leading to peak broadening and difficulties in accurate mass measurement. McKay et al. demonstrated that it is possible to estimate the observed mass increase of a species as a function of peak broadening and thus to predict the actual mass of the species, peak broadening being independent of the mass analyzer.[137]

   The possibility of detecting supramolecular complexes, with molecular weights of over 40 MDa, was demonstrated with whole viruses by Siuzdak et al. in 1996.[141,142] Viral ions were generated in the gas phase by ESI from two different viruses: rice yellow mottle virus (RYMV) and tobacco mosaic virus (TMV). RYMV (MW 6.2 MDa) consists of ssRNA surrounded by a homogenous protein shell made up from 180 copies of a single coat protein, and TMV (MW 40.5 MDa), which is a rod-shaped virus formed from approximately 2140 identical proteins and an ssRNA. In both cases, viral capsids that contain RNA are stabilized by noncovalent interactions between multiple protein subunits. In an initial report, it was only demonstrated that viral ions RYMV and TMV could be generated by ESI, and then transmitted up to a collector placed in front of the MS detector.[141] The experiment was conducted in positive polarity on a standard ESI-triple quadrupole instrument with a maximum detection of *m/z* 2400. As *m/z* ratios of viral ions were expected far above this mass limit, ions were transmitted by operating the quadrupole analyzers in RF-only mode to allow complete transmission of higher *m/z* without particular mass selection. Viruses collected on a glycerol-coated brass plate (the collector) were then examined by transmission electron microscopy. Structures of the icosahedral RYMV capsid and the rod shape of the helical TMV capsid were clearly recognized.[141] By deflecting the ions produced by ESI so that only neutral viruses could pass, no virus was collected. This demonstrated that collected viruses were charged viral ions and not neutral viruses. Moreover, complementary experiments were conducted by inoculating tobacco plants with TMV collected from the mass spectrometer and consequently, tobacco plants developed lesions characteristic of infection. This demonstrated that even after the ionization process, the transfer through the quadrupole analyzers and the impact on the collector plate, viruses conserved their native biomolecular structure and viability.[141]

Mass measurement of these intact viruses was then reported in 2001 from a TOF analyzer by using a detector capable of measuring the charge state and velocity of ions simultaneously.[143] It was shown that RYMV and TMV ions produced by ESI possessed a charge distribution of between +300 and +1000 with higher charge states observed for TMV. The recording of a mass spectrum from RYMV led to the detection of a maximum signal centered around 6–7 MDa, which agreed well with the theoretical mass of an intact RYMV (6.5 MDa). In the case of TMV, an average molecular weight was detected between 39 and 42 MDa, which was also consistent with the known mass of TMV (40.5 MDa). In both cases, peak broadening and tailing toward the high masses were attributed to incomplete desolvation and adduct formation.[143]

The coupling of ESI to ion mobility spectrometry has also been successfully applied to the estimation of virus diameters, providing further evidences that no disruption of the tertiary or quaternary structure of the capsid occurred during desolvation and ionization. However, charge-reduced electrospray size spectrometry applied recently to the analysis of bacteriophage viruses with molecular masses ranging from 3.6 MDa to the gigadalton range defined the new limits of the ESI technique.[144] It was found that the "small" bacteriophage virus MS2 (MW 3.6 MDa) survived the ESI process, whereas bacteriophages T2 and T4 with larger sizes (MW > GDa) than the diameter of the electrospray jet and droplets dissociated during the ESI process, because the viruses could not be fully encapsulated by the small droplets. In this case, no viable T2 or T4 virions were detectable after being electrosprayed. The exact mechanism of fragmentation could not be determined, but it was shown that fragmentation occurred at the virus capsid head–tail noncovalent interface.[144]

10.7   COMPLEMENTARY APPROACHES

Because detection of intact noncovalent complexes is not always an easy task or sufficiently informative, complementary approaches have been developed to gain information about the solution structure of complexes. For example, the location of contact points between DNA and proteins within the complexes can be estimated by combining enzymatic proteolysis and MALDI-MS.[145] The comparison of the proteolytic digestion products of a protein, both in the presence and in the absence of DNA, can provide information about possible changes in cleavage site accessibility of the protein. In the presence of DNA, a high level of protection against proteolysis in certain region of the protein can clearly indicate regions that come into contact with DNA. Other approaches use chemical reagents or UV radiations to make cross-linking before MS analysis. The proteolysis of the resulting DNA–protein cross-link, followed by MS and MS/MS analysis, can reveal DNA–protein contact points in a similar manner. The main advantage of this strategy is that fragile noncovalent interactions are "frozen" in stable covalent heteroconjugates. The MS analysis of protein–nucleic acid interactions by photochemical cross-linking has been reviewed by Steen et al.[146]

10.8 CONCLUSION AND FUTURE PROSPECTS

MS can provide much more information than just the primary structure of isolated nucleic acids or proteins. Owing to the MALDI and ESI ionization techniques introduced by K. Tanaka and J. B. Fenn (Nobel Prize in Chemistry, 2002), noncovalent complexes between nucleic acids and proteins, the two major components of life, can be transferred into the gas phase and the identity and stoichiometry of the various binding partners can be readily deduced from their mass.[147,148] Numerous reports show that under the appropriate conditions, these complexes conserve, at least in part, their solution structure. The protein foldings of main recognition motifs have been involved in these DNA–protein complex studies: HTH, zinc fingers, and leucine-zipper.[92,93,102,108,109,111,133,149] None of them has been revealed as untransferable into the gas phase.

There has been very significant progress in recent years in ionization techniques and mass analysers. Development of new ionization methods and analyzers makes it possible to think that mass limits and complexity of the studied objects might still evolve considerably. MS has certainly a major role to play in the studies of biomolecular interactions, even if it cannot provide the detail of x-ray or NMR. MS appears as a complementary technique owing to its sensitivity and speed. The few studies on ribosomes and viruses give an idea of the considerable potential of MS in the structural biology field and, in particular, in nucleic acid–protein interactions. What will be the next new mass limit or complexity that MS will be able to reach? What will be the next steps? How to proceed further? A great deal of effort has focused on DNA, while little work has taken place on RNA complexes. Moreover, all the DNA/RNA forms have not yet been explored in conjunction with proteins, DNA triplex, and DNA quadruplex with proteins, for example. Compared to other systems, nucleic acid–protein interaction studies suffer from a lack of attempts to measure solution or gas-phase affinity constants. Tandem MS experiments remain scarce. Even if certain interactions between biomolecules are modified during the transfer from the liquid phase to the gas phase and that MS results still need complementary data from well-established techniques because artifact problems remain, these problems constitute a wide area to investigate and they introduce fascinating questions. The development of ion mobility (completed by molecular modeling) and irradiation techniques (UV and IR) of trapped ions can now improve our limited knowledge of gas-phase structures and help understand the real extent of the structural modifications that can occur both during, and even after, the ionization process. Since ESI and DESI share some characteristics in the ionization process, the DESI technique used for imaging MS might offer the opportunity to capture noncovalent complexes directly from tissues. Even if the development of ESI lagged behind that of MALDI for noncovalent complex studies, perhaps MALDI will make up this delay with the development of AP-IR-MALDI in near physiological conditions. Finally, MS capabilities for nucleic acid–protein noncovalent complex studies seem to remain largely underexploited and this technique should continue to play an increasing role in this field.

REFERENCES

1. Bloomfield, V.A., Crothers, D.M., and Tinoco, J.I., *Nucleic Acids, Structures, Properties, and Functions*, University Science Book, Sausalito, CA, 1999.
2. Boeger, H., Bushnell, D.A., Davis, R., Griesenbeck, J., Lorch, Y., Strattan, J.S., Westover, K.D., and Kornberg, R.D., Structural basis of eukaryotic gene transcription, *FEBS. Lett.* 579(4), 899–903, 2005.
3. Mello, C.C. and Conte, D., Revealing the world of RNA interference, *Nature* 431(7006), 338–342, 2004.
4. Neidle, S., *Nucleic Structure and Recognition*, Oxford University Press, Oxford, 2002.
5. Initial sequencing and analysis of the human genome, *Nature* 291, 860–921, 2001.
6. The sequence of the human genome, *Science* 409, 1304–1351, 2001.
7. Berman, H.M., Westbrook, J., Feng, Z., Gilliland, G., Bhat, T.N., Weissig, H., Shindyalov, I.N., and Bourne, P.E., The protein data bank, *Nucleic Acids Res.* 28(1), 235–242, 2000.
8. Hanson, C.L. and Robinson, C.V., Protein–nucleic acid interactions and the expanding role of mass spectrometry, *J. Biol. Chem.* 279(24), 24907–24910, 2004.

9. Rusconi, F., Guillonneau, F., and Praseuth, D., Contributions of mass spectrometry in the study of nucleic acid-binding proteins and of nucleic acid–protein interactions, *Mass Spectrom. Rev.* 21(5), 305–348, 2002.

10. Garvie, C.W. and Wolberger, C., Recognition of specific DNA sequences, *Mol. Cell* 8(5), 937–946, 2001.

11. Luscombe, N.M., Laskowski, R.A., and Thornton, J.M., Amino acid–base interactions: A three-dimensional analysis of protein–DNA interactions at an atomic level, *Nucleic Acids Res.* 29(13), 2860–2874, 2001.

12. Vazquez, M.E., Caamano, A.M., and Mascarenas, J.L., From transcription factors to designed sequence-specific DNA-binding peptides, *Chem. Soc. Rev.* 32(6), 338–349, 2003.

13. Jones, S., Daley, D.T.A., Luscombe, N.M., Berman, H.M., and Thornton, J.M., Protein–RNA interactions: A structural analysis, *Nucleic Acids Res.* 29(4), 943–954, 2001.

14. Berman, H.M., Olson, W.K., Beveridge, D.L., Westbrook, J., Gelbin, A., Demeny, T., Hsieh, S.H., Srinivasan, A.R., and Schneider, B., The nucleic-acid database—a comprehensive relational database of 3-dimensional structures of nucleic-acids, *Biophys. J.* 63(3), 751–759, 1992.

15. Luscombe, N.M., Austin, S.E., Berman, H.M., and Thornton, J.M., An overview of the structures of protein–DNA complexes, *Genome Biol.* 1(1), 1–10, 2000.

16. Brennan, R.G. and Matthews, B.W., The helix–turn–helix DNA-binding motif, *J. Biol. Chem.* 264(4), 1903–1906, 1989.

17. Harrison, S.C. and Aggarwal, A.K., DNA recognition by proteins with the helix–turn–helix motif, *Ann. Rev. Biochem.* 59, 933–969, 1990.

18. Fraenkel, E., Rould, M.A., Chambers, K.A., and Pabo, C.O., Engrailed homeodomain–DNA complex at 2.2 angstrom resolution: A detailed view of the interface and comparison with other engrailed structures, *J. Mol. Biol.* 284(2), 351–361, 1998.

19. Wolfe, S.A., Nekludova, L., and Pabo, C.O., DNA recognition by Cys(2)His(2) zinc finger proteins, *Ann. Rev. Biophys. Biomol. Struct.* 29, 183–212, 2000.

20. Pavletich, N.P. and Pabo, C.O., Zinc finger DNA recognition—crystal-structure of a Zif268-DNA complex at 2.1-A, *Science* 252(5007), 809–817, 1991.

21. Konig, P. and Richmond, T.J., The x-ray structure of the Gcn4-Bzip bound to Atf Creb site DNA shows the complex depends on DNA flexibility, *J. Mol. Biol.* 233(1), 139–154, 1993.

22. Kim, Y.C., Geiger, J.H., Hahn, S., and Sigler, P.B., Crystal-structure of a yeast Tbp Tata–box complex, *Nature* 365(6446), 512–520, 1993.

23. White, C.L., Suto, R.K., and Luger, K., Structure of the yeast nucleosome core particle reveals fundamental changes in internucleosome interactions, *EMBO J.* 20(18), 5207–5218, 2001.

24. Nagai, K., RNA–protein complexes, *Curr. Opin. Struct. Biol.* 6(1), 53–61, 1996.

25. Draper, D.E., Themes in RNA–protein recognition, *J. Mol. Biol.* 293(2), 255–270, 1999.

26. Perez-Canadillas, J.M. and Varani, G., Recent advances in RNA–protein recognition, *Curr. Opin. Struct. Biol.* 11(1), 53–58, 2001.

27. Williamson, J.R., Induced fit in RNA–protein recognition, *Nat. Struct. Biol.* 7(10), 834–837, 2000.

28. Morozova, N., Allers, J., Myers, J., and Shamoo, Y., Protein–RNA interactions: Exploring binding patterns with a three-dimensional superposition analysis of high resolution structures, *Bioinformatics* 22(22), 2746–2752, 2006.

29. Ellis, J.J., Broom, M., and Jones, S., Protein–RNA interactions: Structural analysis and functional classes, *Proteins-Struct. Function Bioinform.* 66(4), 903–911, 2007.

30. Auweter, S.D., Oberstrass, F.C., and Allain, F.H.T., Sequence-specific binding of single-stranded RNA: Is there a code for recognition? *Nucleic Acids Res.* 34(17), 4943–4959, 2006.

31. Auweter, S.D., Fasan, R., Reymond, L., Underwood, J.G., Black, D.L., Pitsch, S., and Allain, F.H.T., Molecular basis of RNA recognition by the human alternative splicing factor Fox-1, *EMBO J.* 25(1), 163–173, 2006.

32. Lewis, H.A., Musunuru, K., Jensen, K.B., Edo, C., Chen, H., Darnell, R.B., and Burley, S.K., Sequence-specific RNA binding by a Nova KH domain: Implications for paraneoplastic disease and the fragile X syndrome, *Cell* 100(3), 323–332, 2000.

33. Ganem, B., Li, Y.T., and Henion, J.D., Detection of noncovalent receptor ligand complexes by mass-spectrometry, *J. Am. Chem. Soc.* 113(16), 6294–6296, 1991.

34. Ganem, B., Li, Y.T., and Henion, J.D., Observation of noncovalent enzyme substrate and enzyme product complexes by ion-spray mass-spectrometry, *J. Am. Chem. Soc.* 113(20), 7818–7819, 1991.

35. Katta, V. and Chait, B.T., Observation of the Heme Globin complex in native myoglobin by electrospray-ionization mass-spectrometry, *J. Am. Chem. Soc.* 113(22), 8534–8535, 1991.

36. Veenstra, T.D., Electrospray ionization mass spectrometry in the study of biomolecular non-covalent interactions, *Biophys. Chem.* 79(2), 63–79, 1999.

37. Loo, J.A., Electrospray ionization mass spectrometry: A technology for studying noncovalent macromolecular complexes, *Int. J. Mass Spectrom.* 200(1–3), 175–186, 2000.

38. Ganem, B. and Henion, J.D., Going gently into flight: Analyzing noncovalent interactions by mass spectrometry, *Bioorganic Med. Chem.* 11(3), 311–314, 2003.

39. Heck, A.J.R. and van den Heuvel, R.H.H., Investigation of intact protein complexes by mass spectrometry, *Mass Spectrom. Rev.* 23(5), 368–389, 2004.

40. Ashcroft, A.E., Recent developments in electrospray ionisation mass spectrometry: Noncovalently bound protein complexes, *Natural Prod. Rep.* 22(4), 452–464, 2005.

41. Akashi, S., Investigation of molecular interaction within biological macromolecular complexes by mass spectrometry, *Med. Res. Rev.* 26(3), 339–368, 2006.

42. Loo, J.A., Studying noncovalent protein complexes by electrospray ionization mass spectrometry, *Mass Spectrom. Rev.* 16(1), 1–23, 1997.

43. Przybylski, M., Kast, J., Glocker, M.O., Durr, E., Bosshard, H.R., Nock, S., and Sprinzl, M., Mass spectrometric approaches to molecular characterization of protein–nucleic acid interactions, *Toxicol. Lett.* 82–3, 567–575, 1995.

44. Veenstra, T.D., Electrospray ionization mass spectrometry: A promising new technique in the study of protein/DNA noncovalent complexes, *Biochem. Biophys. Res. Commun.* 257(1), 1–5, 1999.

45. Beck, J.L., Colgrave, M.L., Ralph, S.F., and Sheil, M.M., Electrospray ionization mass spectrometry of oligonucleotide complexes with drugs, metals, and proteins, *Mass Spectrom. Rev.* 20(2), 61–87, 2001.

46. Hofstadler, S.A. and Griffey, R.H., Analysis of noncovalent complexes of DNA and RNA by mass spectrometry, *Chem. Rev.* 101(2), 377–390, 2001.

47. Fenn, J.B., Mann, M., Meng, C.K., Wong, S.F., and Whitehouse, C.M., Electrospray ionization for mass-spectrometry of large biomolecules, *Science* 246(4926), 64–71, 1989.

48. Wilm, M. and Mann, M., Analytical properties of the nanoelectrospray ion source, *Anal. Chem.* 68(1), 1–8, 1996.

49. Karas, M. and Hillenkamp, F., Laser desorption ionization of proteins with molecular masses exceeding 10,000 Daltons, *Anal. Chem.* 60(20), 2299–2301, 1988.

50. Farmer, T.B. and Caprioli, R.M., Determination of protein–protein interactions by matrix-assisted laser desorption/ionization mass spectrometry, *J. Mass Spectrom.* 33(8), 697–704, 1998.

51. Bolbach, G., Matrix-assisted laser desorption/ionization analysis of non-covalent complexes: Fundamentals and applications, *Curr. Pharm. Des.* 11(20), 2535–2557, 2005.

52. Daniel, J.M., Friess, S.D., Rajagopalan, S., Wendt, S., and Zenobi, R., Quantitative determination of noncovalent binding interactions using soft ionization mass spectrometry, *Int. J. Mass Spectrom.* 216(1), 1–27, 2002.

53. Cavanagh, J., Benson, L.M., Thompson, R., and Naylor, S., In-line desalting mass spectrometry for the study of noncovalent biological complexes, *Anal. Chem.* 75(14), 3281–3286, 2003.

54. Zamprionio, C.G., Giannakopulos, A.E., Zeller, M., Bitziou, E., Macpherson, J.V., and Derrick, P.J., Production and properties of nanoelectrospray emitters used in Fourier transform ion cyclotron resonance mass spectrometry: Implications for determination of association constants for noncovalent complexes, *Anal. Chem.* 76(17), 5172–5179, 2004.

55. Benkestock, K., Sundqvist, G., Edlund, P.O., and Roeraade, J., Influence of droplet size, capillary-cone distance and selected instrumental parameters for the analysis of noncovalent protein-ligand complexes by nano-electrospray ionization mass spectrometry, *J. Mass Spectrom.* 39(9), 1059–1067, 2004.

56. Gabelica, V., Vreuls, C., Filee, P., Duval, V., Joris, B., and De Pauw, E., Advantages and drawbacks of nanospray for studying noncovalent protein–DNA complexes by mass spectrometry, *Rapid Commun. Mass Spectrom.* 16(18), 1723–1728, 2002.

57. Guilhaus, M., Selby, D., and Mlynski, V., Orthogonal acceleration time-of-flight mass spectrometry, *Mass Spectrom. Rev.* 19(2), 65–107, 2000.

58. Sobott, F., Hernandez, H., McCammon, M.G., Tito, M.A., and Robinson, C.V., A tandem mass spectrometer for improved transmission and analysis of large macromolecular assemblies, *Anal. Chem.* 74(6), 1402–1407, 2002.

59. Sobott, F. and Robinson, C.V., Characterising electrosprayed biomolecules using tandem-MS—the noncovalent GroEL chaperonin assembly, *Int. J. Mass Spectrom.* 236(1–3), 25–32, 2004.

60. Chernushevich, I.V. and Thomson, B.A., Collisional cooling of large ions in electrospray mass spectrometry, *Anal. Chem.* 76(6), 1754–1760, 2004.

61. Xavier, K.A., Eder, P.S., and Giordano, T., RNA as a drug target: Methods for biophysical characterization and screening, *Trends Biotechnol.* 18(8), 349–356, 2000.
62. Van Berkel, G.J., An overview of some recent developments in ionization methods for mass spectrometry, *Eur. J. Mass Spectrom.* 9(6), 539–562, 2003.
63. Doroshenko, V.M., Laiko, V.V., Taranenko, N.I., Berkout, V.D., and Lee, H.S., Recent developments in atmospheric pressure MALDI mass spectrometry, *Int. J. Mass Spectrom.* 221(1), 39–58, 2002.
64. Laiko, V.V., Baldwin, M.A., and Burlingame, A.L., Atmospheric pressure matrix assisted laser desorption/ionization mass spectrometry, *Anal. Chem.* 72(4), 652–657, 2000.
65. Lewis, W.G., Shen, Z.X., Finn, M.G., and Siuzdak, G., Desorption/ionization on silicon (DIOS) mass spectrometry: Background and applications, *Int. J. Mass Spectrom.* 226(1), 107–116, 2003.
66. Takats, Z., Wiseman, J.M., Gologan, B., and Cooks, R.G., Mass spectrometry sampling under ambient conditions with desorption electrospray ionization, *Science* 306(5695), 471–473, 2004.
67. Cody, R.B., Laramee, J.A., and Durst, H.D., Versatile new ion source for the analysis of materials in open air under ambient conditions, *Anal. Chem.* 77(8), 2297–2302, 2005.
68. Takats, Z., Wiseman, J.M., Gologan, B., and Cooks, R.G., Electrosonic spray ionization. A gentle technique for generating folded proteins and protein complexes in the gas phase and for studying ion—Molecule reactions at atmospheric pressure, *Anal. Chem.* 76(14), 4050–4058, 2004.
69. Sakamoto, S. and Yamaguchi, K., Hyperstranded DNA architectures observed by cold-spray ionization mass spectrometry, *Angew. Chem. Int. Ed.* 42(8), 905–908, 2003.
70. Yamaguchi, K., Cold-spray ionization mass spectrometry: Principle and applications, *J. Mass Spectrom.* 38(5), 473–490, 2003.
71. Takamizawa, A., Itoh, Y., Osawa, R., Iwasaki, N., Nishimura, Y., Akashi, S., and Hiraoka, K., Selective dissociation of non-covalent bonds in biological molecules by laser spray, *J. Mass Spectrom.* 39(9), 1053–1058, 2004.
72. Hiraoka, K., Laser spray: Electric field-assisted matrix-assisted laser desorption/ionization, *J. Mass Spectrom.* 39(4), 341–350, 2004.
73. Laiko, V.V., Taranenko, N.I., Berkout, V.D., Yakshin, M.A., Prasad, C.R., Lee, H.S., and Doroshenko, V.M., Desorption/ionization of biomolecules from aqueous solutions at atmospheric pressure using an infrared laser at 3 mu m, *J. Am. Soc. Mass Spectrom.* 13(4), 354–361, 2002.
74. Laiko, V.V., Taranenko, N.I., and Doroshenko, V.M., On the mechanism of ion formation from the aqueous solutions irradiated with 3 mu m IR laser pulses under atmospheric pressure, *J. Mass Spectrom.* 41(10), 1315–1321, 2006.
75. Cooks, R.G., Ouyang, Z., Takats, Z., and Wiseman, J.M., Ambient mass spectrometry, *Science* 311(5767), 1566–1570, 2006.
76. Wiseman, J.M., Ifa, D.R., Song, Q.Y., and Cooks, R.G., Tissue imaging at atmospheric pressure using desorption electrospray ionization (DESI) mass spectrometry, *Angew. Chem. Int. Ed.* 45(43), 7188–7192, 2006.
77. Hirabayashi, A., Sakairi, M., and Koizumi, H., Sonic spray ionization method for atmospheric-pressure ionization mass-spectrometry, *Anal. Chem.* 66(24), 4557–4559, 1994.
78. Hu, Q.Z., Noll, R.J., Li, H.Y., Makarov, A., Hardman, M., and Cooks, R.G., The Orbitrap: A new mass spectrometer, *J. Mass Spectrom.* 40(4), 430–443, 2005.
79. Clemmer, D.E. and Jarrold, M.F., Ion mobility measurements and their applications to clusters and biomolecules, *J. Mass Spectrom.* 32(6), 577–592, 1997.
80. Pringle, S.D., Giles, K., Wildgoose, J.L., Williams, J.P., Slade, S.E., Thalassinos, K., Bateman, R.H., Bowers, M.T., and Scrivens, J.H., An investigation of the mobility separation of some peptide and protein ions using a new hybrid quadrupole/travelling wave IMS/oa-ToF instrument, *Int. J. Mass Spectrom.* 261(1), 1–12, 2007.
81. Loo, J.A., Berhane, B., Kaddis, C.S., Wooding, K.M., Xie, Y.M., Kaufman, S.L., and Chernushevich, I.V., Electrospray ionization mass spectrometry and ion mobility analysis of the 20S proteasome complex, *J. Am. Soc. Mass Spectrom.* 16(7), 998–1008, 2005.
82. Gidden, J., Baker, E.S., Ferzoco, A., and Bowers, M.T., Structural motifs of DNA complexes in the gas phase, *Int. J. Mass Spectrom.* 240(3), 183–193, 2005.
83. Gruic-Sovulj, I., Ludemann, H.C., Hillenkamp, F., Weygand-Durasevic, I., Kucan, Z., and Peter-Katalinic, J., Detection of noncovalent tRNA-aminoacyl-tRNA synthetase complexes by matrix-assisted laser desorption/ionization mass spectrometry, *J. Biol. Chem.* 272(51), 32084–32091, 1997.
84. Lecchi, P. and Pannell, L.K., The detection of intact double-stranded DNA by Maldi, *J. Am. Soc. Mass Spectrom.* 6(10), 972–975, 1995.

85. Little, D.P., Jacob, A., Becker, T., Braun, A., Darnhofer-Demar, B., Jurinke, C., van den Boom, D., and Koster, H., Direct detection of synthetic and biologically generated double-stranded DNA by MALDI-TOF MS, *Int. J. Mass Spectrom.* 169, 323–330, 1997.
86. Kirpekar, F., Berkenkamp, S., and Hillenkamp, F., Detection of double-stranded DNA by IR- and UV-MALDI mass spectrometry, *Anal. Chem.* 71(13), 2334–2339, 1999.
87. Sudha, R. and Zenobi, R., The detection and stability of DNA duplexes probed by MALDI mass spectrometry, *Helv. Chim. Acta* 85(10), 3136–3143, 2002.
88. Juhasz, P. and Biemann, K., Mass-spectrometric molecular-weight determination of highly acidic compounds of biological significance via their complexes with basic polypeptides, *Proc. Natl. Acad. Sci. USA* 91(10), 4333–4337, 1994.
89. Tang, X.D., Callahan, J.H., Zhou, P., and Vertes, A., Noncovalent protein–oligonucleotide interactions monitored by matrix-assisted laser desorption/ionization mass-spectrometry, *Anal. Chem.* 67(24), 4542–4548, 1995.
90. Lecchi, P., Le, H.M.T., and Pannell, L.K., 6-Aza-A-Thiothymine—a matrix for Maldi spectra of oligonucleotides, *Nucleic Acids Res.* 23(7), 1276–1277, 1995.
91. Lin, S.H., Cotter, R.J., and Woods, A.S., Detection of non-covalent interaction of single and double stranded DNA with peptides by MALDI-TOF, *Proteins-Struct. Function Genetics*, 12–21, 1998.
92. Lehmann, E. and Zenobi, R., Detection of specific noncovalent zinc finger peptide-oligodeoxynucleotide complexes by matrix-assisted laser desorption/ionization mass spectrometry, *Angew. Chem. Int. Ed.* 37(24), 3430–3432, 1998.
93. Lin, S.H., Long, S.R., Ramirez, S.M., Cotter, R.J., and Woods, A.S., Characterization of the "helix clamp" motif of HIV-1 reverse transcriptase using MALDI-TOF MS and surface plasmon resonance, *Anal. Chem.* 72(11), 2635–2640, 2000.
94. Luo, S.Z., Li, Y.M., Qiang, W., Zhao, Y.F., Abe, H., Nemoto, T., Qin, X.R., and Nakanishi, H., Detection of specific noncovalent interaction of peptide with DNA by MALDI-TOF, *J. Am. Soc. Mass Spectrom.* 15(1), 28–31, 2004.
95. Ohara, K., Smietana, M., and Vasseur, J.J., Characterization of specific noncovalent complexes between guanidinium derivatives and single-stranded DNA by MALDI, *J. Am. Soc. Mass Spectrom.* 17(3), 283–291, 2006.
96. Horneffer, V., Strupat, K., and Hillenkamp, F., Localization of noncovalent complexes in MALDI-preparations by CLSM, *J. Am. Soc. Mass Spectrom.* 17(11), 1599–1604, 2006.
97. Thiede, B. and von Janta-Lipinski, M., Noncovalent RNA-peptide complexes detected by matrix-assisted laser desorption/ionization mass spectrometry, *Rapid Commun. Mass Spectrom.* 12(23), 1889–1894, 1998.
98. Ganem, B., Li, Y.T., and Henion, J.D., Detection of oligonucleotide duplex forms by ion-spray mass-spectrometry, *Tetrahedron Lett.* 34(9), 1445–1448, 1993.
99. Lightwahl, K.J., Springer, D.L., Winger, B.E., Edmonds, C.G., Camp, D.G., Thrall, B.D., and Smith, R.D., Observation of a small oligonucleotide duplex by electrospray ionization mass-spectrometry, *J. Am. Chem. Soc.* 115(2), 803–804, 1993.
100. Schnier, P.D., Klassen, J.S., Strittmatter, E.E., and Williams, E.R., Activation energies for dissociation of double strand oligonucleotide anions: Evidence for Watson–Crick base pairing in vacuo, *J. Am. Chem. Soc.* 120(37), 9605–9613, 1998.
101. Gabelica, V. and De Pauw, E., Collision-induced dissociation of 16-mer DNA duplexes with various sequences: Evidence for conservation of the double helix conformation in the gas phase, *Int. J. Mass Spectrom.* 219(1), 151–159, 2002.
102. Donald, L.J., Hosfield, D.J., Cuvelier, S.L., Ens, W., Standing, K.G., and Duckworth, H.W., Mass spectrometric study of the Escherichia coli repressor proteins, Iclr and GclR, and their complexes with DNA, *Protein Sci.* 10(7), 1370–1380, 2001.
103. Rostom, A.A., Fucini, P., Benjamin, D.R., Juenemann, R., Nierhaus, K.H., Hartl, F.U., Dobson, C.M., and Robinson, C.V., Detection and selective dissociation of intact ribosomes in a mass spectrometer, *Proc. Natl. Acad. Sci. USA* 97(10), 5185–5190, 2000.
104. Greig, M.J., Gaus, H., Cummins, L.L., Sasmor, H., and Griffey, R.H., Measurement of macromolecular binding using electrospray mass-spectrometry—determination of dissociation-constants for oligonucleotide-serum albumin complexes, *J. Am. Chem. Soc.* 117(43), 10765–10766, 1995.
105. Cheng, X.H., Harms, A.C., Goudreau, P.N., Terwilliger, T.C., and Smith, R.D., Direct measurement of oligonucleotide binding stoichiometry of gene V protein by mass spectrometry, *Proc. Natl. Acad. Sci. USA* 93(14), 7022–7027, 1996.

106. Cheng, X.H., Morin, P.E., Harms, A.C., Bruce, J.E., BenDavid, Y., and Smith, R.D., Mass spectrometric characterization of sequence-specific complexes of DNA and transcription factor PU.1 DNA binding domain, *Anal. Biochem.* 239(1), 35–40, 1996.

107. Veenstra, T.D., Benson, L.M., Craig, T.A., Tomlinson, A.J., Kumar, R., and Naylor, S., Metal mediated sterol receptor-DNA complex association and dissociation determined by electrospray ionization mass spectrometry, *Nat. Biotechnol.* 16(3), 262–266, 1998.

108. Potier, N., Donald, L.J., Chernushevich, I., Ayed, A., Ens, W., Arrowsmith, C.H., Standing, K.G., and Duckworth, H.W., Study of a noncovalent trp repressor: DNA operator complex by electrospray ionization time-of-flight mass spectrometry, *Protein Sci.* 7(6), 1388–1395, 1998.

109. Xu, N.X., Pasa-Tolic, L., Smith, R.D., Ni, S.S., and Thrall, B.D., Electrospray ionization-mass spectrometry study of the interaction of cisplatin-adducted oligonucleotides with human XPA minimal binding domain protein, *Anal. Biochem.* 272(1), 26–33, 1999.

110. Craig, T.A., Benson, L.M., Tomlinson, A.J., Veenstra, T.D., Naylor, S., and Kumar, R., Analysis of transcription complexes and effects of ligands by microelectrospray ionization mass spectrometry, *Nat. Biotechnol.* 17(12), 1214–1218, 1999.

111. Deterding, L.J., Kast, J., Przybylski, M., and Tomer, K.B., Molecular characterization of a tetramolecular complex between dsDNA and a DNA-binding leucine zipper peptide dimer by mass spectrometry, *Bioconjugate Chem.* 11(3), 335–344, 2000.

112. Kapur, A., Beck, J.L., Brown, S.E., Dixon, N.E., and Sheil, M.M., Use of electrospray ionization mass spectrometry to study binding interactions between a replication terminator protein and DNA, *Protein Sci.* 11(1), 147–157, 2002.

113. Cavanagh, J., Thompson, R., Bobay, B., Benson, L.M., and Naylor, S., Stoichiometries of protein–protein/DNA binding and conformational changes for the transition-state regulator AbrB measured by pseudo cell-size exclusion chromatography-mass spectrometry, *Biochemistry* 41(25), 7859–7865, 2002.

114. Benson, L.M., Vaughn, J.L., Strauch, M.A., Bobay, B.G., Thompson, R., Naylor, S., and Cavanagh, J., Macromolecular assembly of the transition state regulator AbrB in its unbound and complexed states probed by microelectrospray ionization mass spectrometry, *Anal. Biochem.* 306(2), 222–227, 2002.

115. Bobay, B.G., Benson, L., Naylor, S., Feeney, B., Clark, A.C., Goshe, M.B., Strauch, M.A., Thompson, R., and Cavanagh, J., Evaluation of the DNA binding tendencies of the transition state regulator AbrB, *Biochemistry* 43(51), 16106–16118, 2004.

116. Jurchen, J.C., Garcia, D.E., and Williams, E.R., Further studies on the origins of asymmetric charge partitioning in protein homodimers, *J. Am. Soc. Mass Spectrom.* 15(10), 1408–1415, 2004.

117. Benesch, J.L.P., Aquilina, J.A., Ruotolo, B.T., Sobott, F., and Robinson, C.V., Tandem mass spectrometry reveals the quaternary organization of macromolecular assemblies, *Chem. Biol.* 13(6), 597–605, 2006.

118. Jones, C.M., Beardsley, R.L., Galhena, A.S., Dagan, S., Cheng, G.L., and Wysocki, V.H., Symmetrical gas-phase dissociation of noncovalent protein complexes via surface collisions, *J. Am. Chem. Soc.* 128(47), 15044–15045, 2006.

119. Terrier, P., Tortajada, J., and Buchmann, W., A study of noncovalent complexes involving single-stranded DNA and polybasic compounds using nanospray mass spectrometry, *J. Am. Soc. Mass Spectrom.* 18(2), 346–358, 2007.

120. van den Heuvel, R.H.H., Gato, S., Versluis, C., Gerbaux, P., Kleanthous, C., and Heck, A.J.R., Real-time monitoring of enzymatic DNA hydrolysis by electrospray ionization mass spectrometry, *Nucleic Acids Res.* 33(10), 2005.

121. Kamadurai, H.B., Subramaniam, S., Jones, R.B., Green-Church, K.B., and Foster, M.P., Protein folding coupled to DNA binding in the catalytic domain of bacteriophage lambda integrase detected by mass spectrometry, *Protein Sci.* 12(3), 620–626, 2003.

122. Shi, X.G., Nishimura, Y., Akashi, S., Takamizawa, A., and Hiraoka, K., Evaluation of binding affinity of protein-mutant DNA complexes in solution by laser spray mass spectrometry, *J. Am. Soc. Mass Spectrom.* 17(4), 611–620, 2006.

123. Akashi, S., Osawa, R., and Nishimura, Y., Evaluation of protein-DNA binding affinity by electrospray ionization mass spectrometry, *J. Am. Soc. Mass Spectrom.* 16(1), 116–125, 2005.

124. Nielsen, P.E., Egholm, M., Berg, R.H., and Buchardt, O., Sequence-selective recognition of DNA by strand displacement with a thymine-substituted polyamide, *Science* 254(5037), 1497–1500, 1991.

125. Egholm, M., Buchardt, O., Nielsen, P.E., and Berg, R.H., Peptide nucleic-acids (Pna)—oligonucleotide analogs with an achiral peptide backbone, *J. Am. Chem. Soc.* 114(5), 1895–1897, 1992.

126. Greig, M.J., Gaus, H.J., and Griffey, R.H., Negative ionization micro electrospray mass spectrometry of oligonucleotides and their complexes, *Rapid Commun. Mass Spectrom.* 10(1), 47–50, 1996.

127. Griffith, M.C., Risen, L.M., Greig, M.J., Lesnik, E.A., Sprankle, K.G., Griffey, R.H., Kiely, J.S., and Freier, S.M., Single and Bis peptide nucleic-acids as triplexing agents—binding and stoichiometry, *J. Am. Chem. Soc.* 117(2), 831–832, 1995.

128. Sforza, S., Tedeschi, T., Corradini, R., Dossena, A., and Marchelli, R., Direction control in DNA binding of chiral D-lysine-based peptide nucleic acid (PNA) probed by electrospray mass spectrometry, *Chem. Commun.* (9), 1102–1103, 2003.

129. SannesLowery, K.A., Hu, P.F., Mack, D.P., Mei, H.Y., and Loo, J.A., HIV 1 Tat peptide binding do to TAR RNA by electrospray ionization mass spectrometry, *Anal. Chem.* 69(24), 5130–5135, 1997.

130. Liu, C.L., Tolic, L.P., Hofstadler, S.A., Harms, A.C., Smith, R.D., Kang, C.H., and Sinha, N., Probing regA/RNA interactions using electrospray ionization Fourier transform ion cyclotron resonance-mass spectrometry, *Anal. Biochem.* 262(1), 67–76, 1998.

131. Cassiday, L.A., Lebruska, L.L., Benson, L.M., Naylor, S., Owen, W.G., and Maher, L.J., Binding stoichiometry of an RNA aptamer and its transcription factor target, *Anal. Biochem.* 306(2), 290–297, 2002.

132. Callaghan, A.J., Grossmann, J.G., Redko, Y.U., Ilag, L.L., Moncrieffe, M.C., Symmons, M.F., Robinson, C.V., McDowall, K.J., and Luisi, B.F., Quaternary structure and catalytic activity of the Escherichia coli ribonuclease E amino-terminal catalytic domain, *Biochemistry* 42(47), 13848–13855, 2003.

133. Hagan, N. and Fabris, D., Direct mass spectrometric determination of the stoichiometry and binding affinity of the complexes between nucleocapsid protein and RNA stem-loop hairpins of the HIV-1 psi-recognition element, *Biochemistry* 42(36), 10736–10745, 2003.

134. Yu, E. and Fabris, D., Direct probing of RNA structures and RNA-protein interactions in the HIV-1 packaging signal by chemical modification and electrospray ionization Fourier transform mass spectrometry, *J. Mol. Biol.* 330(2), 211–223, 2003.

135. Hagan, N.A. and Fabris, D., Dissecting the protein-RNA and RNA–RNA interactions in the nucleocapsid-mediated dimerization and isomerization of HIV-1 stemloop 1, *J. Mol. Biol.* 365(2), 396–410, 2007.

136. Ilag, L.L., Videler, H., McKay, A.R., Sobott, F., Fucini, P., Nierhaus, K.H., and Robinson, C.V., Heptameric (L12)(6)/L10 rather than canonical pentameric complexes are found by tandem MS of intact ribosomes from thermophilic bacteria, *Proc. Natl. Acad. Sci. USA* 102(23), 8192–8197, 2005.

137. McKay, A.R., Ruotolo, B.T., Ilag, L.L., and Robinson, C.V., Mass measurements of increased accuracy resolve heterogeneous populations of intact ribosomes, *J. Am. Chem. Soc.* 128(35), 11433–11442, 2006.

138. Benjamin, D.R., Robinson, C.V., Hendrick, J.P., Hartl, F.U., and Dobson, C.M., Mass spectrometry of ribosomes and ribosomal subunits, *Proc. Natl. Acad. Sci. USA* 95(13), 7391–7395, 1998.

139. Hanson, C.L., Videler, H., Santos, C., Ballesta, J.P.G., and Robinson, C.V., Mass spectrometry of ribosomes from Saccharomyces cerevisiae—implications for assembly of the stalk complex, *J. Biol. Chem.* 279(41), 42750–42757, 2004.

140. Videler, H., Ilag, L.L., McKay, A.R.C., Hanson, C.L., and Robinson, C.V., Mass spectrometry of intact ribosomes, *FEBS. Lett.* 579(4), 943–947, 2005.

141. Siuzdak, G., Bothner, B., Yeager, M., Brugidou, C., Fauquet, C.M., Hoey, K., and Chang, C.M., Mass spectrometry and viral analysis, *Chem. Biol.* 3(1), 45–48, 1996.

142. Bothner, B. and Siuzdak, G., Electrospray ionization of a whole virus: Analyzing mass, structure, and viability, *Chem. Biochem.* 5(3), 258–260, 2004.

143. Fuerstenau, S.D., Benner, W.H., Thomas, J.J., Brugidou, C., Bothner, B., and Siuzdak, G., Mass spectrometry of an intact virus, *Angew. Chem. Int. Ed.* 40(6), 541–544, 2001.

144. Hogan, C.J., Kettleson, E.M., Ramaswami, B., Chen, D.R., and Biswas, P., Charge reduced electrospray size spectrometry of mega- and gigadalton complexes: Whole viruses and virus fragments, *Anal. Chem.* 78(3), 844–852, 2006.

145. Cohen, S.L., Ferredamare, A.R., Burley, S.K., and Chait, B.T., Probing the solution structure of the DNA-binding protein max by a combination of proteolysis and mass-spectrometry, *Protein Sci.* 4(6), 1088–1099, 1995.

146. Steen, H. and Jensen, O.N., Analysis of protein-nucleic acid interactions by photochemical cross-linking and mass spectrometry, *Mass Spectrom. Rev.* 21(3), 163–182, 2002.

147. Fenn, J.B., Electrospray wings for molecular elephants (Nobel lecture), *Angew. Chem. Int. Ed.* 42(33), 3871–3894, 2003.

148. Tanaka, K., The origin of macromolecule ionization by laser irradiation (Nobel lecture), *Angew. Chem. Int. Ed.* 42(33), 3860–3870, 2003.

149. Fisher, R.J., Fivash, M.J., Stephen, A.G., et al. Complex interactions of HIV-1 nucleocapsid protein with oligonucleotides, *Nucleic Acids Res.* 34(2), 472–484, 2006.

11 MALDI-TOF Detection of Specific Noncovalent Complexes of Highly Acidic Biomolecules with Pyrenemethylguanidinium

Michael Smietana, Keiichiro Ohara, and Jean-Jacques Vasseur

CONTENTS

11.1 INTRODUCTION

Noncovalent interactions play an essential role in a vast number of chemical and biological processes. Recognition and selectivity are achieved through noncovalent contacts between nearly all the biopolymers (proteins, DNA, RNA, polysaccharides, and membrane lipids) that are present in living organisms. The formation and dissociation of these weak interactions are crucial in a vast number of biochemical events. For example, glycosidases have developed specifically adjusted active sites optimally configured for transition-state stabilization through noncovalent interactions.[1] Weak interactions between nucleases and DNA or RNA are central to recombination, repair, and replication of these molecules in cells.[2] Protein–protein, protein–DNA, protein–ligand, and antibodies–antigens interactions are other examples that involve noncovalent interactions,[3,4] which are also of fundamental importance for the pharmaceutical industry, in the evaluation of potential drug candidates.

There is a wide variety of noncovalent interactions that differ in their nature and strength. Coulombic interactions (charge–charge interactions) may be attractive or repulsive, depending on the signs of the charges involved. Dipole–dipole interactions that also may be attractive or repulsive depend on the relative angular orientation of the dipoles. All other types of interactions are attractive in nature because permanent or induced dipoles accommodate the forces acting on them.[5,6] All the forces depend on the distance $(1/r^n)$ and the dielectric constant $(1/\varepsilon)$ of the medium. Solvents, such as water, with high dielectric constants will substantially reduce all the interaction energies except for hydrophobic interactions, where the presence of water increases the attraction between two hydrophobic binding partners. This is a particularly important parameter when comparing solution and gas-phase energies. Coulombic and dipole-based interaction energies are therefore expected to be amplified when going from solution into the gas phase. A special case that requires further comment is the hydrogen-bonding interactions that are primarily electrostatic and directional. Hydrogen bonds and van der Waals bonds are special types of dipole-induced interactions. Due to solvation and dielectric conditions, hydrogen bond strengths are much higher in the gas phase than in water or other polar solvents.[7] Finally, systems exhibiting cation–π and π–π interactions also have an active role in the formation of noncovalent complexes (NCXs). Several techniques exist to characterize these interactions: calorimetry, often used to monitor biological recognition events;[8] x-ray diffraction;[9] and nuclear magnetic resonance (NMR) are the most widely reported techniques for the study of NCXs. One of the main advantages of NMR is the possibility of following the formation of a complex through the observation of the chemical shift changes of the host and the guest.[10] As the choice of an analytical technique is often dictated by parameters such as analyte and solvent compatibilities, sensitivity, stability, and specificity of the NCX, mass spectrometry (MS)-based methods have proven to be particularly advantageous compared to other biophysical methods.[11–13] Ganem et al. were the first to describe the use of a soft ionization method for the identification of NCXs.[14] Since then, electrospray ionization-mass spectrometry (ESI-MS) has received the majority of attention in the analysis of noncovalently bound systems and has been extensively reviewed.[13,15,16] The application of matrix-assisted laser desorption/ionization mass spectrometry (MALDI-MS) to the detection of NCXs is considered to be challenging as noncovalent host–guest interactions have to endure both the cocrystallization with the matrix and the absorption of the energy from the laser pulse. Nevertheless, MALDI-MS experiments are attractive, because the phenomenon of multiple charging is not encountered like it is with ESI-MS studies. The introduction of 6-aza-thiothymine (ATT) for the detection of intact noncovalent protein complexes was of crucial importance, and ligand-membrane receptors, protein–protein, peptide–protein, peptide–peptide, protein–DNA, and peptide–DNA NCXs have all been studied by MALDI.[17,18] One of the main motivations for these studies was the possibility to derive information about the cell machinery from the structure, stability, and conformation of NCXs' gas-phase ions. At the heart of these studies stand the interactions of the cationic guanidinium group located at the terminus of the amino acid arginine (R) side chain with various polyanionic biomolecules.[6] Resonance stabilization in the highly basic guanidinium group through a "Y-shaped" delocalization leads to the formation of highly stable complexes with carboxylate, phosph[on]ate, and sulf[on]ate complementary units.

Juhasz and Biemann were the first to report MALDI-MS detection of complexes between highly acidic compounds (such as DNA) and a peptide or a small protein rich in arginine, lysine, and/or histidine.[19] Subsequently, Vertes et al. studied the importance of both the oligonucleotide bases and the amino acid composition of the proteins for the formation of NCXs.[20] In these studies, the nature of the matrix played a key role and acidic matrix solutions were found to protonate A, C, and G, preventing the binding of phosphates to basic residues of peptides. Among the four DNA homo-oligomers, abundant protein–nucleic acid complexes were detected only for $p(dT)_n$. Interactions with bases other than thymine could be observed with the nonacidic (pKa 6.5) ATT matrix. ATT was then used by Pannell and Woods for the observation of intact DNA duplexes and their interactions with various basic peptides, respectively.[21,22] From all these reports focusing on acid–base interactions several conclusions can be drawn for the formation of the NCXs:

1. The oligonucleotide base sequence does not seem to have any influence on peptide–DNA complexes—it should be noted that protein–DNA NCXs were recently found to be sequence dependent.[19,22,23]
2. The proportion and the arrangement of arginine amino acids within the peptides play a key role.[19,22]
3. The choice of the matrix is of crucial importance.

In light of the importance of the guanidinium–phosphate interactions, the structural requirements of the guanidinium unit for successful MALDI detection have not been addressed. We therefore decided to investigate the interaction and ionization of guanidine-derived compounds with highly acidic biomolecules. The goals of the experiments were threefold:

1. To define the complex-forming nature and MALDI behavior of guanidine-derived analytes.
2. To identify the structural elements that drive these interactions.
3. To confirm the specificity and binding stoichiometry of the guanidinium-derived NCXs.

11.2 GUANIDINIUM–PHOSPHATE RECOGNITION[24]

11.2.1 OBSERVATION OF NONCOVALENT COMPLEXES

For the purpose of the discussion below, we have chosen to use the notation introduced by Juhasz and Biemann.[19] That is, $(m:n)^+$ describes the composition and charge state of the complex ion $[mMo + nMg + H]^+$, where Mo and Mg refer to the molecular weights of the oligonucleotide and the guanidinium derivative, respectively. To determine the nature of the interactions involved, we investigated the effect of modifications to the guanidinium structure as well as the influence of the nucleotide base and DNA backbone composition (Table 11.1). The detection and the intensity of guanidinium–DNA complex ions were found to be highly dependent on the choice of matrix. To demonstrate this effect, **DNA1**–guanidinium **9** complex detection was compared among four matrices namely, 3-hydroxypicolinic acid (3-HPA), 2′,4′,6′-trihydroxyacetophenone (THAP), nicotinic acid

---

TABLE 11.1
Oligonucleotides and Guanidinium Derivatives Used in This Study

Oligonucleotide	Sequence	M (Da)[a]
DNA1	5′-CGATCG-3′	1791.2
DNA2	5′-AAAA-3′	1189.9
DNA3	5′-TTTTTTTTTTTT-3′	3587.4
DNA4	5′-AGAATTGGGTGT-3′	3739.5
DNA5	5′-ACACCCAATTCT-3′	3548.4
DNA6	5′-TTTTTTT-3′	2066.4
DNA7	5′-T$_{ps}$T$_{ps}$T$_{ps}$T$_{ps}$T$_{ps}$T$_{ps}$T-3′	2162.8
DNA8	5′-A$_{mp}$C$_{mp}$A$_{mp}$C$_{mp}$C$_{mp}$C$_{mp}$A$_{mp}$A$_{mp}$T$_{mp}$T$_{mp}$C$_{mp}$T -3′	3526.6
DNA9	5′-TA$_{mp}$C$_{mp}$A$_{mp}$C$_{mp}$C$_{mp}$C$_{mp}$A$_{mp}$A$_{mp}$T$_{mp}$T$_{mp}$C$_{mp}$T -3′	3830.8

continued

TABLE 11.1   (continued)
Oligonucleotides and Guanidinium Derivatives Used in This Study

Guanidinium	Structure[b]	M (Da)[a]
1		60
2		75
3		175
4		131
5		144
6		174
7		142
8		156
9		150
10		274

continued

TABLE 11.1   (continued)
Oligonucleotides and Guanidinium Derivatives Used in This Study

Guanidinium	Structure[b]	M (Da)[a]
11		321
12		345

[a] Molecular masses correspond to the fully protonated DNA sequences and guanidinium derivatives.

[b] For the synthesis of these compounds, see Ohara et al.[25] Nps corresponds to a nucleotide phosphorothioate analogue and N_{mp} corresponds to a methylphosphonate analogue.

(NA), and ATT in both positive and negative ionization modes. Although some complexes could be observed in the negative mode, their intensities and abundance were much lower than in the positive ionization mode, especially with a neutral matrix. The MALDI (+ mode) mass spectra of **DNA1** with 10 equivalents of guanidinium **9** recorded in these four matrices are shown in Figure 11.1. In both THAP and ATT matrices the $(1:0)^+$ (m/z 1790.3), $(1:1)^+$ (m/z 1938.2), and, to a lesser extent, $(1:2)^+$ (m/z 2088.8) complex ions were observed. In the 3-HPA matrix, the mass spectrum was dominated by a strong $(1:0)^+$ ion (m/z 1791.0), and there was no signal detected in the NA matrix. These results indicate that the matrix is a key factor that influences the stability of ions formed in the MALDI process. Consequently, guanidinium–oligonucleotide complex formation was then studied in ATT. In all cases, ammonium citrate was used as an additive to the matrix for a better resolution.

Our studies showed that the number of adducts would depend on the relative DNA–guanidinium molar ratio. By probing a 10–30-fold molecular excess of guanidinium **9** while keeping the analyte **DNA1** concentration constant at 10^{-4} M, more complexes could be seen along with an increase of their respective intensity. By using a 30-fold molecular excess of the guanidinium **9**, three adducts ($\Delta = 149$ mu) are detected with a tiny $(1:3)^+$ peak (m/z 2238.6). However, with 10 equivalents of probe, only the $(1:1)^+$ (m/z 1942.0) complex ion is readily observed, albeit at a very low abundance, whereas the $(1:2)^+$ (m/z 2090.1) complex ion can only be presumed to exist (Figure 11.2). We thus decided to use a 10-fold molar excess of guanidinium over the DNA analyte to find the strongest binding derivatives. **DNA1** was chosen to evaluate the structural requirements for abundant complexes formation. The abundance and intensity of complex ions were found to be highly dependent on the guanidinium structure. When multiple ions were detected, the mass difference between two peaks was found to correlate with the molecular weight of the guanidinium derivative, but no signal over the maximum guanidinium–phosphate stoichiometry was ever observed.

Figure 11.3 compares the MALDI mass spectra of **DNA1** obtained with 10 equivalents of guanidinium compounds **1, 3, 6, 10,** and **12**. With compound **1** (Figure 11.3a, inset), a dominant $(1:0)^+$

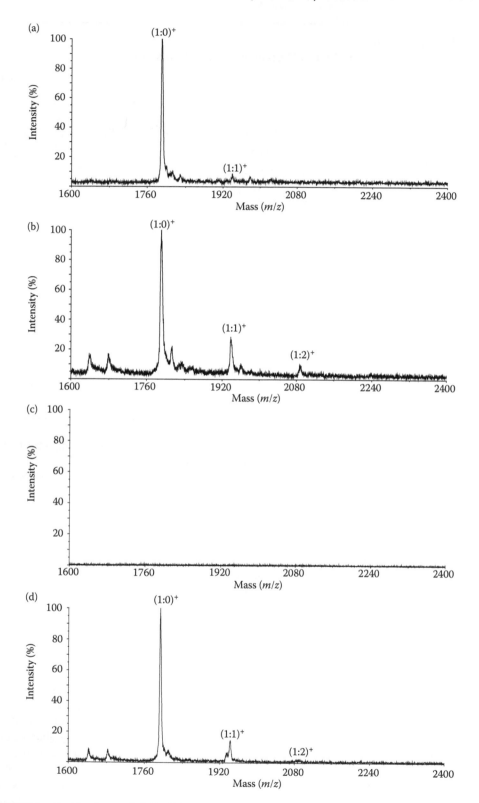

FIGURE 11.1   Positive ion mode MALDI mass spectra of **DNA1** with 10 equivalents of guanidinium **9** using (a) 3-HPA, (b) THAP, (c) NA, and (d) ATT matrices.

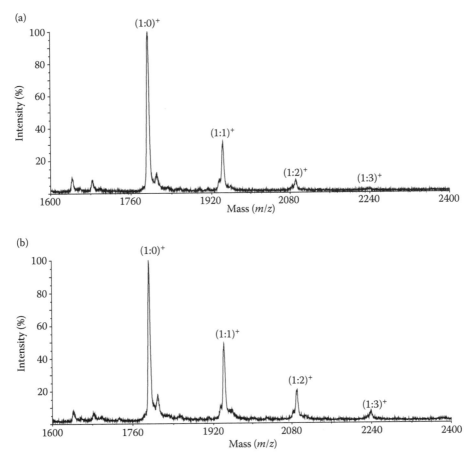

FIGURE 11.2 Positive ion mode MALDI mass spectra of **DNA1** with (a) 20 equivalents and (b) 30 equivalents of guanidinium derivative **9**.

noncomplexed ion is observed along with a slight peak corresponding to the $(1:1)^+$ ion. Surprisingly, with arginine **3** (Figure 11.3b, inset) and with all hydrophilic compounds, only the free oligonucleotide sequence could be observed. A notable exception is seen with *bis*-guanidinium derivative **6**, which is the only hydrophilic compound to display a relative abundance complex ratio >0.4. This difference is attributed to the *bis*-functionalization of compound **6**, as the ratio decreases to 0.2 with 10 equivalents of guanidinium groups. In fact, the $(1:1)^+$, $(1:2)^+$, and $(1:3)^+$ ions start to appear when an alkyl or phenyl group is connected to the guanidinium scaffold. To quantify the relative abundance of the complexes, the ratio of the complex signals over those of the complexed plus the noncomplexed DNA was calculated for each singly charged complex involving **DNA1** (Figure 11.4). The evaluation of Figure 11.4 implies that dipole–dipole and hydrogen-bonding interactions that probably exist in solution are not strong enough to be transferred into the gas phase. An extremely different shape is observed for compounds **10–12**. Indeed, with compounds **10** and **12** (Figure 11.3d and e), the free DNA $(1:0)^+$ ion is no longer the most abundant ion. Instead, a ladder of peaks displaying all possible complexes could be observed with $(1:1)^+$ (*m/z* 2065.8 and 2139.0, respectively) and $(1:2)^+$ (*m/z* 2338.2 and 2481.8, respectively) complex ions as the most intense peaks. Moreover, $(1:3)^+$ (*m/z* 2610.7 and 2824.0, respectively), $(1:4)^+$ (*m/z* 2883.2 and 3166.1, respectively) and, to a lesser extent, $(1:5)^+$ (*m/z* 3156.1 and 3509.2, respectively) complex ions are easily detected. All peaks are separated by the anticipated mass differences of 273 and 344 Da, respectively. In some cases, nonspecific ATT adducts could also be seen in the spectra. If we consider the hypothesis that free DNA came directly from the deposit (and was not produced by gas-phase dissociation of the complex), the higher abundance of complex signals observed with compounds **10–12** indicates a higher degree of stabilization probably induced by π-staking interactions. Differences in the ionization efficiency as a result of the chemical modification of the analytes are clearly marked. Comparison of the spectra obtained with compounds **10** and **12** also permits the evaluation of the importance of the alkyl chain. The butylamine tether present in compound **12** was used to probe the importance of lipophilicity. A lipophilic difference exists between compounds **10** (calculated C log $P = 0.5$) and **12** (calculated C log $P = 0.77$) without inducing a notable difference in the mass spectra that was recorded. We can therefore assume that the formation of abundant and intense complexes between

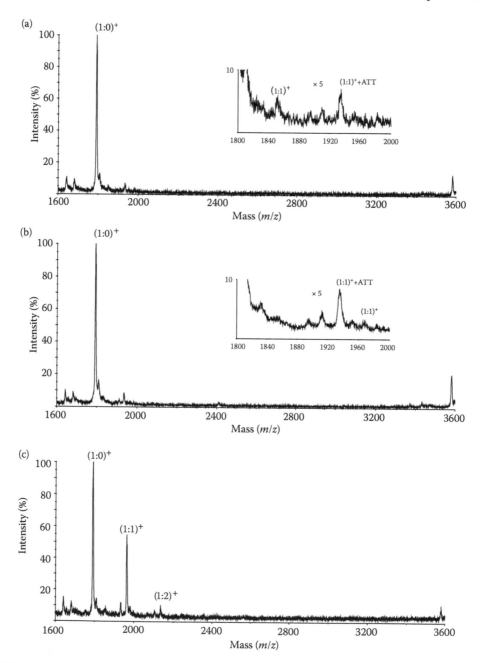

FIGURE 11.3 Positive ion mode MALDI mass spectra of **DNA1** with 10 equivalents of guanidinium functions of derivatives (a) **1**, (b) **3**, (c) **6**, (d) **10**, and (e) **12**.

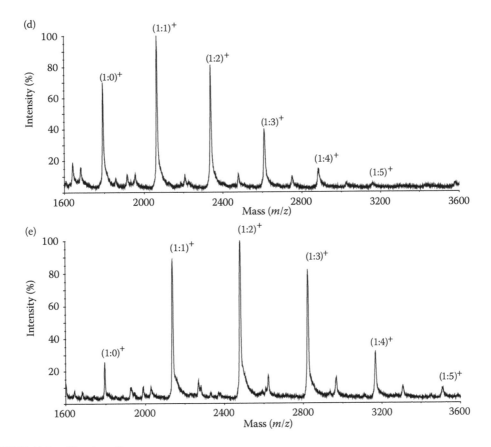

FIGURE 11.3   (Continued).

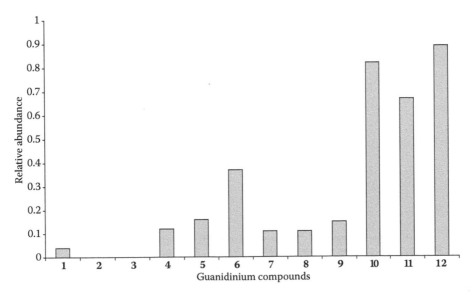

FIGURE 11.4   Relative abundance of the complex signals over the complexed plus noncomplexed **DNA1**.

guanidinium compounds and the phosphate groups of DNA is favored by the inclusion of large, delocalized aromatic structures stabilized by π-stacking interactions.

11.2.2 CONTROL EXPERIMENTS

Since compound **10** showed a high binding affinity for the oligonucleotide phosphate linkage and because its synthesis is straightforward, we decided to use it for control experiments.[25] We first evaluated the importance of the guanidinium function by comparing the results obtained with the ammonium counterparts of derivatives **4** and **10** with **DNA1**. Indeed, no complexes were detected with 10 equivalents of lysine or pyrenemethylamine (data not shown in this chapter). In each case, only the noncomplexed $(1:0)^+$ ion (m/z 1791.3) was observed. These differences can be explained by the highly interactive and complementary functional phosphate and guanidinium units and by the basicity exhibited by amines being insufficient to stabilize salt–bridge interactions, at least in the gas phase. It appears that while the guanidinium group is essential, it is not sufficient for the detection of NCXs.

Nonspecific aggregation in the gas phase during the MALDI process is always a concern when studying noncovalent association.[18] Control experiments need to be carried out to check the specificity of a complex detected in the gas phase. As shown in Figures 11.2 and 11.3, by adding a 10–30-fold molecular excess of guanidinium **10** while keeping the analyte concentration constant at 10^{-4} M, the $(1:4)^+$ and $(1:5)^+$ complex ions could be detected along with an increase in the intensities of less complexed ions, but no adduct signal over the maximum guanidinium–phosphate stoichiometry could be observed. These observations were confirmed when the tetrameric sequence **DNA2** was monitored with 10 equivalents of compound **10**. Under these conditions, only four peaks could be detected with a relative abundance ratio of 0.34 (Figure 11.5). Along with a dominant $(1:1)^+$ complex ion (m/z 1464.1), the $(1:2)^+$ (m/z 1736.4) and $(1:3)^+$ (m/z 2008.4) ions were clearly observed, but no signal over the maximum guanidinium–phosphate stoichiometry (i.e., $(1:4)^+$ ion) was detected. To leave no doubt that these NCXs preexist in solution and are not the result of gas-phase aggregation, Zenobi's dry mixing experiment was carried out.[26] In this experiment, two matrix solutions containing **DNA1** (10^{-4} M) and compound **10** (10^{-3} M) were prepared separately. These solutions were then dried at room temperature and the resulting powders were mixed together and deposited directly on the plate. Upon laser irradiation, only the free sequence ($(1:0)^+$ ion) could be detected. After the addition of 2 µL of a 1:1 acetonitrile:water mixture to these mixed powders and redrying, the ladder of peaks corresponding to the previously described complex ions was recovered (Figure 11.6). It should be noted that the relative abundance ratios of the complexed ions were however different

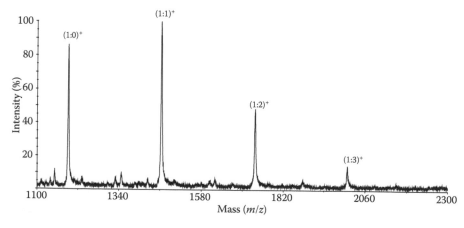

FIGURE 11.5   Positive ion mode MALDI mass spectrum of **DNA2** with 10 equivalents of guanidinium derivative **10**.

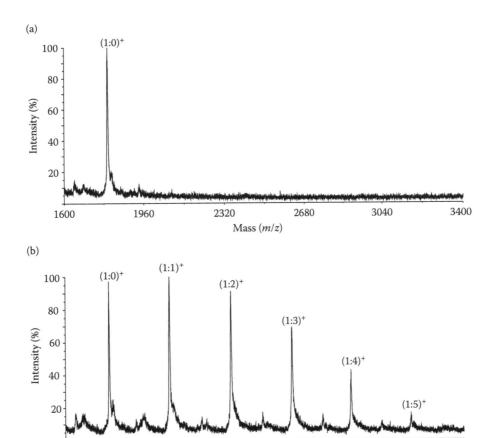

FIGURE 11.6   Positive ion mode MALDI-TOF spectra of the mixed powders of **DNA1** and compound **10** before (a) and after (b) the addition of 2 μL of a 1:1 acetonitrile:water solution.

from those obtained (Figure 11.3d). These results corroborate the formation in solution of complexes that are able to survive their transfer into the gas phase.

11.2.3   EFFECT OF DNA BASE SEQUENCE AND LENGTH

Sensitivity was first assessed, and a series of dilutions of **DNA1** was done. Reproducible analyte complexes from $(1:0)^+$ through $(1:5)^+$ ions could be obtained with 5 pmol of **DNA1** per target spot with 900 equivalents of compound **10** without affecting the spectra qualities.

We then probed the effect of DNA size and composition. Sequence **DNA3**, containing 11 phosphodiester internucleotide linkages, was chosen to evaluate the influence of the sequence length. The addition of 10 equivalents of compound **10** to **DNA3** led, as anticipated, to a ladder of peaks where elevated intensities of $(1:1)^+$ (*m/z* 3857.4) through $(1:6)^+$ ions (*m/z* 5218.7) were observed. Minor peaks corresponding to the $(1:7)^+$ (*m/z* 5490.9), $(1:8)^+$ (*m/z* 5761.6), and $(1:9)^+$ (*m/z* 6034.6) ions were also observed, but a larger amount of guanidinium **10** was needed to observe the fully complexed **DNA3**. Indeed, the addition of 60 equivalents of compound **10** led to a noticeable amount of the $(1:10)^+$ (*m/z* 6328.6) and $(1:11)^+$ (*m/z* 6604.6) ions (Figure 11.7).

The effect of DNA composition was then investigated by studying complexes involving heterosequences **DNA4** and **DNA5** with compound **10**. Observed species were qualitatively the same as in

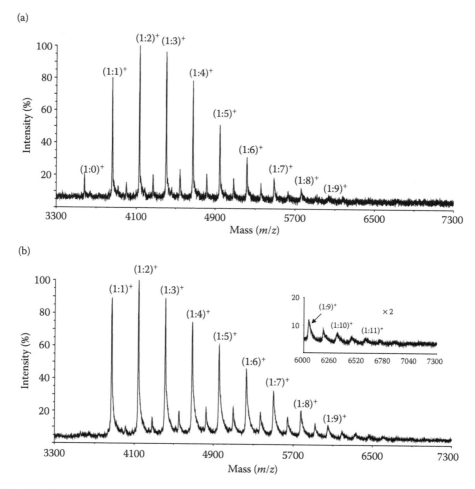

FIGURE 11.7 Positive ion mode MALDI mass spectra of **DNA3** with (a) 10 equivalents and (b) 60 equivalents of guanidinium derivative **10**.

the case of **DNA3** (Figures 11.8 and 11.9). In both cases, 70 equivalents of compound **10** were needed to access the fully complexed sequence, suggesting the absence of a base sequence effect on complex formation. Under these conditions, the relative abundance ratio reached values >0.9, with the free $(1:0)^+$ ion of sequence **DNA4** (m/z 3741.3) almost disappearing (Figure 11.8).

We then shifted our attention toward members of the family of phosphorus-modified nucleic acids. Of the many analogues of oligodeoxynucleotides explored as antisense agents, phosphorothioate analogues have been studied the most extensively.[27] Phosphorothioate analogues, in which a nonbridging oxygen atom in the negatively charged phosphodiester backbone has been replaced by a sulfur atom, have a greater resistance to nuclease degradation than phosphodiesters (Scheme 11.1). These properties have made PS-oligos first-generation antisense oligonucleotides.[28] In phosphorothioate anions, the P–S bond is a single bond with a negative charge localized on the sulfur atom.[29] It is recognized that hydrogen bonds become weaker descending down a given group, and this is usually rationalized on the basis of relative electronegativities.[30] Moreover, it has been demonstrated that the H bonds to sulfur are around 0.5 Å longer than the analogous bonds to oxygen and that complexes involving oxygen are more stable than those with sulfur.[31] A comparison of the homopolymeric sequences of **DNA6** and **DNA7** suggests that electrostatic interactions appear to be more important than hydrogen bond interactions for the ionization of NCXs. Indeed, similar intensities of

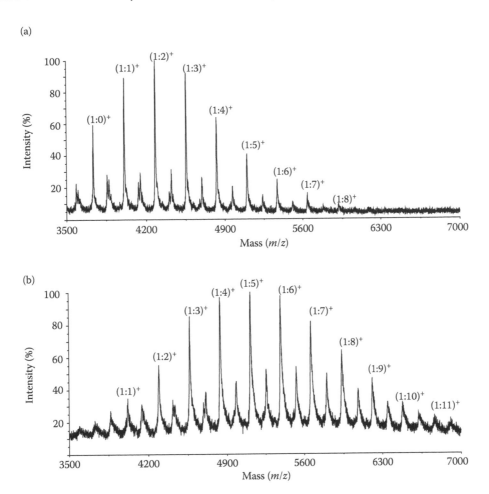

FIGURE 11.8 Positive ion mode MALDI mass spectra of **DNA4** with (a) 10 equivalents and (b) 70 equivalents of guanidinium derivative **10**.

complex ions were obtained with both oligonucleotides analogues (Figure 11.10). The peak distribution goes up to the maximum possible number of adducts with $(1:1)^+$ (m/z 2341.9 and 2438.1, respectively) and $(1:2)^+$ (m/z 2613.3 and 2710.9, respectively) complex ions as the most intense peaks. The fully complexed $(1:6)^+$ ions (m/z 3701.0 and 3799.9, respectively) are equally detected and the relative abundance ratios reach high-level values (0.92 and 0.96, respectively).

The influence of DNA composition was further explored with the use of methylphosphonate analogues in which a nonbridging oxygen atom in the negatively charged phosphodiester backbone has been replaced by a methyl group (Scheme 11.1). Nonionic and hydrophobic methylphosphonate analogues have been found to be effective antisense reagents.[32] The replacement of a nonbridging oxygen atom by a methyl group leads to a difference of only 1 Da per modification between the two species, rendering the methylphosphonate identification sometimes difficult. Since these analogues cannot be engaged in either electrostatic or H-bonding interactions, we decided to evaluate the addition of compound **10** to methylphosphonate-modified sequences **DNA8** and **DNA9**. **DNA8** is an all-modified methylphosphonate 12 mer, whereas **DNA9** is a 13 mer that bears only one standard phosphodiester linkage. The addition of 10 equivalents of compound **10** to **DNA9** leads to the observation of two peaks corresponding to $(1:0)^+$ (m/z 3829.0) and $(1:1)^+$ (m/z 4099.5) ions. The relative abundance ratio was calculated to be 0.17 and no $(1:2)^+$ complex ion could be seen even in the presence of 80 equivalents of compound **10**. With **DNA8**, an overwhelming $(1:0)^+$ (m/z 3521.8)

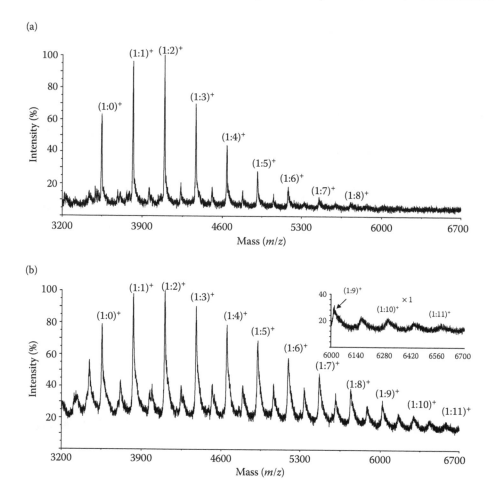

FIGURE 11.9 Positive ion mode MALDI mass spectra of **DNA5** with (a) 10 equivalents and (b) 70 equivalents of guanidinium derivative **10**.

ion was detected (Figure 11.11). An undersized (1:1)⁺ ion can be seen (<4% in intensity), which is thought to be the result of internal hydrolysis during the synthesis. Guanidinium derivatives such as compound **10** can therefore be used as derivatizing agents to monitor the degree of hydrolysis of modified oligonucleotides.

11.2.4 EFFECT OF SALT CONCENTRATION

It is generally accepted that ionic strength significantly affects the kinetics of noncovalent interactions. When NaCl concentration was increased, sodium ions competed with guanidinium cations for the formation of ionic bonds with DNA.[22] When **DNA1** and 10 equivalents of compound **10** were incubated in a saturated NaCl solution, NCXs could still be observed along with sodium adducts (Figure 11.12). Interestingly, all forms of **DNA1** (complexed and noncomplexed) were found to match the number of phosphate groups present in the sequence. Indeed, five adducts (Δ23 Da) are observed for the uncomplexed oligonucleotide. For the (1:1)⁺ complex ions, a series of four peaks (Δ23 Da) are observed. The same trend goes on to the (1:2)⁺ complex ions where only three sodium adducts could be observed. These results demonstrate the unexpected strength of the interactions between hydrophobic guanidinium derivatives and phosphate groups and reinforce the predominance of ionic and π–π interactions over hydrogen bonding.

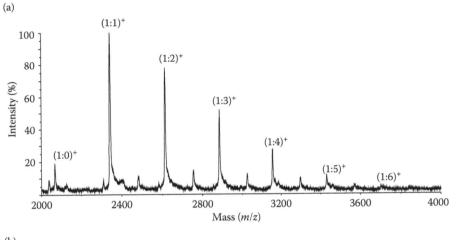

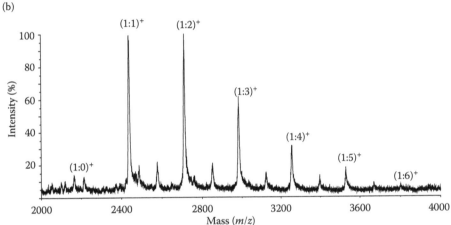

FIGURE 11.10 Positive ion mode MALDI mass spectra of 10 equivalents of guanidinium derivative **10** with (a) **DNA6** and (b) **DNA7**.

11.2.5 OBSERVATION OF COMPLEXES INVOLVING DUPLEX DNA

The analysis of oligonucleotide duplexes by MALDI has been regarded as a tremendous challenge and there have only been a few publications that have reported the detection of oligonucleotide double strands by MALDI-MS.[33–37] It is an important issue, in the MS of NCXs with duplex DNA, to assure

B = base

$X^- = O$: phosphodiester

$X^- = S$: phosphorothioate

$X^- = CH_3$: methylphosphonate

SCHEME 11.1 Chemical structures of modified oligonucleotides.

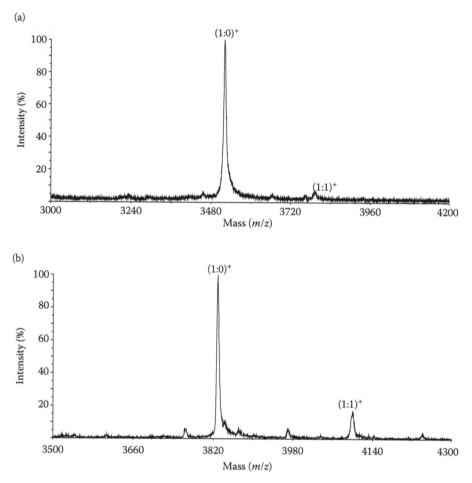

FIGURE 11.11   Positive ion mode MALDI mass spectra of 10 equivalents of guanidinium derivative **10** with (a) **DNA8** and (b) **DNA9**.

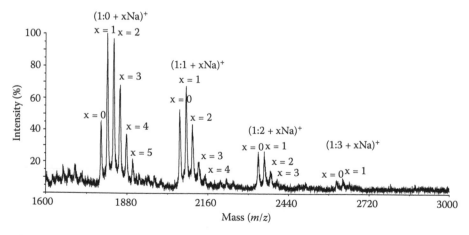

FIGURE 11.12   Positive ion mode MALDI mass spectrum of **DNA1** with 10 equivalents of guanidinium derivative **10** in saturated NaCl.

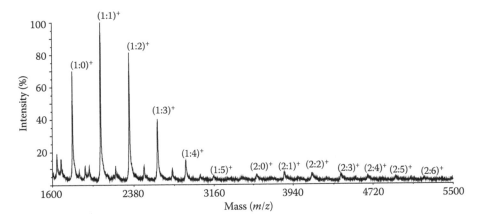

FIGURE 11.13 Positive ion mode MALDI mass spectrum of **DNA1** with 10 equivalents of guanidinium derivative **10** for duplex detection.

that the observed peaks correspond to specifically bound subunits rather than nonspecific aggregates or clusters. Kirpekar et al. reported that a minimum of 12 base pairs is necessary for the detection of DNA duplexes under MALDI conditions.[35] For these experiments, we used the neutral matrix ATT as the matrix solvent in a mixture of water–acetonitrile (1:1), as previously described. Our goal was to find out to what extent the double-stranded sequences are conserved under MALDI conditions in the presence of guanidinium **10**. Self-complementary **DNA1** was chosen, and along with the signals displayed by the single-stranded complexes, a series of smaller peaks could be detected in the duplex mass range of *ca.* 4200 Da (Figure 11.13). These signals go from the free duplex [(2:0)+, *m/z* 3580.3] through the (2:6)+ complexed ion (*m/z* 5219.1). A mass difference of 273 Da is observed between these peaks suggesting a complexation of compound **10** with the double-stranded sequence. It is unlikely that these are the results of intercalation only because **DNA1** displays a maximum of three intercalating sites. To check the specificity of these complexes, **DNA2** was submitted to the same conditions. This non-self-complementary sequence displayed a series of nonspecific **DNA2** dimers complexed to guanidinium **10** (Figure 11.14). The (2:0)+ (*m/z* 2380.6), (2:1)+ (*m/z* 2652.4), (2:2)+ (*m/z* 2924.7), (2:3)+ (*m/z* 3194.5), and (2:4)+ (*m/z* 3466.7) complexed ions are also separated by a mass difference of

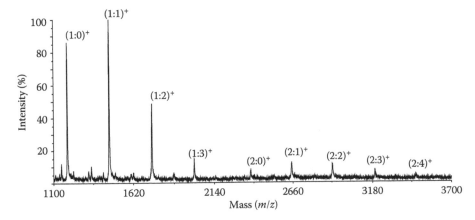

FIGURE 11.14 Positive ion mode MALDI mass spectrum of **DNA2** with 10 equivalents of guanidinium derivative **10** for duplex detection.

273 Da. These nonspecific signals might therefore be a consequence of gas-phase aggregation of specific monomers complexed with compound **10**.

11.2.6 INSTRUMENTATION

MALDI-TOF mass spectra were recorded on a Voyager DE (PerSeptive Biosystems, Framingham, MA) mass spectrometer equipped with an N_2 laser (337 nm). MALDI conditions were as follows: laser power, 2500–2700 (arbitrary units); all spectra have been acquired in the positive linear mode with an acceleration voltage of 24 kV; guide wire, 0.05% of accelerating voltage; grid voltage, 94% of accelerating voltage; and delay extraction time, 550 ns. All spectra reported here were not smoothed. Typically, 40–50 laser shots were averaged for each spectrum.

11.2.7 CONCLUSION

This study was designed and carried out with the aim of studying the interaction between guanidinium derivatives and complementary phosphate units of DNA. By methodically varying the structure attached to the guanidinium moiety, we demonstrated its influence on the efficiency by which it can be transferred as a noncovalent complex with DNA from solution to the gas phase during the MALDI process. We have detected intact specific noncovalent complexes between highly acidic single-stranded DNA and highly basic guanidinium molecules at neutral pH using MALDI-TOF. This interaction is not DNA sequence specific but does require certain structural arrangements of the guanidinium residues. The abundance and intensities of guanidinium–phosphate complexes in a single strand of DNA were found to depend on the concentration of the guanidinium compound, its hydrophobicity, and its ability to form π–π interactions. In addition, we have shown that complexation attributed mainly to ionic interactions could not be disrupted at high ionic strength. These results indicate that screening of small molecule noncovalent interactions by MALDI-MS is an effective way for directly screening and comparing the interactions between complementary functional units. Pyrenemethylguanidinium **10** composed of a forked, Y-shaped, planar guanidine and by a polycyclic aromatic core was shown to display directed hydrogen bonding, as well as nondirected Coulombic and hydrophobic interactions with complementary phosphate units.

11.3 GUANIDINIUM–SULFATE RECOGNITION

11.3.1 INTRODUCTION

Encouraged by the results obtained with compound **10** and complementary phosphate motifs, we decided to shift our attention toward MALDI-MS detection of sulfate–guanidinium interactions.[38]

Among polysulfated biomolecules, polysulfated carbohydrates are highly relevant molecules involved in many biological processes. Naturally occurring polysulfated carbohydrates as well as synthetic polysulfated carbohydrate drugs—such as the antiulcer sucrose octasulfate or the octasulfated pentasaccharide anticoagulant Arixtra—are both highly important to the food and pharmaceutical industries. Unfortunately, because of their chemical heterogeneity and highly acidic character, MS analyses of sulfated molecules are constituently hampered by the loss of sulfo groups and sensitivity problems. Although ESI-MS has been the method of choice for the analysis of polysulfated carbohydrates, the data interpretation is usually complex because ions are detected in multiply charged state.[39–42] On the other hand, the advantage of mainly detecting singly charged ions in MALDI is hampered by the difficulty in generating intact molecules. So far, several methods have been reported that show improvement in this area, including the use of atmospheric pressure (AP)-MALDI,[43] complexation with basic peptides, such as $(Gly–Arg)_n$, where n exceeds the number of sulfate groups by 1,[19,44,45] alkali metal exchange,[46] or the use of ionic liquid matrices (ILM).[47–49] The use of ILM has demonstrated the most effective suppression of dissociation (although not totally) of the sulfate groups but, unfortunately, does not avoid Na/H exchange. The analytes are therefore detected as sodium adducts and not as their protonated form, which complicates the analysis of unknown compounds.

So far, most of the research has been aimed at preventing the loss of SO_3 from the sulfated oligosaccharides. Considering the results obtained between compound **10** and the phosphate groups of DNA, we thought that *specific and complete* fragments, which resulted from the loss of SO_3, could be very useful in providing, in a single experiment, both structural and compositional information.

11.3.2 Analysis of Chondroitin Sulfates

Chondroitin sulfates are major constituents of glycosaminoglycan (GAG). They are formed by a linear polysaccharide chain made of dimeric units composed of an alternating carbohydrate backbone of D-glucuronic acid (GlcA) and 2-acetamido-2-deoxy-D-galactose (GalNAc) and contain, on average, one sulfate group per disaccharide unit (Scheme 11.2).[50] MALDI-TOF analysis (both positive and negative modes) was performed on synthetic samples of chondroitin sulfates with various sulfation patterns.[50,51] Among several matrices that were evaluated, *p*-nitroaniline provided the best shot-to-shot reproducibility and signal stability in the analysis.[52] We found that 40 equivalents of **10** were necessary to detect polysulfated carbohydrates with high sensitivity. Although no signal could be observed without addition of compound **10** (data not shown in this chapter), the addition of 40 equivalents of pyrenemethylguanidinium gave in both ionization modes a ladder of peaks

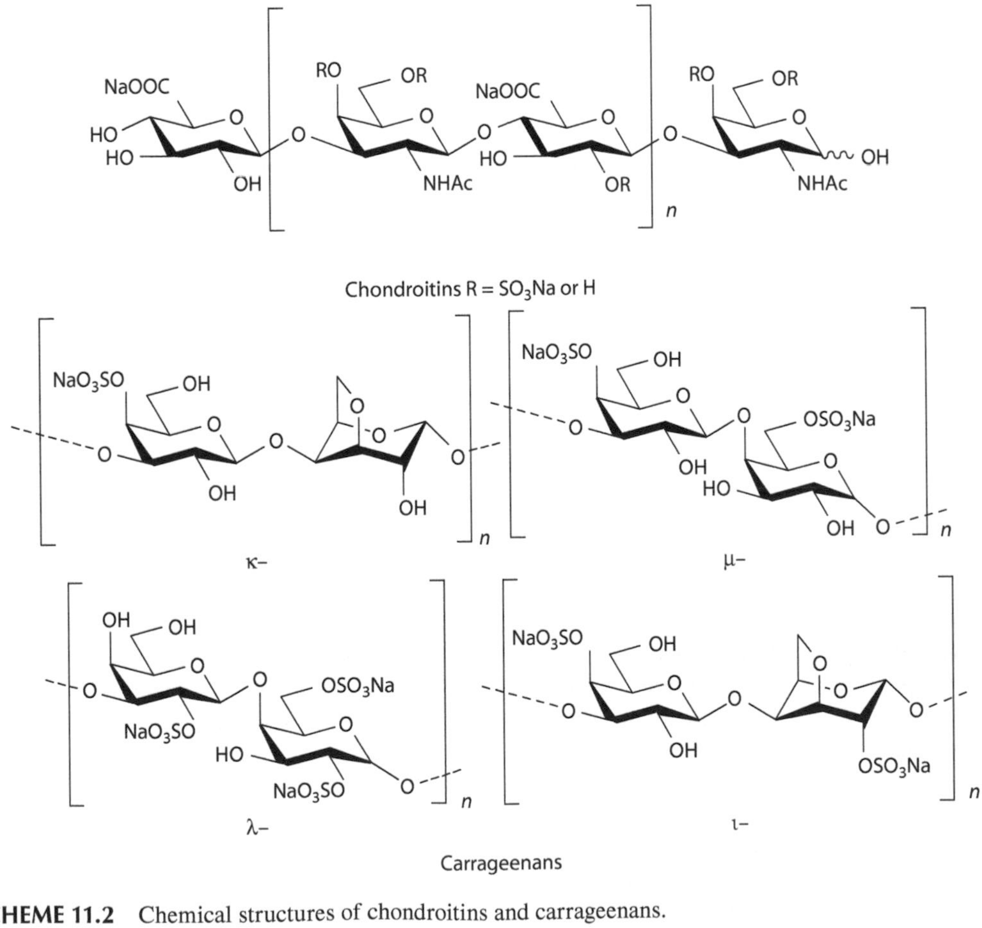

SCHEME 11.2 Chemical structures of chondroitins and carrageenans.

(a)

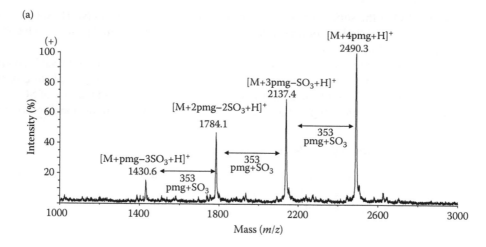

(b)

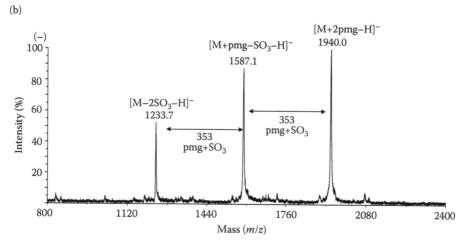

FIGURE 11.15   Positive (a) and negative (b) ion mode MALDI mass spectra of CS–C hexa with 40 equivalents of compound **10**.

dependent on the number of sulfate groups present in the chondroitin backbone. An illustrative example is given in Figure 11.15 with the trisulfated hexasaccharide CS–C hexa. In both ionization modes, a mass difference of 353 Da between each peak is observed; this corresponds to a SO_3–**10** cleavage with an H substituting the SO_3–**10** pair. This observation along with the m/z value of the fully complexed ion indicates that all sulfate groups are complexed by compound **10**. In the positive mode, a ladder of four peaks is obtained corresponding to the complexation of four molecules of compound **10** with CS–C hexa. The fully complexed ion is easily detected at m/z 2490.3, whereas the calculated m/z value is 2488.6. In the negative mode, three well-distinguished peaks can be observed, with the highest m/z value at 1940.0 corresponding to a complexation of the saccharide with two pyrenemethylguanidinium molecules (calculated m/z 1940.4). This specific SO_3–**10** loss (net loss of 353 Da) associated with the highest m/z values observed in both ionization modes allows an unambiguous determination of the degree of sulfation of the molecule (three sulfate groups) as well as the molecular weight of the saccharide. From these spectra, the molecular weight of the fully protonated neutral analyte could be determined as 1395.0 Da in the positive mode and as 1397.3 Da in the negative mode (calculated 1395.2 Da).

Overall seven chondroitin sulfates were analyzed, and from the mass spectra obtained it was possible to draw out the following general rules: (1) for an analyte bearing n sulfate groups, a number of n peaks was detected in the negative mode and $[n + 1]$ peaks in the positive mode; (2) the highest m/z values were found to follow the general formula $m/z = [M_{OS} + (n + 1) \times M_{10} + H]^+$ in the positive mode and $m/z = [M_{OS} + (n - 1) \times M_{10} - H]^-$ in the negative mode (where M_{OS} and M_{10} refer to the molecular weight of the fully protonated oligosaccharide and compound **10**, respectively); and (3) the peaks with the lowest m/z values correspond to the fully desulfated ion in the positive mode and to a monosulfated ion in the negative mode. It is assumed that the $(n + 1)$ peaks observed in the positive mode correspond to a weaker interaction of compound **10** with the carboxylate groups present in the oligosaccharides. Additionally, there were no sodium adducts and no glycosidic cleavage fragments detected.

11.3.3 ANALYSIS OF CARRAGEENANS

Carrageenans are sulfated D-galactans extracted from red algae. The structure is a linear water-soluble chain consisting of disaccharide repeating units of $1 \rightarrow 3$ β-D galactopyranose (G-unit) and $1 \rightarrow 4$ α-D-galactopyranose (D-unit).[53] Kappa-(κ), iota-(ι), and lambda-(λ) carrageenans are constituted by the following repetitive disaccharide units, G4S-DA, G4S-DA2S, and G2S-D2S,6S, respectively.[54] Kappa carrageenans frequently contain variable amounts of their biological precursors, namely mu-(μ) (G4S-D6S) (Scheme 11.2).[55] These natural polymers possess the ability to form gels or viscous solutions, and as such they are extensively used as texturing, thickening, suspending, or stabilizing agents in various industrial applications ranging from food products to pharmaceuticals.[56,57]

As it was the case with chondroitins, no peaks could be detected without addition of compound **10**. However, adding 40 equivalents of compound **10** to various carrageenans led once more to specific ladders of peaks directly relating to the number of sulfate groups and the molecular weight of the protonated analyte. Thirteen carrageenans, differing in both length and sulfation patterns, were analyzed and were shown in both ionization modes to follow the general formula of n peaks for n sulfate groups. As shown in Figure 11.16 with the nonasulfated hexasaccharide λλλ, the highest m/z values (4439.1 in the positive mode and 3889.0 in the negative mode) follow the same general formula found for the chondroitins. The lowest m/z values correspond in both ionization modes to a monosulfated ion, free in the negative mode, and complexed with two molecules of compound **10** in the positive mode. These observations appear to be logical as carrageenans do not bear carboxylic function. In the case of λλλ, a series of nine peaks (Δ353 Da) are observed in both ionization modes, confirming therefore the degree of sulfation (nine sulfo groups) and the molecular weight (1708.1 and 1706.0 Da, calculated 1709.9) from the positive and negative modes, respectively. As this method differs mainly from other approaches by the formation of specific complexes, spectra obtained in both ionization modes allow the unambiguous analysis of unknown compound as well as mixture of compounds.

11.3.4 SENSITIVITY

Usually, in both ionization modes, routine analysis would require 10 pmol of sample material per target spot. However, the detection limit was assessed and we found that 1 pmol of sample material per target spot can be achieved in the negative mode without affecting the quality of the spectra by adding 300 equivalents of compound **10**. Furthermore, the observation of the fully complexed oligosaccharide ions is an important issue for the determination of both the molecular weight and sulfation degree. For the analysis of unknown compounds, the fully complexed and desulfated ions might be observed as small peaks and their intensities might depend on the molecular ratio of the analyte versus compound **10**. An easy way to analyze unknown samples is to increase the concentration of compound **10** until no signal intensity change of the maximum m/z ion is observed. It is therefore important to make sure that spectra obtained in both ionization modes match the general rules developed in Sections 11.3.2 and 11.3.3.

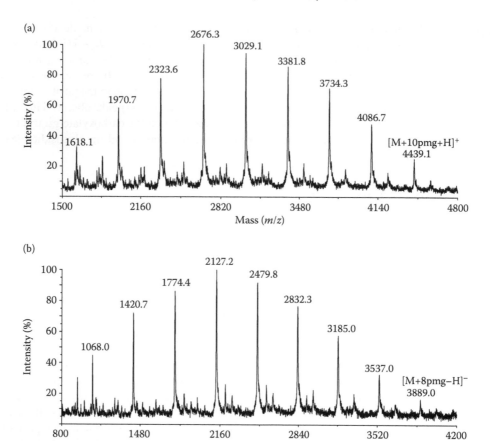

FIGURE 11.16   Positive (a) and negative (b) ion mode MALDI mass spectra of λλλ with 40 equivalents of compound **10**.

11.3.5   Conclusion

The analysis of sulfated polysaccharides, which plays important biological roles, is of primary importance. However, the high acidity of sulfo groups and their liability toward fragmentation have made such analysis challenging. In sharp contrast with other techniques that aim at avoiding fragmentation, our goal was to develop an analytical tool that could reveal, in a single experiment, the degrees of polymerization (molecular weight) and sulfation (number of sulfate groups) of the analyte. Pyrenemethylguanidinium was found to be an effective derivatizing agent and ionization enhancer for MALDI of polyanionic oligosaccharides. Appling this method, the length of the oligosaccharide and the number of sulfate groups are accurately assigned for any given oligosaccharide up to decasaccharides.

11.4   SUMMARY AND OUTLOOK

MALDI-MS usually represents a very powerful technique for the analysis of biomolecules. Accuracy, sensitivity, wide molecular weight range, and singly charged ions are among the main advantages conveyed by this method. However, mainly due to cation adducts, highly acidic compounds are generally difficult to ionize by MALDI. By screening variously substituted guanidines, we found that pyrenemethylguanidine contains highly interactive functional moieties able to complex efficiently phosphate and sulfate complementary units. The formation of specific complexed ions permits the determination of the molecular weight as well as the number of phosphate–sulfate groups of biomolecules otherwise difficult or impossible to ionize by MALDI. The molecular weight could be measured with high sensitivity (picomoles) and good mass accuracy. Pyrenemethylguanidine, which is easy to synthesize, is a particularly strong complexing agent and an ionization enhancer. The utility of this specific complex formation is just beginning to be realized and will serve to extend this concept to other highly acidic biolomolecules or pharmaceutically relevant analytes. Along with the molecular weight of phosphation–sulfation degree, the sequence and the position of acidic groups of various biomolecules should also be possible.

ACKNOWLEDGMENTS

We gratefully acknowledge l'Université de Montpellier 2 and l'Association pour la Recherche sur le Cancer (ARC) for their financial support. The authors thank Dr. W. Helbert and Dr. J.-C. Jacquinet, who provided many of the carbohydrate compounds.

REFERENCES

1. Zechel, D.L. and Withers, S.G., Glycosidase mechanisms: Anatomy of a finely tuned catalyst, *Acc. Chem. Res.* 33, 11–18, 2000.
2. Desai, N.A. and Shankar, V., Single-strand-specific nucleases, *FEMS Microbiol. Rev.* 26, 457–491, 2003.
3. Akashi, S., Investigation of molecular interaction within biological macromolecular complexes by mass spectrometry, *Med. Res. Rev.* 26, 339–368, 2006.
4. Bich, C., Scott, M., Panagiotidis, A., Wenzel, R.J., Nazabal, A., and Zenobi, R., Characterization of antibody–antigen interactions: Comparison between surface plasmon resonance measurements and high-mass matrix-assisted laser desorption/ionization mass spectrometry, *Anal. Biochem.* 375, 35–45, 2008.
5. Daniel, J.M., Friess, S.D., Rajagopalan, S., Wendt, S., and Zenobi, R., Quantitative determination of noncovalent binding interactions using soft ionization mass spectrometry, *Int. J. Mass Spectrom.* 216, 1–27, 2002.
6. Schug, K.A. and Lindner, W., Noncovalent binding between guanidinium and anionic groups: Focus on biological- and synthetic-based arginine/guanidinium interactions with phosph[on]ate and sulf[on]ate residues, *Chem. Rev.* 105, 67–113, 2005.
7. Prins, L.J., Reinhoudt, D.N., and Timmerman, P., Noncovalent synthesis using hydrogen bonding, *Angew. Chem., Int. Ed. Engl.* 40, 2383–2426, 2001.
8. Jelesarov, I. and Bosshard, H.R., Isothermal titration calorimetry and differential scanning calorimetry as complementary tools to investigate the energetics of biomolecular recognition, *J. Mol. Recogn.* 12, 3–18, 1999.
9. Fyfe, M.C.T., Stoddart, J.F., and Williams, D.J., X-ray crystallographic studies on the noncovalent syntheses of supermolecules, *Struct. Chem.* 10, 243–259, 1999.
10. Meyer, B. and Peters, T., NMR Spectroscopy techniques for screening and identifying ligand binding to protein receptors, *Angew. Chem., Int. Ed. Engl.* 42, 864–890, 2003.
11. Smith, D.L., Deng, Y.Z., and Zhang, Z.Q., Probing the non-covalent structure of proteins by amide hydrogen exchange and mass spectrometry, *J. Mass Spectrom.* 32, 135–146, 1997.
12. Brodbelt, J.S., Probing molecular recognition by mass spectrometry, *Int. J. Mass Spectrom.* 200, 57–69, 2000.
13. Loo, J.A., Studying noncovalent protein complexes by electrospray ionization mass spectrometry, *Mass Spectrom. Rev.* 16, 1–23, 1997.
14. Ganem, B., Li, Y.T., and Henion, J.D., Detection of noncovalent receptor ligand complexes by mass-spectrometry, *J. Am. Chem. Soc.* 113, 6294–6296, 1991.
15. Schalley, C.A., Molecular recognition and supramolecular chemistry in the gas phase, *Mass Spectrom. Rev.* 20, 253–309, 2001.
16. Rusconi, F., Guillonneau, F., and Praseuth, D., Contributions of mass spectrometry in the study of nucleic acid-binding proteins and of nucleic acid–protein interactions, *Mass Spectrom. Rev.* 21, 305–348, 2002.

17. Glocker, M.O., Bauer, S.H.J., Kast, J., Volz, J., and Przybylski, M., Characterization of specific noncovalent protein complexes by UV matrix-assisted laser desorption ionization mass spectrometry, *J. Mass Spectrom.* 31, 1221–1227, 1996.
18. Bolbach, G., Matrix-assisted laser desorption/ionization analysis of non-covalent complexes: Fundamentals and applications, *Curr. Pharm. Design* 11, 2535–2557, 2005.
19. Juhasz, P. and Biemann, K., Mass-spectrometric molecular-weight determination of highly acidic compounds of biological significance via their complexes with basic polypeptides, *Proc. Natl. Acad. Sci. USA* 91, 4333–4337, 1994.
20. Tang, X.D., Callahan, J.H., Zhou, P., and Vertes, A., Noncovalent protein–oligonucleotide interactions monitored by matrix-assisted laser desorption/ionization mass-spectrometry, *Anal. Chem.* 67, 4542–4548, 1995.
21. Lecchi, P., Le, H.M.T., and Pannell, L.K., 6-Aza-a-thiothymine—a matrix for MALDI spectra of oligonucleotides, *Nucleic Acids Res.* 23, 1276–1277, 1995.
22. Lin, S.H., Cotter, R.J., and Woods, A.S., Detection of non-covalent interaction of single and double stranded DNA with peptides by MALDI-TOF, *Protein Struct. Funct. Genet. Z*, 12–21, 1998.
23. Luo, S.Z., Li, Y.M., Qiang, W., Zhao, Y.F., Abe, H., Nemoto, T., Qin, X.R., and Nakanishi, H., Detection of specific noncovalent interaction of peptide with DNA by MALDI-TOF, *J. Am. Soc. Mass Spectrom.* 15, 28–31, 2004.
24. Ohara, K., Smietana, M., and Vasseur, J.J., Characterization of specific noncovalent complexes between guanidinium derivatives and single-stranded DNA by MALDI, *J. Am. Soc. Mass Spectrom.* 17, 283–291, 2006.
25. Ohara, K., Smietana, M., Restouin, A., Mollard, S., Borg, J.P., Collette, Y., and Vasseur, J.J., Amine-guanidine switch: A promising approach to improve DNA binding and antiproliferative activities, *J. Med. Chem.* 50, 6465–6475, 2007.
26. Friess, S.D. and Zenobi, R., Protein structure information from mass spectrometry? Selective titration of arginine residues by sulfonates, *J. Am. Soc. Mass Spectrom.* 12, 810–818, 2001.
27. Agrawal, S., Importance of nucleotide sequence and chemical modifications of antisense oligonucleotides, *Biochem. Biophys. Acta Gene Struct. Exp.* 1489, 53–68, 1999.
28. Eckstein, F. and Gish, G., Phosphorothioates in molecular biology, *Trends Biochem. Sci.* 14, 97–100, 1989.
29. Frey, P.A. and Sammons, R.D., Bond order and charge localization in nucleoside phosphorothioates, *Science* 228, 541–545, 1985.
30. Sennikov, P.G., Weak H-Bonding by 2nd-Row (Ph3, H2s) and 3rd-Row (Ash3, H2se) hydrides, *J. Phys. Chem.* 98, 4973–4981, 1994.
31. Platts, J.A., Howard, S.T., and Bracke, B.R.F., Directionality of hydrogen bonds to sulfur and oxygen, *J. Am. Chem. Soc.* 118, 2726–2733, 1996.
32. Crooke, S.T., Progress in antisense therapeutics, *Med. Res. Rev.* 16, 319–344, 1996.
33. Lecchi, P. and Pannell, L.K., The detection of intact double-stranded DNA by Maldi, *J. Am. Soc. Mass Spectrom.* 6, 972–975, 1995.
34. Sudha, R. and Zenobi, R., The detection and stability of DNA duplexes probed by MALDI mass spectrometry, *Helv. Chim. Acta* 85, 3136–3143, 2002.
35. Kirpekar, F., Berkenkamp, S., and Hillenkamp, F., Detection of double-stranded DNA by IR- and UV-MALDI mass spectrometry, *Anal. Chem.* 71, 2334–2339, 1999.
36. Little, D.P., Jacob, A., Becker, T., Braun, A., Darnhofer-Demar, B., Jurinke, C., van den Boom, D., and Koster, H., Direct detection of synthetic and biologically generated double-stranded DNA by MALDI-TOF MS, *Int. J. Mass Spectrom.* 169, 323–330, 1997.
37. Bahr, U., Aygun, H., and Karas, M., Detection and relative quantification of siRNA double strands by MALDI mass spectrometry, *Anal. Chem.* 80, 6280–6285, 2008.
38. Ohara, K., Jacquinet, J.C., Jouanneau, D., Helbert, W., Smietana, M., and Vasseur, J.J., Matrix-assisted laser desorption/ionization mass spectrometric analysis of polysulfated-derived oligosaccharides using pyrenemethylguanidine, *J. Am. Soc. Mass Spectrom.* 20, 131–137, 2009.
39. Chai, W.G., Luo, J.L., Lim, C.K., and Lawson, A.M., Characterization of heparin oligosaccharide mixtures as ammonium salts using electrospray mass spectrometry, *Anal. Chem.* 70, 2060–2066, 1998.
40. Gunay, N.S., Tadano-Aritomi, K., Toida, T., Ishizuka, I., and Linhardt, R.J., Evaluation of counterions for electrospray ionization mass spectral sulfated analysis of a highly sulfated carbohydrate, sucrose octasulfate, *Anal. Chem.* 75, 3226–3231, 2003.
41. Zaia, J., Mass spectrometry of oligosaccharides, *Mass Spectrom. Rev.* 23, 161–227, 2004.

42. Zaia, J., Li, X.Q., Chan, S.Y., and Costello, C.E., Tandem mass spectrometric strategies for determination of sulfation positions and uronic acid epimerization in chondroitin sulfate oligosaccharides, *J. Am. Soc. Mass Spectrom.* 14, 1270–1281, 2003.

43. Zhang, J.H., Zhang, J., LaMotte, L., Dodds, E.D., and Lebrilla, C.B., Atmospheric pressure MALDI Fourier transform mass spectrometry of labile oligosaccharides, *Anal. Chem.* 77, 4429–4438, 2005.

44. Juhasz, P. and Biemann, K., Utility of noncovalent complexes in the matrix-assisted laser-desorption ionization mass-spectrometry of heparin-derived oligosaccharides, *Carbohydr. Res.* 270, 131–147, 1995.

45. Venkataraman, G., Shriver, Z., Raman, R., and Sasisekharan, R., Sequencing complex polysaccharides, *Science* 286, 537–542, 1999.

46. Laremore, T.N. and Linhardt, R.J., Improved matrix-assisted laser desorption/ionization mass spectrometric detection of glycosaminoglycan disaccharides as cesium salts, *Rapid Commun. Mass Spectrom.* 21, 1315–1320, 2007.

47. Laremore, T.N., Murugesan, S., Park, T.J., Avci, F.Y., Zagorevski, D.V., and Linhardt, R.J., Matrix-assisted laser desorption/ionization mass spectrometric analysis of uncomplexed highly sulfated oligosaccharides using ionic liquid matrices, *Anal. Chem.* 78, 1774–1779, 2006.

48. Tissot, B., Gasiunas, N., Powell, A.K., Ahmed, Y., Zhi, Z.L., Haslam, S.M., Morris, H.R., Turnbull, J.E., Gallagher, J.T., and Dell, A., Towards GAG glycomics: Analysis of highly sulfated heparins by MALDI-TOF mass spectrometry, *Glycobiology* 17, 972–982, 2007.

49. Fukuyama, Y., Nakaya, S., Yamazaki, Y., and Tanaka, K., Ionic liquid matrixes optimized for MALDI-MS of sulfated/sialylated/neutral oligosaccharides and glycopeptides, *Anal. Chem.* 80, 2171–2179, 2008.

50. Lopin, C. and Jacquinet, J.C., From polymer to size-defined oligomers: An expeditious route for the preparation of chondroitin oligosaccharides, *Angew. Chem., Int. Ed. Engl.* 45, 2574–2578, 2006.

51. Jacquinet, J.C., Synthesis of a set of sulfated and/or phosphorylated oligosaccharide derivatives from the carbohydrate-protein linkage region of proteoglycans, *Carbohydr. Res.* 341, 1630–1644, 2006.

52. Salih, B. and Zenobi, R., MALDI mass spectrometry of dye-peptide and dye-protein complexes, *Anal. Chem.* 70, 1536–1543, 1998.

53. van de Velde, F., Peppelman, H.A., Rollema, H.S., and Tromp, R.H., On the structure of kappa/iota-hybrid carrageenans, *Carbohydr. Res.* 331, 271–283, 2001.

54. Knutsen, S.H., Myslabodski, D.E., Larsen, B., and Usov, A.I., A modified system of nomenclature for red algal galactans, *Bot. Mar.* 37, 163–169, 1994.

55. van de Velde, F., Knutsen, S.H., Usov, A.I., Rollema, H.S., and Cerezo, A.S., H-1 and C-13 high resolution NMR spectroscopy of carrageenans: Application in research and industry, *Trends Food Sci. Technol.* 13, 73–92, 2002.

56. Renn, D., Biotechnology and the red seaweed polysaccharide industry: Status, needs and prospects, *Trends Biotechnol.* 15, 9–14, 1997.

57. McReynolds, K.D. and Gervay-Hague, J., Chemotherapeutic interventions targeting HIV interactions with host-associated carbohydrates, *Chem. Rev.* 107, 1533–1552, 2007.

12 Quantitative Identification of Nucleic Acids via Signature Digestion Products Detected Using Mass Spectrometry

Mahmud Hossain and Patrick A. Limbach

CONTENTS

12.1   INTRODUCTION

The functional unit of information in living systems is the gene. A gene is defined biochemically as a segment of deoxyribonucleic acid (DNA) (or, in a few cases, ribonucleic acid, RNA) that encodes the information required to produce a functional biological product—a protein. Genetic theory contributed the concept of coding by genes. Physics permitted the determination of molecular structure by x-ray diffraction analysis. Chemistry revealed the composition of nucleic acids and proteins.[1] These and other major advances gave rise to the *central dogma of molecular biology* comprising the three major processes in the cellular utilization of genetic information. The first is *replication*, the copying of parental DNA to form daughter DNA molecules with identical nucleotide

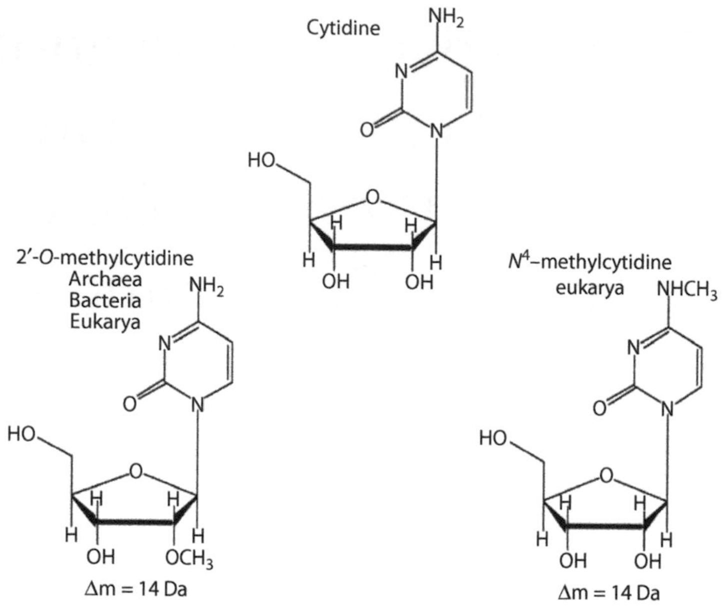

FIGURE 12.1 Example of major nucleoside, cytidine, and common posttranscriptional modifications. Modifications can occur on the ribose sugar (2'-*O*-methylcytidine) or on the heterocyclic base (N^4-methylcytidine). Modifications occur in all kingdoms as noted on these two representative modified nucleosides. MS is particularly useful for identifying modified nucleosides due to the change in mass of the nucleoside upon modification. In these examples, nucleoside methylation increases the mass by 14 Da.

sequences. The second is *transcription*, the process by which parts of the genetic message encoded in DNA are copied precisely into RNA. The third is *translation*, whereby the genetic message encoded in messenger RNA is translated on the ribosome into a polypeptide with a particular sequence of amino acids.

The expression of the information in a gene normally involves production of an RNA molecule transcribed from a DNA template. Strands of RNA and DNA may seem quite similar at first glance, differing only in the hydroxyl group at the 2' position of the pentose and the substitution of uracil (U) for thymine (T). However, unlike DNA, most RNAs exist as single strands. These strands fold back on themselves with the potential for much greater structural diversity than DNA. RNA is thus suited to a variety of cellular functions. RNA is the only macromolecule known to function both in the storage and transmission of information and in catalysis, leading to much speculation about its possible role as an essential chemical intermediate in the development of life on this planet. The discovery of catalytic RNAs, or ribozymes, has changed the very definition of an enzyme, extending it beyond the domain of proteins.[2] RNAs are also unique in that they can be modified by a number of enzymes posttranscriptionally to generate even more diverse structures that are important for a number of biological processes (Figure 12.1).[3]

12.2 NON-PROTEIN-CODING RNAs

A non-protein-coding RNA (ncRNA) is any RNA molecule that is not translated into a protein.[4] ncRNAs include transfer RNAs (tRNAs), ribosomal RNAs (rRNAs), small interfering RNA (siRNAs), small nucleolar RNAs (snoRNAs), micro RNAs (miRNAs), guide RNAs (gRNAs), piwi-interacting RNAs (piRNAs), and transfer-messenger RNAs (tmRNAs). tRNA carries the correct amino acid to a growing polypeptide chain at the ribosomal site of protein biosynthesis during translation, in addition to other recently found cellular roles. rRNAs are synthesized in the nucleolus by RNA polymerase I, and are the central component of the ribosome, the protein manufacturing machinery of all living cells. The function of rRNA is to provide a mechanism for decoding mRNA into polypeptides and to interact with tRNAs during translation by providing peptidyl transferase (PT) activity. siRNA is a class of small RNA molecules (20–25 nt in length) formed through cleavage of long double-stranded RNA molecules. siRNAs are particularly important for taming the activity of transposons and combating viral infection, but they can also regulate protein-coding genes. Synthetic siRNA can also be artificially expressed for experimental purposes.[5] snoRNAs are a class of small RNA molecules that guide chemical modifications (methylation or pseudouridylation) of rRNAs and other RNA genes.[6] Small Cajal Body-specific RNAs (scaRNAs) are a class of small RNA molecules similar to snoRNAs that specifically localize in the Cajal body, a nuclear organelle involved in the biogenesis of small nuclear RNA (snRNA). U85 is the first reported scaRNA. Unlike typical snoRNA, U85 can guide both pseudouridylation and 2'-O-methylation.[7] miRNAs are small RNAs that are the reverse complement of portions of an mRNA transcript and alter the expression of one or several genes through RNA interference. They are around 20–25 nucleotides (nt) long in the mature form, single-stranded, and are integrated into the RNA-induced silencing complex (RISC).[5] gRNAs are RNAs that function in RNA editing. gRNA-mediated RNA editing has been found in the mitochondria of kinetoplastids, in which mRNAs are edited by inserting or deleting stretches of uridylates. The gRNA forms part of the editosome and contains sequences that hybridize to matching sequences in the mRNA, to guide the mRNA modifications. Other types of RNA editing are found in many eukaryotes, including humans.[8] piRNAs are small RNAs (25–30 nt length) that are generated from long single-stranded precursors. They function in association with the piwi subfamily of Argonaute proteins and are essential for the development of germ cells.[9,10] The bacterial tmRNA, also known as 10Sa RNA or SsrA, is so-named for its dual tRNA-like and mRNA-like nature. tmRNA recognizes ribosomes that have trouble translating or reading an mRNA and stall, leaving an unfinished protein that may be detrimental to the cell. tmRNA gene sequences have been identified in all completely sequenced bacterial genomes (in 17 of 20 phyla) and in certain phage, mitochondrial, and plastid genomes, but not yet in archaeal or eukaryotic nuclear genomes.[11] The *signal-recognition particle* (SRP) is an RNA–protein complex present in the cytoplasm of cells that binds to the mRNA of proteins that are destined for secretion from the cell. The RNA component of the SRP in eukaryotes is called 4.5S RNA.[12]

12.3 TRANSFER RNAs

Even before deciphering the genetic code during the 1960s, Francis Crick had postulated in 1956 that protein synthesis is mediated by "adaptor" RNA molecules.[13] Two years later, Hoagland discovered a nucleic acid fraction of low molecular weight that served as a carrier for amino acids, transporting them to the place where polypeptides are synthesized. This fraction was termed soluble RNA (sRNA) and was described as mixture of components, each with adaptor ability for a particular amino acid.[14] Nowadays, we know that the sRNA or adaptor molecules are the tRNA and they are linking factors between the RNA and protein worlds. The specificity of deciphering results because tRNAs contain at one tip of their L-shaped tertiary structure an anticodon complementary to a specific codon and at the other tip the corresponding aminoacyl residue linked by an energy-rich ester bond ($\Delta G^{o'} = \sim\!- 6$ kcal mol^{-1}). Charging of tRNAs is performed by synthetases (aaRS), and all tRNAs that can carry the same amino acid (isoacceptor tRNAs) are usually recognized by the same enzyme.[15]

tRNA constitutes 10–15% of the total RNA in *Escherichia coli*. Each tRNA has a molecular weight of about 25 kDa and a relative sedimentation coefficient of 4S giving rise to the original name "4S RNA." The size of tRNA is variable, but on average they have a length of 76 nt. The longest tRNA identified so far is tRNA^{Ser} from *E. coli* having 91 nt, whereas in nematode mitochondria

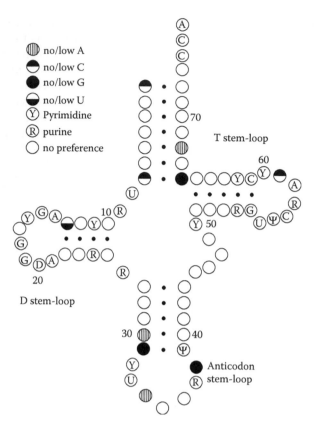

FIGURE 12.2 Cumulative map of bacterial tRNAs showing conserved nucleosides. Similar maps can be drawn for other classes of tRNAs. A = adenosine; C = cytidine; G = guanosine; U = uridine; D = dihydrouridine; and Ψ = pseudouridine. The major structural features are also noted: D stem-loop; anticodon stem-loop (ASL); and T stem-loop.

very short cripple tRNAs (about 56 nt) are found lacking either the D or the TΨC stem-loops.[16] Sequence comparison of various tRNAs reveals that all tRNAs adopt a cloverleaf secondary structure, which is characterized by three stems (acceptor stem, D stem, and anticodon stem) and four loops or hairpins (D loop, anticodon loop, variable loop, and T loop) (Figure 12.2). The variable loop exists between the T loop and the anticodon loop, which can be anywhere between 4 and 24 nts. According to the length of this variable loop, tRNAs have been classified as class I (4–5 nts in variable loops, the vast majority) and class II (10–24 nts in variable loops, tRNA[Leu], tRNA[Ser], and tRNA[Tyr] in eubacteria and some organelles).[17] The 3 Å crystal structures of yeast tRNA[Phe] and tRNA[Asp] confirmed that tRNA molecules adopt an L shape. This shape is the product of a double coaxial stacking between the acceptor stem with its CCA end and the T stem-loop forming the short arm of the L arm, and the ASL and the D stem-loop forming the long arm. In this way, the cloverleaf structure of the tRNA is transformed into two main domains: the acceptor and anticodon arms, respectively, enclosing an angle of about 90°. The extremities of both domains represent the functional "hot spots" of tRNAs.[18,19]

tRNAs are the most modified RNA molecules; almost 25% of the nucleotides of a tRNA are modified. Eighty nucleotide modifications out of more than 100 reported modifications in all RNA molecules have been observed in tRNAs.[3,20] Usually the modification reaction is an alteration of, or addition to, existing bases in the tRNA, an exception being the base queuosine (Q). This base is found 5′ to the anticodon at position 34 of tRNAs that read NAU or NAC codons (where N is any nucleotide), and the modification requires an enzyme that exchanges free Q with guanosine. Many examples of tRNA modifications include ribose/base methylations (2′-*O*-methylguanosine, Gm; 2′-*O*-methylcytidine, Cm; and 5-methylcytidine, m⁵C), base isomerization (U to pseudouridine, Ψ), base reduction (U to D; dihydrouridine), base thiolation (2-thiocytidine, s²C; 2-thiouridine, s²U; and 4-thiouridine, s⁴U), and base deamination (inosine, I). Some modifications are conserved features of all tRNA molecules (D residues that give rise to the name of the D-arm, Ψ found in the TΨC sequence).[15]

In addition to the main role in protein synthesis, tRNAs are also involved in a series of other activities in the cell.[15] Examples include: tRNA^Lys III is used as a primer for DNA synthesis by Human Immunodeficiency Virus (HIV) reverse transcriptase[21]; some tRNAs induce the formation of antitermination structures of the nontranslated region upstream (UTR) of the structural genes of some amino acid operons (*ile-leu*, *his*, and *trp*) and of some tRNA synthetase genes (*thrS*, *tyrS*, *leuS*, and *pheS*)[22]; a deacylated tRNA binds to the ribosome-bound enzyme, RelA, for synthesizing signaling molecule (p)ppGpp in bacterial stringent response[23]; tRNA^Glu is an activated intermediate in the biosynthetic pathway of δ-aminolevulinate,[24] a tetrapyrrole precursor of porphyrins in plants and bacteria; amino acid residues from aminoacyl tRNAs are used in a cross-linking reaction in peptidoglycan synthesis of bacterial cell wall;[25,26] RNA polymerase III activity in silkworm depends on several transcription factors, including tRNA^Ile containing transcription factor TFIIIR, a fraction in which transcription activity is provided by RNA;[27] aminoacyl tRNAs are involved in a proteolytic "N-end rule" pathway,[28] which governs the half-life of a protein in a cell; and in mammals, many diseases are known that are caused by tRNA defects in the mitochondria, including several human neurodegenerative disorders.[29,30]

12.4 RIBOSOMAL RNAs

The rRNAs lie at the core of the protein synthesis machinery. In all organisms the ribosome consists of two subunits. These are designated as the 40S and 60S subunits in eukaryotes and the 30S and 50S subunits in Bacteria, Archaea, and the cytoplasmic organelles of eukaryotes, mitochondria, and chloroplasts. In almost all organisms the small ribosomal subunit contains a single RNA species (the 18S rRNA in eukaryotes and the 16S rRNA elsewhere). In Bacteria and Archaea, the large ribosomal subunit contains two rRNA species (the 5S and the 23S rRNAs); in most eukaryotes the large subunit contains three RNA species (the 5S, 5.8S, and 25S/28S rRNAs). Sequence analysis shows that the 5.8S rRNA corresponds to the 5′ end of the bacterial and archaeal 23S rRNAs, and was presumably generated early in eukaryotic evolution by insertion of a spacer sequence. Chloroplast large ribosomal subunits also contain three RNAs; in this case, the 4.5S rRNA is derived from the 3′ terminus of the bacterial 23S rRNA. Finally, in mitochondria the large subunit rRNA is smaller in size and is designated as the 21S rRNA.[31]

rRNAs were long regarded as mere scaffolds for the ribosomal proteins, but recent work has shown that the rRNAs in fact carry out the key reactions in translation. A major function of ribosomal proteins is ensuring the correct structure of rRNA, allowing its tight packing around the active center of the ribosome. There are a number of significant cellular contributions of rRNA.[32] Inspection of the structure of rRNAs in distant organisms from the three domains of life reveals that despite substantial differences among primary sequence, both the small subunit rRNA (SSU-rRNA) and the large subunit rRNA (LSU-rRNA) display remarkable conservation of their secondary, and probably tertiary, structures. Core structures can be drawn for the SSU- and LSU-rRNAs that can accommodate the secondary structures of all rRNAs. The most conserved elements are presumed to be of functional significance. Notably, almost all the rRNA posttranscriptional modifications (base and ribose methylation as well as conversion of uridine to Ψ) fall within these conserved core regions of the rRNAs. In the primary sequences of the rRNAs, the conserved sequences are separated by variable regions, in which the primary and secondary structures diverge more rapidly in evolution. The overall length of these regions is also poorly conserved and is generally longer in eukaryotes; they are therefore often referred to as expansion segments. The core structures of the

SSU- and LSU-rRNAs contain 10 and 18 such variable regions, respectively. In general, the variable regions are not necessary for ribosome function.

The combination of biochemical approaches, mostly cross-linking experiments and chemical foot-printing of rRNAs bound to tRNAs or to various antibiotics, with the genetic analysis of *E. coli* strains bearing mutations in their rRNAs led to the definition of several functional domains in the rRNAs. Domains were ascribed to the most basic functions of the ribosome (i.e., decoding or codon–anticodon recognition and PT activity) as well as to antibiotic binding and interactions with ribosomal proteins and translational factors. The general picture that emerged from these studies is that the decoding center of the ribosome lies within its small subunit and the PT activity is carried out within the large subunit. The accuracy of translation is determined by components of both subunits, probably reflecting interactions of the tRNAs with both ribosomal subunits.

The identification of the ribosomal components involved in peptide-bond formation has been a long-standing challenge in ribosome research. This area was mostly explored in *E. coli*, making use of systems for *in vitro* reconstitution of the subunits. It is currently accepted that the catalytic activity of the ribosome lies in its RNA components while the ribosomal proteins act as chaperones in ribosome assembly and as cofactors to increase the efficiency of the RNA-mediated PT reaction and the accuracy of translation. Antibiotics have been of great use in ribosome research, particularly where their site(s) of interaction with the ribosome could be correlated with specific translational defects, providing putative functions for particular sections of rRNAs. Most characterized antibiotics appear to bind primarily to rRNA. In some cases, the methylation status of specific rRNA residues correlates with antibiotic resistance or sensitivity.

12.4.1 Classical Methods for Isolating the Major ncRNAs

Research on the characterization of tRNA can be traced back to the 1950s.[33] The closely related physiochemical properties of tRNAs have made their isolation as pure species difficult. Numerous procedures for the fractionation of tRNAs have been reported. Countercurrent distribution was one of the first techniques used to purify tRNA from the soluble fraction of a rat liver homogenate. The solvent system described by Warner and Vaimberg[34] was used with slight modification, and the distribution was carried out in the 100-tube countercurrent apparatus described by Raymond.[35] Countercurrent distribution of RNA gives a broad distribution pattern, measured by absorption at 260 nm. Redistribution of materials from different parts of the first distribution pattern establishes that there has been actual fractionation of the RNA. While this technique was most useful for early studies of tRNAs, it was somewhat unwieldy and separation of tRNAs, while effective, was quite time consuming.[36]

Countercurrent distribution was followed by separation techniques based on the use of benzoylated diethylaminoethyl (DEAE) cellulose (BD cellulose) by Tener et al.,[37] DEAE–Sephadex by Nishimura et al.,[38] reversed-phase chromatography (RPC) by Kelmers et al.,[39] and Sepharose 4B by Holms.[40] The BD-cellulose column separates tRNAs based on the interaction of exposed hydrophobic bases to the matrix and provides essentially a one-step purification of yeast tRNA^Phe starting from total yeast tRNA. This tRNA uniquely contains the hydrophobic base yW (wybutosine) that is exposed and is located next to the anticodon sequence. Because of this, the yeast tRNA^Phe binds tightly to the column even in 1 M NaCl and can be eluted off the column only in the presence of 10% ethanol. DEAE-Sephadex and Sepharose 4B columns also proved quite useful for large-scale purification of many tRNAs. RPC columns, which could be run at different pHs and temperatures and at high pressure, were quite versatile. Because the columns are run at high pressure, chromatography is quite rapid allowing for fractionation of not just tRNAs but aminoacyl tRNAs.

Other procedures include chemical modification of uncharged or charged tRNA by the use of monoclonal anti-adenosine monophosphate (AMP) antibody affinity columns and by polyacrylamide gel electrophoresis. A combination of hydrophobic chromatography on phenyl-Sepharose and reversed-phase high performance liquid chromatography (HPLC) was used to purify individual tRNAs with high specific activity. The efficiency of chromatographic separation was enhanced by biochemical manipulations of the tRNA molecule, such as aminoacylation, formylation of the amin-oacyl moiety, and enzymatic deacylation.[41] A murine monoclonal anti-AMP antibody affinity matrix was used for isolation of individual species of tRNAs. The antibodies were prepared using 5′-AMP covalently attached to bovine serum albumin as the antigen and exhibited high affinity for 5′-AMP but greatly reduced affinity for 3′-AMP. Native uncharged tRNAs that terminate in a 5′-AMP group on the amino acid acceptor arm of the molecule bind tightly to the anti-AMP affinity matrix, whereas aminoacylated tRNAs are not retained. This allows separation of a particular tRNA species as its aminoacyl derivative from a complex mixture of uncharged tRNAs under very mild conditions.[42] Elongation factor Tu from *Thermus thermophilus* containing six histidine residues on its C-terminus, EF-Tu(C-His6), was used for purification of aminoacyl-tRNA isoacceptors, from aminoacylated bulk tRNA, by affinity chromatography.[43] Another method employed the Elongation factor Tu from *E. coli*, immobilized on activated CH Sepharose 4B, to purify aminoacylated tRNA isoacceptors from bulk tRNAs.[44] tRNAs from the posterior silk gland and carcass tissues of the silk worm *Bombyx mori* were fractionated by high-resolution polyacrylamide gel electrophoresis. Nonlabeled tRNA of the posterior silk gland, purified by benzoylated DEAE-cellulose column chromatography and by countercurrent distribution, was used to aid in identification of tRNA^{Ala}, tRNA^{Gly}, and tRNA^{Ser} isoacceptor species. The high resolution of tRNA separation on polyacrylamide gels thus provided a quantitative estimate of the posterior silk gland isoacceptor tRNA distribution which is adapted to produce large amounts of proteins, silk fibroin, during the fifth larval instar.[45]

These methods have been and are of great value for both analytical and preparative experiments. However, they have not permitted the complete separation of the extremely complex mixture represented by the set of cellular tRNAs. More than one chromatographic procedure must be used to resolve and purify only a few tRNAs, which is time consuming and usually results in low recovery values. Analogously, chemical modification of tRNA molecules allows for only the separation of charged from uncharged tRNAs.[46] Thus, these aforementioned methods are less than ideal for routine analytical research involving individual tRNAs in complex biological mixtures.

12.4.2 CONTEMPORARY METHODS FOR THE CHARACTERIZATION OF MAJOR ncRNAs

Over the last couple of decades, several methods based on chromatography, electrophoresis, and mass spectrometry (MS) have been developed to characterize tRNAs. Filter and solution hybridization assays were used for the characterization of yeast cytosolic tRNA^{Phe}, and bovine mitochondrial tRNA^{Ser}.[47–49] The hybridization efficiencies varied considerably among probes, which are complementary to different regions of the tRNAs, although there was little efficiency variation in the probes toward DNA substrates including the same nucleotide sequence. This efficiency variation was shown to be due to tRNA-specific higher-order structures as well as a hypermodified nucleotide in the anticodon loop.[48] Dong used two-dimensional polyacrylamide gel electrophoresis to fractionate tRNA from *E. coli* and isolated components were identified by hybridization to tRNA-specific oligonucleotide probes.[50] Systematic measurement of the abundance of different isoaccepting tRNAs were measured by using [5,6-³H]uridine and ^{32}P-containing growth media.

There are still several shortcomings hampering the more widespread use of hybridization: (i) slow kinetics of hybridization to a solid phase, and thus the necessity for a large amount of immobilized DNA to achieve efficient hybridization; (ii) conversely, the requirement to limit the DNA amount for active immobilization to a support; (iii) nonspecific interaction of tRNAs with support materials; (iv) deterioration of the tRNA recovery rate when the columns are used repeatedly; and (v) the need to use different immobilized columns for each tRNA species. The major limitation associated with hybridization efficiency of oligonucleotide probes is the tRNA's unpredictability due to secondary/tertiary structures.[48,51] Mir also supported this idea by studying the effects of structure on nucleic acid heteroduplex formation in analyzing hybridization of tRNA^{Phe} to a complete set of complementary oligonucleotides, ranging from single nucleotides to dodecanucleotides.[52] Buvoli improved the sensitivity of northern blotting by alleviating the hindrance to tRNA hybridization.[53] They showed that a combination of oligodeoxynucleotides employed in the prehybridization reaction can reshape the tRNA structure and increase the sensitivity of specific detection of a suppressor tRNA derived from the human serine tRNA up to 222-fold. This approach is complex, time consuming, and overall, not suitable for global analysis of tRNA abundances. Recently, Miyauchi reported an automated parallel isolation of multiple species of noncoding RNAs including tRNAs from *E. coli* by the reciprocal circulating chromatography (RCC) method.[54] This approach requires multiple separation techniques—chromatography, gel electrophoresis, and MS. A comprehensive, rapid, and sensitive detection of sequence variants of human mitochondrial tRNA genes has been developed on the basis of a modification of denaturing gradient gel electrophoresis (DGGE).[55] It has shown to be a reliable and efficient mutation screening technique but not suitable for studying the characterization of tRNA sequence with its myriad posttranscriptional modifications.

12.5   SIGNATURE DIGESTION PRODUCTS

Recently, we proposed the use of signature digestion products (SDPs) as a means of detecting individual ncRNAs from a sample containing one or more ncRNAs.[56] The approach is based on the selective enzymatic digestion of an individual ncRNA by an ribonuclease (RNase) (e.g., RNase T1), which generates a number of specific endonuclease digestion products. A comparison of an organism's complete complement of ncRNA RNase digestion products yields a set of unique or "signature" digestion product(s) that ultimately enable the detection of individual ncRNAs from a total RNA pool. RNase T1 cleaves at the 3′ side of all unmodified guanosine residues; RNase U2 and TA cleave at the 3′ side of unmodified purines, with a preference for unmodified adenosines; and RNase A cleaves at the 3′ side of pyrimidines including pseudouridine. While other RNases may be used, these RNases provide the appropriate balance among selectivity, sequence coverage, and availability.[57] Below, we describe the development of this SDP approach for the identification and quantification of tRNAs.

To illustrate this concept, the theoretical RNase T1, RNase U2, and RNase A digestion products of *E. coli* tRNA[Phe] and tRNA[Trp] are shown in Figure 12.3 and listed in Table 12.1. tRNA[Phe] has 12 mass values that are identical or shared by the RNase T1 digestion products of other *E. coli* tRNAs, whereas three mass values are unique to *E. coli* tRNA[Phe]. For RNase U2 (or TA) digestion, 12 mass values are shared with other *E. coli* tRNAs and two mass values are unique. Similarly, for RNase A, 11 mass values are shared and five mass values are unique. These unique mass values are the RNase SDPs of tRNA[Phe]. The mass spectrometric detection of any one or more of these signature products from a complex mixture will confirm the presence of the corresponding tRNA. Likewise, *E. coli* tRNA[Trp] yields four RNase T1 SDPs, four RNase U2 (or TA) SDPs, and two RNase A SDPs. The detection of any of these SDPs by MS can be used to identify *E. coli* tRNA[Trp].

The SDP approach can be used to identify tRNAs from any particular organism, providing sufficient sequence information is available for each tRNA. At present, only the following organisms have (nearly) complete sequences, including posttranscriptional modifications, for tRNAs: *Haloferax volcanii, E. coli, Mycoplasma caprcolum,* and *Saccharomyces cerevisiae,* although more organisms have a large contingent of their tRNA sequences known at the posttranscriptional level. The theoretical (or expected) SDPs for these organisms are available at the SDP website—RNAccess (http://bearcatms.uc.edu/rnaccess). An examination of all theoretical tRNA SDPs finds that they can be classified into one of three different types:

Type I—a signature product unique to one tRNA.

Type II—a signature product specific to one tRNA family.

Type III—a signature product selective for two or more, but not all, of the members of one tRNA family.

E. coli tRNA^Phe

5'-pGCCCGGA(s⁴U)AGCUCAGDCGGDAGAGCAGGGGAΨUG
AA(ms²i⁶A)AΨCCCCGU(m₇G)(acp₃U)CCUUGGTΨCGAUUC
CGAGUCCGGGCACCA-3'

RNase T1	pGp CCCGp Gp **A(s⁴U)AGp** CUCAGp DCGp Gp DAGp
	AGCAGp Gp Gp Gp AΨUGp **AA(ms²i⁶A)AΨCCCCGp**
	U(m⁷G)(acp³U)CCUUGp Gp TΨCGp AUUCCGp AGp
	UCCGp Gp Gp CACCA-OH

RNase U2	pGCCCGGAp (s⁴U)Ap GCUCAp GDCGGDAp GAp GCAp
	GGGGAp ΨUGAp AP (ms²i⁶A)Ap
	ΨCCCCGU(m⁷G)(acp³U)CCUUGGTΨCGAp UUCCGAp
	GUCCGGGCAp CCA-OH

RNase A	pGCp Cp Cp **GGA(s⁴U)AGCp** Up Cp AGDCp
	GGDAGAGCp AGGGGAΨUp GAA(ms²i⁶A)AΨCp Cp Cp
	Cp GUp **(m⁷G)(acp³U)Cp** Cp Up Up GGTΨCp GAUp
	Up Cp Cp GAGUp Cp Cp GGGCp ACp Cp A-OH

E. coli tRNA^Trp

5'-pAGGGGCG(s⁴U)AGUUCAADDGGDAGAGCACCGGU
(Cm)UCCA(ms²i⁶A)AACCGGGU(m⁷G)UUGGGAGTΨCGAG
UCUCUCCGCCCCUGCCA-3'

RNase T1	pAGp Gp Gp Gp CGp (s⁴U)AGp **UUCAADDGp** Gp
	DAGp AGp CACCGp Gp **U(Cm)UCCA(ms²i⁶A)AACCGp**
	Gp Gp **U(m⁷G)UUGp** Gp Gp AGp TΨCGp AGp
	UCUCUCCGp CCCCUGp CCA-OH

RNase U2	pAp **GGGGCG(s⁴U)Ap** GUUCAp Ap **DDGGDAp** GAp
	GCAp **CCGGU(Cm)UCCAp** (ms²i⁶A)Ap Ap
	CCGGGU(m⁷G)UUGGGAp GTΨCGAp
	GUCUCUCCGCCCCUGCCA-OH

RNase A	pAGGGGCp G(s⁴U)AGUp Up Cp **AADDGGDAGAGCp**
	ACp Cp GGUp (Cm)Up Cp Cp **A(ms²i⁶A)AACp** Cp
	GGGUp (m⁷G)Up Up GGGAGTΨCp GAGUp Cp Up Cp
	Up Cp Cp GCp Cp Cp Cp Up GCp Cp A-OH

FIGURE 12.3 Representative determination of SDPs. Digestion of individual ncRNAs will generate specific RNase digestion products of unique mass that can be used to identify the original ncRNA. Here, *E. coli* tRNA Phe and Trp are shown as examples. For each tRNA, the expected, typical RNase digestion products after RNase T1, RNase U2 (or RNase TA), and RNase A hydrolysis are shown. Bolded RNase digestion products are unique to all *E. coli* tRNAs and can be used to identify the tRNA from which they were generated. See Table 12.1 for *m/z* values that would be detected during mass spectral analysis of these digestion product mixtures.

To illustrate the use of SDPs for the identification of tRNAs, both tRNA families (tRNAs specific for a particular amino acid) and individual tRNAs, the analyses of a completely sequenced organism, *E. coli*, and a nearly complete one, *Bacillus subtilis*, are presented.

12.6 IDENTIFICATION OF *E. COLI* tRNA FAMILIES

For *E. coli*, there are 22 tRNA families each of which is specific for a particular amino acid in protein synthesis, and these families result in 48 unique tRNAs due to the presence of isoaccepting tRNAs in 13 of the tRNA families. Table 12.2 provides a listing of the *E. coli* tRNAs that can be

TABLE 12.1
SDPs for *E. Coli* tRNAs Phe and Trp

E. coli tRNA Phe

5′-pGCCCGGA(s^4U)AGCUCAGDCGGDAGAGCAGGGGAΨUGAA(ms^2i^6A)AΨ
CCCCGU$(m^7G)$$(acp^3U)$CCUUGGT$\Psi$CGAUUCCGAGUCCGGGCACCA-3′

RNase Product	Average Mass	RNase
$(m^7G)(acp^3U)$Cp	1089.71	A
A(s^4U)AGp	1343.88	T1
GGA(s^4U)AGCp	2339.48	A
GAA(ms^2i^6A)AYCp	2405.62	A
U$(m^7G)(acp^3U)$CCUUGp	2658.61	T1
AGGGGAΨUp	2669.61	A
GGDAGAGCp	2670.64	A
GUCCGGGCAp	2949.78	U2/TA
AA(ms^2i^6A)AΨCCCCGp	3321.17	T1
ΨCCCCGU$(m^7G)(acp^3U)$CCUUGGTΨCGAp	6481.90	U2/TA

E. coli tRNA Trp

5′-pAGGGGCG(s^4U)AGUUCAADDGGDAGAGCACCGGU(Cm)UCCA(ms^2i^6A)AA
CCGGGU(m^7G)UUGGGAGTΨCGAGUCUCUCCGCCCCUGCCA-3′

U(m^7G)UUGp	1640.97	T1
A(ms^2i^6A)AACp	1754.24	A
CCCCUGp	1810.15	T1
DDGGDAp	1962.21	U2/TA
UCUCUCCGp	2502.47	T1
UUCAADDGp	2555.54	T1
GGGGCG(s^4U)Ap	2700.69	U2/TA
CCGGU(Cm)UCCAp	3189.92	U2/TA
AADDGGDAGAGCp	3945.44	A
U(Cm)UCCA(ms^2i^6A)AACCGp	3946.55	T1
CCGGGU(m^7G)UUGGGAp	4306.59	U2/TA

detected by SDPs generated from RNases T1, TA, and A. The purine-selective endonucleases (RNase T1 and RNase TA) yield more predicted SDPs for *E. coli* tRNAs, 104 and 150 respectively, than does the pyrimidine-selective endonuclease RNase A, which yields only 82. Individually, RNase A generates predicted SDPs to identify 13 tRNA families, and RNase TA generates predicted SDPs to identify 14 tRNA families. RNase T1 is the most efficient endonuclease for this organism, as it generates predicted SDPs to identify all tRNA families except Thr and Sec6.

Figure 12.4 and Table 12.3 present representative matrix-assisted laser desorption/ionization (MALDI)-MS data obtained from the RNase digestion of tRNAs isolated from an *E. coli* MRE 600 cell lysate. Analysis of the RNase T1 digestion yielded seven Type I signature products (for tRNAs Arg III, Gln II, Leu I, Leu II, Leu III, Tyr II, and Val III), 13 Type II signature products (for tRNAs Asn, Asp, Cys, Glu, Gly, His, Ini, Lys, Met, Phe, Pro, Trp, and Tyr), and seven Type III signature products (for tRNAs of Ala I/II, Glu II/III, Ile I/II, Leu II/III, Ser I/IV, Ser III/IV/V, and Val II/III). With RNase A, 10 Type 1 signature products (for tRNAs Arg V, Gln II, Gly I, Gly II, Gly III, Ser I, Ser II, Ser III, Ser V, and Val III), seven type II signature products (for tRNAs Asp, Ini, Phe, Sec6, Thr, Trp, and Tyr), and two Type III signature products (for tRNAs of Ala II/III and Arg III/IV) were identified. Analysis of the RNase TA digestion revealed 10 Type I signature products (for tRNAs Ala I, Arg III, Gly III, Ile I, Ile II, Pro II, Ser I, Ser IV, Thr II, and Val II), six Type II signature products

TABLE 12.2
Predicted Identification of *E. coli* tRNAs by SDPs

Type of SDP	Ribonuclease	Identified tRNAs
Type I	RNase T1	Ala I, Ala II, Arg III, Arg IV, Arg V, Asn, Asp, Cys, Gln I, Gln II, Gly I, Gly II, Gly III, His, Ile III, Leu I, Leu II, Leu III, Lys, Met, Phe, Pro I, Pro II, Pro III, Sec6, Ser I, Ser II, Ser V, Thr I, Thr II, Trp, Tyr II, Val II, Val III
	RNase A	Ala I, Ala II, Arg I, Arg II, Arg III, Arg IV, Arg V, Asn, Asp, Cys, Gln I, Gln II, Glu I, Gly I, Gly II, Gly III, Ile I, Ile II, Ile III, Leu I, Leu III, Met, Phe, Pro III, Sec6, Ser I, Ser II, Ser III, Ser IV, Ser V, Trp, Val I, Val III
	RNase U2/TA	Ala I, Ala II, Arg I, Arg II, Arg III, Arg IV, Arg V, Asn, Asp, Cys, Gln I, Gln II, Glu I, Gly I, Gly II, Gly III, His, Ile I, Ile II, Ile III, Ini I, Ini II, Leu I, Leu II, Leu III, Lys, Met, Phe, Pro I, Pro II, Pro III, Sec6, Ser I, Ser II, Ser III, Ser IV, Ser V, Thr I, Thr II, Trp, Val I, Val II, Val III
Type II	RNase T1	Asn, Asp, Cys, Gln, Glu, Gly, His, Ini, Lys, Met, Phe, Pro, Sec6, Thr, Trp, Tyr
	RNase A	Asn, Asp, Cys, Gln, Glu, Ini, Met, Phe, Sec6, Thr, Trp, Tyr
	RNase U2/TA	Asn, Asp, Cys, Gln, Glu, His, Ini, Lys, Met, Phe, Sec6, Thr, Trp, Tyr
Type III	RNase T1	Ala I/III, Ala II/III, Arg I/II, Glu II/III, Ile I/II, Leu II/III, Leu I/III, Ser I/IV, Ser III/V, Ser II/IV/V, Pro I/II, Val II/III
	RNase A	Ala II/III, Arg III/IV, Ile I/II, Leu II/III, Pro I/II, Pro II/III, Ser III/IV, Val I/II
	RNase U2/TA	Ala I/III, Ala II/III, Arg I/II, Glu I/II, Glu II/III, Gly I/II, Ile I/II, Leu II/III, Ser II/IV, Ser III/IV, Ser II/III/IV, Val I/II

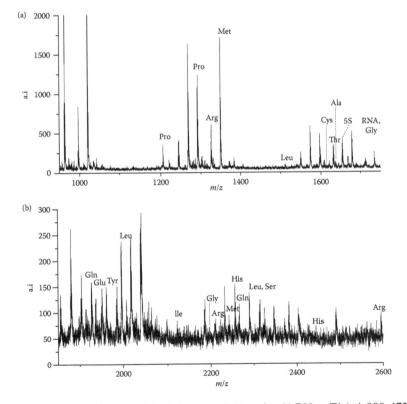

FIGURE 12.4 SDP analysis with MALDI-MS of *E. coli* tRNAs using (a) RNase T1 (*m/z* 900–1750), (b) RNase T1 (*m/z* 1750–2600), (c) RNase A, and (d) RNase TA. Identified tRNAs are noted on each mass spectrum. A summary of *E. coli* tRNA identifications is listed in Table 12.3. (Reproduced from Hossain, M. and Limbach, P. A., *Anal. Bioanal. Chem.* 394, 1125–1135, 2009. With permission.)

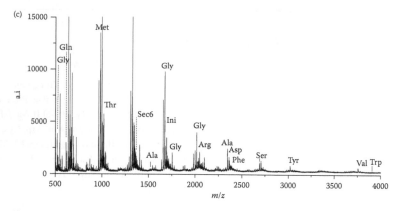

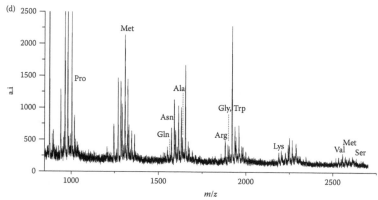

FIGURE 12.4 Continued.

TABLE 12.3
Summary of MALDI-MS Detection of SDPs for
***E. coli* tRNAs**

tRNAs	RNase T1	Rnase A	RNase TA
Ala	✓	✓	—
Ala I	—	—	✓
Ala II	—	—	—
Ala III	—	—	—
Arg	—	✓	✓
Arg I	—	—	—
Arg II		—	—
Arg III	✓	—	✓
Arg IV	—	—	—
Arg V	—	✓	—
Asn	✓	—	✓
Asp	✓	✓	—
Cys	✓	—	—
Gln	—	—	✓
Gln I	—	—	—
Gln II	✓	✓	—

continued

TABLE 12.3 (continued)
Summary of MALDI-MS Detection of SDPs for
E. coli **tRNAs**

tRNAs	RNase T1	Rnase A	RNase TA
Glu	✓	—	—
Glu I	—	—	—
Glu II	—	—	—
Glu III	—	—	—
Gly	✓	—	—
Gly I	—	✓	—
Gly II	—	✓	✓
Gly III	—	✓	—
His	✓	—	—
Ile	✓	—	—
Ile I	—	—	✓
Ile II	—	—	✓
Ile III	—	—	—
Ini	✓	✓	—
Ini I	—	—	—
Ini II	—	—	—
Leu	✓	—	—
Leu I	✓	—	—
Leu II	✓	—	—
Leu III	✓	—	—
Lys	✓	—	✓
Met	✓	—	✓
Phe	✓	✓	✓
Pro	✓	—	—
Pro I	—	—	—
Pro II	—	—	✓
Pro III	—	—	—
Sec6	—	✓	—
Ser	✓	—	—
Ser I	—	✓	—
Ser II	—	✓	✓
Ser III	—	✓	—
Ser IV	—	—	✓
Ser V	—	✓	—
Thr	—	✓	—
Thr I	—	—	—
Thr II	—	—	✓
Trp	✓	✓	✓
Tyr	✓	✓	—
Tyr I	—	—	—
Tyr II	✓	—	—
Val	✓	—	✓
Val I	—	—	—
Val II	—	—	✓
Val III	✓	✓	—

Source: Reproduced from Hossain, M. and Limbach, P. A., *Anal. Bioanal. Chem.* 394, 1125–1135, 2009. With permission.

(for tRNAs Asn, Gln, Lys, Met, Phe, and Trp), and three Type III signature products (for tRNAs of Arg I/II, Ile I/II, and Val I/II). All of the major peaks in the mass spectra are easily assignable to expected digestion products, and all expected products generate distinct signals. No prior fractionation is needed for this analysis. Combining the experimentally detected signature products of any type from all three endonucleases identifies all 22 tRNA families from *E. coli*. While RNase T1 was most effective for identifying tRNA families, these results do not appear to correlate strongly with the number of predicted signature products.

12.7 IDENTIFICATION OF INDIVIDUAL *E. COLI* tRNAs

Although the signature digestion approach is relatively efficient at identifying tRNA families, a more informative identification is one made at the level of individual tRNAs. As noted above, 13 tRNA families contain a total of 36 *E. coli* tRNA isoacceptors resulting in 48 unique tRNAs. Experimentally, 35 of these 48 were detected through either RNase T1, A, or U2 digestions. Five tRNA isoacceptors (Arg III, Gln II, Gly II, Ser I, and Val III) can be detected by more than one endonuclease, and six tRNAs without isoacceptors (Asn, Asp, Lys, Met, Phe, and Trp) can be detected by more than one endonuclease.

There is no specific preference if only two endonucleases were taken in combination. As shown in Figure 12.5, RNase T1 plus RNase A identified 30 tRNAs, RNase T1 plus RNase TA identified 27 tRNAs, and RNase A plus RNase TA identified 29 tRNAs. For *E. coli*, the use of multiple endonucleases generally results in a 10–15% increase in the identification of tRNAs. All of the major (>2% of total tRNAs) tRNAs (Ala II, Ala III, Arg I, Arg II, Asp, Cys, Glu I, Glu II, Glu III, Gly I, Gly II, Gly III, Ile I, Ile II, Ile III, Leu I, Leu II, Leu III, Lys, Ser II, and Val III) are detected by at least one ribonuclease.

12.8 ANALYSIS OF *BACILLUS SUBTILIS* tRNAs

Similarly, *Bacillis subtilis* has been analyzed for the number of predicted and experimentally detected SDPs. There are 21 tRNA families in *Bacillus subtilis* containing a total of 28 unique

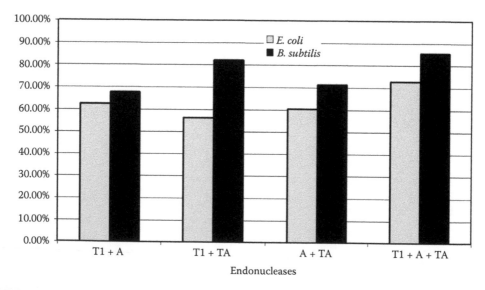

FIGURE 12.5 Effectiveness of endonuclease combinations at identifying individual tRNAs from mixtures of *E. coli* (gray bars) and *B. subtilis* (black bars). Three endonucleases in combination yield greater coverage than any two endonucleases. (Reproduced from Hossain, M. and Limbach, P. A., *Anal. Bioanal. Chem.* 394, 1125–1135, 2009. With permission.)

TABLE 12.4
Predicted Identification of *Bacillus subtilis* tRNAs by SDPs

Type of SDP	Ribonuclease	Identified tRNAs
Type I	RNase T1	Ala, Arg, Asn I, Asn II, Asp, Cys, Gln, Glu II, Gly, Ile, Ini, Leu, Met, Phe, Pro, Ser I, Ser II, Ser III, Thr, Trp, Tyr I, Tyr II, Val
	RNase A	Ala, Arg, Asn II, Asp, Cys, Gln, Glu II, Gly, Ile, Ini, Leu, Lys I, Lys II, Met, Phe, Pro, Ser I, Ser II, Ser III, Thr, Trp, Tyr I, Tyr II, Val
	RNase U2/TA	Ala, Arg, Asn I, Asn II, Asp, Cys, Gln, Glu I, Glu II, Gly, Ile, Ini, Leu, Met, Phe, Pro, Ser I, Ser II, Ser III, Thr, Trp, Val
Type II	RNase T1	Ala, Arg, Asn, Asp, Cys, Gln, Glu, Gly, His, Ile, Ini, Leu, Lys, Met, Phe, Pro, Thr, Trp, Tyr, Val
	RNase A	Ala, Arg, Asp, Cys, Gln, Gly, Ile, Ini, Leu, Met, Phe, Pro, Thr, Trp, Tyr, Val
	RNase U2/TA	Ala, Arg, Asn, Asp, Cys, Gln, Glu, Gly, His, Ile, Ini, Leu, Lys, Met, Phe, Pro, Thr, Trp, Tyr, Val
Type III	RNase T1	Ser I/III, Ser II/III
	RNase A	None
	RNase U2/TA	Ser I/III, Ser II/III

sequenced tRNAs, 13 of which are tRNA isoacceptors. Table 12.4 lists the *Bacillus subtilis* tRNAs that are detectable by SDPs. Reviewing this information, all tRNA families yield predicted SDPs, regardless of the endonuclease used, except for tRNA His, which does not have any RNase A-generated SDPs. RNase T1 generates distinguishable signature products from all isoacceptors except Glu I, His I, His II, Lys I, and Lys II. RNase A generates distinguishable signature products from all isoacceptors except Asn I, Glu I, His I, and His II. RNase TA generates distinguishable signature products from all isoacceptors except His I, His II, Lys I, Lys II, Tyr I, and Tyr II. RNase TA generates the largest number of SDPs (107), with RNase T1 generating 80 and RNase A generating 55. Further, all tRNA families of *Bacillus subtilis* have multiple signature products, ranging from six for tRNAs Cys and Gln to a high of 43 for Ser.

As for *E. coli*, a mixture of *Bacillus subtilis* tRNAs was digested individually with RNase T1, RNase A, and RNase TA. The digestions were analyzed by MALDI-MS and the results are presented in Figure 12.6 and Table 12.5. Analysis of the RNase T1 digestion yielded three Type I signature products (for tRNAs Asn II, Glu II, and Ser I), 15 Type II signature products (for tRNAs Ala, Arg, Gln, Glu, Gly, His, Ile, Ini, Leu, Met, Pro, Thr, Trp, Tyr, and Val), and one Type III signature product (for tRNA Ser I/II). With RNase A, two Type I signature products (for tRNAs Glu II and Lys I) and eight Type II signature products (for tRNAs Arg, Asp, Cys, Leu, Met, Pro, Trp, and Val) were identified. Analysis of the RNase TA digestion revealed three Type I signature products (for tRNAs Asn I, Ser II, and Ser III), 13 Type II signature products (for tRNAs Ala, Asp, Gln, Gly, Glu, His, Ile, Ini, Leu, Lys, Phe, Trp, and Val), and one Type III signature product (for tRNAs of Ser II/III).

12.8.1 IDENTIFICATION OF *BACILLUS SUBTILIS* tRNA FAMILIES

Combining the experimentally detected signature products of any type from all three endonucleases identifies all 21 tRNA families from *Bacillus subtilis*. Individually, RNase T1 was most effective, identifying 17 tRNA families, with RNase TA identifying 15 and RNase A identifying only 10. Similar to the results found for *E. coli*, the purine-selective endonucleases (RNase T1 and RNase TA) yield the greatest number of predicted SDPs and are most effective at identifying tRNA families

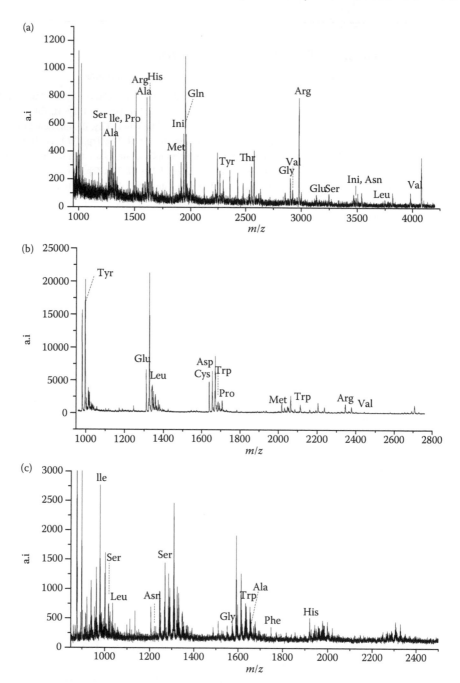

FIGURE 12.6   SDP analysis with MALDI-MS of *Bacillus subtilis* tRNAs using (a) RNase T1, (b) RNase A, and (c) RNase TA. Identified tRNAs are noted on each mass spectrum. A summary of *Bacillus subtilis* tRNA identifications is listed in Table 12.5. (Reproduced from Hossain, M. and Limbach, P. A., *Anal. Bioanal. Chem.* 394, 1125–1135, 2009. With permission.)

present in the sample. There appears to be no preference for the identification of tRNA families if only two endonucleases were used, as RNase T1 plus RNase A identified 20 tRNA families as did RNase T1 plus RNase TA. RNase A plus RNase TA was, again, least effective in identifying only 19 of 21 tRNA families.

TABLE 12.5
Summary of MALDI-MS Detection of SDPs for
***Bacillus subtilis* tRNAs**

TRNAs	RNase T1	RNase A	RNase TA
Ala	✓	—	✓
Arg	✓	✓	—
Asn	✓	—	✓
Asn I	—	—	✓
Asn II	✓	—	—
Asp	—	✓	✓
Cys	—	✓	—
Gln	✓	—	✓
Glu	✓	✓	✓
Glu I	—	—	—
Glu II	✓	✓	—
Gly	✓	—	✓
His	✓	—	✓
His I	—	—	—
His II	—	—	—
Ile	✓	—	✓
Ini	✓	—	✓
Leu	✓	✓	✓
Lys	—	✓	✓
Lys I	—	✓	—
Lys II	—	—	—
Met	✓	✓	—
Phe	—	—	✓
Pro	✓	✓	—
Ser	✓	—	✓
Ser I	✓	—	—
Ser II	—	—	✓
Ser III	—	—	✓
Thr	✓	—	—
Trp	✓	✓	✓
Tyr	✓	—	—
Tyr I	—	—	—
Tyr II	—	—	—
Val	✓	✓	✓

Source: Reproduced from Hossain, M. and Limbach, P. A., *Anal. Bioanal. Chem.*
394, 1125–1135, 2009. With permission.

12.8.2 IDENTIFICATION OF INDIVIDUAL *BACILLUS SUBTILIS* tRNAs

Only *Bacillus subtilis* has 28 unique tRNAs, with only six tRNA families having tRNA isoacceptors. Thus, for this organism, Type I signature products would appear to be less important
than for *E. coli*, which has 13 tRNA families generating 36 tRNA isoacceptors. Experimentally,

the use of multiple endonucleases enabled the identification of 24 of the 28 unique tRNAs, which is a significantly higher percentage than found for *E. coli*. Two tRNA isoacceptors (Glu II and Ser III) were detected by more than one endonuclease, and 12 tRNAs without isoacceptors (Ala, Arg, Asp, Gln, Gly, Ile, Ini, Leu, Met, Pro, Trp, and Val) were detected by more than one endonuclease. Again, we did not find any significant preference if only two endonucleases were taken in combination. As shown in Figure 12.5, RNase T1 plus RNase A identified 19 tRNAs, RNase T1 plus RNase TA identified 23 tRNAs, and RNase A plus RNase TA identified 20 tRNAs. As for *E. coli*, the use of multiple endonucleases generally results in 10–15% increase in the identification of tRNAs.

12.9 MULTIPLE ENDONUCLEASE STRATEGY

The similarity of results from two different bacteria led us to generate a strategy for the MALDI-MS identification of tRNAs starting from an initially unseparated mixture. First, if sample size or enzyme availability dictates the use of only a single endonuclease, RNase T1 should be the preferred option. For both *E. coli* and *Bacillus subtilis*, this endonuclease identified the largest percentage of tRNA families and individual tRNAs. These experimental findings are somewhat surprising, as RNase TA generated the largest number of potential signature products for both organisms. This discrepancy may arise because RNase TA (or RNase U2) digestions need to be carried out under limiting conditions (i.e., stopping at the cyclic phosphate) to improve the selectivity of digestion at adenosine residues rather than purine residues in general. Besides the use of RNase T1, there is no significant preference for either RNase A or RNase TA, and it is clear that all three endonucleases should be used whenever possible. Not only does the use of multiple endonucleases increase the number of unique tRNAs detected within the sample, but the use of multiple endonucleases also generates extremely useful confirmatory identifications of tRNAs detected by any other endonuclease.

Because the MALDI-MS analysis of *E. coli* and *Bacillus subtilis* tRNAs yielded comparable numbers of identified SDPs, regardless of the endonuclease used or tRNA source, it appears likely that this MS platform is the limiting factor in the numbers of tRNA SDPs that can be identified. Thus, organisms that have smaller numbers of individual tRNAs (<30) can be directly analyzed by MALDI-MS. Organisms with larger numbers of individual tRNAs are best analyzed through a combination of sample fractionation/separation and subsequent MS analysis, either by LC-MS or by the MALDI-MS analysis of fractionated tRNA mixtures.

12.10 RELATIVE QUANTIFICATION OF ncRNAs

The SDP approach for the detection of ncRNAs is also compatible with an enzyme-mediated isotope labeling approach for the relative quantification of RNA.[58] To obtain relative quantification information, two sample sets are required. One set is digested by an RNase in the presence of ^{16}O-containing water, while the comparison sample is digested by the same RNase in the presence of ^{18}O-containing water. Aliquots from two samples are then mixed in equal proportions and analyzed with MALDI-MS. Relative quantification is achieved by comparing the ion abundance ratios of relevant SDPs for any particular ncRNA (Figure 12.7).

For the determination of relative quantification of a particular digestion product, it is necessary to know the ion abundance ratios of its unlabeled and labeled peaks (**A** and **A + 2**) extracted from the MALDI mass spectrum as shown in Figure 12.8. Peak **A** has a contribution only from the ^{16}O-digestion product whereas peak **A + 2** has contribution from both ^{16}O- and ^{18}O-labeled digestion products. For a single digestion product peak, with no other product ions within ±3 Da, the ion abundance ratio of the ^{18}O- and ^{16}O-labeled digestion products is calculated by Equation 12.1.[58] The ratios of the

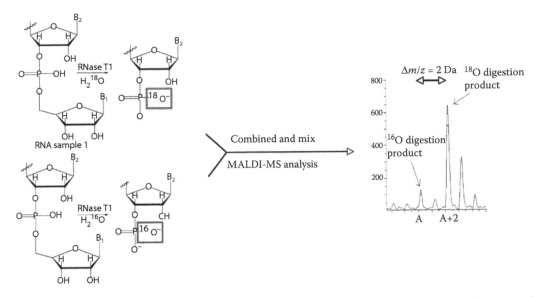

FIGURE 12.7 ¹⁸O- and ¹⁶O-labeling approach for relative quantification. By combining equivalent total amounts of ncRNAs from two experiments, one labeled with ¹⁸O and the other with ¹⁶O, relative abundances can be determined. In this example, the ¹⁸O-labeled RNase digestion product has increased relative to the same RNase digestion product labeled with ¹⁶O. In this way, relative quantification of ncRNAs is possible, and because labeling occurs during enzymatic digestion, this approach can be used with the SDP approach for the simultaneous identification and quantification of individual ncRNAs.

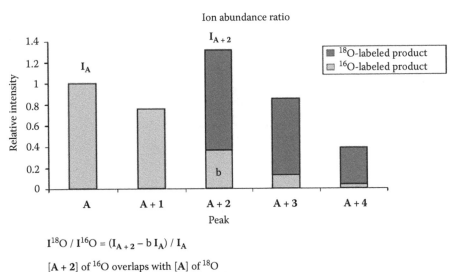

$$I^{18}O \, / \, I^{16}O = (I_{A+2} - b \, I_A) \, / \, I_A$$

[A + 2] of ¹⁶O overlaps with [A] of ¹⁸O
I_A: monoisotopic peak abundance of [A]
I_{A+2}: monoisotopic peak abundance of [A + 2]
b: expected A + 2 contribution from ¹⁶O

FIGURE 12.8 Ion abundance ratio for simple peak. Visual and mathematical expression used to calculate ion abundance ratios for ¹⁶O- and ¹⁸O-labeled RNase digestion products. The I¹⁸O/I¹⁶O ratio can be used to calculate relative abundances of specific RNase digestion products. This formula can only be used when no overlapping RNase digestion products are present. For cases when overlap is present, see Figure 12.9.

isotopic pair ion abundances of digestion products provide a measure of the relative quantity of the ncRNA pairs.

$$\frac{I^{18}O}{I^{16}O} = \frac{[(I_{A+2}) - b*I_A]}{I_A},$$ (12.1)

where I_A represents the monoisotopic peak abundance of the ^{16}O product, I_{A+2} represents the combination of the monoisotopic peak abundance of the ^{18}O digestion product and the $A + 2$ isotopic peak abundance of the ^{16}O digestion product, and b represents the $A + 2$ isotopic peak abundance contribution from the ^{16}O digestion products.

For overlapping peak pairs, for example, when two digestion product peaks are located 1 Da away from each other in the spectrum, the ion abundance ratio calculated from Equation 12.1 is not accurate. For example, digestion products of CCAAAGp (*m/z* 1960.3) and UAACAGp (*m/z* 1961.3) are separated by 1 Da. In this situation, the isotopic peak abundances of A, $A + 1$, $A + 2$, and $A + 3$ have different contributions from different digestion products (Figure 12.9). Peak A has contribution only from the lower-molecular-weight (first) ^{16}O-labeled digestion Peak, $A + 1$ has contributions from both the first ^{16}O- and higher-molecular-weight (second) ^{16}O-labeled products, peak $A + 2$ has contributions from the first ^{16}O-, second ^{16}O-, and first ^{18}O-labeled products, and peak $A + 3$ has contributions from the first ^{16}O- and ^{18}O-, and second ^{16}O- and ^{18}O-labeled products.

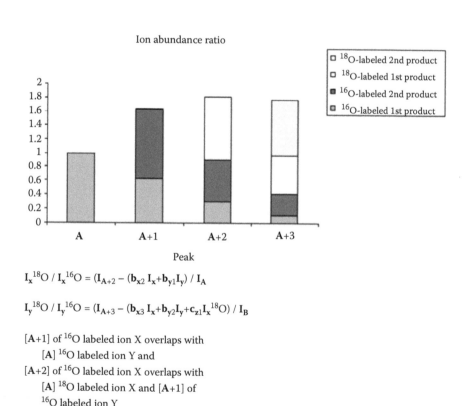

$I_x^{18}O / I_x^{16}O = (I_{A+2} - (b_{x2} I_x + b_{y1}I_y)) / I_A$

$I_y^{18}O / I_y^{16}O = (I_{A+3} - (b_{x3} I_x + b_{y2}I_y + c_{z1}I_x^{18}O)) / I_B$

[A+1] of ^{16}O labeled ion X overlaps with
    [A] ^{16}O labeled ion Y and
[A+2] of ^{16}O labeled ion X overlaps with
    [A] ^{18}O labeled ion X and [A+1] of
    ^{16}O labeled ion Y

FIGURE 12.9 Ion abundance ratios when overlapping digestion products are present. Visual and mathematical expression used to calculate ion abundance ratios for ^{16}O- and ^{18}O-labeled RNase digestion products when two different RNase digestion products are present, differing by 1 Da, such as a sequence change from a C to U. This formula allows one to determine the relative abundances for both RNase digestion products by correcting for isotope overlaps.

To determine the ion abundance ratio of the lower-molecular-weight digestion product ($I_X{}^{18}O/$ $I_X{}^{16}O$) of the overlapping pair (e.g., m/z 1960.3), it is also necessary to know the ion abundances ($I_Y{}^{16}O$) of the second product (e.g., m/z 1961.3), where $I_Y{}^{16}O=(I_{X+1} - b_{X1} * I_X)$ and b_{X1} represents $A + 1$ isotopic peak abundance contribution from the first ^{16}O-labeled product. The ion abundance ratio of first digestion products (e.g., m/z 1960.3) with ^{18}O and ^{16}O labels can be determined by

$$\frac{I_X{}^{18}O}{I_X{}^{16}O} = \frac{[I_{A+2} - (b_{X_2} * I_X + b_{Y_1} * I_Y)]}{I_A}, \tag{12.2}$$

where I_A represents the monoisotopic peak abundance of the ^{16}O-labeled first digestion product, I_{A+2} represents the combination of the monoisotopic peak abundance of the ^{18}O digestion product, $A + 1$ isotopic peak abundance of the ^{16}O-labeled second digestion product, and the $A + 2$ isotopic peak abundance of the ^{16}O-labeled first digestion product, b_{X1} and b_{X2} represent the $A + 1$ and $A + 2$ isotopic peak abundance contribution from the ^{16}O-labeled first digestion products, and b_{Y1} represents the $A + 2$ isotopic peak abundance contribution from the ^{16}O-labeled second digestion products.

To determine the ion abundance ratio of second digestion product ($I_B{}^{18}O/I_B{}^{16}O$) of the overlapping pair (e.g., m/z 1961.3), it is also necessary to know that the ion abundance contributions of the ^{16}O- and ^{18}O-labeled first and second digestion products on peak $A + 3$ position. Peak $A + 3 = (I_B{}^{18}O + a_3 * I_A + b_{Y2} * I_B + c_{Z1}*{}^{18}O)$. Rearranging this equation, $I_B{}^{18}O = (I_A - (a_3 * I_A + b_{Y2} * I_B + c_{Z1} * {}^{18}O))$. The ion abundance ratio of second digestion product (e.g., m/z 1961.3) with their ^{18}O and ^{16}O labels can be determined by

$$\frac{I_Y{}^{18}O}{I_Y{}^{16}O} = \frac{[I_{A+3} - (b_{X_3} * I_X + b_{Y_2} * I_Y + c_{Z1} * I_X{}^{18}O)]}{I_B}, \tag{12.3}$$

where I_A represents the monoisotopic peak abundance of the ^{16}O-labeled first digestion product, I_{A+3} represents the combination of the monoisotopic peak abundance of the ^{18}O digestion product and the $A + 2$ isotopic peak abundance of the ^{16}O digestion product, a_3 represents the $A + 3$ isotopic peak abundance contribution from the ^{16}O-labeled first digestion products, b_2 represents the $A + 3$ isotopic peak abundance contribution from the ^{16}O-labeled second digestion products and c_1 represents the $A + 3$ isotopic peak abundance contribution from the ^{18}O-labeled first digestion product, which is similar to a_1.

12.11 RELATIVE QUANTIFICATION OF LARGE rRNAs OF *E. COLI* BY THEIR SDPs

While the SDP approach is particularly useful for identifying specific ncRNAs, such as tRNAs, in a relatively complex mixture, it is also applicable to larger ncRNAs, such as rRNAs. Moreover, when combined with the isotopic labeling for relative quantification, this approach provides a rapid and sensitive technique to monitor changes in rRNA levels. To conduct such an analysis, *E. coli* cells were grown in 3-(*N*-morpholino)propanesulfonic acid (MOPS) minimal medium (control sample) and EZ-rich defined medium (experimental sample). Cells were harvested and total RNAs were first separated with acid phenol–chloroform extraction. Large rRNAs (16S and 23S rRNAs) were separated from smaller ones by ethanol precipitation onto a glass bead cartridge. After determining the purity, large rRNA mixtures were digested with ribonucleases followed by analysis using MALDI-MS without any prior purification. rRNAs from MOPS minimum media and EZ-rich defined medium were labeled with ^{16}O and ^{18}O during enzymatic digestion.

Twenty-seven SDPs from *E. coli* 16S rRNA and 68 SDPs from *E. coli* 23S rRNA are possible when digested with RNase T1. These signature products do not overlap with any tRNA digestion products. Of those, there are five 16S rRNA signature products that can serve as Q-SDPs, using the criteria established above, and 13 23S rRNA signature products that can serve as Q-SDPs. Similarly, RNase A digestion results in 5 Q-SDPs from 16S rRNA and 10 Q-SDPs from 23S rRNA.

12.12   SDP APPLICATIONS AND DIRECTIONS

In this chapter, we have outlined the concept of SDPs for identification of ncRNAs, have illustrated the utility of this approach for identifying tRNAs without requiring separation of individual tRNAs, shown the steps required to establish quantifiable SDPs, and presented the application of Q-SDPs to quantify changes in rRNAs. While these data were obtained using MALDI-MS, there is no specific requirement for MALDI or electrospray ionization for such experiments. This technique thus opens up new opportunities to study biological processes that can impact the type and distribution of ncRNAs within an organism. Below, we briefly describe but a select few of the processes that should be amenable to this approach.

Mitochondria, though containing their own genome, import the vast majority of their macromolecular components from the cytoplasm.[59] RNAs that function in the mitochondria, in contrast to the majority of mitochondrial proteins, are generally encoded by the mitochondrial genome. However, it is well established today that the transport of nucleus-encoded tRNAs into mitochondria is occurring in a number of evolutionary distinct organisms such as plants, the yeast *S. cerevisiae* and many protozoans.[60] In the yeast *S. cerevisiae*, two nuclear DNA-encoded tRNAs were reported as mitochondrially targeted, tRNA[Gln 61] and tRNA.[Lys 162] Mitochondrial tRNA import might have some exciting practical applications. However, the identification of numerous specific tRNAs simultaneously in a complex mixture is challenging due to their almost similar secondary and tertiary structures. Contemporary methods are able to identify tRNA import in smaller extent,[61,62] but global identification of imported tRNAs are rare. On the other hand, tRNA import to mitochondria from the cytosol in diverse living systems makes it a semiuniversal phenomenon. Identifying the SDPs that differentiate cytosolic from mitochondrial tRNAs would enable this approach to monitor tRNA import and export phenomena, including identification of editing and modification sites related to these transport processes.

Although not yet explored experimentally, SDP identification of specific siRNAs or miRNAs would provide an approach that complements or enhances microarray-based strategies for examining the distribution of these small, regulatory RNAs within biological systems. In particular, because the SDP approach preserves information about modification status of any individual ncRNA, this approach could be used to track any changes in miRNA methylation throughout various steps in gene expression/protein translation.[63–66] However, because the miRNA levels within a cell are significantly lower than the levels of tRNAs or rRNAs, further work is necessary to improve the sensitivity of the SDP approach.

ACKNOWLEDGMENTS

Financial support for this work was provided by the National Institutes of Health, the National Science Foundation and the University of Cincinnati.

REFERENCES

1. Lehninger, A., Nelson, D., and Cox, M., *Principles of Biochemistry*, W.H. Freeman & Company, New York, 2000.
2. Johnston, W., Unrau, P., Lawrence, M., Glasner, M., and Bartel, D., RNA-catalyzed RNA polymerization: Accurate and general RNA-templated primer extension. *Science* 292, 1319–1325, 2001.
3. Limbach, P., Crain, P., and McCloskey, J., Summary: The modified nucleosides of RNA. *Nucleic Acids Res.* 22, 2183–2196, 1994.
4. Eddy, S., Non-coding RNA genes and the modern RNA world. *Nature Rev. Genet.* 2, 919–929, 2001.

5. Großhans, H. and Filipowicz, W., The expanding world of small RNAs. *Nature* 451, 414–416, 2008.

6. Mattick, J. and Makunin, I., Non-coding RNA. *Hum. Mol. Genet.* 15, R17–R29, 2006.

7. Jàdy, B. and Kiss, T., A small nucleolar guide RNA functions both in 2′-O-ribose methylation and pseudouridylation of the U5 spliceosomal RNA. *EMBO J.* 20, 541–551, 2001.

8. Blum, B. and Simpson, L., Guide RNAs in kinetoplastid mitochondria have a nonencoded 3′-oligo(U) tail involved in recognition of the preedited region. *Cell* 62, 391–397, 1990.

9. Klattenhoff, C. and Theurkauf, W., Biogenesis and germline functions of piRNAs. *Developement* 135, 3–9, 2008.

10. Lau, N., Seto, A., Kim, J., Kuramochi-Miyagawa, S., Nakano, T., Bartel, D., and Kingston, R., Characterization of the piRNA complex from rat testes. *Science* 313, 363–367, 2006.

11. Gillet, R. and Felden, B., Emerging views on tmRNA-mediated protein tagging and ribosome rescue. *Mol. Microbiol.* 42, 879–885, 2001.

12. Walter, P., Ibrahimi, I., and Blobel, G., Translocation of proteins across the endoplasmic reticulum I. signal recognition protein (SRP) binds to in-vitro-assembled polysomes synthesizing secretory protein. *Cell Biol.* 1981, 91, 545–550, 1981.

13. Crick, F. and Crook, E., ed., *The Structure of Nucleic Acids and Their Role in Protein Synthesis*, Cambridge University Press, London, 1957.

14. Hoagland, M., Stephenson, M., Scott, J., Hecht, L., and Zamecnik, P.J., A soluble ribonucleic acid intermediate in protein synthesis. *J. Biol. Chem.* 231, 241–257, 1958.

15. Marquèz, V. and Nierhaus, K., tRNA and synthetases, in *Protein Synthesis and Ribosome Structure*, K. Nierhaus and D. Wilson, eds, Wiley-VCH Verlag GmbH & Co., KGaA, Weinheim, pp. 145–167, 2004.

16. Watanabe, Y., Tsuruit, H., Ueda, T., Furushima, R., Takamiya, S., Kita, K., Nishikawa, K., and Watanabe, K., Primary and higher order structures of Nematode (*Ascaris Suum*) mitochondrial tRNAs lacking either the T or D stem. *J. Biol. Chem.* 269, 22902–22906, 1994.

17. Auffinger, P. and Westhof, E., Effects of pseudouridylation on tRNA hydration and dynamics: A theoretical approach, in *Modification and Editing of RNA*, H. Grosjean and R. Benne, eds, ASM, Washington, DC, 1998.

18. Kim, S., Suddath, F., Quigley, G., McPherson, A., Sussman, J., Wang, A., Seeman, N. and Rich, A., Three-dimensional tertiary structure of yeast phenylalanine transfer RNA. *Science* 185, 435–440, 1974.

19. Moras, D., Comarmond, M., Fischer, J., Weiss, R., Thierry, J., Ebel, J., and Giegé, R., Crystal structure of yeast tRNA^{Asp}. *Nature* 288, 669–674, 1980.

20. Grosjean, H., Modification and editing of RNA: An overview, in *Fine-tuning of RNA Functions by Modification and Editing*, H. Grosjean, ed., Springer, Heidelberg, Berlin, pp. 1–22, 2005.

21. Das, A., Klaver, B., and Berkhout, B., Reduced replication of human immunodeficiency virus type 1 mutants that use reverse transcription primers other than the natural tRNA$_3^{\text{Lys}}$. *J. Virol.* 69, 3090–3097, 1995.

22. Gollnick, P. and Babitzke, P., Transcription attenuation. *Biochim. Biophys. Acta* 1577, 240–250, 2002.

23. Wendrich, T., Blaha, G., Wilson, D., Marahiel, M., and Nierhaus, K., Dissection of the mechanism for the stringent factor RelA. *Mol. Cell* 10, 779–788, 2002.

24. Kannangara, C., Gough, S., Bruyant, P., Hoober, J., Kahn, A., and Wettstein, D., tRNAGlu as a cofactor in δ-aminolevulinate biosynthesis: Steps that regulate chlorophyll synthesis. *Trends Biochem. Sci.* 13, 139–143, 1988.

25. Hegde, S. and Shrader, T., FemABX family members are novel nonribosomal peptidyltransferases and important pathogen-specific drug targets. *J. Biol. Chem.* 2001, 6998–7003, 2001.

26. Plapp, R. and Strominger, J., Biosynthesis of the peptidoglycan of bacterial cell wall. XVII. Biosynthesis of peptidoglycan and of interpeptide bridge in lactobacillus viridescens. *J. Biol. Chem.* 245, 3667–3674, 1970.

27. Dunstan, H., Young, L., and Sprague, K., TFIIIR is an isoleucine tRNA. *Mol. Cell Biol.* 14, 3588–3595, 1994.

28. Varshavsky, A., The N-end rule: Functions, mysteries, uses. *Proc. Natl. Acad. Sci. USA* 93, 12142–12149, 1996.

29. Florentz, C., Molecular investigations on tRNAs involved in human mitochondrial disorders. *Biosci. Rep.* 22, 81–98, 2002.

30. Graeber, M. and Müller, U., Recent developments in the molecular genetics of mitochondrial disorders. *J. Neurol. Sci.* 153, 251–263, 1998.

31. Garrett, R., Douthwaite, S., and Liljas, A., *The Ribosome: Structure, Function, Antibiotics and Cellular Interaction*, ASM, Washington, DC, 2000.

32. Lafontaine, D. and Tollervey, D., Ribosomal RNA, in *Encyclopedia of Life Sciences*, Nature Publishing Group, London, pp. 1–7, 2001.

33. Holley, R. and Merril, S., Countercurrent distribution of an active ribonucleic acid. *J. Am. Chem. Soc.* 81, 753, 1959.
34. Warner, R. and Vaimberg, P., Countercurrent fractionation of ribonucleic acids. *Fed. Proc.* 17, 331, 1958.
35. Raymond, S., Compact countercurrent distribution apparatus. *Anal. Chem.* 1958, 30, 1214–1216.
36. RajBhandary, U. and Köhler, C., Early days of tRNA research: Discovery, function, purification and sequence analysis. *J. Biosci.* 31, 439–451, 2006.
37. Gillam, I., Millward, S., Blew, D., Tigerstrom, M., and Tener, G., The separation of soluble ribonucleic acids on benzoylated diethylaminoethylcellulose. *Biochemistry* 6, 3043–3056, 1967.
38. Nishimura, S., Harada, F., Narushima, U., and Seno, T., Purification of methionine-, valine-, phenylalanine- and tyrosine-specific tRNA from *Escherichia coli*. *Biochem. Biophys. Res. Commun.* 142, 133–148, 1967.
39. Pearson, R., Weiss, J., and Kelmers, A., Improved separation of transfer RNA's on polychlorotrifuroethylene-supported reversed-phase chromatography columns. *Biochim. Biophys. Acta* 228, 770–774, 1971.
40. Holms, W., Hurd, R., Reid, B., Rimerman, R., and Hatfield, G., Separation of transfer ribonucleic acid by sepharose chromatography using reverse salt gradients. *Proc. Natl. Acad. Sci. USA* 72, 1068–1071, 1975.
41. Cayama, E., Yepez, A., Rotondo, F., Bandeira, E., Ferreras, A., and Triana-Alonso, F., New chromatographic and biochemical strategies for quick preparative isolation of tRNA. *Nucleic Acids Res.* 28, e64, 2000.
42. Zhu, R., Ching, W., Chung, H., Rhee, S., and Stadman, T., Purification of individual tRNAs using a monoclonal anti-AMP antibody affinity column. *Anal. Biochem.* 161, 460–466, 1987.
43. Ribeiro, S., Nock, S., and Sprinzl, M., Purification of aminoacyl-tRNA by affinity chromatography on immobilized *Thermus thermophilus* EF-Tu-GTP. *Anal. Biochem.* 228, 330–335, 1995.
44. Chinali, G., Isolation of tRNA isoacceptors by affinity chromatography with immobilized elongation factor Tu from *Escherichia coli*. *J. Biochem. Biophys. Methods* 34, 1–10, 1997.
45. Garel, J., Garber, R., and Siddiqui, M., Transfer RNA in posterior silk gland of *Bommbyx mori*: Polyacrylamide gel mapping of mature transfer RNA, identification and partial structural characterization of major isoacceptor species. *Biochemistry* 16, 3618–3624, 1997.
46. Kanduc, D. Fractionation of rat liver tRNA by reversed-phase high performance liquid chromatography: Isolation of Iso-tRNAs(Pro). *Prep Biochem* 24, 167–174, 1994.
47. Kumazawa, Y., Yokogawa, T., Hasegawa, E., Miura, K., and Watanabe, K., The aminoacylation of structurally variant phenylalanine tRNAs from mitochondria and various nonmitochondrial sources by bovine mitochondrial phenylalanyl-tRNA synthetase. *J. Biol. Chem.* 264, 13005–13011, 1989.
48. Kumazawa, Y., Yokogawa, T., Tsurui, H., Miura, K., and Watanabe, K., Effect of the higher-order structure of tRNAs on the stability of hybrids with oligodeoxyribonucleotides: Separation of tRNA by an efficient solution hybridization. *Nucleic Acids Res.* 20, 2223–2232, 1992.
49. Yokogawa, T., Kumazawa, Y., Miura, K., and Watanabe, K., Purification and characterization of two serine isoacceptor tRNAs from bovine mitochondria by using a hybridization assay method. *Nucleic Acids Res.* 17, 2623–2638, 1989.
50. Dong, H., Nilsson, L., and Kurland, C., Co-variation of tRNA abundance and codon usage in *Escherichia coli* at different rates. *J. Mol. Biol.* 260, 649–663, 1996.
51. Petyuk, V., Zenkova, M., Giege, R., and Vlassov, V., Hybridization of antisense oligonucleotides with the 3′ part of tRNA[Phe]. *FEBS Lett.* 444, 217–221, 1999.
52. Mir, K., Southern, E., Determining the influence of structure on hybridization using oligonucleotide arrays. *Nat. Biotechnol.* 17, 788–792, 1999.
53. Buvoli, A., Buvoli, M., and Leinwand, L., Enhanced detection of tRNA isoacceptors by combinatorial oligonucleotide hybridization. *RNA* 6, 912–918, 2000.
54. Miyauchi, K., Ohara, T., and Suzuki, T., Automated parallel isolation of multiple species of non-coding RNAs by the reciprocal circulating chromatography method. *Nucleic Acids Res.* 35, e24, 2007.
55. Michikawa, Y., Hofhaus, G., Lerman, L., and Attardi, G., Comprehensive, rapid and sensitive detection of sequence variants of human mitochondrial tRNA genes. *Nucleic Acids Res.* 25, 2455–2463, 1997.
56. Hossain, M. and Limbach, P. A., Mass spectrometry-based detection of transfer RNAs by their signature endonuclease digestion products. *RNA* 13, 295–303, 2007.
57. Hossain, M. and Limbach, P. A., Multiple endonucleases improve MALDI-MS signature digestion product detection of bacterial transfer RNAs. *Anal. Bioanal. Chem.* 394, 1125–1135, 2009.
58. Meng, Z. and Limbach, P. A., Quantitation of ribonucleic acids using [18]O labeling and mass spectrometry. *Anal. Chem.* 77, 1891–1895, 2005.

59. Entelis, N., Kolesnikova, O., Martin, R., and Tarassov, I., RNA delivery into mitochondria. *Adv. Drug Deliv. Rev.* 49, 199–215, 2001.

60. Schneider, A., Import of RNA into mitochondria. *Trends Cell Biol.* 4, 282–286, 1994.

61. Rinehart, J., Krett, B., Rubio, M., and Alfonso, J., Soll, D., *Saccharomyces cerevisiae* imports the cytosolic pathway for Gln-tRNA synthesis into the mitochondrion. *Genes Dev.* 19, 583–592, 2005.

62. Martin, R., Schneller, J., Stahl, A., and Dirheimer, G., Import of nuclear deoxyribonucleic acid coded lysine-accepting transfer ribonucleic acid (anticodon C-U-U) into yeast mitochondria. *Biochemistry* 18, 4600–4605, 1979.

63. Boutet, S., Vazquez, F., Liu, J., Beclin, C., Fagard, M., Gratias, A., Morel, J. B., Crete, P., Chen, X., and Vaucheret, H., Arabidopsis HEN1: A genetic link between endogenous miRNA controlling development and siRNA controlling transgene silencing and virus resistance. *Curr. Biol.* 13, 843–848, 2003.

64. Katiyar-Agarwal, S., Gao, S., Vivian-Smith, A., and Jin, H., A novel class of bacteria-induced small RNAs in Arabidopsis. *Genes Dev.* 21, 3123–3134, 2007.

65. Li, J., Yang, Z., Yu, B., Liu, J., and Chen, X., Methylation protects miRNAs and siRNAs from a 3'-end uridylation activity in Arabidopsis. *Curr. Biol.* 15, 1501–1507, 2005.

66. Yu, B., Yang, Z., Li, J., Minakhina, S., Yang, M., Padgett, R. W., Steward, R., and Chen, X., Methylation as a crucial step in plant microRNA biogenesis. *Science* 307, 932–935, 2005.

13 Electrospray Ionization Mass Spectrometry for the Direct Analysis of Modified Nucleosides in Small RNAs

C. Callie Coombs and Patrick A. Limbach

CONTENTS

13.1 INTRODUCTION

Ribonucleic acid (RNA) has long been known to contain modified nucleosides in addition to the four major bases. Modified nucleosides, which are present in all types of RNA, are particularly prevalent in tRNAs where there are 70 different nucleoside modifications that only appear within these small RNAs. The different modifications were originally summarized by Limbach et al.[1] and a continuously updated database can be found at http://library.med.utah.edu/RNAmods/. Mass spectrometry (MS) is a valuable tool for their study because all modified nucleosides, with the exception of pseudouridine, result in a change in mass from the parent major nucleoside.

Nucleosides are modified posttranscriptionally at precise positions by the action of RNA-modifying enzymes.[2] A number of these enzymes have been identified; for example, 31 different enzymes have been identified that modify cytosolic tRNAs in *Saccharomyces cerevisiae*. Further, 18 enzymes are required for modifying ribosomal RNA in *Escherichia coli*.[3] Modifying enzymes in different organisms often share conserved sequences, suggesting evolutionary significance.[2] While a precise role for each modification has not yet been discovered, there have been reports elucidating the functional significance of specific modified nucleosides in the RNA in which they reside.[4–7] For example, Ofengand hypothesized that pseudouridine, the most abundant modified nucleoside, serves as "molecular glue" that stabilizes RNA conformations through an additional hydrogen bond donor.[7]

There have been numerous studies regarding a phenotypic change in an organism based solely on a difference in the modification of a single nucleoside. The majority of known nucleoside-related diseases are caused by the absence of a modified nucleoside in mitochondrial tRNAs. Mitochondrial myopathy, encephalopathy, lactic acidosis, and stroke (MELAS) patients are missing the modified nucleoside 5-taurinomethyluridine at the wobble position of the tRNA, while the non-MELAS cohort has this modification.[8] Furthermore, MERRF, or myoclonic epilepsy and ragged-red fiber disease, is a disease that results when a mutation occurs in the DNA coding for mitochondrial tRNA^{Lys}. Specifically, it is a point mutation (A8344G), which corresponds to the wobble position of the mature tRNA.[9] As a result, instead of uridine, cytidine is present, which is unable to be modified into the 5-taurinomethyl-2-thiouridine found in wild-type cells.[10] While mature tRNA molecules always contain modifications, there are other types of RNA, such as 5S ribosomal RNA (rRNA), that may not contain a modification, depending on the organism from which they originate. For example, 5S rRNA in *E. coli* contains no modified nucleosides, whereas 5S rRNA in *Sulfolobus solfataricus* and *Pyrodictium occultum* each contains two modified nucleosides.[6,11] Some microRNAs (miRNAs) and short interfering RNA (siRNA) contain methylated nucleosides, such as those enzymatically processed by the enzyme *HEN1*.[12–14]

Most MS methods for determining posttranscriptional modifications of RNA at the nucleoside level are based on the use of a separation technique, primarily via on-line chromatography [e.g., liquid chromatography (LC)-MS].[15]

In this chapter, we describe a direct-infusion electrospray ionization (ESI)-MS approach that can serve as a screening technique for the presence of modified nucleosides from small RNAs.

13.2   EXPERIMENT

13.2.1   MATERIALS

E. coli tRNA^{Tyr1}, adenosine, cytidine, nuclease P1, and venom phosphodiesterase I were purchased from Sigma (St. Louis, MO, USA). Alkaline phosphatase was purchased through Worthington (Lakewood, NJ, USA). All solvents were high performance liquid chromatography (HPLC) grade and purchased from Tedia Company (Fairfield, OH, USA). Nanopure water (18 MΩ) was obtained from a Barnstead (Dubuque, IA, USA) System and was autoclaved prior to use.

13.2.2   ENZYMATIC DIGESTION

RNA samples were digested to nucleosides according to a previously described protocol.[16] The concentration of RNA was estimated by measuring at 260 nm using a Shimadzu UV-visible spectrophotometer (Columbia, MD, USA). Samples were heated using a Thermolyne 17600 (Dubuque, Iowa, USA) dri-bath. Subsequent incubations with different enzymes (nuclease P1, venom phosphodiesterase, and alkaline phosphatase) were performed using either an Innova 4000 (Edison, NJ. USA) incubator shaker at 100 rpm or the Thermolyne dri-bath. Solutions used during the digestion (0.1 M ammonium acetate, pH 5.3, and 1.0 M ammonium bicarbonate) were filter-sterilized using 0.20 µm filters. Once the digestion was complete, RNA samples were stored at −80°C.

The completeness of a digestion was verified using a previously described procedure, with minor variations.[17] A 4.6 mm × 25.0 cm Supelcosil LC-18S column, with 5 µm particle size and 120 Å pore size, and an LC-18-S guard column were used (Supelco, Bellefonte, PA, USA). A nucleoside test mix, also purchased through Supelco, was used to verify column performance. A Hitachi D-7000 with an L-7100 pump and L-7400 UV detector (San Jose, CA, USA) was used at a flow rate of 2.0 mL/min with UV detection at 254 nm. In the case of tRNA^{Val}, all unmodified and modified nucleosides yielded peaks near predicted retention times, within ±1 min, with the exception of 5,6-dihydrouridine, which does not possess a significant chromophore.[18]

13.2.3 Direct-Infusion Electrospray MS

Samples were analyzed using a Perseptive Biosystems (Foster City, CA, USA) Mariner electrospray time-of-flight (TOF) mass spectrometer in the positive ion mode. Adenosine and cytidine standards were used to optimize instrumental parameters. Products of the enzymatic digest, dissolved in methanol, autoclaved water, and formic acid (50:49.9:0.1, v/v) were directly injected at 0.500 μL/min using a Harvard Apparatus (Holliston, MA, USA) syringe pump. Control experiments were performed in the same manner.

13.3 RESULTS AND DISCUSSION

13.3.1 Direct-Infusion Analysis of Modified Nucleosides

The motivation for this study was to determine the analytical characteristics of a direct-infusion ESI approach for screening for modified nucleosides from small RNAs. To characterize this approach, standard tRNA digests were generated from known isoaccepting tRNAs. First, a tRNA^{Tyr} nucleoside digest was directly infused into the ESI-TOF mass spectrometer. After the ESI-MS conditions for analyzing nucleosides were optimized, tRNA digestion mixtures were analyzed at various concentrations all dissolved in 50% methanol, 49.9% water, and 0.1% formic acid. From a concentration range of 0.85–17 μg/mL of tRNA, five different modified nucleosides could be detected above a signal-to-noise ratio of 3:1 (Figure 13.1). Control experiments were performed to verify the origin of non-nucleoside peaks found in the mass spectrum. As expected, these peaks could be components of the digestion or electrospray buffers. None of the peaks detected during control experiments interfered with the peaks arising from known nucleoside modifications in tRNA^{Tyr}. However, because interferences from the buffer or digestion may hamper detection of nucleoside modifications, it is essential to verify all detected ions before analysis of unknown samples .

The known modifications in *E. coli* tRNA^{Tyr} are pseudouridine (Ψ), 5-methyluridine (m^{5}U), 4-thiouridine (s^{4}U), 2′-*O*-methylguanosine (Gm), queuosine (Q), and N6-(3-methyl-2-butenyl)-2-methylthioadenosine (ms^{2}i^{6}A). The limit of detection for ms^{2}i^{6}A is nearly 10 times lower than the

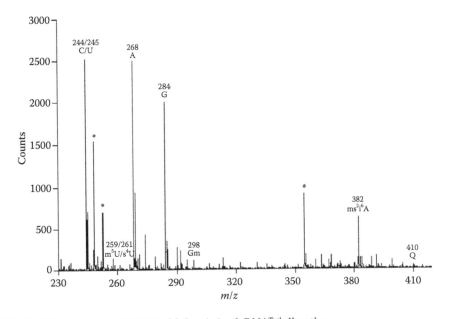

FIGURE 13.1 Direct-infusion ESI-MS of 8.5 μg/mL of tRNA^{Tyr} digestion.

TABLE 13.1
The Nucleosides Found in tRNA^Tyr1, the Number of Times They Occur, the Relative Mole Ratios, and the Detection Limits (pg) Determined in this Study

Nucleoside	Abundance	Relative Mole Ratio	Detection Limit (pg)
Cytidine	27	1	—
Uridine	12	0.44	—
Guanosine	12	0.44	—
Adenosine	18	0.67	—
Pseudouridine	2	0.07	—
4-Thiouridine	1	0.04	400
2-O-methylguanosine	1	0.04	460
Queuosine	1	0.04	640
2-Methylthio-N^6-isopentenyladenosine	1	0.04	59
Ribosylthymine (5-methyluridine)	1	0.04	410

limits of detection for m^{5}U, s^{4}U, Gm, and Q (Table 13.1). In all cases, these modified nucleosides can be detected even in the presence of the more concentrated major (i.e., unmodified) nucleosides. Further, these nucleosides generated a linear response over all concentrations investigated, except in the most concentrated mixture (42.5 μg/mL), when the detector had become saturated (Figure 13.2).

The data revealed a correlation between the sensitivity of each nucleoside and its pK_a value as expected because the ESI process occurs in solution.[19] The pH of the electrospray buffer was approximately 2.5. The nucleoside with the greatest slope, ms^{2}i^{6}A, has a pK_a of ~2.9, whereas the nucleoside with the lowest slope, m^{5}U, has a pK_a of −3.0.[20,21] Thus, nucleosides with higher pK_as yielded the best response for these analyses. These data also reveal that, generally, uridine-derived nucleosides will be more difficult to detect by this approach because uridine nucleosides are more difficult to protonate than other nucleosides. Their detection may be improved by use of electrospray buffers at lower pH to favor the formation of the positive ion.

Pseudouridine, the only mass-silent-modified nucleoside, requires derivatization or HPLC separation for confirmation. This was the only modified nucleoside present in tRNA^Tyr1 that could not be unambiguously identified. In general, any nucleoside modifications resulting in the same shift in molecular weight from the major nucleoside would not be distinguishable by this method. For example, a common nucleoside modification is methylation. By direct-infusion ESI-MS, the three modified adenosine isomers 1-methyladenosine, 2-methyladenosine, and 2′-O-methyladenosine would not be immediately distinguishable. If isomers yielded different major fragment ions, the approach characterized here would not preclude an tandem mass spectrometry (MS/MS) step for differentiation of the isomers present.[22,23]

13.3.2 COMPARISON TO LC-MS

Direct-infusion MS allowed detection of all mass-shifting-modified nucleosides from a tRNA sample. The modified nucleosides could be detected when the concentration of the tRNA from which they were digested was 0.85 μg/mL or greater. The molecular weight of tRNA^Tyr1 is 27,352 Da, so at a modified nucleoside molecular weight of 280 Da (chosen to represent an average modified nucleoside), the concentration in terms of a single modified nucleoside corresponds to 9 ng/mL. This concentration can be used to find a detection limit for nucleoside detection by taking into account

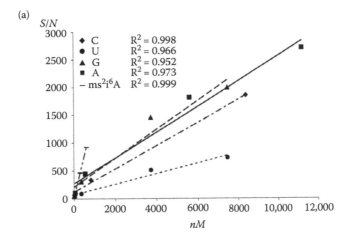

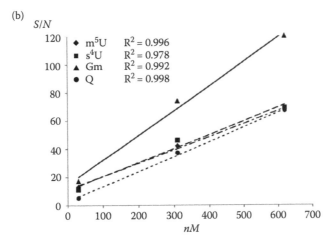

FIGURE 13.2 Response curve for the following nucleosides detected in the nucleoside digestion mixture of tRNA^{Tyr1} (a) C (◆), U (●), G (▲), A (■), ms^{2}i^{6}A (–) and (b) m^{5}U (◆), s^{4}U (■), Gm (▲), and Q (●).

the flow rate (0.500 μL/min) and the experimental time for analysis (10 min). The limit of detection for finding a modified nucleoside in a tRNA molecule by the approach described here is approximately 0.04 ng, starting with 4 ng of purified tRNA. The starting material dynamic range of the direct-infusion method is 4–200 ng.

As developed here, modified nucleosides can be detected among major nucleosides for tRNA samples. tRNAs are relatively small RNA molecules, ranging from 70 to 90 nucleotides. A modified nucleoside would be more abundant, and thus easier to detect among the major nucleosides in a tRNA (1 out of 80) compared to a larger RNA, such as a rRNA, where there may be only one modification among several hundreds of unmodified nucleosides (Table 13.2).

Takeda et al. have characterized modified nucleosides using Frit-fast atom bombardment (FAB) LC-MS starting with 3–10 μg of undigested tRNA,[24] and Banks et al. achieved a detection limit of 4 ng of undigested RNA in the detection of nucleosides using ultrasonically assisted LC-ESI-MS.[25] Although the direct ESI-MS approach described here has similar limits of detection, the potential advantage and application of this approach would be in higher throughput screening of samples for the presence or absence of modified nucleosides using an instrumental approach like the Advion Nanomate™.[26]

TABLE 13.2
Modified Nucleosides in *E. coli* 16S rRNA and Mole Abundances, Relative to Guanosine

Nucleoside	Relative Mole Ratio
Guanosine	1
Pseudouridine	0.002
7-Methylguanosine	0.002
2-Methylguanosine	0.006
5-Methylcytidine	0.004
N^4,2'-O-dimethylcytidine	0.002
3-Methyluridine	0.002
N^6,N^6-dimethyladenosine	0.004

13.4 CONCLUSION

This work has characterized a direct-infusion ESI-MS approach for detection of modified nucleosides. Potential clinical applications to this method arise for any disease that is linked to the presence or absence of a nucleoside modification. Experiments would involve the direct-infusion analysis of small RNA hydrolyzates isolated from speculated diseased cells, which could be compared against wild-type cells. Such an approach may allow rapid diagnostic screening, especially if modified nucleosides are found to be appropriate biomarkers for specific disease state differentiation. This approach would best be applicable in the determination of nucleoside modifications in relatively simple systems, while LC-MS remains the preferential technique for more complex characterizations. In addition, when complicating isomers of modified nucleosides are present, MS-MS approaches could be developed for greater specificity.

ACKNOWLEDGMENTS

Financial support for this work was provided by the National Institutes of Health (GM58843) and the University of Cincinnati.

REFERENCES

1. Limbach, P.A., Crain, P.F., and McCloskey, J.A., Summary: The modified nucleosides of RNA, *Nucleic Acids Res.* 22, 2183, 1994.
2. Ferré-D'Amaré, A., RNA-modifying enzymes, *Curr. Opin. Struct. Biol.* 13, 49, 2003.
3. Grosjean, H., Fine-tuning of RNA functions by modification and editing, *Topics Curr. Genet.* 12, 1–6, 2005.
4. Del Campo, M., Recinos, C., Yanez, G., Pomerantz, S.C., Guymon, R., Crain, P.F., McCloskey, J.A., and Ofengand, J., Number, position, and significance of the pseudouridines in the large subunit ribosomal RNA of *Haloarcula marismortui* and *Deinococcus radiodurans*, *RNA* 11, 210, 2005.
5. Helm, M., Post-transcriptional nucleotide modification and alternative folding of RNA, *Nucleic Acids Res.* 34, 721, 2006.
6. Kowalak, J.A., Dalluge, J.J., McCloskey, J.A., and Stetter, K.O., Role of posttranscriptional modification in stabilization of transfer RNA from hyperthermophiles, *Biochemistry* 33, 7869, 1994.
7. Ofengand, J.F.L., Ribosomal RNA pseudouridines and pseudouridine synthases, *FEBS Lett.* 514, 17, 2002.
8. Kirino, Y., Goto, Y., Campos, Y., Arenas, J., and Suzuki, T., Specific correlation between the wobble modification in mutant tRNAs and clinical features of a human mitochrondrial disease, *PNAS* 102, 7127, 2005.

9. Shoffner, J.M., Lott, M.T., Lezza, A.M.S., Seibel, P., Ballinger, S.W., and Wallace, D.C., Myoclonic epilepsy and ragged-red fiber disease (MERRF) is associated with a mitochondrial DNA tRNALys mutation, *Cell* 61, 931, 1990.
10. Yasukawa, T., Kirino, Y. Ishii, N., Holt, I.J., Jacobs, H.T., Makifuchi, T., Fukuhara, N., Ohta, S., Suzuki, T., and Watanabe, K., Wobble modification deficiency in mutant tRNAs in patients with mitochondrial diseases, *FEBS Lett.*, 579, 2948, 2005.
11. Bruenger, E., Kowalak, J.A., Kuchino, Y., McCloskey, J.A., Mizushima, H., Stetter, K.O., and Crain, P.F., 5S rRNA modification in the hyperthermophilic archaea *Sulfolobus solfataricus* and *Pyrodictium occultum*, *FASEB J.* 7, 196, 1993.
12. Li, J., Yang, Z. Yu, B., Liu, J., and Chen, X., Methylation protects siRNAs and miRNAs from a 3′-end uridylation activity in arabidopsis, *Curr. Biol.* 15, 1501, 2005.
13. Yang, Z.E., Y.W., Yu, B., and Chen, X., HEN1 recognizes 21–24 small RNA duplexes and deposits a methyl group onto the 2′ OH of the 3′ terminal nucleotide, *Nucleic Acids Res.* 34, 667, 2006.
14. Yu, B., Yang, Z. Li, J., Minakhina, S., Yang, M., Padgett, R.W., Steward, R., and Chen, X., Methylation as a crucial step in plant microRNA Biogenesis, *Science* 307, 932 2005.
15. McCloskey, J.A., Constituents of nucleic acids: Overview and strategy, *Methods Enzymol.* 193, 771, 1990.
16. Crain, P.F., Preparation and enzymatic hydrolysis of DNA and RNA for mass spectrometry, *Methods Enzymol.* 193, 782, 1990.
17. Pomerantz, S.C. and McCloskey, J.A., Analysis of RNA hydrolyzates by liquid chromatography-mass spectrometry, *Methods Enzymol* 193, 796, 1990.
18. Dalluge, J.J., Hashizume, T., and McCloskey, J.A., Quantitative measurement of dihydrouridine in RNA using isotope dilution liquid chromatography, *Nucleic Acids Res.* 24, 3242, 1996.
19. Smith, R.D., Loo, J.A., Edmonds, C.G., Barinaga, C.J., and Udseth, H.R., New developments in biochemical mass spectrometry: Electrospray ionization, *Anal. Chem.* 62, 882, 1990.
20. Solaris V 4.67, Scifinder Scholar™, American Chemical Society, 2006.
21. Poulter, C.D. and Frederick, G.D., Uracil and its 4-hydroxy-1(*H*) and 2-hydroxy-3(*H*) protomers, *Tetrahedron Lett.* 26, 2171, 1975.
22. Gregson, J.M. and McCloskey, J.A., The dissociation chemistry of permethylated guanosines as articulated by MS/MS, *Tetrahedron Lett.* 34, 6665, 1993.
23. Zhou, S., Sitaramaiah, D., Pomerantz, S.C., Crain, P.F., and McCloskey, J.A., Tandem mass spectrometry for structure assignments of Wye nucleosides from transfer RNA, *Nucleos. Nucleot. Nucleic Acids* 23, 41, 2004.
24. Takeda, N., Masashi, N., Yoshizumi, H., and Tatematsu, A., structural characterization of modified nucleosides in tRNA hydrolysates by frit-fast atom bombardment liquid chromatography/mass spectrometry, *Biol. Mass Spectrom.* 23, 465, 1994.
25. Banks, J.F., Shen, S., Whitehouse, C.M., and Fend, J., Ultrasonically assisted electrospray ionization for LC/MS determination of nucleosides from a transfer RNA digest, *Anal. Chem.* 66, 406, 1994.
26. Van Pelt, C.K., Zheng, S., and Henion, J., Characterization of a fully automated nanoelectrospray system with mass spectrometric detection for proteomic analyses, *J. Biomol. Tech.* 13, 72, 2002.

14 LC-MS/MS for the Examination of the Cytotoxic and Mutagenic Properties of DNA Lesions *In Vitro* and *In Vivo*

Chunang Gu, Haizheng Hong, and Yinsheng Wang

CONTENTS

14.1 INTRODUCTION

The human genome is frequently assaulted by endogenous and exogenous sources.[1] To gain insights into the biological consequences of DNA damage, it is important to ask how effectively a DNA lesion inhibits DNA replication and how frequently it causes mutations. These questions are often addressed by first constructing oligodeoxyribonucleotides (ODNs) harboring a structurally defined DNA lesion. The ODNs can then be treated with purified DNA polymerases and the resulting replication products can be assessed by using the traditional gel electrophoresis techniques.

Pre-steady-state and steady-state kinetic assays are commonly employed for determining the polymerase stalling and miscoding properties of DNA lesions *in vitro*.[2–8] While the former time scale provides detailed mechanistic information on individual steps in the polymerization and proofreading pathways, the latter is used to determine nucleotide misincorporation rates for polymerases acting on substrates with a structurally defined DNA lesion in different sequence contexts.[9] In these experiments, the primer–template complex is incubated with a DNA polymerase in the presence of an individual dNTP at a concentration allowing for less than 20% of the primer to be extended. The efficiency and fidelity of nucleotide incorporation are then determined by measuring the velocities for the incorporation of one type of nucleotide at a time and fitting the velocity data with the Michaelis–Menton equation[5,10]:

$$V_{obs} = \frac{V_{max}*[dNTP]}{K_m + [dNTP]}.$$

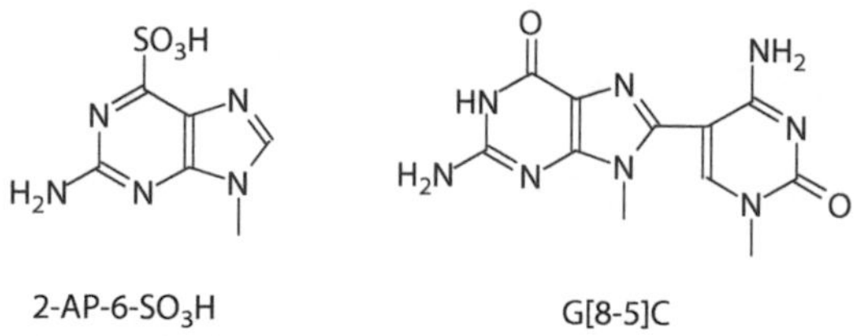

SCHEME 14.1   Structures of 2-AP-6-SO₃H and G[8-5]C discussed in the chapter.

Alternatively, the lesion-carrying ODNs can be inserted into a plasmid, and the plasmid can be transfected into host cells that support its replication. The progeny emanating from the replication of the plasmid can then be isolated from the host cells and subsequently assayed for the change in DNA sequence at or near the original lesion site. The alteration in DNA sequence is often determined by phenotypic selection followed by DNA sequencing, though a new technique has been developed where no selection is needed and the nature and the frequency of the mutation can be determined by ^{32}P-post-labeling and thin-layer chromatography (TLC) analysis.[11]

The advent of sensitive mass spectrometric techniques and the understanding of the fragmentation chemistry of ODNs facilitate the sequencing of ODNs of reasonable lengths by using mass spectrometry (MS). These developments also open up the possibility of employing mass spectrometry (MS) for interrogating the products arising from the replication of lesion-containing ODNs *in vitro* or from a lesion-carrying plasmid genome *in vivo*.[12-14] In this context, Guengerich et al.[12-18] and we recently employed an liquid chromatography-tandem mass spectrometry (LC-MS/MS) method to investigate the multiple bypass mechanisms of polymerases toward DNA lesions *in vitro*.[15-22] The LC-MS/MS method offers efficient determination of the sequences and distributions of various replication products resulting from the polymerase reaction with the mutual presence of all four natural dNTPs.

In this chapter, we will discuss the LC-MS/MS method[22] for the *in vitro* replication studies on damage-containing DNA substrates, and we will employ the yeast polymerase η (pol η)-mediated replication of a 2-aminopurine-6-sulfonic acid (2-AP-6-SO₃H)-containing DNA substrate as an example to illustrate the method (Scheme 14.1). We will then describe the combination of shuttle vector technology with LC-MS/MS for examining the cytotoxic and mutagenic properties of DNA lesions in cells.[12-14]

14.2   LC-MS/MS FOR *IN VITRO* REPLICATION STUDIES

14.2.1   PREPARATION OF 2-AP-6-SO₃H-CONTAINING ODN SUBSTRATES

Construction of ODN substrates housing a site specifically inserted and structurally defined lesion constitutes the first step toward examining the cytotoxic and mutagenic properties of the DNA lesion at the molecular level. Two methods are frequently used for the preparation of lesion-carrying ODN substrates. In one method, the phosphoramidite building block for a modified nucleoside is first synthesized, and modified ODNs can be assembled by automated solid-phase DNA synthesis.[23] The other method involves postsynthetic modification, in which an ODN carrying a reactive group is first prepared and then subsequently treated with a reactive chemical to induce modification of DNA bases in the ODN.[24-26] The first method is often preferred because the lesion of interest can be incorporated at any position, and generally the lesion-carrying ODN can be readily isolated from the synthetic mixture by employing high performance liquid chromatography (HPLC). Because the chemical reaction may be inefficient and/or occur at more than one site, it is sometimes difficult to obtain sufficient and pure products with the second method. In addition, the sequences of the modified ODNs constructed by the second method can be restricted.

To obtain pure ODNs containing a 2-AP-6-SO$_3$H at a specific site, we first synthesized a 6-thioguanine (6-TG)-containing ODN by using the commercially available phosphoramidite building block of 6-thio-2′-deoxyguanosine (Glen Research). After cleavage from the control pore glass support and the removal of protecting groups, the synthetic mixture is purified by employing reversed-phase HPLC. The 6-TG in ODNs is then selectively oxidized to 2-AP-6-SO$_3$H following previously published procedures.[27] The desired 2-AP-6-SO$_3$H-containing substrate is subsequently isolated from the oxidation mixture by HPLC.

14.2.2 IDENTIFICATION OF *IN VITRO* REPLICATION PRODUCTS BY LC-MS/MS

With the availability of the lesion-carrying ODN substrate, the bypass and miscoding properties of the 2-AP-6-SO$_3$H can be assessed by using LC-MS/MS. It is worth noting that the primer–template complex has to be long enough to facilitate the binding of DNA polymerases to the lesion-containing substrate. This constraint, however, introduces some difficulty for examining directly the replication products by MS because these products are often too long for sequencing by MS/MS. A strategy to over come this difficulty is via the use of a uracil-containing primer, that is, d(GCTAGGATCAUAGCG) (Scheme 14.2). After *in vitro* replication and cleavage with uracil DNA glycosylase (UDG), the portion of the primer strand extended by the polymerase can be liberated in short ODNs (Scheme 14.2), which are more amenable to sequencing by MS/MS.[15]

For the *in vitro* polymerase reaction, the uracil-containing primer and the 2-AP-6-SO$_3$H-bearing template (100 pmol each) are annealed and incubated at 37°C in the presence of a mixture of four dNTPs (1 mM each) and a buffer containing 10 mM Tris-HCl (pH 7.5), 5 mM MgCl$_2$, and 7.5 mM DTT. Yeast pol η (1.6 μg) is added and the reaction is continued overnight. The replication reaction is stopped by deactivating the enzyme via heating to 65°C for 10 min, and the resultant mixture is subsequently incubated with UDG (4 units) in a buffer containing 20 mM Tris-HCl (pH 8.0), 1 mM DTT, and 1 mM EDTA at 37°C for 5 h. The UDG cleavage reaction is subsequently quenched by adding piperidine until its final concentration reaches 0.25 M. The mixture is then incubated at 60°C for 1 h, the proteins in the mixtures are removed by chloroform extraction, and the aqueous layer is dried by using a Speed Vac. The hot piperidine treatment induces the formation of strand breaks at the UDG-produced abasic site (Scheme 14.2). The dried residue is redissolved in 100 μl H$_2$O for LC-MS/MS analysis, and a 25-μl aliquot is injected in each LC-MS/MS run.

The replication mixture is separated by using a 0.5 × 150 mm Zorbax SB-C18 column (particle size 5 μm, Agilent Technologies). The 1,1,1,3,3,3-hexafluoro-2-propanol (HFIP) buffer is adopted

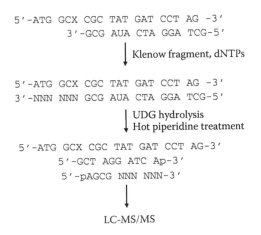

SCHEME 14.2 LC-MS/MS method for sequencing prime extension products of 2-AP-6-SO$_3$H. "X" and "N" represent 2-AP-6-SO$_3$H and random nucleotide incorporated, respectively.

to improve the efficiency in both the HPLC separation and the electrospray ionization of ODNs.[28] The gradient program for HPLC elution is as follows: 0–5 min, 0–20% methanol in 400 mM HFIP (pH is adjusted to 7.0 with the addition of triethylamine); 5–40 min, 20–50% methanol in 400 mM HFIP. The flow rate is set at 6.0 µL/min, which is delivered by using an Agilent 1100 capillary HPLC pump. The effluent from the LC column is directed to an LTQ linear ion-trap mass spectrometer (Thermo Fisher Scientific, San Jose, CA) operated in the negative ion mode. The capillary temperature needs to be maintained at a relatively high temperature, that is, 300°C, to minimize the formation of HFIP adducts of ODNs.

To identify replication products, the mass spectrometer is first set to operate in a full-scan mode and subsequently data-dependent scan mode, where the four most abundant ions found in MS are chosen sequentially for fragmentation to give MS/MS by the instrument control software. The MS/MS data are manually analyzed to determine the sequences of the ODNs. In this regard, understanding the fragmentation chemistry of ODNs is very important for identifying the sequences of the replication products based on MS/MS results. Upon collisional activation in a mass spectrometer, an ODN undergoes cleavages at the N-glycosidic bond and the 3' C–O bond of the same nucleotide to render the $[a_n$–Base$]$ and w_n ions.[29] Mechanistic studies on ODN fragmentation revealed that the proton transfer from the backbone phosphate group to nucleobases and the subsequent loss of the nucleobase are important for the rupture of the 3' C–O bond.[30–32] Because of the low proton affinity of thymine, chain cleavage on the 3' side of thymidine is frequently not observed.[30–32] The presence of the w_n and $[a_n$–Base$]$ ions in the product ion spectrum facilitates the determination of the sequence of an ODN.

A drawback with data-dependent acquisition is that the tandem mass spectra of some replication products are of low quality. In most cases, the system picks up the most abundant ions for MS/MS analysis. In this respect, some tandem mass spectra ions in low abundances are not of high enough quality for complete sequence assignments. These MS/MS data are acquired again by monitoring specifically the fragmentation of the precursor ions. After identifying the ODNs from the replication mixture, the quantification of replication products is carried out by operating the instrument in such a way that the precursor ions for the ODNs housing the extended portions of the primer strand are chosen for fragmentation in the product ion scan mode. The latter mode of analysis provides an improved signal-to-noise ratio for the measurement of replication products.

To illustrate, we use the analysis of the yeast pol η-catalyzed primer extension products of the 2-AP-6-SO₃H-containing substrate. We identified, by LC-MS/MS, 10C, 10A, 10G, 11C, 11T, 11C_T, and 11A_T, which correspond to the 3' components of the full-length replication products. The sequences for the identified products are listed in Scheme 14.3. Other than these full-length products, we can also find the unextended primer (4-mer), one-nucleotide insertion product (5C), and frameshift mutation products (6C, 6A, 8A, and 8C).

```
5'- ATG GCX CGC TAT GAT CCT AG -3'
          3'- GCG Ap - 5'   4mer    22%  ± 2
        3'- C GCG Ap - 5'   5C      1.3% ± 0.0
      3'- C   C GCG Ap - 5'   6C      2.7% ± 0.0
      3'- C   A GCG Ap - 5'   6A      0.6% ± 0.0
    3'- TAC   C GCG Ap - 5'   8C      29%  ± 3
    3'- TAC   A GCG Ap - 5'   8A      6.9% ± 0.7
    3'- TAC CGC GCG Ap - 5'   10C     20%  ± 2
    3'- TAC CGA GCG Ap - 5'   10A     3.9% ± 0.4
    3'- TAC CGG GCG Ap - 5'   10G     0.6% ± 0.0
  3'- ATAC CGC GCG Ap - 5'   11C     1.5% ± 0.0
  3'- ATAC CGT GCG Ap - 5'   11T     0.5% ± 0.0
  3'- TTAC CGC GCG Ap - 5'   11C_T   6.8% ± 0.6
  3'- TTAC CGT GCG Ap - 5'   11T_T   0.4% ± 0.0
  3'- TTAC CGA GCG Ap - 5'   11A_T   2.7% ± 0.3
```

SCHEME 14.3 Replication products of yeast pol η-mediated reaction of 2-AP-6-SO₃H-containing ODNs.

14.2.3 QUANTIFICATION OF REPLICATION PRODUCTS BY LC-MS/MS

Because the hydrophobicity and free energy of solvation vary with the sequences and lengths of the ODNs, which affect their signal intensities in ESI-MS,[33] accurate LC-MS/MS quantification of replication products necessitates the consideration of the difference in ionization efficiencies for these products. This can be achieved by using a relative-ratio method.[22] In this method, one needs to first obtain the standard phosphorylated ODNs. We typically obtain these ODNs by treating 20-nmol unphosphorylated ODNs with 20 U of T4 polynucleotide kinase in a buffer containing 50 mM Tris-HCl (pH 7.5), 10 mM $MgCl_2$, 1 mM ATP, 10 mM DTT, and 25 µg/ml BSA at 37°C for 1 h. Immediately after the reaction, we extract the mixture with chloroform to remove the enzymes, dry the aqueous layer, and isolate the desired 5′-phosphorylated ODNs by using HPLC, in which a 60-min gradient program with 0–40% methanol in 50 mM phosphate buffer (pH 6.8) can be employed. The phosphate buffer is chosen to minimize the loss of terminal phosphate groups from the 5′-phosphorylated ODNs. When necessary, the 5′ phosphate groups in the reaction products can be removed by phosphatase treatment (see *in vivo* replication studies discussed in Section 14.3).

The relative ionization efficiencies of these standard ODNs can then be assessed by using LC-MS/MS. A mixture comprising of 5 pmol each of the standard phosphorylated ODN substrates, which are identified in the mixture of the pol η-mediated replication reaction of the 2-AP-6-SO_3H-containing ODNs (Scheme 14.3), is dispersed in the same buffer as that used in the extension assays. The ODN mixture is then subjected to LC-MS and MS/MS analyses with the same experimental setup as that used for the analysis of the replication mixture. The peak area, found in the total-ion chromatogram (TIC) for the most abundant deprotonated molecular ion for each standard ODN or in the selected-ion chromatogram (SIC) for the formation of three abundant fragment ions of the ODN, was normalized against that of one specific ODN. As depicted in Figure 14.1, the 13 5′-phosphorylated ODNs

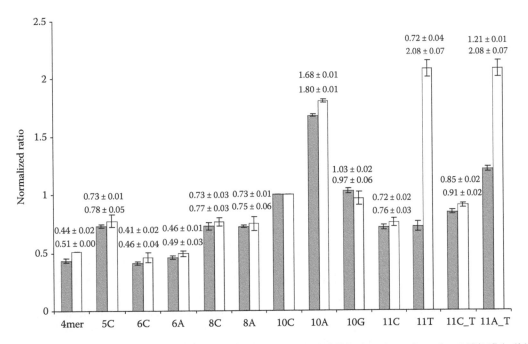

FIGURE 14.1 Ratios of peak areas of individual ODNs to that of 10C in the mixture based on MS/MS (solid bar) and MS (open bar). All authentic ODNs were identified from the replication mixture of 2-AP-6-SO_3H-containing substrate induced by yeast pol η. Error bars represent the standard deviations of results from three independent experiments. The top and bottom numbers listed above the bars are data obtained from the SICs for the formation of three abundant fragment ions in MS/MS and based on the SICs for the formation of the most abundant molecular ion in MS, respectively.

identified from the pol η-mediated replication mixture of the 2-AP-6-SO$_3$H-bearing substrate indeed exhibit substantially different efficiencies in producing the most abundant molecular ions. This result highlights the importance of considering the varied ionization efficiencies of different ODNs in the LC-MS/MS quantification of replication products.

The relative efficiencies for the formation of product ions need to be examined to determine if MS/MS can provide an enhanced signal-to-noise ratio for measuring the relative amounts of different ODNs present in the replication mixture. The formation of three abundant fragment ions from the LC-MS/MS analysis of 5 pmol each of the 13 authentic ODNs is monitored. The ratios obtained for most ODNs are similar to those found for the measurement of molecular ions, and the efficiencies for the formation of three abundant product ions are again significantly different for different ODNs (Figure 14.1). It is worth noting that two isomeric ODN products are found in the pol η-catalyzed replication mixture of 2-AP-6-SO$_3$H-containing substrates (Scheme 14.3), that is, 11T or 11A_T, which exhibit very similar retention times. Thus, MS/MS has to be employed for the quantification of these two products.

In the yeast pol η-mediated reaction mixture of 2-AP-6-SO$_3$H-containing ODNs, 13 replication products can be found (Scheme 14.3). To determine the percentage of the unextended primer (4-mer) in the reaction mixture, one can calculate first the ratio of the peak area for the 4-mer [$A°(4\text{-mer})$] over that for ODN 10C [$A°(10C)$] found in the SICs for the analysis of the mixture of standard ODNs, that is, $R°$ (4-mer):

$$R°(4\text{-mer}) = \frac{A°(4\text{-mer})}{A°(10C)}$$

The corresponding ratio for the analysis of the replication mixture can also be calculated (the SICs for monitoring the replication products are depicted in Figure 14.2):

$$R(4\text{-mer}) = \frac{A(4\text{-mer})}{A(10C)}.$$

$A°$ and A represent the peak areas found in SICs for the analysis of standards and replication samples, respectively. The ratio determined for the replication mixture can then be normalized against that determined for the standards, which gives normalized ratio for the 4-mer, that is, RR(4-mer):

$$RR(4\text{-mer}) = \frac{R(4\text{-mer})}{R°(4\text{-mer})}.$$

The normalized ratio is then calculated for each identified replication product, and the percentage of the 4-mer in the replication mixture can be calculated from the ratio of the normalized ratio for the 4-mer over the sum of the normalized ratios for all the replication products:

$$P(4\text{-mer}) = \frac{RR(4\text{-mer})}{\begin{array}{l}[RR(4\text{-mer}) + RR(5C) + RR(6C) + RR(6A) + RR(8C) + RR(8A) + RR(10C) \\ + RR(10A) + RR(10G) + RR(11C) + RR(11T) + RR(11C\_T) + RR(11A\_T)]\end{array}}.$$

Other replication products present in the reaction mixture can be calculated in the same manner. The percentage of each product in the replication mixture is summarized in Scheme 14.3.

If we consider only the full-length products, the respective frequencies for the insertions of dCMP, dTMP, and dAMP opposite 2-AP-6-SO$_3$H are 29%, 1.2%, and 5.8% (Scheme 14.3). Other than nucleotide misincorporation opposite the lesion sites, we also find a substantial amount of −2 frameshift products in the replication mixtures for 2-AP-6-SO$_3$H-containing substrates (6C, 6A, 8C, and 8A, a total of ~39%, Scheme 14.3). The results from this *in vitro* replication study highlight the

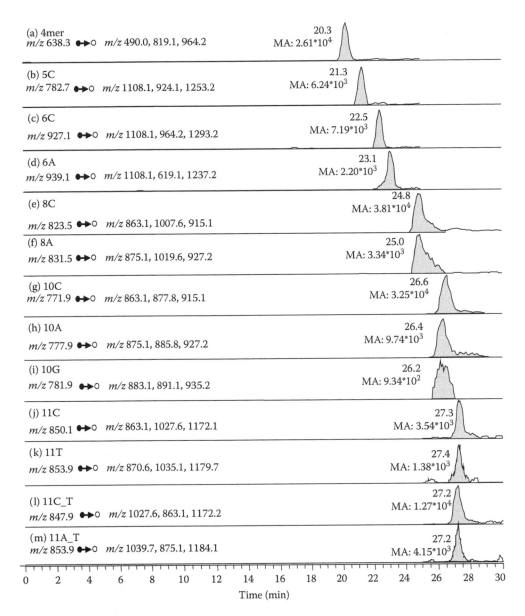

FIGURE 14.2 SICs from the analysis of the replication mixture of the substrate containing 2-AP-6-SO$_3$H by yeast pol η. The integrated peak areas and a summary of the sequences and percentages of replication products are also shown in Scheme 14.3.

usefulness of the LC-MS/MS method in identifying and quantifying replication products resulting from single-base substitutions or from frameshift mutations.

Compared with the traditional gel electrophoresis assays in the DNA adducts mutagenesis study, the LC-MS-based approach is efficient and high-throughput. A single LC-MS/MS experiment facilitates one to gain insights into both the nucleotide incorporation opposite the lesion and the primer extension products beyond the lesion site. Meanwhile, the polymerase reaction with the incubation of all four dNTPs and lesion-containing ODNs, which, relative to the conditions used for steady-state kinetic measurements, mimics better the real replication conditions *in vivo*.

14.3 RESTRICTION ENDONUCLEASE AND MASS SPECTROMETRY (REAMS) ASSAY FOR *IN VIVO* DNA REPLICATION STUDIES

The above LC-MS/MS method can also be extended to examine how a particular DNA lesion affects replication by hindering DNA polymerase and inducing mutations *in vivo*.[12] This method is adapted from the REAP (restriction endonuclease and postlabeling) and CRAB (competitive replication and adduct bypass) assays, which are shuttle vector-based methods first reported by Essigmann et al.[11,34–35] The LC-MS/MS method was developed based on these assays because the entire population of progeny is assayed for bypass efficiency and mutation frequency in these two assays. Thus, the REAMS method can offer statistically important conclusions about how a particular lesion is replicated *in vivo*. In addition, the method is high-throughput and it does not require phenotypic selection.

Since the REAP and CRAB assays were discussed in detail in the original research articles[34,35] and in a recent review,[11] we choose to describe briefly the experimental procedures and discuss our results with the study on the G[8–5]C lesion (structure shown in Scheme 14.1), in which the C8 of guanine is covalently linked with the C5 of its neighboring 3′ cytosine.[8,12,25] For more detailed information about the shuttle vector technology, the readers should consult a recent review by Delaney and Essigmann.[11]

The REAMS assay begins with the construction of a replicable single-stranded genome, that is, M13mp7(L2), housing a site specifically incorporated lesion. We obtained a G[8–5]C-containing ODN from the UVB irradiation of 5-bromocytosine-containing duplex DNA[12,25,36] and ligated the damage-carrying ODN into the EcoRI-linearized M13mp7(L2) genome by using T4 DNA ligase.[11] The ligation scaffolds and unligated linear plasmid DNA can then be degraded by the 3′ → 5′ exonuclease activity of the T4 DNA polymerase (Figure 14.3).[37]

To examine the bypass efficiency with the CRAB assay, a control competitor genome is also prepared by inserting a 19-mer unmodified ODN into the EcoRI-cleaved M13 genome as described

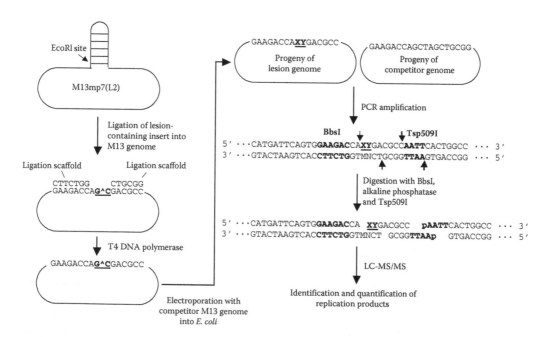

FIGURE 14.3 REAMS assay for assessing the mutagenicity and cytotoxicity of the G[8-5]C lesion *in vivo*. For the two ligation scaffolds, only the portions of the sequence that are complementary to the 16-mer lesion insert are shown. The PCR products and restriction fragments for the competitor genome are not listed.[12]

in CRAB assay (Figure 14.3). The damage-containing genome is then mixed with the competitor genome at a known molar ratio, transformed into wild-type AB1157 *Escherichia coli* cells by electroporation, and allowed for propagation in the host *E. coli* cells for 6 h.[11,34,35] The AB1157 cells are then pelleted and the progeny phages in the supernatant can be collected. Because some lesion-containing genomes are not electroporated into cells, the isolated supernatant also carries some unreplicated genomes. To minimize the effect of residual lesion-containing genomes in subsequent analysis, the progeny/lesion–genome ratio is further increased, by orders of magnitude, via infecting SCS110 *E. coli* with the viable progeny phage.[11,29,30] After culturing the cells at 37°C for 5 h, the progeny phage is isolated and the single-stranded DNA from the progeny phage is extracted.

The region in the resulting progeny genome of interest is amplified by polymerase chain reaction (PCR).[11,34,35] The PCR products are then digested with BbsI, the terminal phosphate groups in the restriction fragments are removed by treatment with shrimp alkaline phosphatase, and the resulting DNA fragments are subjected to further digestion with Tsp509I (Figure 14.3). After the above enzymatic treatments, the DNA fragment of interest from the full-length replication product is liberated as an 8-mer ODN, d(MNGACGCC), where "M" and "N" represent the nucleobases present at the G[8-5]C lesion site after *in vivo* DNA replication (Figure 14.3). Meanwhile, an 11-mer ODN, d(GCTAGCTGCGG), is produced from the PCR product of the competitor genome (Figure 14.3).

After removing the enzymes by chloroform extraction, the restriction digestion mixture is then subjected directly to LC-MS/MS analysis as described above for the analysis of *in vitro* replication products. In this particular example, LC-MS/MS facilitates the identification of three replication products from the propagation of the lesion-bearing genome: d(<u>GC</u>GACGCC) (**8G**), d(<u>TC</u>GACGCC) (**8T**), and d(<u>CC</u>GACGCC) (**8C**, the underlined nucleobases are those inserted at the original G[8-5]C lesion site during DNA replication). The SIC and tandem mass spectrum for the **8T** ODN are depicted in Figure 14.4.

After identifying the replication products, the relative amounts of these products can then be quantified following the aforementioned relative-ratio method, which facilitates the calculation of mutation frequencies. The bypass efficiency can then be calculated from the ratio of the total amounts of **8G**, **8T**, and **8C** over the amount of d(GCTAGCTGCGG) (**11-mer**) with the consideration of the lesion–competitor genome ratio used in the initial transformation experiment. The results showed that the G[8-5]C lesion can block considerably DNA replication *in vivo* as represented by a 20% bypass efficiency. Whereas no detectable mutation was found for the 3′ cytosine portion of the G[8-5]C lesion, 8.7% G → T and 1.2% G → C transversion mutations were observed for the 5′ guanine portion of the lesion.

By using *E. coli* strains that are defective in translesion synthesis DNA polymerases,[38] the REAMS method can also allow one to identify the role of a specific DNA polymerase in the translesion synthesis of the DNA lesion *in vivo*. For instance, to determine which translesion synthesis DNA polymerase is involved in the mutagenic bypass of the G[8-5]C lesion in *E. coli* cells, we can examine the bypass efficiencies and mutation frequencies of the lesion in the isogenic AB1157 *E. coli* strains that are depleted in one of the three translesion synthesis DNA polymerases, pol II, pol IV, and pol V.[38]

The results with polymerase-deficient *E. coli* strains showed that the bypass efficiencies dropped from 20% for the wild-type strains to 14, 13 and 8% for the strains that are defective in pol II, pol IV, and pol V, respectively. Although no significant difference in mutation frequency is found for the wild-type strains and the strains that are deficient in either pol II or pol IV, the G → T and G → C mutations are almost completely abolished in the pol V-deficient strain. This result demonstrates conclusively that mutagenic bypass of the G[8-5]C lesion in *E. coli* cells requires pol V.[12]

The novel LC-MS/MS method for the study of the cytotoxicity and mutagenicity of DNA lesions obviates the use of ^{32}P-labeling and facilitates the simultaneous monitoring of bypass efficiencies and mutation frequencies of DNA lesions. In addition, the REAMS method interrogates a short piece of ODN carrying the original lesion site rather than examining only the nucleotide at the

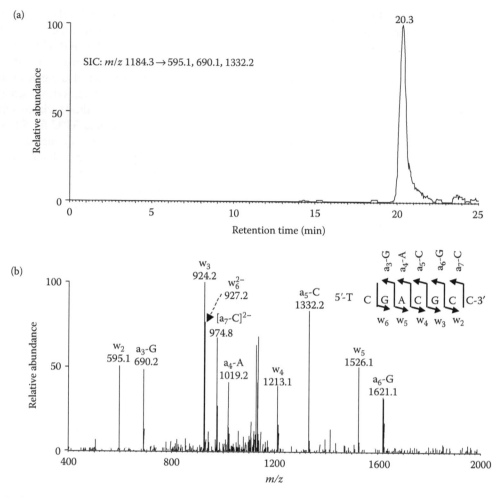

FIGURE 14.4 LC-MS/MS for monitoring the restriction fragments of interest with a G–T mutation at the original guanine portion of the lesion, that is, d(TCGACGCC). Shown in (a) is the SIC for the formation of indicated fragment ions of the ODN, and illustrated in (b) is the MS/MS of the $[M-2H]^{2-}$ (m/z 1184.3) of the ODN.

initial lesion site as in the traditional REAP assay. Therefore, the sequence information offered by MS/MS provides unambiguous identification of the replication products.

It is important to discuss the sensitivity and throughput of the method for *in vivo* replication studies. In this context, the large ion capacity, its associated high sensitivity, and the fast scan rate of the linear ion-trap mass spectrometer make this instrument well suited for this type of analysis. By using the HFIP mobile phase for the on-line HPLC separation and the LTQ linear ion-trap mass spectrometer, one can obtain tandem mass spectra of sufficient quality for determining the sequence of an ODN from an injection of about 10 fmol of the ODN. Thus, with restriction fragments from 2.5 pmol of PCR products, one can determine reliably a mutation frequency that is 0.5% or above. With larger amounts of the PCR product, even lower levels of mutation frequency can be detected by this method. In addition, in this LC-MS/MS experiment, the mass spectrometer was configured to fragment sequentially five different ODNs in a single analysis. Therefore, the REAMS assay is high-throughput; the mutation frequencies and lesion bypass efficiencies of single-nucleobase lesions can be monitored in one LC-MS/MS experiment.

14.4 CONCLUSIONS

LC-MS/MS, because of its high efficiency and its capacity in offering sequence information for ODNs, has been developed as a new tool for assessing the cytotoxic and mutagenic effects of DNA lesions *in vitro*.[15–22] Different from the pre-steady-state and steady-state kinetic assays, the LC-MS/MS method can allow for the analysis of the products emerging from the *in vitro* replication reactions with the mutual presence of all four dNTPs, instead of in the existence of one type of dNTP at a time as in the kinetic assays. It may represent better the real polymerization reaction conditions *in vivo*, thereby providing more accurate measurements.

The quantitative analysis of individual ODNs in the reaction mixture by LC-MS/MS is compounded by the difference in ionization efficiencies of ODNs. By assessing quantitatively the ionization and detection efficiencies of different ODNs through the analysis of authentic compounds, one can determine accurately the distributions of replication products by LC-MS/MS.

The LC-MS/MS method is also proven to be useful for assessing the genotoxic effects of DNA lesions *in vivo*.[12–14] The *in vivo* mutation frequency and bypass efficiency can be determined in a single experiment. With the availability of host cells that are deficient in one of the translesion synthesis DNA polymerases, one can also gain important insights into the role of various translesion synthesis DNA polymerases in the bypass of a structurally defined lesion *in vivo*. It can be envisaged that the method can be readily exportable to the replication studies of other host cells including human cells with appropriately designed shuttle vector systems.

REFERENCES

1. Lindahl, T., Instability and decay of the primary structure of DNA, *Nature* 362, 709–715, 1993.
2. Bloom, L.B., Otto, M.R., Beechem, J.M., and Goodman, M.F., Influence of 5′-nearest neighbors on the insertion kinetics of the fluorescent nucleotide analog 2-aminopurine by Klenow fragment, *Biochemistry* 32, 11247–11258, 1993.
3. Bloom, L.B., Otto, M.R., Eritja, R., Reha-Krantz, L.J., Goodman, M.F., and Beechem, J.M., Pre-steady-state kinetic analysis of sequence-dependent nucleotide excision by the 3′-exonuclease activity of bacteriophage T4 DNA polymerase, *Biochemistry* 33, 7576–7586, 1994.
4. Bloom, L.B., Chen, X., Fygenson, D.K., Turner, J., O'Donnell, M., and Goodman, M.F., Fidelity of *Escherichia coli* DNA polymerase III holoenzyme. The effects of beta, gamma complex processivity proteins and epsilon proofreading exonuclease on nucleotide misincorporation efficiencies, *J. Biol. Chem.* 272, 27919–27930, 1997.
5. Goodman, M.F., Creighton, S., Bloom, L.B., and Petruska, J., Biochemical basis of DNA replication fidelity, *Crit. Rev. Biochem. Mol. Biol.* 28, 83–126, 1993.
6. Johnson, K.A., Conformational coupling in DNA polymerase fidelity, *Annu. Rev. Biochem.* 62, 685–713, 1993.
7. Kuchta, R.D., Mizrahi, V., Benkovic, P.A., Johnson, K.A., and Benkovic, S.J., Kinetic mechanism of DNA polymerase I (Klenow). *Biochemistry* 26, 8410–8417, 1987.
8. Gu, C. and Wang, Y., LC-MS/MS identification and yeast polymerase eta bypass of a novel gamma-irradiation-induced intrastrand cross-link lesion G[8-5]C, *Biochemistry* 43, 6745–6750, 2004.
9. Goodman, M.F. and Fygenson, K.D., DNA polymerase fidelity: From genetics toward a biochemical understanding, *Genetics* 148, 1475–1482, 1998.
10. Creighton, S., Bloom, L.B., and Goodman, M.F., Gel fidelity assay measuring nucleotide misinsertion, exonucleolytic proofreading, and lesion bypass efficiencies, *Methods Enzymol.* 262, 232–256, 1995.
11. Delaney, J.C. and Essigmann, J.M., Assays for determining lesion bypass efficiency and mutagenicity of site-specific DNA lesions in vivo, *Methods Enzymol.* 408, 1–15, 2006.
12. Hong, H., Cao, H., and Wang, Y., Formation and genotoxicity of a guanine cytosine intrastrand cross-link lesion *in vivo*, *Nucleic Acids Res.* 35, 7118–7127, 2007.
13. Yuan, B., Cao, H., Jiang, Y., Hong, H., and Wang, Y., Efficient and accurate bypass of N^2-(1-carboxyethyl)-2′-deoxyguanosine by DinB DNA polymerase *in vitro* and *in vivo*, *Proc. Natl. Acad. Sci. USA* 105, 8679–8684, 2008.
14. Yuan, B. and Wang, Y., Mutagenic and cytotoxic properties of 6-thioguanine, S^6-methylthioguanine, and guanine-S^6-sulfonic acid. *J. Biol. Chem.* 283, 23665–23670, 2008.

15. Zang, H., Goodenough, A.K., Choi, J.Y., Irimia, A., Loukachevitch, L.V., Kozekov, I.D., Angel, K.C., Rizzo, C.J., Egli, M., and Guengerich, F.P., DNA adduct bypass polymerization by *Sulfolobus solfataricus* DNA polymerase Dpo4: Analysis and crystal structures of multiple base pair substitution and frameshift products with the adduct $1,N^2$-ethenoguanine, *J. Biol. Chem.* 280, 29750–29764, 2005.

16. Choi, J.Y., Zang, H., Angel, K.C., Kozekov, I.D., Goodenough, A.K., Rizzo, C.J., and Guengerich, F.P., Translesion synthesis across $1,N^2$-ethenoguanine by human DNA polymerases, *Chem. Res. Toxicol.* 19, 879–886, 2006.

17. Eoff, R.L., Angel, K.C., Egli, M., and Guengerich, F.P., Molecular basis of selectivity of nucleoside triphosphate incorporation opposite O6-benzylguanine by *Sulfolobus solfataricus* DNA polymerase Dpo4: Steady-state and pre-steady-state kinetics and x-ray crystallography of correct and incorrect pairing, *J. Biol. Chem.* 282, 13573–13584, 2007.

18. Eoff, R.L., Irimia, A., Egli, M., and Guengerich, F.P., *Sulfolobus solfataricus* DNA polymerase Dpo4 is partially inhibited by "wobble" pairing between O6-methylguanine and cytosine, but accurate bypass is preferred, *J. Biol. Chem.* 282, 1456–1467, 2007.

19. Choi, J.Y., Angel, K.C., and Guengerich, F.P., Translesion synthesis across bulky N2-alkyl guanine DNA adducts by human DNA polymerase kappa, *J. Biol. Chem.* 281, 21062–21072, 2006.

20. Choi, J.Y., Chowdhury, G., Zang, H., Angel, K.C., Vu, C.C., Peterson, L.A., and Guengerich, F.P., Translation synthesis across O6-alkylguanine DNA adducts by recombinant human DNA polymerases, *J. Biol. Chem.* 281, 38244–38256, 2006.

21. Choi, J.Y., Stover, J.S., Angel, K.C., Chowdhury, G., Rizzo, C.J., and Guengerich, F.P., Biochemical basis of genotoxicity of heterocyclic arylamine food mutagens: Human DNA polymerase eta selectively produces a two-base deletion in copying the N2-guanyl adduct of 2-amino-3-methylimidazo[4,5-f]quinoline but not the C8 adduct at the NarI G3 site, *J. Biol. Chem.* 281, 25297–25306, 2006.

22. Gu, C., and Wang, Y., *In vitro* replication and thermodynamic studies of methylation and oxidation modifications of 6-thioguanine, *Nucleic Acids Res.* 35, 3693–3704, 2007.

23. Beaucage, S.L. and Iyer, R.P., The synthesis of modified oligonucleotides by the phosphoramidite approach and their applications, *Tetrahedron* 49, 6123, 1993.

24. Harris, C.M., Zhou, L., Strand, E.A., and Harris, T.M., New strategy for the synthesis of oligodeoxynucleotides bearing adducts at exocyclic amino sites of purine nucleosides. *J. Am. Chem. Soc.* 113, 4328–4329, 1991.

25. Zeng, Y. and Wang, Y., Facile formation of an intrastrand cross-link lesion between cytosine and guanine upon pyrex-filtered UV light irradiation of d(^{Br}CG) and duplex DNA containing 5-bromocytosine, *J. Am. Chem. Soc.* 126, 6552–6553, 2004.

26. Zhang, Q. and Wang, Y. Generation of 5-(2'-deoxycytidyl)methyl radical and the formation of intrastrand cross-link lesions in oligodeoxyribonucleotides. *Nucleic Acids Res.* 33, 1593–1603, 2005.

27. Xu, Y., Post-synthetic introduction of labile functionalities onto purine residues via 6-methylthiopurines in oligodeoxyribonucleotides, *Tetrahedron* 52, 10737–10750, 1996.

28. Apffel, A., Chakel, J., Fischer, S., Lichtenwalter, K., and Hancock, W., Analysis of oligonucleotides by HPLC-electrospray ionization mass spectrometry, *Anal. Chem.* 69, 1320–1325, 1997.

29. McLuckey, S.A., Van Berkel, G.J., and Glish, G.L., Tandem mass spectrometry of small, multiply charged oligonucleotides, *J. Am. Soc. Mass Spectrom.* 3, 60–70, 1992.

30. Wang, Z., Wan, K.X., Ramanathan, R., Taylor, J.S., and Gross, M.L., Structure and fragmentation mechanisms of isomeric T-rich oligodeoxynucleotides: A comparison of four tandem mass spectrometric methods, *J. Am. Soc. Mass Spectrom.* 9, 683–691, 1998.

31. Wan, K.X., and Gross, M.L., Fragmentation mechanisms of oligodeoxynucleotides: Effects of replacing phosphates with methylphosphonates and thymines with other bases in T-rich sequences, *J. Am. Soc. Mass Spectrom.* 12, 580–589, 2001.

32. Wang, Y., Taylor, J.S., and Gross, M.L., Fragmentation of electrospray-produced oligodeoxynucleotide ions adducted to metal ions, *J. Am. Soc. Mass Spectrom.* 12, 550–556, 2001.

33. Null, A.P., Nepomuceno, A.I., and Muddiman, D.C., Implications of hydrophobicity and free energy of solvation for characterization of nucleic acids by electrospray ionization mass spectrometry, *Anal. Chem.* 75, 1331–1339, 2003.

34. Delaney, J.C. and Essigmann, J.M., Mutagenesis, genotoxicity, and repair of 1-methyladenine, 3-alkylcytosines, 1-methylguanine, and 3-methylthymine in alkB *Escherichia coli*, *Proc. Natl. Acad. Sci. USA* 101, 14051–14056, 2004.

35. Delaney, J.C. and Essigmann, J.M., Context-dependent mutagenesis by DNA lesions, *Chem. Biol.* 6, 743–753, 1999.

36. Gu, C., Zhang, Q., Yang, Z., Wang, Y., Zou, Y., and Wang, Y., Recognition and incision of oxidative intrastrand cross-link lesions by UvrABC nuclease, *Biochemistry* 45, 10739–10746, 2006.
37. Moriya, M., Single-stranded shuttle phagemid for mutagenesis studies in mammalian cells: 8-Oxo-guanine in DNA induces targeted G.C– > T.A transversions in simian kidney cells, *Proc. Natl. Acad. Sci. USA* 90, 1122–1126, 1993.
38. Jarosz, D.F., Beuning, P.J., Cohen, S.E., and Walker, G.C., Y-family DNA polymerases in *Escherichia coli*, *Trends Microbiol.* 15, 70–77, 2007.

15 Influence of Metal Ions on the Structure and Reactivity of Nucleic Acids

Janna Anichina and Diethard K. Bohme

CONTENTS

15.1 INTRODUCTION

It is not possible to overestimate the role of metal ions in determining the three-dimensional architecture of nucleic acids, their helices, and higher-order molecular assemblies. Magnesium and potassium cations, for instance, are the key components in stabilizing secondary and tertiary elements of nucleic acid structure in ribozymes and telomeres. Also, metal ions can alter the acid–base and hydrogen-bonding properties of DNA in a most exciting manner: In some cases, a drop in a nucleic acid pK_a value up to 7 log units may be achieved.[1] The presence of metal ions is crucial for many biochemical reactions involving phosphate group transfer and hydrolysis, where the metal–nucleotide complex may function as a substrate, cofactor, and/or modulator of enzymatic activity.

Interesting are the properties of the recently synthesized M-DNA,[2] a novel DNA–metal ion complex in which a divalent metal ion (Zn^{2+}, Ni^{2+}, or Co^{2+}) is embedded into the center of the DNA duplex (Figure 15.1). The addition of metal ions to DNA results in the deprotonation of the N3 or N1 nitrogen of thymine or guanine, respectively.[2]

M-DNA has been demonstrated to have unusual conductive properties that might be potentially utilized for the development of novel complex nanoarchitectures. Fluorescence quenching experiments have indicated that M-DNA exhibits the behavior of a molecular wire.[3]

Metal ions possess properties particularly important in the nanoworld. Nanoscopic metal wires may allow the conduction of electrons through self-assembled networks, whereas magnetic coupling of metal ions may create nanomagnets with a defined orientation and strength of the magnetic field.

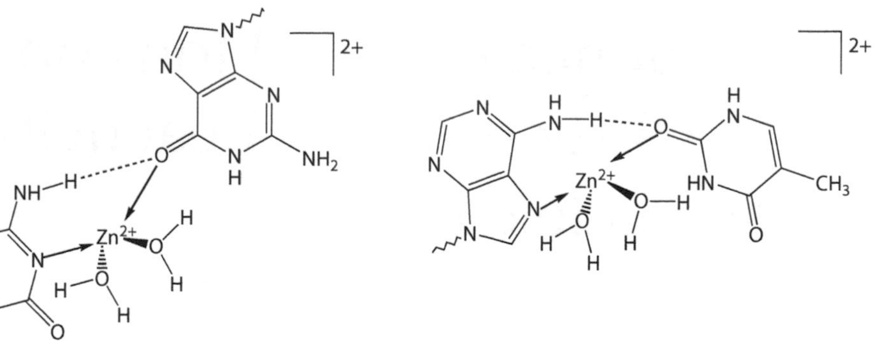

FIGURE 15.1 Proposed structure of Zn^{2+}-coordinated AT and GC complexes formed in DNA. (From Wettig, S.D., Wood, D.O., and Lee, J.S., *J. Inorg. Biochem.* 94, 94, 2003. With permission.)

The incorporation of metal ions into DNA with its self-assembling properties may result in the development of complex functional nanostructures with a range of desired characteristics.[4]

Another substantial impact of metal ions upon structure and functions of nucleic acids is their role as mediators of interactions with small molecules and drugs. *In vivo* DNA is wrapped tightly in a sphere-like structure rather than stretched out. At the same time, at a neutral pH, all the phosphate groups of the polyanionic backbone of nucleic acids are fully deprotonated, resulting in the need for a cationic species to neutralize the Coulomb repulsion present in the tightly packed structure. In cell nuclei, the reduction of the Coulomb repulsion is achieved by means of highly positively charged histone proteins and smaller polyamine molecules. Metal ions and their complexes also play a significant part in this process.

Current research in the field of metal–DNA interactions is dedicated to the formation of the noncanonical base pairs from natural nucleobases in the presence of metal ions, the exchange of hydrogens in Watson–Crick-type base pairing by metal ions, and the reversible and irreversible binding of metal ions to sites other than on the nucleobases.[5–8]

The recognition of DNA by metallocomplexes with different binding modes results in significant changes in physical properties of the DNA and, as a consequence, in its *in vivo* chemistry. There may also be interference with gene expressions, the inhibition of translation, transcription, and replication processes, and the enhancement or reduction in the stability of duplexes, triplexes, and higher-order structures of DNA.[9]

One of the major driving forces in the study of sequence-selective DNA binding is a desire to understand, on the molecular level, the nature of the interactions of antitumor and metalloantibiotics with nucleic acids. The anticancer activity of *cis*-diaminedichloroplatinum (II), most commonly known as *cis*-platin, was discovered in 1965 when inhibition of cell division was observed upon electrolysis using a Pt electrode[10,11] Since the discovery of the anticancer properties of *cis*-platin, an array of platinum complexes, resembling the structure of *cis*-platin, has been synthesized and shown to be therapeutically efficient (Scheme 15.1).

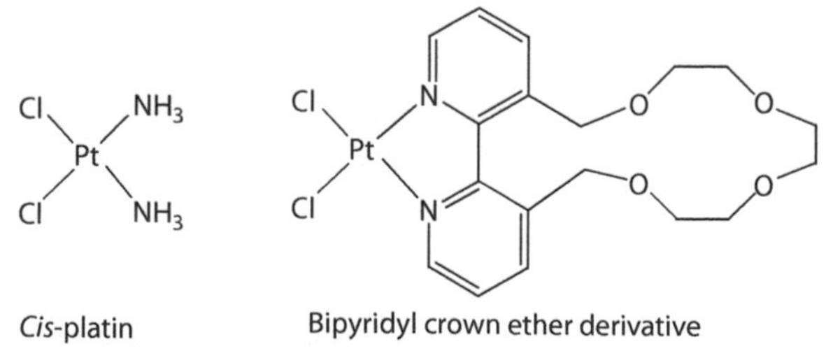

Cis-platin   Bipyridyl crown ether derivative

SCHEME 15.1 *Cis*-platin and its crown ether derivative.

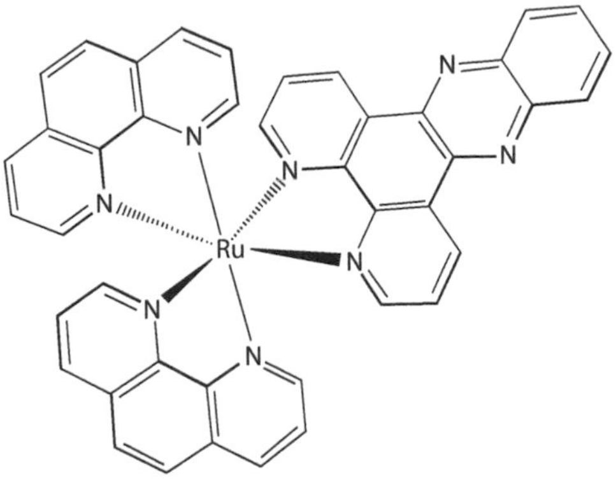

SCHEME 15.2 DNA intercalator $[Ru(Phen)_2(Dppz)]^{2+}$.

All antitumor platinum drugs are known to add to DNA in 1:1 ratio and to form simultaneously intra- and interstrand cross-links.[12,13] The formation of these lesions results in the inhibition of DNA replication, which is the main reason behind the cytotoxicity of the platinum complexes.[14]

Related to the platinum complexes are the Ru species coordinated with aromatic heterocyclic planar ligands; for example, 1,10-phenanthroline or dipyridophenazine (Scheme 15.2).[15]

Finally, because nucleic acids are chiral molecules, chiral metal complexes may interact enantioselectively with DNA.[15] All of this is possible due to the plethora of binding sites for metal ions/complexes in the structure of nucleic acids, such as phosphates, hydroxyl groups of the deoxyriboses, and nucleobases.

The most comprehensive review of metal ion–DNA interactions in both the solution and the crystal phase has drawn attention to the preferences expressed by different types of metal ions for different DNA sites.[16] Some of these preferences are 13, 15, and 16.

- *Phosphate groups* have been shown to be good ligands for all types of metal ions and form salt-like complexes between positively charged metal ions and negatively charged phosphate groups.
- *Sugar hydroxyl groups*, being "hard" bases, are generally attractive to alkali and alkali-earth ions but not to the transition metal ions that cannot be classified as either hard or soft exhibiting instead the propensies of both hard and soft types.
- *Ring nitrogens of bases* are considered to be good ligands for all transition metals, since nitrogen donors are usually higher than oxygen donors in the spectrochemical series. The crystal field splitting energy favors the formation of bonds to nitrogen. For N-9-substituted adenine and guanine bases, the N-7 site is the preferred metal-binding site, followed by the N-1 site for adenine. The N-3 site is sterically blocked from the metal coordination by the sugar moiety connected to the N-9 substituent (Scheme 15.3). The dashed lines indicate available binding sites of common purines and pyrimidines. "S" represents the deoxyribose moiety.

The advent of electrospray ionization (ESI)[17,18] and electrospray ionization mass spectrometry (ESI-MS) has generated tremendous advances in the structural characterization of biomolecules such as DNA and oligodeoxynucleotides (ODNs), and their metallated adducts and noncovalent complexes with proteins and drugs. Collision-induced dissociation (CID) has proven to be a powerful tool in providing important information regarding the sequences of ODNs and probable sites for the binding of metal ions. Even though ESI tandem mass spectrometry (ESI-MS/MS) has been used for the bulk of such studies, ESI/ion mobility MS combined with molecular dynamics simulations also has been utilized to understand the gas-phase conformation of double-stranded DNA adducts

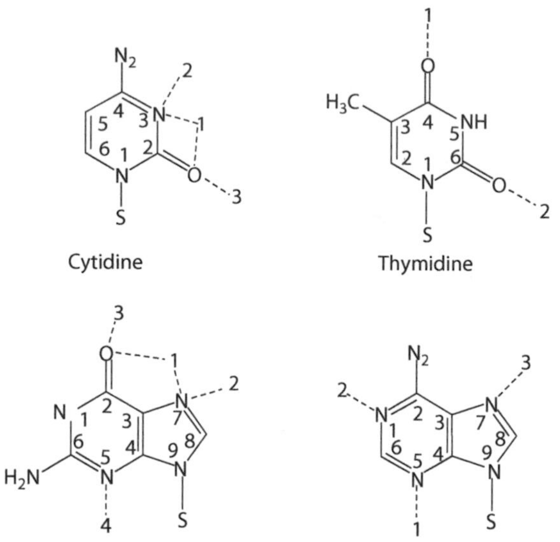

SCHEME 15.3   Numbering scheme for preferential sites of metal ion coordination with DNA nucleobases.

with metal ions, the influence of metal ions upon Watson–Crick base pairing, and other important information.[19–22] There are a number of factors to be taken into account when using ESI and ESI-MS/MS to study ionic metal–oligonucleotide complexes. First, sensitivity may be reduced because of difficulties in removing excess metal salts, and because of the distribution of available ion current among species with different metal isotopes. High resolution may be required for accurate mass measurements, because many metals of interest have complex isotope patterns. Another problem might arise from the chemical reactivity of metal complexes since they may be more labile compared to alkylators, making detection of intact complexes difficult and/or resulting in less informative MS/MS spectra.

Another mass spectrometric technique emerging as an efficient tool in the analysis of nucleic acids is matrix-assisted laser desorption/ionization time-of-flight mass spectrometry (MALDI-TOF-MS).[23–27] There are two major problems encountered in the MALDI-TOF-MS analysis of nucleic acids: the fragmentation of ODNs, especially larger species, that occurs upon MALDI, and the formation of ODN adducts with alkali metal cations leading to a significant decrease in resolution and detection sensitivity.

Table 15.1 provides an overview of the studies dedicated to the interaction of metal ions with DNA and oligonucleotides ions utilizing a mass spectrometric approach.

15.2   ESI-MS OF OLIGONUCLEOTIDE COMPLEXES WITH METAL IONS

15.2.1   BARE ODNs: MAIN TRENDS IN FRAGMENTATION

With the advent of ionization techniques such as ESI and MALDI that are utilized to transfer large biological ions into the gas phase, CID or collision-activated dissociation (CAD) began to be utilized for structural investigations of proteins, peptides, and nucleic acids. The fragmentation of ODN ions has been a major endeavor in biological MS for several decades. An excellent review on the trends in the gas-phase fragmentation of ODNs appeared in 2004. A summary of the key points of this review is provided here as a benchmark for the study of fragmentation of metallated oligonucleotide ions.

We mention at the outset the fragmentation nomenclature for the nucleic acids, now commonly adopted, suggested by McLuckey et al.[48] This nomenclature is illustrated in Scheme 15.4. The four

TABLE 15.1

Overview of the Mass Spectrometric Studies Dedicated to Metal Ion–DNA Interactions

ODN	Metal Ion	Instrumentation	Results	Reference
dAMP; dGMP; dCMP; dTMP	Cu^{2+}, Pt^{2+} complexed with diethylenetriamine or 2.2′:6′, 2″-terpyridine	ESI-LCQ quadrupole ion trap *in the positive ion mode*	Formation of radical cations of Cu(II) and nucleotides Correlation the CID results with vertical ionization energies	28
CG, GC, AT, TA, TC	Shiff base complex of Cr(III) $[Cr(salprn)(H_2O)_2]ClO_4$ salprn = 1,3 bis(salicyl-ideneamino)propane	ESI-QqQ	Cr(III) compound was found to attach all dinucleotides CID revealed that Cr^{3+} attaches to the phosphate and a base	29
AA, GG, TT, CC, AT, TA, TC, CT, GC, CG, GT, TG, AC, CA	Na^+	MALDI-TOF, ion mobility Molecular dynamics	Formation of multiply metallated, clusters of dinucleotides Structural motifs are suggested	30
$(GG)_2$; $(TT)_2$; $(GC)_2$	Li^+, Na^+, K^+, Mg^{2+}, Ca^{2+}, Cr^{2+}, Mn^{2+}, Fe^{2+}, Co^{2+}, Ni^{2+}	MALDI-TOF ion mobility	Influence of metal ions upon Watson–Crick pairing	31
$(GC)_2$	Cu^+, Ag^+, Cd^+, Pt^{2+}, Cu^{2+}, Zn^{2+}, Fe^{2+}, Co^{2+}, Ni^{2+}, Li^+, Na^+, K^+, Cs^+, Mg^{2+}, Ca^{2+}, Cr^{2+}, Cr^{3+}, Mn^{2+}, Fe^{3+}	MALDI-TOF ion mobility	Influence of metal ions upon Watson–Crick base pairing	32
ACGT, AGCT, TGT, GGT, ACA, AGCTAG, AG, TG, CA, CT, GG	Ca^{2+}, Ce^{4+}, Ce^{3+}, Co^{2+}, Cu^{2+}, Fe^{3+}, Mg^{2+}, Zn^{2+}	MALDI-FT-ICR	Evaluation of sequential metal attachment Determination of metal-binding site	33
ACGD, AATT, AGCT AGCA, TGT, GGT	Fe(III) and Fe(II) salts mixed with the matrix compound	MALDI-FT-ICR	Presence of Fe(II) and Fe(III) was found to be controlled by number of proton abstracted from the phosphates MS/MS showed that the presence of Fe alters the CID pattern	34
TTCAT, TTGAT, TTGCT, TTGGT	Fe^{2+}, Fe^{3+}, Zn^{2+}	ESI-Q-TOF *in the positive ion mode*	Attachment of three Zn^{2+}, one Fe^{3+}, and three Fe^{2+} observed in ESI spectra Stabilizing effect of metal ions was demonstrated in MS/MS experiments	35
CACGTG	$[cis\text{-}Pt(NH_3)_2Cl_2]$	ESI-EBE-TOF	Formation of $Pt(NH_3)_2^{2+}$ adducts of the ODN was demonstrated MS/MS was used to elucidate Pt coordination	36
CTTAAG, CTAAAG	Na^+, K^+	MALDI-TOF	Using a new matrix enabled transfer of oligos shorter than 30 bp and reduction of alkali metal adducts	37

continued

TABLE 15.1 (continued)
Overview of the Mass Spectrometric Studies Dedicated to Metal Ion–DNA Interactions

ODN	Metal Ion	Instrumentation	Results	Reference
(CGCGAATTCGCGCGAAATCGCG) (TTACTCTGTTAAT GTCCTTG) (TCCACCATTAACA CCCAAAGCTA) (TCCACCATTTGCA CCCAAAGCTA) (TCCACCATTAGCA CCCAAAGCTA) (CATGTATAACCGC ATTATGCTGAGTGATAT)	Tetramethyl pyridine porphirin-4 of Cu, Fe and Mn [Ru12S4dppz]Cl$_2$	ESI-LCQ ion trap	Relative affinities of the ODNs for the metallocomplexes were studied	38
GCTTGCAT; TTGGCCCTCCTT	UO$_2^{2+}$, Na$^+$, Mg^{2+}	ESI-FT-ICR	CID of 1:1 metal ion: ODN complexes showed that UO$_2^{2+}$ bind preferentially to the thymine region of nucleic acids	39
GGCTAGCC	cis-[Ru(bpy)$_2$Cl$_2$], [Ru(tpy)Cl$_3$], [Pd(en) Cl$_2$], [RuCl$_2$(DMSO)$_4$]	ESI-MS	Formation of adducts between ruthenium and palladium compounds was demonstrated in the ESI spectra	40
GCGAATTCGC	K$^+$, Rb$^+$, Cs$^+$, Ca^{2+}, Ba^{2+}, Sr^{2+}, Mn^{2+}, Ni^{2+}, Co^{2+}, Ag$^+$, Cd^{2+}, Hg^{2+}, Pb^{2+}	ESI-LCQ Duo quadrupole ion trap	Comparison of the CID patterns of 1:1 ODN: metal complexes in different charge states as a function of the nature of the attached metal ion	41
ATACATGGTACATA	Ru(II) and Os(II) complexes of the type [M(η^6-p-cumene)Cl$_2$(L)] L = pta, pta-Me$^+$	ESI Q-TOF	Ru(II) and Os(II) complexes of 1:1 and 1:2 stoichiometry were shown to form in the solution	42
D1 = CCTCGGCCGGCCGACC/ GGTCGGCCGGCCGAGG D2 = CCTCATGGCCATGACC/ GGTCATGGCCATGAGG D3 = CCTCAAAATTTTGACC/ GGTCAAAATTTTGAGG.	[Ru(Phen)$_3$]$^{2+}$ [Ru(Phen)$_2$(dpq)]$^{2+}$ [Ru(Phen)$_2$(dppz)]$^{2+}$	Micromass QTOF2 with Z-spray probe	Competition experiments for mixed binding indicated that the largest number of additions was 3:3:1 [Ru(Phen)$_3$]$^{2+}$: [Ru(Phen)$_2$(dpq)]$^{2+}$ DNA	43

2+

[Ru(Phen)$_2$dppz]$^{2+}$

2+

[Ru(Phen)$_2$dpq]$^{2+}$

2+

[Ru(Phen)$_3$]$^{2+}$

Trombin-binding aptamer—GCTTGGTGTGGTTGG AATTAATGTAATTAA AGTTAGTGTAGTTAG GATTAGTGTGATTA	K^+, Sr^{2+}, Pb^{2+}, Ba^{2+}	ESI-MS	Comparison of the relative affinity of metal ions for the aptamer and the control ODNs indicated much greater affinity for the aptamer	44
A number of ODNs of different base composition and length, single- and double-stranded	cis-Platin and its derivatives	ESI-MS	A review summarizing the data on the interactions of ODNs with Pt complexes	21
27-mer CCGGATCCT- AATACGACT- CACTATAGGAGG 34-mer TAATACGACTCA CTATAGGCAGAA TTCCCGCGGC 47-mer CCGAATTCTGTA CCGTCAGCGTCA GCGTCATTGACG CTGCGCCCAGGA TCCGG	Stainless steel probe provided the source for Fe adducts as well as iron powder added to the matrix	MALDI-TOF	Formation of Fe-containing adducts was demonstrated upon utilization of the steel probe. Fe adduction was found to depend on the length of oligodeoxynucleotide	45
CAGCGTGCGCA TCCTACCC, GGGAAGGATGG GGCACGCTG BBR3464	BBR3464 (structure is shown below)	ESI–FT-ICR	Stoichiometry of binding of cis-platin analog with the ODN	46

$[\quad]^{4+}$

Cl—Pt(NH$_3$)(H$_3$N)—N(H$_2$)— ... —Pt(H$_2$N)(H$_3$N)(NH$_3$)— ... —Pt(H$_2$N)(H$_3$N)(NH$_3$)—Cl

continued

TABLE 15.1 (continued)
Overview of the Mass Spectrometric Studies Dedicated to Metal Ion–DNA Interactions

ODN	Metal Ion	Instrumentation	Results	Reference
Duplexes of d(5'GCGGGGATGGGGCG) d(5'CGCCCCATCCCCGC) d(5'GCGGGAATTGGGCG) d(5'CGCCCAATTCCCGC) d(5'GCGGAAATTTGGCG)	Mg(II), Ni(II), Cu(II) with novel benzoxazole and benzimidazole ligans (structures are shown below)	ESI-LCQ ion trap	Evaluation of binding of the model ODN with metallocomplexes	47

Note: Entries are ordered according to the length of the ODN.

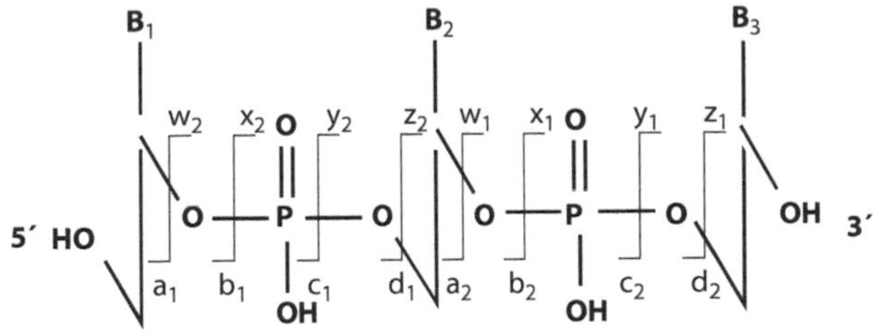

SCHEME 15.4　Nomenclature for fragmentation of synthetic oligonucleotides.[48]

possible cleavages along the phosphodiester backbone are denoted by the lowercase letters a, b, c, and d for the fragments containing the 5′-terminus and w, x, y, z for the fragments containing the 3′-terminus. The numerical subscripts indicate the number of bases from the respective termini. The uppercase letter B represents a base with the numerical subscript indicating its position in the sequence starting from the 5′-end. The main advantage of the McLuckey nomenclature over others is that it addresses the cleavages of the P–O bonds in addition to those of the 5′- and 3′-phosphoester linkages. This nomenclature is used widely in the ESI and MALDI literature.

Fragmentations of monodeoxynucleotides, dideoxynucleotides, and trideoxynucleotides have been extensively studied by ESI-MS/MS. Habibi-Goudarzi and McLuckey[49] systematically studied the fragmentation of the singly deprotonated ions of the four 2′-deoxyribonucleoside 3′-monophosphates and their corresponding 5′-monophosphates, and the 16 possible 2′-deoxyribodinucleoside 3′,5′-monophosphates as a function of the resonance excitation amplitude utilizing ESI-ion-trap-MS^n. Boschenok and Sheil[50] investigated the fragmentation patterns of deprotonated, protonated, and cationized by sodium and potassium 2′-deoxyribodinucleoside 3′,5′-monophosphates with a quadrupole–hexapole–quadrupole mass spectrometer and on a magnetic sector-TOF-MS. Ho and Kebarle[51] determined the energy thresholds for the formation of the major product ions from the CID of 10 deprotonated mononucleotides with a modified triple quadrupole mass spectrometer and found good correlation with values determined using semiempirical PM3 calculations. The CID of mononucleotides indicated that loss of the neutral base constitutes the major fragmentation channel for all the mononucleotides. Formation of PO_3^- is a minor channel for the 2′-deoxyribonucleoside 3′-monophosphates, whereas for the corresponding 5′-monophosphates, it tends to be relatively abundant. The 3′-phosphate group facilitates base loss relative to loss of PO_3^-. For the two sets of 2′-deoxyribonucleoside monophosphates, the tendency for neutral base loss follows the order of dAp, dTp > dCp > dGp and pdA, pdT > pdC ≈ pdG.[49] The CID of several single-stranded ODN anions was investigated using quadrupole ion traps.[19] The main trends in the dissociation of larger multiply charged ODNs are as follows:

　i. The first dissociation step involves the loss of a base. The relative loss of a charged base versus neutral base is strongly dependent on the charge on the parent ion. For low charge states, neutral base loss is favored without any obvious base preference. At high charge states, loss of a charged base is favored.

　ii. Loss of the base at the 3′-terminus tends to be disfavored.

　iii. The subsequent dissociation event involves cleavage of the 3′ C–O bond of the sugar from which the base was lost. As a result, complementary w-type and (a–B)-type ions are being formed. Cleavages at other locations along the phosphodiester chain are significantly less abundant.

　iv. Base loss from the first- and second-generation fragments promotes further backbone cleavages to yield internal fragment ions. Multiple base loss increases with increasing activation and increasing charge state.

Despite the fact that the MS of ODNs has matured significantly over the last decade, more information is needed about the influence of the nature of the base, its location, the charge state of the ODN, and its length. Computational modeling combined with appropriate mass spectrometric measurements could provide essential information on the mechanisms of the gas-phase dissociation of ODNs and could shed light on the intrinsic gas-phase reactivity of the deprotonated nucleic acids. In both quantum theory and ESI-MS, experiments were performed to explore the gas-phase structures,[52] energies, and stabilities against CID of an oligodeoxydeoxynucleotide with the sequence CCC, in its singly and doubly deprotonated forms, $(CCC-H)^-$ and $(CCC-2H)^{2-}$, respectively.[52] Reactions of deprotonated CCC with hydrogen bromide were investigated utilizing a novel chemical reactor within an ESI-selected ion flow tube (SIFT)-QqQ configuration developed for the investigation of molecule reactions with electrosprayed ions. The results of these measurements allowed an experimental assessment of the proton affinities of the singly and doubly deprotonated trideoxynucleotide that were predicted by computations.

The minimum energy structure of $(CCC-2H)^{2-}$ is presented in Figure 15.2. It is evident that the preferred conformation to a large degree is dictated by the two phosphate groups. The phosphate group from the 5'-end accepts two hydrogen bonds, one from the 5'-terminal hydroxyl group and the other from one of the amino hydrogens of the 3'-terminal cytosine; in addition, a weak coordination from the slightly acidic 6-H of the central cytosine ring to one of the slightly basic ether oxygens, (P)–O–(C), of the phosphate is observed; the phosphate group of the 3'-end accepts one H bond, from one of the amino hydrogens of the 5'-terminal; in addition a weak coordination from 6-H of the 3'-terminal cytosine ring to one of the ether oxygens of the phosphate is also observed in this case. CID performed on the selected dianion $(CCC-2H)^{2-}$ indicates that this ion dissociates by charge separation via two competing channels (Figure 15.3): (a) formation of a deprotonated cytosine $(C-H)^-$ and its complementary fragment and (b) backbone cleavage leading to the formation of either w_1^- and a_2^- or d_1^- and z_2^- ions. The latter two pairs of ions cannot be distinguished in this experiment since they have identical sets of m/z values. The relative abundances of channels (a) and (b) are $48 \pm 2\%$ and $52 \pm 2\%$, respectively.

CID experiments performed on the selected dianion $(CCC-2H)^{2-}$ indicate that this ion dissociates by charge separation via two competing channels: formation of a deprotonated cytosine and its complementary fragment and the backbone cleavage leading to the formation of either w_1^- and a_2^- or d_1^- and z_2^-. On the basis of the computed minimum energy structure, it is possible to identify the location of the cytosine that is lost as $(C-H)^-$. Since the 5'-end phosphate group accepts two hydrogen bonds, whereas the 3'-end phosphate group accepts only one, the overall negative charge

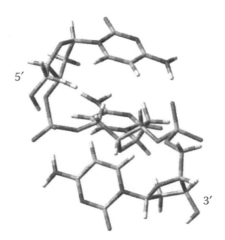

FIGURE 15.2 Computed minimum energy structure of $(CCC-2H)^{2-n}$. (From Anichina, J., et al., *J. Am. Soc. Mass Spectrom.* 19, 987, 2008. With permission.)

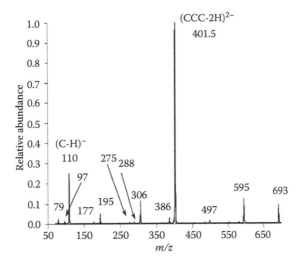

FIGURE 15.3   MS/MS spectrum for the dissociation of m/z 401.4 (CCC–2H)$^{2-}$ averaged over the region from −1 to −30 V (on the left) and CID profiles for this dissociation (on the right). The profiles of the product ions with relative intensities lower than 0.02, for example, those for m/z = 693, 595, 497, 386, 288, and 275, are not shown. (From Anichina, J., et al., *J. Am. Soc. Mass Spectrom.* 19, 987, 2008. With permission.)

on the 5′-end phosphate is reduced compared to that at the 3′-end, so the probability of the proton transfer from the departing cytosine to the more negative 3′-end phosphate is higher (Figure 15.3).

The two monoprotonated structures, in which the proton dissociated from the phosphate adjacent to the 5′- and 3′-end respectively, are both dominated by a strong hydrogen bond from the protonated phosphate group to the unprotonated (*vide infra*) (Figure 15.4). This interaction limits the effective conformational space for both structures. In addition to this central hydrogen bond, the nonprotonated phosphate effectively accepts hydrogen bonds from suitably situated hydrogen bond donors. In the case of protonation at the phosphate group closest to the 5′-terminus—the corresponding compound is denoted (5′)—the 3′-phosphate accepts hydrogen bonds from the protonated phosphate group as well as the two terminal OH groups. The same occurs for the 5′-phosphate in 3′(CCC–H)⁻ (3′) (see Figure 15.4). The CID of the monoanion, (CCC–H)⁻, was found to lead to the loss of cytosine as the primary dissociation event. In addition, there is a signal corresponding to loss

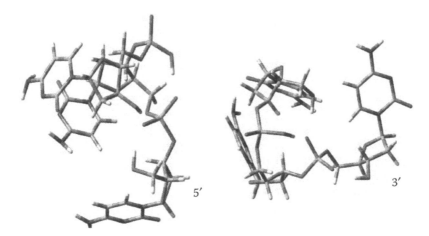

FIGURE 15.4   Computed minimum-energy structures of CCC trideoxynucleotide deprotonated at either the 5′-end (left structure) or the 3′-end (right structure). (From Anichina, J., et al., *J. Am. Soc. Mass Spectrom.* 19, 987, 2008. With permission.)

of furfuryl alcohol at slightly higher onset voltage (OV). The availability of a nearby acidic proton for transfer to (C–H)$^-$, most likely the one located at the neighboring protonated phosphate group, is also a critical factor. H/D exchange experiments performed in solution indicated a deuteron rather than a proton transferred to the departing neutral nucleobase confirming the involvement of the phosphate proton in the mechanism of the loss of a neutral base in CID of ODNs.

A brief summary of the main trends in the CID of higher-order DNA adducts, such as double-stranded species, is also useful. Gabelica and De Paw[53] examined the CID of a series of 12-mer and 16-mer duplexes in a quadrupole ion trap with different activation times and energies. Upon CID of the selected duplexes, two types of dissociations were observed: noncovalent charge separation producing two single strands sharing the charge of the double-stranded parent ion and covalent fragmentation, associated with the loss of a neutral base. This work also showed that for complexes containing a high number of contributing interactions organized at the interface of the ligands, the noncovalent dissociation is likely to be a multistep process. This has important implications for the interpretation of dissociation kinetics. The overall (observed) rate constant will not reflect a one-step dissociation with a single activation energy. Rather, the apparent rate constant is the resultant of consecutive small dissociation steps with little energy barriers. For the observed dissociation rate to reflect the contribution of all noncovalent interactions, the dissociation process must not have a rate-limiting step.[53]

Pan et al.[54] also systematically studied a series of 31 non-self-complementary matched DNA duplexes ranging in size from 5- to 12-mers. They showed that the branching ratio of covalent fragmentation to noncovalent dissociation is dependent on the so-called charge level of the duplex under investigation. By "charge level," the authors mean the ratio of the actual charge of the duplex to the total possible charge of the duplex. They determined that higher duplex "charge level" values resulted in more covalent fragmentation compared to noncovalent dissociation. Essentially, using "charge level" as a descriptive parameter, the authors could compare duplexes that differ in sequence, size, and charge state.

Another important factor defining the branching between covalent fragmentation and noncovalent dissociation channels is the base composition of the duplexes under investigation. It was demonstrated in a number of studies, both experimental and computational, that most of the noncovalent interactions that lead to the bonding of two or more DNA strands, such as electrostatic interactions, Watson–Crick hydrogen bonding, π-staking, and so on, are not only preserved but also are enhanced in the gas phase.[55–60] Sequences with a higher GC content are known to have larger melting temperatures (T_m) in solution since a GC base pair forms three hydrogen bonds whereas AT forms only two. In the gas phase, they produce better yields in their ESI spectra, whereas upon CID, they exhibit a higher propensity toward covalent fragmentation compared with AT-rich sequences.

15.2.2   ESI-MS/MS of ODN Complexed with Metal Ions

Most of the studies of the interactions between ODNs and metal ions, utilizing ESI-MS and CID-MS/MS, have been carried out in the negative ion mode. However, for shorter ODN sequences, the negative ion mode is not always suitable. For example, the addition of a monocation to a dimer results in a neutral complex that cannot be selected for in the negative ion mode. The constituent parts of the nucleic acids, for example, individual nucleobases, nucleosides, and nucleotides, are also problematic to study in the negative ion mode since they do not form anions in solution and do not transfer protons in the electrospray process. Most of the reported ESI-MS/MS studies of metal complexes with nucleosides and mononucleotides, therefore, have been performed in the positive ion mode.

Studies of interactions between alkali metal ions and nucleobases have been summarized in a review that appeared in reference.[61] The focus of this review is the application of ESI-MS to an examination of the role of alkali metal ions in the aggregation of nucleobases and nucleosides and the binding energies between alkali metal ions and nucleobases. The affinity of an alkali metal ion for a nucleobase is defined as the variation in enthalpy in the following chemical reaction:

$$BM^+ \rightarrow B + M^+.$$

Affinities of nucleobases for alkali metal cations have been determined utilizing two types of measurements: measurement of the threshold collision-induced dissociation (TCID) and the kinetic method.

In TCID measurements with a guided ion-beam mass spectrometer,[62] the threshold is determined as the minimum energy required for the dissociation into products without internal excitation at 0 K. The complex of the nucleobase with an alkali metal ion is collisionally activated and dissociated by collisions with xenon according to the following equation:

$$BM^+ + Xe \rightarrow M^+ + B + Xe.$$

Product ions and unreacted parent ions from the original beam are then focused into a quadrupole mass analyzer. The kinetic energies of the ions are converted into center-of-mass (CM) energies according to the following relation:

$$E_{CM} = E_{lab} \times m \times (m + M)^{-1},$$

where m and M are the masses of the neutral and the ionic species. Ion abundances are converted into absolute cross sections.

The kinetic method is a more approximate method for the determination of thermochemical properties and is based on rates of the competitive dissociation of chosen cluster complexes, namely, a heterodimer containing a nucleobase of interest, an alkali metal ion, and a reference ligand, whose affinity for the chosen alkali metal ion is known. The products of the two dissociation channels monitored are shown as follows:

$$B \; ... \; M^+ \; ... \; R \quad \overset{k_1}{\underset{k_2}{\nearrow \searrow}} \quad \begin{matrix} BM^+ + R \\ \\ RM^+ + B. \end{matrix}$$

Here R designates the reference ligand and k_1 and k_2 represent the rate constants for the competitive dissociation channels. Both the reverse activation energies and the difference in the required entropy for the competitive channels are assumed to be negligible and only one isomer is assumed to be dissociating. In this case, the ratio of the ion intensities (abundances), BMR^+ and RM^+, is related to the difference in the affinities for the alkali metal ion of the nucleobase and the reference ligand according to the following relationship:

$$\ln\left(\frac{k_1}{k_2}\right) = \ln\left(\frac{[BMR^+]}{[RM^+]}\right) \approx \frac{\Delta(\Delta H)}{RT_{eff}},$$

where T_{eff} is the effective temperature of the heterodimer complex. Table 15.2 provides a comparison of the alkali metal ion affinities obtained by the threshold collision induced dissociation (TCID) and the kinetic method.

Neither the kinetic method nor TCID provides any information about the binding sites for the alkali metal cations, but these can be predicted by theory. For example, using DFT calculations (MP2(full)/6-311G(2d2p)/1MP2/6-31G), Rogers and Armentrout[62] have predicted that the energetically favored sites for Li$^+$, Na$^+$, and K$^+$ on thymine and uracil are the exocyclic O$_4$ oxygen atoms and

TABLE 15.2
Alkali Metal Ion Affinities of Nucleobases Determined Using the Kinetic Method and TCID (in kcal mol^{-1})

Complex	Kinetic Method at 298 K	Kinetic Method Adjusted to 0 K	TCID at 0 K
LiU$^+$	50.4	50.0	50.55
NaU$^+$	33.7	33.5	32.17
KU$^+$	24.1	23.9	24.93
LiT$^+$	51.4	50.9	50.22
NaT$^+$	34.4	34.2	32.34
KT$^+$	54.0	53.0	54.04
LiA$^+$	54.0	53.0	54.04
NaA$^+$	41.1	39.7	39.37
KA$^+$	25.3	24.9	22.73
LiC$^+$	55.4		
NaC$^+$	42.3		
KC$^+$	26.3		
LiG$^+$	57.1		
NaG$^+$	43.5		
KG$^+$	28.0		

Source: Wang, Y., *The Encyclopedia of Mass Spectrometry*, Elsevier, Amsterdam, Boston, Heidelberg, London, New York, Oxford, Paris, San Diego, San Francisco, Singapore, Sydney, Tokyo, Vol. 4, p. 577, 2005.

the N_7 nitrogen of adenine. Possible tautomerization of the nucleobases can, however, influence the coordination of the metal ions with nucleobases and therefore has to be taken into account.

An earlier study with an octamer containing two terminal phosphate groups, pGCTTGCATp, and a 12-mer with no phosphate at either of the termini, TTGGCCCTCCTT, was dedicated to the interaction of these deprotonated ODNs with UO_2^{2+} dications.[39] The ESI spectra of both bare ODNs indicated that the extent of deprotonation under the chosen operating conditions was similar for both species, for example, 3−, 4−, and 5− charge states were observed with 4− ions being the most abundant. ESI spectra that were obtained also demonstrated the presence of sodium cation impurities suppressing the signals of the bare species. However, addition of an equimolar amount of $UO_2(NO_3)_2$ resulted in the formation of a complex of UO_2^{2+} and the ODNs with 1:1 stoichiometry. Addition of uranyl cations resulted in a dramatic decrease in the intensity of the sodiated ODNs. CID measurements with the uranyl adduct of the octamer indicated that the primary dissociation event involved a facile elimination of the terminal phosphate group. Indirectly, this is considered to be a proof of the nonterminal uranyl cation binding with the octamer. The sodium adduct showed a similar fragmentation pattern but exhibited a much lower specificity of binding than the uranyl dication adduct. Overall, ESI-FT-ICR-MS^n (Fourier transform ion cyclotron resonance) measurements indicated that the metal cations are localized in the TT-rich region of the octamer. The work represents one of the first dedicated to obtaining information about the metal ion-binding site from CID measurements.

Keller and Brodbelt[21] employed CAD to assess the gas-phase fragmentation behavior of a series of 1:1 ODN:metal complexes over a range of charge states using several 10-residue ODNs and a wide array of alkali, alkaline earth, and transition metals. A ThermoQuest LCQ Duo quadrupole ion-trap mass spectrometer was utilized in these experiments operating in the negative ion mode. The dissociation of the lower charge states was found to proceed via the loss of a neutral base similar to the bare ODNs yielding $(a_n–B)$-type ions that retain the intact phosphodiester backbone. Secondary dissociation of these products yields an array of backbone cleavage products. CID data obtained for the adducts of GCCAATTCGC and K^+, Ba^{2+}, Ca^{2+}, and Ni^{2+} indicate that metal ion–ODN interactions survive the electrospray process. The preferred localization of the metal cation can be estimated by examining which of the sequence (secondary fragmentation) products retain the metal. Also, CAD mass spectra of the ODN metal complexes produced different base loss patterns than the free deprotonated ODN (Figure 15.5).

The loss of cytosine was most strongly favored for the low charge states of ODNs metallated with complexes with alkali and alkali-earth ions. In order to rationalize these base losses, additional experiments were performed to determine what, if any, sequence dependence might exist. CAD results obtained for the ODNs GAATTCGCGC, GCGCGAATTC, CGCTTAAGCG, and AAGCGCGCTT were quite similar to those shown in Figure 15.5: loss of guanine dominated for the $[M-2H]^{2-}$ precursors and loss of cytosine dominated for the $[M-4H+Ba^{2+}]^{2-}$ precursors. According to the authors, this uniform behavior indicates that the variation of the position of cytosine rings along the sequence has very little influence on the fragmentation behavior that was observed. In higher charge states, CAD of some metal adducts of GCGAATTCGC produced intense fragment ions corresponding to metallated a_6^{2-} ions. These species result from direct cleavage of the

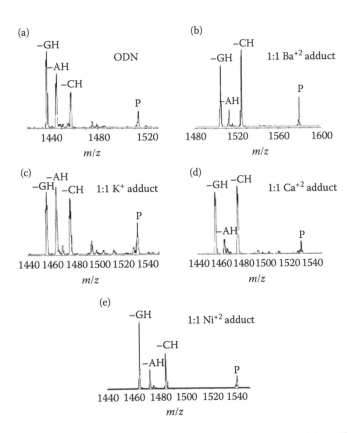

FIGURE 15.5 Base loss region of CAD spectra for GCGAATTCGC and several 1:1 metal complexes in the –2 charge state: (a) ODN alone (0.75 V/30 ms trapping time); (b) 1:1 adduct with Ba^{2+} (0.85 V/30 ms); (c) 1:1 adduct with K^+ (0.95 V/30 ms); (d) 1:1 adduct with Ca^{2+} (0.9 V/30 ms), and (e) 1:1 adduct with Ni^{2+} (0.78 V/30 ms). (From Keller, K.M. and Brodbelt, J.S., *J. Am. Soc. Mass Spectrom.* 16, 28, 2005. With permission.)

phosphate backbone and are not commonly observed for deprotonated ODNs. The formation of these ions from metallated ODNs is highly dependent on the identity of the metal, being most favored for large divalent cations with empty valence shells such as Ba^{2+} and Sr^{2+}.

The dissociation behavior of several oligonucleotides such as GGCTATAA and TA$\underline{X}$ATAT (where $\underline{X}$ is one of the possible nucleobases) adducted with Na^+, Ca^{2+}, Li^+, K^+, and Mg^{2+} has been investigated by Wang et al.[63] utilizing ESI-MS/MS. A very special feature of the metallated species subjected to CID was the absence of the labile protons on any of the present phosphate groups of the ODNs that were studied. These protons were completely replaced by the metal ions. This substitution resulted in a dramatic change in the fragmentation pattern of the metallated adducts compared to the bare ODNs in the same charge state (Figure 15.6). Similar to the observation summarized in Section 15.2.1, bare ODNs in a lower charge state (2−) were found to fragment via the loss of a neutral base. This dissociation pattern was suggested to be directed by the proton affinity of the departing nucleobase. The protons transferred to the bases were found to originate from the protonated phosphate groups (as demonstrated in a series of CIDs performed with the deuterated ODNs). In comparison with the bare ODNs, their metallated analogs exhibited a much greater stability against CID (reflected in the breakdown curves of bare and metallated species) and also yielded many more $[w+M^{n+}]^{-/2-}$ types of ions in their MS/MS spectra compared to $[a-base]^{2-}$ ions characteristic of the bare deprotonated ODNs. Results of the study provided further evidence for the involvement of an acidic phosphate protons in the mechanism of the loss of a neutral nucleobase from the partially deprotonated oligonucleotides in low charge states.

Anichina et al.[64] investigated the influence of a sodium cation on the stability against CID of a deprotonated trideoxynucleotide with a CCC sequence. The minimum energy structure of the sodiated dianion of the trinucleotide was computed in a hierarchical fashion employing four methods, which in reverse order of complexity and accuracy are AMBER8 (molecular mechanics), PM3 (semiempirical), RI-BP-86/SVP (density functional and small basis set), and RI-MP2/TZVPP

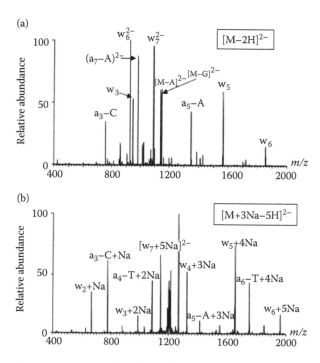

FIGURE 15.6 Product ion spectra of the ESI-produced doubly deprotonated GGCTATAA (a) and its sodiated analog (b). (From Wang, Y., Taylor, J.-S., and Gross, M.L., *J. Am. Soc. Mass Spectrom.* 12, 550, 2001. With permission.)

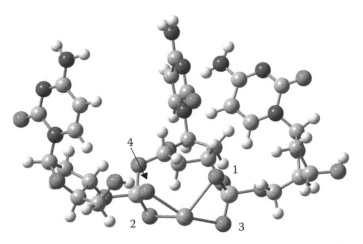

FIGURE 15.7 Minimum energy structure of the sodiated monoanion of CCC. 1–4 refers to the four oxygen atoms (O1, O2, O3, O4). (Adapted from Anichina, J., Uggerud, E., and Bohme, D.K., unpublished results.)

(Møller–Plesset perturbation theory to the second order). The binding energy of a sodium ion (sodium cation affinity) to $(CCC-2H)^{2-}$ was computed to be 694 kJ mol^{-1} (RI-MP2) (Figure 15.7). The sodium cation–oxygen bond lengths are 2.29 Å (Na–O1), 2.36 Å (Na–O2), 2.43 Å (Na–O3), and 2.59 Å (Na–O4). The bond lengths indicate that the sodium cation is not located in the same plane with four oxygen atoms. The dissociation of $[Na(CCC-2H)]^-$ was found to proceed via the loss of a neutral cytosine as a primary dissociation event (Figure 15.8). The ensuing loss of the adjacent furfuryl alcohol indicates that the initial loss of a neutral nucleobase takes place at one of the ends of the monoanion rather than in the middle. As demonstrated in Figure 15.8, the sodiated monoanion exhibits a greater stability in the gas phase, for example, at the same value of the laboratory collision voltage, the dissociation of the protonated species is more "pronounced." The onset voltages for both monoanions were reported to be (-21.6 ± 0.4) and (-29.4 ± 0.3) V for protonated and sodiated monoanions, respectively.

Tandem mass spectrometric measurements performed with a deuterated analog of the sodium complex indicated that, unlike deuterated $(CCC-D)^-$, the metallocomplex dissociates via the loss of a neutral base formed as a result of the transfer of a proton, not a deuteron. Therefore, a different

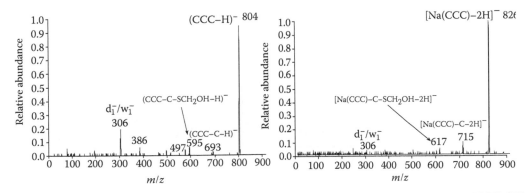

FIGURE 15.8 MS/MS spectra for the dissociation of 804 $(CCC-H)^-$ (on the right) and 826 $[Na(CCC-2H)]^-$ (on the left) at the laboratory collision voltage -30 V. (Adapted from Anichina, J., Uggerud, E., and Bohme, D.K., unpublished results.)

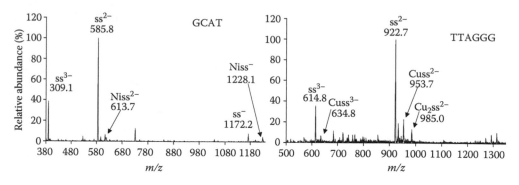

FIGURE 15.9 ESI spectra obtained for 20 μM solution of GCAT mixed with 5-fold excess Ni(II) salt and TTAGGG mixed with 5-fold excess of Cu(II) nitrate in the 20:80 methanol–water mixture at DP (declustering potential) = −30 V. (Adapted from Anichina, J. and Bohme, D.K., unpublished results.)

mechanism is involved in the primary loss that leads to an increase in the onset voltage of the sodium complex compared to the protonated species.

Anichina et al.[65] also have investigated metallocomplexation of single- and double-stranded ODNs that vary in length and nucleobase composition such as GCGC (**I**), ATAT (**II**), GCAT (**III**), CATAC (**IV**), GCATGC (**V**), AGTCTG (**VI**), and TTAGGG (**VII**). Addition of the salts of the first-row transition metal dications to the solutions of the selected ODNs resulted in the formation of multiply metallated single- and double-stranded anions in a range of charge states (Figure 15.9). Formation of these metal ion complexes was found to occur by the displacement of the corresponding number of protons from the ODNs. The composition of an ODN complexes of the observed metal dications can be represented by the following formula: $[M_k^{2+}(ODN_m-nH)]^{(n-2k)-}$, where k is the number of metal dications in the complex, n is the number of protons lost, and m is the number of strands in the ODN.

A shift in the distribution of the intensities of the species with different numbers of attached metal dications was observed upon an increase in the concentration of the corresponding metal salt. The latter observation suggests that the yields of the ions in the ESI spectra of the metal–ODN systems may reflect their distribution in solution. However, nonspecific clustering in the electrospray process cannot be excluded (Figure 15.10).

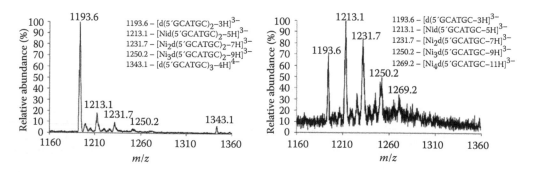

FIGURE 15.10 ESI spectra of the solutions containing previously annealed GCATGC (20 μM) with 10-fold excess (right) and 5-fold (left) of nickel nitrate in 20:80 methanol–water. (Adapted from Anichina, J. and Bohme, D.K., unpublished results.)

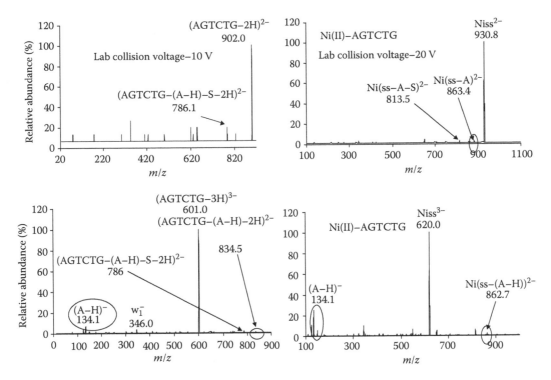

FIGURE 15.11 MS/MS spectra of doubly (top) and triply charged (bottom) AGTCTG (left) and its Ni(II) complexes (right) acquired at the laboratory collision voltage −10 V (for the bare ODN) and −20 V (for the NI(II) adducts). The products of the primary dissociation are circled. (Adapted from Anichina, J. and Bohme, D.K., unpublished results.)

The CID of the complexes of the single-stranded ODNs with metal ions that were studied was found to be similar to the bare ODNs in that it was charge-directed (Figure 15.11).

Lower charge states tend to eliminate a neutral nucleobase, whereas higher-charged species were shown to charge separate producing a monoanion of a nucleobase and a complementary multiply charged, metal-containing product. However, dissociation of the metallocomplexes was found to occur at elevated values of laboratory collision voltage which is indicative of a stabilizing effect of the metal ions against CID. In order to further explore the stabilizing influence of a metal ion, we compared the stability of the bare deprotonated ODNs with that of the same species with a metal dication attached. For example, Figure 15.12 shows such a comparison for a hexadeoxynucleotide in a charge state of −5 in the absence and presence of Zn^{2+}.[65] The dissociation patterns of the triply charged complexes of **V**, **VI**, and **VII** hexadeoxynucleotides with the first row transition metal dications were investigated by MS/MS. The primary dissociation of these ions was found to be similar to the penta-anions of the corresponding bare ODNs. The deprotonated nucleobases lost upon dissociation were the same for the bare penta-anions and the triply charged metallocomplexes, for example, adenine in the case of AGTCTG and GCATGC and thymine in the case of TTAGGG and their metallocomplexes. The presence of metal ions, however, results in stabilization of the ODNs, as revealed in the elevated values of the tangent values (TVs) of the metallocomplexes (Table 15.3).

The relative gas-phase stabilities of $(ODN-3H)^{3-}$ and $[M(ODN-5H)]^{3-}$ were compared utilizing the value of the intercept of the linear portion of the parent ion disappearance curve with the applied collision voltage axis (TV). The TV values of the triply charged AGTCTG, GCATGC, and TTAGGG were found to be (-37.2 ± 0.7) V (1.71 eV), (-34.7 ± 0.7) V (1.61 eV), and (-39.2 ± 0.8) V (1.76 eV),

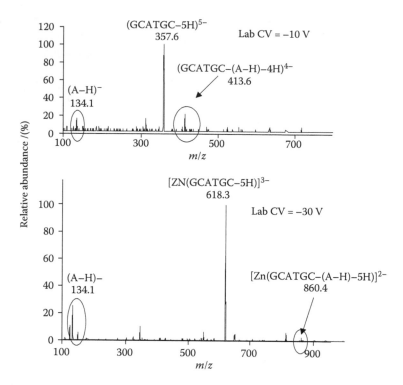

FIGURE 15.12 MS/MS spectra of (GCATGC–5H)$^{5-}$ (top) and its triply charged Zn(II) complex (bottom) acquired at the laboratory collision voltage –10 and –30 V, respectively. The products of the primary dissociation are circled.

respectively. These numbers indicate that the triply charged metallocomplexes exhibit greater gas-phase stability than the trianions of the bare ODNs. In the structures of protonated and metallated versions of the triply charged ODNs, the main difference will probably be in the sites of the binding of two protons versus one metal dication. While a metal dication is shared, the two protons

TABLE 15.3

Tangent Voltages Acquired for the Dissociation of (ODN–5H)$^{5-}$ and [M(ODN–5H)]$^{3-}$ (in V) Along with Apparent Centre of Mass Dissociation Energies (in eV) Calculated According to the Equation: $TE_{CM}^{app} = TV \times z \times m(N_2) \times (m_{N2} + m_{ion})$

Metal Ion	AGTCTG	GCATGC	TTAGGG
Bare ODN^{5-}	(–22.1 ± 0.7) (1.69)	(–20.6 ± 0.5) (1.59)	(–23.3 ± 0.6) (1.74)
Mn^{2+}	(–44.3 ± 0.5) (1.98)	(–41.4 ± 0.5) (1.86)	(–45.1 ± 0.6) (1.97)
Fe^{2+}	(–42.7 ± 0.8) (1.90)	(–40.7 ± 0.7) (1.83)	(–44.3 ± 0.8) (1.96)
Co^{2+}	(–39.4 ± 0.7) (1.75)	(–40.0 ± 0.6) (1.79)	(–43.4 ± 0.7) (1.88)
Ni^{2+}	(–38.7 ± 0.4) (1.72)	(–37.8 ± 0.8) (1.69)	(–40.3 ± 0.5) (1.76)
Cu^{2+}	(–38.0 ± 1.0) (1.69)	(–36.3 ± 0.7) (1.62)	(–39.3 ± 0.7) (1.75)
Zn^{2+}	(–41.3 ± 0.8) (1.83)	(–40.7 ± 0.3) (1.81)	(–39.6 ± 0.6) (1.74)

Source: Anichina, J. and Bohme, D.K., unpublished results.
Note: The apparent CM values are given in parentheses.

are localized on the phosphate groups that they belong to; therefore, the effect of "sharing" rather than localization should explain an increase in the gas-phase stability of the metallocomplexes.[65]

A systematic study of the CID of multiply metallated duplexes was conducted for a duplex with composition d(5′GCATGC).[66] MS/MS experiments performed with the double-stranded species indicate that all metallated duplexes dissociate noncovalently forming two single strands with metal ions still attached to one of them or to both depending on the extent of metallation. Reactions (1)–(5) summarize the dissociation pathways for the observed complexes of the first row transition metal dications with deprotonated d(5′GCATGC)$_2$. The singly charged ions that are underlined were not observed due to the limitations in the available range in m/z.

$$[Md(5′GCATGC)_2 - 5H]^{3-} \nearrow \underline{[Mss-3H]^-} + [ss-2H]^{2-} \tag{15.1}$$

$$\searrow [Mss-4H]^{2-} + [ss-H]^- \tag{15.2}$$

$$[M_2d(5′GCATGC)_2 - 7H]^{3-} \nearrow \underline{[Mss-3H]^- + [Mss-4H]^{2-}} \tag{15.3}$$

$$\searrow \underline{[M_2ss-5H]^-} + [ss-2H]^{2-} \tag{15.4}$$

$$[M_3d(5′GCATGC)_2 - 9H]^{3-} [M_2ss-5H] + [Mss-4H]^2 \tag{15.5}$$

The dissociation of singly metallated species was found to proceed via separation into two strands by two competing channels (1) and (2). Pathway (1) dominates with M = Mn, Fe, Co, and Ni, whereas both pathways (1) and (2) occur with M = Cu and Zn and in relatively equal amounts (Figure 15.13). Doubly metallated ODNs also fragment via two channels with reaction (3) being predominant for all the metal dications. Double-stranded d(5′GCATGC) trianions with three metal dications attached exhibited only one fragmentation pathway, reaction (5). MS/MS experiments

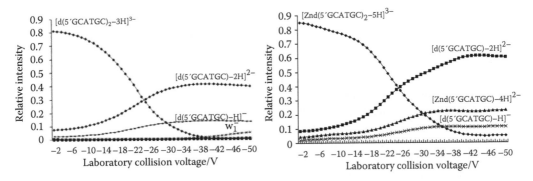

FIGURE 15.13 CID profiles for the dissociation of [Znd(5′GCATGC)$_2$–5H]$^{3-}$ (right) and [d(5′GCATGC)$_2$–3H]$^{3-}$ (left). (Adapted from Anichina, J. and Bohme, D.K., unpublished results.)

TABLE 15.4
Tangent Voltages (in V) and Corresponding Apparent CM Energies (in eV) Given in Parentheses of Singly, Doubly, and Triply Metallated Duplexes of the First Row Transition Metal Dications

	$[Md(5'GCATGC)_2-5H]^{3-}$	$[M_2d(5'GCATGC)_2-7H]^{3-}$	$[M_3d(5'GCATGC)_2-9H]^{3-}$
Mn	-46.7 ± 0.3 (1.07)	-44.0 ± 0.4 (0.99)	-43.2 ± 0.4 (0.96)
Fe	-43.7 ± 0.3 (1.00)	-41.6 ± 0.7 (0.94)	-39.2 ± 0.4 (0.87)
Co	-40.0 ± 0.4 (0.92)	-34.7 ± 0.5 (0.78)	-36.5 ± 0.4 (0.81)
Ni	-38.7 ± 0.4 (0.87)	-36.1 ± 0.4 (0.81)	-36.1 ± 0.4 (0.80)
Cu	-29.4 ± 0.3 (0.68)	-34.1 ± 0.4 (0.77)	-28.6 ± 0.3 (0.63)
Zn	-35.6 ± 0.4 (0.81)	-33.8 ± 0.4 (0.76)	-33.1 ± 0.3 (0.73)

Source: Anichina, J. and Bohme, D.K., unpublished results.

Note: The values of the apparent CM tangent energies were calculated according to the following equation: $E_{CM}^{ap} = z \times TV \times m_{N2} \times (m_{N2} + m_{ion})^{-1}$, where z represents the charge of the corresponding parent ion.

could not be performed with quadruply metallated trianions due to the very low yields of these ions. The relative gas-phase stabilities of the different trianionic metallocomplexes of the double-stranded ODN compared are presented in Table 15.4.

The tangent voltage value for the bare trianion of the duplex was found to be -30.7 ± 0.2 V (0.71 eV in the apparent CM frame). Table 15.4 indicates that metallocomplexation increases the gas-phase stability of the double-stranded ODN with the exception of singly and triply metallated copper adducts. The presence of metal ions in the structure of the double-stranded ODN is expected to relieve Coulombic strain by interacting with its polyanionic phosphodiester backbone and so increases the gas-phase stability. The TV values that were observed for the metallated ODN also indicate that the interaction of the metal ion(s) and the nucleic acid takes place with the deprotonated phosphate groups of both strands; otherwise we could expect the stabilities of the metallated species against CID to be similar to that of the bare $(d(5'GCATGC)_2-3H)^{3-}$.

15.2.3   ESI-MS/MS of ODN Adducts with Metallocomplexes

The first reports on ESI-MS/MS of metallocomplexes added to model DNA were almost exclusively restricted to platinum compounds, given the importance of the coordination compound of Pt(II) in cancer treatment.[46,67,68] These mass spectrometric measurements were able to show that $Pt(NH_3)_2$ binds to DNA relatively slowly (around 5 days) and incompletely, but in an irreversible manner. The relative binding affinity of $Pt(NH_3)_2$ to different sequences was derived from the relative intensities of $Pt(NH_3)_2$–oligonucleotide complexes in their ESI spectra. Also, the *cis*-platin adduct with 5'-CACGTG-3' was subjected to CID. The MS/MS experiments indicated that the most abundant product ions arose from the loss of two amine groups from the molecular ions. Ions corresponding to losses of neutral guanine and cytosine from the precursor ion were also observed in the MS/MS spectrum. The low mass range ($m/z < 600$) of the negative ion MS/MS spectrum was comprised entirely of nucleotide, nucleobase, and sugar and phosphate fragments, with no product ions observed that contained bound platinum. Again the formation of these species would be unfavorable in the negative ion mode owing to Pt dication.[36] In contrast to the negative ion spectrum, the positive ion spectrum of the platinated ternary adduct was found to be dominated by platinated sequence ions and platinated nucleotide and nucleobase fragments. The major ions observed arise from the loss of one or more bases, followed by cleavage of the adjacent 3' C–O bond in the ribose residue from which the base is lost to yield the 3′ sequence (i.e., *w* ions, see McLuckey nomenclature), accompanied by loss of one amine group from the bound platinum resulting in the species $[Pt^{2+}(NH_3)(w_n)_2]^+$. Analysis of the MS/MS spectra acquired in both positive and negative ion modes as a function of the applied collision energy allowed identification of the binding sites of *cis*-platin on the oligonucleotide. Taken together, these data confirmed that platinum coordination occurs at two guanines, G4 and G6 (viz. 1,3 intrastrand cross-link). The site of platinum coordination was also identified in a separate set of experiments that involved major product ions formed via collisional activation in the source.[36] The next large group of studies on the interactions of metallocomplexes with DNA and ODNs appeared after the discovery that Ru(II) triphenanthrolines and related complexes can noncovalently bind DNA in a shape-selective manner.[69] Photophysical and nuclear magnetic resonance (NMR) studies led to the proposal that the cationic tris(phenanthroline) complexes bind to DNA through three non-covalent modes: (i) electrostatically; (ii) hydrophobically against the minor groove; and (iii) by par-tial intercalation of one of the phenanthroline ligands into the DNA base stack from the side of the major groove. Early experiments indicated a preference for the right-handed Δ isomers upon interca-lation into the right-handed DNA, while a small preference for the Λ isomers could be observed for binding in a complementary fashion against the right-handed groove.[70–72] Studies with these simple complexes provided a basis for conceptualizing how octahedral complexes might interact noncova-lently with DNA and for exploring how the properties of the metal complexes, most notably their photophysical and redox characteristics, might be utilized in developing novel probes for DNA.[73]

Resonating with the work of bioinorganic synthetic chemists, a series of ESI-MS studies dedicated to the investigation of noncovalent interactions between ruthenium complexes and DNA demon-strated the full potential of ESI-MS in deriving relative binding affinities of metallocomplexes for DNA.[42,73–76] Several adducts of tetrahedral polyaminocomplexes of ruthenium(II) with three differ-ent non-self-complementary duplex oligonucleotides containing 16 base pairs were examined utiliz-ing ESI-MS. The structures of the corresponding ligands are presented in Figure 15.14.[43] The sequences of the corresponding oligonucleotides were

D1: d(CCTCGGCCGGCCGACC/GGTCGGCCGGCCGAGG)
D2: d(CCTCATGGCCATGACC/GGTCATGGCCATGAGG)
D3: d(CCTCAAAATTTTGACC/GGTCAAAATTTTGAGG).

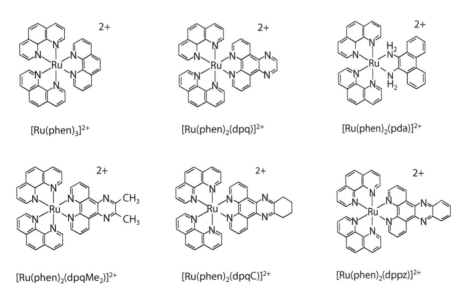

$[Ru(phen)_3]^{2+}$ $[Ru(phen)_2(dpq)]^{2+}$ $[Ru(phen)_2(pda)]^{2+}$

$[Ru(phen)_2(dpqMe_2)]^{2+}$ $[Ru(phen)_2(dpqC)]^{2+}$ $[Ru(phen)_2(dppz)]^{2+}$

FIGURE 15.14 Structures of ruthenium compounds utilized in an ESI-MS comparative study. (From Urathmakul, T., et al., *Dalton Trans.* 17, 2683, 2004. With permission.)

The binding affinity of ruthenium compounds was found to be dependent on both the nature of the ligands in the coordination sphere of the metal and the nucleobase sequence of the duplexes. D2 was shown to possess the largest affinity for the metallocomplexes. The relative affinities were estimated on the basis of a comparison of relative abundances of different DNA–ruthenium complex species in the ESI spectra of mixtures of different metallocomplexes with one selected duplex. The relative affinities of ruthenium compounds toward D2 follow the order $[Ru(phen)_2dppz]^{2+} > [Ru(phen)_2 dpqMe_2]^{2+} > [Ru(phen)_2dpqMe_2]^{2+} > [Ru(phen)_2(dpqC)]^{2+} > [Ru(phen)_2(dpq)]^{2+} > [Ru(phen)_2(pda)]^{2+} > [Ru(phen)_3]^{2+}$. Similar trends in relative affinity were observed for the ruthenium compounds toward D1 and D3 when relative abundance data from reaction mixtures containing these duplexes were examined.

The same research group conducted comparative experiments to study binding interactions of two series of ruthenium complexes, $[Ru(phen)_2L]^{2+}$ and $[RuL'_2(dpqC)]^{2+}$, to a double-stranded DNA hexadecamer and to derive orders of relative binding affinity. The results that were generated utilizing exclusively ESI-MS were later compared to those obtained by absorption spectrophotometry, by CD spectroscopy, and from DNA melting curves. The data generated by techniques other than MS support the use of ESI-MS to assess the binding affinity of metallocomplexes to DNA.[73]

Anichina and Bohme[76] utilized tandem mass spectrometric measurements to test not only the binding affinity of ruthenium trisphenanthrolines to duplex oligonucleotides but also to investigate the influence of the presence of a mismatching base pair in the oligonucleotide sequence on MS/MS measurements. Ruthenium trisphenanthrolines have sparked intensive discussions throughout the 1990s about the possible binding mode of the metallocomplex with DNA and model ODNs.[77,78] The consensual binding model that was finally proposed involves the formation of an externally bound adduct of the ruthenium complex with DNA. Essentially, ruthenium trisphenanthrolines interact with the major groove of DNA by partial insertion. This conclusion was reached as the result of theoretical computations, experimental measurements of sedimentation coefficients, solution viscosity measurements of bulk DNA in the absence and presence of the metallocomplex, and UV-VIS measurements. CID experiments performed with both complexes indicated that their dissociations occur in a similar fashion via charge and strand separation yielding a bare doubly charged single strand of ODN and a singly charged complex of the other strand with the attached ruthenium complex (Figure 15.15). A comparison of the TVs that provide a measure of the relative gas-phase stabilities of ionic species against CID for $[Ru(phen)_3(ODN)_2-5H]^{3-}$ (with ODN = d(5'GCATGC) and d(5'GCTTGC)) indicated that charge and strand separation is not affected by the presence of a mismatching base pair (Figure 15.15).

This result supports an external, nonselective binding of the ruthenium complex to both duplexes and is consistent with other experimental and theoretical data.[77,78] Since the main process that is observed in the CID of both complexes involves noncovalent strand separation and since the number of hydrogen bonds that are broken upon dissociation is the same, the close values for the relative gas-phase stabilities of the noncovalent adducts that were obtained imply that the ruthenium complex is not involved in the binding with π-system of the stacked base pairs. The gas-phase stabilities expressed as TVs (converted into apparent CM) were found to be 1.06 eV for $[Ru(phen)_3d(5'GCATGC)_2-5H]^{3-}$ and 1.08 eV for $[Ru(Phen)_3d(5'GCTTGC)_2-5H]^{3-}$. The corresponding values for the bare triply charged ODNs were found to be 0.72 and 0.73 eV, respectively. Therefore, ruthenium trisphenanthroline stabilizes the duplex via electrostatic interactions with the polyanionic phosphodiester backbone of ODNs.

Relative binding affinities of different metallocomplexes for oligonucleotides were determined by Wan et al.,[38] who investigated competitive binding of three metalloporphyrins, such as CuTMpyP-4, FeTMpyP-4, and MnTMpyP-4, and $[Ru(II)12S4dppz]Cl_2$ with the duplex of d(5'ATATAT). Copper and iron porphyrins exhibited a similar affinity for the duplex that was larger than that for Mn porphyrin and the ruthenium complex. The two latter compounds were found to be equal in the binding affinity for the selected duplex.

A similar set of experiments that involved an evaluation of DNA binding with metallocomplexes of benzoxazole ligands was performed by Mazittelli et al.[47] The binding of the selected

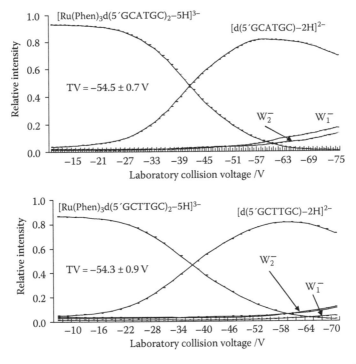

FIGURE 15.15 CID profiles obtained for d(5′GCATGC) and d(5′GCTTGC) complexes with ruthenium tris-phenanthroline. The products whose intensities were lower than 0.05 are not shown. (Adapted Anichina, J. and Bohme, D.K., unpublished results.)

oligonucleotides to metal ions, ligands, and ligands in the presence of metal ions was evaluated separately. The structures of the ligands utilized in this study are presented in Table 15.1. The ligands with the shorter side chains only formed DNA complexes in the presence of metal cations, most notably for 7 and 8 binding to DNA in the presence of Cu^{2+}. The binding of long side-chain ligands was enhanced by Cu^{2+} and to a lesser degree by Ni^{2+} and Zn^{2+}. The authors also assessed cytotoxicities of all of the ligands against the A549 lung cancer and MCF7 breast cancer cell lines. It was found that the ligands whose binding to DNA was enhanced in the presence of metal ions also demonstrated the greatest cytotoxic activity.

Anichina et al.[66] employed the ESI-CID-MS/MS approach for deriving the mode of binding of metallocomplexes to the double-stranded ODN with a d(5′GCATGC) sequence. The metallocomplexes selected for study included singly ligated coordination compounds of the first row transition metal dications with bleomycin A_2 and triethylenetetramine as well as singly, doubly, and triply ligated 1,10-phenanthroline species. The structures of bleomycin A_2 abd triethylenetetramine are depicted in Figure 15.16.

The bleomycins (BLMs) are natural products clinically employed in combination chemotherapy in the treatment of lymphomas, cervical cancers, and squamous cell carcinomas of head and neck.[79] The structure of bleomycin A_2 (Figure 15.16) can be viewed as consisting of four distinct domains: (i) the metal-binding domain responsible for the coordination with a metal cation (four circled N atoms are believed to coordinate a central ion in the equatorial plane, while the amino group belonging to the β-aminoalanine moiety provides one axial ligand leaving the sixth coordination site available for a solvent molecule (or some other ligand)); (ii) the disaccharide domain; (iii) the linker-peptide region; and (iv) the DNA-binding domain. The bithiazole tail is responsible for multiple modes of DNA binding, including probable partial intercalation and binding within the minor groove, suggested in some *in vitro* experiments. Anichina et al.[66] demonstrated the formation of the ternary

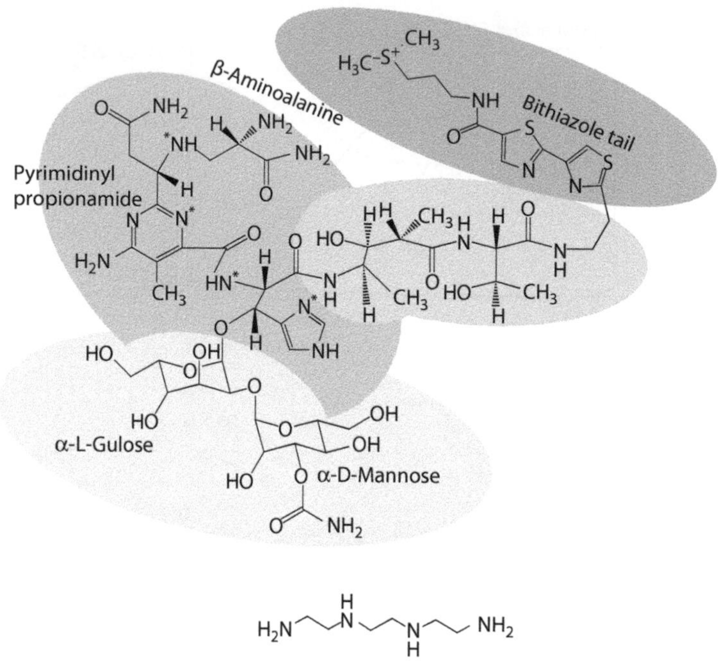

FIGURE 15.16 Structures of bleomycin A$_2$ (top) and triethylenetetramine (bottom). (From Anichina, J. and Bohme, D.K., *J. Phys. Chem. B* 113, 328, 2009. With permission.)

adducts of metallated bleomycins A$_2$ with double-stranded d(5′GCATGC) upon addition of an equimolar (with respect to the concentration of single-stranded hexanucleotide) amount of blenoxane (a mixture of several forms of BLM with 70% content of A$_2$ form) to the solution of a metal ion mixed with the annealed ODN. Selected [MBLMds−6H]$^{4−}$ (with BLM = bleomycin A$_2$ and ds = d(5′GCATGC)$_2$) species were then subjected to CID that was found to proceed via charge and strand separation according to reaction 15.6:

$$[MBLMds−6H]^{4−} \rightarrow [ss−2H]^{2−} + [MBLMss−4H]^{2−}. \tag{15.6}$$

Comparison of the relative gas-phase stabilities of the bare double-stranded oligonucleotide with its metallated BLM adducts indicated that binding of the metallocomplexes did not significantly affect gas-phase strand separation. The authors therefore suggested a groove binding mode for the metallated BLMs with the selected duplex rather than intercalation into the stacked base pairs. They also concluded that metal ions were not directly involved in binding to the nucleic acid that remained complexed with the ligand. Also, the adducts of the duplex d(5′GCATGC) with singly, doubly, and triply ligated phenanthroline complexes of the first row transition metal dications were found to dissociate in a similar fashion with the loss of a neutral ligand being the primary dissociation event (Figure 15.17).

In terms of the relative gas-phase stability, triply ligated metallocomplexes were found to form the most stable complexes with the duplex that was attributed to possible stereo-selective interactions of the metal species and the double-stranded oligonucleotide.

Interestingly, formation of the ternary adduct of the d(5′GCATGC)$_2$ with triethylenetetramine was observed only in the presence of Cu(II). On the basis of the MS/MS of [CuTrien d(5′GCATGC)$_2$−5H]$^{2−}$, the authors suggested that the Trien ligand chelates the copper cation in this noncovalent assembly rather than interacting electrostatically with the polyanionic phosphodiester backbone of the duplex.

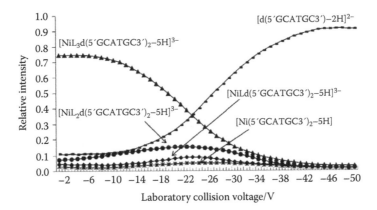

FIGURE 15.17 CID profiles for the dissociation of $[NiL_3d(5'GCATGC3')_2-5H]^{3-}$ with L = 1, 10-phenanthroiline. (From Anichina, J. and Bohme, D.K., *J. Phys. Chem. B* 113, 328, 2009. With permission.)

15.3 MALDI-TOF OBSERVATIONS OF METAL–NUCLEIC ACID ADDUCT FORMATION

In contrast to ESI, MALDI is not routinely utilized for gas-phase studies of the complexation of metal ions with oligonucleotides. Usually, formation of alkali metal ion adducts causes a significant reduction in the yields of oligonucleotides in MALDI spectra.[37] A few studies, however, are dedicated to MALDI of mixtures of oligonucleotides with metal ions. For instance, Hettich[34] demonstrated that negative oligonucleotide ions containing iron can be generated by MALDI from a stainless steel target disk, either by defocusing the laser beam or by mixing iron salts such as $FeCl_3$ with the matrix compound during sample preparation. High-resolution mass measurements revealed the presence of Fe^{3+} and Fe^{2+} in the metal–oligonucleotide ions. The presence of Fe^{3+} is unexpected, and must involve replacement of protons from the nucleic bases or ribose groups as well as the phosphate groups of the oligonucleotides. Inspection of a range of small oligonucleotides and mononucleotides reveals that the presence of both Fe^{2+} and Fe^{3+} in the iron–biomolecule complexes is dependent on the number of acidic hydrogens that can be replaced in the oligonucleotide or nucleotide. Collisional dissociation of several metal–tetranucleotide ions revealed that the presence of the iron ion alters the fragmentation observed. The iron atom was observed to be present in all of the fragment ions, and, whenever possible, seemed to enhance the abundance of fragment ions containing both iron and a guanine nucleic base.

The formation of metalloadducts of oligonucleotides as the result of matrix–substrate–oligonucleotide was first demonstrated in 1995.[45] A stainless steel MALDI probe was utilized in the experiments that resulted in the detection of adducts of singly charged iron-containing complexes of 27-, 34-, 47-, and 68-mers in the corresponding MALDI spectra. Later, Hettich[33] demonstrated the application of MALDI-FT-ICR-MS[n] to the study of the fundamental interactions of metal ions with small, single-stranded oligonucleotides.[33] Of special interest was an evaluation of sequential metal ion attachment and the determination of binding site(s). Whereas the incorporation of Cu(I) into dinucleotides revealed the maximum number of protons that can be replaced by singly charged metal ions, the incorporation of iron(II, III) into the same biomolecules was sensitive to steric binding factors as well. The authors proposed a mechanism for the formation of metallocomplexes according to which the complexes of the metal ion and the laser matrix compound (2,5-dihydroxybenzoic acid, DHB) initially coordinate with the oligonucleotides, and in these complexes, the attached metal ions enhance the elimination of neutral DHB molecules. Ions of divalent (Zn^{2+} and Fe^{2+}), trivalent (Fe^{3+} and Ce^{3+}), and tetravalent (Ce^{4+} and Th^{4+}) metals were found to bind strongly to oligonucleotides and were detected even in the fragment ions. Gas-phase structures of the species containing the metal ion coordinated to both phosphate and nucleobase were proposed based on high-resolution mass measurements and information about their fragmentation.

MALDI studies provide molecular-level information about interactions of metal ions with oligo-nucleotides, and may be the basis for determining how certain metal ions can be exploited as selective probes for the interrogation of biomolecular structure. However, more systematic and comparative studies are needed to relate the metallocomplexes observed in MALDI spectra with their solution counterparts.

15.4   ION MOBILITY MS IN THE EXPLORATION OF THE GAS-PHASE STRUCTURES OF METAL ION–NUCLEIC ACID COMPLEXES

The mobility of an ion is simply a measure of how fast the ion drifts through a buffer gas under the influence of a weak, uniform electric field.[80] This mobility depends on the mass of the ion and, for larger species, also on the geometric shape of the ion. Compact ions with small collision cross sections undergo fewer collisions with the buffer gas and hence drift faster than more extended ions. Conformational information about the ions is obtained by comparing the cross sections determined from arrival time distributions (ATDs) with the cross sections calculated using theoretical models. Essentially, ion mobility MS is the only mass spectrometric approach that provides information about the gas-phase shapes of the ions. Currently, there are very few publications dedicated to the ion mobility of metallocomplexes of nucleic acids.[31,32] Baker et al.[32] analyzed the metal-binding properties of the dinucleotide duplex, dCG–dCG, in the gas phase with ion mobility MS. Both MALDI and ESI were used to generate [M(dCG–dCG)]$^+$ complexes with M = Cu$^+$, Ag$^+$, Cd^{2+}, Pt^{2+}, Cu^{2+}, Zn^{2+}, Fe^{2+}, Co^{2+}, Ni^{2+}, Li$^+$, Na$^+$, K$^+$, Cs$^+$, Mg^{2+}, Ca^{2+}, Cr^{2+}, Cr^{3+}, Mn^{2+}, and Fe^{3+}. The collision cross section of each complex was measured in helium using ion mobility-based methods and was compared with calculated cross sections of theoretical structures.

The main conclusion of this study is that the [M(dCG–dCG)]$^+$ complex formed from metal cations (classified as hard acids) combining with dCG–dCG organized into a globular structure, whereas dCG–dCG and the soft acid metal cations examined, formed a structure in which the two C–G base pairs were Watson–Crick bound.[32]

MALDI-TOF mass spectra obtained for a series of dinucleotides cationized by Na$^+$ indicated that up to 7 Na$^+$ could bind to a given duplex and up to 10 Na$^+$ could bind to triplexes, with n Na$^+$ ions replacing ($n-1$) hydrogens.[31] A combined ion mobility and molecular modeling approach indicated that the sodium ions initially cluster around the deprotonated phosphates, but as more Na$^+$ are added, they become more dispersed and bind to multiple sites and groups in the duplex/triplex. Examples of duplex and triplex structures incorporating multiple sodium cations are shown in Figure 15.18. Overall, ion mobility comprises a promising mass spectrometric approach that contributes toward understanding of the gas-phase geometry of nucleic acids complexed with metal ions.

15.5   THEORETICAL MODELING OF METAL ION–DNA ADDUCTS

Computational modeling of metal ion–nucleic acid complexes represent a serious challenge due to the large size of oligonucleotides, the large number of atoms they are comprised of, and therefore the large number of degrees of freedom that should be taken into account. Frequently, the computations are performed in an hierarchical fashion that employ several methods that in reverse order of complexity and accuracy usually are molecular mechanics approach (the most popular software package is AMBER), semiempirical, and density functional computations with different basis sets.[31,64] First, the molecular mechanic modeling force fields are constructed, and then, this geometry is subjected to energy minimization. After that several nuclear configurations are sampled and subjected to full geometry optimization using DFT. If there are several possible stereoisomers, the same procedure is repeated several times for each stereoisomer.

Theoretical modeling has shown to be extremely useful in the interpretation of MS/MS and reactivity measurements.[64]

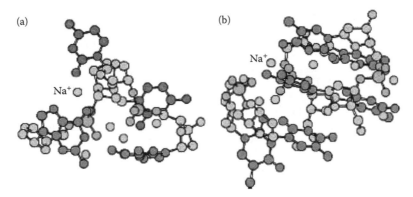

FIGURE 15.18   Lowest energy structures of (a) $[(dCG–dCG) + 5Na – 4H]^+$ and (b) $[(dTT·dTT·dTT) + 10Na – 9H]^+$ obtained by molecular mechanics calculations. Cytosine is shown in red, guanine in blue, and thymine in green. Na^+ ions are yellow and the phosphate groups are orange. The Na^+ ions bind preferentially to oxygen atoms on the phosphate groups and bases. (From Gidden, J., et al., *Int. J. Mass Spectrom.* 240, 183, 2005. With permission.)

15.6   CONCLUSION AND FUTURE DIRECTIONS

The plethora of information on metal ion–nucleic acid interactions obtained by means of versatile mass spectrometric approaches described in this chapter demonstrates the essentially infinite potential of MS to explore and solve various important and unanswered problems associated with the *in vivo* and *in vitro* chemistry of nucleic acids with metal ions.

Two very promising mass spectrometric approaches have become available recently, which allow the exploration of the properties of DNA and its adducts with metal ions and complexes in terms of their intrinsic reactivity in ion–molecule and ion–ion reactions occurring in pristine gas-phase environments.[81,82]

Feil et al.[81] have been able to transfer single-stranded deprotonated hexadeoxynucleotide $[AGTCTG–nH]^{n–}$ and both single- and double-stranded $[GCATGC–nH]^{n–}$ anions into the gas phase and expose them to chemical reactions with various agents using a novel ESI-qQ-SIFT-QqQ instrument. The DNA anions that were investigated proved to be remarkably unreactive with strong oxidants (O_3, O_2, and N_2O) and potential intercalators (benzene, pyridine, toluene, and quinoxaline). Hydration also was observed to be inefficient. However, $[AGTCTG–nH]^{n–}$ anions with $n = 5$, 4, 3, and 2 were seen to be sequentially protonated and/or hydrobrominated with HBr (but not damaged) and displayed an interesting "end effect" against protonation. Measurements were achieved for rate coefficients of reaction and the efficiencies of protonation. These experimental results point toward the exciting prospect of measuring the intrinsic chemistry of other bare and metallated DNA-like anions, including double-stranded oligonucleotide anions, in the gas phase at room temperature.

Ion–ion reactions with the participation of oligonucleotides and metallocomplexes were investigated by Barlow et al.[83] by subjecting multiply deprotonated hexadeoxyadenylate anions, $(A_6–nH)^{n–}$, where $n = 3–5$ to reaction with bis- and tris-1,10-phenanthroline complexes of Cu(II), Fe(II), and Co(II), as well as with the tris-1,10-phenanthroline complex of Ru(II). Competing reaction channels were found to be involved in electron transfer from the ODN anion to the cation, the formation of a complex between the anion and cation, and the incorporation of the transition metal into the ODN.

REFERENCES

1. Lippert, B., Alterations of nucleobase pKa values upon metal coordination: Origins and consequences. *Prog. Inorg. Chem.* 54, 385, 2005.
2. Wettig, S.D., Wood, D.O., and Lee, J.S., Thermodynamic investigation of M-DNA: A novel metal ion-DNA complex. *J. Inorg. Biochem.* 94, 94, 2003.

3. Aich, P., Skinner, R.J.D., Wettig, S.D., Steer, R.P., and Lee, J.S., Long range molecular wire behaviour in a metal complex DNA. *J. Biomol. Struct. Dynam.* 20, 1, 2002.

4. Clever, G.H., Kaul, C., and Carell, T., DNA–metal base pairs, *Angew. Chem. Int. Ed. Engl.* 46, 6226, 2007.

5. Egli, M., DNA-cation interactions quo vadis? *Chem. Biol.* 9, 277, 2002.

6. Hud, N.V. and Polak, M., DNA-cation interactions: the major and minor grooves are flexible ionophores. *Curr. Opin. Struct. Biol.* 11, 293, 2001.

7. Schliepe, J., Berghoff, U., and Lippert, D.C., Automated solid phase synthesis of platinated oligodeoxy-nucleotides via nucleoside phosphates. *Angew. Chem. Int. Ed. Engl.* 35, 646, 1996.

8. Wahl, M.C., and Sundaralingam, M., A-DNA duplexes in the crystal. In *Oxford Handbook of Nucleic Acids Structure.*, S. Neidle, ed., Oxford University Press, Oxford, p. 135, 1999.

9. Giovannageli, C., Sun, J.-S., and Helene, C., Nucleis acids: Supramolecular structures and rational design of sequence-specific ligands. In *Comprehensive Supramolecular Chemistry*, Y. Murakami, ed., 4 ed., Pergamon, New York, p. 193, 1996.

10. Rosenberg, B., VanCamp, L., and Krigas, T., Inhibition of cell division in *E. coli* by electrolysis products from a platinum electrode. *Nature* 205, 698, 1965.

11. Rosenberg, B., VanCamp, L., Trosko, J.E., and Mansour, V.H., Platinum compounds: A new class of potent antitumour agents. *Nature* 222, 385, 1969.

12. Elmroth, S.K.C. and Lippard, S.J., Platinum binding to d(GPG) target sequences and phosphorothioate linkages in DNA occurs more rapidly with increasing oligonucleotide length. *J. Am. Chem. Soc.* 116, 3633, 1994.

13. Ling, E.C.H., Allen, G.W., and Hambley, T.W., DNA binding of a platinum(II) complex designed to bind intrastrand and not intrastrand. *J. Am. Chem. Soc.* 116, 2673, 1994.

14. Greene, M.H., Is cisplatin a human carcinogen? *J. Natl. Cancer Inst.* 84, 306, 1992.

15. Dupureur, C.M. and Barton, J.K., Use of selective deuteration and 1H NMR in demonstrating major groove binding of Δ-[Ru(phen)$_2$dppz]$^{2+}$ to d(GTCGAC)$_2$. *J. Am. Chem. Soc.* 116, 10286, 1994.

16. Eichhorn, G.L., Metal complexes of nucleic acid derivatives and nucleotides. *Metal Ions in Genetic Information Transfer*, G.L. Eichhorn, and L.G. Marzilli, eds, Elsevier, Amsterdam, 340 pp., 1984.

17. Yamashita, M. and Fenn, J.B., Electrospray ion source. Another variation on the free-Jet theme. *J. Phys. Chem.* 88, 4451, 1984.

18. Fenn, J.B., Mann, M., Meng, C.K., and Wong, S.F., Electrospray ionization: Principles and practice. *Mass Spectrom. Rev.* 9, 37, 1990.

19. Wu, J. and McLuckey, S.A., Gas-phase fragmentation of oligonucleotide ions. *Int. J. Mass Spectrom.* 237, 197, 2004.

20. Beck, J.L., Colgrave, M.L., Ralph, S.F., and Sheil, M.M., Electrospray ionization mass spectrometry of oligonucleotide complexes with drugs, metals, and proteins. *Mass Spectrom. Rev.* 20, 61, 2001.

21. Keller, K.M. and Brodbelt, J.S., Charge state-dependent fragmentation of oligonucleotide/metal complexes. *J. Am. Soc. Mass Spectrom.* 16, 28, 2005.

22. Baker, E.S., Bernstein, S.L., Gabelica, V., De Paw, E., and Bowers, M.T., G-quadruplexes in telomeric repeats are conserved in a solvent-free environment. *Int. J. Mass Spectrom.* 253, 225, 2006.

23. Lecchi, P., Le, H.M.T., and Pannel, L.K., 6-Aza-2-thiothymine: A matrix for MALDI spectra of oligo-nucleotides. *Nucleic Acids Res.* 23, 1276, 1995.

24. Nordhoff, E., Luebbert, C., Thiele, G., Heiser, V., and Lehrach, H., Rapid determination of short DNA sequences by the use of MALDI-MS. *Nucleic Acids Res.* 28, e86, 2000.

25. Schatz, P., Dietrich, P., and Schuster, M., Rapid analysis of CpG methylation patterns using RNase T1 cleavge and MALDI-TOF. *Nucleic Acids Res.* 32, e167, 2004.

26. Spottke, B., Gross, J., Galla, H.J., and Hillencamp, F., Reverse sanger sequencing of RNA by MALDI-TOF mass spectrometry after solid phase purification. *Nucleic Acids Res.* 32, e97, 2004.

27. Fu, Y., Xu, S., Pan, C., Ye, M., Zou, H., and Guo, B., A matrix of 3,4-diaminobenzophenone for the analy-sis of oligonucleotides by matrix-assisted laser desorbtion/ionization time-of-flight mass spectrometry. *Nucleic Acids Res.* 34, e94, 2006.

28. Wee, S., O'Hair, R.A.J., and McFadyen, D., Can radical cations of the constituents of nucleic acids be formed in the gas phase using ternary transition metal complexes? *Rapid Commun. Mass Spectrom.* 19, 1797, 2005.

29. Madhusudanan, K.P., Katti, S.B., Vijayalakshmi, R., and Nair, B.U., Chromium (III) interactions with nucleosides and nucleotides: A mass spectrometric study. *J. Mass Spectrom.* 34, 880, 1999.

30. Baker, E.S., Gidden, J., Ferzoco, A., and Bowers, M.T., Sodium stabilization of dinucleotide multiplexes in the gas phase. *Phys. Chem. Chem. Phys.* 6, 2786, 2004.

31. Gidden, J., Baker, E.S., Ferzoco, A., and Bowers, M.T., Structural motifs of DNA complexes in the gas phase. *Int. J. Mass Spectrom.* 240, 183, 2005.

32. Baker, E.S., Manarel, M.J., Gidden, J., and Bowers, M.T., Structural analysis of metal interactions with the dinucleotide duplex, dCGxdCG using ion mobility mass spectrometry. *J. Phys. Chem. B Lett.* 109, 4808, 2005.

33. Hettich, R.L., Investigating the effect of transition metal ion oxidation state on oligodeoxynucleotide binding by matrix-assisted laser desorbtion/ionization fourier transform ion cyclotron resonance mass spectrometry. *Int. J. Mass Spectrom.* 204, 55, 2001.

34. Hettich, R.L., Formation and characterization of iron-oligonucleotide complexes with matrix-assisted laser desorbtion/ionization fourier transform ion cyclotron resonance mass spectrometry. *J. Am. Soc. Mass Spectrom.* 10, 941, 1999.

35. Monn, S.T.M. and Schurch, S., Investigation of metal-oligonucleotide complexes by nanoelectrospray tandem mass spectrometry in the positive ion mode. *J. Am. Soc. Mass Spectrom.* 16, 370, 2005.

36. Ianniti-Tito, P., Weimann, A., Wickham, G., and Sheil, M.M., Structural analysis of drug-DNA adducts by tandem mass spectrometry. *Analyst* 125, 627, 2000.

37. Fu, Y., Xu, S., Pan, C., Ye, M., Zou, H., and Guo, B., A matrix of 3,4-diaminobenzophenone for the analysis of oligonucleotides by matrix-assisted laser desorbtion/ionization time-of-flight mass spectrometry. *Nucleic Acid Res.* 34, e94, 2006.

38. Wan, K.X., Shibue, T., and Gross, M.L., Gas phase stability of double—stranded oligodexynucleotides and their noncovalent complexes with DNA-binding drugs as revealed by collisional activation in an ion trap. *J. Am. Chem. Soc.* 122, 300, 2000.

39. Wu, Q., Cheng, X., Hofstadler, S.A., and Smith, R.D., Specific metal-oligonucleotide binding studied by high resolution tandem mass spectrometry. *J. Mass Spectrom.* 31, 669, 1996.

40. Beck, J.L., Humphries, A., Sheil, M.M., and Ralph, S.F., Electrospray ionization mass spectrometry of ruthenium and palladium complexes with oligonucleotides. *Eur. J. Mass Spectrom.* 5, 489, 1999.

41. Keller, K.M. and Brodbelt, J.S., Charge state-dependent fragmentation of oligonucleotide/metal complexes. *J. Am. Soc. Mass Spectrom.* 16, 28, 2005.

42. Dorsier, A., Dyson, P.J., Gossens, G., Rothlisberger, U., Scopelliti, R., and Tavernelli, T., Binding of organometallic ruthenium(II) and osmium (II) complexes to an oligonucleotide: A combined mass spectrometric and theoretical study. *Organometallics* 24, 2114, 2005.

43. Urathmakul, T., Beck, J.L., Sheil, M., Aldrich-Wright, J.R., and Ralph, S.F., A mass spectrometric investigation of non-covalent interaction between ruthenium complexes and DNA. *Dalton Trans. 17,* 2683, 2004.

44. Vairamani, M. and Gross, M.L., G-quadruplex formation of thrombin-binding aptamer detected by electrospray ionization mass spectrometry. *J. Am. Chem. Soc.* 125, 42, 2003.

45. Christian, N.P., Colby, S.M., Giver, L., Houston, C.T., Arnold, R.J., Ellington, A.D., and Reilly, J.P., High resolution matrix-assisted laser desorbtion/ionization time-of-flight analysis of single-stranded DNA of 27 to 68 nucleotide in length. *Rapid Commun. Mass Spectrom.* 9, 1061, 1995.

46. Kloster, M.B.G., Hannis, J.C., Muddiman, D.C., and Farell, N., Consequences of nucleic acid conformation on the binding of trinuclear platinum drug. *Biochemistry* 38, 14731, 1999.

47. Mazitelli, C.L., Rodriguez, M., Kerwin, S., and Brodbelt, J.S., Evaluation of metal-mediated DNA binding of benzoxazole ligands by electrospray ionization mass spectrometry. *J. Am. Soc. Mass Spectrom.* 19, 209, 2008.

48. McLuckey, S.A., Van Berkel, G.J., and Glish, G.I., Tandem mass spectrometry of small multiply-charged oligonucleotides. *J. Am. Soc. Mass Spectrom.* 3, 60, 1992.

49. Habibi-Goudarzi, S. and McLuckey, S.A., Ion trap collisional activation of the deprotonated deoxymononucleoside and deoxydinucleoside monophosphates. *J. Am. Soc. Mass Spectrom.* 6, 102, 1995.

50. Boschenok, J. and Sheil, M.M., Electrospray tandem mass spectrometry of oligonucleotides. *Rapid Commun. Mass Spectrom.* 10, 144, 1996.

51. Ho, Y.H. and Kebarle, P., Studies of the dissociation mechanisms of deprotonated mononucleotides by energy resolved collision-induced dissociation. *Int. J. Mass Spectrom.* 165, 433, 1997.

52. Anichina, J., Feil, S., Uggerud, E., and Bohme, D.K., Structures, fragmentation and protonation of trideoxynucleotide CCC mono- and dianions. *J. Am. Soc. Mass Spectrom.* 19, 987, 2008.

53. Gabelica, V. and De Paw, E., Comparison between solution-phase stability and gas-phase kinetic stability of oligodeoxynucleotide duplexes. *J. Am. Chem. Soc. Mass Spectrom.* 13, 910, 2002.

54. Pan, S., Sun, X., and Lee, J.K., DNA stability in the gas versus solution phases: A systematic study of thirty-one duplexes with varying length, sequence, and charge level. *J. Am. Soc. Mass Spectrom.* 17, 1383, 2006.

55. Gao, O., Cheng, X., Smith, R.D., Yang, C.F., and Goldberg, I.H., Binding specificity of post-activated neocarzinostatin chromophore drug-bulged DNA complex studied using electrospray ionization mass spectrometry. *J. Mass Spectrom.* 31, 31, 1996.

56. Rogniaux, H., Van Dorselaer, A., Barth, P., Biellmann, J.F., Brabanton, J., Van Zandt, M., Chevrier, B., et al., Binding of aldose reductase inhibitors: Correlation of crystallographic and mass spectrometric studies. *J. Am. Soc. Mass Spectrom.* 10, 638, 1999.

57. Gao, J. Wu, O., Carbeck, J.D., Lei, Q.P., Smith, R.D., and Whitesides, G.M., Probing the energetics of the dissociation of carbonic anhydrase-ligand complexes in the gas phase. *J. Biophys.* 76, 3253, 1999.

58. Gabelica, V. and De Pauw, E., Collision-induced dissociation of 16-mer DNA duplexes with various sequences: Evidence for conservation of the double helix conformation in the gas phase. *Int. J. Mass Spectrom.* 219, 151, 2002.

59. Robertazzi, A. and Platts, J.A., Gas-phase DNA oligonucleotide structures. A QM/MM and atoms in molecules study. *J. Phys. Chem. A* 110, 3992, 2006.

60. Baker, E.S., Bernstein, S.L., Gabelica, V., De Pauw, E., and Bowers, M.T., G-quadruplexes in telomeric repeats are conserved in a solvent-free environment. *Int. J. Mass Spectrom.* 253, 225, 2006.

61. Wang, Y., Interactions between alkali metal ions and nucleic acid bases. *The Encyclopedia of Mass Spectrometry*, Elsevier, Amsterdam, Boston, Heidelberg, London, New York, Oxford, Paris, San Diego, San Francisco, Singapore, Sydney, Tokyo, Vol. 4, p. 577, 2005.

62. Rogers, M. and Armentrout, P.B., Noncovalent interactions of nucleic acid bases (uracil, thymine, and adenine) with alkali metal ions. Threshold collision-induced dissociation and theoretical studies. *J. Am. Chem. Soc.* 122, 8548, 2000.

63. Wang, Y., Taylor, J.-S., and Gross, M.L., Fragmentation of electrospray-produced oligodeoxynucleotide ions adducted to metal ions. *J. Am. Soc. Mass Spectrom.* 12, 550, 2001.

64. Anichina, J., Uggerud, E., and Bohme, D.K., Sodium cation influence upon stability and reactivity of a trideoxynucleotide, CCC: Experiment and theory (unpublished results).

65. Anichina, J. and Bohme, D.K., General trends in collision induced dissociation of metallated single- and double-stranded oligonucleotides (unpublished results).

66. Anichina, J. and Bohme, D.K., Mass-spectrometric studies of the interactions of selected metallo-antibiotics and drugs with deprotonated hexadeoxynucleotide GCATGC. *J. Phys. Chem. B.* 113, 328, 2009.

67. Gonnet, F., Kocher, F., Plais, J., Bplbach, G., Chottard, J., and Tabet, Kinetic analysis of the reaction between d(TTGGCCAA) and [Pt(NH₃)₃(H₂O)]²⁺ by enzymatic degradation of the products and ESI and MALDI mass spectrometries. *J., Inorg. Chem.* 35, 1653, 1996.

68. Lowe, G., McClosckey, J.A., Ni, J., and Vilaivan, T.A., A Mass spectrometric investigation of the reaction between 4,4'-vinylenedipyridinebis[2,2':6',2'-terpyridine platinum(II)] and self-complementary oligo-nucleotide d(CpGpTpApCpG). *Bioinorg. Med. Chem.* 4, 1007, 1996.

69. Barton J.K., Metals and DNA: Molecular left-handed compliments. *Science* 233, 727, 1986.

70. Satyanarayana, S., Dabrowiak, J.C., and Chaires, J.B., Neither Δ- nor Λ-tris(phenanthroline)ruthenium(II) binds to DNA by classical intercalation. *Biochemistry* 31, 9319, 1992.

71. Satyanarayana, S., Dabrowiak, J.C., and Chaires, J.B., Tris(phenanthroline)ruthenium(II) enantiomer interactions with DNA: Mode and specificity of binding. *Biochemistry* 32, 2573, 1993.

72. Eriksson, M., Leijon, M., Hiort, C., Norden, B., and Graslund, A., Binding of delta- and lambda-[Ru(phen)₃]²⁺ to [d(CGCGATCGCG)₂] studied by NMR. *Biochemistry* 33, 5031, 1994.

73. Erkkila, K.E., Odom, D.T., and Barton., J.K., Recognition and reaction of metallointercalators with DNA. *Chem. Rev.* 99, 2777, 1999.

74. Urathamakul, T., Waller, D.J., Beck, J.L., Aldrich-Wright, J.R., and Ralph, S.F., Comparison of mass spectrometry and other techniques for probing interaction between meta complexes and DNA. *Inorg. Chem.* 47, 6621, 2008.

75. Talib, J., Green, C., Davis, K.J., Urathamakul, T., Beck, J.L., Aldrich-Wright, J.R., and Ralph, S.F., A comparison of the binding of metal complexes to duplex and quadruplex DNA. *Dalton Trans. 8,* 1018, 2008.

76. Anichina, J. and Bohme D.K., Mass spectrometric evidence of external binding of Ru(II) trisphenanthro-line to double-stranded hexadeoxynucleotides (unpublished results).

77. Rehmann, J.P. and Barton, J.K., Proton NMR studies of tris(phenanthroline) metal complexes bound to oligonucleotides: Characterization of binding modes. *Biochemistry* 29, 1701, 1990.

78. Coggan, D.Z., Haworth, I.S., Bates, P.J., Robinson, A., and Rodger, A., DNA Binding of ruthenium tris(1,10-phenanthroline): Evidence for the dependence of binding mode on metal complex concentration. *Inorg. Chem.* 38, 4486, 1999.

79. Chen, J., and Stubbe, J., Bleomycins: New methods will allow reinvestigation of old issues. *Curr. Opin. Chem. Biol.* 8, 175, 2004.

80. Mason, E.A. and McDaniel, E.W., *Transport Properties of Ions in Gases*, Wiley, New York, 1988.

81. Feil, S.W., Koyanagi, G.K., Anichina, J., and Bohme, D.K., Chemical stability and reactivity of deprotonated oligonucleotides (DNA) in the gas phase: Protonation and solvation with hydrogen bromide. *J. Phys. Chem. B* 112, 10375, 2008.

82. Stephenson, J.L. and McLuckey, S.A., Counting basic sites in oligopeptides via gas-phase ion chemistry. *Anal. Chem.* 69, 281, 1997.

83. Barlow, C.B., Hodges, B.D.M., Xia, Y., 'Hair, R.A.J., and McLuckey, S.T., Gas-phase ion/ion reactions of transition metal complex cations with multiply charged oligodeoxynucleotide anions. *J. Am. Soc. Mass Spectrom.* 19, 281, 2008.

Index

Milton Keynes UK
Ingram Content Group UK Ltd.
UKHW052024071024
449327UK00027B/2415